NONLINEAR DYNAMICS AND CHAOS IN SEMICONDUCTORS

Series in Condensed Matter Physics

Other titles in the series include:

Modern Magnetooptics and Magnetooptical Materials
A K Zvedin and V A Kotov

Permanent Magnetism
R Skomski and J M D Coey

SERIES IN CONDENSED MATTER PHYSICS

Series Editors
J M D Coey and D R Tilley

NONLINEAR DYNAMICS AND CHAOS IN SEMICONDUCTORS

Kazunori Aoki

Department of Electronic and Electrical Engineering,
Kobe University,
Japan

IoP
Institute of Physics Publishing
Bristol and Philadelphia

British Library Cataloguing-in-Publication Data
A catalogue record of this book is available from the British Library.

ISBN 0 7503 0514 2

Library of Congress Cataloging-in-Publication Data are available

Commissioning Editor: James Revill
Production Editor: Simon Laurenson
Production Control: Sarah Plenty
Cover Design: Victoria Le Billon, Frederique Swist
Marketing Executive: Colin Fenton

Published by the Institute of Physics Publishing, wholly owned by The Institute of Physics, London

Institute of Physics Publishing, Dirac House, Temple Back, Bristol BS1 6BE, UK

US Office: The Institute of Physics, The Public Ledger Building, Suite 1035, 150 South Independence Mall West, Philadelphia, PA 19106, USA

Produced and typeset by Gray Publishing, Tunbridge Wells
Printed in the UK by Bookcraft, Midsomer Norton, Somerset

To my family, Kazuko, Mariko, Minako and Sayaka

Contents

Preface

In natural science and technology, there is a universality class of dynamical systems that show complex chaotic behaviors. These chaotic behaviors are strongly related to the nonlinear characteristics of each dynamical system. As a function of control parameter, periodic oscillation bifurcates to nonperiodic and irregular motion, namely "deterministic chaos". Nonperiodicity or chaos stems from deterministic property hidden behind nonlinear characteristics of dynamical systems. It is surprising that deterministic chaos can be seen in numerous dynamical systems: biological systems, chemical reaction systems (Belousov–Zhabotinski reaction), lasers, semiconductor systems, fluid flow systems, and so on. Many different nonlinear systems show universal rules of bifurcation routes to chaos.

Very often, spatially-extended dynamical systems give rise to a pattern formation as a result of nonequilibrium phase transition, where the spatial pattern will be static or evolve to an oscillatory mode (periodic or chaotic). While a deep understanding about the pattern formation process in a specified physical system or chemical system is necessary for a full comprehension of spatiotemporal behaviors, it is quite interesting to note that the temporal evolutions are governed by *universality rules* of the chaos theory which arise essentially from the nonlinear character of the dynamical systems.

In semiconductors physics, studies on nonlinear carrier transport have been ongoing since the 1960s. At the early stage in semiconductor physics, many authors studied the nonlinear carrier transport associated with the impact ionization of neutral impurities in Ge and GaAs, the Gunn oscillations in GaAs and InP, and the acousto-electric effects in CdS. Many devices have been proposed in order to develop new, modern electronic devices of oscillators and switching devices such as the "oscillistor", the "cryosar", and the Gunn diode. However, many of these proposed devices, except for the Gunn diode, were found not to be on the application base due to unstable oscillations and unreliable switching. The failures could have been partly attributed to the unavoidable occurrence of chaos. Two decades later, chaos in semiconductor systems was discovered by the author in 1982. After this discovery, many researchers have studied the complex chaotic behaviors in GaAs, InSb, InP, Si, and Ge. The discovery of chaos brought us a *paradigm shift* in the field of semiconductor physics.

The aim of this book is to describe chaos experiments in semiconductors and to characterize the chaos by means of various analytical methods of chaos theory. The book is intended for students studying master's degrees, doctoral students, engineers and scientists, especially semiconductor physicists and researchers in electrical and electronics engineering. The book provides an elementary introduction to chaos by logistic map in Chapter 1; nonlinear carrier transport, nonequilibrium phase transition, SNDC and NNDC, and an example of chaos in the Gunn diode in Chapter 2. Basic theory of impact ionization of shallow impurities and pattern formation of current density filament is given in Chapter 3. Basic ideas of linear stability analysis and examples of periodically-driven chaos are given in Chapter 4. In this chapter, catastrophe theory is supplemented; the breakdown of perfect delay in the cusp catastrophe which plays a crucial role for the occurrence of chaos in semiconductors is especially stressed. Observation of chaos, experimental methods and model simulations in bulk semiconductors and semiconductor heterostructures are described in Chapters 5–8. Bifurcation routes to chaos, fractal dimension of strange attractor, circle map behaviors, controlling chaos and spatiotemporal dynamics of the current density filaments are discussed in detail.

I wish to thank E Schoell, W Prettl, F-J Niedernostheide, J Parisi, J Peinke, J Hirschinger and U Rau for their useful comments and discussions of early drafts of this book. I would express special thanks to Roger Cooper, commissioning editor of IOP, who gave me the opportunity to write this book. I also acknowledge the support of the editorial staff of IOP; KM Cantley, Victoria Le Billon and Ann Berne.

Kazunori Aoki
Kobe

Chapter 1

Introduction

1.1 What is chaos?

1.1.1 Logistic map

In the natural sciences there is a wide diversity of dynamical systems, ranging from the nano-scale, e.g., hydrogen atoms [1], to macroscopic size, e.g., the solar system [2], which exhibit instabilities leading to chaos or nonperiodic time evolutions. In condensed matter physics, typical examples which show chaotic behavior are turbulence in fluid dynamics [3–5], the Belousov–Zhabotinsky (B-Z) reaction in chemistry [6, 7], charge density wave in low-dimensional conductors [8, 9] and nonlinear carrier transport in semiconductors [10–12]. It can be quite surprising to know and recognize that the dynamical behaviors among different physical systems are unified to a universal class of chaos theory [13–17]. It is also surprising to note that even the common feature of the dynamical behaviors is characterized by a simple mathematical map (the so-called logistic map [18–20]).

The question may be asked: what is chaos? The word "chaos" was first used in science by Li and Yorke, who are mathematicians [18]. The Greek word "chaos" was originally a verb meaning to gape open and was often used to refer to the primeval emptiness of the universe before things came into being (*Encyclopedia Britannica*, Vol. 5, p. 276 [21]). More simply, it means: (i) utter disorder and confusion, and (ii) the unformed original state of the universe. In science, the word chaos means a nonperiodic oscillatory state (or solution) which stems from the nonlinear nature of the deterministic physical systems. The term deterministic chaos means that the system behaviors, which can be adequately modeled by a set of nonlinear differential equations, reveal seemingly stochastic time evolution (chaos) without any microscopic fluctuation or external noise embedded in the system. Nonlinear chaotic current in semiconductors, for instance, shows quite irregular oscillations in long-term evolution in an unpredictable

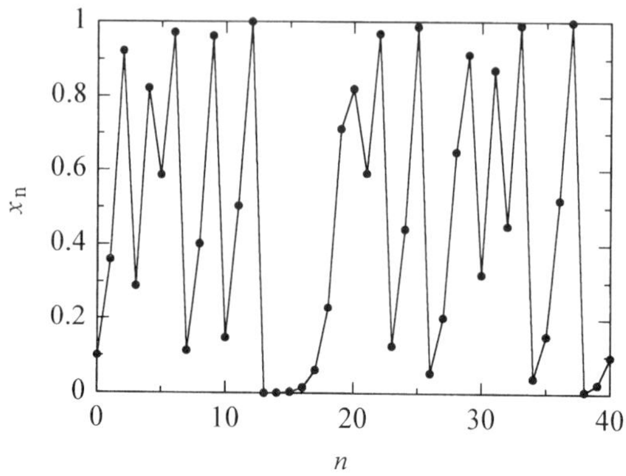

Figure 1.1 Chaotic output x_n vs n in the logistic map. The initial condition is $x_0 = 0.1$.

manner. At first sight, the current signals $x(t)$ like noise; but, on analyzing the signals by the methods used in chaos theory [13–17], they are frequently found to be deterministic chaos. Furthermore, a chaotic orbit reconstructed in the phase space forms a strange attractor, which topologically has a fractal structure [13–17].

One of the bifurcation routes from periodic oscillations to chaos is **period doubling bifurcation**, known as the **Feigenbaum scenario**. For a periodic oscillation, the peak current x_n takes a fixed value x^* for each period $(n = 1, 2, 3, \ldots)$, namely $\{x_n\} = \{x^*, x^*, x^*, \ldots\}$. The oscillatory state is called **period 1** with the period T. Increasing the applied voltage, for instance, period 1 bifurcates to **period 2**, where the peak current oscillates with the period $2T$ as $\{x_n\} = \{x_1^*, x_2^*, x_1^*, x_2^*, \ldots\}$. With a further increase in the applied voltage, period 2 bifurcates to **period 4** with the period $4T$, and so on. After the successive bifurcations, chaos finally emerges. Period doubling bifurcation is universal among many nonlinear dynamical systems.

As a prototypical example of mathematics, one may start with a one-dimensional (1D) iteration of the logistic map:

$$\begin{aligned} x_{n+1} &= f_u(x_n), \\ &= \mu x_n(1 - x_n), \quad (x_n \in [0, 1],\ n = 1, 2, 3, \ldots), \end{aligned} \tag{1.1}$$

where x_{n+1} (or x_n) is called the **order parameter** at discrete time $n + 1$(or n), and μ is the **control parameter**. The map is formally written as:

$$f_\mu : \quad I \to I, \tag{1.2}$$

where $I = [0, 1]$. Equation (1.1) was introduced by May for the population dynamics in biology [22] and become highly popular through Li and Yorke's literature [18], "Period three implies chaos". Feigenbaum presented a universal scaling law characterized by Feigenbaum constants [13, 16, 17] via the renormalization group theory [16, 17, 19]. Bifurcation routes to chaos (period doubling cascade, intermittency) predicted by the logistic map have been evidenced in many physical systems [13–17] (and also in nonlinear carrier transport in semiconductors [10–12]. The details are not discussed for those readers who are unfamiliar with chaos theory; instead, some essential features deduced by the logistic map are described. The numerical iteration of (1.1) is quite simple. Using a simple computer language, the map function $y = \mu x(1-x)$ can be calculated for a given initial value $x_0(y = x_1 = \mu x_0(1-x_0))$. For the second iteration, one puts $x = y$ and calculates $y(= x_2 = \mu x_1(1-x_1))$. The procedure is repeated up to an arbitrary number of iterations. Figure 1.1 shows the chaotic output x_n vs n for $x_0 = 0.1$. The mapping procedure is shown schematically in Fig. 1.2. First, an intersection between the parabolic curve $y = \mu x(1-x)$ and a vertical line $x = x_0$ (initial value) gives $y = x_1$. As the second step a horizontal line is drawn from the point (x_0, x_1) to a point (x_1, x_1) which is an interaction between $y = x$ and $y = x_1$. The mapping process is repeated again and again. The arrows in Fig. 1.2 show the direction of the forward mapping (1.1). The reader must exactly distinguish the directions between forward (Fig. 1.2) and reverse mapping (introduced later in Fig. 1.7). In Fig. 1.2 the intersections between $y = \mu x(1-x)$ and $y = x$ are $x^* = 0, 1 - 1/\mu$ which are called **fixed points**, where $\mu > 1$. A fixed point is stable if $|f'(x^*)| < 1$, and otherwise unstable, where the prime denotes the derivative $(\mathrm{d}/\mathrm{d}x)$. Since $f'(0) > 1$ for $\mu > 1$, $x^* = 0$ is an unstable fixed point. The fixed point $x^* = 1 - 1/\mu$ is stable if $1 < \mu \leq 3$, namely, $\lim_{n\to\infty} |x_n - x^*|$

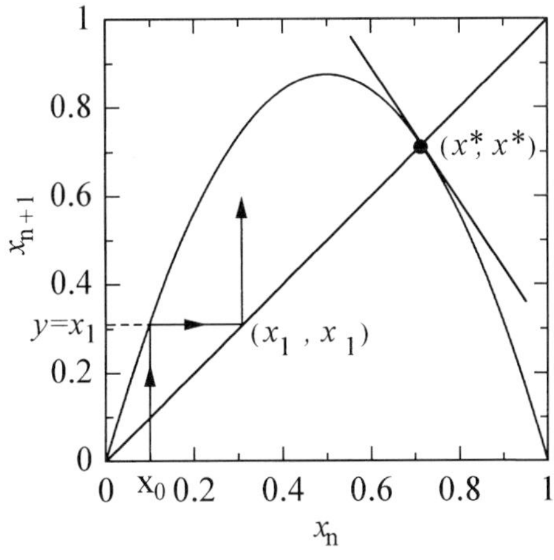

Figure 1.2 The mapping procedure for the logistic map. The point (x^*, x^*) is a fixed point, at which the slope of the logistic map is given by the solid line.

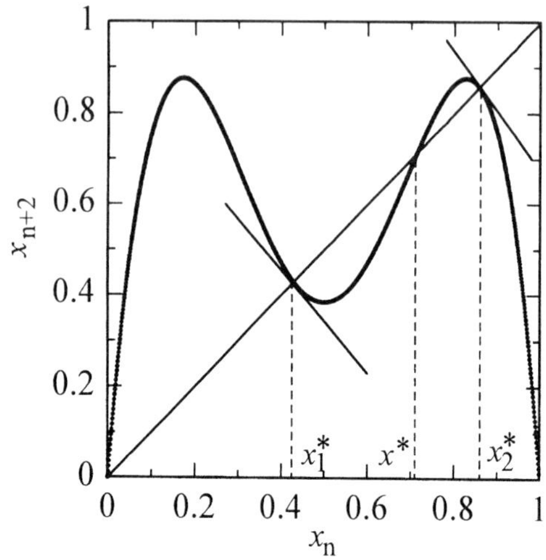

Figure 1.3 The mapping function $y = f(f(x))$, where $\mu = 3.5$. x^* is an unstable fixed point, and x_1^* and x_2^* are stable fixed points.

$\to 0(x^* = f(x^*), |f'(x^*)| < 1)$. Starting from an initial value x_0 in Fig. 1.2, the steady-state sequence $\{x_n\}_{n\to\infty}$ for (1.1) is $\{x^*x^* \ldots\}$, the sequence of which is called the "period 1 cycle". What happens if the control parameter is increased further ($\mu > \mu_1 = 3$)? Note that $x^* = 1 - 1/\mu$ is no longer a stable fixed point for $\mu > 3$ since now $|f'(x^*)| > 1$ Consider the map function $y = f(f(x)) = f^2(x)$, as given in Fig. 1.3. Then, the logistic map (1.1) is converted automatically to $x_{n+2} = f^2(x_n)$. The intersections with the line $y = x$ give four fixed points, which are solutions of

$$x^4 - 2x^3 + \frac{\mu + 1}{\mu}x^2 + \frac{1 - \mu^2}{\mu^3}x = 0. \tag{1.3}$$

Since $x^* = 0, 1 - 1/\mu$ are already the solutions of $x = f^2(x)$, (1.3) is decomposed to

$$x(x - 1 + 1/\mu)(x^2 + ax + b) = 0, \tag{1.4}$$

where $a = -(1 + 1/\mu)$ and $b = (\mu + 1)/\mu^2$. Then, from $x^2 + ax + b = 0$, one obtains the remainder solutions x_1^*, x_2^*:

$$x_{1,2}^* = \frac{1 + \mu \pm \sqrt{\mu^2 - 2\mu - 3}}{2\mu}, \tag{1.5}$$

where $x_1^* < x^* < x_2^*$. The fixed points $x_{1,2}^*$ are stable if $|f^{2'}(x_{1,2}^*)| < 1$. Using the relation $x_{1,2}^{*2} + ax_{1,2}^* + b = 0$, and after some calculations one has:

$$f^{2'}(x_{1,2}^*) = -\mu^2 + 2\mu + 4. \tag{1.6}$$

The condition $|f^{2'}(x^*_{1,2})| < 1$ imposes

$$3 < \mu \le 1 + \sqrt{6}. \tag{1.7}$$

With the parameter regime (1.7), one obtains stable steady-state iterations $x_1^* = f^2(x_1^*)$ and $x_2^* = f^2(x_2^*)$. Correspondingly, $x_2^* = f(x_1^*)$ and $x_1^* = f(x_2^*)$ in the original logistic map (1.1), and the sequence of steady-state iterations is $\{x_1^* x_2^* x_1^* x_2^* \ldots\}$, which is called the "period 2 cycle". For $\mu \ge \mu_1 = 3$, it is said that period 1 bifurcates to period 2. The period doubling bifurcation is shown schematically in Fig. 1.4. With $\mu = 1 + \sqrt{5}$, $f^{2'}(x^*_{1,2}) = 0$ and the period 2 cycle is called a superstable cycle [17], where

$$x_1^* = 1/2,\ x_2^* = \frac{3 + \sqrt{5}}{2(1 + \sqrt{5})}.$$

For $\mu > \mu_2 = 1 + \sqrt{6}, x^*_{1,2}$ become unstable, leading to the second period doubling. It gives rise to four stable fixed points $x^*_{\pm 1}, x^*_{\pm 2}$, where $x^*_{-1} < x^*_{+1} < x^*_{-2} < x^*_{+2}$. In the map function $y = f^4(x)$, the iterations are given by $x^*_{\pm 1} = f^4(x^*_{\pm 1})$, $x^*_{\pm 2} = f^4(x^*_{\pm 2})$. Correspondingly, $x^*_{+1} = f^2(x^*_{-1})$, $x^*_{+2} = f^2(x^*_{-2})$, and further $x^*_{-2} = f(x^*_{-1}), x_{+1} = f(x^*_{-2}), x_{+2} = f(x^*_{+1})$, $x_{-1} = f(x^*_{+2})$. The sequence $\{x^*_{-1} x^*_{-2} x^*_{+1} x^*_{+2} \ldots\}$ is termed the period 4 cycle. The period 4 cycle is stable in the range $\mu_2 < \mu \le \mu_3$. For $\mu_3 < \mu \le \mu_4$ the period 8 cycle appears and so on. The scheme of successive period doubling is called the Feigenbaum scenario [23]. The period 2^k cycle approaches asymptotically to an emergence of chaos with $k \to \infty$ at $\mu_\infty \sim 3.57$. Above μ_∞, chaos and many (or, strictly, an infinite number of) periodic windows appear. Here, a period-m window means a periodic orbit

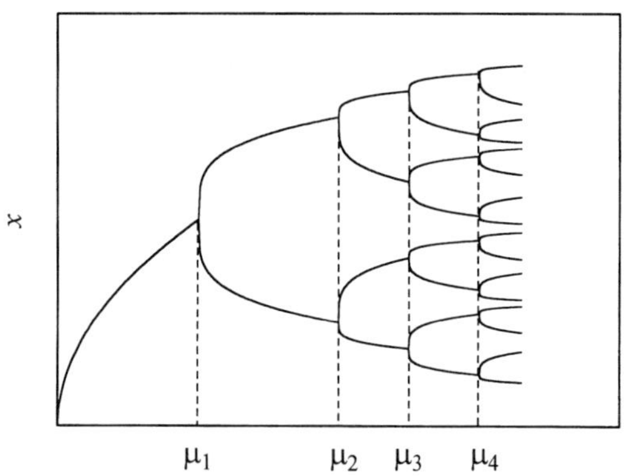

Figure 1.4 Pitchfork bifurcation (schematic).

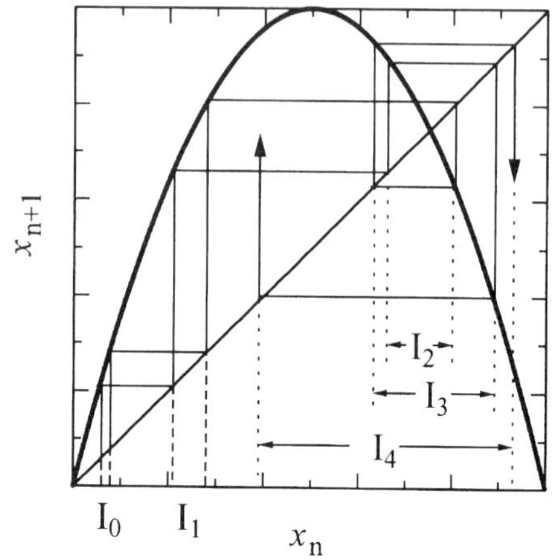

Figure 1.5 Sensitive dependence on the initial conditions.

which satisfies $x^* = f^m(x^*)$, where

$$\underbrace{f^m = f_\mu \circ f_\mu \circ \ldots \circ f_\mu}_{m \text{ times}} : I \to I. \tag{1.8}$$

Finally, the end point of chaos is reached at $\mu = 4$, where the chaos is fully developed in the unit interval I. Hereafter, fully developed chaos is considered for $\mu = 4$. Chaos is well characterized by sensitive dependence on initial conditions. Starting from the nearby initial values x_0, x'_0, the distance $|x - x'|$ diverges by successive iterations (see Fig. 1.5) because $|\mathrm{d}f(x)/\mathrm{d}x| > 1$ for $x \in [0, (1 - 1/\mu)/2)$ and $x \in ((1 + 1/\mu)/2, 1]$. As a result, x_n and x'_n have quite different values after several iterations even if the initial values are too close (e.g., $|x_0 - x'_0| = 10^{-10}$). So, after many iterations, x_n and x'_n, which were closely located at the beginning, lose their correlation completely. Namely, the chaotic orbit is **unpredictable**. Two chaotic orbits that started from two nearby initial values are shown in Fig. 1.6. Let I_n be an interval defined by $I_n = [\inf(x_n, x'_n), \sup(x_n, x'_n)]$ for the nth iteration, then iterations $I_{n+1} = f(I_n)$ experience overlaps among I_n and $I_m, I'_m, I''_m, \ldots$. For instance, in Fig. 1.5, note that $I_2 \subset I_3 \subset I_4$. The overlapping among intervals is termed mixing. Chaotic orbit loses its passed memory successively during its itinerancy in the unit interval, showing unpredictability.

More precisely, consider loss of memory. Assume that a system (or computer) is allowed to have memory of initial value x_0 and the latest i values of iterated points $x_{n-i}, x_{n-i+1}, \ldots, x_{n-1}$ (where i is fixed). Then, the computer will have lost the memory of previously iterated points $x_1, x_2, \ldots, x_{n-i-1}$ owing to the memory limit. Now, ask: is it always possible to reach the initial value of x_0, starting from x_{n-i} by the reverse map: $x = f^{-1}(y)$? The answer is apparently "no". As drawn schematically in

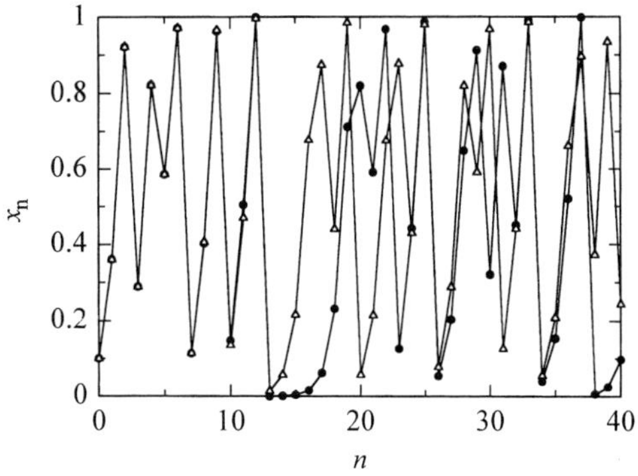

Figure 1.6 Two chaotic orbits initiated from two nearby initial values. The orbits begin to deviate from each other at $n = 13$.

Fig. 1.7, there are two kinds of reverse map:

$$f_R^{-1}: \quad x_j = f_R^{-1}(x_{j+1}) = (1 + \sqrt{1 - 4x_{j+1}/\mu})/2, \tag{1.9}$$

$$f_L^{-1}: \quad x_j = f_L^{-1}(x_{j+1}) = (1 - \sqrt{1 - 4x_{j+1}/\mu})/2, \tag{1.10}$$

where f_R^{-1} and f_L^{-1} are, respectively, right and left reverse mapping, and $j = n - i,\ n - i - 1, \ldots, 0$. The time reversal processes are apparently probabilistic since the memory has been lost: one may choose f_R^{-1} or f_L^{-1} (or simply R or L) at random (like a coin toss) for each reverse mapping with an equal probability of 1/2. There are 2^{i+1} different branches reaching

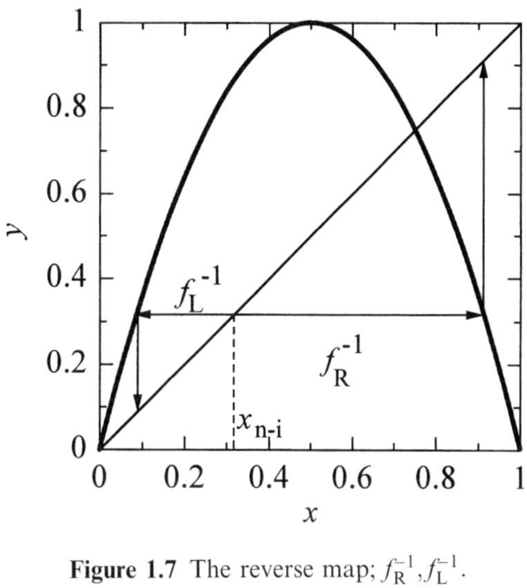

Figure 1.7 The reverse map; f_R^{-1}, f_L^{-1}.

2^{i+1} different initial values x_0^*, among which only one reverse mapping, say RLLRLRRRL ..., hits the specified correct value x_0. The probability of reaching the correct value is thus $1/2 \times \ldots \times 1/2 = 2^{-(i+1)}$. The time reversal process is just like a **Markov process** [24].

1.1.2 Ergodic theorem

From logistic map (1.1), chaos can be seen to be quite irregular and unpredictable, although it is deterministic by (1.1), showing sensitive dependence on initial conditions (cf. Fig. 1.6). A chaotic orbit loses memory about its passed trajectory, as evidenced by the reverse mapping. Nonlinear currents in semiconductors exhibit such chaotic behavior. It should be remembered that this discussion is about dissipative nonlinear processes (or dissipative physical systems). Current in semiconductors is a prototypical example of dissipative flow via resistive electronic conduction (i.e., joule heating). A counter example is conservative classical mechanics [17, 25], e.g., a particle in a stadium. Turbulence in the fluid system, chemical turbulence and electronic turbulence in semiconductors may be a direct result of hierarchical complexity of the "**dissipative structures**" introduced by Nicolis and Prigogine [26] and are certainly related to loss of information in chaos theory [17]. Concerning the stochasticity of chaos, there is a universal rule, i.e., an ergodic theorem which asserts that a time average is equal to a spatial average. The theorem has been used traditionally in statistical thermodynamics [27]: the ensemble average of a physical quantity at time t over a whole system (phase space) which is composed of numerous subsystems equals the time average of the microscopically fluctuating quantity at a specified subsystem. Let $\rho(x)$ be probability density (the **invariant probability measure**, as given below), then applying this to the chaos system, the ergodic theorem for every continuous function can be read as [28]:

$$\lim_{T\to\infty} \frac{1}{T}\int_0^T \phi(f^t(x(0)))\mathrm{d}t = \int \rho(x')\phi(x')dx' \tag{1.11}$$

for any initial conditions $x(0)$. In the logistic map, one may choose ϕ as:

$$\phi(x') = \begin{cases} 1 & x' \in [x - \frac{1}{2}\varepsilon, x + \frac{1}{2}\varepsilon], \\ 0 & \text{otherwise,} \end{cases} \tag{1.12}$$

where ε is a small quantity ($\varepsilon \ll 1$). For the iterative map, the time integral $\int \mathrm{d}t$ should be replaced by the sum $\sum_i$ over the discretized time. By taking an arbitrarily chosen initial value x_0 and applying theorem (1.11), one obtains:

$$\rho(x) = \lim_{\varepsilon \to 0} \lim_{N \to \infty} \frac{1}{N} \sum_{j=0}^{N} \phi(f^j(x_0)). \tag{1.13}$$

Thus, the invariant probability measure is a probability of how many times the trajectory visits a specified small bin $I_\varepsilon = [x - \frac{1}{2}\varepsilon, x + \frac{1}{2}\varepsilon]$ during long-term iterations. If $\phi(x) = 1$ is taken for any $x \epsilon [0, 1]$, then $\int \rho \, dx = 1$ from (1.13). The ergodic theorem holds also for random reverse mapping [replace $f^i(x_0)$ by $f^{-j}(x_{n+1})$ and apply (1.9), (1.10) in the right-hand side (r.h.s.) of (1.11)], where n is an arbitrarily large integer. In Fig. 1.8, the invariant probability measure is shown for both the forward and reverse mapping.

1.1.3 Chaos include many unstable periodic states

If $f^m(x_n) = x_n$ for $n = 0, 1, 2, 3 \ldots$, then the orbit $\{x_n\}$ shows periodic and regular motion with period m. Chaos appears as a result of bifurcations from periodic states by varying the control parameter μ. Chaos does not show any periodic motions stably, but careful examination suggests the existence of many unstable periodic orbits. Using regular reverse mapping, any kind of unstable periodic orbits can be constructed artificially in the chaotic regime, which are asymptotically stable only for the reverse mapping. For instance, a regular reverse mapping of

$$f_{\mathrm{R}}^{-m}(f_{\mathrm{L}}^{-n}(x)) : \overbrace{\mathrm{LLL} \ldots \mathrm{L}}^{n} \overbrace{\mathrm{RRR} \ldots \mathrm{R}}^{m}$$

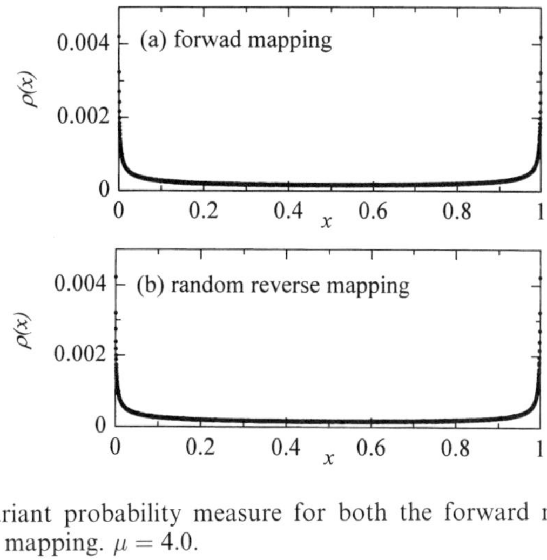

Figure 1.8 Invariant probability measure for both the forward mapping and the random reverse mapping. $\mu = 4.0$.

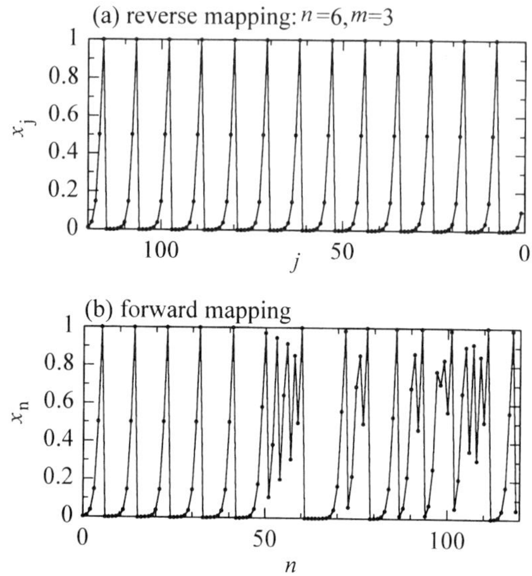

Figure 1.9 (a) Reverse mapping $f_R^{-m}(f_L^{-n}(x))$ for $n = 6$ and $m = 3$, and (b) the orbit by forward mapping, using an initial value which was the final value obtained by reverse mapping.

(repetition of n times f_L^{-1} mapping plus m times f_R^{-1} mapping) approaches asymptotically to a periodic state of period $m + n$, for any integers m and n. An example for $n = 6, m = 3$ is shown in Fig. 1.9(a). The asymptotic orbit is never stable for the forward mapping $x_{n+1} = f(x_n)$, even when the final value obtained by the reverse mapping is used as an initial value for the forward mapping (Fig. 1.9b). This results from its sensitive dependence on the initial conditions (note that computer round-off errors occur even with double precision).

The appearance of unstable orbits by reverse mapping reflects all of the histories of bifurcations that the logistic map has experienced for $\mu < 4$. The finding may be related to Sarkovskii's famous theorem [17, 29], which will be mentioned briefly in chapter 4.

1.1.4 Entropy positive

Since chaos is unpredictable in its future evolution, it may be described as seemingly stochastic, although it is an exactly deterministic phenomenon. Since the future cannot be predicted exactly from present data or from a set of previous data, a probabilistic concept such as an invariant probability measure is highly acceptable. So, it is naturally conjectured that some kinds of entropy, if defined, may be positive for chaotic behavior. By analogy with thermodynamics [24, 27], periodic states correspond to a completely ordered state at zero temperature (entropy zero) and chaos corresponds to a thermal equilibrium of a canonical ensemble with a definite positive entropy. Starting from an unstable periodic state as an initial condition (approximately zero entropy, cf. Fig. 1.9a), the orbit by forward mapping soon becomes chaotic with a definite positive entropy, which is reminiscent of the "**law of increasing entropy**". Thus, even when starting from an ordered state, the orbit begins quickly to itinerate in the chaotic world with positive entropy (Fig. 1.9b). The entropy in chaos theory [Kolmogorov (K) entropy] is related to a loss of information. For the logistic map in Fig. 1.10, one can see more precisely how the information loss can be related to the divergence rate of the chaotic trajectories or the Lyapunov exponent. By dividing a small interval $[x, x + \Delta x]$ into n bins, one obtains $K_1^* = \ln z_1$ ($z_1 = n$, density of states). After one iteration, the small interval is mapped onto an interval $[f(x), f(x + \Delta x)]$ of which the density of states is given by $z_2 = n|\mathrm{d}f/\mathrm{d}x|$, then $K_2^* = \ln z_2 = \ln z_1(|\mathrm{d}f/\mathrm{d}x|)$. Therefore, the loss of information is obtained as $K^* = \ln(|\mathrm{d}f/\mathrm{d}x|)$. From the ergodic theory (1.11), the K-entropy is given by:

$$\int_0^1 \rho(x)(\ln(|\mathrm{d}f/\mathrm{d}x|)\mathrm{d}x = \lim_{N\to\infty}(1/N)\sum_{j=0}^{N} \ln|\mathrm{d}f(x_j)/\mathrm{d}x|, \tag{1.14}$$

$$= \langle \ln|\mathrm{d}f(x)/\mathrm{d}x|\rangle = \langle\lambda_j\rangle = \lambda, \tag{1.15}$$

where $x_{j+1} = f(x_j)(j = 0, 1, 2, \ldots)$ and

$$\mathrm{d}f(x_j)/\mathrm{d}x = \lim_{\Delta x\to 0}[f(x_j') - f(x_j)]/\Delta x \quad (\Delta x = |x_j' - x_j|).$$

Since the distance between two nearby trajectories $|x_j' - x_j|$ diverges as

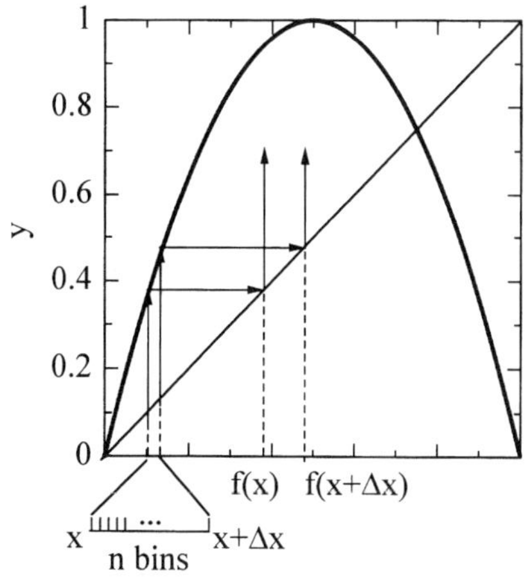

Figure 1.10 Divergence rate and information loss in the logistic map.

$|f(x'_j) - f(x_j)| = \exp(\lambda_j)|x'_j - x_j|$ after one iteration, (1.15) indicates the mean of the divergence rate during long-term iterations. In (1.15), λ is a Lyapunov exponent in the one-dimensional case.

1.2 Universality and individuality

1.2.1 From universality

In Section 1.1, the logistic map was introduced as a simple mathematical model to show some essential features of chaos. The behaviors of the logistic map have been observed in numerous nonlinear physical systems, as given in Table 1.1. Let $X_1, X_2, \ldots, X_m$ be order parameters (or simply independent variables) and $\mu_1, \mu_2, \ldots, \mu_k$ be control parameters, then the dynamical process of a physical system is given by partial differential equations (PDEs):

$$\partial \mathbf{X}/\partial t = \mathbf{F}(\mathbf{X}; \{\nabla X_i\}; \{\nabla^2 X_i\}; \{\partial X_i/\partial t\}; r; t; \{\mu_j\}), \qquad (1.16)$$

where $i = 1, 2, \ldots, m$, $j = 1, 2, \ldots, k$, $\mathbf{X} = \{X_1, X_2, \ldots, X_m\}$, $\mathbf{F} = \{F_1, F_2, \ldots, F_m\}$, ∇ and ∇^2 are spatial differential operators (nabla and Laplacian) and r is a spatial coordinate (e.g., Cartesian coordinates $\{x, y, z\}$). The notations used in (1.16) follow [26]. If the functions $F_i s$ are independent of spatial coordinates and the operators (namely, $F(\mathbf{X}; \{\partial X_i/\partial t\}; t; \{\mu_j\})$), then degrees of freedom d_f of the system are $m + 1$ (note: m independent variables plus time variable). When the system depends explicitly on time t, e.g., via periodic driving force $A\cos(\omega t)$, the system is said to be non-autonomous (the system functions are given by

Table 1.1 Classification of nonlinear dynamical systems by categories A and B

Category A	Ref.	Category B	Ref.
Forced pendulum	[30, 31]	Fluid dynamics	[3–5, 36]
Mechanical vibration	[21, 32]	Semiconductor carrier transport	[10–12]
Nonlinear circuits (varactor diode, Chua circuit, etc.)	[33–35]	Laser, optical bistability	[37, 38]
		Chemical reaction (e.g., B-Z reaction)	[6, 7]
		Josephson junction	[39, 40]
		Charge density wave	[8, 9]

$F(\mathbf{X}; \{dX_i/dt\}; t; \{\mu_j\})$; $d_f = m + 1$), otherwise the system is an autonomous one, where $A = 0$, for which functions are given by $F(\mathbf{X}; \{dX_i/dt\}; \{\mu_j\})$ with the degrees of freedom of $d_f = m$.

Dynamical systems of category A in Table 1.1 are nonautonomous, with $d_f = m + 1$, of which dynamical behaviors are not spatially extended. Category B lists the dissipative physical systems which show spatially extended dynamical behaviors and chaos, and/or spatially extended nonequilibrium phase transition. A large part of the dynamical systems in category B exhibits spatial patterns which may evolve to turbulent states (Fig. 1.11) [36]. The degrees of freedom are essentially quite large (e.g., $\approx 10^{23}$ in the fluid flow system) for the spatially extended dissipative systems because of the spatial degrees of freedom. The turbulent patterns may be simulated by a computer. However, in order to study the bifurcation phenomenon to chaos, the diffusion term(s) are often neglected for economical reasons;

$$d\mathbf{X}/dt = \mathbf{F}(\{X_i\}; \{dX_i/dt\}; t; \{\mu_j\}). \tag{1.17}$$

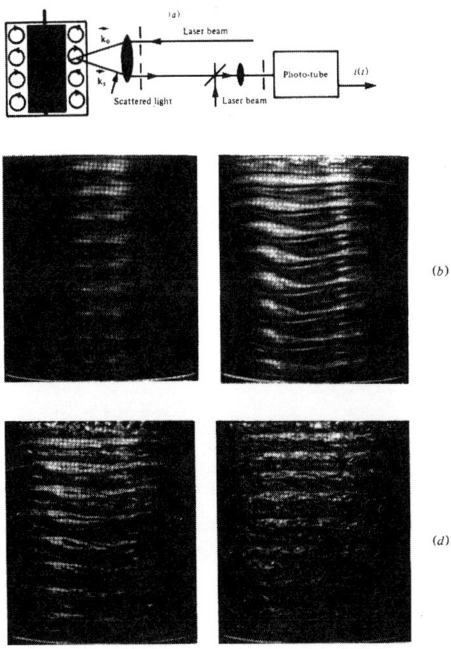

Figure 1.11 Regular and turbulent pattern in the Taylor vortex flow. (a) The Reynolds number: $R/R_c = 1.1$. Time-independent Taylor vortex flow with 18 vortices; (b) $R/R_c = 6.0$. (c) $R/R_c = 16.0$; and (d) $R/R_c = 23.5$. Parts (b) and (c) illustrate wavy vortex flow, while in the turbulent pattern in (d) the waves have disappeared. [36].

In semiconductors, for instance, filamentary current flows between ohmic contacts, and the output current frequently shows chaos under a certain parameter regime $\{\mu_j\}$. A reduced model (impact ionization model [10, 11] which neglects carrier diffusion) can reproduce well the measured time series of the nonlinear currents in semiconductors. Similar situations appear in other physical systems of category B.

All of the dissipative physical systems listed in Table 1.1 show universality of bifurcation and chaos as partly predicted by the logistic map (1.1). Among them, the dynamical behaviors uncover: (i) anharmonic vibrations (far from harmonic vibrations); (ii) nonequilibrium states far from thermodynamical equilibrium; (iii) instability around an unstable saddle point; and (iv) the existence of homoclinic points. The dynamical behaviors of many physical systems (driven pendulum, nonlinear circuits, charge density wave and Josephson junction in Table 1.1) are ruled by the anharmonic equation of motions:

$$\ddot{X} + \gamma(X)\dot{X} + \partial V/\partial X = f(t), \tag{1.18}$$

where X is the order parameter, $\gamma(X)$ is the friction coefficient, V is the potential energy function ($-\partial V/\partial X$: the restoring force), $f(t)$ is the external driving force (e.g., $A\sin\omega t$) and dots indicate the time derivatives ($\dot{X} = \mathrm{d}X/\mathrm{d}t$, $\ddot{X} = \mathrm{d}^2X/\mathrm{d}t^2$).

A driven pendulum is the most famous example used to demonstrate how the unstable saddle point plays an important role in chaotic motion. The equation of motions is given by:

$$\dot{\theta} + \gamma\dot{\theta} + \omega_0^2 \sin\theta = A\cos\omega t, \tag{1.19}$$

where θ is the angle of the pendulum and ω_0 is the natural frequency of the pendulum. By putting $x = \theta,\ y = \dot{\theta}$ and $z = \omega t$, the nonautonomous system is formally converted to the autonomous system with $d_f = 3$;

$$\begin{aligned} \dot{x} &= y, \\ \dot{y} &= -\gamma y - \omega_0^2 \sin x + A\cos z, \\ \dot{z} &= \omega \end{aligned} \tag{1.20}$$

A necessary condition for the chaotic oscillation is $A \neq 0$ and $\omega \neq 0$. Figure 1.12 (a–c) shows trajectories (phase portraits) in the phase space $(\theta, \dot{\theta})$ for the different parameter regimes: (a) $\gamma = \mathrm{A} = 0$ (undamped case without driving force), (b) $\gamma > 0,\ A = 0$ (damped case without driving force), and (c) $\gamma \neq 0,\ A \neq 0$ (damped case with driving force). For parameter regime (a) the system is conservative with the constant energy $H = -\omega_0^2\cos\theta + \dot{\theta}^2/2 = -\omega_0^2\cos x + y^2/2 = \text{const.}$ (note: $\dot{H} = 0$). The trajectories are determined uniquely by the initial conditions $(\theta_0, \dot{\theta}_0)$: (i) harmonic (or anharmonic) oscillations, e.g., for $\dot{\theta}_0 = 0,\ |\theta_0| \ll \pi$ (or $0 \lesssim |\theta_0| \lesssim \pi$), (ii) clockwise rotation for $\dot{\theta}_0 > 0, |\theta_0| = \pi$, and (iii) anticlockwise rotation for

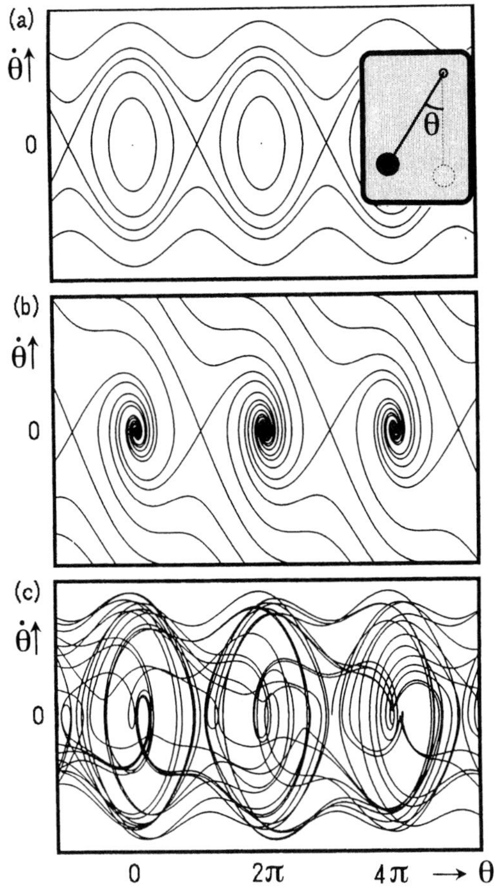

Figure 1.12 Phase portraits for the driven pendulum with the different parameter regimes: (a) $\gamma = A = 0$, (b) $\gamma > 0,\ A = 0$, and (c) $\gamma \neq 0,\ A \neq 0$.

$\dot{\theta}_0 < 0, |\theta_0| = \pi$. Trajectories which pass through the unstable saddle points ($\dot{\theta} = 0, \theta = n\pi$; n is odd) are called separatrices. For the parameter regime of $\gamma > 0$ and $A = 0$, the system is dissipative with the total energy $H = -\omega_0^2 \cos\theta + \dot{\theta}^2/2$ $(\dot{H} = -\gamma(\dot{\theta})^2 < 0)$. Any trajectories which start from arbitrarily chosen initial values $(\theta_0, \dot{\theta}_0)$ fall into one of the steady states ($\dot{\theta} = 0,\ \theta = \pm\ n\pi$; $n = 0, 2, 4\ldots$), as shown in Fig. 1.12b. By driving the pendulum with a periodic force $A \cos \omega t$, one can observe chaotic oscillation (Fig. 1.12c). The trajectory in Fig. 1.12c is explained by the random phase space diffusion of the chaotic trajectory, or strongly anharmonic oscillation from a confined domain of $(n-1)\pi \lesssim \theta \lesssim n\pi$ (n is odd) to the adjacent domain of either $n\pi \lesssim \theta \lesssim (n+1)\pi$ or $(n-2)\pi \lesssim \theta \lesssim (n-1)\pi$. The diffusion occurs near the saddle points (Fig. 1.12c) at which the pendulum seems to cease its motion (remember that $\dot{\theta} = \ddot{\theta} = 0$ at the saddle point when $A = 0$).

Table 1.2 summarizes the friction coefficients $\gamma(X)$, the restoring forces $-\partial V(X)/\partial X$ and the potential energy functions of the model equation (1.18) for various physical systems. In Fig. 1.13 a simulated chaotic trajectory is shown in the phase space $(X, \dot{X})$ for a particle moving in a double-well potential, where the equation of motion is called the Duffing equation (Table 1.2) [41, 42]. Many nonlinear electronic circuits exhibit (or might potentially exhibit) chaos, the most well known of which are the Van der Pol oscillator [43], varactor diode [33, 34] and Chua circuit [35].

For instance, a nonlinear circuit which is composed of a Si p-n junction diode, an inductor L and a resistor R shows chaos when it is driven sinusoidally by an ac bias $V_{ac} \sin \omega t$ (Fig. 1.14a). The equation of motion of charge q in the p-n junction driven damped oscillator is given by

$$\ddot{q} + \gamma(q)\dot{q} - f(q) = A_{ac} \sin \omega t, \tag{1.21}$$

where the nonlinear damping coefficient $\gamma(q)$ is given by $\gamma(q) = \mathrm{R/L} + (\partial I_d/\partial v)/C(q)$, the restoring force by $f(q) = -[v(q) + \mathrm{R}I_d(v(q))]/\mathrm{L}$, I_d is the conduction current of the diode, and $v(q)$ is the junction voltage [34]. Buskirk and Jeffries [34] simplified (1.21) by assuming $\gamma(q) = \gamma = \text{const.}$, and $f(q) = -e^q + 1$:

$$\ddot{q} + \gamma\dot{q} + e^q - 1 = A \sin t. \tag{1.22}$$

According to the model equation (1.22) after Buskirk and Jeffries, one can easily reproduce the chaotic phase trajectories (Fig. 1.14b).

Table 1.2 Friction coefficients, restoring force and potential functions in various dissipative physical systems

Nonlinear system	$\gamma(X)$	$-\partial V/\partial X$	V
Driven pendulum Josephson junction Charge density wave	γ	$-\omega_0^2 \sin X$	$-\omega_0^2 \cos X$
Van del Pol oscillator	$\gamma(X^2 - 1)$	$-X$	$X^2/2$
Varactor diode, p-n junction	$\approx \gamma$	$\cong -\exp(X) + 1$	$\exp(X) - X$
Mechanical vibrations particle in a double-well potential	γ	$-X^3 + X$	$X^4/4 - X^2/2$

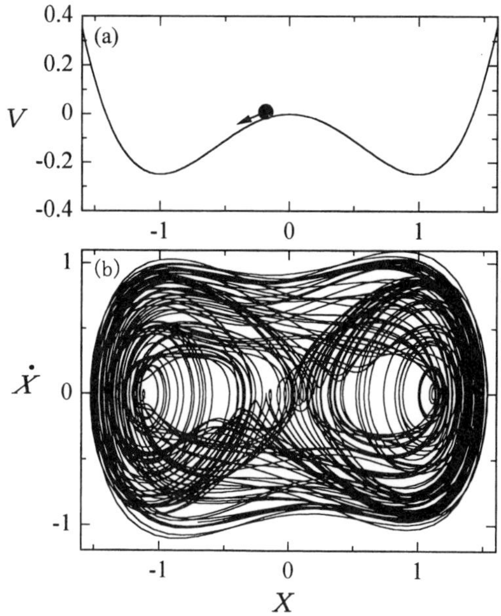

Figure 1.13 Chaotic trajectory in the phase space $(X, \dot{X})$ for a particle moving in a double well potential.

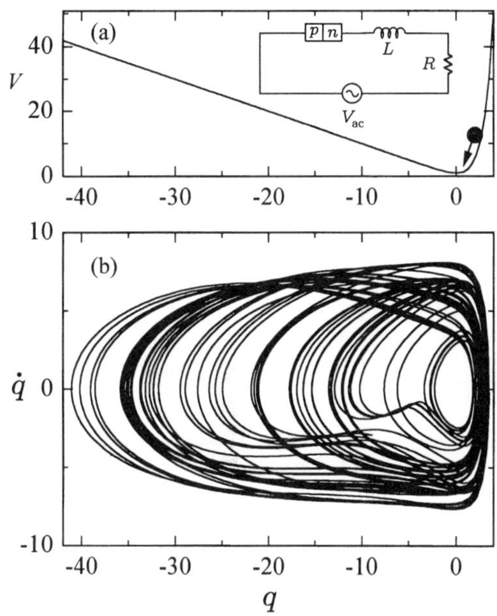

Figure 1.14 (a) The potential function V as a function of q. The inset shows the nonlinear circuit which is composed of a Si p-n diode, an inductor L and a resistor R: (b) chaotic response in the phase space.

Here, the system equation modeled for the driven Josephson junctions (the resistivity shunted junction model [39, 40]) is simply noted:

$$C\dot{V} + V/R + I_c \sin\ \phi = I_0 + I_r \cos\ \omega t, \tag{1.23}$$

where I_c is the critical current, I_0 the dc bias, I_r the rf current, C the capacitance and R the resistance, and the phase difference ϕ across the junction is given by the Josephson equation $\dot{\phi} = 2eV/\hbar$. When $I_0 = 0$, the model equation coincides exactly with the equation of motion for the driven pendulum (1.19).

In the semiconductor carrier transport, a truncated model for the impact-ionization avalanche [10] can be written as

$$\dot{n} + \frac{\partial V_E}{\partial n} = 0, \tag{1.24}$$

$$\dot{E} = -\gamma(E - E_0 + CJ), \tag{1.25}$$

$$\frac{\partial V_E}{\partial n} = an^3 + b(E)n^2 + c(E)n + d, \tag{1.26}$$

where n is the carrier density, E the electric field, E_0 the dc bias, γ the inverse relaxation time, J the current density, a–d are the coefficients due to the generation–recombination process between free carriers and the impurity ground state, and C is a constant. The potential function V_E is field dependent and shows double well potential in some range of E. Consequently, the J–E characteristic is S-shaped, as shown in Fig. 1.15. Since the term $\ddot{n}$ is absent in (1.24), the system is strongly dissipative or overdamped compared with (1.18). For a load line as given in Fig. 1.15, the system shows a spontaneous current oscillation but in absence of chaos because of $d_f = 2$. The multilevel impact ionization model presented by

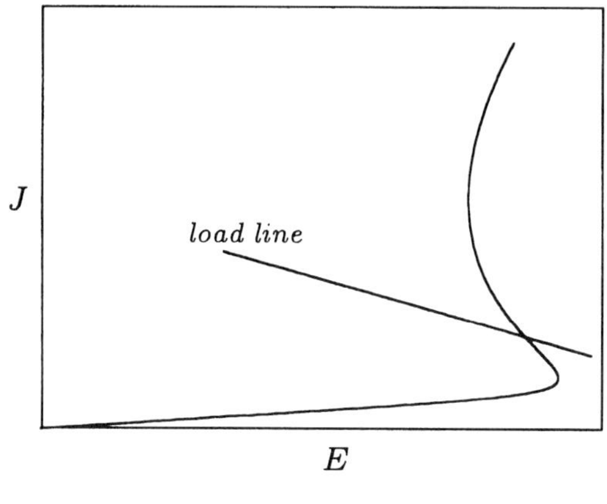

Figure 1.15 S-shaped $J - E$ characteristic in the semiconductor system.

Schöll [11] gives chaotic solutions, where $d_f = 3$. In general, any autonomous system can be converted to a nonautonomous system simply applying a driving force. In (1.24)–(1.26), one can find chaotic solutions replacing the dc bias E_0 by $E_0 + E_{ac} \sin(2\pi f_0 t)$.

Why does universality appear among different physical systems? This questions may not be answered easily from the model equations alone since the dynamical behaviors of certain physical systems are simulated only by computation and examined by the experiments. One may say that the nonlinearity itself guarantees the logistic map behaviors. Alternatively, the Poincaré surface of a section of the phase space trajectory forms topologically a round maximum function or unimodal function like in the logistic map. But why is this so? The simplest orbit is a closed orbit (period 1 cycle) or a sinusoidal oscillation. When the system bifurcates smoothly to a new oscillatory state how is this possible?

Suppose that a periodic orbit bifurcates smoothly to a new oscillatory state when a physical parameter μ exceeds a critical value μ_1. From an argument of the topology in the phase space, the possible orbit after the bifurcation may lie on either a Möbius band or a torus, as shown in Fig. 1.16, where the former gives rise to period doubling and the latter a quasi-periodic orbit. The smooth bifurcation imposes that the phase space topologies both approach asymptotically to the closed orbit with $\mu \to \mu_1$, where $\mu > \mu_1$. The Poincaré surface of a section is a set of intersections between the orbit and a fixed plane in Fig. 1.16, and it constructs a two-dimensional map: $u_{n+1} = g(u_n, v_n)$; $v_{n+1} = h(u_n, v_n)$, where the origin of the conventional coordinate (u, v) is placed at the fixed point $\overrightarrow{X}^*$ on the plane. Note that $\overrightarrow{X}^*$ is measured by the original coordinate system. Stability of the

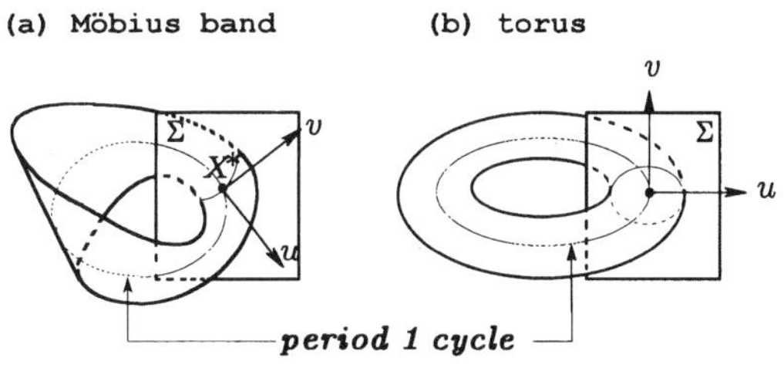

Figure 1.16 (a) The Period-1 cycle lies on the Möbius band. The unique orbit penetrates the generalized surface Σ by a unique point which is called the Poincaré surface of a section. The origin of the coordinates u, v is settled on the fixed point X^*. Note that the u-axis is perpendicular to the surface of the Möbius band, while the v-axis directs along the surface, where both the u- and v-axes lie on Σ. Suppose that the orbit of the period-2 cycle formed by the period doubling bifurcation lies at the edge of the Möbius band. (b) Period-1 cycle inside the torus. The quasi-periodic orbit lies on the surface of the torus.

fixed point $\overrightarrow{X}^* = f_\mu^T(\overrightarrow{X}^*)$ is discussed first by the linear map:

$$\begin{pmatrix} u_{n+1} \\ v_{n+1} \end{pmatrix} = \begin{pmatrix} -1 & 0 \\ 0 & -1 \end{pmatrix} \begin{pmatrix} a & 0 \\ 0 & d \end{pmatrix} \begin{pmatrix} u_n \\ v_n \end{pmatrix}, \tag{1.27}$$

where $a > 0$, $d > 0$. For $a < 1$, $d < 1 (\mu < \mu_1)$, $\overrightarrow{X}^*$ is a stable fixed point, giving a period 1 cycle. Note that for the iterations $u_{n+1} = -au_n$; $v_{n+1} = -dv_n$ the phase space point (u_n, v_n) approaches the origin (0, 0) for any initial conditions (u_0, v_0). If $a < 1$, $d \geq 1 (\mu \geq \mu_1) \overrightarrow{X}^*$ becomes an unstable saddle point (see Fig. 1.17a) since $u_n \to 0$, $v_n \to \pm\infty$. In the nonlinear dynamical systems, one can introduce the nonlinear map instead of (1.27):

$$\begin{pmatrix} u_{n+1} \\ v_{n+1} \end{pmatrix} = \begin{pmatrix} -1 & 0 \\ 0 & -1 \end{pmatrix} \begin{pmatrix} a & 0 \\ 0 & d - \alpha v_n^2 \end{pmatrix} \begin{pmatrix} u_n \\ v_n \end{pmatrix} \tag{1.28}$$

where $\alpha > 0$. By doing so, there is an unstable saddle $(0, 0)$ and two stable fixed points $(0, v_{1,2}^*)$ for $a < 1, d > 1$, as schematically shown in Fig. 1.17b, where

$$v_1^* = \sqrt{\frac{d-1}{\alpha}}, \quad v_2^* = -\sqrt{\frac{d-1}{\alpha}}.$$

The iterations $v_2^* = h(v_1^*), v_1^* = h(v_2^*)$ mean a period 2 cycle, and subsequently there is period doubling:

$$\overrightarrow{X}_2^* = f_\mu^T(\overrightarrow{X}_1^*), \quad \overrightarrow{X}_1^* = f_\mu^T(\overrightarrow{X}_2^*).$$

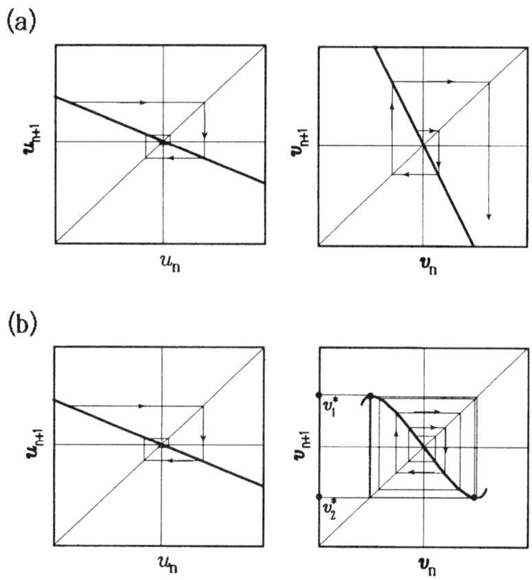

Figure 1.17 (a) The linearized maps for u_n and v_n. (b) The map function for (1.28).

In (1.28) the smooth bifurcation is confirmed since

$$v_1^*, v_2^* \to 0 \quad (\overrightarrow{X}_1^*, \overrightarrow{X}_2^* \to \overrightarrow{X}^*)$$

when $d \to 1$.

The above discussion may be summarized as follows. The nonlinear dynamical system modeled by a set of continuous differential equations (1.17) can be converted to the map system $u_{n+1} = g(u_n, v_n)$, and $v_{n+1} = h(u_n, v_n)$ by taking the Poincaré surface of a section. The map functions g, h can be expanded to the Taylor series around the fixed point $(0, 0)$:

$$g(u, v) = \sum_{k=1}^{\infty} \frac{1}{k!} \sum_{i=0}^{i=k} \frac{\partial^k g}{\partial u^i \partial v^{k-i}} |u^i v^{k-i}, \tag{1.29}$$

$$h(u, v) = \sum_{k=1}^{\infty} \frac{1}{k!} \sum_{i=0}^{i=k} \frac{\partial^k h}{\partial u^i \partial v^{k-i}} |u^i v^{k-i}. \tag{1.30}$$

If the dynamical system contains the significant nonlinearity of the order u^k or $v^k (k \geq 3)$ in the map functions, the map behavior near the bifurcation point may be well characterized by an approximated map behavior (1.28). Thus, many nonlinear dynamical systems may be likely (potentially) to exhibit period doubling as a universal bifurcation scheme. The topology of the Möbius band is confirmed in (1.28) by the π-rotation matrix:

$$\begin{pmatrix} -1 & 0 \\ 0 & -1 \end{pmatrix}. \tag{1.31}$$

Another bifurcation scheme is the Hopf bifurcation (Fig. 1.16b). The iteration map is described by the polar coordinate (r, θ):

$$r_{n+1} = ar_n - br_n^3, \tag{1.32}$$

$$\theta_{n+1} = \theta_n + \Omega, \tag{1.33}$$

where $r^2 = u^2 + v^2, \theta = \tan^{-1}(v/u)$, $a, b > 0$, and Ω is assumed to be an irrational number. When $a < 1$ $(\mu < \mu_1)$ then $r^* = 0$ $(u^* = v^* = 0)$ from $r^*(1 - a + br^{*2}) = 0$, and the solution is a stable period 1 cycle. If $a \geq 1$ $(\mu \geq \mu_1)$, then $r_1^* = 0$ (unstable) and

$$r_2^* = \sqrt{\frac{a-1}{b}}.$$

Then, the steady-state solution is given by:

$$r_2^* = \sqrt{\frac{a-1}{b}}, \tag{1.34}$$

$$\theta_{n+1} = \theta_n + \Omega \tag{1.35}$$

The coordinate (r_2^*, θ_n) now indicates rotation on a circle with radius r_2^*

giving the Poincaré surface of a section of a torus. Again, the smooth bifurcation is apparent, as $r_2^* \to 0$ when $a \to 1$. For the map behavior of the Hopf bifurcation route to chaos, the dissipative circle map is traditionally used (see chapter 6). Various experimental observations have proved the universality:

- **period doubling bifurcation:** the fundamental period of oscillation doubles repeatedly by varying a control parameter. The bifurcation sequence is period $1 \to$ period $2 \to$ period $4 \to \ldots \to$ period $2^n (n \to \infty)$. Chaos appears above a critical value of the control parameter at which the oscillation period is 2^∞
- **quasi-periodic route to chaos:** a periodic oscillation with the fundamental frequency f_0 bifurcates to a torus (quasi-periodic oscillation) with two incommensurate frequencies f_0 and $f_1 (f_1/f_0$: irrational). The torus further bifurcates to chaos by varying the control parameter
- **intermittency:** at a certain value of the control parameter, periodic oscillation experiences sudden bursts.

This universality has had an impact on the fields of condensed-matter physics. Furthermore, chaos theory now prevails in every area of the "nonlinear world", in which chaotic behaviors were previously felt to be intractable problems.

1.2.2 Towards individuality

The physics of spatially extended systems (e.g., fluid systems, chemical reaction systems, and carrier transport systems in semiconductors) have traditionally been described by theoretical frames based on thermodynamics, classical mechanics and statistical physics (electromagnetism and quantum mechanics are also included in semiconductor physics [44–47] and other areas of physics). Within these frames, nonequilibrium phase transitions (first-order and second-order phase transitions) are well characterized in condensed matter physics. It has been traditional since Newton's classical mechanics, for determinism in the equations of motion, to predict physical quantities $X(t)$ in their future evolutions. However, once the intractable problem or chaos happens in a physical system, the traditional analytical approach is unable to solve it. Chaos theory is transcendent in the sense that it can solve such an intractable problem which is related to the nonequilibrium phase transition. Following the pioneering work by Li and York [18], May [22] and Feigenbaum [19], many scientists in the 1980s showed universality of chaos in their experiments and models [13–17]. Their aim was unification from individuality to universality. Chaos theory is now moving on to the second stage, from universality to individuality, in order to determine precisely the complex dynamical behaviors in individual physical systems. In each of the spatially extended physical (or chemical) systems, it will be a great challenge to investigate the fully developed turbulent states which stem from the spatiotemporal dynamical behaviors (e.g., Fig. 1.11). In semiconductors, many theoretical approaches have been taken to understand the inhomogeneous spatial patterns of the nonlinear currents [11, 12]. However, many unresolved problems remain in our understanding of electronic turbulence in semiconductors. There may be some criticism that chaos is simply a mathematical product, not physics. Elaborations to solve the problems are important in order to place the chaotic phenomena and move into an area (nonlinear carrier transport theory) of semiconductor physics.

The physical systems of category B in Table 1.1 are strongly dissipative, except for the Josephson junction and the spatially extended instabilities occur via diffusion and nonlinearity. For instance, fluid dynamics is described by the Navier–Stokes equation. It may be surprising that the complex pattern of turbulence for Newtonian fluid (e.g., the Karman vortex and turbulence) is fundamentally characterized by the basic equation. In experiments on Rayleigh–Bénard convection [48, 49] and Taylor–Couette flow [50], a period doubling route to chaos [49] and the Hopf-bifurcation route to chaos [48, 50] have been observed. Since the turbulence is apparently a cooperative phenomenon of a huge number of molecules, one can imagine the irreversibility of chaos and turbulence, where the energy dissipation occurs at the microscopic level.

In chemical reaction systems, spatiotemporal behaviors are modeled generally as:

$$\dot{n}_i = D_i \nabla^2 n_i + f(\{n_i\}; t; \{\mu\}), \quad i = 1, 2, \ldots, \tag{1.36}$$

where n_i is the concentration of molecule i, and D_i is a diffusion constant. Typical examples are seen in the Brusselator and Oregonator models [24, 51]; it is also known as the inhibitor–activator model, in which typically diffusion-induced instability is likely to occur. The PDEs (1.36) may reproduce target patterns and spiral patterns of the B-Z reaction [53] in Fig. 1.18.

There is a strong analogy between the chemical reaction system and the semiconductor system. There are many autocatalytic processes in the chemical reaction systems [6, 7, 26, 51–55], among which the B-Z reaction [48–50] is a typical example showing chaotic behavior. The key words here include: open flow, autocatalytic, self-organization and far from thermal equilibrium. An analogy is taken from an autocatalytic reaction scheme [26]:

$$\mathrm{A} + 2\mathrm{X} \underset{k_2}{\overset{k_1}{\rightleftharpoons}} 3\mathrm{X} \tag{1.37}$$

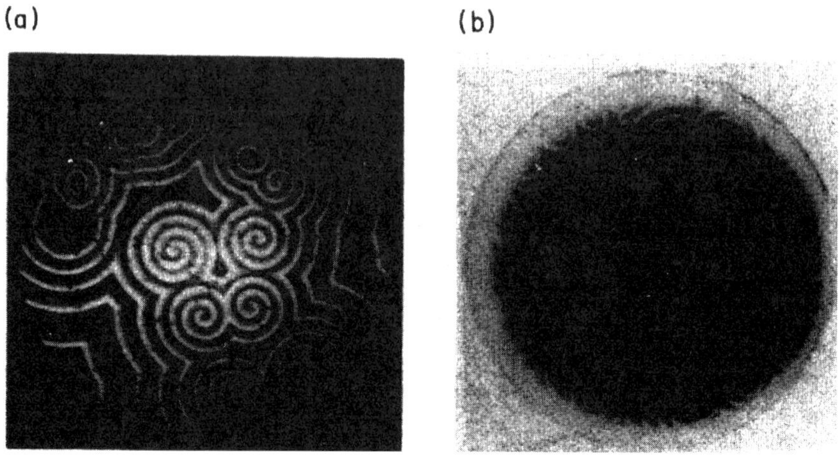

Figure 1.18 Spiral patterns in excitable media. (a) Belousov–Zhabotinsky reaction photographed in blue light. regions of maximum concentration of the blue indicator ferroin appear white. The spirals rotate with a period of about 1 min. The dish is 90 mm diameter with 1 mm depth of reagent. (b) Dark field photograph of a monolayer of slime mold *Dictyostelium discoideum* cells. The contrast is produced by different refractive properties of the cells responding to a pulse of the diffusing chemical adenosine monophosphate (cAMP). Spirals are evident. They rotate every 5 min. The dish is also 90 mm in diameter.

$$X \underset{k_4}{\overset{k_3}{\rightleftharpoons}} B \tag{1.38}$$

where the reactant bath is composed of molecules A and B, and molecule X is an intermediate reactant. Reactions (1.37) and (1.38) are totally equivalent to:

$$A \rightleftharpoons B. \tag{1.39}$$

The rate equation for the concentration of the molecule X is given by:

$$\dot{n} = -k_2 n^3 + k_1 n_A n^2 - k_3 n + k_4 n_B, \tag{1.40}$$

where $[X] = n$, $[A] = n_A$, and $[B] = n_B$ ($[\cdot]$ is concentration). When each of the reactions (1.37) and (1.38) are under equilibrium states, the ratio $(n_A/n_B)_{eq}$ is given by:

$$\left(\frac{n_A}{n_B}\right)_{eq} = \frac{k_4 k_2}{k_1 k_3}, \tag{1.41}$$

since $k_1 n_A n_{eq}^2 = k_2 n_{eq}^3$ from (1.37) and $k_3 n_{eq} = k_4 n_B$ from (1.38), where n_{eq} is the concentration of molecule X. The nonequilibrium state can be shown

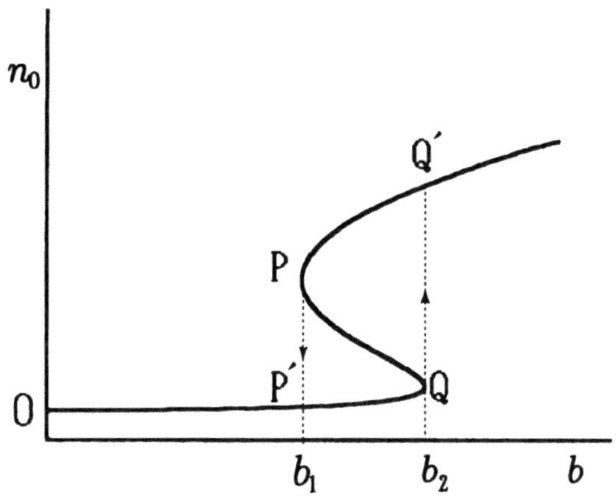

Figure 1.19 Concentration n_0 of the molecule X as a function of a control parameter b. Redrawn from [26].

by putting $\dot{n} = 0$ in (1.40);

$$n_0^3 - an_0^2 + kn_0 - b = 0, \qquad (1.42)$$

where $a = k_1 n_A / k_2$, $b = k_4 n_B / k_2$, $k = k_3 / k_2$ and $n_0 = [X]$ under the nonequilibrium state. If a and b are considered as two control parameters, the solution of (1.42) gives bistability (multistability) as a function of b under some fixed value of a, as given in Fig. 1.19 [26]. When both of the control parameters are varied simultaneously, the bistability region is shown in the parameter space (a, b) by a cusp-shaped region in Fig. 1.20, where a unique solution outside the cusp region indicates the thermal equilibrium or near-

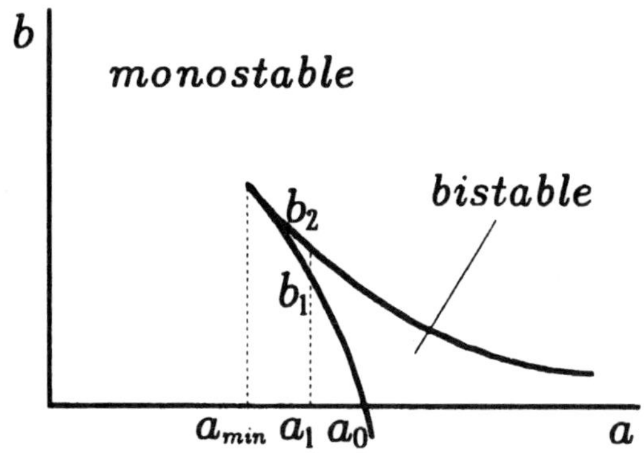

Figure 1.20 The bistability region in the (a, b) parameter space, which is shown by the cusp-shaped area. Redrawn from [26].

thermal equilibrium. The bistability region indicates the occurrence of a nonequilibrium phase transition (first-order phase transition) beyond the thermal equilibrium. The state equation (1.42) is reminiscent of van der Waals equation [56, 57]. A remarkable point is that an explosive reaction occurs at a critical point (Q or P in Fig. 1.19), in contrast to the coexistence of liquid/gas phases in van der Waals liquid [56, 57]. The reaction system (1.37), (1.38) exhibits several paths of possible thermodynamical bifurcations: (i) explosive bifurcation and hysteresis ($Q \rightarrow Q' \rightarrow P \rightarrow P' \rightarrow Q$); and (ii) smooth bifurcation from the thermal equilibrium to the nonequilibrium states (or vice versa). Such an explosive reaction scheme is a fascinating problem which can be discussed using catastrophe theory [58].

Similar behaviors appear in semiconductor carrier transport. The generation–recombination process of free carriers between the conduction band and donor ground state for n-type semiconductors is expressed by the autocatalytic reaction scheme:

$$D^0 \underset{B_T}{\overset{A_T}{\rightleftharpoons}} D^+ + e^-, \tag{1.43}$$

where e^-, D^0 and D^+ denote the free electron, the neutral donor and the ionized donor, respectively, and A_T and B_T are the thermal ionization rate and the thermal recombination rate. By applying an electric field E, the impact ionization proceeds above a threshold level $E > E_c$;

$$D^0 + e^- \xrightarrow{A_I} D^+ + 2e^-, \tag{1.44}$$

where A_I is the impact ionization rate. The impact ionization is quite similar to the ionization process of a hydrogen atom by an electron, except that the impurity atom is embedded in a dielectric medium with the dielectric constant $\kappa_0\kappa$ (namely, a semiconductor). Similarly, the impact ionization of neutral acceptor can be explained by holes in p-type semiconductors. In the carrier transport theory, one must explain various types of scattering process which are not included explicitly in (1.43) and (1.44) and which make the problem much more complex than those in the chemical reaction system. The screening effect of ionized-impurity scattering in high-purity n-GaAs induces an impact ionization avalanche and S-shaped negative differential resistivity at 4.2 K [59], which can be modeled by the above reaction scheme (1.43) with an approximated expression $A_I^0 \simeq A_I(E) + \gamma n$ in (1.44) [10]. The approximation leads to (1.24–1.26). Thus, scattering problems in the carrier transport theory impose the individuality of semiconductor chaos among spatially extended dynamical systems. There is an important difference between a chemical reaction system and a semiconductor system. In the autocatalytic processes (1.43) and (1.44), the free electrons must obey the Fermi–Dirac statistics with the distribution function $f(\epsilon^*)$, where ε^* is the microscopic energy of free carriers. An impact-ionization process (1.44) occurs with $\varepsilon^* > \varepsilon_D$ (ε_D is the ionization energy of a neutral donor), while free carriers with $\varepsilon^* < \varepsilon_D$ mainly participate in the thermal process (1.43). Therefore, (1.43), (1.44) cannot be handled by the individual (single-particle) excitation regime with ε^*. This complication does not appear in the explosive reaction (1.37), (1.38). From the collective excitation regime with mean kinetic energy $\langle \varepsilon^* \rangle = \varepsilon$, one can argue the hydrodynamical equations of motion, and in the regime of non-equilibrium phase transition, entropy production and excess entropy are also discussed (chapter 4) [26, 51, 60]. The most remarkable point to be stressed is that chaos in semiconductors can be discussed and viewed from the aspect of the cusp catastrophe [58] and chaos.

1.3 Motivations

1.3.1 From the viewpoint of semiconductor chaos

Nonlinear science includes enormous fundamental aspects of statistical physics, e.g., chaos in nonlinear dissipative systems as introduced above, structures and growth process of fractals [61], neural networks [62, 63] and cellular automata [64]. Many powerful numeric tools have been devised in nonlinear science. In semiconductor chaos systems, many new findings are to be expected by developing various measurement techniques in experiments, which may be analyzed from the viewpoint of nonlinear science and which may lead to a full understanding of electronic turbulence.

Ideas of synergetics [24] and catastrophe theory [58] increase our understanding of nonequilibrium phase transition and chaos in semiconductors. Current motivation may be summarized by the following questions.

1. Can the universality of chaos be confirmed in semiconductor carrier transport?
2. How is the structural stability of the nonequilibrium phase transition broken under the driven and spontaneous regimes? This question is equivalent to: how does chaos emerge in semiconductor carrier transport related to the impact-ionization avalanche?
3. How can we understand electronic turbulence?

References

[1] J.N. Bardsley and B. Sundaram, Phys. Rev. **A32** (1985) 689; K.A.H. Van Leeuwen, G.V. Oppen, S. Rewicik, J.B. Bowlin and P.M. Koch, Phys. Rev. Lett. **55** (1985) 2231.

[2] J. Wisdom, Proc. of a Royal Society Discussion Meeting, Dynamical Chaos, eds. H.V. Berry, I.C. Percival and N.O. Weiss (Princeton University Press, Princeton, NJ, 1987) p. 109.

[3] H.L. Swinney and J.P. Gollub, *Hydrodynamic Instabilities and the Transition to Turbulence* (Springer, Berlin, 1981).
[4] M. Dubois, M.A. Rubio and P. Berge, Phys. Rev. Lett. **51** (1983) 1446.
[5] A. Libchaber, Proc. R. Soc. London **A413** (1987) 633.
[6] J.C. Roux, A. Ross, S. Bachelart and C. Vidal, Physica **2D** (1981) 395.
[7] P. Richetti, J.C. Roux, F. Argoul and A. Arneodo, J. Chem. Phys. **86** (1987) 3339.
[8] P. Monceau, J. Richard and M. Renard, Phys. Rev. **B25** (1982) 931.
[9] E. Brown, G. Mozurkewich and O. Gruner, Phys. Rev. Lett. **54** (1984) 2272.
[10] K. Aoki, Proc. of a NATO Advanced Workshop on Negative Differential Resistance and Instabilities in 2-D Semiconductors, eds. N. Balkan, B.K. Ridley and A.J. Vickers (Plenum Press, London, 1993) p. 393; see also Y. Abe ed., Appl. Phys. **A48**, A special issue on Nonlinear Chaotic Transport Phenomena in Semiconductors.
[11] E. Schöll, *Nonequilibrium Phase Transitions in Semiconductors – Self-organization Induced by Generation and Recombination Processes* (Springer, Berlin, 1987).
[12] J. Peinke, J. Parisi, O.E. Rössler and R. Stoop, *Encounter with Chaos – Self-Organized Hierarchical Complexity in Semiconductor Experiments* (Springer, Berlin, 1992).
[13] P. Cvitanovic ed., *Universality in Chaos – An Introduction and Reprints Volume* (Adam Hilger, Bristol, 1984).
[14] P. Berge, Y. Pomeau and Ch. Vidal, *Order within Chaos – Towards a Deterministic Approach to Turbulence* (Wiley, New York, 1986).
[15] J. Guckenheimer and P. Holmes, *Nonlinear Oscillations, Dynamical Systems and Bifurcations of Vector Fields* (Springer, Berlin, 1983).
[16] Hao Bai-Lin, *Chaos – An Introduction and Reprints Volume* (World Scientific, Singapore, 1984); *Chaos II – An Introduction and Reprints Volume* (World Scientific, Singapore, 1990) and references therein.
[17] H.G. Schuster, *Deterministic Chaos – An Introduction* (Physik-Verlag, Weinheim, 1988).
[18] T.Y. Li and J.A. Yorke, Am. Math. Monthly **82** (1975) 985.
[19] M.J. Feigenbaum, J. Stat. Phys. **19** (1978) 25.
[20] N. Metropolis, M.L. Stein and P.R. Stein, J. Comb. Theor. **15** (1973) 25.
[21] F.C. Moon, *Chaotic Vibrations – An Introduction for Applied Scientists and Engineers* (John Wiley & Sons, New York, 1987); *Chaotic and Fractal Dynamics – An Introduction for Applied Scientists and Engineers* (John Wiley & Sons, New York, 1992).
[22] R.M. May, Nature **261** (1976) 459.
[23] J.-P. Eckmann, Rev. Mod Phys. **53** (1981) 643.
[24] H. Haken, *Synergetics – An Introduction, Nonequilibrium Phase Transitions and Self-Organization in Physics, Chemistry and Biology*, 2nd ed. (Springer, Berlin, 1982).
[25] M.V. Berry, Proc. of a Royal Society Discussion Meeting, Dynamical Chaos, eds. M.V. Berry, I.C. Percival and N.O. Weiss (Princeton University Press, Princeton, NJ, 1987) p. 183.
[26] G. Nicolis and I. Prigogine, *Self-Organization in Nonequilibrium Systems – From Dissipative to Order Through Fluctuations* (John Wiley & Sons, New York, 1977).
[27] C. Kittel, *Elementary Statistical Physics* (John Wiley & Sons, New York, 1958).
[28] J.-P. Eckmann and D. Ruelle, Rev. Mod. Phys. **57** (1985) 617.
[29] A.N. Sarkovskii, Ukr. Mat. Z. **16** (1964) 61 (in Russian).
[30] R.W. Leven and B.P. Koch, Phys. Lett. **A86** (1981) 71.

[31] J.B. McLaughlin, J. Stat. Phys. **24** (1981) 375.
[32] F.C. Moon, Phys. Rev. Lett. **53** (1984) 962.
[33] P.S. Linsay, Phys. Rev. Lett. **47** (1981) 1349.
[34] R.V. Buskirk and C. Jeffries, Phys. Rev. **A31** (1985) 3332.
[35] T. Matsumoto, IEEE Trans. Cir. & Sys. **CAS-31** (1984) 1055; T. Matsumoto, L.O. Chua and M. Komuro, IEEE Trans. Circ. & Sys. **CAS-32** (1985) 798.
[36] P.R. Fenstermacher, H.L. Swinney and J.P. Golub, J. Fluid Mech. **94** (1979) 103.
[37] F.T. Arecchi, in *Instabilities and Chaos in Quantum Optics*, eds. F.T. Arecchi and R.G. Harrison (Springer, Berlin, 1987) pp. 9–47.
[38] K. Ikeda and O. Akimoto, Phys. Rev. Lett. **48** (1982) 617; F.A. Hopf, D.L. Kaplan, H.M. Gibbs and R.L. Shoemaker, Phys. Rev. **A25** (1982) 2171.
[39] A.B. Huberman, J.P. Crutchfield and N.H. Packard, Appl. Phys. Lett. **37** (1980) 750.
[40] M. Octavio and C.R. Nasser, Phys. Rev. **B30** (1984) 1586.
[41] Y. Ueda, J. Stat. Phys. **20** (1979) 181.
[42] R. Seydel, Physica **17D** (1985) 308.
[43] U. Parlitz and W. Lauterborn, Phys. Rev. **A36** (1987) 1428.
[44] J. Pozhela, *Plasma and Current Instabilities in Semiconductors* (Pergamon, Oxford, 1981).
[45] B.R. Nag, *Theory of Electrical Transport in Semiconductors* (Pergamon, Oxford, 1972).
[46] S.M. Sze, *Physics of Semiconductor Devices* (John Wiley & Sons, New York, 1969).
[47] K. Seeger, *Semiconductor Physics – An Introduction* (Springer, Berlin, 1989).
[48] M. Sano and Y. Sawada, in *Chaos and Statistical Methods*, ed. Y. Kuramoto (Springer, Berlin, 1984) p. 226.
[49] A. Libchaber, C. Laroche and S. Fauve, J. Physique Lett. **43** (1982) L211.
[50] H.L. Swinney and J.P. Gollub, Phys. Today **31** (1978) 41.
[51] G. Nicolis and I. Prigogine, *Exploring Complexity – An Introduction* (R. Riper, KG Verlag, Muchen, 1989).
[52] M.L. Smoes, in *Dynamics of Synergetic Systems,* ed. H. Haken (Springer, Berlin, 1980) p. 80.
[53] A.N. Zaikin and A.M. Zhabotinsky, Nature **225** (1970) 535.
[54] R.J. Field and R.M. Noyes, J. Chem. Phys. **60** (1970) 1877.
[55] J.L. Hudson, M. Hart and D. Marinko, J. Chem. Phys. **71** (1979) 1601.
[56] L.D. Landau and E.M. Lifshitz, *Statistical Physics* (Pergamon, Oxford, 1957).
[57] A.B. Pippard, *Response and Stability – An Introduction to the Physical Theory* (Cambridge University Press, Cambridge, 1985).
[58] R. Thom, *Structural Stability and Morphogenesis – An Outline of a General Theory of Models* (Benjamin/Cummings, London, 1975).
[59] R.S. Crandall, J. Phys. Chem. Solids **31** (1970) 2069.
[60] P. Gransdorff and I. Prigogine, *Thermodynamic Theory of Structure and Fluctuations* (Wiley-Interscience, London, 1971).
[61] H.O. Peitgen and P.H. Richter, *The Beauty of Fractals* (Springer, Berlin, 1986).
[62] J.J. Hopfield, Proc. Natl. Acad. Sci. USA **79** (1982) 2554.
[63] I. Tsuda, E. Koerner and H. Shimizu, Prog. Theor. Phys. **78** (1987) 51.
[64] S. Wolfram, Physica **10D** (1984) 1–35.

Chapter 2

Nonlinear carrier transport in semiconductors

2.1 Carrier transport theory

2.1.1 Statistics of carriers

In physics, various types of nonequilibrium phase transitions are encountered, which can be classified as either **first order** or **second order** [1]. In a first-order phase transition, the physical variable (state variable) shows a discontinuous transition when the control parameter is varied continuously, while for a second-order phase transition the state variable is continuous but its derivative shows discontinuity at the transition point. This chapter is concerned with the **nonequilibrium phase transition of first order**, which can be seen in the semiconductor carrier transport [2]. The steady-state solution of the current density **J** at an electric field **E** is given by a well-known formula:

$$\mathbf{J} = \sigma \mathbf{E}. \tag{2.1}$$

Here, electrical conductivity σ is given by

$$\sigma = \begin{cases} en\mu_{\mathrm{e}} & \text{for n-type semiconductors,} \quad (2.2) \\ ep\mu_{\mathrm{h}} & \text{for p-type semiconductors,} \quad (2.3) \end{cases}$$

where $n(p)$ is free electron (free hole) density, $\mu_{\mathrm{e}}(\mu_{\mathrm{h}})$ is the electron (hole) mobility, and e is the magnitude of elementary charge. For the double injection diode [3], $\sigma = e(n\mu_{\mathrm{e}} + p\mu_{\mathrm{h}})$. There are typically two types of J–E characteristics (Fig. 2.1). As is apparent from Fig. 2.1, there is an unstable region $\partial J/\partial E < 0$ ($\partial\sigma/\partial E < 0$), followed by S-shaped negative differential conductivity **(SNDC)** in Fig. 2.1a, and N-shaped negative differential

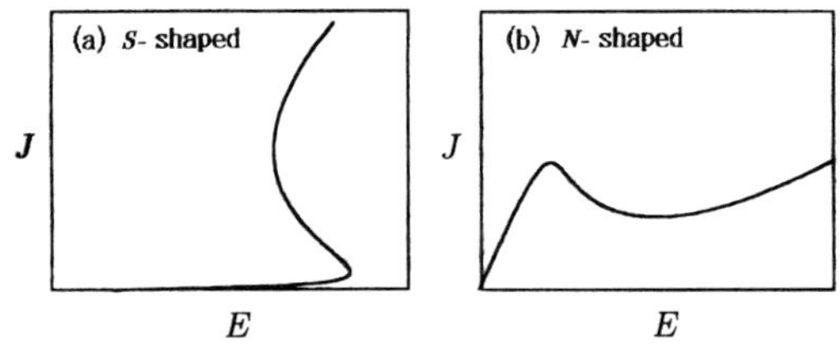

Figure 2.1 (a) S-shaped and (b) N-shaped J–E characteristics.

conductivity **(NNDC)** in Fig. 2.1b. For n-type semiconductors,

$$\frac{\partial \sigma}{\partial E} = e\left(\mu_e \frac{\partial n}{\partial E} + n\frac{\partial \mu_e}{\partial E}\right) < 0. \tag{2.4}$$

The NDC means $\partial n/\partial E$ and/or $\partial \mu_e/\partial E < 0$. The SNDC is typical in the impact ionization avalanche of neutral impurities, leading to spatial formation of a **current filament.** The NNDC is seen in the Gunn diode, leading to the formation of a **high-field domain,** where intervalley scattering of free electrons as well as polar optical phonon scattering and acoustic phonon scattering play an important role [4, 5].

Both NDCs exhibit oscillatory instability. In order to understand the nonequilibrium phase transition and the dynamical behaviors, it is necessary to know the fundamental properties of the carrier transport theory [6, 7]. This chapter will cover the important parts of the theory that may be helpful for the present interests. Readers who are familiar with the carrier transport theory may skip the first section of this chapter. In sections 2.2 and 2.3, the experimental and theoretical studies of NNDC and the SNDC will be introduced.

2.1.1.1 Fermi-Dirac distribution function

In semiconductor physics, the statistics of charge carriers (electrons or holes) will first be studied under thermal equilibrium [4, 6, 7], where free electrons are supplied by the donor impurities in n-type semiconductors and free holes by the acceptor impurities in p-type semiconductors. The probability that electrons (or holes) have the energy $\tilde{\varepsilon}$ with the wave vector **k** is given by the **Fermi–Dirac distribution function** $f_0(\tilde{\varepsilon})$:

$$f_0(\tilde{\varepsilon}) = \{\exp[\mathrm{sgn}(-q)(\tilde{\varepsilon} - \zeta)/k_B T_L] + 1\}^{-1}, \tag{2.5}$$

$$\tilde{\varepsilon} = \begin{cases} \varepsilon_c + \hbar^2 k^2/2m_e & \text{for electrons} \\ \varepsilon_v + \hbar^2 k^2/(-2m_h) & \text{for holes,} \end{cases} \tag{2.6}$$

where $\mathrm{sgn}(-q) = +1$ for electrons $(q = -e)$, $\mathrm{sgn}(-q) = -1$ for holes

$(q = e)$, ζ is the Fermi level, k_B is the Boltzmann constant, T_L is the lattice temperature, ε_c and ε_v are the energy edges of conduction band and valence band, respectively (see Fig. 2.2), m_e and $-m_h$ are the effective masses, and $\hbar$ is Planck's constant divided by 2π $(\hbar = h/2\pi)$. In (2.6), the effective mass of free holes is taken as negative $(-m_h)$. Since the effective mass is given by

$$m^* = \hbar^2(\partial^2\tilde{\varepsilon}/\partial k^2)^{-1} \tag{2.7}$$

the convexity of the parabolic valence band in Fig. 2.2 gives the negative value $-m_h$. The hole is a quasi-particle which is a positive charge carrier generated by the acceptor impurity. When one of the neighboring covalent electrons moves to the vacant site (acceptor) with a displacement $-\delta\mathbf{r}$, the positive hole moves in the opposite direction with a displacement $+\delta\mathbf{r}$. Namely, transport of positive (negative) mass occurs in the direction of $-\delta\mathbf{r}$ $(+\delta\mathbf{r})$. Therefore, the hole carries the negative mass. However, in the carrier transport theory, one must conventionally treat the **positively charged hole as if it behaves like a particle of positive effective mass** m_h. The convention is apparent on rewriting (2.6), treating $\varepsilon_v - \tilde{\varepsilon}$ as the kinetic energy of the hole with mass m_h:

$$\hbar^2\mathbf{k}^2/2m_h = \varepsilon_v - \tilde{\varepsilon}. \tag{2.8}$$

The semiconductor is called degenerate if $\zeta > \varepsilon_c$ for n-type and $\zeta < \varepsilon_v$ for p-type, while it is nondegenerate when $\varepsilon_v < \zeta < \varepsilon_c$. It is convenient to measure

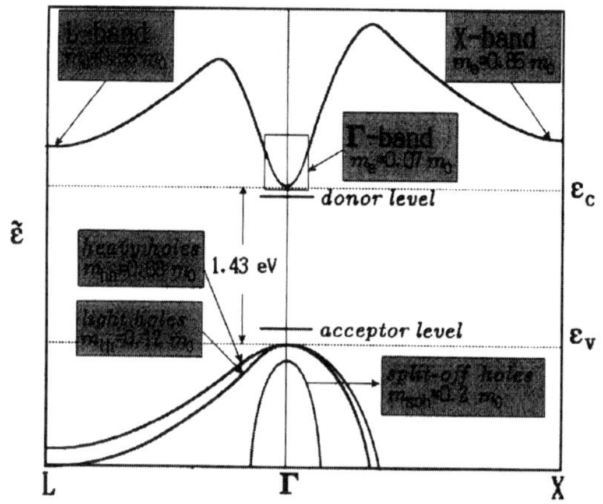

Figure 2.2 Band structure in n-GaAs. Values of the effective mass at each band are given. Note that the band-gap energy 1.43 eV $(\varepsilon_G = \varepsilon_c - \varepsilon_v)$ is the value at room temperature.

the carrier energies from the bottoms of the band edges:

$$\varepsilon^* = \begin{cases} \tilde{\varepsilon} - \varepsilon_c = \hbar^2\mathbf{k}^2/2m_e & \text{for free electrons} \\ \varepsilon_v - \tilde{\varepsilon} = \hbar^2\mathbf{k}^2/2m_h & \text{for free holes.} \end{cases} \tag{2.9}$$

From now on, n-type semiconductors are considered in a self-contained manner. Extension of the carrier transport theory to the p-type materials is easily done without losing any generality (one can apply the statistics to both light and heavy hole bands).

Since the conduction band has the density of states $D(\varepsilon^*)d\varepsilon^*$, which is the number of electrons allowed to populate in the range $[\varepsilon^*, \varepsilon^* + d\varepsilon^*]$, the number of electrons $n(\varepsilon^*)d\varepsilon^*$ is given by a product:

$$n(\varepsilon^*)d\varepsilon^* = D(\varepsilon^*)f(\varepsilon^*)d\varepsilon^*, \tag{2.10}$$

$$D(\varepsilon^*)d\varepsilon^* = 2(V/4\pi^2)(2m_e/\hbar^2)^{3/2}\sqrt{\varepsilon^*}d\varepsilon^*, \tag{2.11}$$

where $f(\varepsilon^*)$ is the Fermi-Dirac distribution function (2.5) and V is the crystal volume. By integrating over the whole energy range, the total concentration is given by [4]:

$$n = \frac{1}{V}\int_0^\infty D(\varepsilon^*)f(\varepsilon^*)\mathrm{d}\varepsilon^*, \tag{2.12}$$

$$= N_c F_{1/2}((\zeta - \varepsilon_c)/k_B T_L), \tag{2.13}$$

where the effective density of states N_c and the Fermi-integral $F_{1/2}$ are [4]:

$$N_c = 2\left(\frac{m_e k_B T_L}{2\hbar^2}\right)^{1/2}, \tag{2.14}$$

$$F_j = (1/j!)\int_0^\infty \frac{x^j \mathrm{d}x}{\exp(x-\eta)+1}, \quad (j = 1/2), \tag{2.15}$$

respectively, $x = \varepsilon^*/k_B T_L$, and $\eta = (\zeta - \varepsilon_c)/k_B T_L$. By applying electric field, the electron gas system deviates from the thermal equilibrium. It is convenient to introduce the effective temperature of electrons T_e, which is normally higher than T_L. For a while, T_L is replaced by T_e in the distribution function.

2.1.1.2 Maxwell–Boltzmann distribution function

For nondegenerate semiconductors, the Fermi–Dirac distribution function is approximated by the **Maxwell–Boltzmann (MB) distribution function**:

$$f(\varepsilon^*) \sim \exp[-(\varepsilon^* - \varepsilon_F)/k_B T_e], \tag{2.16}$$

where ε_F is the Fermi energy measured from the conduction band. Under the steady state slightly deviated from thermal equilibrium, current density $\mathbf{J}$ vs electric field $\mathbf{E}$ is almost linear and obeys the ohmic law $\mathbf{J} = \sigma_0\mathbf{E}$

($\sigma_0 = en\mu_e$), where the electron density n and the mobility μ_e are constant under a fixed effective temperature T_e. For a nondegenerate semiconductor, assuming the MB distribution function (2.16), then the carrier distribution in the parabolic Γ-band (Fig. 2.2) is given by [4]

$$n(\varepsilon^*)\mathrm{d}\varepsilon^* = 2N_c(k_B T_e)^{-2/3} \exp(\eta)\sqrt{\varepsilon^*} \exp(-\varepsilon^*/k_B T_e)\mathrm{d}\varepsilon^*, \tag{2.17}$$

$$= 2N_c \exp(\eta)\sqrt{x} \exp(-x)\mathrm{d}x. \tag{2.18}$$

Hence,

$$n = 2N_c \exp(\eta) \int_0^\infty \sqrt{x} \exp(-x)\mathrm{d}x. \tag{2.19}$$

The mean kinetic energy $\epsilon = \langle \epsilon^* \rangle$ per carrier is given by:

$$\varepsilon = \langle \varepsilon^* \rangle = \frac{1}{n} \int_0^\infty \varepsilon^* D(\varepsilon^*) f(\varepsilon^*)\mathrm{d}\varepsilon^* \tag{2.20}$$

$$= 3k_B T_e/2. \tag{2.21}$$

In (2.20), the integration of $\varepsilon^{*3/2} \exp(-\varepsilon^*/k_B T_e)$ is done by partial integration and by using (2.19). Alternatively,

$$\varepsilon = \langle \varepsilon^* \rangle = \left\langle \frac{1}{2} m_e (v_{x^2} + v_{y^2} + v_{z^2}) \right\rangle = 3(k_B T_e/2), \tag{2.22}$$

by the equi-partition rule in analogy with the molecular gas. For the MB distribution function, the electrons carry the crystal momentum $\mathbf{p} = \hbar\mathbf{k}$ of which the mean value is a priori zero:

$$\langle p_j \rangle = \langle \hbar k_j \rangle = \frac{1}{nV} \int_{-\infty}^{\infty} \hbar k_j D(\varepsilon^*) f(\varepsilon^*)\mathrm{d}x\mathrm{d}y\mathrm{d}z\mathrm{d}p_x\mathrm{d}p_y\mathrm{d}p_z, \tag{2.23}$$

$$= 0,$$

where $j = x, y, z$ and $\varepsilon^* = (\hbar^2/2m_e)(k_{x^2} + k_{y^2} + k_{z^2})$. The solution is quite apparent by the odd integrand function and by the fact that Brownian motion of thermal particles has zero-mean momentum. A a general remark, the effective temperature T_e is proportional to the variance σ_v of velocity fluctuation or momentum fluctuation:

$$\varepsilon = \frac{3}{2} k_B T_e = \frac{1}{2} m_e \sigma_v = \frac{1}{2} m_e \langle (\mathbf{v} - \langle \mathbf{v} \rangle)^2 \rangle$$

$$= \frac{1}{2} m_e (\langle \mathbf{v}^2 \rangle - \langle \mathbf{v} \rangle^2). \tag{2.24}$$

Since $\langle \mathbf{v} \rangle = 0$ from (2.23), then

$$\varepsilon = \frac{1}{2} m_e \langle \mathbf{v}^2 \rangle = \frac{\hbar^2}{2m_e} \langle \mathbf{k}^2 \rangle.$$

2.1.1.3 *Shifted Maxwell–Boltzmann distribution function*

Under the applied electric field, the mean velocity $\langle \mathbf{v} \rangle$ (the mean momentum $m_e\langle \mathbf{v} \rangle$) should be equal to the drift velocity $\mathbf{v}_d$ (the drift momentum $m_e\mathbf{v}_d = \hbar\mathbf{k}_d$). In the momentum space, the drift momentum shifts towards the opposite direction to the electric field (Fig. 2.3) by a displacement $\hbar\mathbf{k}_d$, then free electrons repopulate inside the shifted conduction band. For the microscopic energy $\tilde{\epsilon}$ of the electrons under the shifted regime,

$$\tilde{\varepsilon} = \varepsilon_c + \frac{1}{2}m_e(\mathbf{v} - \mathbf{v}_d)^2, \tag{2.25}$$

$$= \varepsilon_c + \frac{\hbar^2}{2m_e}(\mathbf{k} - \mathbf{k}_d)^2. \tag{2.26}$$

Since $\epsilon^* = \tilde{\epsilon} - \epsilon_c$, one obtains the **shifted MB distribution function** form (2.5) [4, 8]:

$$f(\varepsilon^*) \sim \exp\left\{-\left[\frac{\hbar^2}{2m_e}(\mathbf{k} - \mathbf{k}_d)^2 - \varepsilon_F\right]/k_B T_e\right\}, \tag{2.27}$$

$$= \exp(\eta)\ \exp\left\{-\frac{\hbar^2}{2m_e}(\mathbf{k} - \mathbf{k}_d)^2/k_B T_e\right\}. \tag{2.28}$$

For the microscopic kinetic energy ε^*,

$$\varepsilon^{*\prime} = \frac{\hbar^2}{2m_e}(\mathbf{k} - \mathbf{k}_d)^2 = \frac{\hbar^2}{2m_e}\mathbf{k}'^2. \tag{2.29}$$

With (2.11), (2.28) and (2.29), the same expressions as in (2.19), (2.21) and

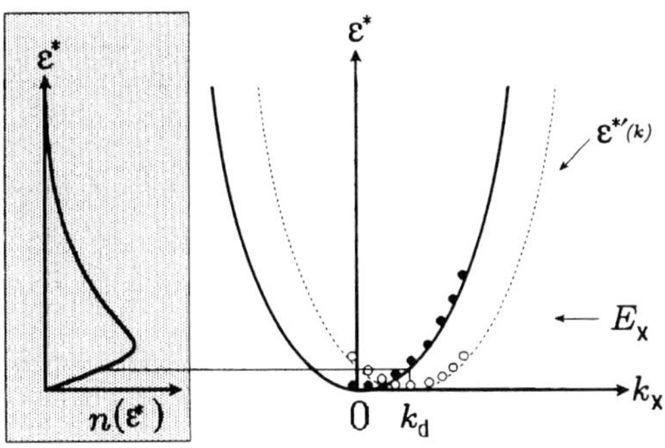

Figure 2.3 Distribution of the electrons (solid circles) at the bottom of the parabolic Γ-band, drawn schematically for the shifted MB distribution function. The broken parabolic curve indicates the energy dispersion curve given by $\epsilon^{*\prime}(\mathbf{k}) = \hbar^2/2m_e(\mathbf{k} - \mathbf{k}_d)^2$. All of the electrons carry the crystal momentum $\hbar\mathbf{k}_d$ under the applied electric field. Suppose that the electrons are virtually thermalized at the bottom of the parabola (broken line, open circles). These electrons lie on the solid line of the parabola $(\hbar^2/2m_e)\mathbf{k}^2$ (solid circles).

(2.23) are obtained for the total carrier concentration n, the mean kinetic energy $\langle\varepsilon^{*\prime}\rangle = 3k_{\mathrm{B}}T_{\mathrm{e}}/2$ and $\langle\mathbf{p}'\rangle = 0$, respectively. For $\langle\varepsilon^{*}\rangle$ and $\langle\mathbf{p}'\rangle = \langle\hbar\mathbf{k}'\rangle$, one can rewrite from (2.29) as:

$$\varepsilon = \frac{\hbar^2}{2m_{\mathrm{e}}}\langle\mathbf{k}^2\rangle = \frac{3}{2}k_{\mathrm{B}}T_{\mathrm{e}} + \frac{\hbar^2}{2m_{\mathrm{e}}}\mathbf{k}_{\mathrm{d}}^2, \tag{2.30}$$

and

$$\hbar\langle\mathbf{k}\rangle = \hbar\mathbf{k}_{\mathrm{d}}. \tag{2.31}$$

From (2.30), the effective temperature is again proportional to the variance of the velocity fluctuation:

$$\frac{3}{2}k_{\mathrm{B}}T_{\mathrm{e}} = \frac{1}{2}m_{\mathrm{e}}\sigma_{\mathrm{v}}, \tag{2.32}$$

where

$$\sigma_{\mathrm{v}} = \frac{\hbar^2}{2m_{\mathrm{e}}}(\langle\mathbf{k}^2\rangle - \mathbf{k}_{\mathrm{d}}^2).$$

The effective temperature T_{e} in (2.16) and (2.27) is called the **electron temperature.** The physical meaning of the electron temperature is apparent by the steady-state solution of the energy balance equation:

$$\varepsilon = \varepsilon_{\mathrm{L}} + e\tau(\varepsilon)\mu_{\mathrm{e}}\mathbf{E}^2, \tag{2.33}$$

where ε is given by (2.30), and $\varepsilon_{\mathrm{L}} = 3k_{\mathrm{B}}T_{\mathrm{L}}/2$. From (2.33), one has [9]:

$$\tau(\varepsilon)e\mu_{\mathrm{e}}\mathbf{E}^2 = \frac{3}{2}k_{\mathrm{B}}\delta T + \frac{\hbar^2}{2m_{\mathrm{e}}}{\mathbf{k}_{\mathrm{d}}}^2, \tag{2.34}$$

where $\delta T = T_{\mathrm{e}} - T_{\mathrm{L}}$. Namely, energy supplied by the electric field is partitioned into two parts, heating the electron gas and drift energy. Since $m_{\mathrm{e}}\mathbf{v}_{\mathrm{d}} = \hbar\mathbf{k}_{\mathrm{d}} = m_{\mathrm{e}}\mu_{\mathrm{e}}\mathbf{E}$ and $\mu_{\mathrm{e}} = e\tau_{\mathrm{m}}(\varepsilon)/m_{\mathrm{e}}$, from (2.34):

$$\frac{3}{2}k_{\mathrm{B}}\delta T = e\mu_{\mathrm{e}}\mathbf{E}^2\left(\tau(\varepsilon) - \frac{1}{2}\tau_{\mathrm{m}}(\varepsilon)\right) \approx e\mu_{\mathrm{e}}\mathbf{E}^2\tau(\varepsilon), \tag{2.35}$$

where $\tau(\varepsilon) \gg \tau_{\mathrm{m}}(\varepsilon)$ (e.g., $\tau(\varepsilon) \sim 10^{-9}$ s, $\tau_{\mathrm{m}}(\varepsilon) \sim 10^{-12}$ s in high-purity semiconductors at low fields and low temperature [4]). Electrons are termed **warm** when $T_{\mathrm{e}} > T_{\mathrm{L}}$ and **hot** when $T_{\mathrm{e}} \gg T_{\mathrm{L}}$ [4, 8]. In (2.33)–(2.35), the electron temperature T_{e} cannot be determined simply from the r.h.s. of the equations because $\tau_{\mathrm{m}}(\varepsilon)$ and $\tau(\varepsilon)$ are also functions of T_{e}.

In order for the approximated forms of the MB and the shifted MB distribution functions (2.16), (2.28) to be valid, one must postulate carrier–carrier scattering, which plays an important role for quick randomization of the distributions of the energy and momentum gained from the field [4, 8]. When the electric field $E(t) = E_0$ $(t > 0)$ is applied in the x-direction, for instance, the electron distribution deviates far from the initial one under the thermal equilibrium ($E = 0$), as shown schematically in Fig. 2.4. However, after a few picoseconds ($\sim$ momentum relaxation time), the electron distribution with the inhomogeneous energies and momenta is rapidly randomized in all directions of the phase space $\mathbf{k} = (k_x, k_y, k_z)$ via electron–electron (e-e) scattering, as well as ionized impurity scattering and polar optical phonon scattering if significant. During the momentum relaxation process via the e-e scattering and the ionized impurity scattering, the electron gas does not lose any part of the total energy because of the elastic scatterings. Energy loss due to acoustic phonon scattering and piezoelectric scattering occurs at a much later stage than during the period of momentum relaxation. The **electron temperature model** using the MB (2.16) or the shifted MB distribution functions (2.28) may be justified for semiconductor materials whose electron concentrations are relatively high [8, 10] to allow e-e scattering. The above discussions are limited to the low-field regime. With high electric fields ($\gtrsim$ a few kV/cm), electrons are excited high above the conduction band edge, making their polar optical phonon scattering [4] and/or transfer to the satellite valleys (X or L valleys) [4, 5, 11] extremely important. For the high-field regime, the electron temperature model is no longer valid.

The MB distribution function and the shifted MB distribution function have been extensively used to characterize the carrier transport phenomenon under relatively low electric fields ($<$ 100 V/cm) [4, 8] mainly because both of the distribution functions (2.16) and (2.28) are simple forms. However, the MB and shifted MB distributions may be incorrect when the carrier density is relatively low, since the e-e scattering is insignificant. Despite this, the MB or shifted MB distribution functions may be used for the analysis of carrier transport properties and for modeling nonlinear dynamical behaviors because of the simple treatment. Measurements using various experimental techniques provide much information about electron tem-

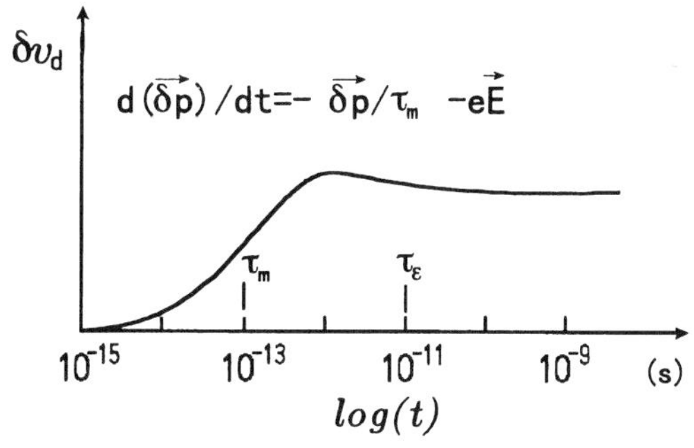

Figure 2.4 Drift velocity as a function of time. Note that the momentum relaxation time is much faster than the energy relaxation time ($\tau_m = 0.1$ ps, $\tau_\varepsilon = 10$ ps). Redrawn from [4].

perature and also precise hot electron distributions, which slightly deviate from simple forms of the MB and shifted MB distributions. From electrical measurements (Hall measurements [8, 12] in n-InAs, Shubnikov de Haas (SdH) effects in n-InSb [13] and n-InAs [12], noise measurements in Ge [14], etc.), and optical measurements (photoluminescence spectra in GaAs [15, 16], optical absorption (Burstein shift) in GaSb [17], intervalence absorption in p–Ge [18], cyclotron resonance in n- and p-Ge [19, 20], etc.), a huge number of carrier-transport characteristics has been investigated, concerning electron temperature [8, 12, 13, 17], electron distribution [15, 16], electron mobility [12] and energy loss rate [12, 13]. From the measurements of photoluminescence spectra [15, 16], the electron (or hole) distribution can be obtained directly.

By the Hall measurements [8, 12], the carrier concentration and the mobility can be measured. From the lattice temperature dependence of the carrier mobility come the well-known power laws of momentum relaxation time $\tau_{\mathrm{m}}^{-1} \propto \varepsilon^{r}$ [4] ($r = 1/2$ for acoustic phonon scattering and $r = -3/2$ for ionized impurity scattering). As shown in Fig. 2.5, from the SdH measurements in InSb [13], information is obtained about the energy loss rates $\langle \partial\varepsilon/\partial t \rangle_{\mathrm{coll}}$ via screened acoustic deformation potential scattering at low lattice temperatures, and polar optical phonon scattering at high lattice temperatures. The measurements by microwave and infrared techniques in n-GaAs [19, 20] have shown how the energy relaxation time $\tau(\varepsilon)$ depends on the lattice temperature. A wide variety of measurements indicates the importance of the scattering processes which seriously affect the momentum and the energy relaxation time $\tau_{\mathrm{m}}(\varepsilon)$ and $\tau(\varepsilon)$. In particular, power laws $\tau_{\mathrm{m}}(\varepsilon)^{-1} \propto \varepsilon^{r}$ and $\tau(\varepsilon)^{-1} \propto \varepsilon^{r'}$ are fundamentally important in order to describe the carrier transport phenomena under the low-field regime. The total energy loss rate $\langle \partial\varepsilon/\partial t \rangle_{\mathrm{tot}}$ is given by the sum of above-mentioned energy loss rate $\langle \partial\varepsilon/\partial t \rangle_{j}$ as:

$$\left\langle \frac{\partial\varepsilon}{\partial t} \right\rangle_{\mathrm{tot}} = \sum_{j} \left\langle \frac{\partial\varepsilon}{\partial t} \right\rangle_{j}, \tag{2.36}$$

where j is the type of scattering (polar optical phonon scattering, etc.). Since $\langle \partial\varepsilon/\partial t \rangle_{j} \propto \tau_{j}^{-1}(\varepsilon)$ and $\langle \partial\varepsilon/\partial t \rangle_{\mathrm{tot}} \propto \tau^{-1}(\varepsilon)$, the **inverse sum rule** for the total energy relaxation time $\tau(\varepsilon)$ is given by

$$\tau(\varepsilon)^{-1} = \sum_{j} \tau_{j}^{-1}(\varepsilon). \tag{2.37}$$

2.1.2 The Boltzmann equation

In semiconductors, detailed descriptions of carrier transport are rather complicated, depending on the doping concentration (degenerated or

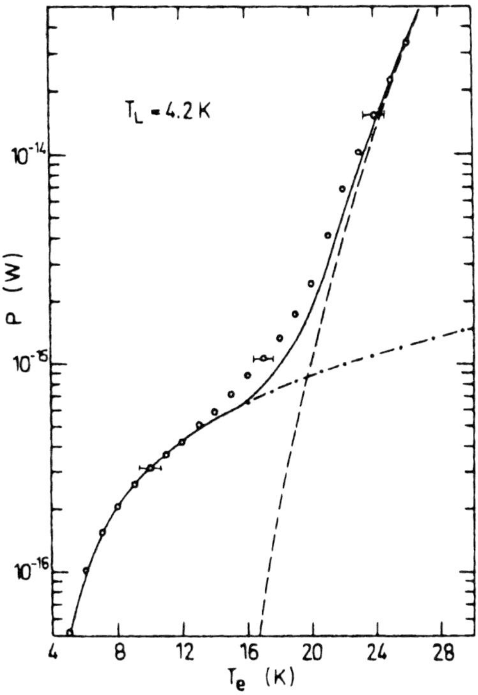

Figure 2.5 Dependence of energy-loss rate P on electron temperature in n-InSb. Dash–dotted curve: energy-loss rate due to the sum of both types of acoustic scattering; dashed curve: energy-loss rate due to polar optical phonon scattering; solid curve: sum of acoustic and polar optical energy-loss rates; circles: experimental data [13].

nondegenerated), the degree of compensation by acceptors (or donors) in n-type (or p-type) materials, the experimental parameter regimes (low- or high-field regimes under various lattice temperatures) and the materials themselves (covalent semiconductors such as Ge and Si or II–VI or III–V polar semiconductors). The **Boltzmann equation** (BE) is required in order to describe accurate formulae, under a low-field regime, of carrier mobility μ_j, momentum relaxation time τ_{mj}, and energy relaxation time τ_j via individual scattering processes j (j stands for ionized impurity scattering, acoustic deformation potential scattering and so on [4]). Alternatively, one can develop the BE under a high-field regime to find out the carrier distribution function by means of the Monte Carlo procedure [5, 10, 21, 22] and the iterative method [23, 24].

However, for economical reasons, accurate analysis in terms of the Boltzmann equation should be made only within the frame of stationary steady states (or static regimes) which are either slightly deviated or far from the thermal equilibrium. For long-term behaviors of the oscillatory instabilities (periodic oscillations and chaos), the BE cannot usually be used because of the difficulty in solving the evolution of the distribution function under the dynamical regimes. For the low-field regime, for instance, one must assume the MB or shifted MB distribution functions with T_e as well as use the scattering formulae which have been obtained from the theory [4].

For the purpose of obtaining the scattering formulae [4], the analytical approach for the BE and the approximated solution are presented. Borrowing the expression from Ziman [25], the distribution function of the carriers $f_k(r)$ in the neighborhood of the space coordinate r varies from place to place owing to the diffusion, the drift motion and the collision (scattering) processes. The time derivative is simply given by:

$$\frac{\mathrm{d}f_k}{\mathrm{d}t} = \left(\frac{\partial f_k}{\partial t}\right)_{\mathrm{diff}} + \left(\frac{\partial f_k}{\partial t}\right)_{\mathrm{drift}} + \left(\frac{\partial f_k}{\partial t}\right)_{\mathrm{coll}}. \tag{2.38}$$

For the static regime, $\mathrm{d}f_k/\mathrm{d}t = 0$. In the r.h.s. of (2.38), the diffusion term $(\partial f_k/\partial t)_{\mathrm{diff}}$ and the drift term $(\partial f_k/\partial t)_{\mathrm{drift}}$ are:

$$\begin{aligned}\left(\frac{\partial f_k}{\partial t}\right)_{\mathrm{diff}} &= -\left(\frac{\partial f_k}{\partial x}\frac{\partial x}{\partial t} + \frac{\partial f_k}{\partial y}\frac{\partial y}{\partial t} + \frac{\partial f_k}{\partial z}\frac{\partial z}{\partial t}\right)\\ &= -\mathbf{v}_k \cdot \nabla_{\mathbf{r}} f_k,\end{aligned} \tag{2.39}$$

$$\begin{aligned}\left(\frac{\partial f_k}{\partial t}\right)_{\mathrm{drift}} &= -\left(\frac{\partial f_k}{\partial k_x}\frac{\partial k_x}{\partial t} + \frac{\partial f_k}{\partial k_y}\frac{\partial k_y}{\partial t} + \frac{\partial f_k}{\partial k_z}\frac{\partial k_z}{\partial t}\right)\\ &= -\dot{\mathbf{k}} \cdot \nabla_k f_k.\end{aligned} \tag{2.40}$$

In (2.40), $\dot{\mathbf{k}}$ is given by the Lorentz force $\mathbf{F}$ as:

$$\dot{\mathbf{k}} = \mathbf{F}/\hbar = \frac{q}{\hbar}(\mathbf{E} + \mathbf{v}_k \times \mathbf{B}), \tag{2.41}$$

where $q = -e$ for the electrons and $\mathbf{B}$ is the magnetic induction. By inserting (2.39)–(2.41) into (2.38) and $\mathrm{d}f_k/\mathrm{d}t = 0$ in (2.38):

$$\left(\frac{\partial f_k}{\partial t}\right)_{\mathrm{coll}} = \mathbf{v}_k \cdot \nabla_{\mathbf{r}} f_k - \frac{e}{\hbar}(\mathbf{E} + \mathbf{v}_k \text{ x } \mathbf{B}) \cdot \nabla_k f_k. \tag{2.42}$$

The minus sign in the r.h.s. of (2.39) and (2.40) appears for the following reason [25]. Let $f_k(\mathbf{r})$ be the probability of finding the electrons with the velocity $\mathbf{v}_k$ in the neighborhood of $\mathbf{r}$ at time $t = 0$. Then, there is the probability $f_k(\mathbf{r} + \mathbf{v}_{k'}\delta t)$ of finding the electrons in the neighborhood of $\mathbf{r} + \mathbf{v}_{k'}\delta t$ at time $t = \delta t$. By the flow of electrons from $\mathbf{r}$ to $\mathbf{r} + \mathbf{v}_{k'}\delta t$, the probability $f_k(\mathbf{r})$ at time $t = \delta t$ decreases by $\delta f_k \approx f_k(\mathbf{r}) - f_k(\mathbf{r} + \mathbf{v}_{k'}\delta t) \approx -v_k(\delta f_k/\delta v_k)$, where $\delta \mathbf{v}_k = \mathbf{v}_{k'} - \mathbf{v}_k$. Similarly, there is a minus sign in (2.40).

For a simple treatment which is available to the low-field regime of semiconductors, the **relaxation time approximation** for $(\partial f_k/\partial t)_{\mathrm{coll}}$ assumed to be [4]:

$$\left(\frac{\partial f_\mathrm{k}}{\partial t}\right)_\mathrm{coll} = -\frac{f_\mathrm{k} - f_\mathrm{k}^0}{\tau_m}. \tag{2.43}$$

If one neglects the diffusion term $\mathbf{v}_\mathrm{k} \cdot \nabla_\mathrm{r} f_\mathrm{k}$ and use $B = 0$ in (2.42), the diffusion approximation of the distribution function from (2.42) and (2.43) is:

$$f_\mathrm{k} = f_\mathrm{k}^0 + \frac{e}{\hbar}\,\tau_m \mathbf{E} \cdot \nabla_\mathrm{k} f_\mathrm{k}, \tag{2.44}$$

where one can approximate $\nabla_\mathrm{k} f_\mathrm{k} \sim \nabla_\mathrm{k} f_\mathrm{k}^0$ in (2.44) under the weak electric field. Since

$$f_\mathrm{k}^0 \sim \exp\left(-\frac{\hbar^2}{2m_\mathrm{e}}\mathbf{k}^2/k_\mathrm{B} T_\mathrm{L}\right),$$

the distribution function

$$f_\mathrm{k} \sim f_\mathrm{k}^0 [1 - (\hbar e\,\tau_m/m_\mathrm{e} k_\mathrm{B} T_\mathrm{L})\mathbf{k} \cdot \mathbf{E}], \tag{2.45}$$

is obtained, where T_L can be replaced by T_e under the hot electron regime. Here, note that distribution function (2.45) is consistent with the shifted MB distribution function (2.28). For a physical quantity X, the average value is given by:

$$\langle X \rangle = \int_0^\infty X\, f_\mathrm{k}\, \mathrm{d}^3\mathbf{k} \Big/ \int_0^\infty f_\mathrm{k}\, \mathrm{d}^3\mathbf{k}. \tag{2.46}$$

In the shifted MB regime using (2.44), the drift velocity is given by [4]:

$$\begin{aligned} \mathbf{v}_\mathrm{d} &= \int_{-\infty}^\infty \mathbf{v}\,(f_\mathrm{k}^0 + (e\,\tau_m/\hbar)\mathbf{E} \cdot \nabla_\mathrm{k} f_\mathrm{k}^0)\, \mathrm{d}^3\mathbf{k} \Big/ \int_{-\infty}^\infty (f_\mathrm{k}^0 + (e\,\tau_m/\hbar)\mathbf{E} \cdot \nabla_\mathrm{k} f_\mathrm{k}^0)\, \mathrm{d}^3\mathbf{k}, \\ &= \int_{-\infty}^\infty \mathbf{k}(e\,\tau_m/m_\mathrm{e})\mathbf{E} \cdot \nabla_\mathrm{k} f_\mathrm{k}^0\, \mathrm{d}^3\mathbf{k} \Big/ \int_{-\infty}^\infty f_\mathrm{k}^0\, \mathrm{d}^3\mathbf{k}, \end{aligned} \tag{2.47}$$

where the integrand $\mathbf{v} f_\mathrm{k}^0$ in the numerator and $\nabla_\mathrm{k} f_\mathrm{k}$ in the denominator are eliminated because of the odd functions. Since $\mathbf{v}_\mathrm{d} = -\mu_\mathrm{e}\mathbf{E} = -(e\langle\tau_m\rangle/m_\mathrm{e})\mathbf{E}$ from (2.47), then:

$$\langle\tau_m\rangle = -\int_{-\infty}^\infty \tau_m \mathbf{k}\, \nabla_\mathrm{k} f_\mathrm{k}\, \mathrm{d}^3\mathbf{k} \Big/ \int_{-\infty}^\infty f_\mathrm{k}^0\, \mathrm{d}^3\mathbf{k}. \tag{2.48}$$

For the relatively low-field regime with the shifted MB distribution function, it is noted that (2.45) is equivalent to

$$f_\mathrm{k} \sim f_\mathrm{k}^0 + f_\mathrm{k}^1 \cos\theta, \tag{2.49}$$

where $\mathbf{kE} = |\mathbf{k}|\,|\mathbf{E}| \cos\theta$, $f_\mathrm{k}^1 = [|\mathbf{k}|\,|\mathbf{E}|\,(e\hbar\,\tau_m/m_\mathrm{e} k_\mathrm{B} T_\mathrm{L})] f_\mathrm{k}^0$, and $f_\mathrm{k}^1 \ll f_\mathrm{k}^0$. For the high-field regime, attempts to find f_k may be made based on an expansion of f_k similar to (2.50). In this regime, f_k^0 may no larger than f_k^1

and may deviate from the simple MB distribution function [8]. Thus, additional terms in the expansion of $f_{\mathbf{k}}$ are needed in a series of Legendre polynomials $P_n(\cos\theta)$ [8]:

$$f_{\mathbf{k}} = \sum_{n=0} f_n\,(\varepsilon^*) P_n(\cos\theta). \tag{2.50}$$

Various methods to find the solution of the high-field Boltzmann transport equation are given in Table 2.1 [8]. The Monte Carlo method is described in Section 2.2.

2.2 Nonequilibrium phase transitions

2.2.1 High-field regime: the Gunn effect

Nonlinear electrical transport in semiconductors exhibits a variety of nonequilibrium phase transitions and instabilities [2, 4, 9, 26], among which many of the nonlinear phenomena are related with negative differential resistance (NDR) [2, 27]. For high- or modest-field transport behaviors in the bulk semiconductors, many instabilities including field-enhanced capture and emission [9, 27–29], acousto-electric effect [4, 30–32], recombination instability [33] and plasma instabilities [26] have been found experimentally or modeled theoretically. The most well-known and widely investigated transport behavior is the **transferred-electron (TE) effect** which exhibits an N-shaped $J-E$ characteristic (or voltage-controlled NNDC; see Fig. 2.1b) and subsequently high-frequency instability. The mechanism of the NNDC was first proposed by Ridley and Watkins [34] and Hilsum [35], and called the RHW mechanism, and soon afterwards oscillatory instability was found in n-GaAs by Gunn [36, 37]. The effect has been applied to the transferred-electron device (TED, also known as the Gunn oscillator) in GaAs and InP and is well established; operating frequencies reach 100 GHz and the pulse powers attain 6 kW (for a review see Ref. [27]). Nevertheless, owing to its nonlinear character unreliable chaotic behavior has been also predicted numerically [38–40]. Here, as the most typical example of the nonequilibrium phase transition, the TE effect is summarized. In Section 2.2 some numerical studies [38–40] are introduced as a precaution of the complex dynamical behaviors.

Qualitatively, the NNDC in the III–V compounds such as GaAs and InP is characterized by electron transfer from the Γ-valley to the L-valley in the conduction bands, as shown schematically in Fig. 2.6 (see also Fig. 2.2), where the energy separation between the bottoms of the Γ-valley and L-valley is $\nabla_{\Gamma L} = 0.29$ eV. Applying a modest electric field to the sample, the electron gas is well heated above the bottom of the Γ-conduction band, thereby electrons suffer various types of scattering (polar optical, piezoelectric and acoustic-deformation potential scattering). If the electron–

Table 2.1 Solution of high-field Boltzmann equation [8]

Analytical methods (assumptions concerning f)		Numerical methods (no assumptions concerning f)	
(a) Electron temperature model	(b) Diffusion approximation	(a) Monte Carlo method	(b) Iterative method: integrodifferential
$f(\mathbf{k}) \propto \exp[-\hbar^2(\mathbf{k} - \mathbf{k}_d)/2mk_B T_e]$.	$f(\mathbf{k}) = f_0(\varepsilon) + f_1(\varepsilon)\cos\theta$	Path of single electron in $\mathbf{k}$ – space	BE is reduced to a differential equation by
Balance equations:	Solution of BE	$\mathbf{k} = \mathbf{k}_0 + e\mathbf{E}(t - t_0)/\hbar$	$g_{n+1}(\mathbf{k}) = \int f_n(\mathbf{k}')W(\mathbf{k}',\mathbf{k})d^3\mathbf{k}'$
$\int k\{(\partial f/\partial t)_\tau + (\partial f/\partial t)_c\}d^3\mathbf{k} = 0$ $\int \varepsilon\{(\partial f/\partial t)_\tau + (\partial f/\partial t)_c\}d^3\mathbf{k} = 0$	$(\partial f/\partial t)_\tau + (\partial f/\partial t)_c = 0$ for f_0 and f_1.	terminated according to three probability distribution functions: (i) for scattering event to occur, (ii) for particular scattering process, (iii) for choosing $\mathbf{k}'$ after scattering event. $f(\mathbf{k})$ obtained from time which the electron spends in each cell of $\mathbf{k}$-space using the ergodic theorem. Method equivalent to a solution of the BE.	then $(e\mathbf{E}/\hbar)(\partial f_{n+1}(\mathbf{k})\partial(\mathbf{k}) + f_{n+1}(\mathbf{k})\lambda(\mathbf{k}) = g_{n+1}(\mathbf{k})$ where $\lambda(\mathbf{k}) = \int W(\mathbf{k},\mathbf{k}')d^3\mathbf{k}$
or $f = f_0 + f_1$; τ exists $f_0 \propto \exp - (\varepsilon/k_B T_e)$ $f_1 = (e\tau(\varepsilon)E/\hbar)(\partial f_0(\mathbf{k}/\partial\mathbf{k})$.	Analytically only possible if further approximations concerning the collision term are made, e.g., equivpartition approximation for phonon occupation number and restriction on scattering mechanisms.		Method converges to $f(\mathbf{k})$
$E(T_e)$ from energy balance $e\mu E^2 = (2/8\pi^2 n)\int \hbar_q W(\mathbf{k},\mathbf{k}+\mathbf{q})d^3\mathbf{q}$			

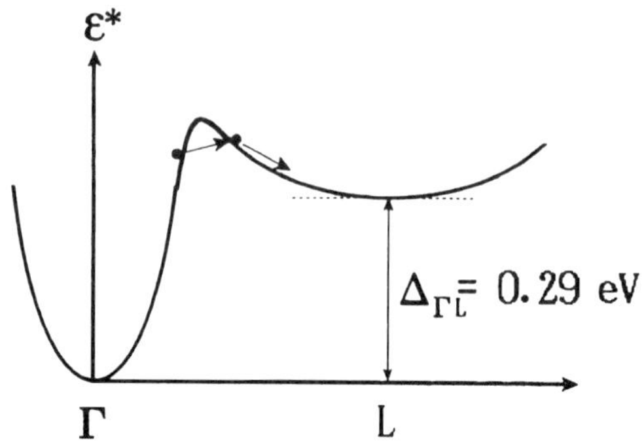

Figure 2.6 Electron transfer from the Γ- valley to the L-valley in the conduction bands.

electron scattering is insignificant (for material with a low carrier density), the electron distribution deviates from the MB or shifted MB distribution. Under the high field (~ 3 kV/cm in GaAs) the electrons gain sufficient energy to be transferred from the high-energy part of the lower lying Γ-valley to the bottom of the higher lying L-valley. It is well known that the effective mass of the electron is inversely proportional to the convexity of the parabolic conduction bands, and in GaAs, $m_1^* = 0.07m_0$ for the Γ-valley and $m_2^* = 0.55m_0$ for four equivalent satellite (L-) valleys [4], so that the transferred electrons in the L-valleys have lower mobility μ_2 than the mobility μ_1 in the Γ-valley. Consequently, the electrical conductivity is effectively given by [4]:

$$\sigma = e(n_1 \mu_1 + n_2 \mu_2). \tag{2.51}$$

If the fraction n_2/n_1 of the transferred electrons to the Γ-electrons become significantly large, the effective electron mobility becomes totally small and thereby nonequilibrium phase transition of first order occurs from the high-mobility to the low-mobility state. By applying a sufficiently high field to cause the intervalley transfer, a high-field domain can be formed at some inhomogeneous region of the cathode. Since the drift mobility of the transferred electrons in the L-valley is small, they tend to accumulate behind the domain front, while the carriers are depleted at the domain front, giving the electron profile and correspondingly the electric field profile in Fig. 2.7 [4]. Since the high-field domain forms a dipole layer, it is also called a **dipole domain**. The domain nucleated at the cathode runs towards the anode with a drift velocity v_d of 2×10^7 cm/s (during the transit the domain amplitude grows), and finally disappears after reaching the anode (see Fig. 2.8 [4, 41]). After the annihilation, a new domain can be formed at the cathode. The repetitive process shows the Gunn oscillation. The oscillation frequency is

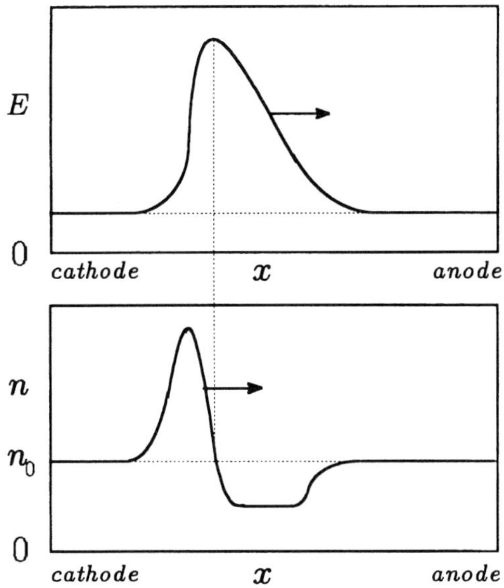

Figure 2.7 Electron profile of the dipole domain (bottom), and the electric field profile (top). Redrawn from [4].

expected to be improved by shortening the sample length L since the transit time is approximately given by $t_r = L/v_d$ ($f_I \sim 5$ GHz for $L = 0.1$ mm, $v_d \sim 2 \times 10^7$ cm/s). This type of the oscillation mode is known as the **transit**

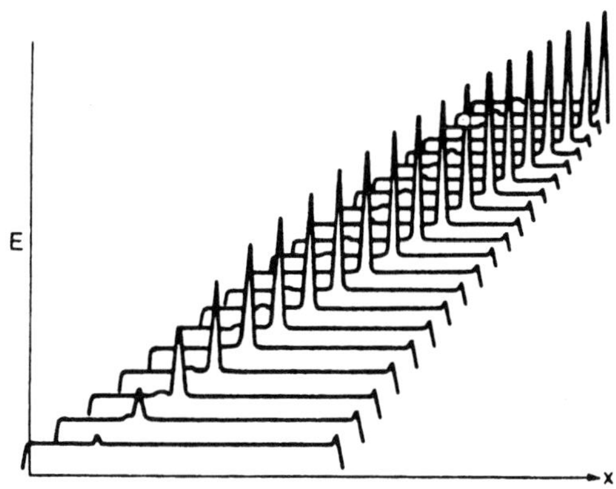

Figure 2.8 Propagation of the dipole domain. After [4]; For the original figure, see [41].

time mode (see next section). However, for a given sample with the doping density n_0, there exists a criterion that the n_0L product should roughly exceed 10^{12} cm^{-2} for oscillatory behavior [4, 27, 42], otherwise the stationary domain may be formed stably at the anode with a shorter sample length. If the sample length is much longer than the minimum length, the unlikely situation of simultaneous generation of multidomains may occur because the new domain may be emitted from the cathode before a domain arrives at the anode [4]. For the above-mentioned domain instability Ridley has discussed from nonequilibrium thermodynamical arguments (from a viewpoint of the **minimum entropy production**) [43]. For the TE effects in Ge and Si see Ref [4].

Numerical studies of the hot electron effects have also been intensively investigated by means of the **Monte Carlo method** [5, 10, 21, 22] or by the **iterative method** [23, 24]; both methods give direct numerical solutions of the BE. For application to devices such as the TED or Gunn oscillator, and also from the interest of nonlinear carrier transport in high-field regimes, accurate knowledge on the distribution function and the drift velocity and field characteristic is required. The Monte Carlo method was first introduced by Kurosawa, who calculated the average energy, drift velocity and energy distribution function for the problem of the streaming effect in p-Ge [44]. Many authors subsequently developed the Monte Carlo method for the TE effect [5, 10, 21, 22, 45, 46]. In particular, Fawcett *et al.* [5] calculated the stationary electron transport properties of GaAs by the Monte Carlo method. They evaluated accurately the distribution function and the drift velocity as a function of the electric field, under the influence of various scattering processes including intravalley scattering, equivalent intervalley scattering for electrons in the L-valleys and nonequivalent scattering of electrons between the Γ-valley and the L-valleys. In their calculation, the nonparabolicity of the conduction-band minimum was taken into account and wave-vector dependence of the periodic part of the Bloch functions for electrons in this minimum was also included [5]. By the Monte Carlo method, the motion of one single electron in momentum space is calculated for a long time, instead of the collected motions of the electron gas. Such a numerical procedure using one electron is allowed for the homogeneous stationary state due to the ergodic property. An essential part of the simulation is to generate random numbers in order to determine the time of free flight, type of various scattering, and wave vector $\mathbf{k}'$ after the scattering. For a number of free flights and scatterings (random events), the wave vector $\mathbf{k}'$ and the corresponding energy are recorded. Then the drift velocity (Fig. 2.9c) and the distribution function $f_{\mathbf{k}}(\varepsilon)$ can be calculated from the data recorded, where the $\mathbf{k}$-space of interest which covers the Γ-band and the equivalent satellite valleys must be divided into many cells with a preferable accuracy in order to find the probability distribution in each cell.

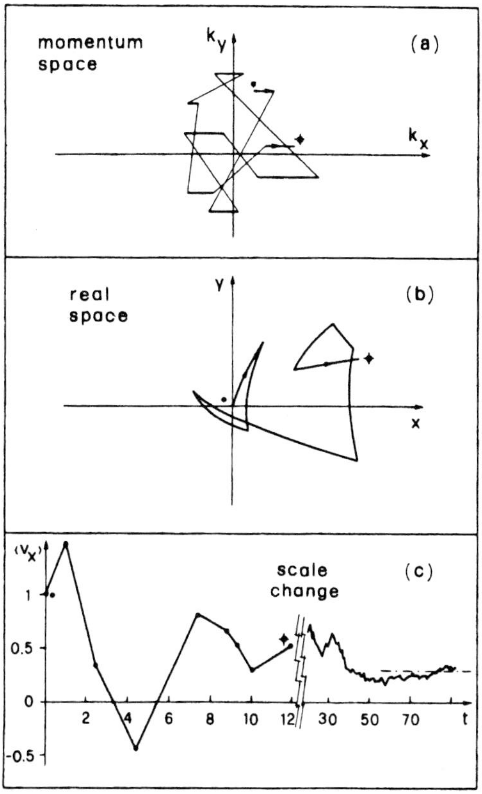

Figure 2.9 Principles of the Monte Carlo method. (a) Simulation of the sampling particle, in the wave–vector space, subject to an acceleration field oriented along the positive x direction. The heavy segments are due to the effect of the field during free flights; curves represent discontinuous variations of **k** due to the scattering process. (b) Path of the particle in real space. (c) Average velocity of the particle obtained as a function of simulation time. The horizontal dashed curve represents the "exact" drift velocity with a very long simulation time [22].

By applying a high-field E in the x direction, the trajectory of the free flight can be calculated by:

$$\mathbf{k}(t) = \mathbf{k}_0 + q\mathbf{E}t/\hbar, \tag{2.52}$$

where $\mathbf{k}_0$ is the wave vector at the beginning of the flight $(t = 0)$, t is the time duration of the free flight, $\mathbf{k}(t)$ is the wave vector at the end of the flight and q is the charge of the carrier ($q = -e$ for electrons, $q = e$ for holes). An example of the trajectory is schematically given in Fig. 2.9 [22] and the trajectory in the real space is shown in Fig. 2.9b. In Fig. 2.9c, the averaged quantity of the drift velocity approaches a well-defined value with increasing simulation time. Denoting scattering rate from a state of the wave vector $\mathbf{k}$ to $\mathbf{k}'$ by $S_j(\mathbf{k}, \mathbf{k}')$, total scattering rate over the possible $\mathbf{k}'$ is given by

$$\lambda_j(\mathbf{k}) = \int d\mathbf{k}' S_j(\mathbf{k}, \mathbf{k}'). \tag{2.53}$$

The probability per unit time that the electron will drift for a time t in the electric field E and then be scattered is given by [5, 22, 26]:

$$P(t) = \lambda[\mathbf{k}(t)] \ \exp\left\{-\int_0^t [\mathbf{k}(t')]dt'\right\}, \tag{2.54}$$

where

$$\lambda(\mathbf{k}) = \sum_{j=1}^{N} \lambda_j(\mathbf{k}).$$

The time of the free flight between collisions can be determined by generating a random number r and solving the equation:

$$r = 1 - \exp\left\{-\int_0^t [\mathbf{k}(t')]dt'\right\}. \tag{2.55}$$

However, this procedure is intricate since $\lambda[\mathbf{k}(t')]$ is a complex function of time [5, 22, 26]. By introducing a fictitious process [5, 22, 26, 45, 46], namely, **self-scattering process**

$$S_0(\mathbf{k}, \mathbf{k}') = \lambda_0(\mathbf{k}) \ \delta(\mathbf{k} - \mathbf{k}'), \tag{2.56}$$

the problem becomes simple. Since the self-scattering is of no physical significance and the delta function does not allow the electron wave vector to change, one can take the function $\lambda_0(\mathbf{k})$ arbitrarily. Replacing $\lambda[\mathbf{k}(t')]$ in (2.54) by $\lambda_0[\mathbf{k}(t')] +\ \lambda[\mathbf{k}(t')]$ and putting $\lambda_0[\mathbf{k}(t')] = \Gamma - \lambda[\mathbf{k}(t')]$, (2.54) becomes a simple form [5, 22, 26]:

$$P(t) = \Gamma \ \exp(-\Gamma t) \tag{2.57}$$

Correspondingly (2.55) becomes

$$r = 1 - \exp(-\Gamma t), \tag{2.58}$$

or $t = \tau_0 \ \ln(1 - r)$, where $\tau_0 = 1/\Gamma$. For a given value of Γ, t can be determined by generating the random number r from (2.58).

The next step is to select the type of scattering processes from polar optical scattering, acoustic phonon scattering, intravalley scattering, equivalent intervalley scattering between the (100) valleys (e.g., in GaAs), and non-equivalent intervalley scattering between the (000) and (100) valleys [5]. Since the probability of the electron being scattered by a scattering process j is proportional to $\lambda_j(\mathbf{k})$, and since

$$\sum_{j=0}^{N} \lambda_j(\mathbf{k}) = \Gamma$$

(being a fixed value), it is only necessary to generate a random number s between 0 and Γ and test the inequality

$$s < \sum_{j=0}^{m} \lambda_j(\mathbf{k}) \tag{2.59}$$

for all m [5, 26]. The scattering process m can be selected if (2.59) is satisfied. If $m = 0$, it is a fictitious scattering and does not count as a real scattering event. After determining m, one must further select the final state $\mathbf{k}'(t)$ of the scattering, where the scattering event is considered to be instantaneous. The procedure to find $\mathbf{k}(t)$ by generating random numbers depends on the type of scattering. For acoustic scattering in the (100) valley and all intervalley scattering, the wave vector is decomposed as $k'_\rho = |\mathbf{k}'| \sin\theta$ and $k'_x = |\mathbf{k}'| \cos\theta$, where the components k'_ρ and k'_x are perpendicular and parallel to the electric field, respectively. By generating a random number r, for a given relation $r = 1/2(1 - \cos\theta)$ is obtained. The absolute value $|\mathbf{k}'|$ is found from the energy conservation which appears in the formula of the scattering rate $S_j(\mathbf{k}, \mathbf{k}')$ [5]. For polar and acoustic scattering in the (000) valley and polar scattering in the (100) valleys, the procedure is rather more complicated; for details see Ref. [5].

From the above procedures, a scattering event is found with the final wave vector $\mathbf{k}'$ and the free flight time t. The wave vector $\mathbf{k}_0$ in (2.52) is replaced by $\mathbf{k}'$ and the free flight trajectory of the electron is simulated again. The random scattering event and free flight are simulated repeatedly for a long time, where many self-scattering events do not contribute to the transport property. Various physical quantities $X[\mathbf{k}(t)]$ can be evaluated by collecting the data at each free flight [22]:

$$\langle X \rangle_T = \frac{1}{T_j} \sum_{i=0} \int_0^{t_i} X[\mathbf{k}(t')]\mathrm{d}t', \tag{2.60}$$

where T_j is the sum of the free flight time in the valley j and t_i is the duration of free flight ($T_j = \sum t_i$). In order to obtain the distribution functions in the (000) and (100) valleys, one must divide (k_z, k_ρ) space into cells with the preferable accuracy, and count the time that the electron spends in each cell of the mesh. Let $X[\mathbf{k}(t)] = n_{\ell\mathrm{m}}$ in (2.60), where ℓm denotes the ℓmth element of the mesh, and let $n_{\ell m} = 1$ if the electron resides inside the ℓmth element and 0 otherwise, then the probability $f_{\mathbf{k}}(\varepsilon^*)$ that the electron spends in the cell is given from (2.60) by

$$f_{\mathbf{k}}(\varepsilon) = \frac{1}{T_j}\sum t_{\ell m}, \tag{2.61}$$

where $t_{\ell m}$ is the residence time in the cell and the sum is taken every time when the electron visits the ℓmth element. The drift velocity $v_{\mathrm{d}j} = \langle \hbar k_x/m_j \rangle$ in valley j (Fig. 2.9c) can be determined by inserting $X[\mathbf{k}(t)] = \hbar k_x/m_j$ in (2.60):

$$v_{\mathrm{d}j} = \frac{1}{T_j}\sum_i \int_0^{t_i} (\hbar k_x/m_j)\,\mathrm{d}t', \tag{2.62}$$

where m_j is the effective mass in the valley j. Since $\mathrm{d}k_x/\mathrm{d}t = qE/\hbar$, one can convert the integral over the time of free flight into the integral over the momentum space, and since

$$k_x = (m_j/\hbar^2)\frac{\partial \epsilon}{\partial k_x},$$

then [5]:

$$v_{\mathrm{d}j} = \frac{1}{K_j}\int_{k_{x\mathrm{i}}}^{k_{xf}} \frac{1}{\hbar}\frac{\partial \epsilon}{\partial k_x}\,\mathrm{d}k_x = \frac{1}{\hbar K_j}(\varepsilon_f - \varepsilon_{\mathrm{i}}), \tag{2.63}$$

where k_{xi} and k_{xf} are the initial and final values of k_x for each free flight, K_j is the total length of the **k**-space trajectory, given by $K_j = qET_j/\hbar$, and ε_{i} and ε_f are the initial and final energies for the flight, respectively.

Numerical results of the distribution function (the spherically symmetrical part for the Γ-valley) in GaAs by Fawcett *et al.* [5] are shown in Fig. 2.10, where the energy in Fig. 2.10a is measured from the bottom of the Γ-band. For GaAs, the electron transfer is considered only from the Γ-valley to the L-valley. At low field (~ 1 kV/cm), the distribution function is approximately Maxwellian (broken curve in Fig. 2.10a), as expected. For the high-field regime, the structure becomes significant as a result of various scattering processes; it becomes flatter owing to the hot electron and it shows a rapid fall-off above the energy 0.36 eV owing to the intervalley scattering, where the energy difference between the Γ-valley and L-valley minima is given by $\Delta_{\Gamma\mathrm{L}} = 0.29$ eV (see Fig. 2.6) [4]. For the fields at 15 and 25 kV/cm, the distribution function exhibits a population inversion between 0.2 and 0.36 eV in Fig. 2.10a. In Fig. 2.10b the distribution function in the wave-vector space is plotted along lines through $\mathbf{k} = 0$ parallel (full curve) and perpendicular (broken curve) to the electric field of 15 kV/cm [5]. The transverse distribution in Fig. 2.10b is apparently symmetrical around $\mathbf{k} = 0$ as expected. The population inversion in Fig. 2.10 is a consequence of a build-up of electrons below 0.36 eV in the transverse profile. The population inversion results from the combination of strong momentum radomization above 0.36 eV due to intervalley scattering and the relatively ineffective

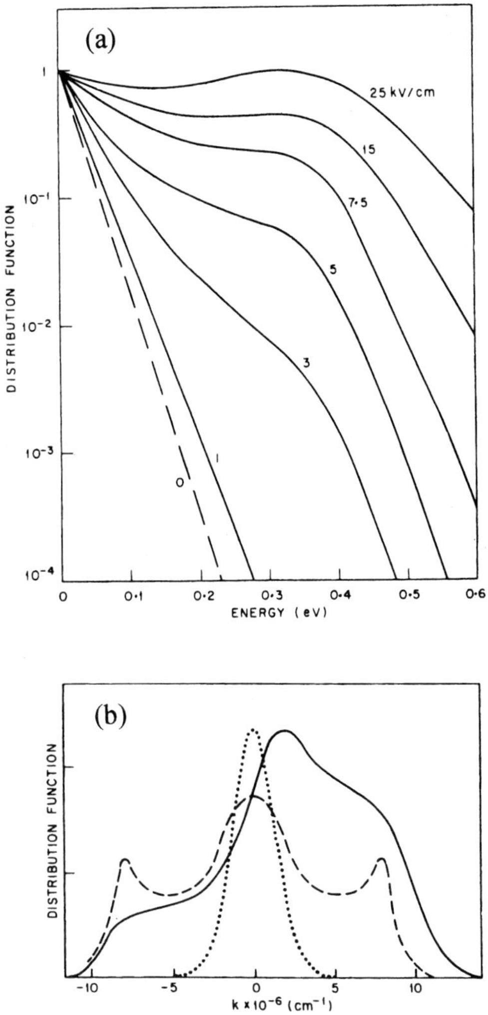

Figure 2.10 (a) Energy dependence of the spherically symmetric part of the distribution function in the (000) valley (Γ – *valley*) for an intervalley deformation potential of 1×10^9 eV/cm for the field strenghts indicated. The broken line is the zero-field Maxwellian. (b) Plots against k of the distribution function in the (000) valley along lines through $k = 0$ parallel (full curve) and perpendicular (broken curve) to a field of 15 kV/cm. The zero-field distribution function is shown by a dotted curve [5].

polar scattering below this energy. For details see Ref. [5]. In Fig. 2.11 the velocity-field characteristics are compared between the experimental results (full lines) and the numerical result by the Monte Carlo method with an

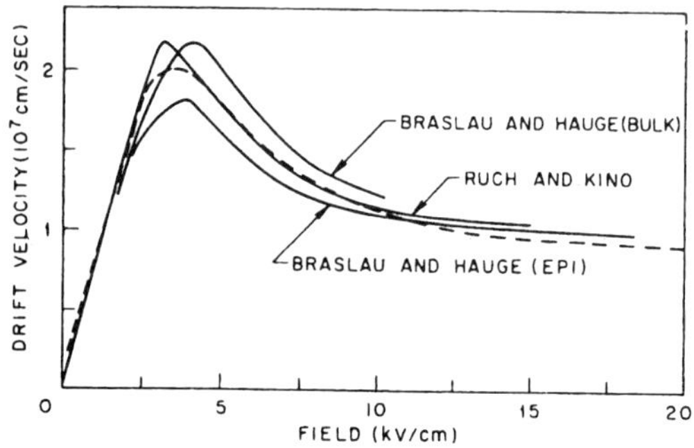

Figure 2.11 Drift velocity calculated by the Monte Carlo method for an intervalley deformation potential of 1×10^9 eV/cm (broken line). The result is compared with the experimental velocity-field characteristics (solid lines) for GaAs, investigated by the authors indicated [5].

intervalley deformation potential Ξ_{ij} of 1×10^9 eV/cm (broken line) [5]. The numerical curve is in good agreement with the experimental curves up to the highest field strength investigated by Ruch and Kino [47] for insulating GaAs, and also by Braslau and Hauge for bulk and epitaxial materials [48]. For InP or GaInSb, three-level electron transfer among the $\Gamma-$, X- and L-valleys must be considered carefully [26, 49].

2.2.2 High-field instability and chaos

Because of its applicability to electron devices, a huge number of studies has been conducted on high-field transport in compound semiconductors such as GaAs and InP [4, 26, 27, 50]. Today, applications of the TED are almost completely established based on the highly reliable and high-speed operations [4, 50]. Recent developments including the real-space transfer device [51–56] even use TED materials with microstructures such as GaAs/AlGaAs heterostructures [56–58] and quantum-well structures [59, 60]. Nevertheless, many interesting problems remain for the TED system, which arise from an interest in the nonequilibrium phase transition and, potentially, its complex nonlinear dynamical behaviors including chaos, as predicted numerically [38–40, 61–63]. In the experiments, some observations have exhibited irregular domain motion and the related current noise [64–66]. Here some theoretical aspects for the propagating domain [67, 68] and the complex dynamical behaviors in the model simulations [38–40, 61–63] are introduced.

The stability of the propagating domain was first discussed by Ridley [43], from a viewpoint of nonequilibrium thermodynamics, discussing the principle of minimum entropy production. Later, Butcher [67] and Butcher *et al.* [68] developed their arguments on stability in terms of an **equal-areas rule**, as introduced below. Suppose that the electrons with charge $-e$ move in the positive x direction so that the field F points in the x-direction (the field is usually denoted by E in this book). The velocity-field characteristic $v(F)$ is given in Fig. 2.12. Although the field dependence of the diffusion coefficient $D(F)$ may be highly important for the precise description [69], the coefficient is ultimately set to be a constant. The fundamental equations to be described are Poisson's equation and the equation of conservation of total current density J:

$$\frac{\partial F}{\partial x} = \frac{e}{\epsilon}(n - n_0), \tag{2.64}$$

$$J = env(F) - e\frac{\partial}{\partial x}\{D(F)n\} + \epsilon\frac{\partial F}{\partial t}, \tag{2.65}$$

where n is the electron density, n_0 is the density of the ionized donors (assumed to be spatially uniform), and ϵ is the static dielectric constant ($\epsilon = \epsilon_r\epsilon_0$). In the r.h.s. of (2.65) the total current density is composed of the conduction current (the first term), the diffusion current (the second term) and the displacement current (the third term). The conservation of the total current density requires $\partial J/\partial x = 0$, which is a special case of the continuity equation $\dot{\rho} + \mathrm{div}\,\mathbf{J} = 0$, where $\rho = e(n_0 - n)$, and $\dot{\rho} = \dot{n} = 0$ for the stationary domain and even for the stationary pattern in the moving frame (see Fig. 2.13). A solution for (2.64), (2.65) may take a pattern of a high-field

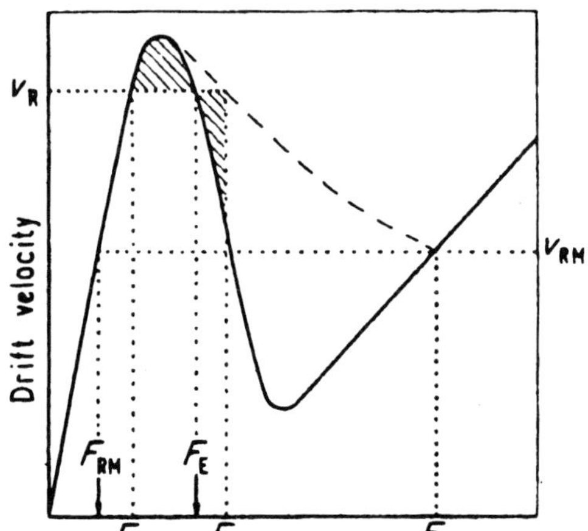

Figure 2.12 Schematic plots of drift velocity against field (full line) and maximum domain field against drift velocity outside the domain (dashed line). The two hatched areas are equal in magnitude [67, 68].

domain, like in Fig. 2.13, which propagates without change in shape with the drift velocity v_D. From Fig. 2.13, n is a two-valued function of F. The domain consists of a depletion layer with $n < n_0$, which is the leading edge of the domain, and an accumulation layer with $n > n_0$, which is the trailing edge.

In order to find the equal-areas rule in the $v(F) - F$ characteristic in Fig. 2.12, one can make use of the single variable $y = x - v_D t$ for the moving frame in Fig. 2.13. Since $\partial J/\partial x = 0$, one can safely put $J = en_0 v_R$, which is the conduction current density outside the domain, where $n = n_0$ and $v_R = v(F_R)$. Noting that $\partial F/\partial x = \mathrm{d}F/\mathrm{d}y$ and $\partial F/\partial t = -v_D \mathrm{d}F/\mathrm{d}y$, (2.64) and (2.65) are converted to:

$$\frac{\mathrm{d}F}{\mathrm{d}y} = \frac{e}{\epsilon}(n - n_0), \tag{2.66}$$

$$\frac{\mathrm{d}}{\mathrm{d}y}\{D(F)n\} = n\{v(F) - v_D\} - n_0(v_R - v_D). \tag{2.67}$$

Dividing these equations gives a differential equation for $\{D(F)n\}$ as a function of the field,

$$\frac{e}{\epsilon}\frac{\mathrm{d}}{\mathrm{d}F}\{D(F)n\} = \frac{n\{v(F) - v_D\} - n_0(v_R - v_D)}{n - n_0} \tag{2.68}$$

Equation (2.68) is generally solved only by numerical methods [69]. By making $D(F)$ a constant, however, the problem becomes quite simple, with

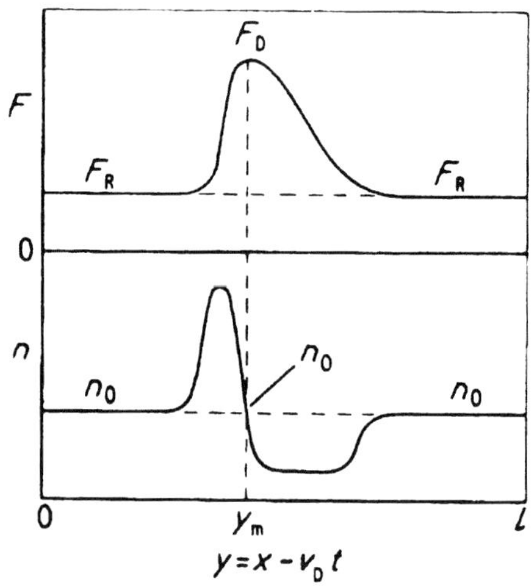

Figure 2.13 Schematic plots of the field F and the electron density n against $y = x - v_D t$ for a stable high-field domain propagating with velocity v_D [68].

the solution:

$$\frac{n}{n_0} - \ln\frac{n}{n_0} - 1 = \frac{\epsilon}{en_0 D}\int_{F_R}^{F} \mathrm{d}F\,\{v(F) - v_D\} - \frac{n_0}{n}\{v_R - v_D\}, \tag{2.69}$$

where the boundary condition $n = n_0$ at $F = F_R$ has been imposed [67, 68]. The solution (2.69) is easily verified by differentiation. As can be seen from Fig. 2.13, $n = n_0$ at $F = F_D$ where F_D is a maximum field of the domain between the depletion and accumulation layers, so that the l.h.s. of (2.69) vanishes when $F = F_D$. Then the integration over the range $[F_R, F_D]$ must also vanish. However, if $v_D \neq v_R$, it is impossible because the second integration $\int(1/n)\mathrm{d}F$ has two different values depending on the two different integration paths including the depletion layer or the accumulation layer (see Fig. 2.13). Therefore, in order for the acceptable solution of (2.69) to exist, it can be concluded that $v_D = v_R$. This relation is physically important, indicating that the domain velocity v_D is equal to the drift velocity of the electrons in the region outside the domain. The important formula of the equal-areas rule is given by the equation:

$$\int_{F_R}^{F_D} \mathrm{d}F\{v(F) - v_D\} = 0. \tag{2.70}$$

Since $v(E)$ is the N-shaped velocity-field characteristic represented by Fig. 2.12, integration in (2.70) clearly indicates that the shaded areas in Fig. 2.12 are equal in magnitude. The broken line in Fig. 2.12 is a schematic plot of F_D against v_R which is determined by the equal-areas rule [67, 68]. It begins at the peak of the velocity-field characteristic where the field equals the threshold field F_T, and ends at the point (F_{DM}, v_{DM}). As F_R decreases from F_T, $v_D(= v_R)$ decreases and F_D increases. So, if we change the field F_R as a control parameter in Fig. 2.12, there is a minimum value F_{RM} for the equal-areas rule to hold, corresponding to the end-point of the broken line. With a further decrease in F_R, the equal-areas rule (2.70) can be no longer satisfied and stable domain propagation is impossible [67, 68]. The field F_R outside the domain is directly related to a bias potential φ by

$$\varphi = F_R L + \varphi_D(F_R), \tag{2.71}$$

where L is the sample length, and $\varphi_D(F_R)$ is the domain potential, which is the area of the domain-field profile in Fig. 2.13 (integrated above the horizontal broken line in Fig. 2.13).

The equal-areas rule seems to guarantee the stable periodic Gunn oscillation with the transit-time mode (Gunn mode) by the mechanism described briefly in Section 2.2.1, and many microwave experiments in GaAs and InP have established the stable transit-time mode operation [50]. However, recent numerical studies [40, 61–63] have strongly suggested that the stable Gunn oscillation may lose its stability and may reveal complex chaotic behaviors when the Gunn diode (GaAs) is coupled with a resonant circuit of the microwave cavity [40, 61, 62] or when the diode is fabricated by a narrow-gap semiconductor such as InSb [63]. The pioneering work was done by Nakamura in 1977 [38] who, for the first time, paid attention to the occurrence of chaos in the Gunn oscillator from a viewpoint of nonlinear dynamical theory. Nakamura [38] has studied the following fundamental equation:

$$\frac{\partial E}{\partial t} = -n_0 v(E) + D\frac{\partial^2 E}{\partial x^2} - v(E)\frac{\partial E}{\partial x} + J(t), \tag{2.72}$$

where all of the notations are the same as in (2.64), (2.65) except that the electric field is replaced by E. Equation (2.72) is found easily by differentiating (2.64) as a function of x and substituting it into (2.65). Then, $E(x, t)$ undergoes fourier analysis in the form:

$$E(x, t) = E_0 + \sum_{k=0} c_k(t)e^{ik(x - v_0 t)}, \tag{2.73}$$

where E_0 is the applied electric field and $v_0 = v(E_0)$. Imposing the periodic boundary condition on $E(x, t)$, the wave number k is given by $k = (2\pi/L)m$, where L is the sample length as usual and m are positive or negative integers. Substituting (2.73) into (2.72) and expanding $v(E)$ into a power series of E, a set of differential equations for c_k can be obtained [38]. The set of system equations is not described here; readers interested in these equations may refer to see Ref. [39]. By setting the system equations with 40 modes ($m = 40$), Nakamura [38] found many complex chaotic solutions, one of which is shown in Fig. 2.14, where two trajectories show sensitive dependence on the initial conditions.

In actual diode systems, various operation modes must be considered when the Gunn diode is operated in the microwave cavity [50, 70–72], where the whole system can be described by an equivalent resonant circuit as given in the inset of Fig. 2.15 [73]. The modes of operation are characterized, in Fig. 2.16, in the parameter space of fL (frequency × length) and n_0L [50],

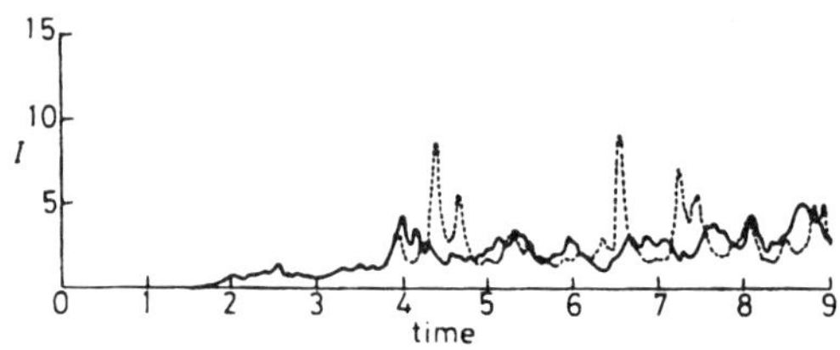

Figure 2.14 Simulated result of time behavior of the current density in the Gunn instability, starting with slightly different initial conditions [38].

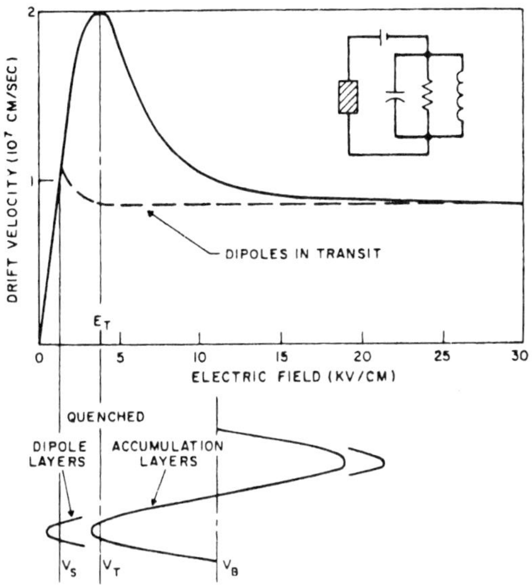

Figure 2.15 Drift velocity of electrons vs electric field in n-GaAs. The broken curve is the dynamic characteristic and applies when dipoles are in transit. The inset shows the equivalent oscillator circuit. A plot of voltage vs time (bottom) illustrates the principle of the quenched modes. The voltage is sinusoidal with a frequency determined by oscillator circuit shown [73].

where f is the resonant frequency. The vertical line at $n_0L = 10^{12}$ cm^{-2} indicates the borderline for the basic **Krömer's criterion** $n_0L \gtrsim 10^{12}$ cm^{-2} concerning the stably propagating domain [42]. With n_0L smaller than 10^{12} cm^{-2} the dipole growth rate is reduced to a point that a dipole domain is not able to grow fully during its transit through the diode [73]. Such a mode is called the **accumulation layer mode** [73]. The diode can be used as a small signal amplifier in this region if the load resistance is sufficiently small [50, 73] (see the region denoted by amplification in Fig. 2.16). In contrast to the resonant circuit, a bulk GaAs diode with $n_0L > 10^{12}$ cm^{-2} supports essentially a stable transit-time mode [50] or, in other words, a stably traveling dipole layer mode [73]. In the resonant circuit, however, the two basic modes can be further classified into more modes **(LSA mode, quenched domain mode, transit-time mode** and **delayed domain mode)**, as shown in Fig. 2.16, where the transit-time mode is also denoted by the dipole layer mode [73]. The delayed domain mode is also called the inhibited domain mode. The LSA in Fig. 2.16 stands for limited space-charge accumulation [70, 72]. Both single and multiple domains are possible; the latter has been investigated numerically by Thim [73] considering doping fluctuation and experimentally by Thim and Barber [74]. The characteristics of the LSA

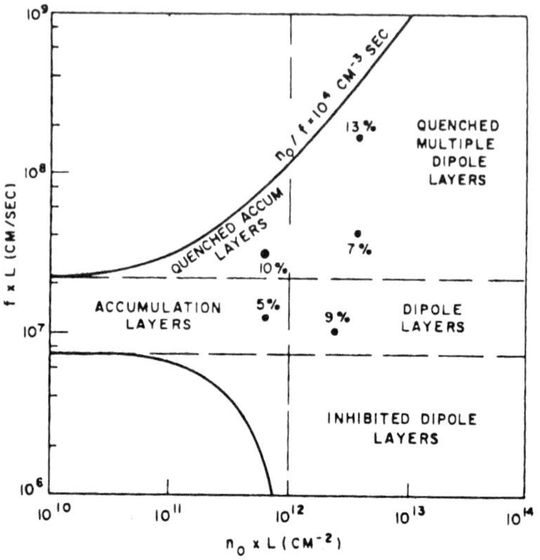

Figure 2.16 Oscillator mode chart for n-GaAs diodes with 11% one-dimensional random doping fluctuations [73].

mode are: (i) the frequency of operation is higher than the reciprocal of the carrier transit time ($fL \gtrsim 2 \times 10^7$ cm/s; see Fig. 2.16); (ii) the power output and efficiency are equal to or higher than when the same device is operated in the transit-time mode or the quenched dipole mode [70]; and consequently (iii) there is high dc-to-rf conversion efficiency in the 50–120 GHz regime. In this mode, the current and the internal field would be almost uniform, proportional to the applied voltage, so that the entire device would follow a current–voltage characteristic of the same relative shape as the velocity-field characteristic (Fig. 2.15). For this mode, the voltage swing is so quick with a relatively high resonant frequency that the space-charge distribution does not have sufficient time to form the high-field domain [50], where the bias voltage drops below the threshold value V_T during the negative half period (see Fig. 2.15). The primary accumulation layer can only be formed near the cathode, while the rest of the diode remains fairly homogeneous [50, 73]. The quenched domain mode in Fig. 2.16 was proposed by Carroll [71]. With the resonant frequency larger than the reciprocal of the carrier transit time ($fL \lesssim 2 \times 10^7$cm/s) and with the voltage swing of relatively large amplitude, the dipole domain is quenched before arriving the anode contact, since the system is driven well below the threshold by the rf voltage (Fig. 2.15). An example of the quenched domain mode has been presented in the numerical simulation by Oshio and Yahata [40] (Fig. 2.17). For the delayed domain mode in Fig. 2.16, where $fL \lesssim 1 \times 10^7$ cm/s with the frequency less than the transit time, the voltage across the

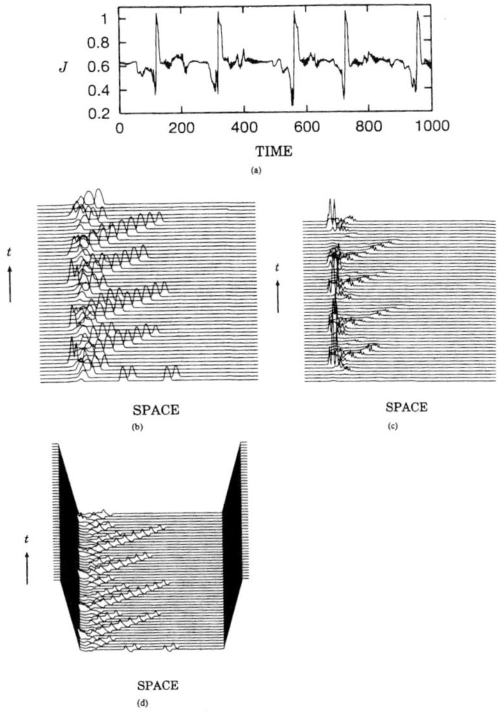

Figure 2.17 Numerical result of quenched mode of the domain propagation. (a) Current density as a function of time. (b, c and d) Spatiotemporal evolution of the electric field $E(x, t)$, the hole density $p(x, t)$ and the electron density $n(x, t)$, respectively. Note that the domain is quenched before reaching the anode. The anode contact is shown by the right black wall in (d) [40].

Gunn diode is below the threshold at the instant when a dipole domain arrives the anode. Then, formation of the subsequent new domain is delayed (inhibited) until the rf voltage rises above the threshold. The current waveform is a square wave 180° out of phase with respect to the voltage, and the frequency of the resonant circuit is tuned exactly to half the transit-time frequency [50, 73].

The above-mentioned system with the microwave cavity can be regarded as a periodically driven Gunn diode system [40, 61, 62]. Numerical investigations by Mosekilde *et al.* [61, 62] and Oshio and Yahata [40] have suggested complex dynamical behaviors and chaos including a **devil's staircase** of frequency-locked oscillations [61, 62]. (chapter 5) and **period doubling bifurcation** [40, 61, 62]. In a one-dimensional model system of GaAs diode presented by Mosekilde *et al.* [61, 62], the carrier density $n_1(x, t)$ and $n_2(x, t)$, respectively, in the high- and low-mobility state carry the current density when the electric field is applied in the x-direction [62]:

$$J_1 = en_1 v_1(E) - eD_1 \frac{\partial n_1}{\partial x}, \tag{2.74}$$

and

$$J_2 = en_2 v_2(E) - eD_2 \frac{\partial n_2}{\partial x}, \tag{2.75}$$

where the electrons are assumed to carry a positive charge e conventionally (no generality is lost by doing so), $v_1(E)$, $v_2(E)$ are the drift velocity, D_1, D_2 are the diffusion constants ($D_1 = 200\ \text{cm}^2/\text{s}$, $D_2 = 8\ \text{cm}^2/\text{s}$) in the lower-lying and higher-lying valleys, respectively, and $E = E(x, t)$ is the local field is determined by the Poisson equation:

$$\frac{\partial E}{\partial x} = \frac{e}{\epsilon}(n_1 + n_2 - n_0), \tag{2.76}$$

where ϵ is the static dielectric constant for GaAs ($\epsilon = \epsilon_r \epsilon_0$, $\epsilon_r = 12.5$), and n_0 is the doping concentration ($n_0 = 2 \times 10^{15}/\text{cm}^3$).

The drift velocity $v_1(E)$ in (2.74) is assumed to follow an analytical function:

$$v_1(E) = \begin{cases} \mu_1 E(1 - aE^2) & \text{for } E < E_{\text{sat}}, \quad (2.77) \\ v_{1,\text{sat}} & \text{for } E \geq E_{\text{sat}}, \quad (2.78) \end{cases}$$

where $\mu_1 = 5000\ \text{cm}^2/\text{Vs}$, $E_{\text{sat}} = 4.3 \times 10^3\ \text{V/cm}$, and $v_{1,\text{sat}} = 2.3 \times 10^7\ \text{cm/s}$ is the high-field saturation drift velocity in the Γ-band. By adjusting parameter a, the expression (2.77) may be fitted fairly well to the Monte Carlo result of the velocity-field characteristic $v_1(E)$ obtained by Fawcett *et al.* [5]. The drift velocity $v_2(E)$ is assumed to be proportional to the electric field, namely $v_2(E) = \mu_2 E$, where $\mu_2 = 320\ \text{cm}^2/\text{Vs}$. From the equations of continuity, the rate equations for the electron densities n_1, n_2 are given by:

$$\frac{\partial n_1}{\partial t} = -\frac{1}{e}\frac{\partial J_1}{\partial x} - \frac{n_2 - \bar{n}_2}{\tau} \tag{2.78}$$

and

$$\frac{\partial n_2}{\partial t} = -\frac{1}{e}\frac{\partial J_2}{\partial x} + \frac{n_2 - \bar{n}_2}{\tau} \tag{2.79}$$

where $\tau = 2$ ps is the nonequivalent intervalley scattering time, and $n_2(E)$ is the steady-state electron distribution in the L-valley corresponding to the applied field E [61, 62]. In (2.79), Mosekilde *et al.* [61, 62] used the simplified analytical approximation for $\bar{n}_2(E)$:

$$\bar{n}_2 = \frac{n_0}{1 + \alpha \exp[\Delta/(k_B T_L + bE^2)]}, \tag{2.80}$$

where T_L is the lattice temperature, and $\Delta = 0.35$ eV is the effective energy difference between the set of Γ-valley and L-valley slightly larger than $\Delta_{\Gamma L} = 0.29$ eV. Here, $\alpha = 0.33$ and $b = 60.0\ \mathrm{eV}/(\mathrm{V/cm})^2$ are the adjusting parameters such that the analytical function for the fraction $n_2(E)/n_0$ of the transferred electrons is as close as possible to the Monte Carlo results calculated with the deformation potential $\Xi_{ij} = 1 \times 10^9$ eV/cm by Fawcett *et al.* [5] (see Section 2.2.1). For the boundary conditions, the thermal equilibrium condition

$$n_1(0,t) = n_0, \qquad n_2(0,t) = 0 \tag{2.81}$$

at the cathode end (region of doping notch [40]) and the neutrality condition

$$n_1(L,t) + n_2(L,t) = n_0 \tag{2.82}$$

at the anode end are imposed [61, 62]. The driving force (rf voltage) is a time-dependent boundary condition for the electric potential $\varphi(x,t)$, and given by the normalization condition

$$\int_0^L E(x,t)\mathrm{d}x = V_{\mathrm{dc}}[1 + A\sin(2\pi f t)], \tag{2.83}$$

where L is the sample length as usual, V_{dc} is the applied dc voltage ($V_{\mathrm{dc}}/L = 4.0$ kV/cm), and f and A (control parameters) denote the frequency and the relative amplitude of the applied microwave, respectively. Since $E(x,t) = \partial\varphi(x,t)/\partial x$, (2.83) is equivalent to [40]:

$$\varphi(L,t) = 0, \qquad \varphi(0,t) = V_{\mathrm{dc}}[1 + A\sin(2\pi f t)]. \tag{2.84}$$

The equations of motion (2.78) and (2.79) and the Poisson equation (2.76) are solved by dividing the full sample length into 150 sections, each of length $\Delta x = 80$ nm [62]. Thus, the equations of motion provide a total of 300 dynamical variables and consequently the set of 300 locally coupled ordinary differential equations, where the spatial derivatives are all approximated by central differences, while time derivatives are approximated by forward differences [62, 75]. Mosekilde *et al.* [61, 62] solved an eight-stage Runge–Kutta pair of order five and six [62, 76]. They introduced the doping notch, the 1% of the crystal closest to the cathode end, where there is a slight (0.1%) enhancement of the specific resistance to facilitate the formation of domains. Oshio and Yahata [40] investigated the same problem of instability by introducing the total carrier density $n(x,t)$ instead of $n_1(x,t)$ and $n_2(x,t)$. They solved the Poisson equation for $\varphi(x,t)$ corresponding to (2.76) as well as the equation of motion for $n(x,t)$ [40], in which the drift term was discretized by using the second-order Lax–Wendroff method [75] and the diffusion term was transformed into the difference form using the Crank–Nicholson method [75].

Mosekilde *et al.* [61, 62] have discussed spatiotemporal behaviors of the carrier densities and the local field as well as temporal behaviors of the total current density:

$$J(t) = J_1(x,t) + J_2(x,t) + \epsilon \frac{\partial E}{\partial t} \tag{2.85}$$

where, as a result of the displacement current density $\epsilon(\partial E/\partial t)$, the total current density becomes independent of the space coordinate and therefore proportional to the external current. A typical example of the numerical results of Mosekilde *et al.* [61, 62] is shown in Fig. 2.18, where $A = 0.2$ and $f = 12.5$ GHz. In the upper panel of Fig. 2.18a, the total current density is given in the upper trace, and the average electric field $E_{av} = (V_{dc}/L)[1 + A \sin(2\pi ft)]$ in the lower trace. The lower panel of Fig. 2.18a is the contour curve of the propagating domain, the time axis is common to both panels and the vertical axis in the lower panel represents the position along the

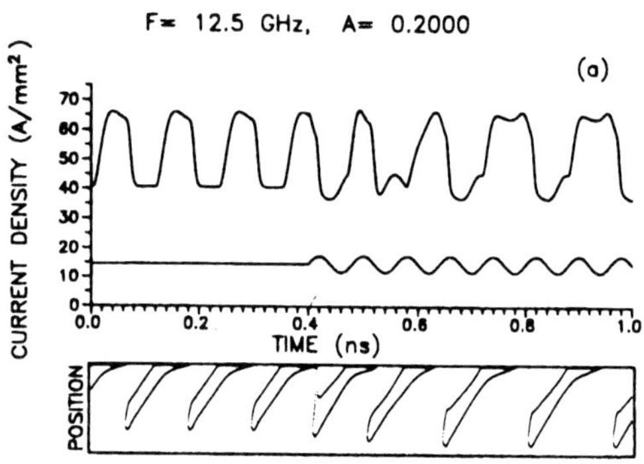

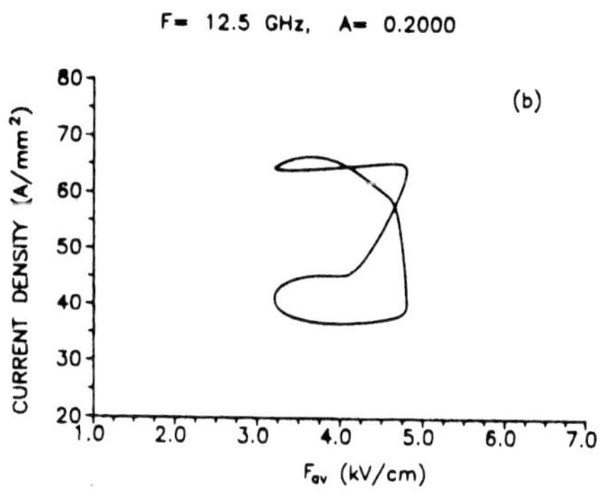

Figure 2.18 Onset of a 2:1 frequency-locked solution upon application of a microwave signal of frequency $f = 12.5$ GHz and relative amplitude $A = 0.2$. (a) Temporal variation of current density, applied field and internal field distribution (contour curve). (b) Phase plot for the stationary 2:1 solution [61].

crystal with the cathode end at the bottom. When the microwave signal is applied at $t = 0.4$ ns, the transit-time mode of the Gunn oscillation turns to the 2:1 frequency-locked solution, where the period of the Gunn oscillation doubles to the driving period $1/f$. Namely, during every two cycles of the microwave signal a dipole domain propagates only once from the cathode end to the anode. By comparing the total current density with the contour curve for $t \gtrsim 0.63$ ns in Fig. 2.18a, it is found that a dipole domain is formed at around the maximum field and propagates towards the anode, and that the current density drops during the formation of the domain. The current rises as the domain leaves the crystal ($t \sim 0.74$ ns). After that, the average field E_{av} becomes sufficiently high for a new domain to be formed after one driving period, at $t \sim 0.76$ ns, but it is not actually formed and the current density remains high during a second negative swing of the microwave signal ($t \sim 0.78$ ns). The transit-time mode (for $t < 0.4$ ns) thus adjusts to the microwave signal by delaying the formation of a new domain, resulting in the 2:1 frequency-locked mode [61]. Figure 2.18b shows the stationary phase-space trajectory, called the phase portrait, which is obtained by plotting simultaneous values of the current density and the averaged field E_{av}.

By increasing the microwave frequency f, the 2:1 frequency-locked mode is maintained up to 18.4 GHz. With further increases, many intervals of the driving frequencies exist in which the domain mode entrains into the frequency-locked modes with 3:1, 4:1, 5:1, ... [61], where 3:1 means one domain running through the crystal for every three microwave cycles, 4:1 one domain for every four microwave cycles, and so on. The 6:1 frequency-locked solution is shown in Fig. 2.19, where $f = 50.0$ GHz and $A = 0.2$. In the upper panel of Fig. 2.19a, the current density shows three oscillations around its high level during the period of domain formation, and subsequently three oscillations around the low level during the period of domain propagation (see also Fig. 2.19b). In the lower panel of Fig. 2.19a, it is seen that the contour curve, namely, the width of the domain field, is modulated by the microwave signal. Surprisingly, the period of the domain propagation and the spatiotemporal pattern are approximately the same as those in the transit-time mode (or Gunn mode; see the unperturbed pattern in lower panel of Fig. 2.18a), except for the width modulation. The 6:1 frequency-locked mode occurs when the driving period $1/f$ is tuned at values roughly equal to 1/6 of the period of the domain propagation in the transit mode.

The control parameters f and A maybe varied simultaneously, and an overview of the more complex frequency-locked solutions is shown in the phase diagram of Fig. 2.20. The $m : 1$ frequency-locked solutions ($m =$ 1, 2, ...) appear in the cusp-shaped regions called **Arnold tongues** (see Chapter 5). The width of each Arnold tongue increases on increasing the driving amplitude A. An infinite number of Arnold tongues is embedded in

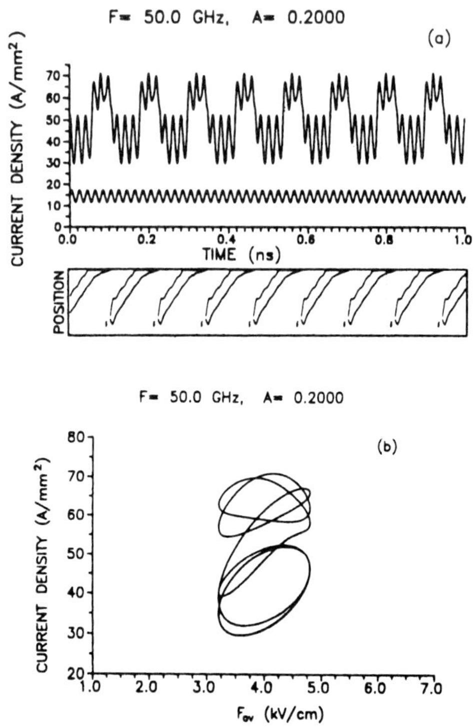

Figure 2.19 Stationary 6:1 frequency-locked solution existing for $f = 50.5$ GHz and $A = 0.2$. (a) Temporal variation of current density, applied field and internal field distribution (contour curve); (b) corresponding phase-space trajectory [61].

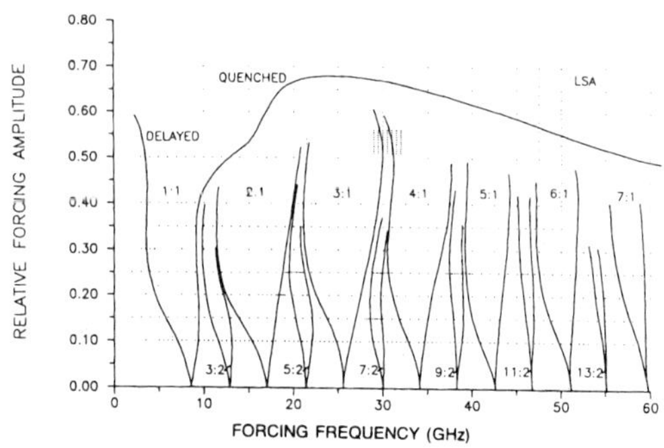

Figure 2.20 Phase diagram showing the distribution of periodic modes over the microwave frequency/microwave amplitude parameter plane. Interspersed with the periodic solutions are solutions with quasi-periodic or chaotic behavior. At high microwave amplitudes, transitions to delayed, quenched and limited space charge accumulation modes take place [61, 62].

the phase diagram. Between the two main tongues of the 1:1 and 2:1 frequency-locked states in Fig. 2.20, for instance, there is an Arnold tongue of the 3:2 frequency-locked state. Between 3:2 and 2:1, there is the 5:3 Arnold tongue (not shown in Fig. 2.20). Between the 5:3 and 2:1 Arnold tongues, there exists a narrower 7:4 Arnold tongue. Here, the 7:4 frequency-locked solution means four domains for every seven microwave cycles [61, 62]. In general, for any two adjacent Arnold tongues of $(m : n)$ and $(m' : n')$, the $(m + m') : (n + n')$ Arnold tongue exists in between. In Fig. 2.20, main (parents) and the submain (daughters) tongues are shown. There are other solutions, namely, the quasi-periodic states for which the ratio f'/f of the response frequency f' of the Gunn diode to the microwave frequency f is irrational and the phase space trajectory of the current density vs field does not show any closed orbits. In the parameter space of Fig. 2.20, with relatively low values of amplitudes A ($\lesssim 0.5$), the set of quasi-periodic solutions forms a sea. Therefore, every (or an infinite number of) Arnold tongue(s) would be embedded in the quasi-periodic sea. If the microwave frequency is increased from its low value, while the amplitude remains fixed, say $A = 0.2$, in Fig. 2.20, the bifurcation scheme is thus an alternating frequency-locked/quasi-periodic sequence. Whenever the control parameter (A,f) is fixed at a point outside any Arnold tongues, the solution is quasi-periodic. By increasing the microwave amplitude, the width of the Arnold tongues broadens, finally leading to an overlap between two Arnold tongues, and thereby resulting in the emergence of chaos and mode converting bifurcations [61, 62]. Mosekilde *et al.* [61, 62] have performed a large number of simulations in the parameter space of Fig. 2.20. In the region between the 3:1 and 4:1 tongues, where $A \gtrsim 0.5$ and $f \sim 30$ GHz, they have precisely investigated and found 13:4, 10:3, 14:4, 7:2, 11:3 and 8:2 frequency-locked solutions. With $A = 0.56$ and $f = 30.0$ GHz, chaotic solutions appear due to the overlapping of the Arnold tongues (Fig. 2.21). The spatiotemporal pattern of the dipole domain exhibits some irregular behavior, in which the domains are quenched and generated many times during running through the crystal. As the microwave amplitude is increased further, the mode converting bifurcations take place.

Numerical studies [61, 62] have also shown the solutions of the well-known modes of operation (quenched, delayed and LSA) as observed in the microwave experiments, where the locations of each mode are denoted in Fig. 2.20. The spatiotemporal pattern of the domain for the quenched mode is similar to the pattern in Fig. 2.17, which has been simulated by Oshio and Yahata [40]. As an ideal case of the LSA mode, Mosekilde *et al.* [61] have shown a uniform distribution of the high field without any formation of the domain. Also examined is the period doubling bifurcation to chaos, which was obtained by a nearly vertical scan of $A(0.57 \lesssim A \lesssim 0.6)$ for $f \sim 12$ GHz in the phase diagram of Fig. 2.20. By increasing A from 0.57, the oscillatory solutions escape from the 3:2 tongue and enter the main 1:1

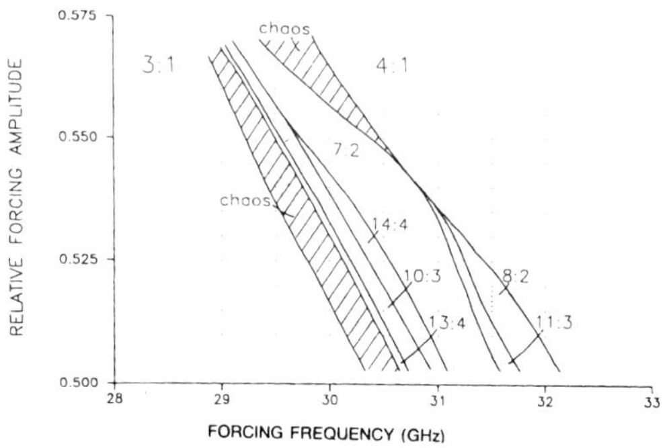

Figure 2.21 Detailed mode distribution in the region between the 3:1 and the 4:1 tongues [62].

Arnold tongue through a narrow 7:6 tongue, then the 1:1 frequency-locked state shows the successive period doubling via the 2:2, 4:4 and 8:8 solutions, where $0.586 \lesssim A \lesssim 0.588$, and finally a chaotic solution appears at $A = 0.585$. With a further increase in A, the 3:3 solution (period 3 window) and again the 1:1 solution are obtained. The period doubling bifurcation has been also observed by Oshio and Yahata [40]. The numerical studies [40, 61, 62] have reproduced well the modes of operation that have been observed in many microwave experiments [50, 70–72]. In contrast, to the authors' knowledge, no experiments aiming to verify the deterministic chaos from a viewpoint of nonlinear dynamics have been conducted.

2.2.3 Low-field regime: the SNDC

The electrical properties of lightly doped or nondoped semiconductors in the low-field regime are mainly determined by the generation and recombination processes of charge carriers between the conduction band and trapping centers. At room temperature, most of the semiconductor materials doped with shallow impurity centers show ohmic conduction in the low-field regime, where free charge carriers are supplied from the impurity centers by thermal ionization. At low temperatures, a large part of the charge carriers is trapped (frozen) at the impurity centers while only a small fraction of the charge carriers is thermally ionized and mobile, so that the semiconductor materials behave like an insulator. However, by increasing the electric field within the low-field regime, the small fraction of the mobile carriers gains sufficient energy to impact-ionize the impurity centers, leading to the non-ohmic transport properties. Above a critical field, the non-ohmic transport becomes dramatic due to the **impact-ionization avalanche**, which induces the nonequilibrium phase transition from a low-conductivity state to a high-conductivity state followed by the SNDC. Thus, the J–E characteristic is typically S-shaped, as in Fig. 2.1a. Concerning the SNDC, many interpretations and models [77-82] were proposed during the 1960s in order to explain the observed negative differential conductivity in Ge [77, 78, 83–86], GaAs [87–90] and CdSe [91], and in order to explain numerically the chaotic behaviors induced by the impact-ionization avalanche in semiconductors [2, 9, 92, 93]. In this section, interpretations and models for the impact-ionization mechanism and the SNDC mechanism are introduced.

Many of the models consider the impact ionization of the impurity ground state (here called the one-level impact-ionization model). The generation and recombination (g-r) processes for n-type semiconductors are shown in Fig. 2.22. The rate equation for the carrier density is described by the equation [86]:

$$\frac{dn}{dt} = A_I n(N_D - N_A - n) + A_T(N_D - N_A - n) - B_T n(n + N_A) - B_I n^2(n + N_A), \qquad (2.86)$$

where n is the carrier density, N_D and N_A are the concentrations of shallow donors and acceptors, respectively, A_I is the impact-ionization rate in units of [cm^3 s^{-1}], A_T [s^{-1}] is the thermal ionization rate, B_T [cm^3 s^{-1}] is the thermal recombination rate and B_I [cm^6 s^{-1}] is the impact recombination rate due to an Auger process. Normally the Auger process is neglected. The Lucky Electron Model for the impact ionization avalanche was proposed theoretically by Shockley [94], who found a formula for the impact-ionization rate:

$$A_I \propto \exp(-\varepsilon_i / e\,\lambda E), \qquad (2.87)$$

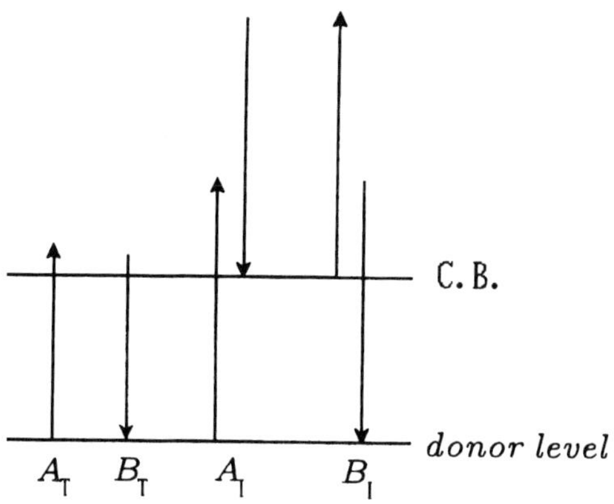

Figure 2.22 Generation and recombination (g-r) processes via ground state of shallow donors in n-type semiconductors. G-r processes via the excited donor states are not included in the simple model (one-level impact-ionization model).

where ε_i is the ionization energy of the impurity and λ is the mean free path of the free carriers. An important observation for the SNDC in p-type Ge at 4.2 k was made by McWhorter and Rediker [77, 78], who found that the SNDC can be seen for compensated materials of In-doped p-type Ge at the breakdown field $E_b \sim 100$ V/cm necessary for the impact-ionization avalanche, and that the SNDC was absent for uncompensated materials which show an abrupt increase in the current at the much lower breakdown field $E_b \sim 10$ V/cm. Here, compensated material is material that also contains unintentionally doped donors with the donor concentration N_D and acceptor concentration N_A, where $N_D < N_A$ for p-type semiconductors. The density of neutral acceptors is $N_A - N_D - p \sim N_A - N_D (p \ll N_A)$ in the prebreakdown regime and the density of ionized acceptors is equal to $P + N_D$, where p is the density of free holes. Similarly, if the discussion is confined to n-type semiconductor materials at low temperatures and in the prebreakdown regime, the density of neutral donors is $N_D - N_A - n \sim N_D - N_A (n \ll N_D)$ and the density of the ionized donors is equal to $n + N_A$, where n is the density of free electrons. McWhorter and Rediker [78] suggested that the origin of the SNDC would be an inelastic scattering of free carriers with a **hydrogen molecule ion**. They thought it possible that the neutral donors (for n-type semiconductors) will tend to share electrons with the neighboring ionized ones forming the hydrogen molecule ion D_2^+, where activation (ionization) energy ε_i of the molecule D_2^+ is supposed to be larger than that of a neutral donor D^0 and thus the breakdown field to ionize the D_2^+ molecule is expected to be larger than that of D^0. The molecule ion may only be possible for compensated materials since donor impurities are partially ionized by compensation even at zero electric field. Free electrons in the conduction band gain energy from an external electric field at a rate $e\,\mu E^2$ and lose their energy in two ways, via the acoustic phonon scattering and the molecular excitation processes. If the energy loss via the molecular excitation is significant, the breakdown field may be much higher than in uncompensated materials. Once the impact ionization of D_2^+ proceeds at an increased electric field, the energy loss via the molecular excitation becomes less significant and the heated carriers may easily impact-ionize the D_2^+ molecules. Positive feedback works on the breakdown field, whereby the molecular excitation disappears owing to the decrease in the density of the D_2^+ molecules, and consequently the impact ionization avalanche induces the SNDC. Regarding this problem, Callaway and Cummings have argued theoretically and calculated the energy loss rate numerically [79].

Zylbersztejn [80] suggested that the energy loss due to a strong mutual interaction (collision) between electrons in the conduction band and electrons in the **impurity band** may be a possible mechanism for the occurrence of the SNDC in strongly compensated Ge. Yamashita [81] assumed that a mutual interaction between the conduction electrons and the impurity system was so strong as to keep an equilibrium between them at the same electron temperature T_e. He also assumed that the current density **J** is carried by the impurity electrons as well as the conduction electrons, namely,

$$\mathbf{J} = e(n_c\mu_c + n_i\mu_i)\mathbf{E}, \tag{2.88}$$

where the subscripts c and i denote the contributions from the conduction electrons and the impurity electrons, respectively, and note that $\mu_i \ll \mu_c$. When the doping concentration is relatively low ($\approx 10^{14} \sim 10^{15}/\mathrm{cm}^3$), the current is carried by the **hopping conduction** of a bound electron from an impurity center to its adjacent empty center rather than by the impurity conduction in the impurity band [81]. By mutual interaction, the impurity system dissipates to the phonon system some of the excess energy of the conduction electrons supplied by the external electric field, where the energy loss rate is denoted by $(\partial\epsilon/\partial t)_{c,i}$. Yamashita considers the energy loss rates of $(\partial\epsilon/\partial t)_{c,i}$, $(\partial\epsilon/\partial t)_{ac}$ due to acoustic phonon scattering, and $(\partial\epsilon/\partial t)_x$ due to one of the possible mechanisms such as exchange interaction between conduction electrons and bound electrons in the singlet state, leading to the ionization of the bound electron and recombination of the conduction electron at the triplet donor state. The sum of the energy loss rates is then balanced with the energy gain $e\mu_c E^2$:

$$e\mu_c E^2 = \left(\frac{\partial\varepsilon}{\partial t}\right)_{ac} + (N_D - N_A)\left(\frac{\partial\varepsilon}{\partial t}\right)_x + n_i\left(\frac{\partial\varepsilon}{\partial t}\right)_{c,i}, \tag{2.89}$$

where $(\partial\varepsilon/\partial t)_{ac} \propto T_e - T_L$ (T_L is lattice temperature). Since the last term is very efficient, the electron temperature does not rise to the breakdown temperature, even when the applied field is larger than the field, which can lead to electrical breakdown for an uncompensated sample [77, 81]. By increasing the electric field in the prebreakdown regime, the density n_c of the conduction electrons and the electron temperature increase simultaneously by impact ionization and the current $J(E)$ becomes non-ohmic because $\mu_c \gg \mu_i$ in (2.88), where the last term of the energy loss rates in (2.89) is reduced but still important. Finally, the energy loss $(\partial\varepsilon/\partial t)_{c,i}$ disappears completely at a critical field (breakdown field) and the impact ionization avalanche occurs instantaneously with positive feedback in such a way that an increase in n_c by impact ionization reduces $(\partial\varepsilon/\partial t)_{c,i}$ and increases T_e, which induces more impact ionization.

An alternative model was proposed by Kurosawa [82], who showed numerically that the distribution function of the free electrons plays an essential role in the occurrence of the SNDC. In the prebreakdown regime, the electron density is quite low and electron–electron scattering is disregarded, so that the electron distribution takes a form of the BE distribution. During breakdown, however, the electron density increases and subsequently electron–electron scattering becomes more important, thereby the electron distribution may shift from the BE to the MB or shifted MB distribution with the electron temperature T_e. The latter can be described by the electron temperature model. The BE distribution function $f_{BE}(\varepsilon)$ is larger at small energy $\sim k_B T_L$ and smaller at $\varepsilon \sim \varepsilon_i$ (donor ionization energy) than the MB distribution function $f_{MB}(\varepsilon)$.

Thus, such a shift in the distribution function during impact ionization may cause the SNDC. All of the above models [77–82] have been proposed in order to explain the SNDC in compensated Ge [77, 78, 83–85]. For the precise measurements of impact ionization and generation–recombination processes, see Ref. [86]. In Fig. 2.23, typical examples of the carrier density, Hall mobility and current density are shown as a function of the electric field for weakly compensated n-Ge. Note that the carrier density and the current density are increased abruptly by the impact ionization at the breakdown field, while the Hall mobility decreases, in contrast to the case in n-GaAs, as stated below. Koenig *et al.* [86] measured the recombination time by pulse measurements and estimated various generation–recombination coefficients (Table 2.2), where mainly Sb-doped n-Ge samples were used with $N_D = 2.5 - 4.6 \times 10^{13}/\text{cm}^3$, $N_A = 0.6 - 1.4 \times 10^{12}/\text{cm}^3$ and the high electron mobilities $\mu_e = 1.2 - 3.2 \times 10^6\ \text{cm}^2/\text{V s}$. The breakdown field was typically $E_b \sim 4\ \text{V/cm}$.

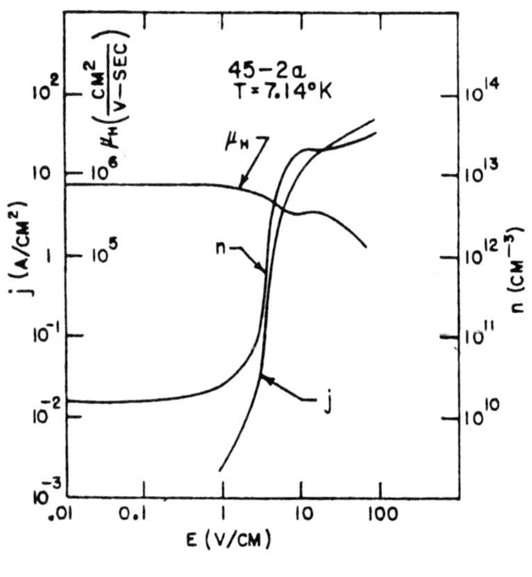

Figure 2.23 Variation in current density, carrier density and Hall mobility vs electric field for a weakly compensated n-Ge sample at a lattice temperature of 7.14 K. The carrier density n has been obtained from the Hall constant assuming a value of unity for the Hall to drift mobility ratio. Data for $j > 5\ \text{A/cm}^2$ were obtained using 3-μs pulses; the other data obtained by dc measurements [86].

Table 2.2 Generation–recombination coefficients in n-Ge

A_I [cm^3/s]	A_T [s^{-1}]	B_T [cm^3/s]	B_I [cm^6/s]
$> 5 \times 10^{-8}$	1.6×10^{-2}	8.4×10^{-4}	0.7×10^{-16}

The value of A_I was taken in the prebreakdown regime.

Especially worth noting is a model proposed by Crandall [95], who attributed the observed SNDC in n-GaAs to the **screening of acoustic phonon scattering and ionized impurity scattering**. He starts from the rate equation of free carriers:

$$\frac{dn}{dt} = (nA_I + A_T)(N_D - N_A - n) - nB_T(N_A + n), \tag{2.90}$$

$$nA_I = \int_0^\infty \sqrt{\varepsilon^*}\, f(n, \varepsilon^*, E)\ \sigma_I(n, \varepsilon^*) d\varepsilon^*, \tag{2.91}$$

$$nB_T = \int_0^\infty \sqrt{\varepsilon^*}\, f(n, \varepsilon^*, E)\ \sigma_T(n, \varepsilon^*) d\varepsilon^*, \tag{2.92}$$

where $f(n, \varepsilon^*, E)$ is the distribution function at a given electric field, and σ_I and σ_T are the collision cross-sections for the impact ionization and the electron-donor recombination, respectively. The Auger process in (2.86) was neglected. Crandall assumes A_T to be a constant. The terms $(N_D - N_A - n)$ and $(n + N_A)$ in (2.90) are the densities of neutral and ionized donors, respectively. During the impact ionization, screening of the acoustic phonon scattering and the ionized impurity scattering are considered to play a key role in the occurrence of the SNDC, and thus the electron distribution function should be important in this problem. One must solve precisely the Boltzmann equation (2.38) under the steady-state condition [81, 82, 90, 96] $(df_k/dt = 0$ with $(\partial f_k/\partial t)_{diff} = 0)$. However, since the precise determination is difficult and inconvenient, the electron temperature model with the MB distribution function may be used [97]:

$$f(n, \varepsilon^*, E) \propto nT_e^{-3/2}\ \exp(-\varepsilon^*/k_B T_e). \tag{2.93}$$

The current-field characteristic is expressed by

$$J = en\ v_d, \tag{2.94}$$

where v_d is the electron drift velocity and e is the electron charge.

From Fig. 2.1a, the SNDC means that $|dI/dE| \to \infty$ at a breakdown field E_b and a sustained field E_s (or holding field E_h). Since the drift velocity, as determined from the solution of the BE, is a function of both n and E, and n calculated from the rate equation (2.90) is a function of E only, the derivative $\mathrm{d}I/\mathrm{d}E$ is given by

$$\frac{\mathrm{d}I}{\mathrm{d}E} = e\left(v_\mathrm{d} + n\frac{\partial v_\mathrm{d}}{\partial n}\right)\frac{\mathrm{d}n}{\mathrm{d}E} + en\frac{\partial v_\mathrm{d}}{\partial E}. \tag{2.95}$$

If any of the derivatives in (2.95) happen to diverge, then $\mathrm{d}I/\mathrm{d}E$ diverges. Crandall [95] suggested that $\partial v_\mathrm{d}/\partial E$ was positive and finite since a situation where $\partial v_\mathrm{d}/\partial E$ was infinite has not observed experimentally. Then, for the SNDC to occur, $\mathrm{d}n/\mathrm{d}E$ must diverge. Since the rate equation (2.90) is a function of both n and E, namely $n = n(n, E)$ and also $\dot{n} = 0$ under the steady-state condition, then $\mathrm{d}\dot{n} = (\partial\dot{n}/\partial E)\mathrm{d}E + (\partial\dot{n}/\partial n)\mathrm{d}n = 0$. The relation $\mathrm{d}n/\mathrm{d}E$ is rewritten as:

$$\frac{\mathrm{d}n}{\mathrm{d}E} = -\frac{\partial\dot{n}}{\partial E} \Big/ \frac{\partial\dot{n}}{\partial n}. \tag{2.96}$$

Since $\partial\dot{n}/\partial E$ is slowly varying, $\mathrm{d}n/\mathrm{d}E$ becomes infinite only when $\partial\dot{n}/\partial n \to 0$. From (2.90):

$$\begin{aligned} -\frac{\partial\dot{n}}{\partial n} &= n(A_\mathrm{I} + B_\mathrm{T}) + A_\mathrm{T}\frac{N_\mathrm{D}}{n}(1 - K) \\ &\quad - nN_\mathrm{D}\left[\left(1 - K - \frac{n}{N_\mathrm{D}}\right)\frac{\partial A_\mathrm{I}}{\partial n} - \left(K + \frac{n}{N_\mathrm{D}}\right)\frac{\partial B_\mathrm{T}}{\partial n}\right], \end{aligned} \tag{2.97}$$

where K is the compensation ratio $N_\mathrm{A}/N_\mathrm{D}$. The steady-state condition $\dot{n} = 0$ in (2.90) has been used for the derivation of (2.97). If A_I and B_T are independent of n, then $-\partial\dot{n}/\partial n$ and from (2.95) $\mathrm{d}I/\mathrm{d}E$ are positive, showing no negative differential resistivity. Therefore, the impact ionization is not a sufficient condition for a SNDC. For simplicity, B_T is assumed to be a constant [95]. In order for the SNDC to occur, $\partial A_\mathrm{I}/\partial n$ must be positive and an increasing function of n when the electric field approaches the critical field. With the electron temperature model (2.93), the impact ionization rate is given by:

$$A_\mathrm{I} \propto (\varepsilon_\mathrm{i}/k_\mathrm{B}T_\mathrm{e})^{1/2}\exp(-\varepsilon_\mathrm{i}/k_\mathrm{B}T_\mathrm{e}), \tag{2.98}$$

for the condition $\varepsilon_\mathrm{i} \gg k_\mathrm{B}T_\mathrm{e}$. Taking the derivative $\partial A_\mathrm{I}/\partial n$, one obtains:

$$\frac{\partial\dot{n}}{\partial n} \propto -\left[1 - (\varepsilon_\mathrm{i}/k_\mathrm{B}T_\mathrm{e} - 0.5)F\frac{\partial T_\mathrm{e}}{\partial n}\frac{n}{T_\mathrm{e}}\right], \tag{2.99}$$

where the function F is positive and given by

$$F = \frac{(KN_\mathrm{D}/n + 1)(1 - K - n/N_\mathrm{D})}{1 + [(KN_\mathrm{D}/n + 1) + 1 - K - n/N_\mathrm{D}]B_\mathrm{T}/nA_\mathrm{I}}. \tag{2.100}$$

For the derivation of (2.100) $\dot{n} = 0$ in (2.90) was used. Since $\varepsilon_\mathrm{i}/k_\mathrm{B}T_\mathrm{e} > 1$ in (2.98) and $F > 0$ in (2.100), for a SNDC to occur (i.e., $\partial\dot{n}/\partial n \to 0$) the electron temperature must be an increasing function of n ($\partial T_\mathrm{e}/\partial n > 0$). Crandall considers the electron mobility $\mu(n, T_\mathrm{e})$ to be determined by ionized impurity scattering and the energy loss to be due to acoustic phonon interaction via the deformation potential [95]. Then the energy balance equation is given by:

$$\frac{\varepsilon - \varepsilon_L}{\tau(\varepsilon)} = e\mu(n, T_e)E^2, \tag{2.101}$$

where $\tau(\varepsilon)$ is the energy relaxation time, $\varepsilon = \frac{3}{2}k_B T_e$ and $\varepsilon_L = \frac{3}{2}k_B T_L$. The electron mobility is given by the Brooks–Herring expression:

$$\mu(n, T_e) = \frac{a(k_B T_E)^{3/2}}{(N_I + n)g(\zeta)}, \tag{2.102}$$

where

$$\zeta = bT_e^2/n, \tag{2.103}$$

and

$$g(\zeta) = \ln(1 + \zeta) - \zeta/(1 + \zeta). \tag{2.104}$$

In (2.102) N_I is the density of the ionized impurity when $n = 0$. The electron mobility by the Brooks–Herring formula describes that it increases by an increase in the carrier density via the screening of the ionized impurity scattering. Thus, an increase in n by impact ionization results in an increase in μ and consequently an increase in T_e from the energy balance (2.101). Inclusion of the screening of electron–phonon interactions would enhance the possibility of a SNDC, because the screening reduces the rate of energy loss due to the acoustic phonon scattering via an increased energy relaxation time $\tau(\varepsilon)$, where $\varepsilon = \varepsilon_L + e\,\mu\tau(\varepsilon)E^2$ from (2.101). For simplicity, this mechanism is neglected here.

From (2.101) and (2.102), the derivative $\partial T_e/\partial n$ is obtained as:

$$\frac{\partial T_e}{\partial n} = \frac{T_e}{2n}\left(1 - \frac{Rn}{N_I + n}\right) \Big/ \left(1 + \frac{2R}{1 - t}\right), \tag{2.105}$$

where $t = T_e/T_L$, and $R = (1 + 1/\zeta)^2 g(\zeta)$. The quantity R is a slowly varying function of n and T_e, and $R > 1$ with $\zeta \gg 1$ for undoped or lightly doped samples [4]. From (2.105), $\partial T_e/\partial n$ is positive when $N_I \gg n$ in the prebreakdown regime, and negative when $n > N_I$ in the postbreakdown regime. If $n \ll N_I$, the main effect of increasing the carrier density by the impact ionization is to screen the ionized impurity scattering, whereby the electron temperature increases as stated before. In the prebreakdown regime the positive feedback of such a mechanism induces the SNDC. With the approximations $n/N_D \ll K$, $t \gg 1$, $n \ll N_I$ and $n^2 A_I/A_T N_D \ll 1$, Crandall has shown that the derivative $\partial \dot{n}/\partial n$ in (2.99) is rewritten as:

$$\frac{\partial \dot{n}}{\partial n} \propto (n(E) - n_1), \tag{2.106}$$

where

$$n_1 = \frac{4(1-K)^2 k_B^2 A_T}{(KB_T\varepsilon_i)^2 b} \exp(CE_b^2/N_I). \tag{2.107}$$

In (2.107) C is a proportionality constant. The approximated form (2.106) together with (2.95) and (2.96) clearly indicates the occurrence of the phase transition at the first turning point $E = E_b$ when $n(E) \to n_1$ with $E \to E_b$ in the prebreakdown regime. For the second turning point in the postbreakdown regime at $E = E_s$,

$$\frac{\partial \dot{n}}{\partial n} \propto \left(\frac{N_I}{R-1} - n(E) \right). \tag{2.108}$$

Corresponding to (2.106), the carrier density may behave like

$$\dot{n} = -\gamma(n - n_1)^2 (n - n_3), \tag{2.109}$$

where γ is a constant, and $n_1 \ll n_3$. If $n = n_1$ at $E = E_b$, then $\dot{n} = 0$. Introducing a small fluctuation ξ_n, where $n = n_1 + \xi_n$, then $\dot{n} \sim \xi_n^2(n_3 - n_1) > 0$ at $t = 0$, and for $t > 0$ the generation rate $\dot{n}$ keeps positive values until n rapidly reaches a high value n_3. Similarly to (2.109), one may have:

$$n = -\gamma(n - n_3)^2 (n - n_1), \tag{2.110}$$

where $n = n_3$ at $E = E_s$. By a small fluctuation the transition of the carrier density occurs from n_3 to n_1 since $\dot{n} < 0$. The screening model by Crandall [95] seems to be highly acceptable for GaAs since a rapid increase in the Hall mobility during the impact ionization [90, 95] can be explained only by such a screening mechanism as described by the Brooks–Herring expression (2.102)–(2.104) [4, 90]. In Fig. 2.24, the Hall mobility (solid circles) increases significantly at the breakdown field (~ 2.0 V/cm) corresponding to a rapid increase in the carrier density (open circles), where the solid line is the Hall mobility calculated using the experimental values of the carrier density. Crandall suggested that the SNDC was absent for uncompensated materials [95].

Khosla *et al.* [91] observed similar behavior of the SNDC in CdSe. In Fig. 2.25, the *J–E* characteristics observed at various temperatures indicate that the SNDC can be seen at lower temperatures, less than 8 K. Correspondingly, the Hall mobility increases rapidly by one order of magnitude during the impact ionization avalanche and even exhibits negative differential mobility ($\partial\mu/\partial E < 0$) in the SNDC region. Khosla *et al.* [91] analyzed their experimental results in terms of the screening model [95] in an approximated form. Considering the steady-state condition $\dot{n} = 0$

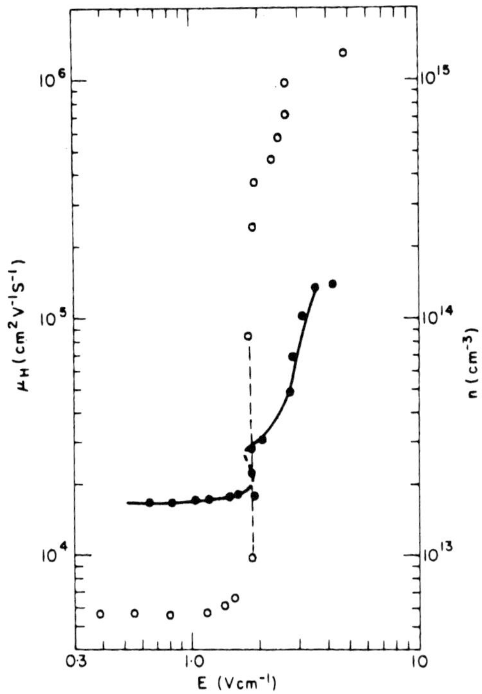

Figure 2.24 Electric field dependence of the Hall mobility μ_H (solid circles) and the carrier density (open circles). The solid curve is the calculated value of the mobility. The dashed curve shows the SNDC region [95].

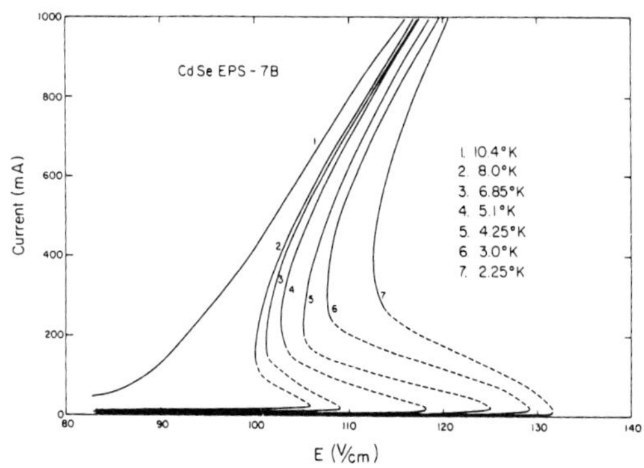

Figure 2.25 Temperature dependence of the current vs applied field in CdSe [91].

for the rate equation (2.89), the carrier density can be solved as

$$n = -\frac{A_T + N_A B_T - (N_D - N_A)A_I}{2(A_I + B_T)} + \frac{\{[A_T + N_A B_T - (N_A - N_D)A_I]^2 + 4(A_I + B_T)(N_D - N_A)A_I\}^{1/2}}{2(A_I + B_T)} \tag{2.111}$$

The carrier density is a nonlinearly increasing function of the impact ionization rate A_I, as given in Fig. 2.26, where three curves are drawn for three different temperatures T_1, T_2 and T_3 ($T_1 < T_2 < T_3$). At very low values of A_I, the carrier density takes its value of thermal equilibrium, $n = (A_T/B_T)(1 - K^{-1})$, where $A_I \sim 0$, $N_D - N_A - n \sim N_D - N_A$, and $n^2 B_T \sim 0$ in (2.90). For very large A_I, neutral donors are fully impact-ionized so that the carrier density saturates at a level given by $n = N_D - N_A$. If A_I is a monotonically increasing function of E and independent of n, say $A_I(E)$, then the SNDC will not occur [91]. If A_I is an increasing function $A_I(n, E)$ of both n and E, as expected from the screening model by Crandall [95], then the SNDC may occur. Now the steady-state solutions can be obtained by finding the intersections between two curves $A_I = A_I(n, E)$ and $n = n(A_I)$ (see Fig. 2.26). For a range $E_2 < E < E_1$ at a given temperature

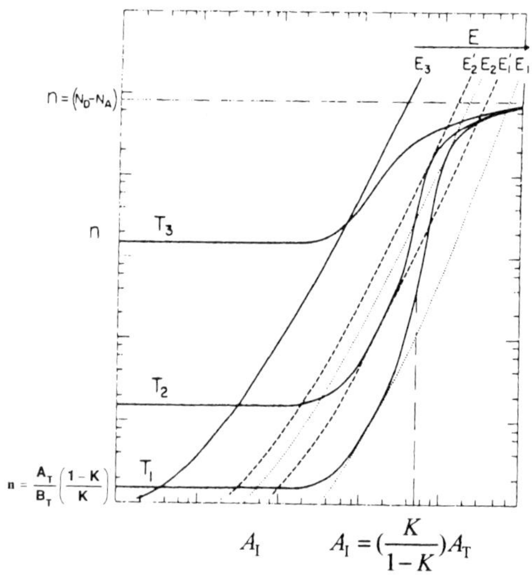

Figure 2.26 n vs. $A_I(n = n(A_I))$ from (2.111) in the text for three temperatures T_1, T_2 and T_3, where $T_1 < T_2 < T_3$. The impact-ionization rate is an increasing function of both n and E; $A_I = A_I(n, E)$. $A_I(n, E_i)$ is also shown at various electric fields, where $E_1 < E_2 < E_3$ [91].

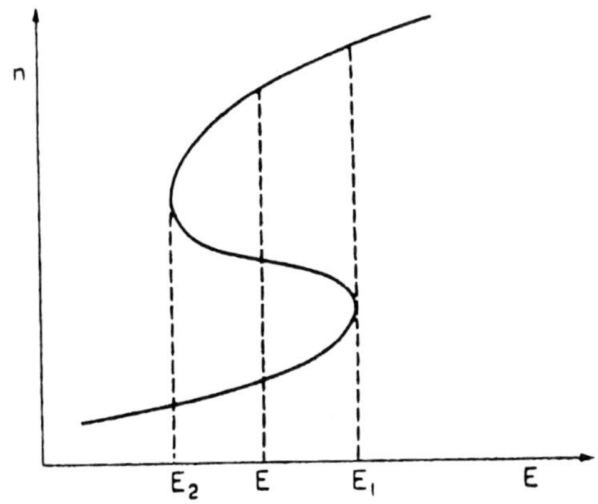

Figure 2.27 S-shaped $n - E$ characteristic, which can be obtained from the steady-state solutions of the rate equation (2.86) if the impact-ionization rate is an increasing function of n (see Fig. 2.26). The Auger recombination process is neglected ($B_1 = 0$) [91].

T_1, there are three solutions in Fig. 2.26. Then, as a function of E, one can expect an S-shaped $n - E$ characteristic, as shown in Fig. 2.27 [91]. Aoki *et al.* [98, 99] adopted the screening model in order to analyze nonlinear dynamical behaviors and chaos in n-GaAs (see chapters 3–5).

2.3 Two-level impact-ionization model

The section introduces the two-level impact-ionization model presented by Schöll [2, 100]. The Schöll model, which is supported by many experimentalists, suggests that the impact ionization processes of the impurity multilevels may induce the SNDC for any n-type or p-type semiconductor material whether the impurity states are deep or shallow. The generation and recombination processes in n-type semiconductors are shown in Fig. 2.28, where the impact ionization rates are given by X_1^* for the donor-excited state and X_1 for the donor-ground state, X_1^S and X^* are the thermal ionization coefficients, and T_1^S and T^* are the recombination coefficients. Denoting the free electron density by n, the ionized donor density by p_t, and the trapped electron densities at the donor-ground and donor-excited states, respectively, by n_{t1} and n_{t2}, a set of rate equations is given by [2, 100]:

$$\dot{n} = X_1^S n_{t2} - T_1^S n p_t + X_1^* n n_{t2} + X_1 n n_{t1}, \tag{2.112}$$

$$\dot{n}_{t1} = -X^* n_{t1} + T^* n_{t2} - X_1 n n_{t1}, \tag{2.113}$$

$$\dot{n}_{t2} = -\dot{n} - \dot{n}_{t1}, \tag{2.114}$$

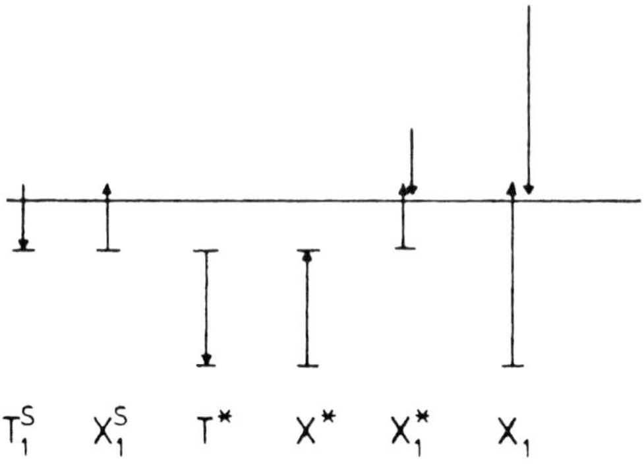

Figure 2.28 Generation and recombination processes for the two-level impact-ionization model in n-type semiconductors [2, 100].

where

$$p_t = n + N_A, \; p_t + n_{t1} + n_{t2} = N_D. \tag{2.115}$$

The first part of (2.115) describes the charge neutrality condition, that the number of positively charged particles (p_t) must be equal to the number of negatively charged particles ($n + N_A$), where the acceptors are all negatively ionized in n-type materials, while the latter part is given from the condition of conservation of impurities. Since $p_t = N_D - n_{t1} - n_{t2}$ from (2.115), the charge neutrality condition in (2.115) can be rewritten as

$$n_{t1} + n_{t2} = N_D - N_A - n, \tag{2.116}$$

where $N_D - N_A$ is the effective doping density since the donor impurities are partially compensated by the acceptors. Eliminating p_t and n_{t2}, two rate equations (2.112) and (2.113) are converted to

$$\dot{n} = a_0 + a_1 n + a_2 n_{t1} + a_3 n n_{t1} + a_4 n^2, \tag{2.117}$$

$$\dot{n}_{t1} = b_0 + b_1 n + b_2 n_{t1} + b_3 n n_{t1}, \tag{2.118}$$

where

$$\begin{aligned} a_0 &= X_1^S(N_D - N_A), & b_0 &= T^* N_D^*, \\ a_1 &= X_1^*(N_D - N_A) - X_1^S - T_1^S N_A, & b_1 &= -T^*, \\ a_2 &= -X_1^S, & b_2 &= -(T^* + X^*), \\ a_3 &= X_1 - X_1^*, & b_3 &= -X_1, \\ a_4 &= -(T_1^S + X_1^*). \end{aligned} \tag{2.119}$$

Under the steady-state condition $\dot{n}_{t1} = 0$ one obtains the solution

$$n_{t1} = \frac{T^*(N_D - N_A - n)}{T^* + X^* + X_1 n}. \tag{2.120}$$

With $\dot{n} = 0$, from (2.117) one obtains

$$an^3 + bn^2 + cn + d = 0, \tag{2.121}$$

where

$$\begin{aligned}
a &= -(T_1^S + X_1^*)X_1 < 0 \\
b &= X_1^S X_1 N_D^* - [X_1 + T_1^S N_A + T^*]X_1 - X^* X_1^* - T_1^S(T^* + X^*) \\
c &= [(X_1^S + T^*)X_1 + X^* X_1](N_D - N_A) - X_1 X^* - T_1(T^* + X^*)N_A \\
d &= X_1 X^*(N_D - N_A) > 0.
\end{aligned} \tag{2.122}$$

Since (2.121) is a cubic function of n and the impact ionization rates X_1 and X_1^* are field dependent, (2.121) may have three distinct solutions n_1, n_2 and n_3 $(0 < n_1 < n_2 < n_3)$ under a certain parameter regime [2, 100], n_1 and n_3 are stable and n_2 is unstable. Then, as a function of the applied field E_0 the $n - E_0$ characteristic may show an S-shaped curve. Schöll has used Shockley's Lucky Electron Model [94] for the impact-ionization rates X_1 and X_1^* [2, 100]:

$$X_1 = X_1^0 \exp(-\varepsilon_D/e\lambda E), \qquad X_1^* = X_1^{*0} \exp(-\varepsilon_D^*/e\lambda E), \tag{2.123}$$

where ε_D, ε_D^* are the ground state and excited donor energy levels, λ is the

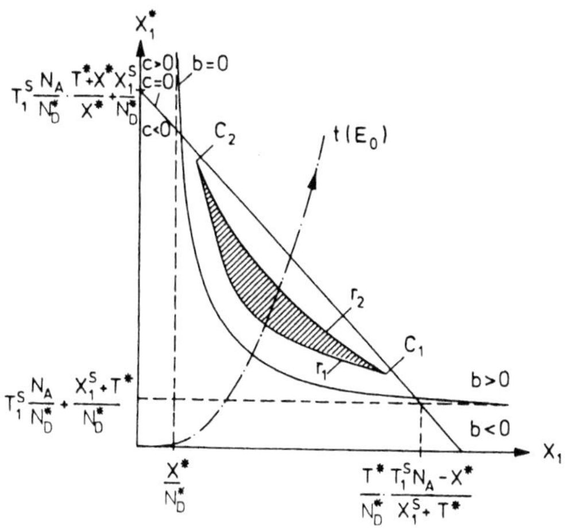

Figure 2.29 Control parameter plane of the impact-ionization coefficients X_1, X_1^* (schematic). The hatched area represents the bistability domain. The notations correspond to donor kinetics in n-type semiconductor materials [2].

Table 2.3 Generation–recombination coefficients used for the two-level impact-ionization model [2, 100]

Figure	$T_1^S(N_D - N_A)$ [μs^{-1}]	T^* [μs^{-1}]	X_1^S [μs^{-1}]	X^* [μs^{-1}]	$N_A/(N_D - N_A)$
2.30(a)	10^3	1	0.5–10	1	0.5
2.30(b)	10^3	1	0.5	1-2	0.5
2.30(c)	10^3	1	0.5	0.5	0.2–0.4

A typical value of $N_D - N_A$ is $10^{15}/cm^3$[100].

mean free path of the electrons, and X_1^0, X_1^{*0} are approximately constant. Since X_1 and X_1^* are functions of E_0 only in (2.123), the bistability region can be investigated by solving (2.121) in the control parameter plane (X_1, X_1^*). For values of X_1 and X_1^* taken from the hatched region in Fig. 2.29, there are three solutions, n_1, n_2 and n_3, indicating bistability [100]. Outside the hatched region there is one steady state only. As a function of E_0, one can obtain the S-shaped current-field characteristics in Fig. 2.30, using the various parameters as given in Table 2.3 and with

$$X_1(N_D - N_A) = 50\ \exp(-6/E_0), \qquad X_1(N_D - N_A) = 10^3 \exp(-1.5/E_0).$$

In Fig. 2.30(a–c) the different curves correspond to the indicated parameter values for (a) X_1^*, (b) X^*, and (c) $N_A/(N_D - N_A)$, respectively (see also Table 2.3). In the two-level impact-ionization model, the electric field effects of electron distribution are not considered and thus one cannot argue about

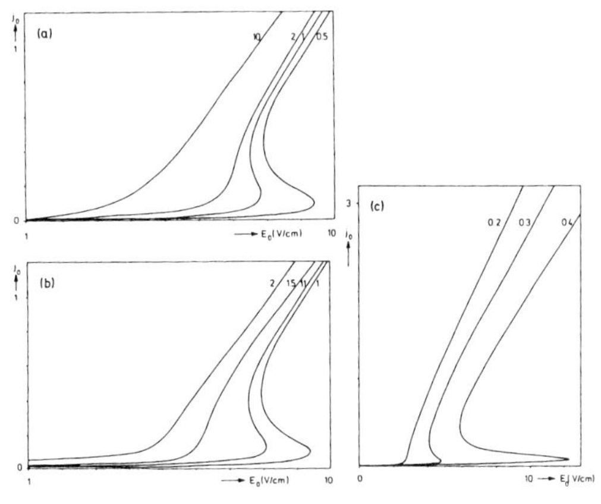

Figure 2.30 Current density-field characteristics, calculated with the numerical parameters of Table 2.3. The current density j_0 is plotted in units $e\mu_e N_D$. The parameters of the different curves are (a) X_1^* in μs^{-1}, (b) X^* in μs^{-1}, (c) $N_A/(N_D - N_A)$. For details, see [2].

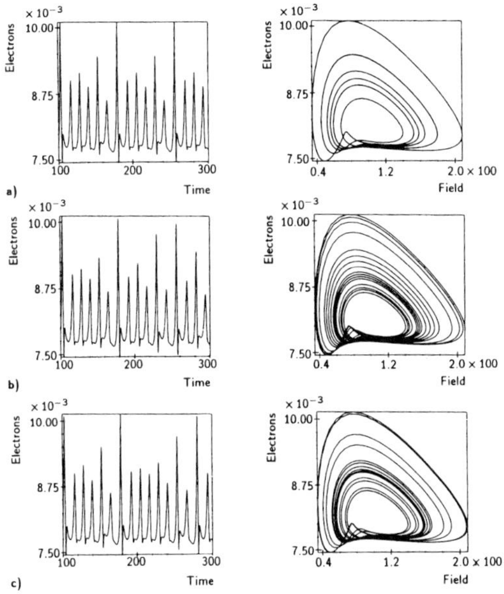

Figure 2.31 Numerical chaotic solution of the two-level impact-ionization model. Left: electron density (in units $10^{-3}N_D$) as a function of time (in units $2 \times 10^6 \tau_m$); right: phase portrait of electron density vs electric field [2].

the electron temperature as well as the electron mobility. However, by the additional inclusion of the impact-ionization process at the excited donor level the model seems to have succeeded in explaining easily the SNDC. A typical example of the phase portrait for a chaotic attractor is shown in Fig. 2.31.

References

[1] A.B. Pippard, *Response and Stability – An Introduction to the Physical Theory* (Cambridge University Press, Cambridge, 1985).

[2] E. Schöll, *Nonequilibrium Phase Transitions in Semiconductors – Self-Organization Induced by Generation and Recombination Processes* (Springer, Berlin, 1987).

[3] N.A. Lampert and R.B. Schilling, in *Semiconductor and Semimetals*, Vol. 6, *Current Injection in Solids* (Academic Press, New York, 1970) pp. 1–96.

[4] K. Seeger, *Semiconductor Physics – An Introduction*, 4th ed. (Springer, Berlin, 1989).

[5] W. Fawcett, A.D. Boardman and S. Swain, J. Phys. Chem. Solids **31** (1970) 1963.

[6] C. Kittel, *Introduction to Solid State Physics*, 5th ed. (John Wiley & Sons, New York, 1976) pp. 207–248.
[7] W. Shockley, *Electrons and Holes in Semiconductors – With Application to Transistor Electronics*. (D. Van Nostrand Company, Princeton, 1950).
[8] G. Bauer, in *Solid-State Physics*, Springer Tracts in Modern Physics, Vol. 74, ed. G. Hohler, *Determination of Hot Electron Distribution Functions in Semiconductors* (Springer, Berlin, 1974) pp. 1–106.
[9] E. Schöll, in *Nonlinear Dynamics in Solids*, ed. H. Thomas (Springer, Berlin, 1992) pp. 31–50.
[10] L.B. Montefusco and C. Jacoboni, Solid State Commun. **10** (1972) 71.
[11] C. Hilsum and H.D. Rees, Electron. Lett. **6** (1970) 277; B.K. Ridley, J. Appl. Phys. **48** (1977) 754.
[12] G. Bauer and H. Kahlert, Phys. Rev. **B5** (1972) 566.
[13] H. Kahlert and G. Bauer, Phys. Rev. **B7** (1973) 2670; see also R.A. Isaacson and F. Bridges, Solid State Commun. **4** (1966) 635.
[14] J.P. Nougier and M. Rolland, Phys. Rev. **B8** (1973) 5728; Solid State Electron. **16** (1973) 1399.
[15] P.D. Southgate and D.S. Hall, Appl. Phys. Lett. **16** (1970) 280.
[16] P.D. Southgate, D.S. Hall and A.B. Dreeben, J. Appl. Phys. **42** (1971) 2868.
[17] H. Heinrich and W. Jantsch, Phys. Rev. **B4** (1971) 2504.
[18] O. Christensen, Phys. Rev. **B7** (1973) 763.
[19] E.M. Conwell, Phys. Rev. **143** (1966) 657.
[20] K. Hess and H. Kahlert, J. Phys. Chem. Solids **32** (1971) 2262.
[21] W. Fawcett and H.D. Rees, Phys. Lett. **11** (1969) 731.
[22] C. Jacoboni and L. Reggiani, Rev. Mod. Phys. **55** (1983) 645.
[23] H. Budd, Proc. Int. Conf. on Phys. Semicond., Kyoto, 1966, J. Phys. Soc. Jpn Suppl. **21** (1966) 420.
[24] H.D. Rees, J. Phys. Chem. Solids **30** (1969) 643.
[25] J.M. Ziman, *Electrons and Phonons – The Theory of Transport Phenomena in Solids* (Oxford University Press, Glasgow, 1960).
[26] J. Pozhela, *Plasma and Current Instabilities in Semiconductors* (Pergamon Press, Oxford, 1981) and references therein.
[27] B.K. Ridley, in Proc. of a NATO Advanced Research Workshop on Negative Differential Resistance and Instabilities in 2-D Semiconductors, eds. N. Balkan, B.K. Ridley and A.J. Vickers (Plenum Press, New York, 1993) pp. 1–21.
[28] B.K. Ridley and R.G. Bratt, Phys. Lett. **4** (1963) 300.
[29] K.W. Böer and W.E. Wilhelm, Phys. Stat. Sol. **3** (1963) 1718.
[30] A.R. Huston and D.L. White, J. Appl. Phys. **33** (1962) 40.
[31] D.L. White, J. Appl. Phys. **33** (1962) 2547.
[32] J.H. McFee, J. Appl. Phys. **34** (1963) 1548.
[33] J.S. Moore, C.M. Penchina, N. Holonyak, M.D. Sirkis and T. Yamara, J. Appl. Phys. **37** (1966) 2009.
[34] B.K. Ridley and T.B. Watkins, Proc. Phys. Soc. London **78** (1961) 293.
[35] C. Hilsum, Proc. IRE **50** (1962) 185.
[36] J.B. Gunn, Solid State Commun. **1** (1963) 88.
[37] J.B. Gunn, IBM J. Res. Develop. **8** (1964) 141.
[38] K. Nakamura, Prog. Theor. Phys. **57** (1977) 1874.
[39] H. Haken, *Synergetics – An Introduction, Nonequilibrium Phase Transitions and Self-Organization in Physics, Chemistry and Biology*, 2nd ed. (Springer, Berlin, 1982) pp. 254–258.

[40] K. Oshio and H. Yahata, J. Phys. Soc. Jpn **62** (1993) 3639; J. Phys. Soc. Jpn **64** (1995) 1823.
[41] P.N. Butcher, W. Fawcett and C. Hilsum, Br. J. Appl. Phys. **17** (1966) 841.
[42] H. Krömer, Proc. IEEE **52** (1964) 1736.
[43] B.K. Ridley, Proc. Phys. Soc. London **82** (1963) 954.
[44] T. Kurosawa, Proc. 8th Int. Conf. on Phys. Semicond., Kyoto, 1966, J. Phys. Soc. Jpn Suppl. **21** (1966) 424.
[45] H.D. Rees, Phys. Lett. **26A** (1968) 416.
[46] H.D. Rees, J. Phys. Chem. Solids **30** (1969) 643.
[47] J.G. Ruch and G.S. Kino, Appl. Phys. Lett. **10** (1967) 40; Phys. Rev. **174** (1968) 921.
[48] N. Braslau and P.S. Hauge, IEEE Trans. Electron Devices **ED-17** (1970) 616.
[49] C. Hilsum and H.D. Rees, Proc. 10th Int. Conf. on Phys. Semicond., Cambridge, 1970, p. 45.
[50] S.M. Sze, *Physics of Semiconductor Devices* (John Wiley & Sons, New York, 1969) pp. 731–787.
[51] K. Hess, H. Morkoc, H. Shichijo and B.G. Streetman, Appl. Phys. Lett. **35** (1979) 469.
[52] M. Keever, H. Shichijo, K. Hess, S. Banerjee, L. Witkowski and H. Morkoc, Appl. Phys. Lett. **38** (1981) 36.
[53] K. Aoki, K. Yamamoto, N. Mugibayashi and E. Schöll, Solid State Electron. **32** (1989) 1149.
[54] E. Schöll and K. Aoki, Appl. Phys. Lett. **58** (1991) 1277.
[55] E. Schöll, in Proc. of a NATO Advanced Research Workshop on Negative Differential Resistance and Instabilities in 2-D Semiconductors, eds. N. Balkan, B.K. Ridley and A.J. Vickers (Plenum Press, New York, 1993) pp. 37–51.
[56] S. Lurgi and A. Kastalsky, Physica **134B/C** (1985) 453.
[57] S. Lurgi and M. Mastrapasqua, in Proc. of a NATO Advanced Research Workshop on Negative Differential Resistance and Instabilities in 2-D Semiconductors, eds. N. Balkan, B.K. Ridley and A.J. Vickers (Plenum Press, New York, 1993) pp. 37–51.
[58] W.I. Masselink, N. Breslan, D. La Tulipe, N.I. Wang and S.L. Wright, Solid State Electron. **31** (1988) 337.
[59] R.A. Höpfel, J. Shah, A.C. Gossard and W. Wiegmann, Physica **134B** (1985) 509.
[60] N. Balkan and B.K. Ridley, Semicond. Sci. Technol. **3** (1988) 507.
[61] E. Mosekilde, R. Feldberg, C. Knudsen and M. Hindsholm, Phys. Rev. **B41** (1990) 2298.
[62] E. Mosekilde, J.S. Thomsen, C. Knudsen and R. Feldberg, Physica **D66** (1993) 143.
[63] A. Čenys, G. Lasiene and K. Pyragas, Solid State Electron. **35** (1992) 975.
[64] K. Matsumoto, Phys. Lett. **31A** (1970) 335.
[65] S. Kabashima, H. Yamazaki and T. Kawakubo, J. Phys. Soc. Jpn **40** (1976) 921.
[66] S. Porokowski, W. Paul, J.C. McGrody, M.I. Nathan and J.E. Smith, Solid State Commun. **7** (1969) 905.
[67] P.N. Butcher, Phys. Lett. **19** (1965) 546.
[68] P.N. Butcher, W. Fawcett and C. Hilsum, Br. J. Appl. Phys. **17** (1966) 841.
[69] J.A. Copeland, IEEE Trans. Electron Devices **ED-13** (1966) 189.
[70] J.A. Copeland, Proc. IEEE **54** (1966) 1479.

[71] J.E. Carroll, Electron. Lett. **2** (1966) 141.
[72] R.H. Chilton, Proc. IEEE **56** (1968) 1124.
[73] H.W. Thim, J. Appl. Phys. **39** (1968) 3897.
[74] H.W. Thim and M.R. Barber, Proc. IEEE **56** (1968) 110.
[75] W.H. Press, B.P. Flannery, S.A. Teukolsky and W.T. Vettering, *Numerical Recipes in C–The Art of Scientific Computing* (Cambridge University Press, Cambridge, 1988).
[76] W.H. Enright, K.R. Jackson, S.P. Norsett and P.G. Thomsen, ACM Trans. Math. Software **12** (1986) 193.
[77] A.L. McWhorter and R.H. Rediker, Proc. IEEE **47** (1959) 1207.
[78] A.L. McWhorter and R.H. Rediker, in Proc. Int. Conf. on Phys. Semicond. Prague, 1960 (Academic Press, New York, 1961) p. 134.
[79] J. Callaway and F.W. Cummings, Phys. Rev. **126** (1962) 5.
[80] A. Zylbersztejn, J. Phys. Chem. Solids **23** (1962) 297.
[81] J. Yamashita, J. Phys. Soc. Jpn **16** (1961) 720.
[82] T. Kurosawa, J. Phys. Soc. Jpn **20** (1965) 1405.
[83] N. Sclar and E. Burstein, J. Phys. Chem. Solids **2** (1957) 1.
[84] S.H. Koenig and G.R. Gunther-Mohr, J. Phys. Chem. Solids **2** (1957) 268.
[85] S.H. Koenig, Phys. Rev. **110** (1958) 986; Phys. Rev. **110** (1958) 988.
[86] S.H. Koenig, R.D. Brown III and W. Schillinger, Phys. Rev. **128** (1962) 1668.
[87] D.J. Oliver, Phys. Rev. **127** (1962) 1045.
[88] D.J. Oliber, in Proc. Int. Conf. on Phys. Semicond., Exeter, 1962 (The Institute of Physics and The Physical Society, London, 1962) p. 133.
[89] R.A. Reynolds, Solid State Electron. **11** (1968) 385.
[90] R.S. Crandall, Phys. Rev. **B1** (1970) 730.
[91] R.P. Khosla, J.R. Fischer and B.C. Burkey, Phys. Rev. **B7** (1973) 2551.
[92] K. Aoki, Proc. of a NATO Advanced Workshop on Negative Differential Resistance and Instabilities in 2-D Semiconductors, eds. N. Balkan, B.K. Ridley and A.J. Vickers (Plenum Press, London, 1993) pp. 393–407.
[93] J. Peinke, J. Parisi, O.E. Rössler and R. Stoop, *Encounter with Chaos – Self-Organized Hierarchical Complexity in Semiconductor Experiments* (Springer, Berlin, 1992).
[94] W. Shockley, Solid State Electron. **2** (1961) 35.
[95] R.S. Crandall, J. Phys. Chem. Solids **31** (1970) 2069.
[96] R.S. Crandall, Phys. Rev. **169** (1968) 585.
[97] R. Stratton, Proc. R. Soc. London **A242** (1957) 355.
[98] K. Aoki, N. Mugibayashi and K. Yamamoto, Proc. 18th Int. Conf. on Phys. Semicond., ed. O. Engstrom (World Scientific, Singapore, 1987) p. 1543; see also K. Aoki and N. Mugibayashi, Appl. Phys. **A48** (1989) 127.
[99] K. Aoki, Solid State Commun. **77** (1991) 87; see also K. Aoki, Phys. Lett **A152** (1991) 485.
[100] E. Schöll, Proc. 3rd Int. Conf. on Hot Carriers in Semicond., in Suppl. J. Phys. (France) **C-7** (1981) 57.

Chapter 3

S-shaped negative differential conductivity and the nonequilibrium phase transition

3.1 Scattering processes

Many models for the S-shaped negative differential conductivity (SNDC) mechanism were introduced in chapter 2. Among these models, it is worth noting that the two-level impact-ionization model by Schöll [1] has prevailed [2–4] as a universal interpretation of the SNDC and chaos caused by the impact-ionization avalanche in semiconductors. However, the SNDC mechanism may depend on the nature of the material. In particular, the negative differential mobility observed during the impact-ionization avalanche may be an important factor in understanding the SNDC mechanism in compound semiconductors such as GaAs [5] and CdSe [6]. Then, as described in chapter 2, the screening effect for electron mobility may play an essential role in the occurrence of the SNDC in GaAs [5, 7–10] or CdSe [6]. This book adopts the screening model [5] for the SNDC mechanism in compound semiconductors. In section 3.2, the screening model is introduced in such a way as to be available for numerical studies of nonlinear dynamics and chaos. Thermodynamical consideration of the SNDC is given in section 3.3, and pattern formation of a current filament induced by the impact-ionization avalanche is argued based on an equal-areas rule [1] in section 3.4. Finally, measurement techniques of the filamentary current are presented in section 3.5. Before describing the model, the chapter starts with the scattering processes which participate in the generation and recombination processes and the SNDC mechanism.

3.1.1 Ionized-impurity scattering: Brooks–Herring formula

For simplicity, consider first the ionized-impurity scattering in n-type semiconductors as usual, with free electron concentration n and donor concentration N_D. A similar method is also used for p-type materials. The charge neutrality is given by $n = N_D{}^+$, where $N_D{}^+$ is the ionized donor concentration. For a compensated material, the relation is $n = N_D{}^+ - N_A$, where N_A is the acceptor concentration. The ionized-impurity scattering is an elastic process. Electrons accelerated by an electric field collide frequently with the ionized donors, and thereby their momenta $\hbar\mathbf{k}$ are altered from collision to collision during the zigzag motion. Thus, the momentum relaxation time is $\tau_m(\varepsilon^*)$ which depends on the microscopic carrier energy ε^*. Similarly to the Brownian motion of classical particles, the momentum relaxation time is defined by [11]:

$$1/\tau_m = N_D{}^+ Q_m v = N_D{}^+ v 2\pi \int_0^\pi Q(\theta)(1 - \cos\theta)\sin\theta \, \mathrm{d}\theta, \qquad (3.1)$$

where $v = \hbar\mathbf{k}/m_e$ is the electron velocity, $Q(\theta)$ is the differential collision cross-section, and θ is the deflection angle in Fig. 3.1. For the quantum mechanical version [11],

$$1/\tau_m = V(2\pi)^{-2} \int k'^2 \mathrm{d}k' \int_0^\pi S(\mathbf{k}, \mathbf{k}')(1 - \cos\theta)\sin\theta \, \mathrm{d}\theta, \qquad (3.2)$$

where V is the crystal volume, and $S(\mathbf{k}, \mathbf{k}')$ is the transition probability by the ionized-impurity scattering from an initial state of the momentum $\mathbf{k}$ to a final one with $\mathbf{k}'$. The transition probability is given by golden rule no. 2

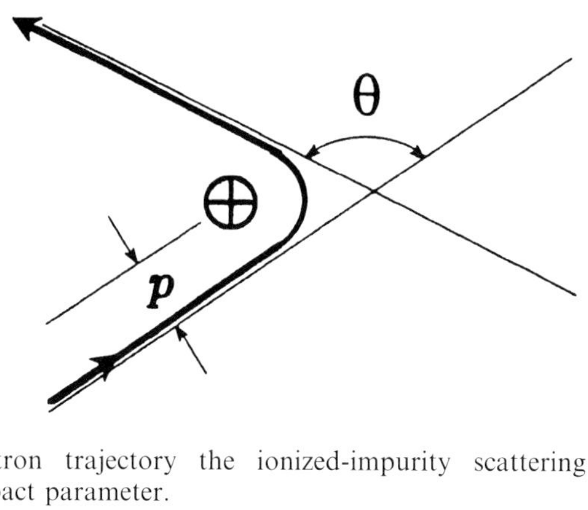

Figure 3.1 Electron trajectory the ionized-impurity scattering (schematic). p indicates the impact parameter.

[11, 12] as:

$$S(\mathbf{k}, \mathbf{k}') = (2\pi/\hbar)\ |H_{kk'}|^2 \delta[\varepsilon^*(\mathbf{k}') - \varepsilon^*(\mathbf{k})], \tag{3.3}$$

$$H_{kk'} = \frac{e}{V} \int V(r) \exp\left[i(\mathbf{k} - \mathbf{k}') \cdot \mathbf{r}\right] \mathrm{d}^3\mathbf{r}, \tag{3.4}$$

where $H_{kk'}$ is the Hamiltonian matrix element, $\varepsilon^*(k) = \hbar^2\mathbf{k}^2/2m_\mathrm{e}$, $\varepsilon^*(\mathbf{k}') = \hbar^2\mathbf{k}'^2/2m_\mathrm{e}$, the Dirac delta means the elastic scattering, namely, $\varepsilon^*(\mathbf{k}) = \varepsilon^*(\mathbf{k}')$ and $|\mathbf{k}| = |\mathbf{k}'|$ with $\mathbf{k} \cdot \mathbf{k}' = |\mathbf{k}|^2 \cos\theta$.

For $V(r)$ in (3.4), one must use the screened Coulomb potential (Debye–Hückel potential) [11, 13–15]:

$$V(r) = -(e/4\pi\epsilon\epsilon_0 r) \exp\left(-r/L_\mathrm{D}\right), \tag{3.5}$$

$$L_\mathrm{D} = \sqrt{\frac{\epsilon\epsilon_0 k_\mathrm{B} T_\mathrm{L}}{ne^2}}. \tag{3.6}$$

The Debye–Hückel potential is obtained by considering the electron concentration $n(r)$ at a distance r from an ionized impurity and by solving the Poisson equation in the linearized form:

$$\Delta V(r) = -e[n(r) - n]/\epsilon\epsilon_0, \tag{3.7}$$

$$n(r) = \frac{1}{\sqrt{2}\hbar^3} \left(\frac{m_e k_\mathrm{B} T_\mathrm{L}}{\pi}\right)^{3/2} F_{1/2}(\eta + eV(r)/k_\mathrm{B} T_\mathrm{L}), \tag{3.8}$$

$$\simeq n \exp\{-eV(r)/k_\mathrm{B} T_\mathrm{L}\} \sim n[1 - eV(r)/k_\mathrm{B} T_\mathrm{L}], \tag{3.9}$$

where n is the uniform electron concentration which is obtained from (3.8)

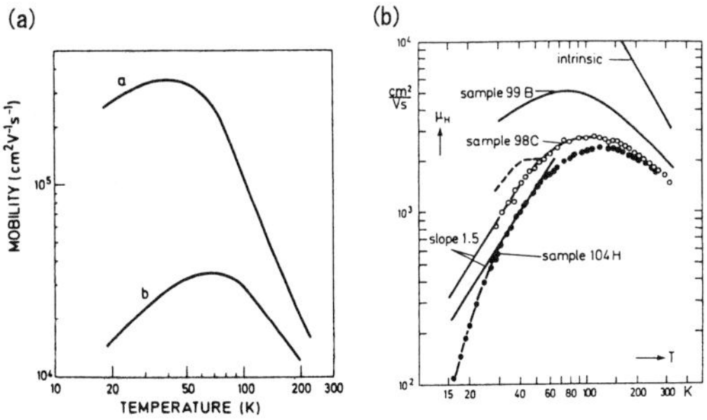

Figure 3.2 (a) Electron mobility vs temperature for n-type GaAs. Curves a and b are for room-temperature carrier concentrations of 3×10^{13} and 1.5×10^{15} cm^{-3}, respectively. After Chattopadhyay and Queisser in Ref. [11]. (b) Hall mobility vs temperature for n-type Zn-doped Ge samples (98 C and 104 H). Mobility data for intrinsic n-type Ge and for uncompensated sample doped with 5×10^{16} Sb atoms/cm^3 are shown for comparison. For sample 98 C, the dashed curve indicates the effect of exposure to light on mobility. After Tyler and Woodbury in Ref. [11].

by putting $V(r) = 0$. The Poisson equation is then rewritten as:

$$d^2(rV(r))/dr^2 = rV(r)/{L_D}^2. \tag{3.10}$$

The general solution is given by $rV(r) = A\exp(-r/L_D) + B\exp(r/L_D)$, which gives (3.5) by using the boundary condition: $V|_{r\to 0} = -e(4\pi\epsilon\epsilon_0 r)^{-1}$ and $V|_{r\to\infty} = 0$.

Now let us return to our discussion for the solution of τ_m via (3.2)–(3.4). Integration of (3.2) over k' is done easily by the sake of the δ-function in (3.3):

$$\int k'^2 S(\mathbf{k}, \mathbf{k}')\mathrm{d}k' = (m_e/\hbar^2)\int k' S(\mathbf{k}, \mathbf{k}')\mathrm{d}\varepsilon^{*\prime} = (2\pi m_e/\hbar^3)k|H_{\mathbf{kk}'}|^2,$$

where the relation

$$\varepsilon^{*\prime} = \frac{\hbar^2}{2m_e}\mathbf{k}'^2$$

was used for the second term. Then, (3.2) is rewritten as:

$$\begin{aligned} 1/\tau_m &= (Vm_e k/2\pi\hbar^3)\int |H_{\mathbf{kk}'}|^2(1 - \cos\theta)\sin\theta\, d\theta \\ &= {N_D}^+(\hbar k/m_e)2\pi(m_e/2\pi\hbar^2)^2(V/{N_D}^+)\int |H_{\mathbf{kk}'}|^2(1 - \cos\theta)\sin\theta\, d\theta \end{aligned} \tag{3.11}$$

Comparing (3.11) with (3.1), the differential collision cross-section is obtained as:

$$Q(\theta) = (m_e V/2\pi\hbar^2)^2|H_{\mathbf{kk}'}|^2. \tag{3.12}$$

The matrix element $H_{kk'}$ is calculated using the screened Coulomb potential (3.5):

$$H_{\mathbf{kk}'} \approx -\frac{e^2}{4\epsilon\epsilon_0 Vk^2}\frac{1}{\sin^2(\theta/2) + (2kL_D)^{-1}}. \tag{3.13}$$

By inserting (3.13) into (3.12), the final form for $Q(\theta)$ is obtained:

$$Q(\theta) = \left[\frac{K/2}{\sin^2(\theta/2) + \beta^{-2}}\right], \tag{3.14}$$

where $K = e^2/(4\pi\epsilon\epsilon_0 m_e\hbar^2 k^2)$ and $\beta = 2kL_D$. From (3.1) and (3.14), τ_m is obtained as:

$$\tau_m = \varepsilon^{*3/2}(1/N_D^+)16\pi\sqrt{2m_e}(\epsilon\epsilon_0/e^2)^2[\ln(1+\beta^2) - \beta^2/(1+\beta^2)]^{-1}, \tag{3.15}$$

where the integration is carried out by putting $z = (1 - \cos\theta)/2 + \beta^{-2}$. Since τ_m is the momentum relaxation time of free electrons with the microscopic energies in the range $[\varepsilon^*, \varepsilon^* + d\varepsilon^*]$, it is necessary to average τ_m

over ε^* by (2.48) in order to obtain the macroscopic value $\langle\tau_m\rangle$ for the collective motion of the electrons and the electron mobility ($\mu_d = -e\langle\tau_m\rangle/m_e$). With the applied electric field $E = (E_x, 0, 0)$, (2.48) is rewritten as

$$\langle\tau_m\rangle = -\int_{-\infty}^{\infty} \tau_m k_x \partial f_k^0/\partial k_x \; 4\pi k^2 \; dk \; / \int_{-\infty}^{\infty} f_k^0 4\pi k^2 \; dk, \tag{3.16}$$

$$= \frac{1}{3}\int_{-\infty}^{\infty} (\hbar^2/m_e k_B T_L) \; \tau_m f_k^0 \; k^4 dk / \int_{-\infty}^{\infty} f_k^0 \; k^2 dk, \tag{3.17}$$

$$= (4/3\sqrt{\pi})(k_B T_L)^{-5/2} \int_0^{\infty} \tau_m \varepsilon^{*3/2} \exp(-\varepsilon^*/k_B T_L) d\varepsilon^*, \tag{3.18}$$

where the relations $f_k^0 = \exp[-(\varepsilon^* - \eta)/k_B T_L], d^3k = 4\pi k^2 dk$, $k_x^2 = k^2/3$ and $\varepsilon^* = \hbar^2 k^2/2m_e$ have been used. The bracket term of τ_m in (3.15) is taken outside the integral of (3.18), replacing k in τ_m by $k = \sqrt{6k_B T_L m_e}/\hbar$, at which the integrand $\varepsilon^{*3}\exp(-\varepsilon^*/k_B T_L)$ in the numerator is a maximum [11]. The final result for the Brooks–Herring (BH) mobility [11, 13] is given by:

$$\mu_{BH} = \frac{128\sqrt{2\pi}(\epsilon\epsilon_0)^2 (k_B T_L)^{3/2}}{e^3 {m_e}^{1/2} {N_D}^+}[\ln(1+\beta_{BH}^2) - \beta_{BH}^2/(1+\beta_{BH}^2)]^{-1}, \tag{3.19}$$

$$\beta_{BH} = 2\left(\frac{6 m_e k_B T_L}{\hbar^2}\right)^{1/2} L_D, \tag{3.20}$$

where β_{BH} is the Brooks–Herring coefficient. The power law is approximately given as ${T_L}^{3/2}$. For the hot electron regime, may be replaced T_L by T_e. Dependence on the carrier concentration comes from ${N_D}^+ = N_D - n$ and β_{BH} via $L_D = \sqrt{\epsilon\epsilon_0 k_B T_L/ne^2}$. As a result, it is expected in the experiments that the power law is less strong than ${T_L}^{3/2}$. Figure 3.2 shows examples of the electron mobility vs lattice temperature in high-purity n-GaAs and Ge [11]. Apparently, the mobility decreases by ionized impurity scattering with decreasing temperature below 50 K.

Prior to the Brooks–Herring theory, Conwell and Weisskopf [11, 16] derived the carrier mobility using a bare Coulomb potential $-e^2/4\pi\epsilon\epsilon_0 r$:

$$\mu_{CW} = \frac{128\sqrt{2\pi}(\epsilon\epsilon_0)^2 (k_B T_L)^{3/2}}{{m_e}^{1/2} {N_D}^+ e^3}\left[\ln\left(1 + \frac{144(\pi\epsilon\epsilon_0)^2 (k_B T_L)^2}{e^4 ({N_D}^+)^{2/3}}\right)\right]^{-1}. \tag{3.21}$$

With the bare Coulomb potential, the differential cross-section $Q(\theta)$ or the inverse relaxation time τ_m^{-1} diverges for zero-angle scattering far from the ionized impurity (i.e. $\theta = 0$, $\mathbf{k} = \mathbf{k}'$, where $|\mathbf{k} - \mathbf{k}'| \sim 2k\sin(\theta/2)$). In order to avoid the divergence, they have assumed the minimum deflection angle θ_{min} and the largest impact parameter p_{max} i.e., the integration (3.11) over θ in the range $[\theta_{\min}, \pi]$:

$$p_{\max} = K\cot(\theta_{\min}/2) = d/2, \tag{3.22}$$

$$d = (N_D{}^+)^{-1/3}, \tag{3.23}$$

where d is the average distance between nearest-neighbor impurity ions and K is the distance defined by $K = e^2/(4\,\pi\,\epsilon\,\epsilon_0 m_e \hbar^2 k^2)$. In the CW approximation, it has been considered that the scattering with $\theta < \theta_{\min}$ ($p > p_{\max}$) may be avoided since the electron is scattered at least by a nearest-neighbor ion with a large deflection angle $\theta' > \theta_{\min}(p' < p_{\max})$ as shown in Fig. 3.3. It is rather surprising to note that both of the formulae μ_{BH} and μ_{CW} give almost identical results (Fig. 3.4).

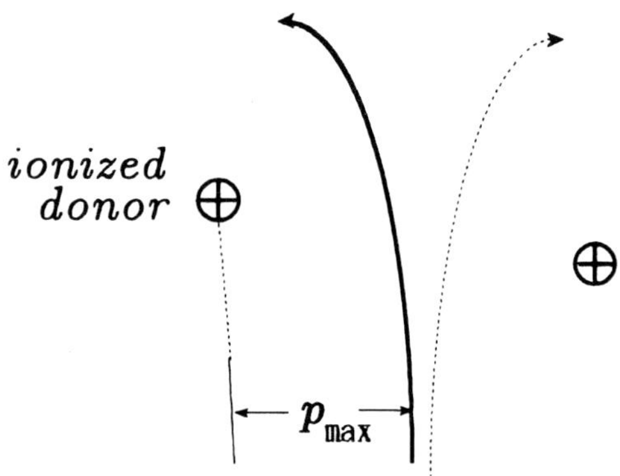

Figure 3.3 Cutoff of the impact parameter due to a nearest-neighbor donor ion.

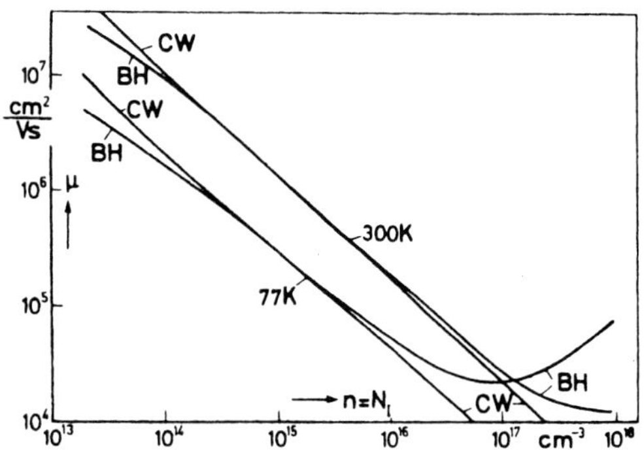

Figure 3.4 Dependence of the mobility on impurity concentration N_I at 77 K and 300 K calculated according to Conwell and Weisskopf (CW) and Brooks and Herring (BH) for a hypothetical uncompensated semiconductor with effective mass equal to the free electron mass, dielectric constant 16 and impurity charge $Ze = e$ [11].

3.1.2 Ionized impurity scattering in compensated semiconductors

So far, ionized impurity scattering has been described for uncompensated materials. For the compensated n-type semiconductors in which acceptors are unintentionally doped ($N_D > N_A; N_A$ = concentration of acceptors), the charge neutrality condition is given by:

$$n + N_A^- = p + N_D^+, \tag{3.24}$$

where N_A^- is the ionized acceptor concentration and p is the excess hole concentration. For the n-type materials, one can put $N_A^- = N_A$ and $p = 0$. Electrons are scattered by the ionized acceptors as well as by the ionized donors. Therefore, N_D^+ must be replaced by $N_D^+ + N_A = n + 2N_A$ for the BH formula and hence μ_{BH} is rewritten as [17].

$$\mu_{BH} = \frac{128\sqrt{2\pi}(\epsilon\,\epsilon_0)^2(k_B T_L)^{3/2}}{m_e^{1/2}(n + 2N_A)e^3}[\ln(1 + b^2) - b^2/(1 + b^2)]^{-1}, \tag{3.25}$$

$$b^2 = \frac{24 m_e \epsilon\epsilon_0 (k_B T_L)^2}{e^2\hbar^2 n_s}, \tag{3.26}$$

where n_s is the modified carrier density given by $n_s = n + (n + N_A)[1 - (n + N_A)/N_D]$ (note: electron density tends to increase near the ionized donors by the attractive Coulomb force [17, 18]; $n_s > n$).

3.1.3 Neutral impurity scattering

At low temperatures, electrons are frozen out at the impurity (donor) levels, and thereby a large part of the impurities are charge neutral. Therefore, the electron-neutral impurity scattering may affect significantly the electron mobility [19–22]. Erginsoy [19] stressed an important role of neutral impurity scattering for the resistivity of semiconductors at low temperatures.

As a general remark, the scattering process in semiconductors is similar to the electron-hydrogen scattering processes [23]. Consider a scattering center (neutral donor) for which an incident electron in the z direction with the wave function ψ feels a potential $V(r)$, as schematically shown in Fig. 3.5. The Schrödinger equation for the neutral impurity scattering is given by:

$$\nabla^2\psi + (k^2 - U(r))\psi = 0, \tag{3.27}$$

where k is the wave vector of the electron and $U(r) = (2m_e/\hbar^2)V(r)$. As well known in the atomic collision theory by Mott and Massey [23] and Schiff [24], asymptomatic form (for sufficiently large r) of the wave function after

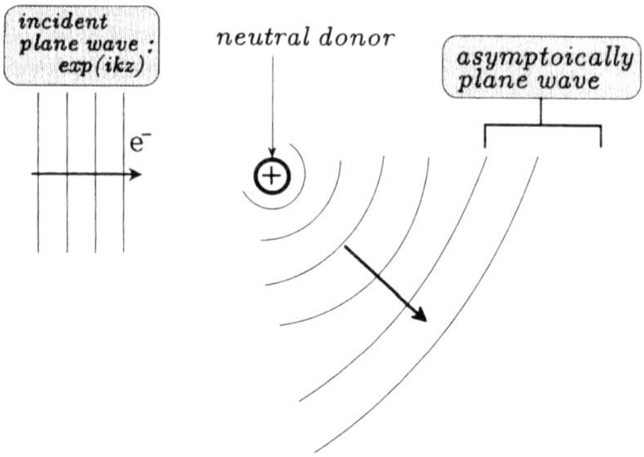

Figure 3.5 Neutral impurity scattering (schematic).

the scattering is expressed as:

$$\psi \sim e^{ikz} + \frac{1}{r}\, e^{ikr} f(\theta). \tag{3.28}$$

Since the plane wave e^{ikz} of the incident electron is given by Legendre expansion $e^{ikz} = \sum_\ell P_\ell(\cos\theta) f_\ell(r)$, where $f_\ell(r) = (kr)^{-1} \sin(kr - \ell\pi/2)$, the wave function after the scattering should be:

$$\psi = \sum_{\ell=0}^{\infty} A_\ell P_\ell(\cos\theta) \mathrm{F}_\ell(r), \tag{3.29}$$

where A_ℓ is the linear combination constant. Function $\mathrm{F}_\ell(r) = Cr^{-1} G_\ell(r)$ is a solution of the equation:

$$\frac{\mathrm{d}^2 \mathrm{G}_\ell}{\mathrm{d}r^2} + \left[k^2 - U(r) - \frac{\ell(\ell+1)}{r^2} \right] \mathrm{G}_\ell = 0, \tag{3.30}$$

With $r \to \infty$, (3.30) is approximated as $\mathrm{d}^2\mathrm{G}_\ell/\mathrm{d}r^2 + k^2 \mathrm{G}_\ell = 0$. Then the asymptomatic solution of (3.30) should be [23]:

$$\mathrm{G}_\ell = \sin(kr - \ell\pi/2 + \eta_\ell), \tag{3.31}$$

$$\mathrm{F}_\ell = (kr)^{-1} \mathrm{G}_\ell, \tag{3.32}$$

where η_ℓ is the phase shift. With $A_\ell = (2\ell+1) i^\ell e^{i\eta_\ell}$ in (3.29), then [23]:

$$\psi = \sum_{\ell=0}^{\infty} (2\ell+1) i^\ell e^{i\eta_\ell} \mathrm{F}_\ell(r) P_\ell(\cos\ \theta). \tag{3.33}$$

In correspondence with (3.28), $\psi - \psi_0$ must be $r^{-1} e^{ikr} f(\theta)$, where

$\psi_0 = \sum (2\ell+1) i^\ell P_\ell(\cos\theta) f_\ell(r)$. From (3.33),

$$\psi - \psi_0 = \sum_{\ell=0}^{\infty} A_\ell P_\ell(\cos\theta) \mathrm{F}_\ell(r) - \sum_{\ell=0}^{\infty} (2\ell+1) i^\ell P_\ell(\cos\theta) f_\ell(r)$$
$$= \frac{e^{ikr}}{r} \frac{1}{2ik} \sum_{\ell=0}^{\infty} (2\ell+1)[e^{2i\eta_\ell} - 1] P_\ell(\cos\theta). \tag{3.34}$$

Therefore,

$$f(\theta) = \frac{1}{2ik} \sum_{\ell=0}^{\infty} (2\ell+1)[e^{2i\eta_\ell} - 1] P_\ell(\cos\theta). \tag{3.35}$$

The probability that the electron is scattered with the solid angle $d\omega$ is given by $PdS = |r^{-1} f(\theta) e^{ikr}|^2 dS = |f(\theta)|^2 d\omega$, where dS is given by $dS = 2\pi r^2 \sin\theta \, \mathrm{d}\theta$. Integrating over θ, one obtains the total cross-section [23]

$$Q_{\mathrm{tot}} = 2\pi \int_0^{2\pi} |f(\theta)|^2 \sin\theta \, \mathrm{d}\theta, \tag{3.36}$$

$$= \frac{4\pi}{k^2} \sum_{\ell=0}^{\infty} (2\ell+1) \sin^2 \eta_\ell. \tag{3.37}$$

For the s-wave scattering ($\ell = 0$),

$$Q_{\mathrm{tot}}^0 = \frac{4\pi}{k^2} \sin^2 \eta_0. \tag{3.38}$$

The problem of neutral impurity scattering is thus solved by determining the phase shift.

The first-order Born approximation [18,23] is a simpler method by which to calculate $|f(\theta)|^2$, but it gives incorrect values of Q_{tot} for the neutral impurity scattering, especially in the low-velocity limit $v \to 0$. Instead, one should use the **partial wave scattering method** as described above. As well as the static field of the impurity atom, one must include the polarization field $\alpha(r)/r^4$ induced by the incident electron [$\alpha(r)$ polarizability of the atom] and the electron exchange effects. Erginsoy [19] approximated the numerical results of the s-wave scattering in hydrogen atoms by Massey and Moiseiwitsch [25], who used Hulthens' variational method [23]:

$$Q_{\mathrm{tot}}^0 \sim \frac{20 \, a_{\mathrm{B}}}{k}, \tag{3.39}$$

where $k = m_0 v/\hbar$ and a_{B} is the Bohr radius of the hydrogen 1s-state ($a_B = \hbar^2/m_0 e^2$); m_0 : electron rest mass). For neutral impurity scattering in semiconductors, m_0 by m_{e} (for effective mass of electrons) and a_{B} by a_0 (effective Bohr radius of the impurity 1s-state in semiconductors). Similar results have been obtained in the electron-exciton scattering problem [26,

Table 3.1 Comparison of ρ_n and ρ_i calculated for Pearson and Bardeen's sample 2 and B: $N_{total} = N_n + N_i$ [cm^{-3}]; ρ_n and ρ_i in ohm·cm [19]

$1/T$	Sample 2 ($N_{total} = 6 \times 10^{17}$)			Sample B ($N_{total} = 1.25 \times 10^{18}$)		
(K)$^{-1}$	$n(N_i)$	ρ_n	ρ_n	$n(N_i)$	ρ_n	ρ_i
12×10^{-3}	3.7×10^{16}	0.0866	0.0526	5.3×10^{17}	0.0077	0.0283
12×10^{-3}	7.4×10^{16}	0.0402	0.0390	5.3×10^{17}	0.0077	0.0250
6×10^{-3}	2.37×10^{17}	0.0086	0.0194	7.4×10^{17}	0.0031	0.0155
5×10^{-3}	3.3×10^{17}	0.0046	0.0153	8.3×10^{17}	0.0029	0.0129

27]. Using formula (3.39) for neutral donor scattering, the momentum relaxation time τ_m is given by:

$$1/\tau_m = N_D{}^0 v \, Q^0_{tot} = 20\hbar N_D{}^0 a_0/m_e, \tag{3.40}$$

where $a_0 = \epsilon\epsilon_0\hbar^2/m_e e^2$. Since τ_m is independent of the electron velocity or wave vector, the electron mobility is immediately obtained as:

$$\mu_n = \frac{e}{20\hbar N_D{}^0 a_0} = \frac{m_e e^3}{20\epsilon\epsilon_0\hbar^3 \; N_D{}^0}. \tag{3.41}$$

Correspondingly, the resistivity is given by:

$$\rho_n = \frac{m_e}{ne^2\tau_m} = \frac{20\epsilon\epsilon_0\hbar^3 N_D{}^0}{m_e e^4 n}. \tag{3.42}$$

Erginsoy [19] has calculated the resistivity ρ_n for Si samples and compared it with the numerical value of ρ_i which is the resistivity due to ionized impurity scattering (Table 3.1), stressing the importance of neutral impurity scattering at low temperatures for nondegenerate samples (sample 2 in Table 3.1). In the experiments, measurements of cyclotron resonance [28] and attenuation of transverse acoustic wave [29] in n-type Ge have shown clear evidences of neutral impurity scattering. For weakly compensated materials whose impurity concentration is not so high, no impurity conduction may occur, but neutral impurity scattering may be an important factor for the carrier transport [30]. For strongly compensated materials, however, ionized impurity scattering may dominate the electron mobility.

A similar scattering process is electron-exciton elastic scattering [26, 27]. Elkomoss and Munschy [26] calculated the electron-exciton elastic scattering cross-sections in the central field and the exchange approximations. They estimated the scattering cross-sections for various semiconductor materials and found a good agreement with the measurements of cyclotron resonance in Si [31]. Approximate polarized-orbital calculations of the electron-exciton elastic scattering cross-sections have been also done by Aoki *et al.* [27] in order to explain broadening of exciton-polariton luminescence observed in n-GaAs as a function of photoexcitation density

[32, 33], where the broadening has been attributed to the lifetime broadening due to shortening of the collision time.

3.1.4 Excitations of neutral impurity

Shallow impurities (donors or acceptors) in semiconductors are treated as hydrogen-like atoms embedded in the continuum medium with the dielectric constant $\epsilon\epsilon_0$. The wave function of the donor ground state is given by:

$$\psi_0 = (\pi {a_0}^3/Z^3)^{-1/2} \exp(-Zr/a_0) \tag{3.43}$$

and the excited states are [23]:

$$\begin{aligned} \psi_{n\ell m} &= N_{n\ell m} r^{\ell} L_{n+\ell}^{2\ell+1}(2rZ/na_0)\ \exp(-Zr/na_0)\ P_\ell^m(\cos\theta) e^{\pm im\phi}, \\ N_{n\ell m}^2 &= \frac{2^{2^\ell}}{n\pi}\ (2\ell+1)\ \frac{(\ell-m)!}{(\ell+m)!}\ \frac{(n-\ell-1)!}{\{(n+\ell)!\}^3} \left(\frac{Z}{na_0}\right)^{2\ell+3}, \end{aligned} \tag{3.44}$$

where r, θ and ϕ are the polar coordinates, n, ℓ and m are the quantum numbers, and L_q^p is the associated Laguerre polynomial. Now let us consider excitation of the donor state from the ith state to the jth state by electron-donor inelastic scattering. The excitation is possible when the kinetic energy of the incident electron is larger than the energy difference $\varepsilon_j - \varepsilon_i$ in Fig. 3.6. The formulae of electron-hydrogen scattering [23] are applied directly, replacing a_B by a_0 and ϵ_0 by $\epsilon\epsilon_0$. The differential cross-section $I_{ij}(K)\,\mathrm{d}K$ is expressed as [23].

$$I_{ij}(K)\mathrm{d}K = \frac{m^2}{2\pi\hbar^4}\ \frac{K\mathrm{d}K}{k^2} \left| \int\int V e^{iKX} \psi_i \psi_j^* \mathbf{dr}\ \mathrm{d}\mathbf{R} \right|^2, \tag{3.45}$$

$$K = |\mathbf{k}_{ij} - \mathbf{k}| = (k_{ij}^2 + k^2 - 2kk_{ij}\cos\theta)^{1/2}, \tag{3.46}$$

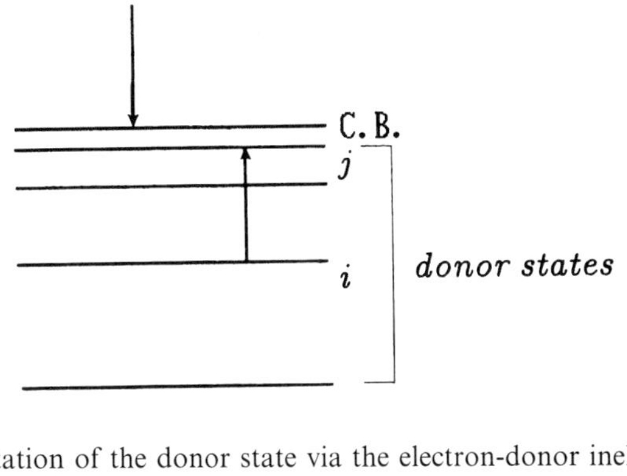

Figure 3.6 Excitation of the donor state via the electron-donor inelastic scattering.

where V is the Coulomb interaction $e^2/|\mathbf{r} - \mathbf{R}|$ between the incident electron and the bound electron, $\mathbf{k}$ and $\mathbf{k}_{ij}$ are the wave vectors of the initial and final states, the subscripts i and j stand for the sets of quantum numbers $(n,\ \ell,\ m)$ and $(n',\ \ell',\ m')$, and $\mathbf{r}$ and $\mathbf{R}$ are the coordinates of the incident electron and the bound electron, respectively. For the scattering, the energy conservation is expressed as:

$$\frac{\hbar^2}{2m_e}(k^2 - k_{ij}^2) = \varepsilon_j - \varepsilon_i. \tag{3.47}$$

After some calculations, (3.45) is rewritten as [23]:

$$I_{ij}(K)dK = \frac{8\pi m^2 e^4}{\hbar^2 k^2}\frac{dK}{K^3}|\gamma_{ij}(K)|^2, \tag{3.48}$$

$$\gamma_{ij}(K) = \sum_{s=1}^{Z}\int \exp(iKX_s)\psi_i\ \psi_j^* \mathbf{dr}, \tag{3.49}$$

where the sum in (3.49) is taken over the atomic number $s = 1, 2, \ldots Z$ ($Z = 1$ for the hydrogenic impurity). For the excitation from the 1s-state to an excited state with the quantum number $(n,\ \ell,\ m)$, (3.49) is calculated directly by using (3.43) and (3.44).

Total cross-section for the ij transition is given by:

$$Q_{ij}(k) = \int_{K_{\min}}^{K\max} I_{ij}(K)\mathrm{d}K, \tag{3.50}$$

where $K_{\min}$ and $K_{\max}$ are given from (3.46) and (3.47) as:

$$K_{\min} = k - \sqrt{k^2 - 2m_e(\varepsilon_j - \varepsilon_i)/\hbar^2},$$

$$K_{\max} = k + \sqrt{k^2 - 2m_e(\varepsilon_j - \varepsilon_i)/\hbar^2}.$$

For the 1s–2p transition, the total cross-section vs k is shown in Fig. 3.7. As expected, k has a threshold value $\sqrt{2m_e(\varepsilon_2 - \varepsilon_1)/\hbar^2}$ for the excitation.

3.1.5 Impact ionization

When the kinetic energy of the incident electron exceeds the binding energy of the neutral impurity ε_D $(= \varepsilon_\infty - \varepsilon_1)$, it can impact-ionize the impurity. The reaction is written as:

$$e^- + D^0 \rightarrow D^+ + 2e^-. \tag{3.51}$$

Let $\mathbf{k}$ and $\mathbf{k}_\kappa$ be the initial and final wave vectors of the incident electron, and $\boldsymbol{\kappa}$ be the wave vector of the impact-ionized electron (see Fig. 3.8).

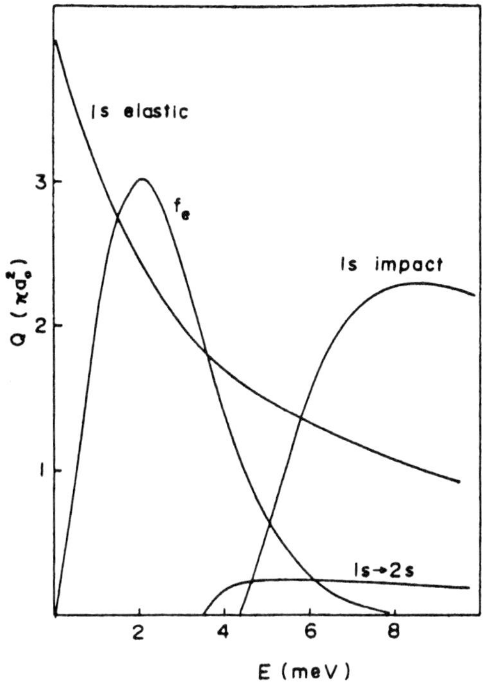

Figure 3.7 Elastic and inelastic collision cross-sections (in units of πa_0^2) for exciton-electron scattering in GaAs as a function of kinetic energy of incident electron. For the calculation, the effective Bohr radius a_0 of 1s exciton and static dielectric constant were chosen to be 115.8 Å and 12.56, respectively. The shifted MB distribution function is also shown by f_e, where the effective electron temperature of 30 K and the drift energy of 2 meV were assumed [34].

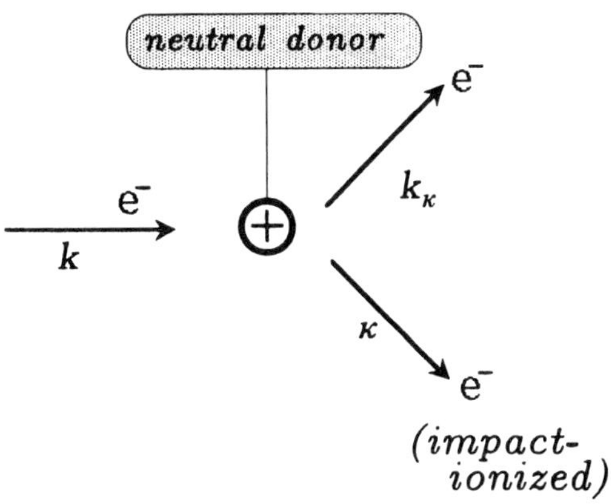

Figure 3.8 Wave vectors of the electrons before and after the impact ionization of a neutral donor.

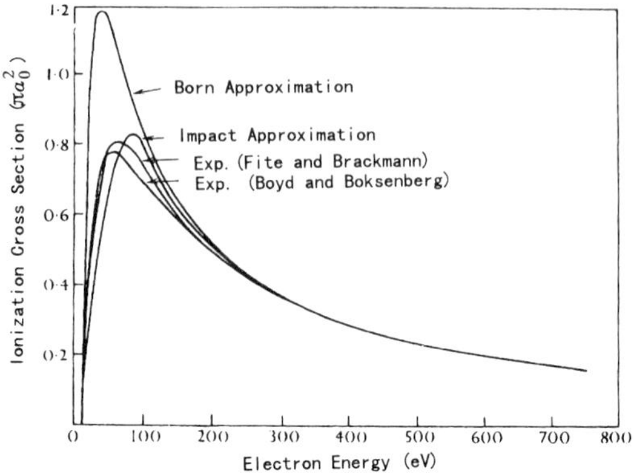

Figure 3.9 Ionization cross-section for the impact ionization of a hydrogen atom. Numerical results by the Born approximation and the impact approximation have been compared with the experimental values of Fite and Brackmann [Phys. Rev. **112** (1958) 1141] and Boyd and Boksenberg [Proc. 4th Int. Conf. on Ionization Phenomena in Gases, Uppsala, 1959 (North Holland, Amsterdam), Vol. 1, p. 529] [23].

Conservation of the energies is given by:

$$\frac{\hbar^2}{2m_e}\mathbf{k}^2 = \frac{\hbar^2}{2m_e}\mathbf{k}_\kappa^2 + \frac{\hbar^2}{2m_e}\boldsymbol{\kappa}^2 + \varepsilon_D. \tag{3.52}$$

With $K = |\mathbf{k} - \mathbf{k}_\kappa|$, the differential cross-section is given by [23]:

$$\begin{aligned} I_{0\kappa}(K)\mathrm{d}\omega\,\mathrm{d}\kappa = {} & \frac{2^{10}\kappa\, k_\kappa}{a_0{}^2 k K^2}\frac{\mu^6\{K^2 + (\mu^2+\kappa^2)/3\}}{\{\mu^4 + 2\mu^2(K^2+\kappa^2) + (K^2-\kappa^2)^2\}^3} \\ & \times \exp\left\{-\frac{2\mu}{\kappa}\arctan\frac{2}{K^2-\kappa^2+\mu^2}\right\} \\ & \times (1-\exp(-2\pi\mu/\kappa)^{-1})\mathrm{d}\omega\,\mathrm{d}\kappa, \end{aligned} \tag{3.53}$$

where $\mu = Z/a_0$, $k = (k_\kappa^2 - \kappa^2 - 2m_e\,\varepsilon_D/\hbar^2)^{1/2}$ from (3.52), and the incident electron is scattered toward the polar angles (θ, ϕ) with the solid angle $\mathrm{d}\omega$. The total cross-section Q^0_{tot} is given by:

$$Q^0_{tot} = \int_0^{\kappa_{max}}\int_{K_{min}}^{K max} I_{0\kappa}(K)\,\mathrm{d}K\,\mathrm{d}\kappa, \tag{3.54}$$

where $K_{min} = k - k_\kappa$, $K_{max} = k + k_\kappa$, and κ_{max} is given by

$$\kappa_{max}^2 = k^2 - 2m_e\,\varepsilon_D/\hbar^2 > 0.$$

$I_{0\kappa}\mathrm{d}\kappa$ is expressed as $(\pi a_0^2)\tilde{I}_{0\kappa}\mathrm{d}\tilde{\kappa}$ by the normalized wave vectors $\tilde{k} = ka_0$, $\tilde{K} = Ka_0$ and $\tilde{k}_\kappa = k_\kappa a_0$. Figure 3.9 shows $Q^0_{tot}/\pi a_0^2$ vs k^2 for impact ionization of hydrogen atom [23], which is a universal relation independent of the binding energy ε_D. One can even apply the theory to the electron-exciton collision processes, as shown in Fig. 3.7.

The impact-ionization theory has been applied to the impact ionization of free excitons in Ge and Si by Smith *et al.* [35], who used Percival's empirical formula for the electron impact-ionization cross-section of the hydrogen atom [36]. Using the empirical formula, they calculated the impact-ionization rate and the exciton density as a function of an applied electric field. In Fig. 3.10 the cross-section is shown by a broken line, while the electron distributions at various electric field intensities are shown by solid lines. At zero electric field, free electrons have insufficient energy to impact-ionize the free excitons. By applying the electric fields, free electrons become more and more energetic and the distribution function broadens towards the higher energies. Consequently, the impact-ionization rate increases abruptly by increasing the electric field, as shown in Fig. 3.11. By considering thermal recombination, thermal dissociation and impact-ionization and -recombination, the generation and recombination process of free excitons is easily described by a rate equation [35], similarly to the case of neutral impurities, from which one can find the exciton density as a function of the electric field in Fig. 3.12. The apparently sudden decrease in the exciton density above a critical field of 1.0 V/cm is due to the impact ionization of free excitons. Smith *et al.* [35] have shown a qualitative agreement with the photoconductivity measurements in Ge [37, 38]. Another example was given by

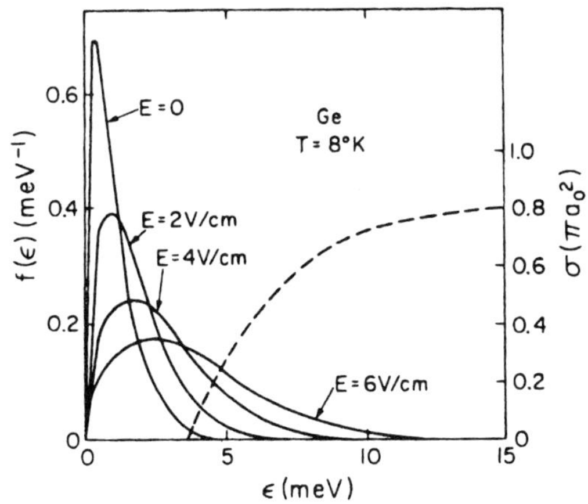

Figure 3.10 Exciton impact-ionization cross-section (dashed line) in units of the geometrical cross-section and electron energy distribution at several applied electric fields (solid lines) vs electron energy in Ge at 8 K. The threshold in the impact-ionization cross-section is the exciton binding energy (3.6 meV). At zero applied electric field, a very small fraction of the electrons is sufficiently energetic to impact-ionize an exciton; as an external field is applied, the number of electrons capable of impact ionization is greatly increased [35].

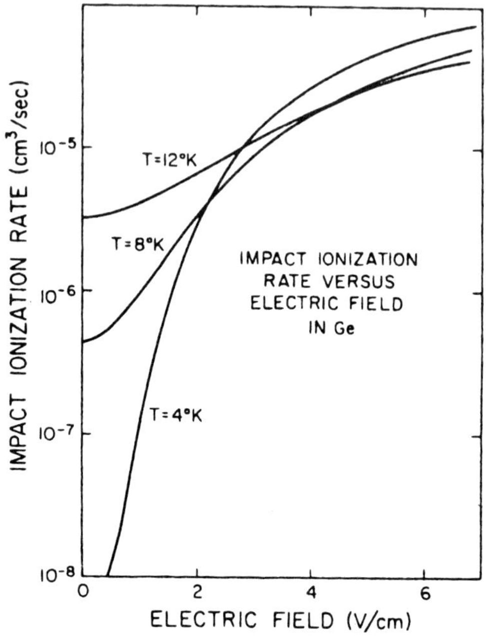

Figure 3.11 Impact-ionization rate vs applied electric field at several temperatures in pure Ge. As the field strength is increased, the fraction of carriers energetic enough to impact-ionize rises rapidly and the impact-ionization rate is greatly increased. The effect is essentially large at low temperature, where the fraction of thermal carriers (zero applied field) sufficiently energetic to impact-ionize is very small and carrier mobilities are high, so that they are easily excited by an electric field [35].

Aoki *et al.* [34], who have also calculated the exciton density as a function of the electric field and argued the measured quenching of the exciton luminescence intensity by the impact-ionization process. In their calculations of the collision cross-sections they used an analytical formula of (3.53), (3.54) instead of Percival's empirical formula.

Precise descriptions of the above-mentioned inelastic scatterings are, however, not useful in investigating the nonlinear dynamical behaviors related to the impact-ionization processes. The total cross-section for the impact-ionization process may be simplified as follows:

$$Q_{\text{tot}}^{0} = \begin{cases} \pi a_0^2 & k \geq \sqrt{2m_e\, \varepsilon_D}/\hbar \\ 0 & k < \sqrt{2m_e\, \varepsilon_D}/\hbar. \end{cases} \qquad (3.55)$$

Using (3.55), the impact-ionization coefficient A_I can be obtained by:

$$A_I = \int_{\varepsilon_D}^{\infty} v Q_{\text{tot}}^{0} f_0(\varepsilon^*) D(\varepsilon^*) \mathrm{d}\varepsilon^*, \qquad (3.56)$$

where v is the electron velocity, and $D(\varepsilon^*)$ is the density of states, which is

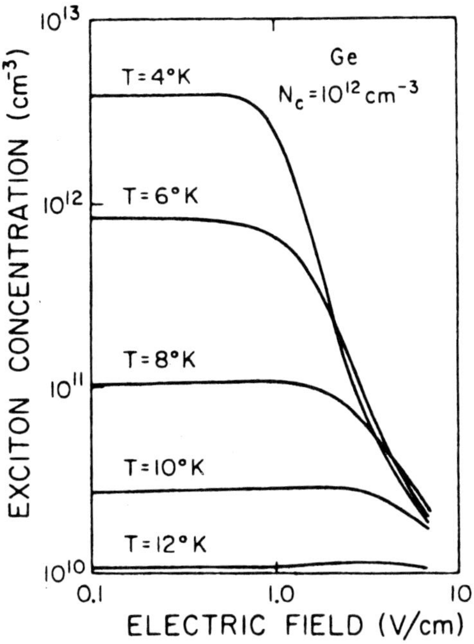

Figure 3.12 Exciton concentration vs applied electric field in Ge at several temperatures; the concentration of electrons and holes has been taken to be 10^{12} cm^{-3}. The sharp drop in the exciton concentration is due to impact ionization [35].

given by:

$$D(\varepsilon^*) = \frac{2}{4\pi^2}\left(\frac{2m_e}{\hbar^2}\right)^{3/2}\sqrt{\varepsilon^*}. \tag{3.57}$$

$f_0(\varepsilon^*)$ is the Maxwell–Boltzmann distribution function. Let the density of states in (3.56) be simply reduced to $\sqrt{\varepsilon^*}$, and multiply the coefficient in (3.57) into $f_0(\varepsilon^*)$, the quantity of which is again denoted by the same symbol f_0. Then (3.56) is rewritten as:

$$A_I = \int_{\varepsilon_D}^{\infty} vQ^0_{tot}\, f_0(\varepsilon^*)\sqrt{\varepsilon^*}\mathrm{d}\,\varepsilon^*. \tag{3.58}$$

For the distribution function, Yamasita [39–41] argued the Boltzmann approach under the hot electron regime and derived

$$f_0(\varepsilon^*) = \frac{2}{\sqrt{\pi}(k_B T_e)^{3/2}} \exp\left(-\varepsilon^*/k_B T_e\right). \tag{3.59}$$

Using (3.59), the impact-ionization coefficient (3.58) is obtained by partial

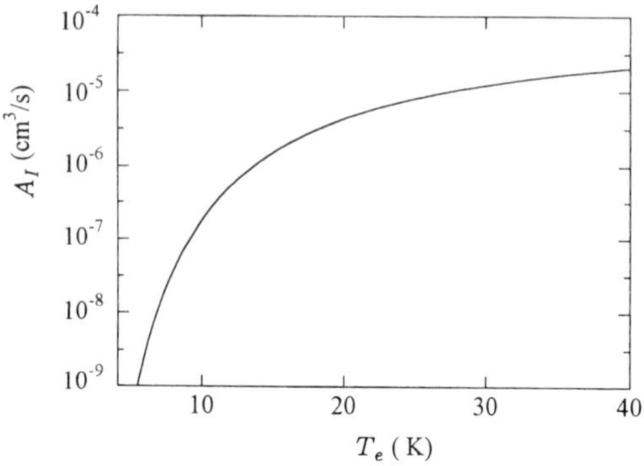

Figure 3.13 Impact-ionization rate (cm^3/s) as a function of electron temperature T_e ($>$ 4.2 K), calculated by (3.60) where $\varepsilon_D = 5.8$ meV, $a_0 = 94.5$ Å for a shallow donor, and $m_e = 0.0675\ m_0$ for n-GaAs. The electron temperature is considered to be an increasing function of applied electric field strength. Cut-off by the lattice temperature (4.2 K) was not taken into account.

integration as:

$$A_I = 2a_0^2(2\pi k_B T_e/m_e)^{1/2}(1 + \varepsilon_D/k_B T_e)\exp(-\varepsilon_D/k_B T_e). \tag{3.60}$$

In Fig. 3.13, the impact-ionization coefficient is plotted as a function of T_e, where the significant dependence on T_e stems from $\exp(-\varepsilon_D/k_B T_e)$. For the shifted MB distribution, (3.58) must be replaced by an approximated expression, as discussed in the following section.

3.2 Impact-ionization model

The system behaviors of the hot carrier transport are generally analyzed by the hydrodynamic balance equations [1, 42–44]. The order parameters for describing the impact-ionization avalanche are the mean momentum per carrier $\mathbf{p}(r,t)$, the mean kinetic energy per carrier $\varepsilon(r,t)$, the carrier density $n(r,t)$ and the local electric field $E(r,t)$ [1, 42, 44]. According to Schöll [1, 42], the system equations are given by:

$$\dot{\mathbf{p}} + (\mathbf{v}\cdot\nabla)\mathbf{p} + \frac{1}{n}\nabla(nk_B T_e) - q\mathbf{E} = -\mathbf{p}\Phi/n - \mathbf{p}/\tau_m(\varepsilon), \tag{3.61}$$

$$\dot{\varepsilon} + (\mathbf{v}\cdot\nabla)\varepsilon + \frac{1}{n}\nabla(nk_B T_e\mathbf{v}) - \frac{\kappa}{n}\Delta T_e - q\mathbf{v}\cdot\mathbf{E} = -\varepsilon\Phi/n - (\varepsilon - \varepsilon_L)/\tau_\varepsilon \tag{3.62}$$

$$\dot{n} + \nabla\cdot(n\mathbf{v}) = \Phi(n,\varepsilon), \tag{3.63}$$

$$\nabla\times\mathbf{H} = \epsilon\epsilon_0\dot{\mathbf{E}} + \mathbf{J} = \mathbf{J}_0. \tag{3.64}$$

(3.61)–(3.64) describe the momentum balance equation, the energy balance equation, the rate equation for the generation–recombination (g-r) process and the rate equation for the dielectric relaxation response, respectively. Here, $\tau_m(\varepsilon)$ is the momentum relaxation time, τ_ε is the energy relaxation time, Φ is the generation–recombination rate which is decomposed to two parts, the generation rate G and the recombination rate R ($\Phi = G - R$), v is the mean electron velocity, q is the carrier charge $q = \pm$e for holes or electrons, respectively), $\mathbf{H}$ is the magnetic field, $\mathbf{J}_0$ is the total external current density, and $\mathbf{J}$ is the sum of the drift and diffusion currents. For the energy balance equation (3.62), the energy-flow density is given by [42]:

$$\mathbf{j}_\varepsilon = n\mathbf{v}\ \varepsilon + n\mathbf{v}k_\mathrm{B}T_\mathrm{e} + \mathbf{j}_Q, \tag{3.65}$$

$$\mathbf{j}_Q = -\kappa\nabla T_\mathrm{e}, \tag{3.66}$$

where κ is the thermal conductivity. Normally the momentum balance equation (3.61) can be eliminated since the time-scale of the momentum relaxation process is much shorter than those for the energy relaxation process (3.62), the g-r process (3.63) and the dielectric relaxation process (3.64). Then, one can put safely $\mathrm{d}\mathbf{p}/\mathrm{d}t \equiv \dot{\mathbf{p}} + (\mathbf{v}\cdot\nabla)\mathbf{p} = 0$ by the slaving principle [45], whereby the carrier mobility can be defined as [42]:

$$\mu(\varepsilon) = \frac{e}{m^*}\ \tau_m'(\varepsilon), \quad \frac{1}{\tau_m'} = \frac{1}{\tau_m} + \frac{\Phi}{n}. \tag{3.67}$$

It is noted that (3.63) is equivalent to the continuity equation

$$\dot{\rho} + \ \mathrm{div}\ \mathbf{J} = 0, \tag{3.68}$$

where ρ is the local charge density given by $\rho = e(N_\mathrm{D} - N_\mathrm{A}) + q(n + n_\mathrm{t})$ (N_D and N_A are donor and acceptor densities, and n_t is the density of trapped electrons), and $\mathbf{J}$ is given by [42]:

$$\mathbf{J} = en\ \mu(\varepsilon)E - qD_\varepsilon\nabla n - \frac{q}{e}\mu(\varepsilon)nk_\mathrm{B}\nabla T_\mathrm{e} \tag{3.69}$$

with the diffusion constant D_ε. In (3.68) n_t equals to the density of neutral impurities and then $\dot{\rho} = q(\dot{n} + \dot{n}_\mathrm{t}) = q(\dot{n} - \Phi)$, which leads to (3.63). For a simplified treatment, disregard the spatial diffusion and gradient in (3.62)–(3.66) and (3.69), then the system equations corresponding to (3.62)–(3.64) become:

$$\dot{\varepsilon} = -\varepsilon\Phi/n - (\varepsilon - \varepsilon_\mathrm{L})/\tau_\varepsilon + q\mathbf{v}\cdot\mathbf{E}, \tag{3.70}$$

$$\dot{n} = \Phi = G - R, \tag{3.71}$$

$$\varepsilon\varepsilon_0\dot{\mathbf{E}} = \mathbf{J}_0 - \mathbf{J}. \tag{3.72}$$

In case when the energy loss rate $-\varepsilon\,\Phi/n$ is negligible, (3.70) becomes:

$$\dot{\varepsilon} = -(\varepsilon - \varepsilon_{\mathrm{L}})/\tau_{\varepsilon} + q\mathbf{v}\cdot\mathbf{E}. \tag{3.73}$$

For the one-level impact-ionization model the g-r process (3.71) is given in an actual form (see section 2.2.3):

$$\dot{n} = nn_{\mathrm{t}}A_{\mathrm{I}} + n_{\mathrm{t}}A_{\mathrm{T}} - nn_{\mathrm{t}}^{+}B_{\mathrm{T}}, \tag{3.74}$$

where $n_{\mathrm{t}}{}^{+}$ is the density of the ionized trapping centers, and the Auger recombination process is neglected as usual. In the following one-level impact-ionization models in n-type semiconductors, $n_{\mathrm{t}} = N_{\mathrm{D}} - N_{\mathrm{A}} - n$, $n_{\mathrm{t}}{}^{+} = n + N_{\mathrm{A}}$. For the SNDC mechanism the screening model [5] will be considered in two different ways, the approximated model [46, 47] and the improved model [48–50], which is equivalent to the original screening model by Crandall [5]. For simplicity all of the spatial diffusion and gradient terms have been dropped.

3.2.1 Approximated one-level impact-ionization model (model A)

The approximated model [46, 47] is similar to the qualitative treatment by Khosla *et al.* [6]. For this model the energy balance equation (3.70) is disregarded. Since the energy relaxation time is still shorter than the inverse g-r rate, the approximation seems to be acceptable [46, 47]. The approximated model takes the impact-ionization coefficient in an approximated form of [46, 47]

$$A_{\mathrm{I}} \sim A_{\mathrm{I}}^{0}(T_{\mathrm{e}}(E)) + \gamma n, \tag{3.75}$$

$$\equiv A\exp(-B/E) + \gamma n, \tag{3.76}$$

where γ is a proportionality constant. The first term of the r.h.s. of (3.75) is approximated by Schokley's formula [1, 46, 47, 51] in (3.76) rather than using (3.60), and the second term is due to the screening effect [5, 6]. By inserting (3.76) into (3.74), then:

$$\dot{n} = G - R = -\gamma n^{3} + a(E)n^{2} + b(E)n + c_{0}, \tag{3.77}$$

where

$$a(E) = \gamma(N_{\mathrm{D}} - N_{\mathrm{A}}) - A\exp(-B/E) - B_{\mathrm{T}}, \tag{3.78}$$

$$b(E) = (N_{\mathrm{D}} - N_{\mathrm{A}})A\exp(-B/E) - A_{\mathrm{T}} - N_{\mathrm{A}}B_{\mathrm{T}}, \tag{3.79}$$

$$c_{0} = (N_{\mathrm{D}} - N_{\mathrm{A}})A_{\mathrm{T}}. \tag{3.80}$$

Taking E_0 as an actual control parameter (E_0: applied electric field), the bias condition is given by

$$E = E_0 - cJ, \tag{3.81}$$

where $c = R_L S/\ell$ for the sample geometry in Fig. 3.14. When $c = 0$ ($R_L = 0$), the bias condition is termed "voltage controlled", while it is "current controlled" when $c \to \infty$ ($R_L \to \infty$). In the dynamical regime (3.81) must be replaced by an equation for the dielectric relaxation response of E in (3.72):

$$\dot{E} = -\gamma'(E - E_0 + cJ), \tag{3.81'}$$

where $1/\gamma'$ is the relaxation time which is given by $R_L\sigma_0/\epsilon\epsilon_0 r$ for an ideal case [1, 47] (σ_0: conductivity; r: the sample resistance in the prebreakdown regime). In the chaos experiments in semiconductors [1, 52], however, the relaxation time seems to be limited by the circuit conditions so that the effective relaxation time τ_{eff} is typically much longer than the ideal values, where τ_{eff} may take its value in the order of or larger than those in the MHz range. In an experiment with a $p^+ - n^+ - p - n^-$ diode [53], the effective relaxation time is limited by a parasitic capacity C_{ex} parallel to the devices and by a load resistor R_L. This may also be true for the chaos experiments in bulk semiconductors [1, 52].

Consider the homogeneous case $\dot{n} = G - R$ under the static regime $E = E_0$ in (3.81), where $c = 0$. Since the r.h.s. of (3.77) is a cubic function of n, one can find three steady-state solutions (n_1, E_0), (n_2, E_0) and (n_3, E_0) for $\dot{n} = 0$ in a certain range of the applied field ($E_h < E_0 < E_b$), otherwise a unique solution (n_1, E_0) for $E_0 < E_h$ or (n_3, E_0) for $E_0 > E_b$ may result. Thus, an S-shaped $n - E$ characteristic can be obtained as a function of the control parameter E_0. The dynamical behaviors and chaos can be modeled by the approximated model, details of which will be described in chapters 4 and 5. After the scale conversions $\nu = n/(N_D - N_A)$ and $t = t/\tau_r(\tau_r$:

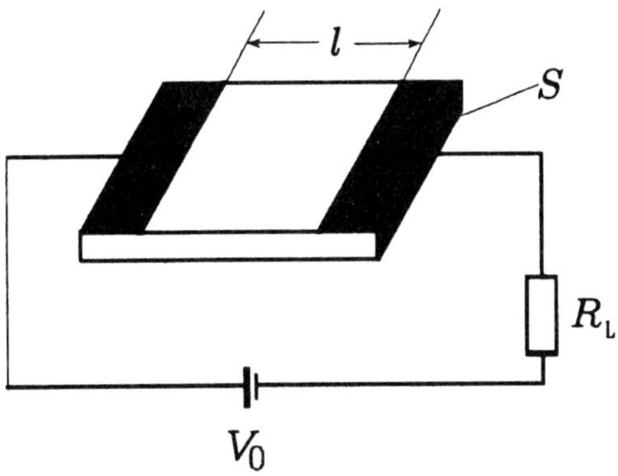

Figure 3.14 Sample geometry with planar-type ohmic contacts.

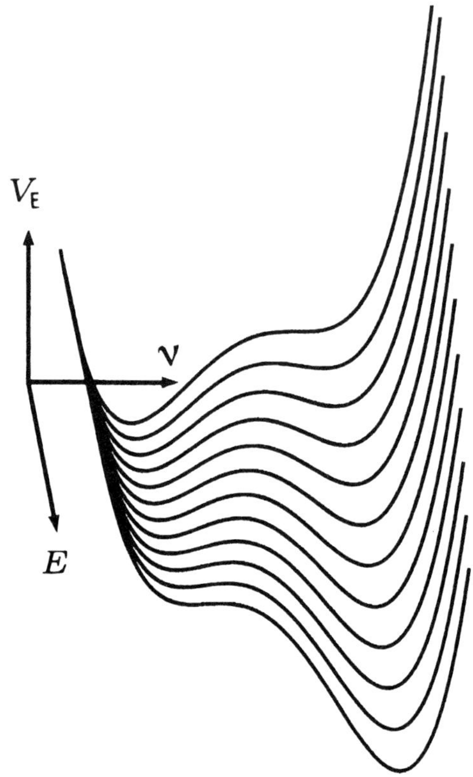

Figure 3.15 Double-well potential V_E in (3.85). The parameters are $\kappa_1 = 1/4\exp(-6/E)$, where $E = 10.947 - 12.12$, $\kappa_2 = 0.0055$, $\kappa_3 = 0.25$ and $C = 0.5$.

dielectric relaxation time) [1], the rate equation (3.77) is rewritten as:

$$\dot{\nu} = -\gamma_1[\nu^3 + (\kappa_1 + \kappa_3 - 1)\nu^2 - (\kappa_1 - \kappa_2 - C\kappa_3/(1-C))\nu - \kappa_2], \quad (3.82)$$

$$\kappa_1 = \tilde{A}\exp(-\tilde{B}/\tilde{E}), \quad (3.83)$$

where $\tilde{A} = A/\gamma(N_D - N_A)$, $\tilde{B}$ and $\tilde{E}$ are the normalized quantities, $\kappa_2 = A_T/\gamma(N_D - N_A)^2$, $\kappa_3 = B_T/\gamma(N_D - N_A)$, $\gamma_1 = \gamma\,\tau_r(N_D - N_A)^2$ and the compensation ratio $C = N_A/N_D$. The dielectric relaxation equation (3.81′) can be converted to:

$$\dot{\tilde{E}} = -\gamma_2(\tilde{E} - \tilde{E}_0 + \tilde{c}\,\nu\tilde{E}), \quad (3.84)$$

where $\gamma_2 = \gamma\,\tau_r$, and $\tilde{E}$, $\tilde{E}_0$ and $\tilde{c}$ are the normalized dimensionless quantities. The model simulation will be given in chapter 4.

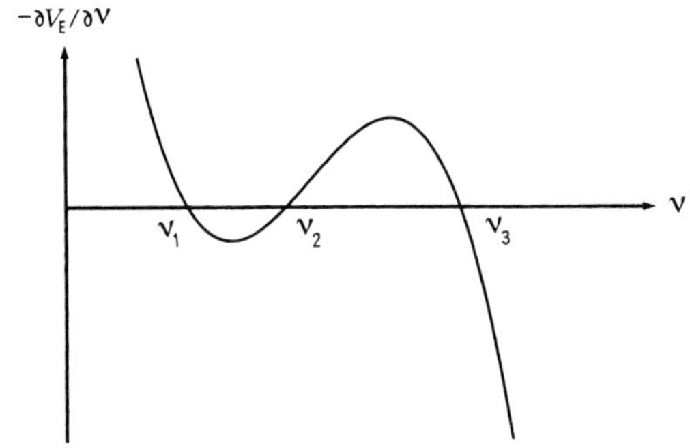

Figure 3.16 An example of the derivative $-\partial V_E/\partial \nu$, where $E = 11.63805$. The steady-state solutions for $\dot{\nu} = 0$ are ν_1, ν_2, and ν_3. The parameters are $\kappa_1 = 1/4 \exp(-/E), \kappa_2 = 0.0055, \kappa_3 = 0.25$ and $C = 0.5$.

For the voltage-controlled condition [$c = 0$ in (3.81)], the control parameter is $E = E_0$. The potential function V_E is of the form of double-well potential in a certain range of E and is apparently given by:

$$V_E = \frac{1}{4}\nu^4 + \frac{1}{3}(\kappa_1 + \kappa_3 - 1)\nu^3 - \frac{1}{2}[\kappa_1 - \kappa_2 - C\kappa_3/(1 - C)]\nu^2 - \kappa_2 \nu. \quad (3.85)$$

Figure 3.15 shows the double-well potential $V_E(\nu)$, where there are two minima (stable equilibrium points) ν_1, ν_3 and a maximum (unstable saddle point) ν_2 for an arbitrary value of E in the range $E_h < E < E_b$. The signs of the derivative $\partial V_E/\partial \nu$ in Fig. 3.16 clearly support that the potential minima (maximum) correspond to the stable (unstable) equilibrium points.

3.2.2 One-level impact-ionization model (model B)

For a more accurate description of the SNDC model (model B) [48–50], one must consider the field-dependent electron mobility (3.25) and the energy balance equation [5]. A set of the differential equations is given by:

$$\dot{\varepsilon} = -\frac{\varepsilon - \varepsilon_L}{\tau_\varepsilon} + e\mu E^2, \quad (3.86)$$

$$\dot{n} = n(N_D - N_A - n)A_I + (N_D - N_A - n)A_T - n(n + N_A)B_T, \quad (3.87)$$

where the electron mobility μ is expressed by the Brooks–Herring formula μ_{BH}, (3.25). The energy relaxation time τ_ε is due to acoustic deformation potential and given by [11]

$$\tau_\varepsilon = \frac{3k_B T_L}{8m_e u_\ell^2 \langle \tau_m^{-1} \rangle}, \tag{3.88}$$

$$\langle {\tau_m}^{-1} \rangle = \frac{2m_e \varepsilon_{ac}^2 k_B T_L}{\pi^{3/2} \hbar^4 \rho u_\ell^2} \left(\frac{2k_B T_e}{m_e} \right)^{1/2} F_1(\eta) F_{1/2}(\eta), \tag{3.89}$$

where $\langle {\tau_m}^{-1} \rangle$ is the inverse momentum relaxation time, u_ℓ is the sound velocity, ε_{ac} is the deformation potential constant, and $F_1(\eta)$ and $F_{1/2}(\eta)$ are the Fermi integral ($\eta = \zeta / k_B T_e$) [11]. For the static regime the bias condition $E = E_0 - cJ$ is given by (3.81).

For the impact-ionization coefficient A_I in (3.87), one can use either (3.60) or the extended expression which can be derived by using the shifted MB distribution function, as discussed below. For the drift motion of electrons in the x direction the distribution function f can be obtained by the diffusion approximation:

$$f = f_0 - \mu^* E \frac{\partial f_0}{\partial v_x} = f_0 \left(1 + \frac{m_e \, \mu^* E}{k_B T_e} \, v_x \right), \tag{3.90}$$

where f_0 is the Boltzman distribution function given by (3.59), $\mu^* = e\tau_m / m_e$ [for τ_m see (3.15)], and v_x is the electron velocity in the x direction. Replacing f_0 in (3.58) by f in (3.90), (3.58) is divided into two parts of the integration, where the former gives the solution (3.60). The latter part of the integration is due to the asymmetric part of the distribution function in (3.90) and given by:

$$A_I = \pi a_0^2 \frac{m_e E}{k_B T_e} \int_{\varepsilon_D}^{\infty} v v_x \mu^* f_0 \sqrt{\varepsilon^*} \, d\, \varepsilon^*. \tag{3.91}$$

By a similar procedure as in (3.18), one can put the microscopic electron mobility μ^* outside the integral by μ_{BH}. In addition, $v_x = \sqrt{v^2/3} = \sqrt{2\varepsilon / 3m_e}$ and $v = \sqrt{2\varepsilon / m_e}$. Then, the integrand in (3.91) takes the form $\varepsilon^{*3/2} f_0$. In order to get an approximated form $\varepsilon^{*3/2}$ is replaced by $\sqrt{\varepsilon_D}\, \varepsilon^*$. The final result for (3.91) is obtained as:

$$A_I = 4a_0^2 \mu_{BH} E \frac{\sqrt{\pi \varepsilon_D}}{\sqrt{3} k_B T_e} \left(\frac{\varepsilon_D}{k_B T_e} - 1 \right) \exp\left(-\varepsilon_D / k_B T_e\right). \tag{3.92}$$

The sum of (3.60) and (3.92) gives the impact-ionization coefficient under the drift regime. With an approximation $\varepsilon_D / k_B T_e \gg 1$, one obtains [48–50]:

$$A_I = 2\sqrt{\frac{2\pi}{m_e}} a_0^2 \, \varepsilon_D \frac{1}{\sqrt{k_B T_e}} \{1 + \sqrt{2m_e \varepsilon_D / 3} \; \mu_{BH} E / k_B T_e\} \exp\left(-\varepsilon_D / k_B T_e\right). \tag{3.93}$$

The static regime of the SNDC can be investigated by (3.86), (3.87) and (3.93), where the electric field E is used as a control parameter. It is convenient to use the scale conversion:

$$\nu = n/(N_D - N_A),\ \nu_I = N_I/(N_D - N_A),\ \tilde{T}_e = T_e/T_L,\ \tilde{\mu} = \mu_{BH}/\mu_0,$$

$$\tilde{E} = \left(\sqrt{\mu_{ac}^0\ \mu_0/u_\ell}\right)E,\ \kappa_1 = A_I/\bar{A}_I,\ \kappa_2 = A_T/(N_D - N_A)\bar{A}_I,$$

$$\kappa_3 = B_T/\bar{A}_I,$$

where μ_{ac}^0 is the zero-field electron mobility due to acoustic deformation potential scattering [11], μ_0 is the zero-field electron mobility by the BH formula, and $\bar{A}_I$ is the impact-ionization coefficient at zero electric field [48–50]. Then, according to (3.86), (3.87) and (3.93), the system equations for the static regime are converted to the dimensionless ones as:

$$\tilde{T}_{e,n+1} = \tilde{T}_{e,n} - \left[\sqrt{\tilde{T}_{e,n}}(\tilde{T}_{e,n} - 1) - \frac{3\pi}{32}\ \tilde{\mu}_n\tilde{E}^2\right], \tag{3.94}$$

$$\nu\ (1-\nu)\kappa_1\ +\ (1-\nu)\kappa_2\ -\ \nu(\nu\ +\ 1/(1-C)\ -\ 1)\kappa_3\ =\ 0, \tag{3.95}$$

$$\kappa_1\ =\ (1\ +\ A\tilde{\mu}_n\tilde{E}/\tilde{T}_{e,n})\sqrt{1/\tilde{T}_{e,n}}\ \exp\left(\alpha(1 - 1/\tilde{T}_{e,n})\right), \tag{3.96}$$

where the energy balance equation (3.86) is converted to the difference equation in (3.94) with the electron temperature $\tilde{T}_{e,n}$, the electron mobility $\tilde{\mu}_n$ and the electron density ν_n at the discrete time $n(n = 0, 1, 2, \ldots)$, the g-r process (3.87) is taken as the steady state ($\dot{n} = 0$) in (3.95), and $C = N_A/N_D$. Corresponding to (3.19), the reduced electron mobility is given by:

$$\tilde{\mu} = \tilde{T}_{e,n}^{3/2}\frac{2C/(1-C)+\nu_0}{2C/(1-C)+\nu}\ \frac{\ln(1+\beta_0^2) - \beta_0^2/(1+\beta_0^2)}{\ln\ u - (u-1)/u}, \tag{3.97}$$

where $u = 1\ +\ \beta_0^2\ \nu_0\tilde{T}_{e,n}^2/\nu$. The numerical procedure is as follows. For given ν and $\tilde{T}_{e,0}$, one can find the electron mobility from (3.97), the impact-ionization coefficient from (3.95) and the electric field $\tilde{E}$ from (3.96), and then $\tilde{T}_{e,1}$ is derived from the energy balance equation (3.94). Repeating the iterative procedure, $\tilde{T}_{e,n}(n = 0, 1, 2, \ldots)$ approaches a steady-state value $\tilde{T}_e$. Then, for a given carrier density, the electron temperature, the electron mobility and the electric field can be found. Thus, one can obtain the S-shaped current-field characteristic $\nu\tilde{\mu}\tilde{E}$ vs $\tilde{E}$.

Figure 3.17 shows the S-shaped characteristics of $\nu - \tilde{E}$, $\tilde{T}_e - \tilde{E}$, and $\tilde{\mu} - \tilde{E}$ curves, which have been calculated using (3.94)–(3.97), where the compensation ratio is $C = 0.526$. Note that an increase in the electron mobility is quite large after the impact-ionization avalanche is completed, while an increase in the electron temperature is relatively small. Such a large

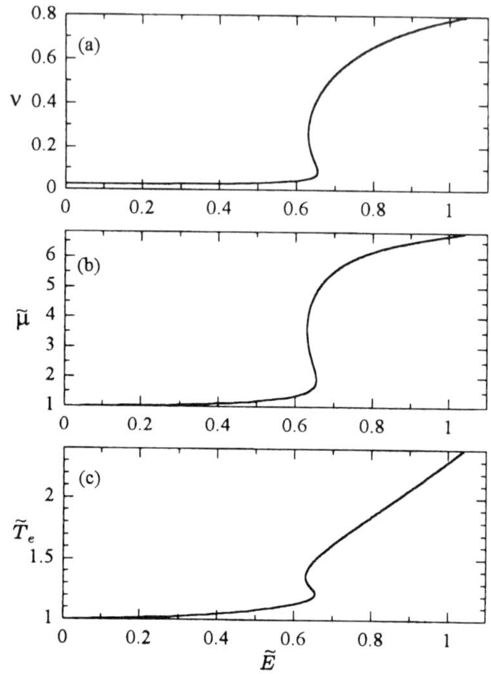

Figure 3.17 S-shaped characteristics of (a) ν vs $\tilde{E}$, (b) $\tilde{\mu}$ vs $\tilde{E}$ and (c) $\tilde{T}_e$ vs $\tilde{E}$. The parameters are $\kappa_2 = 0.05$, $\kappa_3 = 2.6$, $A = 0.11338$, $\alpha = 5.75$, $\beta^2_0 = 30$ and $C = 0.526316$.

increase in the electron mobility is consistent with an experimental result observed by Crandall [5], and in the model simulation it is explained as a direct result of the screening effect of ionized-impurity scattering as supported by the BH formula (3.97). As Crandall [5] has noted and several experiments have shown, the negative differential resistance (NDR) is absent when the compensation ratio of the semiconductor material is small. It is also clearly shown from the present model simulation (Fig. 3.18) that the NDR disappears for $C < 0.488$.

It is also possible to find an S-shaped J–E characteristic by using the impact-ionization coefficient (3.60) which was derived from the Maxwell–Boltzmann distribution function. From (3.60) the impact-ionization coefficient is a unique function of the electron temperature. Then for a given T_e, the carrier density can be obtained from the steady-state solution of the g-r process (3.87). Once $\tilde{T}_e$ and ν are found, the electron mobility can be determined from the BH formula and then the electric field is easily determined from the energy balance equation.

An essential part of the one-level impact ionization model B is summarized below. In a compensated n-type material, acceptors are

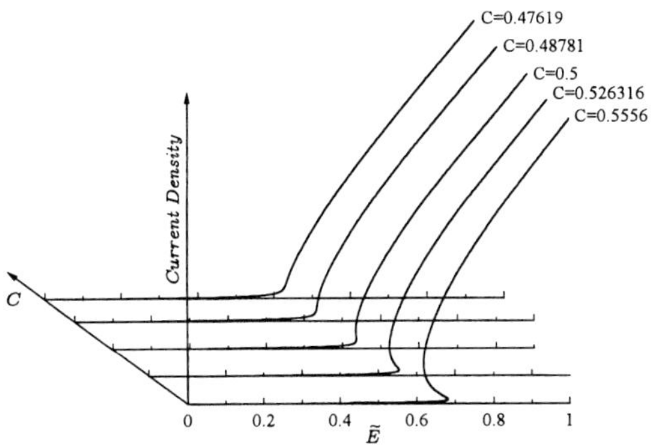

Figure 3.18 $J - \tilde{E}$ characteristics calculated for n-type samples with different compensation ratios $C = N_A/N_D$, where $J = \nu\tilde{\mu}\tilde{E}$. All parameters except for C are the same as in Fig. 3.17.

completely ionized, but a large amount of donor impurities is neutral in the prebreakdown regime, where the electron mobility is determined by the ionized acceptor scattering. In a highly compensated material with a relatively high concentration of acceptor impurities, the electron mobility may be relatively small owing to the ionized acceptor scattering. At the breakdown field E_b, electron density begins to increase by the impact-ionization process. An increase in the electron density Coulomb force between an ionized donor and an incoming electron weakens according to the screened Coulomb potential (3.5), whereby the Debye length in (3.6) contracts slightly by a slight increase in the electron density. Contraction of the Debye length leads to an increase in the electron mobility in (3.97). This leads to an increase in the electron temperature via the energy balance equation (3.94) and leads to more impact ionization. By positive feedback, the hot electrons gain enough energy to complete the impact-ionization avalanche at the breakdown field.

For a computer simulation of chaos by the one-level impact-ionization model B, a complete set of ordinary differential equations (ODEs) is given by:

$$\frac{d\varepsilon}{dt} = -\frac{\varepsilon - \varepsilon_L}{\tau_\varepsilon} + e\mu E^2, \tag{3.98}$$

$$\frac{dn}{dt} = n(N_D - N_A - n)A_I + (N_D - N_A - n)A_T - n(n + N_A)B_T, \tag{3.99}$$

$$\frac{dE}{dt} = -\gamma_d(E - E_0 + cen\mu E), \tag{3.100}$$

where μ is given by the BH formula, γ_d is the inverse dielectric relaxation time, and c is the slope of the load line. Model simulations for chaos will be shown in chapter 7. The spatially extended model will be presented in chapter 8.

3.3 Thermodynamical consideration of the SNDC

The carrier transport systems in semiconductors are apparently open systems since the kinetic energies of the carriers are supplied from a voltage source or a current source of the circuit. Suppose that a carrier transport system is biased near the critical point of the first-order phase transition. When an increment of the mean kinetic energy $\Delta\varepsilon$ is added instantaneously to the system at time $t = 0$, the entropy variation dS at $t = \mathrm{d}t$ is given by [54, 55]

$$dS = d_{\mathrm{e}}S + d_{\mathrm{i}}S, \tag{3.101}$$

where $d_{\mathrm{i}}S$ is the entropy variation of the system and $d_{\mathrm{e}}S$ is the entropy flow from outside (e.g., from a battery of the circuit). The variation $d_{\mathrm{i}}S$ is an entropy production due to irreversible processes of generation and recombination, diffusion, and electrical and thermal conduction. The second thermodynamical law dictates

$$d_{\mathrm{i}}S \geq 0, \tag{3.102}$$

where the equality holds for the thermal equilibrium state. In (3.101) the entropy flow $d_{\mathrm{e}}S$ may be either negative or positive. According to Nicolis and Prigogine [55], a dissipative structure or pattern formation is possible if $dS = 0$, namely,

$$d_{\mathrm{e}}S = -d_{\mathrm{i}}S < 0. \tag{3.103}$$

From the description, the nucleation of a spatial pattern is possible if the system is biased at a critical point and if negative entropy is injected constantly into the system to cancel the positive entropy production $d_{\mathrm{i}}S$. The dissipative structure, which is a fundamental idea of Prigogine, means a spatial coherence as a result of the phase transition to a new nonequilibrium state which is far beyond the thermal equilibrium state. He points out that the coherent structure is a direct result of the destruction of a microscopically disordered state or molecular chaos, namely, a direct result of a self-organized state cooperated among a huge number of molecules [56, 57]. The basic idea can be applied to the SNDC and NNDC in semiconductors [58, 59].

First, the basic idea of the nonequilibrium thermodynamics by Nicolis and Prigogine [55] is introduced and then the idea is applied to semiconductor systems. Let us suppose that an autocatalytic reaction system is composed of molecule X_j $(j = 1, 2, \ldots, n)$ with mass densities ρ_j. The entropy density s_{v} per unit volume is related to the entropy S by the volume integration [55]:

$$S = \int \mathrm{d}V s_{\mathrm{v}}, \tag{3.104}$$

where

$$s_{\mathrm{v}} = s_{\mathrm{v}}(\ \rho_1(\mathbf{r}, t), \ldots, \rho_n(\mathbf{r}, t)). \tag{3.105}$$

The time derivative of s_{v} is given by

$$\frac{\partial s_{\mathrm{v}}}{\partial t} = \sum_j \left(\frac{\partial s_{\mathrm{v}}}{\partial \rho_{\mathrm{j}}}\right) \frac{\partial \rho_j}{\partial t}. \tag{3.106}$$

After [55],

$$\left(\frac{\partial s_{\mathrm{v}}}{\partial \rho_j}\right) = -\frac{\mu_j}{T}, \tag{3.107}$$

where μ_j is the chemical potential. Then, from (3.106),

$$\frac{\partial s_{\mathrm{v}}}{\partial t} = -\sum_j \frac{\mu_j}{T} \frac{\partial \rho_j}{\partial t}. \tag{3.108}$$

The continuity equation for the mass density ρ_j [55] is:

$$\frac{\partial \rho_j}{\partial t} = -\ \mathrm{div}\ \mathbf{J}_j + \sum_\rho \nu_{j\rho} w_\rho, \tag{3.109}$$

where $\mathbf{J}_j = \rho_j \langle \mathbf{v}_j \rangle$ is the current of the mass flow ($\mathbf{v}_j$: the current), $\nu_{j\rho}$ is the stoichiometric coefficient in the autocatalytic process, and w_ρ is the generation rate of molecule X_j.

Substituting (3.109) into (3.108), one obtains the conservation rule for s_{v} [55]:

$$\frac{\partial s_{\mathrm{v}}}{\partial t} = \ \mathrm{div}\ \sum_j \frac{\mu_j}{T} \mathbf{J}_j - \sum_j \mathbf{J}_j \cdot \nabla \frac{\mu_j}{T} + \sum_\rho \left[-\sum_j \frac{\mu_j}{T}\ \nu_{j\rho} \right] w_\rho, \tag{3.110}$$

which used a relation μ_j div $\mathbf{J}_j =$ div $(\mu_j \mathbf{J}_j) - \nabla \mu_j \cdot \mathbf{J}_j$. (3.110) is divided into two parts by

$$\mathrm{div}\ \mathbf{J}_{\mathrm{s}} = -\ \mathrm{div}\ \sum_j \frac{\mu_j}{T} \mathbf{J}_j \tag{3.111}$$

and

$$\sigma = -\sum_j \mathbf{J}_j \cdot \nabla \frac{\mu_j}{T} + \sum_\rho \left[-\sum_j \frac{\mu_j}{T}\ \nu_{j\rho} \right] w_\rho. \tag{3.112}$$

By defining the affinity force $A_\rho \equiv -\sum_j \mu_j \nu_{j\rho}$ for the ρth reaction [55], the entropy production rate σ can be rewritten in a convenient form:

$$\sigma = -\sum_j \mathbf{J}_j \cdot \nabla \frac{\mu_j}{T} + \sum_\rho \frac{A_\rho}{T} w_\rho. \tag{3.113}$$

Further simplification for σ can be done using the generalized forces $\mathbf{X}_j$ as [55]:

$$\sigma = \sum_j \mathbf{J}_j^{\text{diff}} \cdot \mathbf{X}_j^{\text{diff}} + \mathbf{J}^{\text{react}} \mathbf{X}^{\text{react}}, \tag{3.114}$$

where

$$\begin{aligned} \mathbf{J}_j^{\text{diff}} &= \mathbf{J}_j, \quad \mathbf{J}^{\text{react}} = w_\rho \\ \mathbf{X}_j^{\text{diff}} &= -\nabla \frac{\mu_j}{T}, \quad \mathbf{X}^{\text{react}} = \frac{A_\rho}{T}. \end{aligned} \tag{3.115}$$

Then the rate equation (3.110) for the entropy density is summarized as:

$$\frac{\partial s_\text{v}}{\partial t} = -\text{div}\mathbf{J}_\text{s} + \sigma. \tag{3.116}$$

By integrating (3.116) over the system volume, one obtains:

$$\frac{\text{d}S}{\text{d}t} = -\int \mathbf{J}_\text{s} \cdot \text{d}\mathbf{S} + \int \sigma \, \text{d}V. \tag{3.117}$$

Nonequilibrium thermodynamics in SNDC systems are now discussed. Similar discussions are also available for NNDC systems [58]. Ridley [58] derived the basic equation for $\text{d}s_\text{v}/\text{d}t$ available for semiconductor systems, starting from Gibbs' equation:

$$T\frac{\text{d}s_\text{v}}{\text{d}t} = \frac{\text{d}\tilde{U}}{\text{d}t} - \sum_j \tilde{\mu}_j \frac{\text{d}\rho_j}{\text{d}t}, \tag{3.118}$$

where

$$\tilde{U} = U + \sum_j e_j \rho_j \phi, \tag{3.119}$$

$$\tilde{\mu}_j = \mu_j + e_j.$$

In (3.118) $\tilde{U}$ is the total energy, ρ_j and e_j are the density and the charge of the kth component, and ϕ is the electrical potential in the semiconductor system. The energy equation is given by [58]:

$$\frac{\text{d}\tilde{U}}{\text{d}t} = -\text{div}\ \mathbf{J}_u + \sum_j e_j \rho_j \frac{\text{d}\phi}{\text{d}t}, \tag{3.120}$$

where $\mathbf{J}_u$ is the total energy flow, comprising a heat flow $\mathbf{J}_q$ and an electrical energy flow:

$$\mathbf{J}_u = \mathbf{J}_q + \sum_j e_j \mathbf{J}_j \phi, \tag{3.121}$$

where $\mathbf{J}_j$ is the current density of the jth charge carriers. In addition, the continuity equation must be imposed:

$$\frac{\mathrm{d}\rho_j}{\mathrm{d}t} = -\mathrm{div}\ \mathbf{J}_j + \frac{\partial \rho_j}{\partial t}. \tag{3.122}$$

Substituting (3.119)–(3.122) into (3.118) produces the following rate equation for s_v after a small amount of algebra:

$$\begin{aligned}\frac{\mathrm{d}s_\mathrm{v}}{\mathrm{d}t} &= -\mathrm{div}\left[\frac{\mathbf{J}_\mathrm{q} - \sum \mu_j \mathbf{J}_j}{T}\right] \\ &\quad + \frac{1}{T}\Bigg[\Bigg(\mathbf{J}_\mathrm{q} - \sum_j \mu_j \mathbf{J}_j\Bigg)\mathbf{X}_\mathrm{q} + \mathbf{J}\cdot\mathbf{E} + \sum_j \mathbf{X}_j \cdot \mathbf{J}_j \\ &\quad - \sum_j \mu_j \frac{\partial \rho_j}{\partial t} + \sum_j e_j \rho_j \frac{\mathrm{d}\phi}{\mathrm{d}t} - \sum_j e_j \phi\ \frac{\partial \rho_j}{\partial t}\Bigg],\end{aligned} \tag{3.123}$$

where

$$\begin{aligned}&\mathbf{E} = -\mathrm{grad}\ \phi, \qquad \mathbf{J} = \sum_j e_j \mathbf{J}_j, \\ &\mathbf{X}_q = T\ \mathrm{grad}(1/T),\ \mathbf{X}_j = -\mathrm{grad}\ \mu_j.\end{aligned} \tag{3.124}$$

The final result (3.123) is an extended version of (3.110), in the presence of an electric field $\mathbf{E}$. In (3.123) Ridley [58] has neglected the last two terms of the r.h.s., supposing the charge neutrality condition $\sum_j e_j \rho_j = 0$. For an accurate discussion, one cannot neglect these terms because the charge neutrality condition is broken locally in space when the SNDC induces the current filamentation [1]. Nevertheless, for a simple treatment of the nonequilibrium thermodynamics, one may neglect these terms. If joule heating is negligible, one can also drop the term of the thermal force ($\mathbf{X}_q = 0$). Then, (3.123) is approximated to a simpler form:

$$\frac{\mathrm{d}s_\mathrm{v}}{\mathrm{d}t} = -\frac{1}{T}\mathrm{div}(\mathbf{J}_q - \sum_j \mu_j \mathbf{J}_j) + \frac{1}{T}\left[\mathbf{J}\cdot\mathbf{E} + \sum_j \mathbf{X}_j \cdot \mathbf{J}_j - \sum_j \mu_j \frac{\partial \rho_j}{\partial t}\right]. \tag{3.125}$$

Expression (3.125) is equivalent to (3.110), namely equivalent to the sum of (3.111) and (3.112). The last term of the r.h.s in (3.125), may preferably be expressed by $\sum_\rho A_\rho w_\rho / T$ (see chapter 4). The entropy production rate in (3.125) is composed of three terms, energy gain $\mathbf{J}\cdot\mathbf{E}$, energy flow rate and rate of reaction energy ($\sum_\rho A_\rho w_\rho$) during generation and recombination processes (see chapter 4), each divided by the lattice temperature. The main part of the entropy production rate is $\mathbf{J}\cdot\mathbf{E}$, where $\mathbf{J}\cdot\mathbf{E} =$

$J_{\parallel}E_{\parallel} + J_{\perp}E_{\perp} \approx J_{\parallel}E_{\parallel}$. In the steady state $\mathrm{d}s_{\mathrm{v}}/\mathrm{d}t$ must be zero (see (3.103)), whereby the entropy production rate is balanced with the divergence of the flow. Ridley [58] treated the entropy production rate for the SNDC-induced filamentation by an assumption that the steady state of a current filament is near the thermal equilibrium state, where the equilibrium state means $\mathrm{div}\mathbf{J}_{\mathrm{s}} = \sigma = 0$ in (3.111) and (3.112). By this assumption, one can expect the entropy production rate to be as small as possible. This plausible argument is at the basis of the principle of least entropy production [58]. Since $\mathbf{J}\cdot\mathbf{E}/T$ is the main part, the least entropy production may be guaranteed only at a critical point of the sustained field in the S-shaped J–E characteristic [58]. However, it is recognized that this type of argument may be difficult because the steady state of the current filament is actually far beyond the thermal equilibrium state. For this reason further discussion by the principle of least entropy production may not be successful. Argument by the evolution criterion for excess entropy production may be more valuable, and will be presented in chapter 4. For the pattern formation of a filamentary current, description by the equal-areas rule [1] is much more appropriate and is given in the next section.

3.4 Equal area rule and pattern formation

This section introduces the equal-areas rule for the SNDC and discusses the pattern formation of a filamentary current. The first-order phase transition in the SNDC is similar to the gas–liquid transition of a van der Waals gas [1]. A large difference between them is reversibility between two phases, as explained in Fig. 3.19. For the van der Waals gas the reversible process is possible via the formation of droplets (coexistence between gas and liquid phases) without hysteretic behavior, while for the SNDC it is an irreversible process followed by hysteretic behavior of a localized filamentary state, as can be seen in the hysteretic $I - V$ curve. The equal-areas rule for the SNDC [1] is somehow different from that in the van der Waals gas [45, 60].

Let us discuss the equal-areas rule for the one-level impact-ionization model A by introducing the potential densities [59]:

$$\Psi(\tilde{E}(\nu), \nu_2) = \int_{\tilde{E}_0(\nu_1)}^{\tilde{E}(\nu)} (\nu - \nu_2)\mathrm{d}\tilde{E} = \int_{\nu_1}^{\nu} (\nu - \nu_2)\frac{\mathrm{d}\tilde{E}}{\mathrm{d}\nu}\mathrm{d}\nu, \tag{3.126}$$

$$\Phi(\nu, \tilde{E}_0) = \int_{\nu_1}^{\nu} (\tilde{E}_0 - \tilde{E}(\nu))\mathrm{d}\nu, \tag{3.127}$$

where $\tilde{E}_0$ is the applied bias under the voltage-controlled condition, $\nu_i (i = 1, 2, 3)$ are the equilibrium points and $\tilde{E}(\nu)$ is the electric field at which the carrier density takes the value (see Fig. 3.20a). The potential density

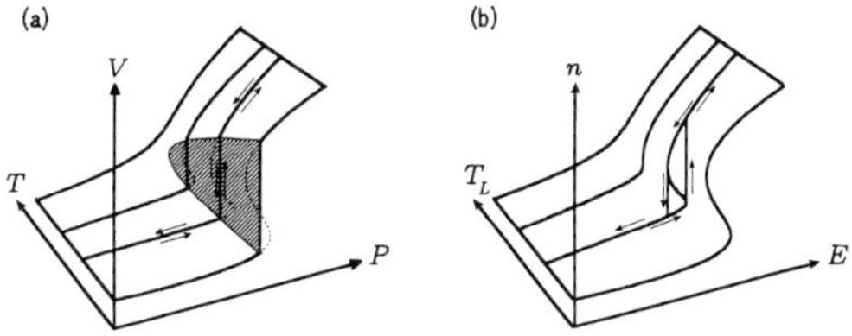

Figure 3.19 (a) Phase diagram for the van der Waals gas, where P is the pressure, V is the volume and T is the temperature. The hatched region illustrates phase coexistence between the gaseous phase and the liquid phase. The thick solid line indicates the first-order phase transition, where reversible switching is possible at the transition point, as shown by arrows. (b) Diagram for the carrier density-field ($n-E$) characteristics, where T_{L} is the lattice temperature.

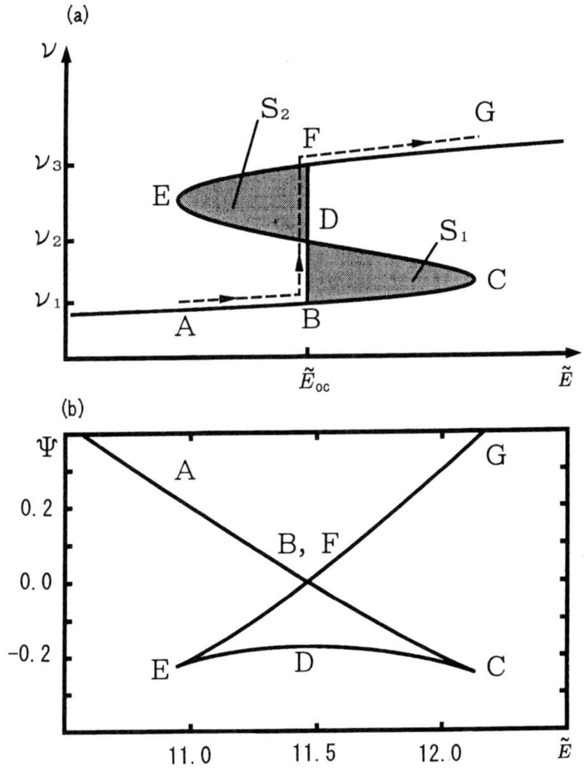

Figure 3.20 (a) Carrier density vs field ($\nu-\tilde{E}$) characteristics, where the applied electric field was chosen to be a special value $\tilde{E}_0 = \tilde{E}_{0c} = 11.462$, at which the equal-areas rule holds $S_1 = S_2$. (b) The potential density Ψ as a function of $\tilde{E}$. ν_2 in (3.126) is one of the equilibrium points at the applied electric field $\tilde{E}_{0c}$.

$\Psi(\tilde{E}(\nu), \nu_2)$ in Fig. 3.20b is easily confirmed by considering the path integral $A \to B \to C \to D \to E \to F \to G$. For the path integral $A \to B \to C$, Ψ decreases monotonically since $\nu_2 > \nu$ and $\mathrm{d}\tilde{E}/\mathrm{d}\nu > 0$, while it increases up to an extremal value at a point D for the path $C \to D$ since $\nu_2 > \nu$ and $\mathrm{d}\tilde{E}/\mathrm{d}\nu < 0$. Similarly, the remaining parts of the path integrals are $D \to E$ and $E \to F \to G$. The areas of the shaded regions are: $S_1 = -\Psi(\tilde{E}(\nu_2), \nu_2)$ by the path integral $B \to C \to D$; $S_2 = S_1 + \Psi(\tilde{E}_0(\nu_3), \nu_2)$ for the path integral $D \to E \to F$, where $\tilde{E}_0(\nu_2) = \tilde{E}_0(\nu_3)$. The equal-areas rule ($S_1 = S_2$) [59] holds for a special value of E_{0c} such that $\Psi(\tilde{E}_{0c}(\nu_3), \nu_2) = 0$. From (3.126), it is also obvious that $\Psi(\tilde{E}_{0c}(\nu_1), \nu_2) = 0$. The degeneracy $\Psi(\tilde{E}_{0c}(\nu_3), \nu_2) = \Psi(\tilde{E}_{0c}(\nu_1), \nu_2) = 0$ allows the reversible switching $A \to B \to F \to G$ or $G \to F \to B \to A$), as shown by a broken line in Fig. 3.20a (see also Fig. 3.20b). Alternatively, there may be irreversible switching $A \to B \to C \to D \to E \to F \to G$. When two different integration paths are formally possible (reversible or irreversible), the system is said to follow **Maxwell's convention** [45, 60]. For $\tilde{E}_{0c}$, the reversible switching means coexistence of two phases (e.g., coexistence of the liquid and gas phases in the van der Waals equation [45, 60]; for the chemical reaction, see the literature by Schlögl [61]). For this problem, coexistence means the current filamentation coexistence between low- and high-conductivity regions. When $S_1 \neq S_2$, reversible switching is unlikely to occur because of the potential difference $\delta\Psi = \Psi(\tilde{E}_0(\nu_3), \nu_2) - \Psi(\tilde{E}_0(\nu_1), \nu_2) \neq 0$, namely, $\Psi_B \neq \Psi_F$ as shown in Fig. 3.21.

The potential density $\Phi(\nu, \tilde{E}_0)$ in (3.127) is similarly argued. Apparently, the areas of the shaded regions in Fig. 3.22 are given by $S_1 = -\Phi(\nu_2, \tilde{E}_0)$

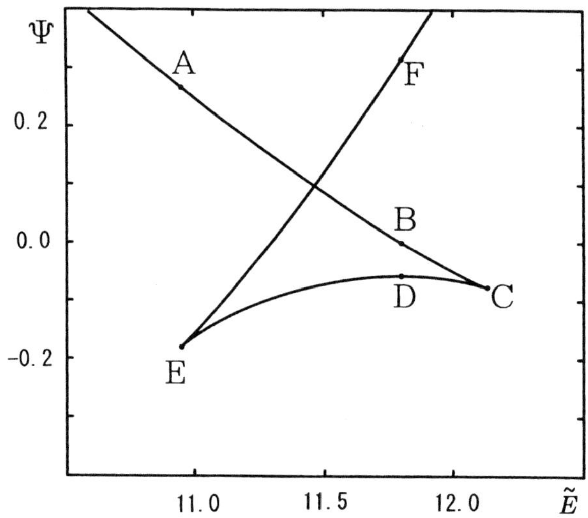

Figure 3.21 Potential density Ψ for an applied electric field $\tilde{E}_0 = 11.8$.

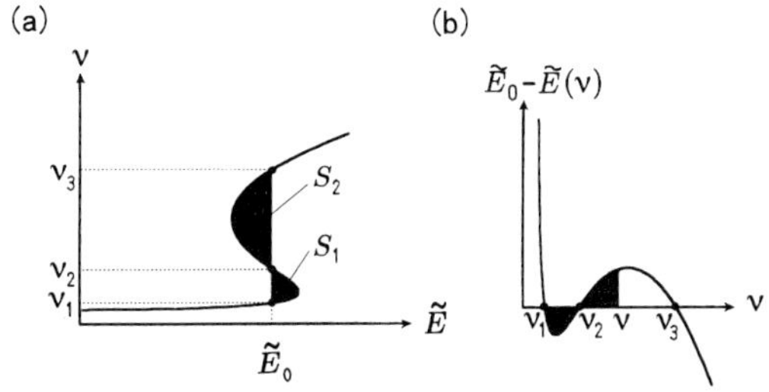

Figure 3.22 (a) Carrier density vs field ($\nu - \tilde{E}$) characteristic for an applied electric field $\tilde{E}_0$, illustrating two areas S_1 and S_2 divided by a load line $\tilde{E} = \tilde{E}_0$ (schematic). (b) $\tilde{E}_0 - \tilde{E}(\nu)$ vs ν (schematic), which is the illustration for the potential density in (3.127).

and $S_2 = S_1 + \Phi(\nu_3, \tilde{E}_0)$. For $\tilde{E}_{0c}$, $\Phi(\nu_3, \tilde{E}_{0c}) = \phi(\nu_1, \tilde{E}_{0c}) = 0$ and thus the equal-areas rule holds ($S_1 = S_2$). The result is shown in Fig. 3.23. Then the potential density $\Phi(\nu, \tilde{E}_{0c})$ supplements the reversible switching more clearly. However, the argument should be extended to the spatially extended problem for current filamentation, as stated below.

Here we shall investigate theoretically how the current filament can be formed stably between the ohmic contacts, on the basis of the extended equal-areas rule presented by Schöll [1]. Let us start from the continuity equation,

$$\dot{\rho} + \text{ div } \mathbf{J}^* = 0, \tag{3.128}$$

$$\rho = e(N_D^+ - n - N_A) \tag{3.129}$$

and a Maxwell equation,

$$\epsilon\epsilon_0 \text{ div } \mathbf{E} = \rho : \tag{3.130}$$

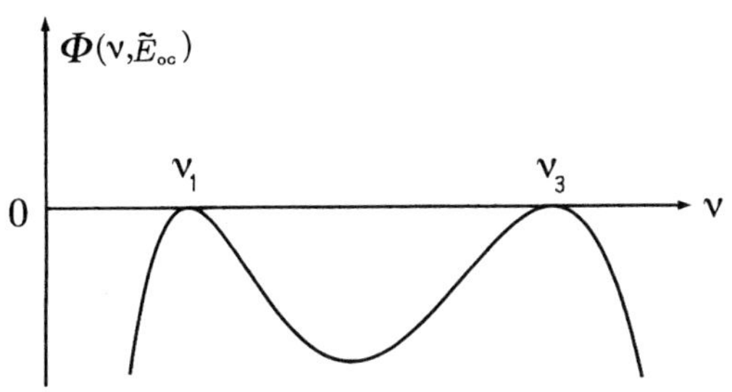

Figure 3.23 Potential density $\Phi(\nu, \tilde{E}_{0c})$ calculated by (3.127), where $\tilde{E}_0 = \tilde{E}_{0c} =$ 11.462.

Under the steady-state condition ($\dot{\rho} = 0$), one has

$$\text{div } \mathbf{J}^* = \text{div } (\sigma\mathbf{E} + eD_e\nabla_n) = 0, \tag{3.131}$$

$$\text{div } \mathbf{E} = \rho/\epsilon\epsilon_0, \tag{3.132}$$

$$\dot{n} = \dot{N}_D^+ = n(N_D - N_D^+)A_I + (N_D - N_D^+)A_T - nN_D^+B_T = 0. \tag{3.133}$$

For the sample geometry in Fig. 3.24, the electric field $\mathbf{E}$ is decomposed into the components of the transverse $\mathbf{E}_\perp$ and the longitudinal ones $\mathbf{E}_0; \mathbf{E} = \mathbf{E}_0 + \mathbf{E}_\perp$. For the current filamentation, the transverse carrier profile should be in an inhomogeneous form, say, $n(y)$. For simplicity, assume $\partial n/\partial x = \partial E_0/\partial x = 0$. Then, from (3.131) and (3.132), one obtains [1]:

$$\partial n/\partial y = -\frac{n\mu}{D_e}E_\perp, \tag{3.134}$$

$$\partial E_\perp/\partial y = \rho/\epsilon\epsilon_0. \tag{3.135}$$

Using new variables; $\nu = n/(N_D - N_A)$, $\tilde{\rho} = \rho/e(N_D - N_A)$, $\tilde{E}_\perp = (eL_D/k_BT_L)E_\perp$, $\tilde{y} = y/L_D$ (L_D: Debye length at lattice temperature, given by $\sqrt{k_BT_L/e^2}\sqrt{\epsilon\epsilon_0/n_0}$), (3.134) and (3.135) are rewritten as

$$\partial\nu/\partial\tilde{y} = -\tilde{E}_\perp\nu, \tag{3.136}$$

$$\partial\tilde{E}_\perp/\partial\tilde{y} = \tilde{\rho}(\nu, \tilde{E}_0). \tag{3.137}$$

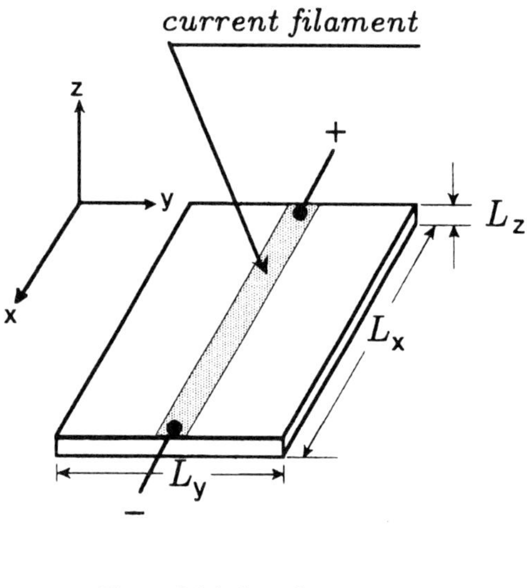

Figure 3.24 Sample geometry.

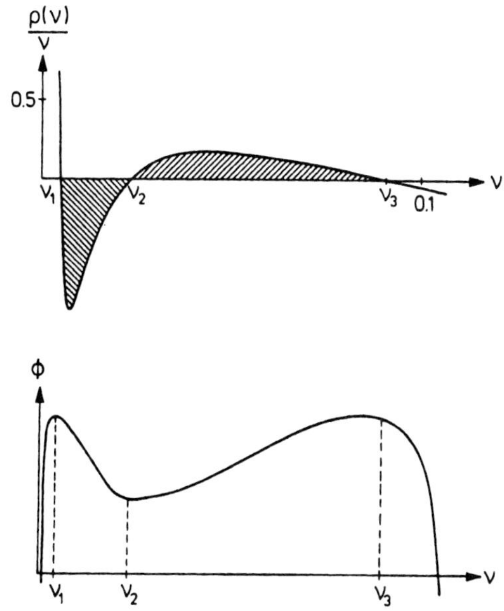

Figure 3.25 Condition for coexistence of the two phases ν_1 and ν_3 with a plane boundary layer (i.e., plane current filament, see Fig. 3.24). The two hatched areas must be equal, in order to secure $\phi(\nu_1) = \phi(\nu_3)$. This is the equal-areas rule for plane current layers [1].

By the logarithmic presentation of $\nu : \zeta = \ln\ \nu$ [1], (3.136) and (3.137) can be unified to:

$$\partial^2\zeta/\partial\tilde{y}^2 + \tilde{\rho}(\zeta, \tilde{E}_0) = 0. \tag{3.138}$$

The equation is analogous to the anharmonic equation of motion (1.18) of the Hamiltonian systems with $\gamma(x) = f(t) = 0$, which is obvious if $\tilde{y}$ is replaced by t and $\tilde{\rho}(\zeta, \tilde{E}_0)$ by $\partial\Phi/\partial\zeta$, where Φ is a potential which will be discussed. Then,

$$\Phi(\nu, \tilde{E}_0) = \int_{\zeta_1}^{\zeta} \tilde{\rho}(\zeta, \tilde{E}_0)\ \mathrm{d}\zeta = \int_{\nu_1}^{\nu} \tilde{\rho}(\nu, \tilde{E}_0)\frac{\mathrm{d}\zeta}{\mathrm{d}\nu}\mathrm{d}\nu, \tag{3.139}$$

where $\zeta_1 = \ln\ \nu_1$ (ν_1: one of the equilibrium points). Since the solution $\tilde{\rho}(\zeta, \tilde{E}_0)$ of (3.138) depends on the transverse position, one should not a priori introduce charge neutrality condition, namely $\tilde{\rho} \neq 0$.

Schöll [1] investigated theoretically, starting from (3.136), (3.137), the spatial structures of the current filaments (plane current layers, cylindrical current filaments and the multicurrent filaments), boundary conditions, stability and instability, oscillatory behaviors, etc. The most important aspect is the breakdown of the local charge neutrality $\tilde{\rho}(\nu, \tilde{y}) \neq 0$. He has treated the charge neutrality problem in a self-contained manner. Figure 3.25 shows one of his typical results for $\tilde{\rho}(\nu)/\nu$ and Φ under the condition of the equal-areas rule. Here, Schöll's theoretical frame will be applied (3.136)–(3.139) [1] to impact-ionization model A [46, 47]. Following Schöll's method [1], the phase portraits or the vector flows in Fig. 3.26 are analyzed in order to understand how the transverse spatial structure of the current filament can be related to the equal-areas rule or the Maxwell convention. The phase portrait can be drawn by using the Runge–Kutta–Gill method. In Fig. 3.26, $(\nu_1, 0)$ and $(\nu_3, 0)$ are the saddle points, while $(\nu_2, 0)$ is the center in the $(\nu, \tilde{E}_\perp)$ phase space. Remember that the ν_is are also the steady-state solutions of the spatially homogeneous rate equation; $\dot{\nu} = -\gamma_1\varphi(\nu) = 0$ (ν_1, ν_3 : stable points, ν_2: unstable saddle point in the temporal evolution), where

$$\varphi(\nu) = \nu^3 + (\kappa_1 + \kappa_3 - 1)\nu^2 - (k_1 - \kappa_2 - \frac{C}{1-C}\kappa_3)\nu - \kappa_2. \qquad (3.140)$$

The charge neutrality holds exceptionally for ν_is ($\tilde{E}_\perp = 0$).

Let $\tilde{E}_{0c}$ be the longitudinal electric field at which the equal-areas rule holds. Then, for $\tilde{E}_0 < \tilde{E}_{0c}$ (Fig. 3.26a), the saddle point $(\nu_3, 0)$ forms a closed orbit (homoclinic orbit), while any vector flows starting from another saddle point $(\nu_1, 0)$ are not localized (unstable orbit of the separatrices goes to infinity). For a special case of $\tilde{E}_0 = \tilde{E}_{0c}$ (Fig. 3.26b), the vector flows starting from both $(\nu_1, 0)$ and $(\nu_3, 0)$ are interconnected to form a closed orbit (heteroclinic orbit). In Fig. 3.26c, similar behavior to Fig. 3.27a is seen in a pairwise manner. The spatial phase portraits in Fig. 3.26 provide information about the transverse profiles of $\nu(\tilde{y})$, $\tilde{E}_\perp(\tilde{y})$, $\nu_D(\tilde{y})$ and $\tilde{\rho}(\tilde{y})$ in the range of $-\infty < \tilde{y} < +\infty$ (note: $\nu_j(-\infty) = \nu_j(+\infty)$ in Fig. 3.26b). In the phase portraits, only the closed orbit seems to be physically meaningful since $\nu(+\infty)$ or $\nu(-\infty)$ should be one of the steady-state solutions ν_1, ν_3. If

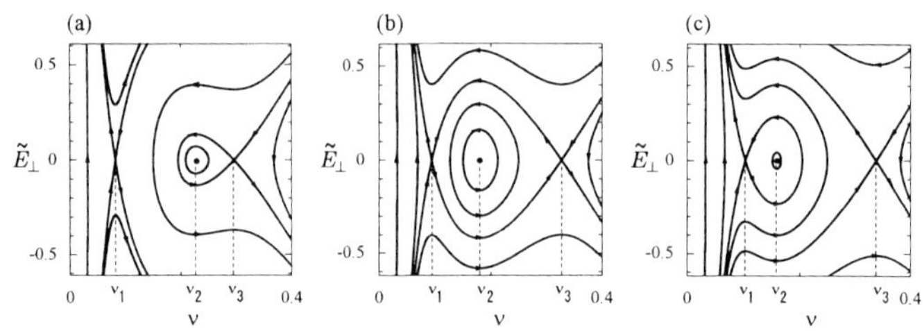

Figure 3.26 $\tilde{E}_\perp$ vs ν for different values of the applied electric fields $\tilde{E}_0$. (a) $\tilde{E}_0 = 11.1$, (b) $\tilde{E}_0 = 11.63805$, (c) $\tilde{E}_0 = 11.9$.

a boundary condition is imposed for a finite size of the sample geometry, the discussions become more complicated [1]. The homoclinic orbit in Fig. 3.26c means a possible nucleation of current filament, while it means a filamentary depletion in Fig. 3.26a. The heteroclinic orbit in Fig. 3.26b means formation of a kink–antikink pair (namely, the current filament). Transverse profiles $\nu(\tilde{y})$, $\nu_D(\tilde{y})$, $\tilde{E}_\perp(\tilde{y})$ and $\tilde{\rho}(\tilde{y})$ are shown in Fig. 3.27, from which it is obvious that the diffusion of the impact-ionized carriers should be counterbalanced to the transverse drift current $(\tilde{E}_\perp \nu)$ and that the gradient of the field $\partial \tilde{E}_\perp / \partial \tilde{y}$ is equal to the total charge density $\tilde{\rho}(\nu, \tilde{E}_0)$.

The equal-areas rule is different from the rule described previously in the homogeneous equation (3.126) (Fig. 3.22). As seen in Fig. 3.25 [1], the areas S_1 and S_2 are now the integrated value of $\tilde{\rho}/\nu$ over ν; $S_1 = -\Phi(\nu_2, \tilde{E}_0)$ and $S_2 = \Phi(\nu_3, \tilde{E}_0) - \Phi(\nu_2, \tilde{E}_0)$, where $\Phi(\nu_1, E_0) = 0$ from (3.139). Apparently,

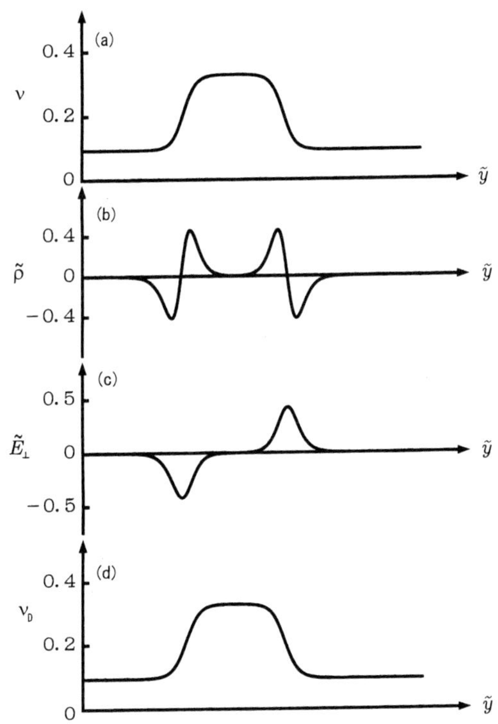

Figure 3.27 Transverse profiles of (a) carrier density, (b) local charge density, (c) transverse electric field, and (d) density of ionized donors. $\tilde{E}_0 = 11.63805$.

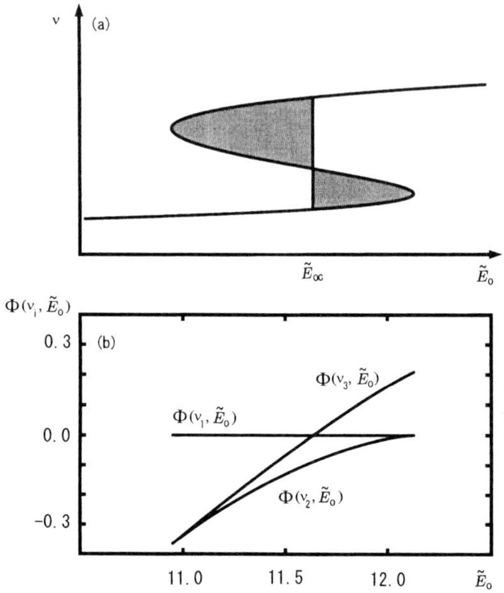

Figure 3.28 Carrier density vs field ($\nu - \tilde{E}_0$) characteristic. (a) At $\tilde{E}_{0c} = 11.638052$, two areas (dark region) divided by the load line no longer equal. However, the areas $S_1 = -\Phi(\nu_2, \tilde{E}_0)$ and $S_2 = \Phi(\nu_3, \tilde{E}_0) - \Phi(\nu_2, \tilde{E}_0)$ defined by (3.139) are equal at $\tilde{E}_{0c}$, where $\Phi(\nu_3, \tilde{E}_0) = 0$. The value of $\tilde{E}_{0c}$ is larger than that in Fig. 3.20. (b) Potential densities $\Phi(\nu_i, \tilde{E}_s 0)(i = 1, 2, 3)$ are given as a function of $\tilde{E}_0$.

$S_1 = S_2$ for $\Phi(\nu_1, \tilde{E}_{0c}) = \Phi(\nu_3, \tilde{E}_{0c}) = 0$. In the S-shaped $\nu - E$ curve of Fig. 3.28a, the two areas (dark regions) divided by the load line $\tilde{E}_0 = \tilde{E}_{0c}$ are no longer equal. Figure 3.28b shows the values $\Phi(\nu_i, \tilde{E}_0)$ $(i = 1, 2, 3)$ as a function of $\tilde{E}_0$, in which the branch $\Phi(\nu_3, \tilde{E}_0)$ approaches the branch $\Phi(\nu_1, \tilde{E}_0)$ by increasing $\tilde{E}_0$ and crosses at $\tilde{E}_0 = \tilde{E}_{0c}$. When compared to the homogeneous case (Fig. 3.20), it is now very clear that the reversible switching (the Maxwell convention) at the crossing point means a smooth spatial transition from a low carrier density ν_1 to a high value ν_3 in the transverse profile $\nu(\tilde{y})$ (coexistence between two laminar phases ν_1 and ν_3). For the behavior, the following is equivalent: a particle located at a potential maximum (at ν_1) can move to another potential maximum (at ν_3) and vice versa (see Fig. 3.25).

The current–voltage (I–V) characteristic is given by [1]:

$$I = e\mu(V/L_x)L_z L_D(N_D - N_A)\int_0^{\tilde{L}_y} \nu(\tilde{y}, \tilde{E}_0)d\tilde{y} \tag{3.141}$$

for the sample geometry in Fig. 3.24, where $\tilde{L}_y = L_y/L_D$. The integration of ν over $\tilde{y}$ should be done for the homoclinic and heteroclinic orbits in Fig. 3.26. Here, some additional important points are given for the results derived by applying Schöll's model [1]. Figure 3.29 shows the I–V curves for the sample dimensions of (a) $L_y = 50L_D$, (b) $L_y = 10L_D$, and (c) $L_y = 5L_D$. For the numerical calculations, no boundary conditions have been imposed. Figure 3.29 is almost identical to the result presented by Schöll [1]. The solid lines are those of the homogeneous solution, while the solid circles are for the inhomogeneous case in question. Under the current-controlled condition, a new nonequilibrium phase transition is seen (see arrows), which was first confirmed by Schöll [1]. The switching behavior are analogous to the NNDC case (formation of the electric field domain in the Gunn diode), where the excess voltage grows into a domain during the constant current bias [1]. At $\tilde{E}_0 = \tilde{E}_{0c}(V = V_c)$, the current in (3.143) exhibits a gap δI [A] (crosses in Fig. 3.29):

$$\delta I = e\mu(V_c/L_x)L_z(L_y - 2W)(N_D - N_A)(\nu_3 - \nu_1), \tag{3.142}$$

where W is the effective width of the current filament (Fig. 3.30). The gap is easily understood by integrating two different carrier profiles in Fig. 3.30, which can be obtained from the heteroclinic orbit starting from ν_1 and ν_3. Therefore, reversible switching may occur across the gap for the current-controlled condition, which really means the Maxwell convention (coexistence between two spatial profiles in space and time; Fig. 3.30). For small sample dimensions, the hysteretic behavior as in Fig. 3.29b disappears (Fig.

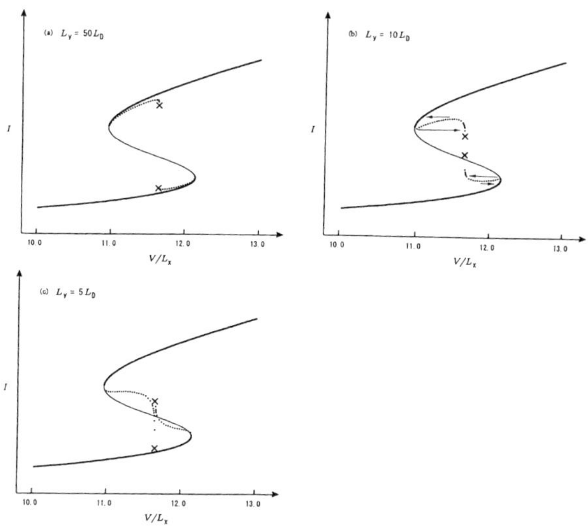

Figure 3.29 The $I - V$ curves for the sample dimensions of (a) $L_y = 50\ L_D$, (b) $L_y = 10\ L_D$, and (c) $L_y = 5\ L_D$. The gap (see crosses) appears at $V/L_x = 11.63805$.

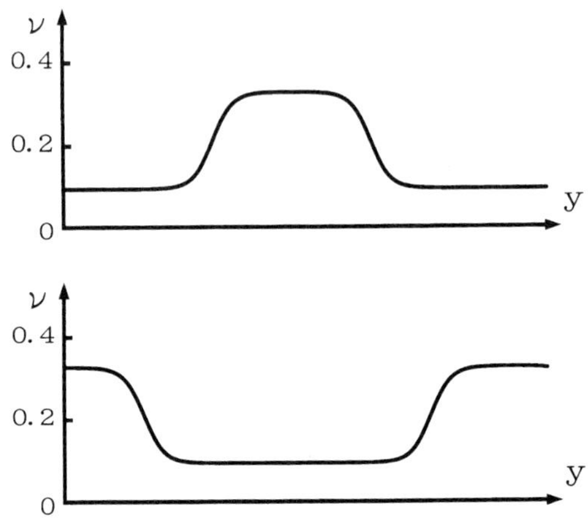

Figure 3.30 Two transverse carrier profiles obtained from the heteroclinic orbit, starting from ν_3 and ending at ν_3 (bottom), starting from ν_1 and ending at ν_1 (top). For the heteroclinic orbit, see Fig. 3.26.

3.29c) and the gap δI becomes negative, assuming the symmetric carrier profiles as in Fig. 3.30.

Under the voltage-controlled condition ($c = 0$) or with the inclined load line ($c \neq 0$, but not very large), however, the hysteretic behavior is standard and almost identical to the homogeneous case.

3.5 Measurement of a localized filamentary current

Since the discovery of chaos in semiconductors [62, 63], a new trend in nonlinear carrier transport research has been to investigate the nonuniform distribution of filamentary currents or field domains [64] from a viewpoint of self-organization and dynamical instability [45, 64]. The first experimental observation of a current filament was made in 1966 by Barnett and Milnes [65, 66] in a Si p-i-n diode. The measurement was made possible by observing the spatial pattern of recombination radiation emitted from a localized region of a space-charge-limited current filament [65–67]. Measurement by recombination radiation became popular [67, 68–71] after the first observation. Later measurement techniques such as low-temperature scanning electron microscopy (LTSEM) [72–77], the potential probe technique [78–80] and scanning laser microscopy [2–4, 81] were developed to study the spatial pattern formation of filamentary currents and the spatiotemporal pattern relating to dynamical behaviors and chaos [1, 64]. The measuring techniques and principles are now surveyed.

For the p-i-n diode with the sample geometry in Fig. 3.31, Barnett and Milnes [65, 66] found a light spot emitted from the surface of p^+ layer due to infrared recombination radiation at 77 K. They attributed the infrared recombination radiation (electroluminescence) to the recombination of the injected carriers inside the high-current-density region of a cylindrical current filament (see Fig. 3.31). The pattern images were recorded by photographs after converting the light spot using the infrared image converter [65, 66].

A recent measurement technique [64, 82–84] is much more sophisticated and quite successful, as shown in Fig. 3.32. Niedernostheide *et al.* have developed the technique by using a streak camera [64, 82, 84]. They investigated spatial and spatiotemporal pattern images by monitoring the electroluminescence distribution in the Si $p^+ - n^+ - p - n^-$ diode. Since the light intensity of the recombination radiation is a good measure of the current density [64–66, 82–84], the electroluminescence distribution gives the pattern image of the current filament. In Fig. 3.32 the electroluminescence light emitted from the $p^+ - n^+ - p$ part of the device is focused into a small stripe oriented parallel to the metallic contacts and converted to the electrons by an infrared-sensitive S1-photocathode [64, 82, 83]. The electrons are accelerated by an acceleration grid, deflected vertically by a deflection plate, and after being multiplied by a microchannel plate they

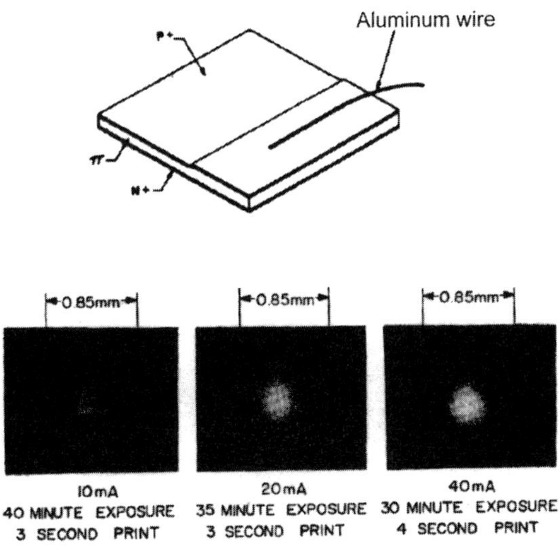

Figure 3.31 Top: Sample geometry of $p^+ - \pi - n^+$ diode for the measurement of recombination radiation from a cylindrical current filament. Bottom: Photographs of 10, 20 and 40 mA filaments, which have been observed as cross-sections of the cylindrical filaments.

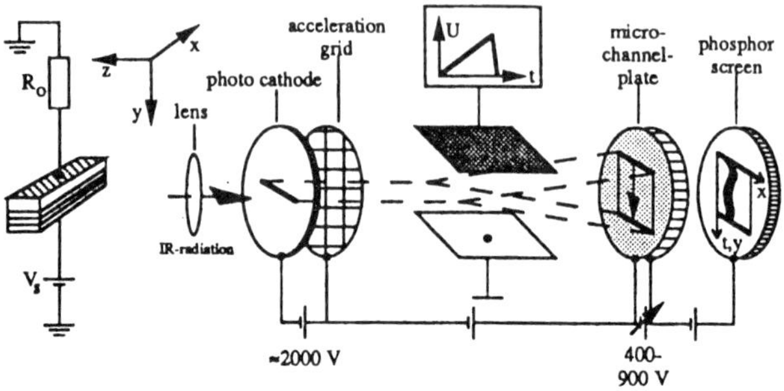

Figure 3.32 Measurement technique of the recombination radiation from a current filament by the streak camera [64].

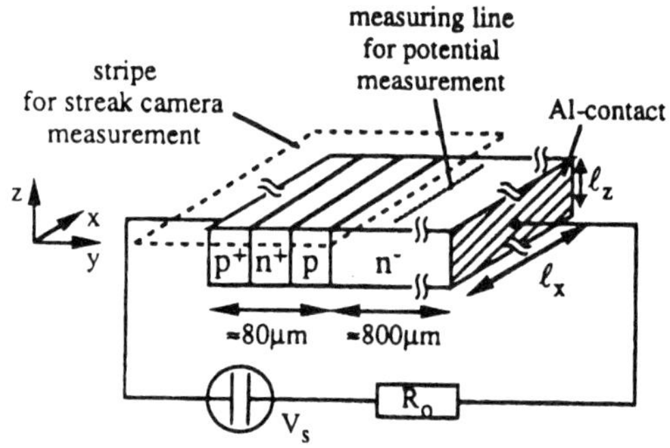

Figure 3.33 Schematic drawing of the electrical circuit and the $p^+ - n^+ - p - n^-$ diode: the dashed rectangle and the dashed line mark the stripe focused to the streak camera and the line along which the potential probe measurements were performed. Typical sample dimensions are $\ell_x = 1$–5 mm, $\ell_z = 0.2$–1 mm. The current filament flows in the y-direction and travels transversely in the x-direction between two sample boundaries [82].

strike a phosphor screen. By using a sawtooth deflection voltage $U(t)$, the vertical axis on the phosphor screen gives the time axis. The image of the spatiotemporal pattern on the phosphor screen is then recorded by a sensitive TV camera. With this technique, temporal resolution to the picosecond range is possible [64]. Niedernostheide *et al.* [64, 82–84] used the $p^+ - n^+ - p - n^-$ diode with the device structure given in Fig. 3.33, where the structure was prepared by a combination of conventional diffusion techniques and ion implantation [82]. In order to provide the n^- layer with a nonblocking electrical contact, an additional n^+ region was implanted before evaporating aluminum on both sides. In Fig. 3.34 a streak image of a traveling filament between the two sample boundaries (see Fig. 3.33) is shown. The vertical lines in Fig. 3.34a indicate the side boundaries of the sample in Fig. 3.33 [82]. Bright regions correspond to regions of intensified infrared radiation and indicate the current filament. Figure 3.34a indicates

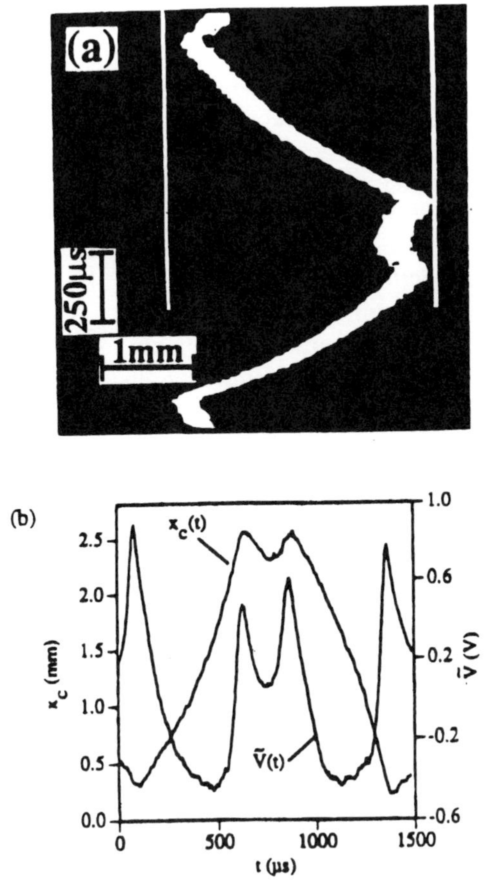

Figure 3.34 Streak camera record of a traveling filament with a superimposed rocking motion (a) and time traces of the ac part $\tilde{V}(t)$ of the device voltage and the position $x_c(t)$ obtained by evaluating the streak camera record (b); the white vertical lines in (a) mark the sample boundaries; parameters: $V_s = 59.2$ V, $V = 29.9$ V, $I =$ 272 mA, $R_0 = 110\Omega$, $T = 293$ K [82].

clearly a traveling mode of the current filament in the transverse direction, where the filament is reflected at both boundaries, but there is a small hesitation of traveling at the right boundary [82]. The transverse profile of the current filament forms a kink–antikink pair [1] and the traveling motion in Fig. 3.34a is similar to the reflection of soliton. In Fig. 3.34b, the ac part $V(t)$ of the sample voltage is compared with the center position $x_c(t)$ of the filament, where the local maxima of $V(t)$ are seen whenever the filament touches the sample boundaries (for a detailed discussion see Ref. [82]). Niedernostheide *et al.* obtained the maximum traveling speed of the traveling filament as 8 μm/μs. The measurement technique is so powerful and the experimental results [64, 82–84] are highly important for understanding spatiotemporal behavior (they will be reintroduced in chapter 6). However, the streak camera method is limited to semiconductor devices such as double-injection diodes, and not normally applicable to single-injection-type bulk materials if recombination radiation is not followed by current filamentation. Niedernostheide *et al.* [64] have also used the potential probe technique (see Fig. 3.33) introduced by Jäger *et al.* [78] and Baumann *et al.* [79].

Jäger *et al.* [78] and Baumann *et al.* [79] used a silicon p-i-n diode in Fig. 3.35, where highly doped p^+ and n^+ layers are formed by conventional B and P diffusion [79]. After forming metallic contacts by aluminum evaporation, the lateral measurement of the surface potential distribution between the contacts of the sample was done using a tungsten probe, as sketched in Fig. 3.35. Scanning of the probe on the sample surface can be done by a computer-controlled manipulator driven by a stepping motor [78–80, 85]. The surface potential is detected by a high-impedance voltmeter. Figure 3.36 is a typical example of the measured potential distribution on the surface, revealing the pattern image of a filament with a width of approximately 300 μm [79]. The probe method may be limited to measurements at room temperature.

The LTSEM method was used by Peinke *et al.* [72, 73], Mayer *et al.* [74, 75] and Rau *et al.* [76, 77]. By measuring electron-beam-induced excess current during two-dimensional beam scanning on the semiconductor surface, one can obtain an image pattern of a current filament, the experimental set-up for which [76] is shown in Fig. 3.37. A special arrangement must be made, in such a way that a sapphire substrate which is

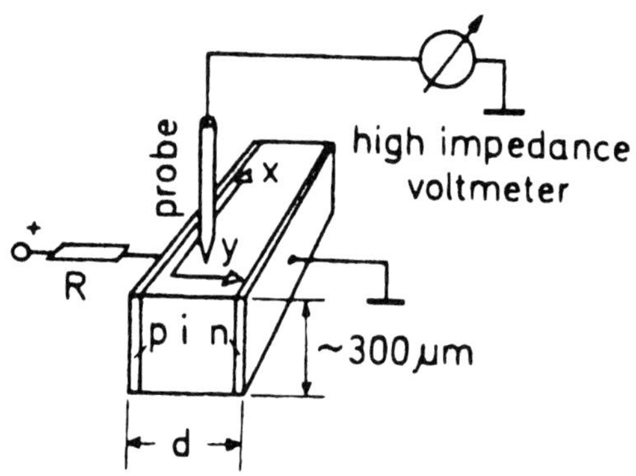

Figure 3.35 Sketch of the sample and electrical network for the potential probe technique. Spatial resolution < 5 μm, potential resolution 10 mV. The resistivity of the compensated silicon is about 100 kΩ [79].

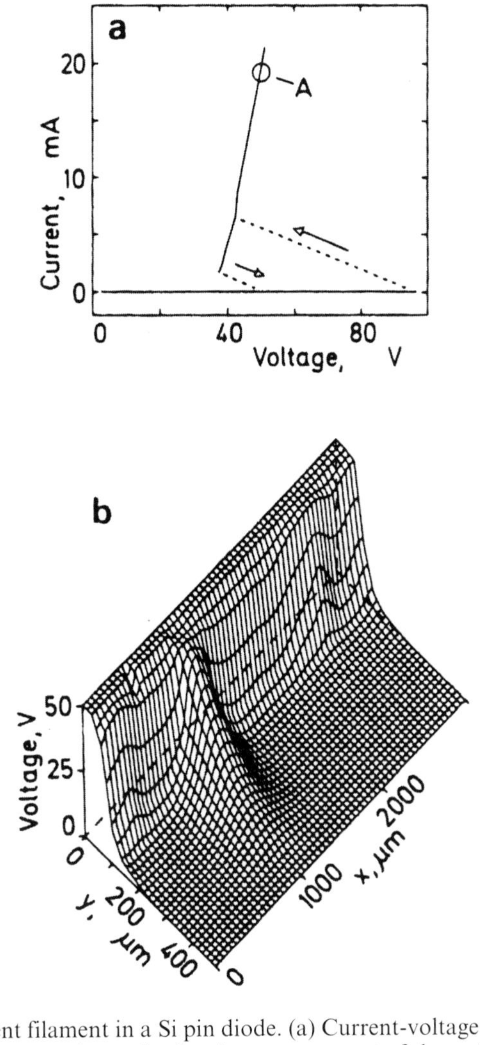

Figure 3.36 Current filament in a Si pin diode. (a) Current-voltage characteristic of a typical device. The operating point for the measurement of the potential distribution is marked by point A. (b) Measured potential distribution on the surface between the contacts of a typical diode at point A of (a) [79].

in direct contact with the liquid-He bath can be mounted in the cryostage of the electron microscope. The samples are glued on to the 1-mm-thick sapphire disc using Stycast cement [74–77]. Beam power is adjusted to be as small as possible, typically 2 μW using a 100 pA beam current. Mayer *et al.* [74, 75] observed the stationary pattern image of a filamentary current filament and also the resonant pattern image of a dynamical filamentary current in In-doped p-Ge with an acceptor concentration of $3 \times 10^{13}/\mathrm{cm}^3$.

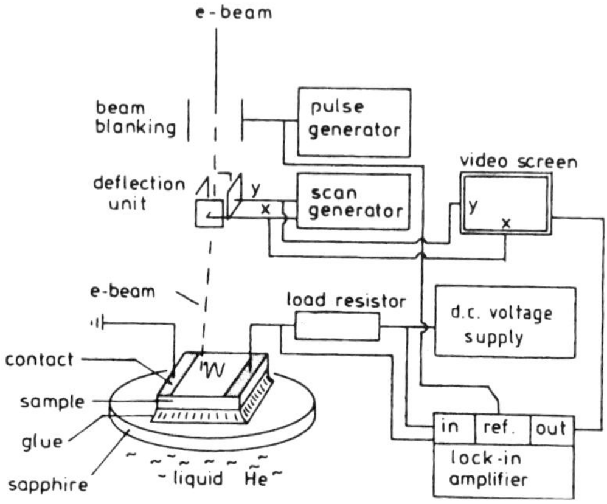

Figure 3.37 Experimental set up of the LTSEM. Taken from U. Rau, PhD thesis, in Ref. [76].

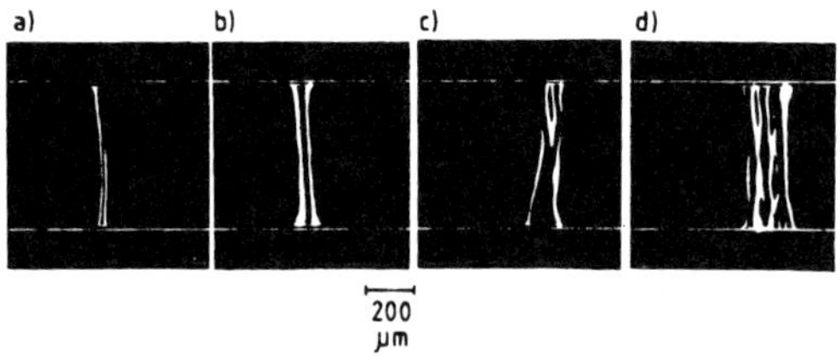

Figure 3.38 Image patterns of the current filament observed by the LTSEM. The current levels are (a) $I = 270\ \mu A$, (b) $430\ \mu A$, (c) $670\ \mu A$, and (d) $1020\ \mu A$. The horizontal white lines are edges of the ohmic contacts. The patterns are obtained by the brightness-modulation images, where two stripes indicate the filament boundaries. At $I = 670\ \mu A$, branching of the current filament is seen in (c). At $I = 1020\ \mu A$, multicurrent filaments can be seen in (d). Taken from U. Rau, PhD thesis, in Ref. [76].

Rau *et al.* [76, 77] investigated the spatiotemporal transport structures in p-Ge and n-GaAs. An example for n-GaAs is shown in Fig. 3.38 [76], where the filamentary region has been given by the brightness-modulated image.

Brandl *et al.* [2], Brandl and Prettl [3, 81] and Novak and Prettl [4] have developed the laser-beam scanning technique (scanning laser microscopy). Besides the study on the nonlinear dynamics of a filamentary current in n-GaAs, the main part of their study focused on the spatial pattern of the current filament under a transverse magnetic field [2–4, 81]. The basic idea of the laser-beam scanning technique is quite similar to the LTSEM [72–77]. When the laser beam touches the filament boundaries during two-dimensional laser-beam scanning on the sample surface, the excess current induced locally by the impact-ionization avalanche can be measured, and thereby one can obtain the spatial pattern image [2–4, 81]. The effect of the transverse magnetic field on the filament was clarified and explained by the Lorentz force. At the filament boundaries, spontaneous current oscillation can be observed, the dynamical mechanism of which was interpreted by the Schöll model [1]. Many observations obtained by the various measurement techniques above include an important part of the instability of the localized current filament and the spatiotemporal patterns, which will be discussed in more detail in Chapter 6.

Readers who are interested in a clear description of the relationship between pattern formation and phase transitions may refer to the review by Cross and Hohenberg [86].

References

[1] E. Schöll, *Nonequilibrium Phase Transitions in Semiconductors – Self-Organization Induced by Generation and Recombination Processes* (Springer, Berlin, 1987); see also E. Schöll, Proc. 3rd Int. Conf. on Hot Carriers in Semicond., in Suppl. J. Phys. (France) **c-7** (1981) 57.
[2] A. Brandl, M. Volcker and W. Prettl, Solid State Commun. **72** (1989) 847.
[3] A. Brandl and W. Prettl, Phys. Rev. Lett. **66** (1991) 3044.
[4] V. Novak and W. Prettl, in *Nonlinear Dynamics and Pattern Formation in Semiconductors and Devices*, ed. F.-J. Niedernostheide (Springer, Berlin, 1995) pp. 145–167.
[5] R.S. Crandall, J. Phys. Chem. Solids **31** (1970) 2069.
[6] R.P. Khosla, J.R. Fischer and B.C. Burkey, Phys. Rev. **B7** (1973) 2551.
[7] N. Holonyak Jr, S.W. Ing Jr, R.C. Thomas and S.F. Bevacqua, Phys. Rev. Lett. **8** (1962) 426.
[8] R.A. Reynolds, Solid-State Electron. **11** (1968) 385.
[9] G.E. Stillman, C.M. Wolfe, I. Melngailis, C.D. Parker, P.E. Tannenwald and J.O. Dimmock, Appl. Phys. Lett. **13** (1968) 83.
[10] T.O. Poehler, Phys. Rev. **B4** (1971) 1223.
[11] K. Seeger, *Semiconductor Physics – An Introduction,* 3rd ed. (Springer, Berlin, 1985); for the mobility data for GaAs in Fig. 3.2, see D. Chattopadhyay and H.J. Queisser, Rev. Mod. Phys. **53** (1981) 745; for the mobility data for Ge in Fig. 3.2, see W.W. Tyler and H.H. Woodbury, Phys. Rev. **102** (1956) 647.
[12] E. Fermi, *Nuclear Physics* (University of Chicago Press, Chicago, 1950) p. 142.
[13] H. Brooks, Phys. Rev. **83** (1951) 879.
[14] R.B. Dingle, Phil. Mag. **46** (1955) 831.
[15] R.D. Rode, *Low-Field Electron Transport* in *Semiconductor and Semimetals,* Vol. **10**, eds. R.K. Willardson and A.C. Beer (Academic Press, New York, 1975) pp. 1–86.
[16] E. Conwell and V.F. Weisskopf, Phys. Rev. **77** (1950) 388.
[17] L.M. Falicov and M. Cuevas, Phys. Rev. **164** (1967) 1025; see also M. Cuevas, Phys. Rev. **164** (1967) 1021.
[18] J.D. Wiley, *Mobility of Holes in III–V Compounds* in *Semiconductor and Semimetals*, Vol. **10**, eds. R.K. Willardson and A.C. Beer (Academic Press, New York, 1975) p. 139.

[19] C. Erginsoy, Phys. Rev. **79** (1950) 1013.
[20] N. Sclar, Phys. Rev. **104** (1956) 1559.
[21] H. Honig, Phys. Rev. Lett. **17** (1966) 186.
[22] R. Maxwell and A. Honig, Phys. Rev. Lett. **17** (1966) 188.
[23] N.F. Mott and H.S.W. Massey, *The Theory of Atomic Collision* (Oxford University Press, Oxford, 1965).
[24] L.I. Schiff, *Quantum Mechanics*, 2nd ed. (McGraw-Hill, New York, 1955).
[25] H.S.W. Massey and B.L. Moiseiwitsch, Phys. Rev. **78** (1949) 865.
[26] S.G. Elkomoss and G. Munschy, J. Phys. Chem. Solids **38** (1977) 557.
[27] K. Aoki, K. Miyamae, T. Kobayashi and K. Yamamoto, J. Phys. Soc. Jpn **21** (1982) 3920.
[28] E. Otsuka, K. Murase and J. Iseki, J. Phys. Soc. Jpn **21** (1966) 1104.
[29] M. Pomeranz, Proc. IEEE **53** (1965) 1438.
[30] B.R. Nag, *Theory of Electrical Transport in Semiconductors* (Pergamon Press, Oxford, 1972); B.R. Nag, *Electron Transport in Compound Semiconductors* (Springer, Berlin, 1980).
[31] T. Ohyama, T. Sanada and E. Otsuka, J. Phys. Soc. Jpn **35** (1973) 822.
[32] R. Ulbrich, Phys. Rev. **B8** (1973) 5719.
[33] K. Aoki, T. Kinugasa and K. Yamamoto, Phys. Lett. **72A** (1979) 63; for exciton polariton structure see also D.D. Sell, S.E. Stokowski, R. Dingle and J.V. DiLorenzo, Phys. Rev. **B7** (1973) 4568.
[34] K. Aoki, T. Kobayashi and K. Yamamoto, Phys. Lett. **79A** (1980) 445.
[35] D.L. Smith, D.S. Pan and T.C. McGill, Phys. Rev. **B12** (1975) 4360.
[36] I.C. Percival, Nucl. Fusion **6** (1966) 182.
[37] M.N. Gurnee, M. Glicksman and P.W. Yu, Solid State Commun. **11** (1972) 11.
[38] T. Yao, K. Inagaki and S. Maekawa, Solid State Commun. **13** (1973) 533.
[39] J. Yamashita, Prog. Theor. Phys. **24** (1960) 357.
[40] J. Yamashita and K. Inoue, J. Phys. Chem. Solids **12** (1959) 1.
[41] J. Yamashita, J. Phys. Soc. Jpn **16** (1961) 720.
[42] E. Schöll, in *Nonlinear Dynamics in Solids*, ed. H. Thomas (Springer, Berlin, 1992) pp. 31–50.
[43] J. Pozhela, *Plasma and Current Instabilities in Semiconductors* (Pergamon Press, Oxford, 1981).
[44] A.F. Volkov and Sh.M. Kogan, Sov. Phys. JETP **25** (1967) 1095.
[45] H. Haken, *Synergetics – An Introduction, Nonequilibrium Phase Transitions and Self-Organization in Physics, Chemistry and Biology*, 2nd ed. (Springer, Berlin, 1978).
[46] K. Aoki, N. Mugibayashi and K. Yamamoto, Proc. 18th Int. Conf. Phys. Semicond., ed. O. Engstrom (World Scientific, Singapore, 1987) p. 1543; K. Aoki, N. Mugibayashi and K. Yamamoto, Phys. Scripta **T14** (1986) 76.
[47] K. Aoki and N. Mugibayashi, Appl. Phys. **A48** (1989) 161.
[48] T. Watanabe, T. Kondo and K. Aoki, Memoirs of the Faculty of Engineering, Kobe University, No. 37 (1990) pp. 199–216.
[49] K. Aoki, Solid State Commun. **77** (1991) 87.
[50] K. Aoki, Phys. Lett. **152A** (1991) 485.
[51] W. Shockley, Solid State Electron. **2** (1961) 35.
[52] K. Aoki and K. Yamamoto, Appl. Phys. **A48** (1989) 111.
[53] F.-J. Niedernostheide, M. Ardes, M. Or-Guil and H.-G. Purwins, Phys. Rev. **B49** (1994) 7370.
[54] P. Gransdorff and I. Prigogine, *Thermodynamic Theory of Structure and Fluctuations* (Wiley-Interscience, London, 1971).

[55] G. Nicolis and I. Prigogine, *Self-Organization in Nonequilibrium Systems – From Dissipative Structures to Order through Fluctuations* (John Wiley & Sons, New York, 1977).
[56] I. Prigogine and I. Stengers, *Order out of Chaos* (Bantam Books, Toronto, 1984).
[57] J. Briggs and F.D. Peat, *Turbulent Mirror – An Illustrated Guide to Chaos Theory and the Science of Wholeness* (Harper & Low, New York, 1990).
[58] B.K. Ridley, Proc. Phys. Soc. **82** (1963) 954.
[59] K. Takeyama and K. Kitahara, J. Phys. Soc. Jpn **39** (1975) 125.
[60] A.B. Pippard, *Response and Stability – An Introduction to the Physical Theory* (Cambridge University Press, Cambridge, 1985).
[61] F. Schlögl, Z. Phys. **253** (1972) 147.
[62] K. Aoki, T. Kobayashi and K. Yamamoto, J. Phys. Soc. Jpn **51** (1982) 2373.
[63] S.W. Teitsworth, R.M. Westervelt and E.E. Haller, Phys. Rev. Lett. **51** (1983) 825.
[64] F.-J. Niedernostheide, ed., *Nonlinear Dynamics and Pattern Formation in Semiconductors and Devices* (Springer, Berlin 1995).
[65] A.M. Barnett and A.G. Milnes, J. Appl. Phys. **37** (1966) 4215.
[66] A.M. Barnett, in *Semiconductor and Semimetals*, Vol. **6**, *Injection Phenomena*, eds. R.K. Willardson and A.C. Beer (Academic Press, New York, 1970) pp. 141–200.
[67] A.M. Barnett and H.A. Jensen, Appl. Phys. Lett. **12** (1968) 341.
[68] A.M. Barnett, H.A. Jensen, V.F. Meiklehan and H.C. Bowers, in *Gallium Arsenide* (Proc. 2nd Symposium, Dallas, Texas, 1968) p. 136.
[69] B.L. Crowder, F.F. Morehead and P.R. Wagner, Appl. Phys. Lett. **8** (1966) 148.
[70] M. Braunstein, A.I. Braunstein and R. Zuleeg, Appl. Phys. Lett. **10** (1967) 313.
[71] G.C. Gerhard and H.A. Jensen, Appl. Phys. Lett. **10** (1967) 333.
[72] J. Peinke, A. Mühlbach, R.P. Huebener and J. Parisi, Phys. Lett. **108A** (1985) 407.
[73] J. Peinke, B. Röhricht, A. Mühlbach, J. Parisi, Ch. Noldeke, R.P. Huebener and O.E. Rössler, Z. Naturforsch. **40A** (1985) 562.
[74] K.M. Mayer, R. Gross, J. Parisi, J. Peinke and R.P. Huebener, Solid State Commun. **63** (1987) 55.
[75] K.M. Mayer, J. Parisi, J. Peinke and R.P. Huebener, Physica **32D** (1988) pp. 306–317.
[76] U. Rau, PhD thesis, University of Tübingen (1991); see also U. Rau, K.M. Mayer, J. Parisi, J. Peinke, W. Clauss and R.P. Huebener, Solid-State Electron. **32** (1989) 1365.
[77] U. Rau, K. Aoki, J. Peinke, J. Parisi, W. Clauss and R.P. Huebener, Z. Phys. B–Condensed Matter **81** (1990) 53.
[78] D. Jäger, H. Baumann, R. Symanczyk, Phys. Lett. **117A** (1986) 141.
[79] H. Baumann, R. Symanczyk, C. Radehaus, H.-G. Purwins and D. Jäger, Phys. Lett. **123A** (1987) 421.
[80] D. Jäger and R. Symanczyk, in *Nonlinear Dynamics in Solids*, ed. H. Thomas (Springer, Berlin, 1992) pp. 68–87.
[81] A. Brandl and W. Prettl, Festkörperproblem, Vol. **30** (Vieweg, Braunschweig, 1990) p. 371.
[82] F.-J. Niedernostheide, M. Kreimer, B. Kukuk, H.-J. Schulze and H.-G. Purwins, Phys. Lett. **191A** (1994) 285.

[83] F.-J. Niedernostheide, B.S. Kerner and H.-G. Purwins, Phys. Rev. **B46** (1992) 7559.
[84] F.-J. Niedernostheide, M. Arps, R. Dohmen, H. Willebrand and H.-G. Purwins, Phys. Stat. Sol. (B) **172** (1992) 249.
[85] R. Symanczyk, S. Knigge and D. Jäger, in *Negative Differential Resistance and Instabilities in 2-D Semiconductors*, eds. N. Balkan, R.K. Ridley and A.J. Vickers, NATO ASI series B, Vol. **307** (Plenum Press, New York, 1993) pp. 269–282.
[86] M.C. Cross and P.C. Hohenberg, Rev. Mod. Phys. **65** (1993) 851.

Chapter 4

Nonlinear dynamics, stability and instability in carrier transport

4.1 Catastrophe theory

Chapter 3 treated theoretically the first-order phase transition and the related S-shaped I–V characteristics in the one-level impact-ionization model. A spatially inhomogeneous solution of the I–V curve was transformed into a spatially homogeneous solution of the I–V curve, leading to our understanding of the formation of a stationary current filament in semiconductors. This chapter covers spatially homogeneous solutions and investigates the nonlinear dynamics of the carrier transport in the impact-ionization avalanche. Dynamical behaviors in the S-shaped I–V characteristics are strongly related to the cusp catastrophe in sections 4.1 and 4.2. Linear stability analysis is introduced in section 4.3, which provides a powerful tool for finding a criterion for stability and instability in a given dynamical system. In order to illustrate the idea of linear stability analysis, some examples (pendulum, Duffing equation and Van der Pol oscillator) are also given in section 4.3. Linear stability analysis is applied to a mathematical model in section 4.4 and a physical model of the impact-ionization avalanche in section 4.5. Some fundamental aspects of chaos will be presented in sections 4.6–4.12.

The following text describes the catastrophe theory. The structural stability of a certain dynamical system can be discussed, under the static regime, in term of the catastrophe manifold $\mathbf{M}$ constructed by the potential function $f(\mathbf{x}; c_1, c_2, \ldots, c_m)$ of an m-parameter family. Catastrophe was first named by a mathematician, Thom [1], and means a sudden change at a critical point in the parameter space. A wave breaking, the diadem of a splash, a swallow's tail, morphogenesis in biology and various types of sudden changes in nature are described by the catastrophe theory [1–3]. The first-order phase transition in semiconductor carrier transport [4] and laser optics [2, 5] are strongly related to an elementary catastrophe (cusp catastrophe), as discussed below. Let $\mathbf{x} = (x_1, x_2, \ldots, x_n)$ be state variables (internal variables, $\mathbf{x} \in \mathbf{R}^n$), c_i $(i = 1, 2, \ldots m)$ be control variables $(c_1, c_2, \ldots c_m) \epsilon \mathbf{R}^m; \mathbf{R}^m$: control space), then catastrophe manifold $\mathbf{M}$ is defined [2] as:

$$\mathbf{M} = \{\mathbf{Y} \in \mathbf{R}^s | \partial f/\partial x_1 = \partial f/\partial x_2 = \ldots = \partial f/\partial x_n = 0\}, \tag{4.1}$$

where $s = n + m, \mathbf{Y} = (x_1, x_2, \ldots, x_n; c_1, c_2, \ldots, c_m)$. The map function f: $\mathbf{R}^n \to \mathbf{R}$ is assumed to be a smooth function of class C^∞. The function f can be noted also as potential function V for dynamical systems. By definition, Y are critical points since $\partial f/\partial x_1 = \partial f/\partial x_2 = \ldots = \partial f/\partial x_\mathrm{n} = 0$[2]. Then, the catastrophe manifold $\mathbf{M}$ is a set of the critical points such that

$$\partial f/\partial x_i|_\mathbf{Y} = 0, \quad i = 1, 2, \ldots. \tag{4.2}$$

Among $\mathbf{Y}$, some locate at local maxima, some at local minima and even at points of inflexion of f. The values of f at $\mathbf{Y}$ are called critical values. Several examples are shown in Fig. 4.1. The critical point $\mathbf{Y}$ is structurally stable on manifold $\mathbf{M}$ if the determinant of the Hessian matrix [2]

$$Hf|_\mathbf{Y} = \left(\frac{\partial^2 \partial^2 f}{\partial x_i \partial x_j}\bigg|_\mathbf{Y}\right) \tag{4.3}$$

is a nonzero (nonsingular) value, namely, det $Hf \neq 0$, while it is structurally unstable when det $Hf = 0$. The critical point is **nondegenerate** if it is structurally stable, and **degenerate** if structurally unstable. For the nondegenerate critical point, the structural stability is not broken by adding an arbitrary, but sufficiently small, function $g(x)$ of at least C^2-topology in the subspace $\mathbf{R}^1$. The determinant of the Hessian matrix satisfies [2]

$$\det H(f + g) \neq 0. \tag{4.4}$$

In contrast, all degenerate critical points are structurally unstable on the manifold $\mathbf{M} \subset \mathbf{R}^s$. As an example, consider $f = x^4, g = \varepsilon x^2$. The critical

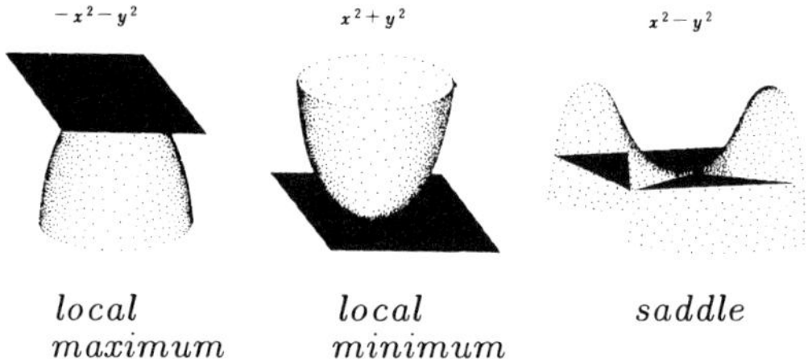

Figure 4.1 Examples of critical points. Local maximum (left), local minimum (middle), and saddle (right), which are tangential to the black sheet. After Ref. [2].

point $x = 0$ is degenerate with Hessian determinant zero. By adding g, the Hessian determinant takes a nonzero value. With $\varepsilon < 0$, the potential function $f + g$ exhibits two local minima and one local maximum, as shown in Fig. 4.2. Thus $f(x)$ is structurally unstable.

The critical point $x = 0$ corresponds to the cusp point in the cusp catastrophe, as seen later. If one chooses $f = x^4 - cx^2$ $(c > 0)$, then there are three critical points $x = 0, x = \pm\sqrt{c/2}$. By adding g such as x or x^2, the structural stability is not broken at two local minima, where the local minima $x = \pm\sqrt{c/2}$ shift slightly against the small perturbation, while the Hessian determinant keeps nonzero values. Then the potential function is structurally stable against a sufficiently small perturbation. If $f = x^4 - cx^3$, there is a local minimum at $x = 3c/4$ (nondegenerate) and a point of inflexion at $x = 0$ (degenerate). Apparently, $x = 0$ is structurally unstable, forming a cliff, while $x = 3c/4$ is stable. A particle on the cliff begins to roll down suddenly against a small perturbation.

Thom [1] described structural stability by considering differential systems, defined by:

- a configuration space that is a finite- or infinite-dimensional manifold **M**.
- a vector field **X** on **M**.

The flow associated with the differential system is obtained by the integration of the vector field on **M**, which gives rise to at least a one-parameter group of diffeomorphism $h_t : \mathbf{M} \to \mathbf{M}$ [1, 2]. If one considers a dynamical system $\varepsilon\dot{\mathbf{x}} = \mathbf{X} = -\text{grad}\, f(\mathbf{x})$, the time evolution on **M** under the **diffeomorphism** of the form $(\mathbf{x}; t, c) \to (\mathbf{x}'; t', c)$, can be discussed, but within an almost static regime. By diffeomorphism, we mean that $h : \mathbf{Y} \to \mathbf{Z}$ is a **homeomorphism**, which is a smooth and differentiable function [2]. By the homeomorphism $h : \mathbf{Y} \to \mathbf{Z}$, we mean that the smooth function h maps **y** in **Y** into **z** in **Z** by one-to-one correspondence. Note that $h : \mathbf{Y} \to \mathbf{Z}(\mathbf{z} = h(\mathbf{y})), h^{-1} : \mathbf{Z} \to \mathbf{Y}(\mathbf{y} = h^{-1}(\mathbf{z})), (h^{-1})^{-1} = h$. ($h$: square $\to$ disk is a homeomorphism. Topologically, the square is equivalent to the

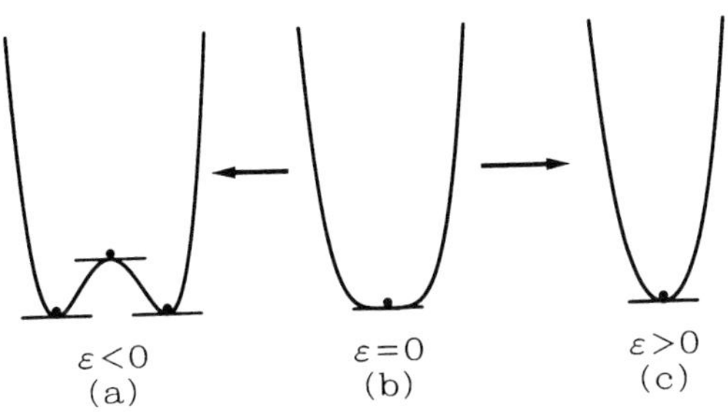

Figure 4.2 The potential function $f + g$, where $f = x^4$, and $g = \varepsilon x^2$.

disk, a cube is equivalent to a sphere and so on.) Thom [1] discusses the set of vector fields **X** on **M** (a vector space of sections of the fiber space of tangent vectors to **M**) by the topology defined by the metric:

$$d(\mathbf{X}_0, \mathbf{X}_1) = \sup{}_{\mathrm{M}}\left\{|\mathbf{X}_0 - \mathbf{X}_1| + \sum \frac{\partial \mathbf{X}_0}{\partial x} - \frac{\partial \mathbf{X}_1}{\partial x}\right\} (\text{the } C^1\text{-topology}). \tag{4.5}$$

According to Thom's expression, a dynamical system **X** on **M** is structurally stable if a sufficiently small (in the C^1-norm) perturbation of the field **X** does not alter the qualitative nature of the system.

Elementary catastrophe, a dynamical system defined by (**M, X**), is treated with the gradient field:

$$\mathbf{X} = -\text{grad } V(\mathbf{x}), \tag{4.6}$$

where $f = V(\mathbf{x})$ is the potential function. Following Thom's classification, there are seven types of famous catastrophe, as listed in Table 4.1. Several types of catastrophe will now be described.

4.1.1 One-parameter family ($m = 1$)

Fold catastrophe is described by the potential and the manifold **M**

$$f = \frac{1}{3}x^3 + ax, \tag{4.7}$$

$$\mathbf{M} = \{(x, a) \in \mathbf{R}^2 | x^2 + a = 0\}. \tag{4.8}$$

The manifold is shown in Fig. 4.3, where $(x, a) = (0, 0)$ is the catastrophe point (singular point) with det $Hf|_{x=0} = 0$. For $a > 0$, the manifold is empty ($x^2 + a = 0$ has no solutions and there are no attractors), while for $a < 0$ there is an attractor at $x = \sqrt{-a}$ and a repeller at $x = -\sqrt{-a}$. In Fig. 4.3, the gradient field **X** is shown by the arrows. The structural stability is intuitively understood by putting a particle on the potential minimum of f (inset figure

Table 4.1 Clasification of elementary catastrophes

m	Center	Universal unfolding	Type
1	x^3	$x^3 + ax$	Folding
2	x^4	$x^4 + ax^2 + bx$	Cusp
3	x^5	$x^5 + ax^3 + bx^2 + cx$	Swallow's tail
4	x^6	$x^6 + ax^4 + bx^3 + cx^2 + dx$	Butterfly
3	$x^3 + y^3$	$x^3 + y^3 + axy - bx - cy$	Hyperbolic umbilic
3	$x^3 - 3xy^2$	$x^3 - 3xy^2 + a(x^2 + y^2) - bx - cy$	Elliptic umbilic
4	$x^2y + y^4$	$x^2y + y^4 + ax^2 + by^2 + cx - dy$	Parabolic umbilic

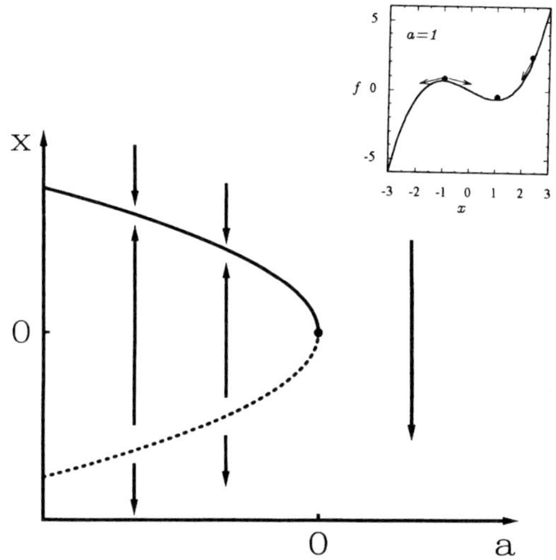

Figure 4.3 Manifold M for the potential function $f = x^3/3 + ax$. The inset shows the potential function f, where $a = 1.0$.

of Fig. 4.3). By increasing a, the attractor and the repeller coalesce, leaving the particle on the cliff at (0,0).

The extended forms of a one-parameter family with n-internal variables $\mathbf{x} = (x_1, x_2, \ldots, x_n)$ are given by:

$$f = x_1^3 + ax_1 + x_2^2 + x_3^2 + \ldots + x_n^2. \tag{4.9}$$

For example, $f = x^3 + ax + y^2; \mathbf{M} = \{(x, y, a) \in \mathbf{R}^3 | \partial f/\partial x = \partial f/\partial y = 0\}$ give the same structure of the bifurcation, forming an attractor at $(\sqrt{-a/3},\ 0,\ a)$ and a repeller at $(-\sqrt{-a/3},\ 0, a)$. If $f = x^3 + ax - y^2$ is chosen, however, there is a saddle point at $(\sqrt{-a/3},\ 0, a)$ and a repeller at $(-\sqrt{-a/3},\ 0, a)$. The equations of motion for $f = x^3 + ax - y^2$ are given by:

$$\dot{x} = -3x^2 - a, \tag{4.10}$$

$$\dot{y} = 2y. \tag{4.11}$$

The linear stability analysis can easily be done at $(\pm\sqrt{-a/3},\ 0, a)$. Thus, there are no stable structures and the bifurcation for the potential $f = x^3 + ax - y^2$.

4.1.2 Two-parameter family ($m = 2$)

Cusp catastrophe is defined by the potential function f, where $f : \mathbf{R}^2 \times \mathbf{R} \to \mathbf{R}$, and the catastrophe manifold $\mathbf{M}$:

$$f = \frac{1}{4}x^4 + \frac{a}{2}x^2 - bx, \tag{4.12}$$

$$\mathbf{M} = \{(x, a, b)\epsilon\mathbf{R}^3 | \partial f/\partial x = 0\}. \tag{4.13}$$

From (4.13), the catastrophe manifold in Fig. 4.4 is constructed by a set of the points (x, a, b) which satisfies

$$x^3 + ax - b = 0. \tag{4.14}$$

The catastrophe map $\chi : \mathbf{M} \to \mathbf{C}$ is the projection of the manifold $\mathbf{M}$ onto the (a, b)-plane:

$$(x, a, b) \mapsto (a, b), \qquad (x, a, b) \in \mathbf{M}. \tag{4.15}$$

A set of the degenerate critical points (x, a, b) on $\mathbf{M}$ satisfies det $Hf = 0$ (the derivative of (4.14) with respect to x):

$$3x^2 + a = 0. \tag{4.16}$$

Eliminating x from (4.14) and (4.16) gives

$$4a^3 + 27b^2 = 0. \tag{4.17}$$

The bifurcation set $\mathbf{K}$ is given by

$$\mathbf{K} = \{(a, b) \in \mathbf{C} | 4a^3 + 27b^2 = 0\}. \tag{4.18}$$

The bifurcation set is cusp shaped, and is the projection of the degenerate critical points (x, a, b) onto the (a, b)-plane, as shown in Fig. 4.4. Origin $P = (0, 0, 0)$, the center of the manifold, is called the cusp point, where the potential function is $x^4/4$ (see Fig. 4.2). The behavior of the variable x is

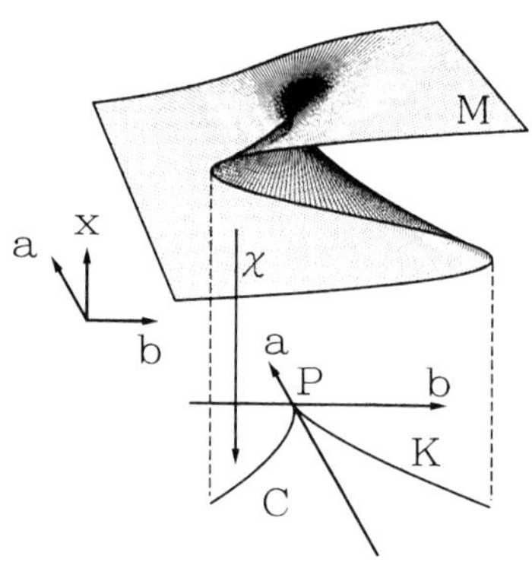

Figure 4.4 Manifold M of the cusp catastrophe and its catastrophe map $M \to C$.

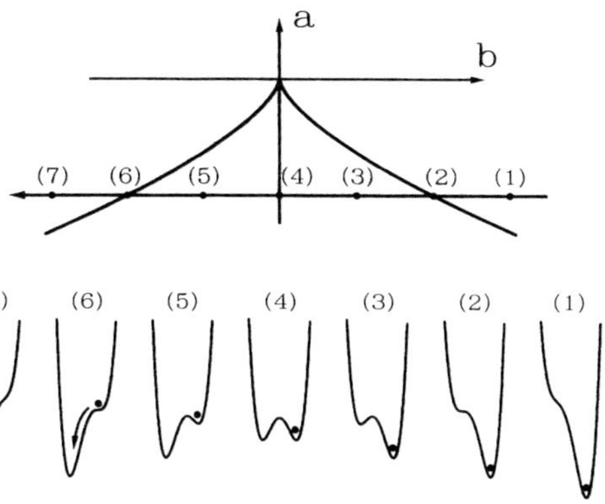

Figure 4.5 Bifurcation set K and the double-well potential on the traverse line of b.

essentially static on the catastrophic manifold, but it shows a catastrophic jump when the control parameter b traverses the cusp region on C, as shown in Fig. 4.5.

The catastrophic jump is understood by considering behavior of a particle on the double-well potential in Fig. 4.5. By decreasing b from above, the particle located in a well (upper sheet of M) shows the catastrophic jump into another well (lower sheet of M) in (6) of Fig. 4.5. The dynamical system (**M, X**) is **quasi-static** whenever the control parameters vary very slowly. That is, the system follows the convention of **perfect delay** [1, 2].

4.1.2.1 Convention of perfect delay

The equilibrium point (x, a, b) on a stable sheet of **M** suffers a catastrophic jump to another sheet whenever it passes through the edge of the folded sheet of **M** (a set of leaving point, line K in the parameter space **C**).

The dynamical system has the following properties: (i) catastrophic jump, (ii) hysteresis, (iii) divergence, and (iv) jump and slow relaxation. Figure 4.6 shows behaviors (ii), (iii) and (iv). In Fig. 4.6b, the divergence is explained as follows. Starting from the same initial point near the cusp point, say

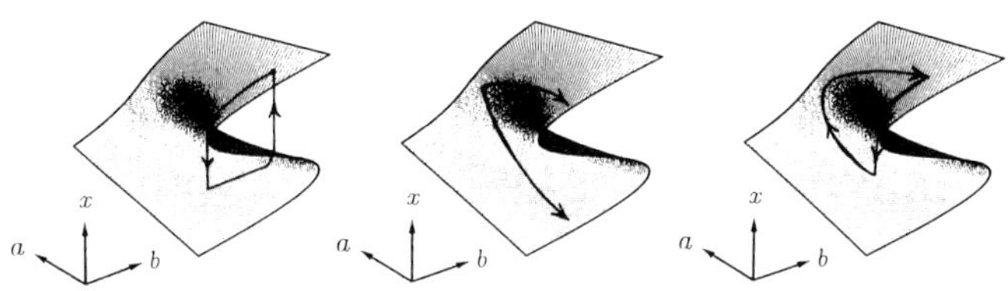

Figure 4.6 (left) Catastrophic jump and hysteresis, (center) divergence, and (right) jump and slow relaxation.

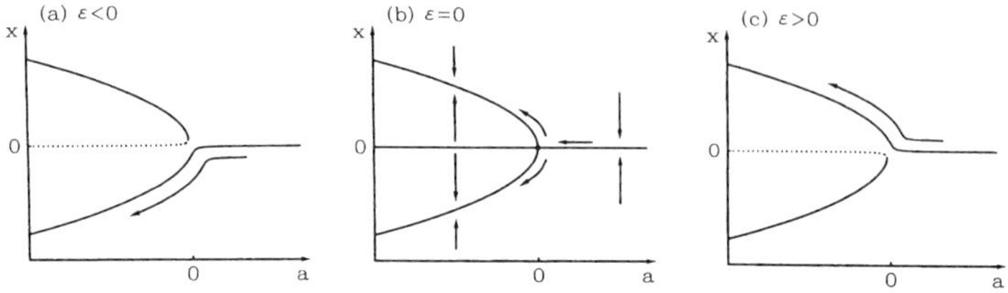

Figure 4.7 Section of the cusp catastrophe manifold for (a) $\varepsilon < 0$, (b) $\varepsilon = 0$, and (c) $\varepsilon > 0$.

(x_0, a_0, b_0), two different paths of the trajectories are possible by small alternation of the control parameters, giving rise to drastic change in the outputs of x. The divergence is related to the structural stability near the cusp point. Figure 4.7 shows the section of the manifold given by

$$x^3 + ax - \varepsilon = 0, \tag{4.19}$$

where $\varepsilon(= b)$ takes small values. With $\varepsilon = 0$, the section of M is a symmetric pitchfork (Fig. 4.7b), where the origin (0,0,0) is structurally unstable since it is degenerate (det $Hf = 0$, two-fold degeneracy). By decreasing a from above, there is no discrimination at the cusp point, about which branch the trajectory should pass. With small perturbations ε, two possible trajectories appear with the help of symmetry breaking (arrows in Fig. 4.7a, c).

Let us consider equations of motion of the dynamical system (**MX**) with $\mathbf{X} = \varepsilon\dot{x}$ (ε: sufficiently small);

$$\varepsilon\dot{x} = -\partial f/\partial x = -(x^3 + ax - b), \tag{4.20}$$

$$\dot{a} = g(x, a, b,), \tag{4.21}$$

$$\dot{b} = h(x, a, b,), \tag{4.22}$$

where point (a, b) varies smoothly on the parameter space **C** such that functions g and h are smooth functions of x, a and b (either linear, sublinear or weakly nonlinear). Functions g and h may include explicitly time t, if necessary. In general, g and h are assumed to be sufficiently many times differentiable. When $\varepsilon = 0$, namely $\partial f/\partial x = 0$, then

$$\partial f/\partial x = x^3 + ax - b = 0, \tag{4.23}$$

$$\dot{a} = g(x, a, b,) \tag{4.24}$$

$$\dot{b} = h(x, a, b,). \tag{4.25}$$

The solution $\{x^*(t), a^*(t), b^*(t)\}$ of the reduced system (4.23)–(4.25) lies exactly on M. For sufficiently small ε, Tikhnov's theorem [6] holds, as below.

Theorem 4.1
Let $\{x^*(t), a^*(t), b^*(t)\}$ be a solution of the reduced system (4.23)–(4.25), which for $t \in [t_0, t_1]$ is contained in M, where the characteristic exponents have negative real parts bounded away from zero (zero means the exponent of the singular points). Then any solution of (4.20)–(4.22) with initial values

$$|x(t_0;\varepsilon) - x^*(t_0)| = O(\varepsilon), \qquad |a(t_0;\varepsilon) - a^*(t_0)| = O(\varepsilon),$$
$$|b(t_0;\varepsilon) - b^*(t_0)| = O(\varepsilon)$$

satisfies

$$x(t;\varepsilon) = x^*(t) + O(\varepsilon),\ a(t;\varepsilon) = a^*(t) + O(\varepsilon),\ b(t;\varepsilon) = b^*(t) + O(\varepsilon)$$

for $t \in [t_0, t_1]$. The theorem supports the intuitive interpretation that the trajectory of the dynamical system (4.20)–(4.22), if ε is small, is close to the case of the reduced system which is governed by the convention of perfect delay. With smaller ε, there is faster relaxation of initial transients of x to the stationary states (equilibrium point or limit cycle). With $\varepsilon \to 0$, the trajectory approaches asymptotically to the solution $(x^*(t), a^*, b^*(t))$. The reduction procedure from (4.20)–(4.22) to (4.23)–(4.25) is related with a well-known adiabatic elimination of fast variable x (slaving principle by Haken [7, 8]). The variable x is slaved on a stable manifold by the slow variables a and b, such that $x = \chi^{-1}(a, b)$ from (4.23).

Quick relaxation of x to an asymptotically stationary state depends on the magnitude of ε. The following theorem will supplement the initial transients, and the distance between two trajectories in the neighborhood of a stable manifold when they are started from two different initial values (see Fig. 4.8).

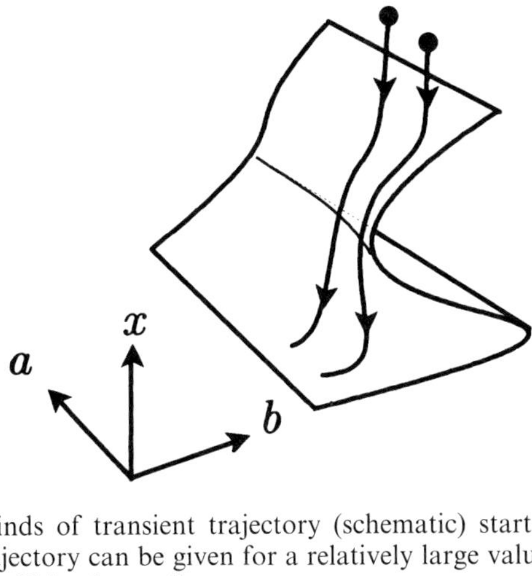

Figure 4.8 Two kinds of transient trajectory (schematic) starting from the initial points. The left trajectory can be given for a relatively large value of ε in (4.20), and the right one for sufficiently small ε.

Theorem 4.2
Given two solutions of (4.20)–(4.22),

$$\{x^{(i)}(t), a^{(i)}(t), b^{(i)}(t)\}, \quad i = 1, 2.$$

Let the initial values satisfy

$$\{x^{(1)}(t_0), a^{(1)}(t_0), b^{(1)}(t_0)\} = (x_0, a_0, b_0)$$

and

$$\{x^{(2)}(t_0), a^{(2)}(t_0), b^{(2)}(t_0)\} = (\bar{x}_0, a_0, b_0)$$

with $(x_0, a_0, b_0) \in \mathbf{M}$, $(\bar{x}_0, a_0, b_0) \notin \mathbf{M}$ and $\bar{x}_0$ such that the solution of $\varepsilon\dot{x} = -x^3 - a_0 x + b_0$ with starting value $\bar{x}_0$ converges to the equilibrium x_0. When the reduced system (4.23)–(4.25) remains outside an ε-independent neighborhood of a point with det $Hf = 0$ for a time interval $[t_1, t_2]$, then for a time interval $[t_1, t_2]$ with $t_1 \geq t_0$

$$|x^{(1)}(t) - x^{(2)}(t)| = 0(\varepsilon), \quad |a^{(1)}(t) - a^{(2)}(t)| = 0(\varepsilon)$$
$$\text{and } |b^{(1)}(t) - b^{(2)}(t)| = 0(\varepsilon).$$

For the behavior in a neighborhood of singularity points, there is another theorem, but no further information about the mathematical viewpoint is given here. Readers who are interested in mathematical treatments, may refer to the book by Grasman [6]. In general, the dynamical system (4.20)–(4.22) has stationary solutions, periodic solutions and chaotic solutions, depending on the nature of the functions $g(x, a, b)$ and $h(x, a, b)$. For sufficient small ε, the dynamical behaviors are quasi-static, quite similar to (4.23)–(4.25), where the convention of perfect delay is an important concept. Oscillatory behavior is given by a discontinuous periodic solution. For relatively large ε, the convection of perfect delay no longer holds, thereby local instability and chaos appear, as presented in section 4.2.

4.1.3 Four-parameter family ($m = 4$)

A four-parameter family of the following type of potential gives the butterfly catastrophe:

$$f = \frac{1}{6}x^6 + \frac{a}{4}x^4 + \frac{b}{3}x^3 + \frac{c}{2}x^2 - dx. \tag{4.26}$$

The manifold $\mathbf{M}$ is a four-dimensional surface embedded in five-dimensional Euclidean space $\mathbf{R}^1 \times \mathbf{R}^4$. By taking a three-dimensional surface of section of the manifold in $\mathbf{R} \times \mathbf{R}^2$, one can visualize the reduced manifold $\mathbf{M}'$ embedded in $\mathbf{R}^3$. For given a and b, one has

$$\mathbf{M}' = \{(x, c, d) \in \mathbf{R}^3 | \partial f/\partial x = 0 \quad \text{for fixed } a, b\}. \tag{4.27}$$

Figure 4.9 shows the manifold $\mathbf{M}'$ and the bifurcation set $\mathbf{K}$ in the (c, d)-plane, where $a = -10, b = 0, c = 0$ to 30 and $d = -99$ to 131. The $x - d$ characteristic (section of $\mathbf{M}'$) is doubly S-shaped in a certain range of c. From $\partial f/\partial x = 0$, one obtains $d = d(x) = x^5 + ax^3 + bx^2 + cx$. Then, the bifurcation set $\mathbf{K}$ is obtained by calculating local extrema of $d(x)$ $(\partial d/\partial x = 0 \leftrightarrow \det Hf = 0)$;

$$\mathbf{K} = \{(c, d) \in \mathbf{R}^2 | \partial d/\partial x = 0\}. \tag{4.28}$$

For the special case of $b = 0$, the bifurcation set is given by

$$d = x(x^4 + ax^2 + c), \tag{4.29}$$

$$x = \pm\sqrt{(-3a \pm \sqrt{9a^2 - 20c})/10}, \tag{4.30}$$

where $c < 0$ for $a > 0$ and $c < 9a^2/20$ for $a < 0$ ($\pm$ signs should be taken such that $-3a \pm \sqrt{9a^2 - 20c} > 0$). For a fixed value of a, one obtains singular point x as a function of c from (4.30) and d from (4.29), giving bifurcation set $\mathbf{K}$ in the (c, d)-plane. In general, for $b \neq 0, d$ is solved directly from $\partial d/\partial x|_c = 0$. The remarkable nature of the butterfly catastrophe is explained in Fig. 4.10. For $a < 0$, there exist three cusps in $\mathbf{K}$, by which a pocket region is formed, as shown by a dark gray region. In the pocket, the potential function f has three local minima, which means tristability in the dynamical system: $\varepsilon\dot{x} = -\partial f/\partial x, \dot{a} = \dot{b} = \dot{c} = \dot{d} = 0$. The light gray region has two local minima, giving bistability. The remaining part of the (c, d)-plane has only one local minimum. As seen in Fig. 4.9, the pocket can be formed by folding an unstable manifold in the cusp catastrophe. Therefore, one may suppose that butterfly catastrophe is an extension of cusp catastrophe. This can be argued semi-quantitatively as follows. For $a > 0$, the term x^6 in f is only additive to ax^4 in its effect. Thus, for $\partial f/\partial x = 0$, one

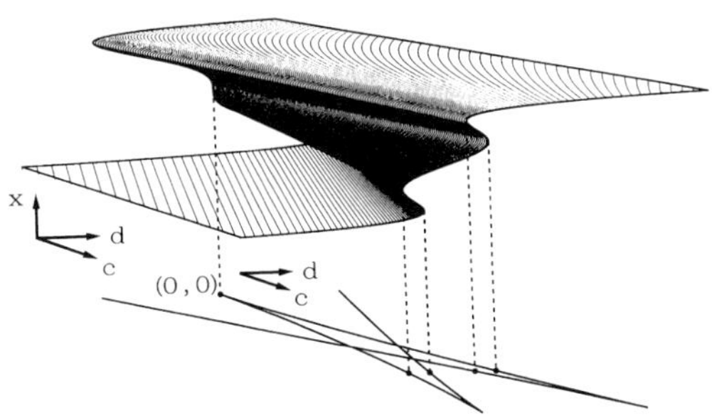

Figure 4.9 Manifold M' of the four-parameter family (butterfly catastrophe), and the bifurcation set K in the (c, d) plane, where $a = -10$, $b = 0$, $c = 0 \sim 30$, $d = -99 \sim 131$.

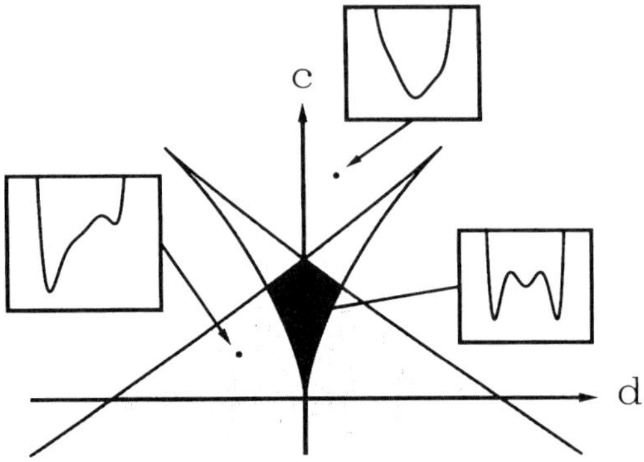

Figure 4.10 Bifurcation set of the butterfly catastrophe. The pocket region is shown by the dark gray region. Symmetric case with $b = 0$.

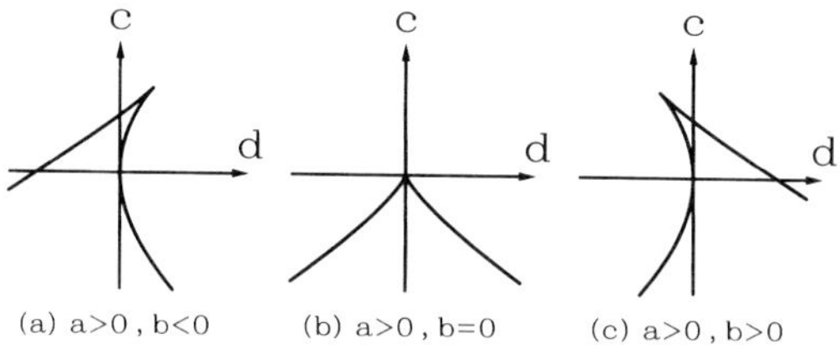

Figure 4.11 Cusp structure by an approximation (4.31) for the four-parameter family potential function (4.26).

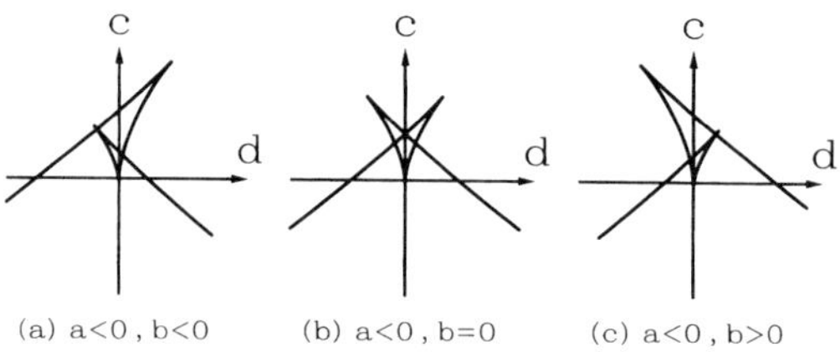

Figure 4.12 Three cusps appear for $a < 0$.

can approximate

$$ax^3 + bx^2 + cx - d \simeq 0. \qquad (4.31)$$

The equation gives a single cusp structure of **K**, where the term bx^2 simply has the effect of shifting and modifying the cusp structure, as shown in Fig. 4.11. For $a < 0$, the term x^6 counterbalances to ax^4, giving rise to a significant structure of **K**. A cusp structure splits into three cusps (Fig. 4.12). A global view has been given by Woodcock and Poston [9] and Bröcker and Lander [10] (see the book by Poston and Stewart [2]), as shown in Fig. 4.13. From Fig. 4.13, the parameter a with large negative values produces the

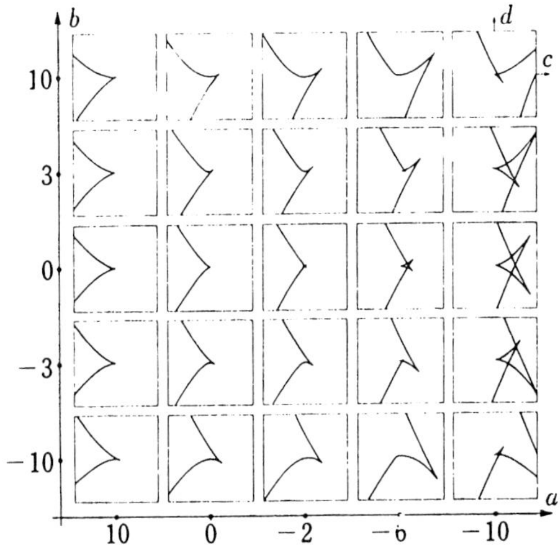

Figure 4.13 Various structures of the bifurcation set for various values of (a, b) [10].

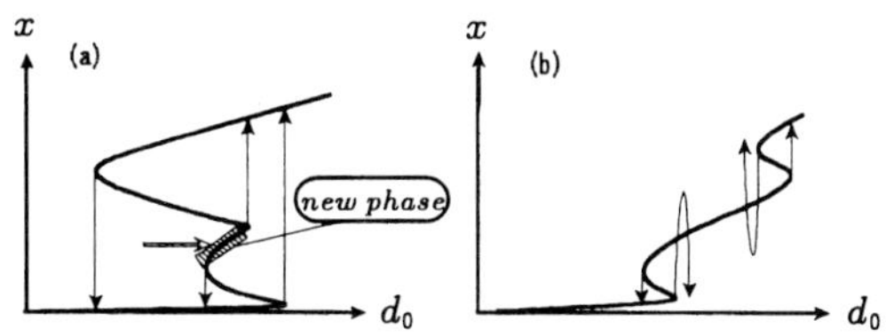

Figure 4.14 $x - d_0$ characteristics by the butterfly catastrophe. (a) The new phase is stable, and (b) the new phase is destabilized.

pocket structure, while, for $a > -2$, the structure is similar to the bifurcation set of cusp catastrophe. The parameter b breaks the symmetry of the bifurcation set.

Now, we shall consider physical aspects of the butterfly catastrophe and relate it to an experiment on the impact-ionization avalanche in GaAs. The pocket region is a new phase, discriminated from two other stable phases. The pocket may be either stable or unstable, depending on the dynamical system $\varepsilon\dot{x} = -\partial f/\partial x, \dot{a} = g(x, a, b, c, d), \dot{b} = h(x, a, b, c, d), \ldots$. Suppose: a dynamical system is given by $\varepsilon\dot{x} = -\partial f/\partial x, \dot{d} = g(x, d)$, where d is a variable parametrized by a control parameter d_0 such that $d = d_0$ when $\dot{d} = 0$, a and b are fixed constants, and c is another control parameter. Given c and d_0, there is a $x - d_0$ curve in Fig. 4.14a, which is a section of the manifold M'. When the new phase is stable, it can be observed. By scanning d_0, the new phase disappears at a singular point and the old one is recovered by switching. After the switching, the new phase is never recurrent without applying a strong external perturbation or varying c. By increasing c, the $x - d_0$ curve looks like Fig. 4.14b, in which one can observe the stable middle branch and two hysteretic behaviors. If the middle branch is destabilized, the new phase is not stationary, but spontaneous oscillation occurs between the two old phases, in a wide range of d_0. Such oscillatory behavior has been observed in an experiment on the impact-ionization avalanche under a longitudinal magnetic field [11], where the magnetic field intensity B corresponds to the control parameter c and the applied bias to d_0. The experimental results have been explained by a mathematical model [12] and by a physical model of the impact-ionization avalanche [13, 14]. Here, the mathematical model is introduced.

Under the longitudinal magnetic field, the current–voltage characteristics show peculiar behavior [11], as shown in Fig. 4.15. With low-field intensity, the I–V_0 curve shows a hysteresis which is caused at 4.2 K by the impact ionization of shallow donors in n-GaAs. By increasing the field intensity B, hysteresis shows a narrowing and current instability sets in after disappearance of the hysteresis, namely for $B > 28$ mT in Fig. 4.15. The frequency of the spontaneous oscillation ranges from 70 to 400 kHz, depending on V_0 [11]. The current instability is termed crossover instability [13, 14]. The experimental results may be explained by a mathematical model of the following type:

$$\dot{x} = -(x-\alpha)(x-\beta)(x-\gamma) + k(\beta_0 - \beta) + \varepsilon, \tag{4.32}$$

$$\dot{\beta} = -\gamma_1(\beta - f_h(x)), \tag{4.33}$$

$$\dot{\varepsilon} = -\gamma_2(\varepsilon - \varepsilon_0), \tag{4.34}$$

where x, β and ε are order parameters,

$$f_h(x) = \beta_0\{1 + h \tanh[c(x - \beta_0)]\} \tag{4.35}$$

is a threshold-like function of x, the parameters $\alpha, \beta_0, \gamma, \gamma_1$ and γ_2 are

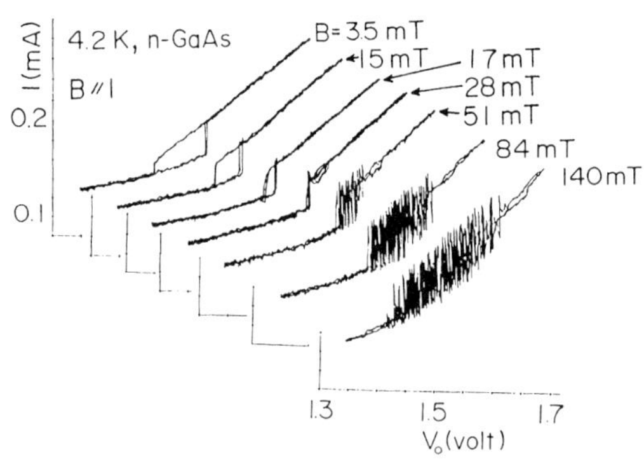

Figure 4.15 A series of $I-V$ curves as a function of B under the longitudinal magnetic field. $R_L = 500\ \Omega$ [11].

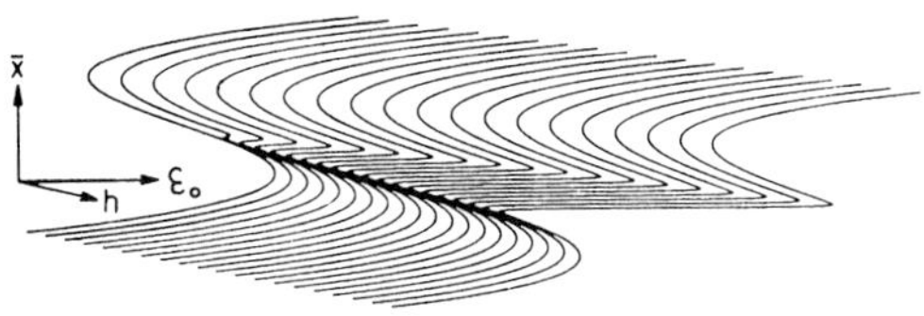

Figure 4.16 A set of $\bar{x} - \varepsilon_0$ curves under steady-state conditions in the parameter space $(h,\ \varepsilon_0,\ \bar{x})$, where $\alpha = 0.1,\ \beta_0 = 0.4,\ \gamma = 0.8,\ k = 0.3$ and $c = 200$ [12].

positive constants, and ε_0, k and h are control parameters. In the model, variables x, β, and ε correspond to the density of the impact-ionized carriers, the magnetoresistance factor and the electric field in the experimental system [11]. A positive constant c in (4.35) is chosen such that the steady-state condition $\bar{\beta} = \bar{x} = f_h(\bar{x})$ has three solutions $\bar{x} = \beta_0, \beta_0 \pm \delta$ for the special case $k = \varepsilon_0 = 0$, where the derivative $\mathrm{d}f_h/\mathrm{d}x|_{\bar{x}} > 1$ is the sufficient condition. Under the steady-state conditions of (4.32)–(4.34), the $\bar{x} - \varepsilon_0$ curve exhibits doubly S-shaped characteristics by increasing $h(h > 0)$. A set of the $x - \varepsilon_0$ curves in the parameter space $(h, \varepsilon_0, \bar{x})$ in Fig. 4.16 shows clearly that the unstable manifold is folded by the function $f_h(x)$, giving rise to a pocket region.

The linear stability analysis of (4.32)–(4.34) is easily done for the folded region (middle branch of the $\bar{x} - \varepsilon_0$ curve). The middle branch is stable if

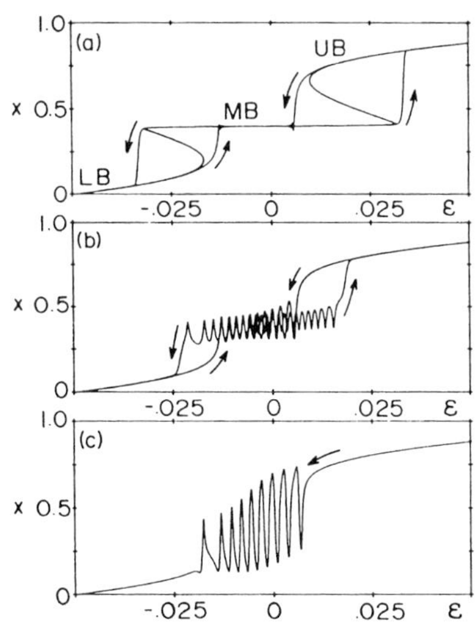

Figure 4.17 Dynamical behavior of x versus ε for (a) slaving mode ($\gamma_1 = 0.4$), (b) breathing mode, ($\gamma_1 = 0.119$), and (c) cross-over mode ($\gamma_1 = 0.03$).

$\gamma_1 > \gamma_0 = (\beta_0 - \alpha)(\gamma - \beta_0)$ and unstable otherwise. Three typical examples are given in Fig. 4.17a–c. The dynamical behaviors of the $x - \varepsilon$ curve are investigated by a slow bias scan of ε_0, where $\alpha = 0.1$, $\beta_0 = 0.4$, $\gamma = 0.8, k = 0.3, h = 0.2, c = 200$, and ε_0 is replaced by $\varepsilon_0 \sin(2\pi f_d t)$, $\varepsilon_0 = 0.05$. With $\gamma_1 = 0.4$ ($> \gamma_0 = 0.12$; see Fig. 4.17a), the system shows simple hysteretic behaviors, while it exhibits oscillatory behaviors; breathing mode for $\gamma_1 = 0.119$ ($\sim \gamma_0$, Fig. 4.17b) and crossover mode for $\gamma_1 = 0.03$ ($\ll \gamma_0$, Fig. 4.17c). The hysteretic behavior still survives for the breathing mode, as indicated by arrows in Fig. 4.17b. Observations of the current instability in Fig. 4.15 are explained phenomenologically by the crossover mode in Fig. 4.17c, where the control parameter h in $f_h(x)$ corresponds to the longitudinal magnetic field B.

4.2 Breakdown of perfect delay

The previous section, described elementary catastrophe and the related dynamical behavior. This section, describes the dynamical process which leads to chaotic behaviors in the cusp catastrophe. An idea of "breakdown of perfect delay" will provide a deep understanding of chaos in the impact-ionization avalanche in semiconductors. The general form of the equations corresponding to a dynamical system (**M, X**) is given by

$$X = \varepsilon \dot{x} = -\frac{\partial V(x, a, b)}{\partial x}, \tag{4.36}$$

$$\dot{a} = g_1(x, a, b, t), \tag{4.37}$$

$$\dot{b} = g_2(x, a, b, t), \tag{4.38}$$

where $V(x, a, b)$ is the potential function of the cusp catastrophe, and g_1, g_2 are smooth functions. The system is autonomous if $g_i = g_i(x, a, b)$, or nonautonomous if $g_i = g_i(x, a, b, t)$ $(i = 1, 2)$. Under the steady-state condition $\dot{x} = \dot{a} = \dot{b} = 0$, there are typically three steady-state solutions (x_1, a_1, b_1), (x_2, a_2, b_2) and (x_3, a_3, b_3), which lie on the folded sheet of the cusp-catastrophe manifold. In the neighborhood of a singular point such that $x_1 = x_2$, the transition time T_{tr} for the switching from $x_1 + \delta$ to $x_3 - \delta$ $(\delta \ll 1)$ is given by [6]:

$$T^1_{tr} = \varepsilon \int_{x_1+\delta}^{x_3-\delta} \left(-\frac{\partial V_{ab}}{\partial x} \right)^{-1} \mathrm{d}x. \tag{4.39}$$

Similarly, T^2_{tr} for the transition in the neighborhood of another singular point is such that $x_2 = x_3$. Then, the transition times T^1_{tr}, T^2_{tr} are proportional to ε, which again supports the convention of perfect delay when $\varepsilon \to 0$. For the oscillatory solution $(x(t), a(t), b(t))$, which may be either periodic or chaotic, one obtains the fundamental period T as evidenced from the Fourier power spectrum. For $\varepsilon \ll 1$,

$$T \sim T_0 + T^1_{\rm tr} + T^2_{\rm tr} + 0(\varepsilon), \tag{4.40}$$

where T_0 is the time spent outside a neighborhood of the singular point [6]. When $T_0 \gg T_{\rm tr}$, the system is almost governed by the rule of perfect delay, where parameters a and b are sufficiently slow variables and x is a fast variable. This is always the case when $\varepsilon \sim 0$. The system is not ruled by the convection of perfect delay if $T_0 \lesssim T_{\rm tr}$, where parameters a and b are not slow variables. For the latter, the perfect delay is broken. ε exists in a wide range, such that the oscillatory solution of (4.36)–(4.38) intersects with the unstable manifold $\mathbf{M}_{\rm u}$ of the cusp catastrophe.

In the vicinity of singular points, there are locally unstable regions, as indicated by Fig. 4.18. Starting from the nearby points in the local region initially at $t = 0$, two orbits diverge remarkably after $t = t_1$ $(t_1 < T)$: one of orbits crosses the unstable manifold $\mathbf{M}_{\rm u}$, while the other one does not (see Fig. 4.18). The expansive nature of the orbits is described as follows. Let T^t be topologically transitive mapping [15] and U a finite volume in $\mathbf{R}^3$. Then, $T^t : U \to U$ is expansive, if for arbitrary points $\mathbf{X}_0, \mathbf{X}'_0 \in U$ there exist t and λ such that $|T^t\mathbf{X}_0 - T^t\mathbf{X}'_0| > \lambda$. The expansive nature is not limited to the chaotic regime, but note that initial transients in the periodic regime show also such behavior in some local volume of U. For the chaotic regime, the requirements are: (i) an expansive nature, (ii) sensitive dependence on initial conditions, and (iii) the existence of an infinite number of unstable periodic orbits (dense in U). The expansive nature is remarkably seen in the Van der Pol-type oscillator and in the impact-ionization avalanche in semiconductors. Several examples are shown here.

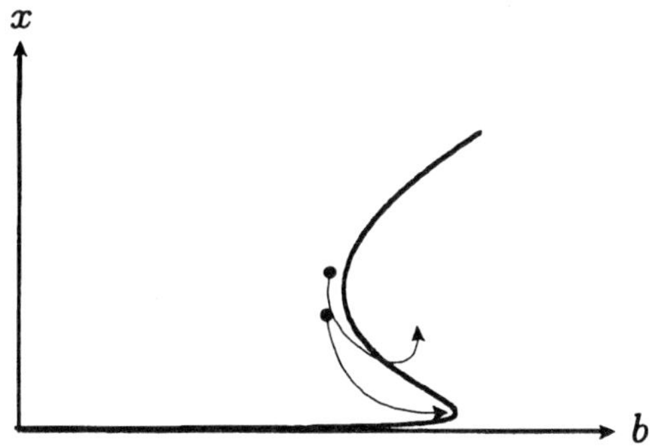

Figure 4.18 Sensitive dependence of two nearby orbits on the initial conditions in the vicinity of singular point.

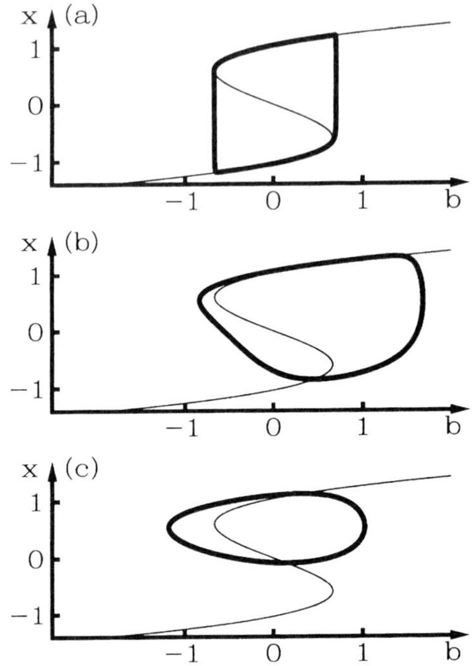

Figure 4.19 Relaxation oscillations simulated with different values of ε for the model equations (4.41) and (4.42) (a) $\varepsilon = 1/800$, (b) $\varepsilon = 1/4$, and (c) $\varepsilon = 1$.

The Van der Pol equation see (4.64) and (4.65), with a constant forcing term c, is given by the Lienard form [6] as:

$$\varepsilon \dot{x} = -\left(\frac{1}{3}x^3 - x - b\right), \tag{4.41}$$

$$\dot{b} = -x + c, \qquad 0 < \varepsilon \ll 1. \tag{4.42}$$

The equations are equivalent to (4.20) and (4.22) by putting $a = -1$. Many types of the Van der Pol equation have been investigated, originally by Van der Pol [16], and by Bonhoeffer [6, 17] and FitzHugh [18] for modeling nerve excitation. The electric potential x across the cell membrane shows a switching behavior (firing) via depolarization. The switching behavior is analogous to that of the current filament in semiconductors [19]. The above equations are similar to an approximated model (4.80), (4.82) of the impact-ionization avalanche in semiconductors, by assuming x as the electron density and b the electric field, where (4.42) may correspond to the dielectric relaxation with the current-control bias condition $\bar{x} = c$. Figure 4.19 shows the switching behaviors by varying ε. With increasing ε, the switching

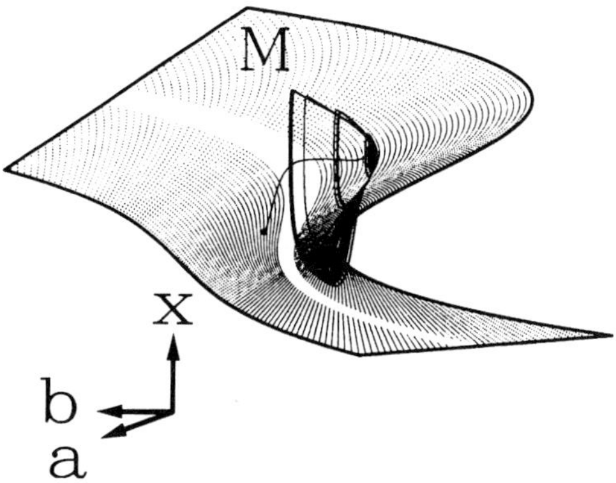

Figure 4.20 Chaotic orbit on the manifold M, simulated for the electical activity of nerve cells. $\varepsilon = 1/15$, $c = 0.12$, and $d = 0.550610737$.

behaviors lose the convention of perfect delay. Another example is given by chaotic relaxation oscillations, which have been studied by Argemi and Rossetto [20] in the modeling of electric activity of nerve cells:

$$\varepsilon\dot{x} = -(x^3 + ax - b), \tag{4.43}$$

$$\dot{a} = -(1 + a - cx), \tag{4.44}$$

$$\dot{b} = -(1 + a - cx)x - x + d, \tag{4.45}$$

where $\varepsilon = 1/15, c = 0.12$ and d is a control parameter. The equilibrium point (x, a, b) locates near singular point of upper sheet of the stable manifold. First period doubling is seen at $d = 0.550687$. By decreasing d $(d < 0.550619)$, one obtains the chaotic solutions or small strange attractors, which are stably trapped around the cliff of the upper sheet of

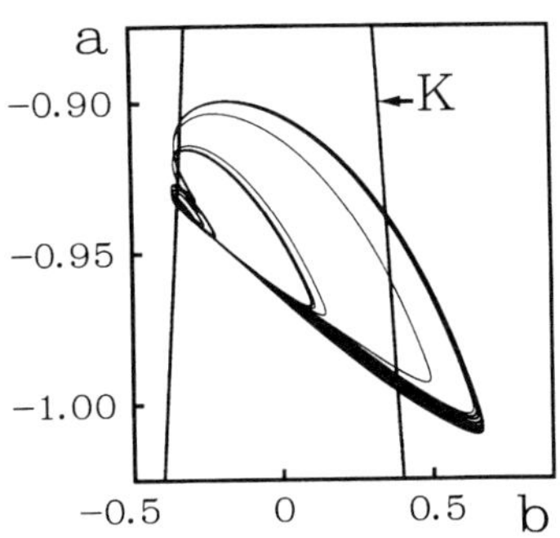

Figure 4.21 Projection of the chaotic orbit on the (a, b) plane. The solid lines K are the bifurcation set. The parameters are the same as those in Fig. 4.20.

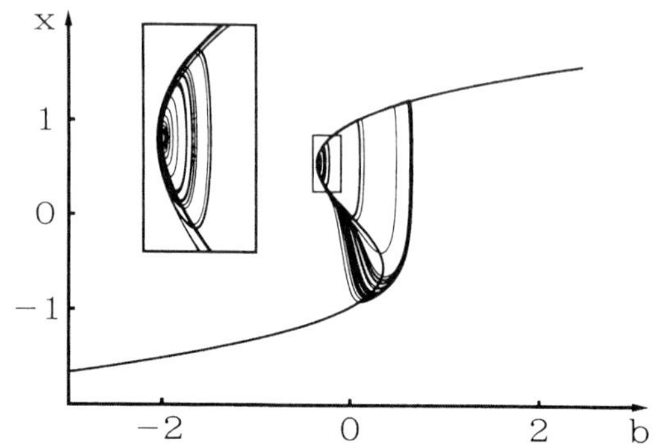

Figure 4.22 Chaotic orbit (x, b), which is the section of the orbit on M at $a = -0.94407$. The inset is the magnified figure of the small attracting region. The S-shaped curve (section of the manifold at $a = -0.94407$) is also shown.

the manifold. With further increase in d ($d \sim 0.55061$), the strange attractor suddenly exhibits a crisis, where the orbit rolls down intermittently onto the lower sheet of the manifold, as shown in Figs. 4.20–4.22, where $d = 0.550610737$. In Fig. 4.20, the chaotic orbits are drawn on the manifold. The projection of the orbits onto the (a, b)-plane is also shown in Fig. 4.21. A small attracting region is seen near the cliff (bifurcation set K of the upper sheet) and it grows suddenly due to the crisis. After switching, the projected orbit (a, b) crosses another edge of K by overshooting, and then it is repelled back to the small region again. The behavior is seen more clearly in Fig. 4.22, where the projected orbit (x, b) is superimposed on the section of the manifold at $a = -0.94407$. From the magnified figure of the small attracting region in Fig. 4.22, the orbit goes along the vicinity of the manifold whenever $\dot{x} < 0$. Readers who intend to study catastrophe theory in more detail may refer to [21–23].

4.3 Linear stability analysis

Consider some nonlinear ODEs:

$$\dot{\mathbf{X}} = \mathbf{F}(\{X_i\}; \{\mu_j\}), \tag{4.46}$$

$$\mathbf{X} = \{X_1, X_2, \ldots, X_i, \ldots, X_m\}, \tag{4.47}$$

$$\mathbf{F} = \{F_1, F_2, \ldots, F_i, \ldots, F_m\}, \tag{4.48}$$

where the system is characterized as an n-parameter family with the control parameters $\{\mu_j\}$ ($j = 1, 2, \ldots, n$), and the degrees of freedom are m with the order parameters X_i ($i = 1, 2, \ldots, m$). Under the steady-state condition $\dot{\mathbf{X}} = 0$, one assumes equilibrium points $\bar{\mathbf{X}}^{(j)} = \{X_1^{(j)}, X_2^{(j)}, \ldots, X_i^{(j)}, \ldots, X_m^{(j)}\}$, $j = 1, 2, \ldots,$ in the m-dimensional phase space. By linearizing (4.46) around an equilibrium point $\bar{\mathbf{X}}^{(j)}$, one obtains the following linearized equations:

$$\dot{\xi}^{(j)} = \begin{pmatrix} \frac{\partial F_1}{\partial X_1} & \frac{\partial F_1}{\partial X_2} & \cdots & \frac{\partial F_1}{\partial X_\mathrm{m}} \\ \frac{\partial F_2}{\partial X_1} & \frac{\partial F_2}{\partial X_2} & \cdots & \frac{\partial F_2}{\partial X_\mathrm{m}} \\ \vdots & & & \\ \frac{\partial F_\mathrm{m}}{\partial X_1} & \frac{\partial F_\mathrm{m}}{\partial X_2} & \cdots & \frac{\partial F_\mathrm{m}}{\partial X_\mathrm{m}} \end{pmatrix} \xi^{(j)} \tag{4.49}$$

$$\xi^{(j)} = \begin{pmatrix} \xi^{(j)}_{X_1} \\ \xi^{(j)}_{X_2} \\ \vdots \\ \xi^{(j)}_{X_\mathrm{m}} \end{pmatrix}, \tag{4.50}$$

where $\xi^{(j)} = X - \bar{X}^{(j)}$ and the Jacobian matrix $[\partial F/\partial X]$ is taken at the jth equilibrium point. Let $\mathbf{X}_0$ be an initial value slightly deviated from $\bar{\mathbf{X}}^{(j)}$ such that $\mathbf{X}_0 = \bar{\mathbf{X}}^{(j)} + \xi_0^{(j)}$, then $\xi^{(j)}(t)$ behaves as $\xi^{(j)}(t) = \xi_0^{(j)} \exp(\lambda t)$. By inserting $\xi^{(j)}(t)$ into (4.49) one obtains the determinant:

$$\begin{vmatrix} \frac{\partial F_1}{\partial X_1} - \lambda & \frac{\partial F_1}{\partial X_2} & \cdots & \frac{\partial F_1}{\partial X_m} \\ \frac{\partial F_2}{\partial X_1} & \frac{\partial F_2}{\partial X_2} - \lambda & \cdots & \frac{\partial F_2}{\partial X_m} \\ \vdots & & & \\ \frac{\partial F_m}{\partial X_1} & \frac{\partial F_m}{\partial X_2} & \cdots & \frac{\partial F_m}{\partial X_m} - \lambda \end{vmatrix} = 0. \tag{4.51}$$

Equation (4.51) is a polynomial of the order m of λ. By solving λ, one can judge whether a specified equilibrium point $\bar{\mathbf{X}}^{(j)}$ is stable or unstable. When all of the real parts of λ_is $(i = 1, 2, \ldots, m)$ are negative, $\bar{\mathbf{X}}^{(j)}$ is stable, while it is unstable if at least one of the real parts of the λ_is is positive. The linear stability analysis is easily done for $m = 2$, as exemplified below.

Example 1. Pendulum without driving force

Consider (1.20) without driving force:

$$\dot{x} = y, \tag{4.52}$$

$$\dot{y} = -\gamma y - \omega_0^2 \sin x. \tag{4.53}$$

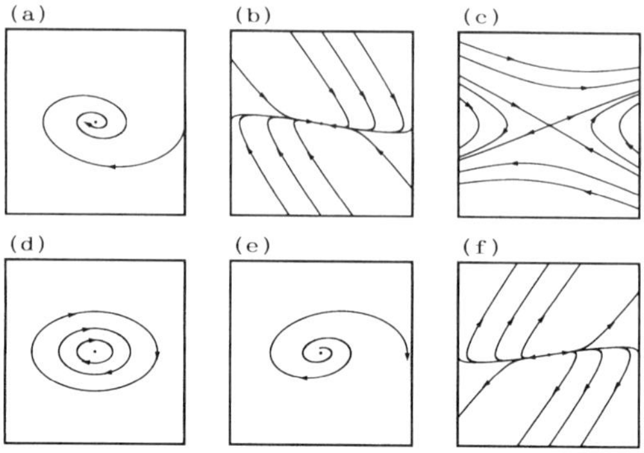

Figure 4.23 Vector flows at an equilibrium point. (a) Stable spiral, (b) stable focus, (c) unstable saddle, (d) center, (e) unstable spiral, and (f) unstable focus.

By putting $\dot{x} = \dot{y} = 0$, one obtains the equilibrium points $(\bar{x}, \bar{y}) = (\pm n\pi, 0)$, $n = 0, 1, 2, \ldots$. The Jacobian matrix is given by:

$$J = \begin{pmatrix} 0 & 1 \\ -\omega_0^2 \cos(n\pi) & -\gamma \end{pmatrix}, \tag{4.54}$$

Then, from (4.51):

$$\lambda(\lambda + \gamma) + \omega_0^2 \cos(n\pi) = 0, \tag{4.55}$$

$$\lambda = \begin{cases} \dfrac{-\gamma \pm \sqrt{\gamma^2 - 4\omega_0^2}}{2} & \text{for } n \text{ even}, \quad (4.56) \\[2ex] \dfrac{-\gamma \pm \sqrt{\gamma^2 + 4\omega_0^2}}{2} & \text{for } n \text{ odd}. \quad (4.57) \end{cases}$$

In the presence of dissipation (damped case, $\gamma > 0$), the equilibrium points of $(\bar{x}, \bar{y}) = (\pm n\pi, 0)$ (n: even) are stable node [25] when $\gamma > 2\omega_0$, as shown in the vector flows of Fig. 4.23b, and they are stable spiral [25] with

$$\lambda_\pm = -\gamma/2 \pm i\sqrt{4\omega_0^2 - \gamma^2}/2$$

when $\gamma < 2\omega_0$, where the vector flows are shown by Fig. 4.23a.

The equilibrium points with n odd are **unstable saddle points** (Fig. 4.23c), where the two characteristic exponents $\lambda_\pm$ are negative and positive in (4.57). With $\gamma = 0$ (undamped case, conservative), the equilibrium points for n even are center ($\lambda_\pm = \pm i\omega_0$), and unstable saddle point for n odd. The vector flows of the unstable spiral in Fig. 4.23e or the unstable focus in Fig. 4.23f are found generally in many dynamical systems, but they are not allowed in the pendulum with either the damped case $\gamma > 0$ or the undamped case ($\gamma = 0$), because $Re(\lambda) \leq 0$).

Example 2. Duffing equation with driving force

The Duffing equation [25–27] with driving force is known to exhibit complex chaotic behaviors. Without the driving force, the Duffing equation which describes the natural motion of a particle in a double-well potential $V = x^4/4 - x^2/2$ is given by

$$\dot{x} = y, \tag{4.58}$$

$$\dot{y} = -\gamma y - x^3 + x, \tag{4.59}$$

where γ is the dissipation coefficient. Under the steady-state condition ($\dot{x} = \dot{y} = 0$), the equilibrium points $(\bar{x}, \bar{y})$ are found to be (0,0), (1,0) and (−1, 0). The Jacobian matrix is given by

$$J = \begin{pmatrix} 0 & 1 \\ -3\bar{x}^2 + 1 & -\gamma \end{pmatrix}. \tag{4.60}$$

From (4.51), the characteristic exponents are obtained as

$$\lambda = \frac{-\gamma \pm \sqrt{\gamma^2 - 4(3\bar{x}^2 - 1)}}{2}. \tag{4.61}$$

For the damped case ($\gamma \neq 0$), the equilibrium points (1,0) and (−1,0) are both stable spirals when $\gamma < 2\sqrt{2}$ (Fig. 4.23a; the particle starting from the initial point x_0 decays and rests at either $\bar{x} = -1$ or $\bar{x} = 1$ after damped oscillation), or stable node when $\gamma > 2\sqrt{2}$ (Fig. 4.23b; the particle decays into either $\bar{x} = -1$ or $\bar{x} = 1$ without damped oscillation). The equilibrium point (0,0) is the saddle point with $\lambda_\pm = -\gamma \pm \sqrt{\gamma^2 + 4}$. For the undamped case ($\gamma = 0$), the equilibrium points (1,0) and (-1,0) are the centers with $\lambda_\pm = \pm i\sqrt{2}$. An example of the chaotic behaviors for the damped case with the driving force will be given in section 4.6.

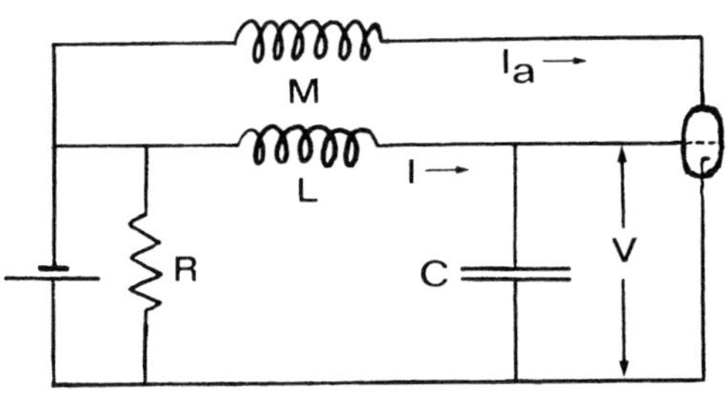

Figure 4.24 Triode circuit exhibiting a self-sustained oscillation. The oscillations are due to the nonlinearity in the triode characteristics [6].

Example 3. The Van der Pol Oscillator

It is well-known that a triode circuit (Fig. 4.24) exhibits relaxation oscillations, which were first observed by Van der Pol (1926). The equation of the Van der Pol oscillator is given by [6, 28]:

$$L\frac{\mathrm{d}I}{\mathrm{d}t} + R\,I + V = M\frac{\mathrm{d}I_\mathrm{a}}{\mathrm{d}t}, \tag{4.62}$$

where V and I are, respectively, the grid voltage and the current, I_a is the anode current, L the self-inductance, and M the mutual inductance. The triode characteristic is idealized as $I_\mathrm{a} = V - \frac{1}{3}V^3$. By the scale conversions [6],

$$x = V/\sqrt{1 - RC/M}; \quad \tilde{t} = t/\sqrt{LC}, \tag{4.63}$$

one obtains

$$\dot{x} = y, \tag{4.64}$$

$$\dot{y} = -\gamma(x^2 - 1)y - x, \tag{4.65}$$

where $\gamma = M/\sqrt{LC} - R\sqrt{C/L}$. With $\dot{x} = \dot{y} = 0$, a unique equilibrium point is apparently $(\bar{x}, \bar{y}) = (0, 0)$. The Jacobian matrix and the characteristic exponents are given by:

$$J = \begin{pmatrix} 0 & 1 \\ -1 & \gamma \end{pmatrix}, \tag{4.66}$$

$$\lambda = \frac{\gamma \pm \sqrt{\gamma^2 - 4}}{2}. \tag{4.67}$$

With $\gamma > 0$, the equilibrium point is an unstable node ($\gamma > 2$) or an unstable spiral ($\gamma < 2$). The orbit $\mathbf{X}(t) = (x(t), y(t))$ starting from a point near the equilibrium point (0,0), say $(\varepsilon, 0)$, diverges as $\|\mathbf{X}(t)\| \sim \varepsilon \exp(\lambda t)$. However, the linear stability analysis does not describe the future trajectory deviating far from the origin. The relaxation oscillation of the Van der Pol oscillator is described qualitatively as follows. Equations (4.64) and (4.65) are rewritten as:

$$\ddot{x} + \gamma(x^2 - 1)\dot{x} + x = 0. \tag{4.68}$$

The dynamical behavior is equivalent to the motion of a particle in a single-potential well of the potential function $x^2/2$ with the effective dissipation coefficient $\gamma_\mathrm{eff} = \gamma(x^2 - 1)$. With $|x| < 1$, the dissipation coefficient is negative, which means that the particle accepts the excess energy from the well. Therefore, the particle keeps going away from the rest point (0,0) until $|x| > 1$. When $|x| > 1$, γ_eff becomes positive and the particle loses the kinetic energy due to the positive dissipation energy. Thus, the particle is injected again inside $|x| < 1$. The process is repeated, followed by the relaxation oscillation.

4.4 Mathematical version of the instability model: an analog for the impact-ionization avalanche

Before covering the physical model of the impact ionization avalanche, this section will discuss simple dynamical behaviors of a mathematical model which shows an S-shaped $x - y$ characteristic (x, y: variables).

Consider a folded manifold **M** which is an algebraic set [1,2]:

$$\mathbf{M} = \{\mathbf{X} \in R^3 | z = -\gamma_1 \partial V/\partial x\}, \tag{4.69}$$

where $\mathbf{X} = (x, y, z)$ and $V(x, y)$ is a double-well potential function with its derivative $\partial V/\partial x$ being a cubic function of x. The manifold can be interconnected with a dynamical system with the variables x and y by a couple of ODEs:

$$\mathrm{d}x/\mathrm{d}t = -\gamma_1\{(x-\alpha)(x-\beta)(x-\gamma) + (y-a)[1 + c(x-\beta)]\}, \tag{4.70}$$

$$\mathrm{d}y/\mathrm{d}t = -\gamma_2\{y - y_0 + cy(x-\beta)\}, \tag{4.71}$$

where $\gamma_1, \alpha, \beta, \gamma$ $(\alpha < \beta < \gamma), a$ and y_0 are constants. Here, (4.69)–(4.71) are treated as a three-parameter family with the control parameters γ_2, c and y_0. The variables x and y correspond to the carrier density n and the electric field E in the impact-ionization avalanche in semiconductors, y_0 and c to the dc bias and the load line, respectively. As a function of y_0, the stationary solution of (4.70) and (4.71) gives an S-shaped $x - y$ characteristic, as shown in Fig. 4.25 where other parameter values except for y_0 are kept constant: $\alpha = 0.1, \beta = 0.2, \gamma = 0.8, \gamma_1 = 1, \gamma_2 = 0.11, a = 0.3$ and $c = 0.6$. A specified load line $y = y_0/\{1 + c(x - \beta)\}$ with $c = 0.6$, in Fig. 4.25, crosses the S-shaped curve by a single equilibrium $(\bar{x}, \bar{y})$ which is stable except for in the

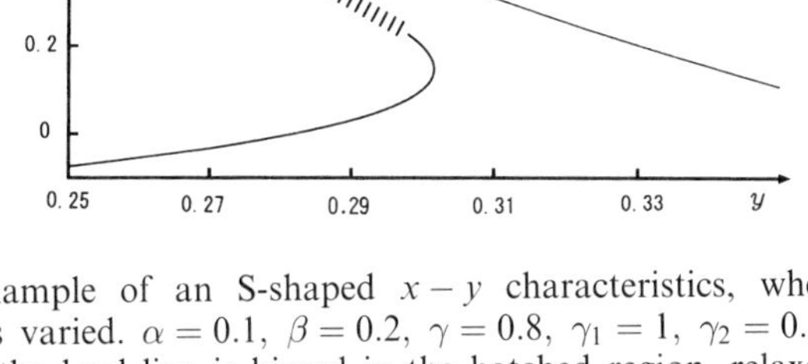

Figure 4.25 Example of an S-shaped $x - y$ characteristics, where the control parameter y_0 is varied. $\alpha = 0.1$, $\beta = 0.2$, $\gamma = 0.8$, $\gamma_1 = 1$, $\gamma_2 = 0.11$, $a = 0.3$ and $c = 0.6$. When the load line is biased in the hatched region, relaxation oscillation occurs.

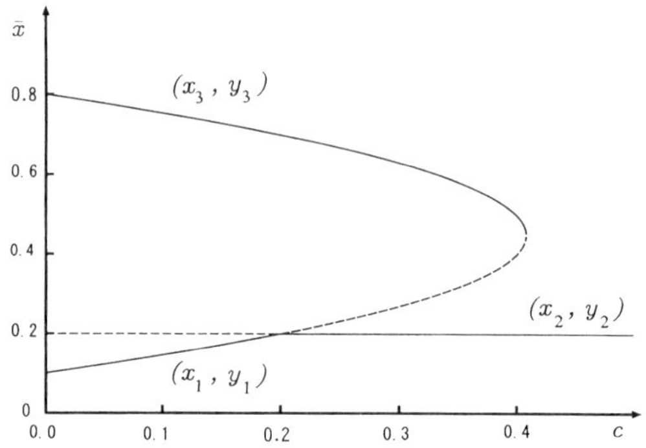

Figure 4.26 Diagram of $\bar{x}$ for the equilibrium points $(x_1, y_1), (x_2, y_2)$ and (x_3, y_3) as a function of c. Dashed lines indicate the unstable saddle points. At $c = 0.40833, x_1$ coincides with x_3. For $c > 0.40833$, only an equilibrium point (x_2, y_2) survives. y_0 was fixed at the value of $y_0 = a = 0.3$.

hatched region. When the load line is biased in the hatched region, relaxation oscillation occurs.

A special case is $y_0 = a = 0.3$, for which one can investigate how the equilibrium points vary as a function of c (Fig. 4.26). For the bias condition of the load line with $0 < c < (\alpha - \gamma)^2/4y_0 = 0.40833$, the steady-state condition ($\dot{x} = \dot{y} = 0$) gives three equilibria (x_i, y_i) $(i = 1, 2, 3)$ in the S-shaped region $(x_1 < x_2 < x_3)$; (x_1, y_1), (x_2, y_2) and (x_3, y_3), where $y_{1,3} = y_0/\{1 + c(x_{1,3} - \beta)\}, x_2 = \beta$ and $y_2 = y_0$. Therefore, one of the equilibria (x_2, y_2) is fixed at (β, y_0) for any values of c. By varying c, the equilibrium points (x_1, y_1) and (x_3, y_3) vary as

$$x_{1,3} = \frac{\alpha + \gamma \pm \sqrt{(\alpha - \gamma)^2 - 4cy_0}}{2}, \tag{4.72}$$

$$y_{1,3} = y_0/\{1 + c(x_{1,3} - \beta)\}. \tag{4.73}$$

Especially with $c = 0$, the equilibria are $(\alpha, y_0), (\beta, y_0)$ and (γ, y_0), as apparent from (4.70)–(4.73). Note that $y_0 = a$. $x_1(x_3)$ increases (decreases) by increasing c and collapse at $c = (\alpha - \gamma)^2/4y_0$. The equilibrium point $x_1(x_3)$ locates in the middle branch $(\partial y/\partial x < 0)$ of the S-shaped $x - y$ curve in Fig. 4.25, for $c > c^* = 0.10449$ $(c > c^{**} = 0.34736)$ (note: $x_1 = \beta$ with $c = 0.2$). The middle branch with the derivative $\partial y/\partial x < 0$ corresponds to the region of the negative differential conductivity (NDC) of the S-shaped I–V curve in the impact ionization avalanche. By linearizing (4.70), (4.71), one obtains the Jacobian matrix (J_{ij}) with the elements:

$$J_{11} = -(x_i - \beta)(x_i - \gamma) - (x_i - \alpha)(x_i - \gamma) - (x_i - \alpha)(x_i - \beta), \tag{4.74}$$

$$J_{12} = 1 + c(x_i - \beta), \tag{4.75}$$

$$J_{21} = -\gamma_2 c y_i, \tag{4.76}$$

$$J_{22} = -\gamma_2[1 + c(x_i - \beta)] \quad (i = 1, 2, 3). \tag{4.77}$$

The characteristic exponents λ are numerically calculated by

$$\lambda = \frac{J_{11} + J_{22} \pm \sqrt{(J_{11} + J_{22})^2 - 4(J_{11}J_{22} - J_{12}J_{21})}}{2}. \tag{4.78}$$

By (4.78), one can investigate the linear stability analysis for each equilibrium point $\bar{\mathbf{X}} = (x_i, y_i)$ $(i = 1, 2, 3)$ in the parameter space $\{c, \gamma_2\}$, e.g., in Figs. 4.27 and 4.28.

For $\bar{\mathbf{X}} = (x_2, y_2) = (\beta, y_0)$, note: $J_{11} = (\beta - \alpha)(\gamma - \beta)$, $J_{12} = 1$, $J_{21} = -\gamma_2 c y_0$, $J_{22} = -\gamma_2$. It is easily shown that the equilibrium point is a saddle point if $J_{11}J_{22} < J_{12}J_{21}$ or $c < (\beta - \alpha)(\gamma - \beta)/y_0 = 0.2$. The saddle point is shown by the broken line in Fig. 4.26. With $c > 0.2$, the equilibrium point is stable if $J_{11} + J_{22} < 0$ or $\gamma_2 > (\beta - \alpha)(\gamma - \beta)$ and unstable if $\gamma_2 < (\beta - \alpha)(\gamma - \beta)$. The spiral condition is given by $c > \{\gamma_2 - (\beta - \alpha)(\gamma - \beta)\}^2/4\ \gamma_2 y_0$. The results for the boundaries are shown in Fig. 4.27. For $\bar{\mathbf{X}} = (x_1, y_1)$, the equilibrium point is a saddle point in the same parameter space as in Fig. 4.27 $(0.2 < c < 0.40833$; $0 < \gamma_2 < 0.5)$. For $\bar{\mathbf{X}} = (x_3, y_3)$, the boundaries are shown in Fig. 4.28.

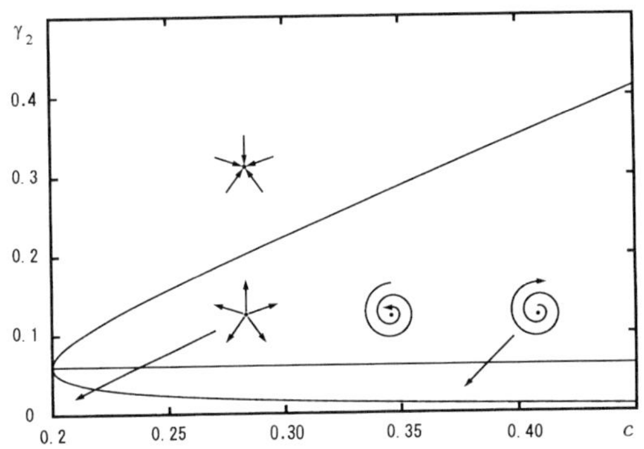

Figure 4.27 Boundaries for the stability of an equilibrium point $(x_2, y_2) = (\beta, y_0)$ in the parameter space (c, γ_2), where $\alpha = 0.1, \beta = 0.2, \gamma = 0.8, \gamma_1 = 1, y_0 = 0.3$ and $a = 0.3$. The horizontal line $\gamma_2 = (\beta - \alpha)(\gamma - \beta) = 0.06$ divides the parameter space into the stable and unstable regions (note: stable spiral for $\gamma_2 > 0.06$, and unstable spiral for $\gamma_2 < 0.06$, respectively).

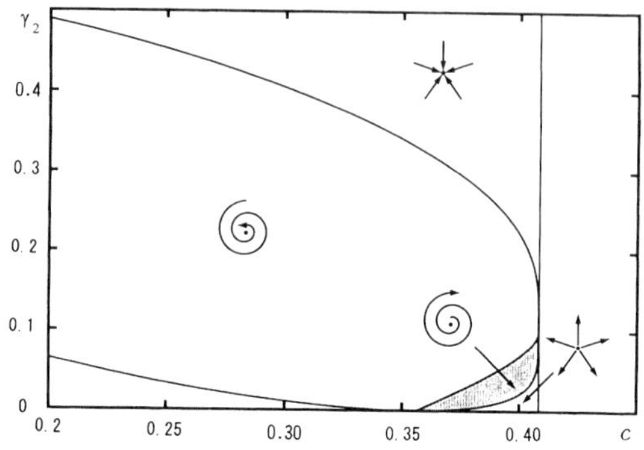

Figure 4.28 Boundaries for the stability of an equilibrium point (x_3, y_3) in the parameter space (c, γ_2), with the same parameters as those in Fig. 4.27. Note that the equilibrium point exists for $c < 0.40833$. The gray region indicates the unstable spiral.

With $c > 0.40833$, only an equilibrium point (β, y_0) survives. As shown in Fig. 4.27, the equilibrium point is destabilized for $\gamma_2 < (\beta - \alpha)(\gamma - \beta)$. Since vector flows starting from the equilibrium point are linked to no other equilibrium points, the orbit initiated from an arbitrarily initial point forms a **limit cycle** [25–27] in the phase space.

The saddle point plays an important role for dynamical behaviors in the presence of multiple equilibrium points. Global features of the vector flows are clarified numerically, not by linear stability analysis. Several types of vector flow related to the saddle point are classified in Fig. 4.29. The saddle point has two main axes in the phase space corresponding to the two characteristic exponents $\lambda_1 (< 0)$ and $\lambda_2 (> 0)$; one of the axes is a pair of the separatrices [29, 30] or the stable orbit (manifold) W_s with λ_1 and the other is the unstable orbit (manifold) W_u with λ_2. When W_s extends to infinity, it is **global**, otherwise it is **local**. A set of the initial points $\{\mathbf{X}_A\}$ which fall asymptotically onto an attractor **A** is called **basin of attraction** [30]. If there exist two basins of attraction $\{\mathbf{X}_A\}$ and $\{\mathbf{X}_B\}$, the separatrices of the saddle point divide the two regions; i.e., W_s is the basin boundary. A typical example of the bistability is schematically shown in Fig. 29a, where the unstable orbits W_u are connected to the stable focus (stable fixed points $\mathbf{X}_2$ and $\mathbf{X}_3$) and W_s divide two regions of the bistability. If the equilibrium points $\mathbf{X}_2$, $\mathbf{X}_3$ are delocalized to form unstable focus or spiral, and linked to W_s (W_s is local; see Fig. 4.29b), there is no basin boundary and the orbits asymptotically form a limit cycle. In Fig. 4.29c, one of the W_s winds and terminates to an unstable limit cycle by the time reversal. The unstable limit cycle is the basin boundary within which (see the gray region) there is a stable spiral $\mathbf{X}_2$. Starting from the outer part of the boundary, the vector

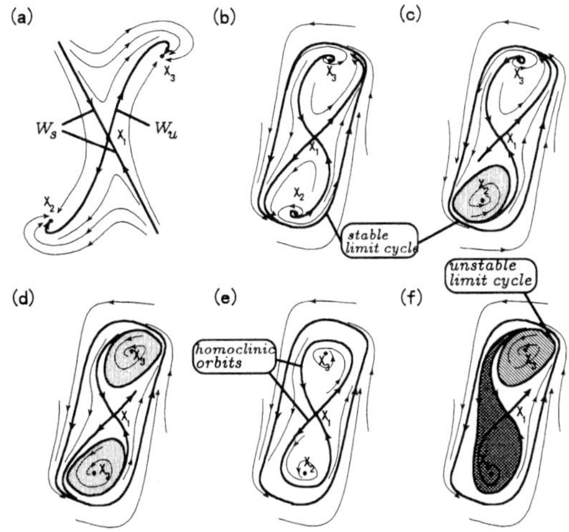

Figure 4.29 Vector flows expected from the linear stability analysis for the dynamical system in (4.70) and (4.71). (a) Vector flows for the bistability, where X_2 and X_3 are the stable focus, and X_1 is the unstable saddle point ($c > 0.2$, see Fig. 4.5). (b) X_2 and X_3 become the unstable spirals and a stable limit cycle is formed, for $0.36 < c < 0.4$ and $\gamma_2 < 0.06$. (c) One of the separatrices winds and terminates to an unstable limit cycle by the time reversal, and the remainder is connected to an unstable spiral X_3. Note that X_2 is the stable spiral. (d) Both of the separatrices wind and terminate to unstable limit cycles by the time reversal. Inside each of the unstable limit cycles (gray region) there is a stable spiral (X_2 or X_3). (e) A special case forming homoclinic orbits. (f) Both of the separatrices wind and terminate to the same unstable limit cycle. (d)–(f) show coexistence of bistability and stable limit cycle.

flows approach asymptotically to a limit cycle. The remainder of the separatrices in Fig. 4.29c is connected to an unstable spiral $\mathbf{X}_3$. Figure 4.29d shows the case in which both of the separatrices W_s are terminated to the unstable limit cycles. The special case is shown in Fig. 4.29e, where the separatrices form the homoclinic orbits via the homoclinic tangency with the unstable manifold W_u. In Fig. 4.29f, the separatrices are both connected to the same unstable limit cycle, and the basins of attraction are indicated by the hatched and cross-hatched regions. From Fig. 4.29d–f, the orbits exhibit bistability and periodic oscillation (limit cycle), depending on the initial conditions. The numerical results for Fig. 4.29c, e, f are given in Figs. 4.30–4.32.

With the same parameter values as those in Fig. 4.31, one can investigate either the bistability behavior or the oscillatory behavior as a function of y_0, depending on the initial conditions. Figure 4.33 shows the oscillatory behavior in (a), and the bistability behavior in (b), where the initial conditions are: (a) $x = 0$, $y = 0$ and (b) final values of x and y at the

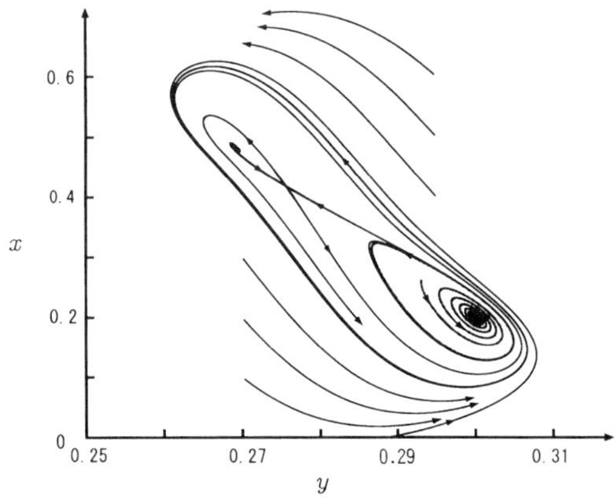

Figure 4.30 Numerical results of the vector flows for $c = 0.405$, and $\gamma_2 = 0.068$ in (4.70) (4.71), corresponding to Fig. 4.29c. Other parameters are $\gamma_1 = 1.0, \alpha = 0.1, \beta = 0.2$ and $\gamma = 0.8$.

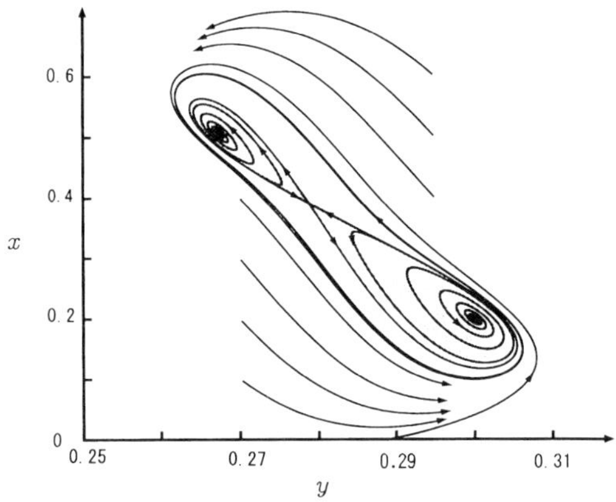

Figure 4.31 Numerical results of the vector flows for $c = 0.3972$, and $\gamma_2 = 0.072$, corresponding to Fig. 4.29e. Other parameters are $\gamma_1 = 1.0, \alpha = 0.1, \beta = 0.2$ and $\gamma = 0.8$.

previous step of the control parameter y_0, respectively. The ODEs (4.70) and (4.71) were integrated numerically by the Runge–Kutta–Gill method [31]. Values of x were plotted at each step of y_0. In Fig. 4.33a, the stationary state becomes unstable and oscillates spontaneously in the range $0.2997 < y_0 < 0.30015$, where vertical bars indicates the dynamical range. In Fig. 4.33b, taking the final values of x, y at the previous step of y_0 as initial values at the next step of y_0, hysteresis is observed with no spontaneous oscillation (vertical bars are due to initial transients at the turning points). The results are explained by the existence of the basin boundary described in Figs. 4.30–

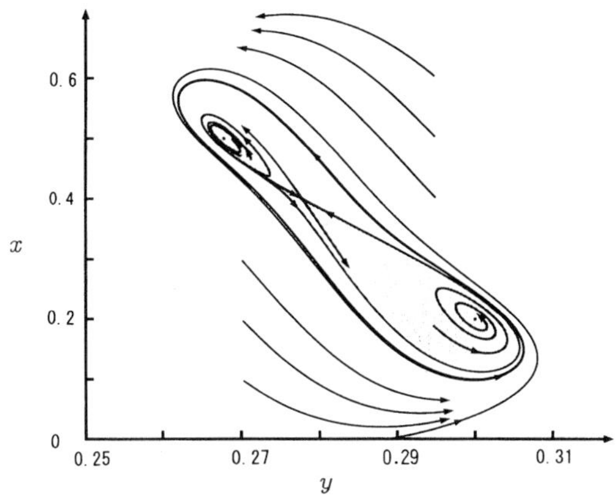

Figure 4.32 Numerical results of the vector flows for $c = 0.4.$ and $\gamma_2 = 0.073$, corresponding to Fig. 4.29f. Other parameters are $\gamma_1 = 1.0, \alpha = 0.1, \beta = 0.2$ and $\gamma = 0.8$.

4.32. By adding external perturbations, switching between the oscillatory state and the bistability state is possible.

4.5 A physical model of the impact-ionization avalanche

Basically, the dynamical behaviors of the impact ionization avalanche are similar to those expected from the ODEs (4.70), (4.71). Here, an approximated model A for the one-level impact-ionization avalanche is presented:

$$M = \{\mathbf{X} \in R^3 | z = -\gamma_1 \partial V_E(\nu)/\partial \nu\}, \tag{4.79}$$

where $\mathbf{X} = (\nu, E, z)$, $z = \dot{\nu}$ and $V_E(\nu)$ is the potential function (3.85). The ODEs are given, as in (3.82)–(3.84), by

$$\dot{\nu} = -\gamma_1[\nu^3 + (\kappa_1 + \kappa_3 - 1)\nu^2 - (\kappa_1 - \kappa_2 - C\kappa_3/(1 - C))\nu - \kappa_2], \tag{4.80}$$

$$\kappa_1 = A \exp(-B/E), \tag{4.81}$$

$$\dot{E} = -\gamma_2[E - E_0 + c\nu E], \tag{4.82}$$

where the notations are given previously and the electron mobility is assumed to be constant. Hereafter, the tildes used in chapter 3 are dropped for simplicity. The manifold $\mathbf{M}$ in (4.79) is constructed in the phase space (ν, E, z) by calculating z simply as functions of ν and E in (4.80). With a fixed value of E, the solution z as a function of ν, e.g., Fig. 3.16, gives a

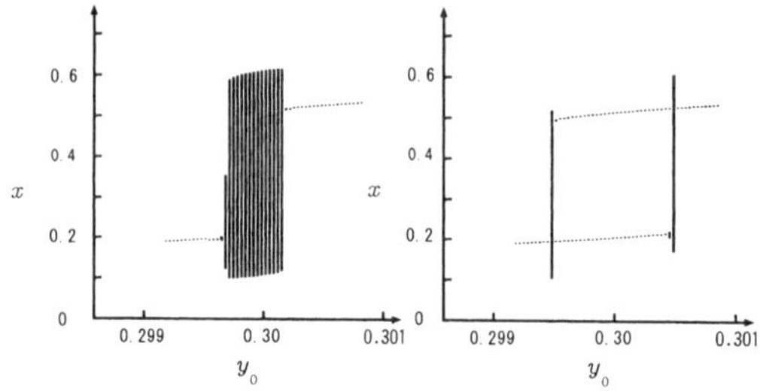

Figure 4.33 (a) Oscillatory behavior as a function of the control parameter y_0; initial condition: $x = 0, y = 0$. (b) Hysteretic behavior as a function of y_0. The final values of x, y at the previous step of y_0 are taken as the initial values at the next step of y_0.

section of **M**. Since $z = \dot{\nu}$ is the total generation rate of the carrier density, $z > 0$ (< 0) means generation (recombination). When the load line is biased in the hysteresis region with the voltage-controlled bias condition [$c = 0$ in (4.82)], there are three equilibria ν_1, ν_2 and ν_3, as shown in Fig. 4.34. At the breakdown field E_b or the holding field E_h, where $\nu_1 = \nu_2$ at $E = E_b$ or $\nu_2 = \nu_3$ at $E = E_h$, the **critical slowing down** [2] is seen in Fig. 4.35. Namely, at the critical points, the total generation rate z is zero. Initially deviated from the critical points by an extremely small perturbation, the orbit starts to switch on at E_b (or switch off at E_h) very slowly at the initial stage. The concept of critical slowing down is quite important in order to interpret the chaotic response in the impact-ionization avalanche, which is described in the next section.

The dynamical behaviors expected from the physical model are essentially the same as those in the mathematical model. Here, some important aspects are summarized and related to experimental observations [4, 19, 32, 33].

(a) With the voltage-controlled bias condition (c: small), hysteretic behavior or bistability is likely to be observed.
(b) With the current-controlled bias condition (c: large), a singly valued solution $\bar{\nu}$ in the negative differential resistance (NDR) region is either stable or unstable (forming a limit cycle), depending on the ratio $\gamma_1/\gamma_2(\gamma_1/\gamma_2 < \mathrm{c}^*$ or $\gamma_1/\gamma_2 > c^*$, respectively; for c^* see below). The latter is typical in semiconductors, known as **spontaneous relaxation oscillation** [6] or **self-sustained oscillation**.
(c) With the intermediate bias condition of c, one can expect a complex situation similar to Fig. 4.33.
(d) Spontaneous chaos is not present as far as the experimental system of the impact ionization avalanche is modeled by the ODEs (4.80), (4.82), where the degrees of freedom are $d_f = 2$.

Remark (b) is important theoretically since many researchers might misread that the system exhibits a spontaneous oscillation whenever the inclined

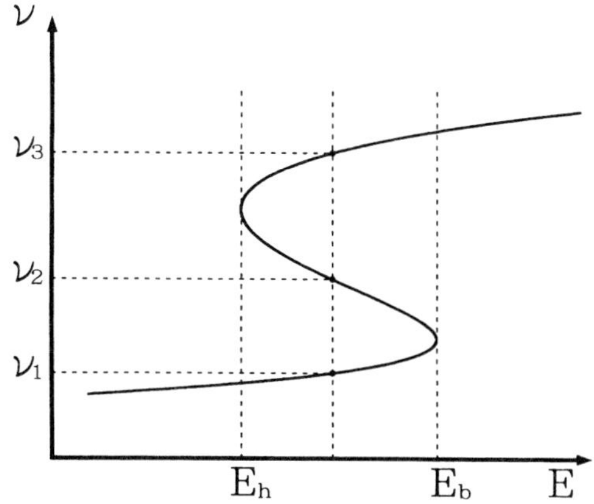

Figure 4.34 $\nu - E$ characteristic.

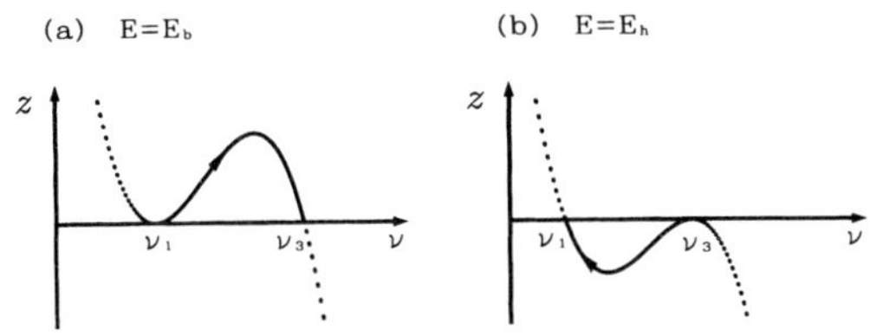

Figure 4.35 z as a function of ν(a) at $E = E_\text{b}$ and (b) $E = E_h$.

load line is biased in the NDR region. Many oscillatory behaviors have been observed in Ge [34, 35]. However, the oscillatory behavior depends on the material nature (the dielectric relaxation time τ_r) and the circuitry condition. For the inclined load line, linear stability analysis around $\bar{\nu}$ (unique steady state solution) can be carried out by the characteristic exponents:

$$\lambda = \frac{a + d \pm \sqrt{(a+d)^2 - 4(ad - bc')}}{2}, \tag{4.83}$$

where

$$\begin{aligned}
a &= -\gamma_1[3\bar{\nu}^2 + 2(\kappa_1 + \kappa_3 - 1)\bar{\nu} - (\kappa_1 - \kappa_2 - C\kappa_3/(1-C))] > 0,\\
b &= -\gamma_1\bar{\nu}(\bar{\nu} - 1)B\kappa_1/\bar{E}^2 > 0,\\
c' &= -\gamma_2 c\bar{E} < 0,\\
d &= -\gamma_2(1 + c\bar{\nu}) < 0.
\end{aligned}$$

Let us confine the present discussion within the condition $ad > bc'$. The real part of the exponents is positive if $a + d > 0$, so that the sufficient condition of γ_1/γ_2 for the spontaneous relaxation oscillation is given by:

$$\gamma_1/\gamma_2 > c^* = -(1 + c\bar{\nu})[3\bar{\nu}^2 + 2(\kappa_1 + \kappa_3 - 1)\bar{\nu} - (\kappa_1 - \kappa_2 - C\kappa_3/(1 - C))]^{-1}. \tag{4.84}$$

When $\gamma_1/\gamma_2 < c^*$, however, the system shows no spontaneous oscillations (the equilibrium point is either focus or spiral for any given control parameters of c and E_0). In other words, when the relaxation time $1/\gamma_2$ of the electric field is relatively fast, the system may show no spontaneous oscillation. Figure 4.36a–c shows an example of the vector flows on the manifold **M** numerically obtained by the ODEs (4.80) and (4.82), where $\gamma_1 = 250$, $\gamma_2 = 1.0$, $E_0 = 14.85$ in (a), $\gamma_1 = 250$, $\gamma_2 = 1.0$, $E_0 = 14.5$ in (b) and $\gamma_1 = 110$, $\gamma_2 = 1.0$, $E_0 = 14.55$ in (c), and the load line is biased in the NDR region with the intermediate bias condition of c ($c = 1.4$). In Fig. 4.36, the S-shaped $\nu - E$ curve is given by the section $z = 0$ (solid line) and the dotted lines are the vector flows with $E =$ const. With $\gamma_1 = 250$, the vector flows starting from an arbitrarily initial point form a limit cycle when the load line is biased in the NDR region (e.g., $E_0 = 14.5$, Fig. 4.36b), while they show damped oscillations when E_0 is biased on the upper branch or lower branch of the ν–E curve (e.g., $E_0 = 14.85$, Fig. 4.36a). By decreasing γ_1, the limit cycle becomes small (Fig. 4.36c, $\gamma_1 = 110$). With further decrease in $\gamma_1 (< 103)$, the limit cycle disappears, showing no spontaneous oscillation.

In the ODEs (4.80), (4.82), there are no chaotic solutions for the following reason. Since the order parameters are ν and E, the degrees of freedom are $d_f = 2$. Since the ODEs are deterministic, the uniqueness theorem of the solutions is required: **the orbits embedded in the d_f-dimensional phase space never intersect each other, otherwise the deterministic property of the ODEs is broken at the intersection point** [26]. With $d_f = 2$, the orbits in the (ν, E) phase space simply form a limit cycle or a fixed point. Namely, chaos-like orbits are never constructed without intersections. If the system is described with $d_f \geq 3$, as in the Schöll model ($d_f = 3$) [4], chaotic solutions can be found without any intersections of the orbits such that the orbits form a **strange attractor** (fractal set with a self-similar structure in the phase space).

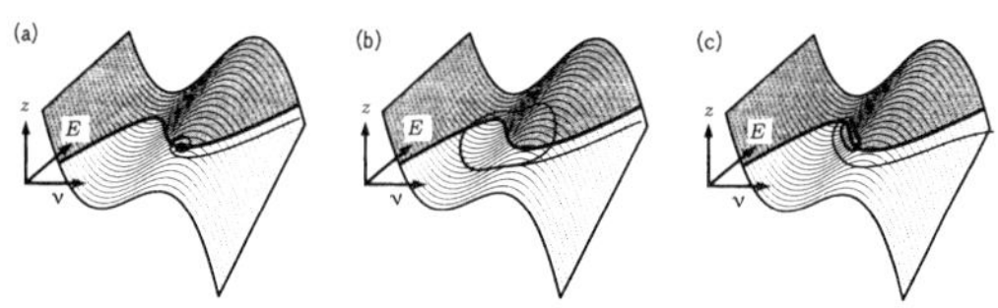

Figure 4.36 Transient trajectory on the manifold M numerically obtained by the ODEs of (4.80) and (4.81). The parameters are (a) $\gamma_1 = 250$, $\gamma_2 = 1.0$, $E_0 = 14.85$; (b) $\gamma_1 = 250$, $\gamma_2 = 1.0$, $E_0 = 14.5$; (c) $\gamma_1 = 110$, $\gamma_2 = 1.0$, $E_0 = 14.55$.

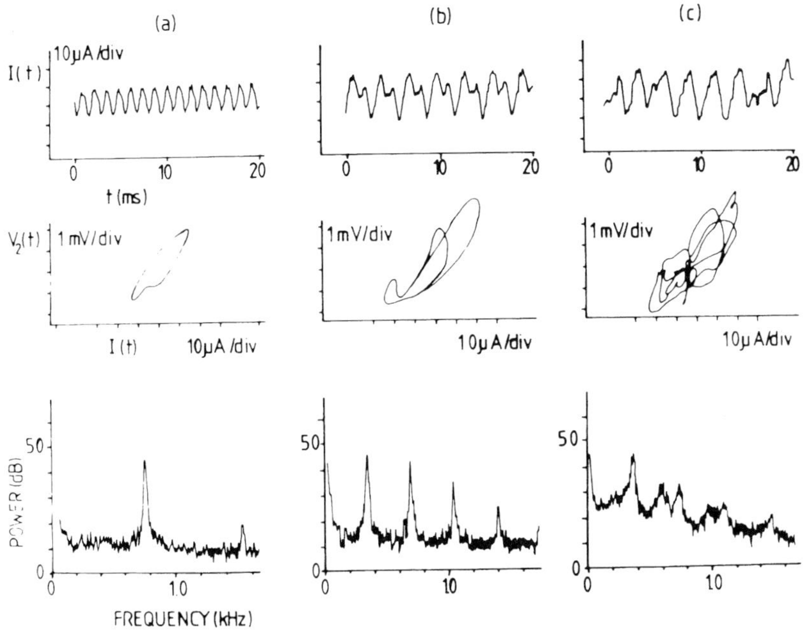

Figure 4.37 Quasi-Ruell–Takens–Newhouse scenario observed with increasing electric fields of about 35 mV/cm in p-Ge at 4.2 K. Top: waveform of the current $I(t)$; middle: phase portraits; bottom: power spectra of $I(t)$ [35].

Many experimental observations [4, 11, 34–39] have shown that spontaneous chaos is often observed under external perturbations (transverse magnetic field [34, 35, 39], longitudinal magnetic field [11], photoexcitation [36]). Several examples have been given in p-Ge by Röhricht *et al.* [35], Ni-compensated Ge by Bumeliène *et al.* [38] and semi-insulating GaAs by Maracas *et al.* [37], as shown in Figs. 4.37 and 4.38. Röhricht *et al.* have mainly investigated chaotic behaviors in the postbreakdown regime caused by impact ionization of shallow acceptors at 4.2 K [35]. The result in Fig. 4.37, measured with zero magnetic field [35], indicates a bifurcation from periodic oscillation with a fundamental frequency f_0 of 0.78 kHz (Fig. 4.37a) to chaos (Fig. 4.37c) via a quasi-Ruelle–Takens–Newhouse scenario [40], where the top view of Fig. 4.37 indicates the current waveforms, the middle the phase portrait of voltage vs current and the bottom the Fourier power spectra in Fig. 4.37. In semi-insulating GaAs [37], chaotic behaviors occur due to the impact ionization of deep defect center(s), presumably related to EL2 centers. The frequency of oscillation is a strong function of temperature, varying by more than four orders of magnitude over the

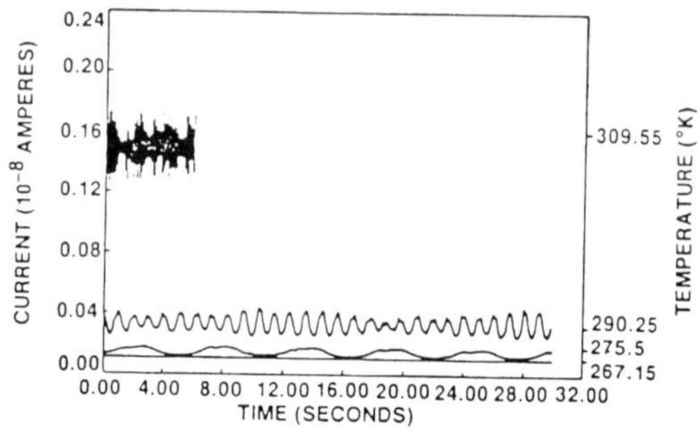

Figure 4.38 Low-frequency oscillations in semi-insulating GaAs. Various waveforms of the current are shown as a function of temperature for a 60-μm-long sample and for a field of 1.7 kV/cm [37].

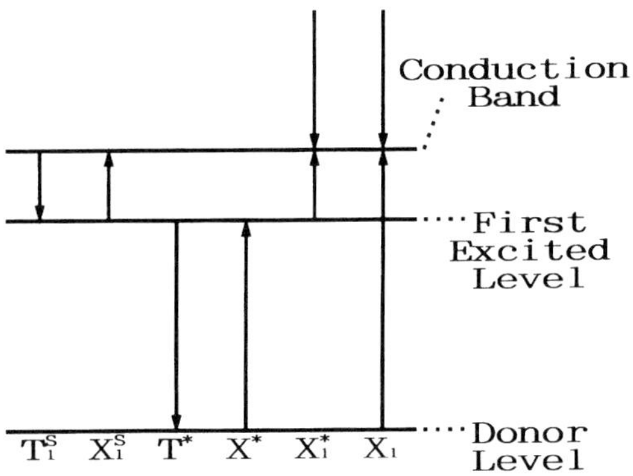

Figure 4.39 Two-level impact-ionization model [4].

temperature range from 250 to 320 K (Fig. 4.38). Fourier transform of the data reveals two main frequency components, which correspond to two deep levels. As a mechanism causing the periodic and chaotic oscillations, Maracas *et al.* have attributed the formation of a space charge region (deep level domain) to the capture of carriers in deep levels. Once formed, the domains traverse the sample length and are collected at the contact [37]. The oscillating mechanism is strongly supplemented by the two-level impact-ionization model by Schöll [4], as shown schematically in Fig. 4.39. The chaotic phenomena described here cannot be explained by the approximated physical model (4.80), (4.82). To the author's knowledge, however, the impact-ionization avalanche by shallow donors in n-GaAs [19, 32, 33] and n-InSb [41] does not reveal spontaneous chaos under unperturbed experimental conditions.

In conclusion, the approximated impact-ionization model (4.80), (4.82) explains phenomenologically how spontaneous relaxation oscillation occurs. The model is consistent with the experimental fact that spontaneous chaos is absent for the impact-ionization avalanche by shallow donors in n-GaAs and n-InSb under the unperturbed regime.

4.6 Periodically driven chaos

The previous section described a simple mechanism of spontaneous relaxation oscillation by the approximated impact-ionization model and noted that the spontaneous chaos is unlikely to happen owing to lack of degrees of freedom ($d_f = 2$). This section will demonstrate that chaotic behaviors appear very easily by applying a dc + ac bias in the NDR region. E_0 in (4.82) is replaced by $E_0 + E_{\rm ac}\sin(2\pi f_0 t)$ as

$$\dot{E} = -\gamma_2(E - E_0 - E_{\rm ac}\sin(2\pi f_0 t) + c\nu E). \qquad (4.85)$$

By applying the ac voltage, the degrees of freedom increase by one ($\mathrm{d_f} = 3$) since the time t is treated as an independent variable. The situation is similar to the Duffing equation [25–27], where a particle in the double-well potential shows no chaotic motion without a driving force. The amplitude $E_{\rm ac}$ and the magnitude of the driving frequency f_0 play important roles in the appearance of chaos [19]. The amplitude should be greater than half of the hysteresis width $\Delta\mathrm{E}$ of the S-shaped J–E curve; $2E_{\rm ac} \gtrsim \Delta E$. The driving frequency should be relatively high and comparable to the inverse relaxation time of the electric field, $\gamma_2 f_0 > 0.1 \times \gamma_2 (\gamma_2 = 1)$.

Figure 4.40 shows the vector flows on the manifold **M** in the periodically driven regime with the voltage-controlled bias condition ($c = 0$). If the driving frequency is extremely low ($f_0 \ll \gamma_2$), the trajectory exhibits hysteresis or quasi-static behavior, as shown in Fig. 4.40a, $f_0 = 0.01$ and $V_{\rm ac} = 0.8$. At the breakdown field or the holding field, critical slowing down can be seen because of the slow driving frequency. By increasing the driving frequency $f_0 = 0.1$ and $f_0 = 0.2$ in Fig. 4.40b, c, the trajectory softens and

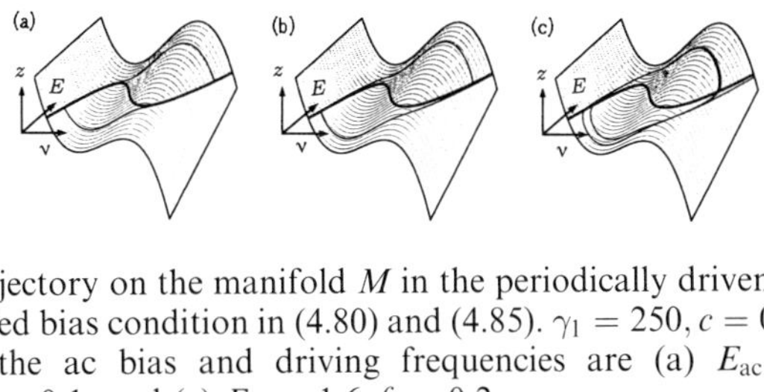

Figure 4.40 Trajectory on the manifold M in the periodically driven regime with the voltage-controlled bias condition in (4.80) and (4.85). $\gamma_1 = 250, c = 0, E_0 = 11.5$. The amplitudes of the ac bias and driving frequencies are (a) $E_{\rm ac} = 0.8, f_0 = 0.01$, (b) $E_{\rm ac} = 1.0, f_0 = 0.1$, and (c) $E_{ac} = 1.6, f_0 = 0.2$.

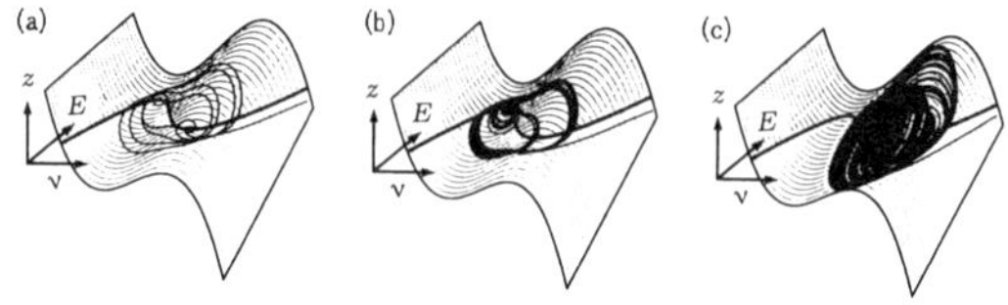

Figure 4.41 Trajectory on the manifold M in the periodically driven regime with the inclined bias condition. $\gamma_1 = 250, c = 1.4$. Other parameters are (a) $E_0 = 14.5$, $E_{ac} = 0.5$, $T_0 = 100$; (b) $E_0 = 14.5$, $E_{ac} = 0.18$, $T_0 = 10$; (c) $E_0 = 15.15$, $E_{ac} = 2.8$, $T_0 = 1.36$, where $T_0 = 1/f_0$. Chaotic behaviors are seen in (b) and (c).

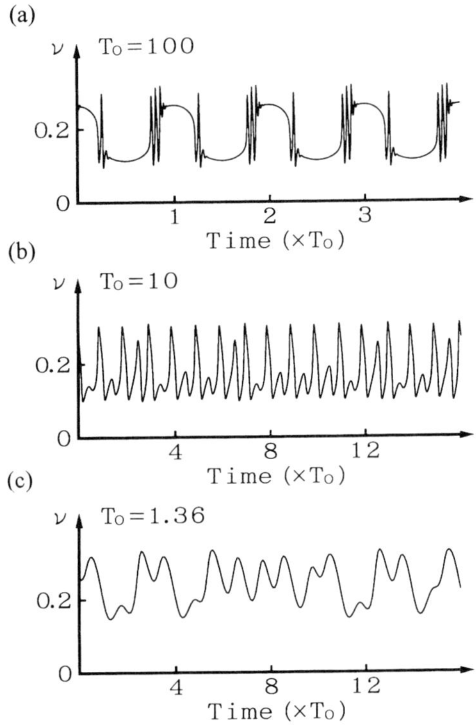

Figure 4.42 Carrier density as a function of time (in units of T_0). (a)–(c) correspond to Fig. 4.41a, c.

deviates from the ideal hysteretic behavior. With $f_0 > 0.1$, the electric field overshoots above the breakdown field (or below the holding field) during the critical slowing down.

For the inclined bias condition ($c > 0$), one can easily find chaotic behaviors, as shown in Figs. 4.41 and 4.42. With $f_0 = 0.01$, the trajectory in Fig. 4.41a looks like a heteroclinic orbit [25, 27, 42] (see also Fig. 4.42a), but it is soon found out that the oscillatory behavior arises from the self-sustained oscillation in the NDR region and that the driving force acts merely by shifting the applied electric field in a quasi-stationary manner, where the frequency of the self-sustained oscillation in Fig. 4.42a is much higher than f_0. No chaotic behaviors are observed. During each period of the driving, the inclined load line moves very slowly from the lower branch to the upper branch via middle branch in the S-shaped ν–E curve, and from the upper to the lower branches. The oscillation occurs when the load line is moving inside the middle branch. With $f_0 > 0.1$, the trajectories begin to show chaotic behaviors, as shown in Figs. 4.41b, c and 4.42b, c. In Fig. 4.42(b, c), the carrier density ν fluctuates in an irregular but deterministic manner. By varying one of the control parameters V_0, V_{ac} or f_0, rich structures of the bifurcation routes to chaos (Feigenbaum scenario [40], Pomeau–Manneville scenario [40], Ruelle–Takens–Newhouse scenario [40] are realized in the model simulation. Details of the bifurcations are described in Chapter 5.

Some fundamental properties of chaos can be investigated by reconstructing the chaotic orbits in the three-dimensional (3D) phase space rather than using the chaotic orbits lying on the 2D manifold **M** in Fig. 4.41. One idea is shown schematically in Fig. 4.43. As described previously, time t is an independent variable which appears explicitly in the ODEs via $E_{ac}\sin(2\pi f_0 t)$. Chaotic trajectory in the vector space $R^3 = \{t, E, \nu\}$ forms an open flow. The vector points (E, ν) at $t_n = n/f_0 (n = 0, 1, 2, \ldots)$ form a Poincaré surface of section or simply a Poincaré map, which is a set of the intersecting points with each plane of $t = t_n$. Figure 4.44 shows a typical result of the Poincaré map of the chaotic trajectory in Fig. 4.42c. By enlarging the squared portion successively, a **self-similar structure** appears, as shown in Fig. 4.44(c,d). The enlarged portion contains finer and finer structures of the original one. The vector points on the Poincaré surface of

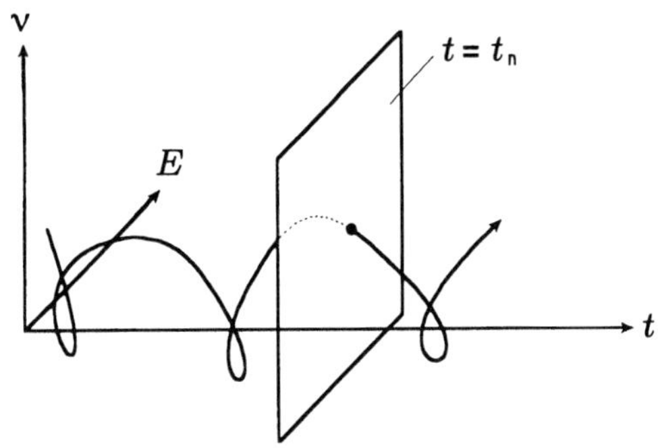

Figure 4.43 Construction of the Poincaré surface of section at $t = t_n$.

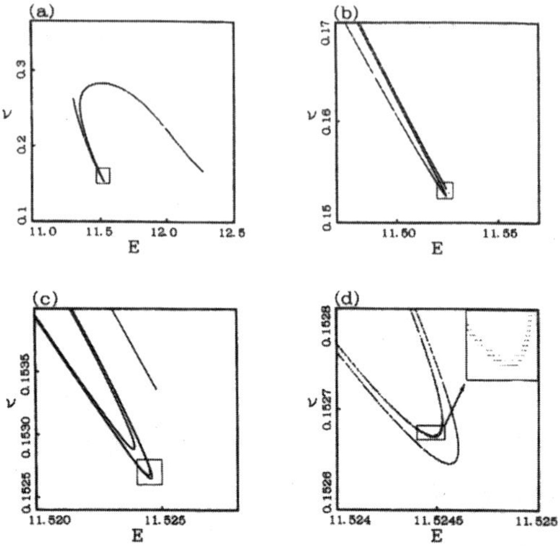

Figure 4.44 Poincaré surface of section at $t = t_n$, for the model simulation in (4.80) and (4.85). $\gamma_1 = 250$, $c = 1.4$, $E_0 = 15.15$, $E_{ac} = 2.8$, and $f_0 = 100/136$. Squared regions are enlarged successively in (b)–(d).

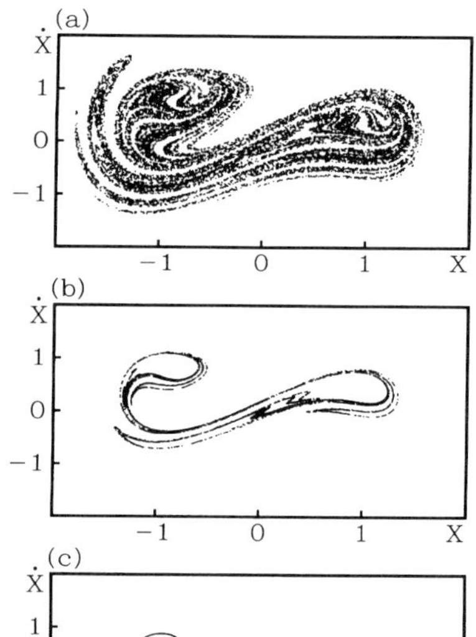

Figure 4.45 Poincaré surface of section for the Duffing equation. The parameters are (a) $F_0 = 0.45$, $\gamma = 0.04$, (b) $F_0 = 0.5$, $\gamma = 0.2$, and (c) $F_0 = 0.7$, $\gamma = 0.5$. $\omega = 1.5$. The attractor in (a) is a fat fractal.

section never coincide with each other since the chaotic orbits never intersect with each other in the 3D phase space (remember the uniqueness theorem of the solution). The self-similar structure in Fig. 4.44 has a Hausdorff dimension or capacity dimension d which is non-integer (fractal) [43, 44].

It is interesting to compare the above result with the fractal distribution of the Poincaré maps for the famous strange attractors of a driven pendulum and Duffing–Ueda attractor of a particle in a double-well potential (Fig. 4.45). Remember the equation of motion (1.18):

$$\ddot{x} + \gamma\dot{x} + \partial v(x)/\partial x = F_0 \cos\omega t, \tag{4.86}$$

where γ is the damping coefficient, $V(x)$ is the potential function and F_0 is the amplitude of the driving force. $V(x) = \omega_0^2(1 - \cos x)$ for the driven pendulum, and $V(x) = x^4/4 - x^2/2$ for the Duffing equation. In Fig. 4.45, the Poincaré map clearly exhibits the fractal distribution of points that fills phase space, while the fractal structure is only recognized by enlarging the Poincaré map for the strange attractor in the present case (Fig. 4.44). The former structure is referred to as a fat fractal and the latter a thin fractal. The main difference between the fat and the thin fractal is due to the absence of an acceleration term in the impact-ionization avalanche model (4.80). By increasing the damping coefficient γ in (4.86), the fat fractal changes its shape to the thin fractal, followed by a decrease in the fractal dimension [43]. Since the impact-ionization process is highly dissipative, it is quite natural and confirmative that the strange attractor is a thin fractal structure with the fractal dimension close to $d = 1$. The fractal dimension D of the attractor in the 3D phase space is given by:

$$D = 1 + d. \tag{4.87}$$

In the experiment of the periodically driven chaos in n-GaAs [19, 32, 33], the fractal dimensions $D \sim 1.9 - 2.2$ have been evaluated.

4.7 Period doubling bifurcation to chaos

The breakdown of perfect delay in the one-level impact-ionization model A′ [45, 46] may be discussed more clearly by:

$$\dot{\nu} = -\gamma_1 \frac{\partial V(\nu, E)}{\partial \nu}, \tag{4.88}$$

$$\dot{E} = -\gamma_2\{E - [E_0 + E_{\mathrm{ac}} \sin(2\pi f_{\mathrm{d}} t)] + C_1 \nu E + \phi\}, \tag{4.89}$$

$$\dot{\phi} = -\gamma_3 \phi + C_2 \nu E, \tag{4.90}$$

where $V(\nu, E)$ is the potential function given by (3.85). The physical interpretation of the model, especially for the order parameter ϕ, is described in chapter 5, but only the dynamical part of the model is discussed

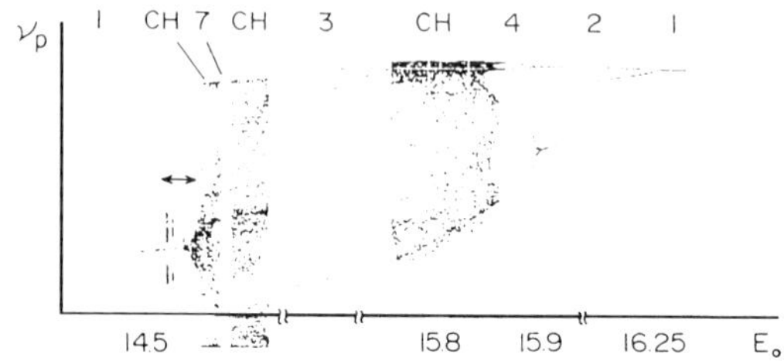

Figure 4.46 Bifurcation diagram as a function of E_0, where $f_d = 10/15$. The numbers indicate the periods of frequency locking, and CH indicates chaos.

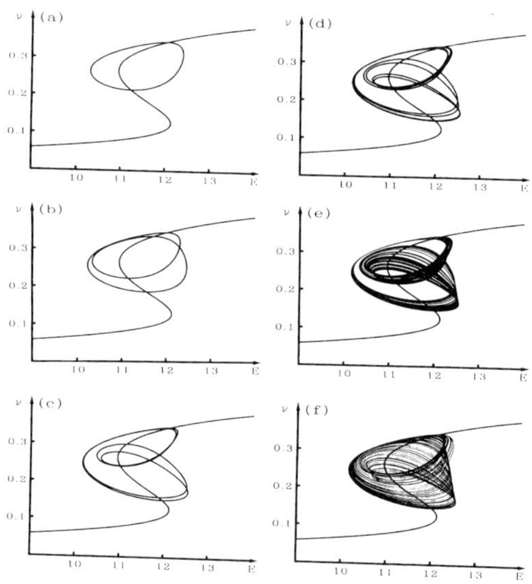

Figure 4.47 Periodic and chaotic orbits obtained as a consequence of period doubling bifurcation, simulated by the one-level impact-ionization model (4.88), (4.89), where $E_{ac} = 4.5$, $f_d = 100/136$; (a)$E_0 = 16.2$, (b)$E_0 = 16.0$, (c)$E_0 = 15.8$, (d)$E_0 = 15.777$, (e)$E_0 = 15.763$, and (f) $E_0 = 15.760$.

here. As a function of E_0, is the rich structure of the bifurcation routes to chaos, as exemplified by the bifurcation diagram in Fig. 4.46 [46]. This section describes how the breakdown of perfect delay induce chaos in the impact-ionization avalanche. For sufficiently large E_0, the solution of (4.88)–(4.90) is given on the upper manifold (postbreakdown regime), which is a periodic oscillation (period 1) with the fundamental period $T = 1/f_d$. By decreasing E_0, the period 1 orbit is destabilized to form period 2 orbit with the period $2T$. With further decrease in E_0, the period 2 orbit is destabilized and period 4 is newly born.

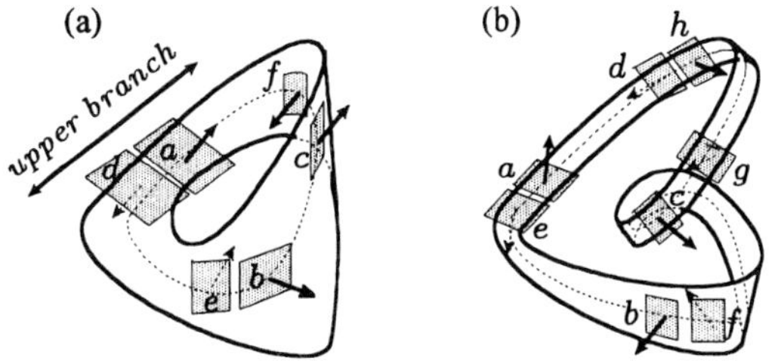

Figure 4.48 Möbius band for (a) period 2 and (b) period 4. For the period 2 orbit in (a), the cycle of the normal bundle is shown as $a \to b \to c \to d \to e \to f \ldots$. Similarity, for the period 4 orbit in (b), $a \to b \to c \to d \to e \to f \to g \to h \to \ldots$.

By successive destabilizations, finer and finer structure of the bifurcations is obtained from period 2^n to period $2^{n+1}(n = 0, 1, 2, \ldots)$. A typical example is shown in Fig. 4.47. The bifurcation is known as period doubling bifurcation (Feigenbaum scenario [40, 47], or sometimes an octave cascade [48]). After the successive bifurcations, the system exhibits the onset of chaos (period 2^∞) at some E_0. A significant characteristic of the periodic and chaotic orbits in Fig. 4.47 is a large overshooting effect of the orbits beyond the singularity point (cliff). By the breakdown of perfect delay, the orbits cross the unstable manifold (NDR branch of the S-shaped ν–E curve), and thereby the system shows the rich structure of the bifurcations.

Period doubling is popular in many dynamical systems, and is explained topologically by Möbius band(s) in the phase space [48]. When period 1 is destabilized (broken line in Fig. 4.48a), a period 2 orbit is newly born, which lies, say, on the edge of the Möbius band. The Möbius band is a closed surface, which is twisted in a region of the upper manifold in Fig. 4.48a for the following reason. The illustration in Fig. 4.48a was drawn with the help of the phase portrait for the period 2 chaotic band in Fig. 4.49, and may be highly accurate. The period 2 orbit in Fig. 4.47b intersects twice with the upper branch during the periodic cycle. After the intersection, the carrier density generated by the complete impact-ionization avalanche takes an almost identical value $\nu(t) \sim \nu(t + T)$ during some overshooting period. Then a normal vector of the tangent plane to the Möbius band (normal bundle) is approximately normal to the upper manifold of the cusp catastrophe. Twisting of the Möbius band occurs after the impact-ionization avalanche. In the center of the Möbius band, there is an unstable period 1 (saddle point orbit). For a period 4 orbit, the Möbius band looks like Fig. 4.48b, where the twisting is seen again at the upper manifold. There is an unstable period 2 orbit (broken line). Among a huge number of topological geometries of dynamical behaviors, Abraham and Shaw [48] illustrated how to form a twisted Möbius band by preparing a folded strip of sheet.

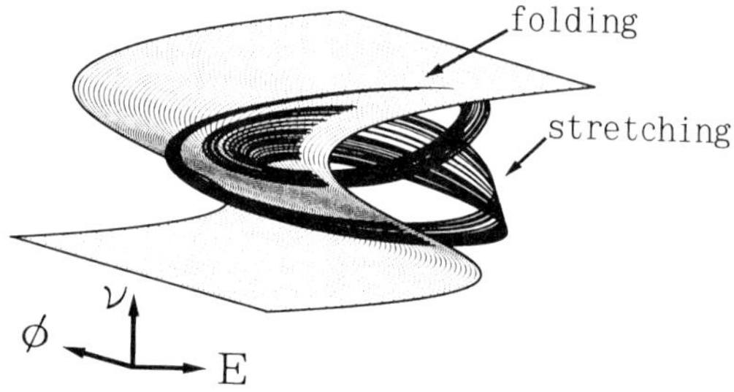

Figure 4.49 Period 2 chaotic band on the catastrophe manifold, where $E_0 = 15.763$, $E_{ac} = 4.5$ and $f_d = 100/136$.

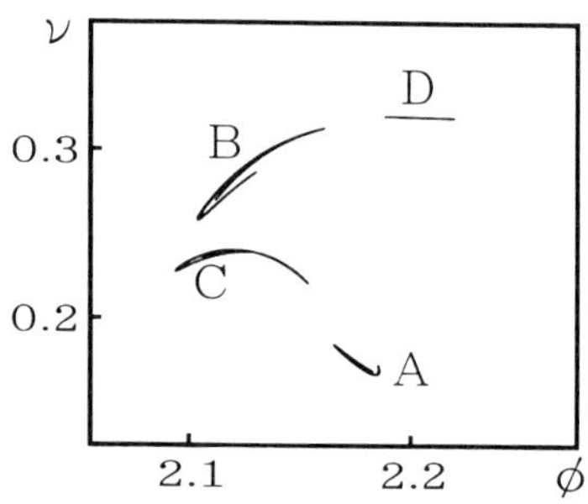

Figure 4.50 Poincaré surface of section at $E = 11.0$, where $E_0 = 15.763$, $E_{ac} = 4.5$ and $f_d = 100/136$.

Combining the inset and outset models [48], they showed that the intersection (known as a homoclinic tangle as an example of Poincaré's solenoid) between stable and unstable manifolds of the combined Möbius bands forms an unstable periodic orbit (saddle point orbit). Note that the unstable manifold is not shown in Fig. 4.48b. By the limit $n \to \infty$, one arrives at edge of chaos, where the 2^∞ periodic orbits lie not on a smooth surface, but on a fractal structure [48], constructed by an infinite number of twistings of the Möbius band.

In the chaotic regime, the orbits form a strange attractor with the fractal structure, which is constructed by stretching and folding the attractor sheet many times.

In Fig. 4.49, a period 2 chaotic band is shown on the catastrophe manifold, where the stretching takes place in the NDR region and the folding on the upper stable manifold. Figure 4.50 shows the Poincaré surface of section which has been taken at each half period at $E = 11.0$ (see Fig. 4.47e). The stretching of the attractor sheet is seen at regions B and C,

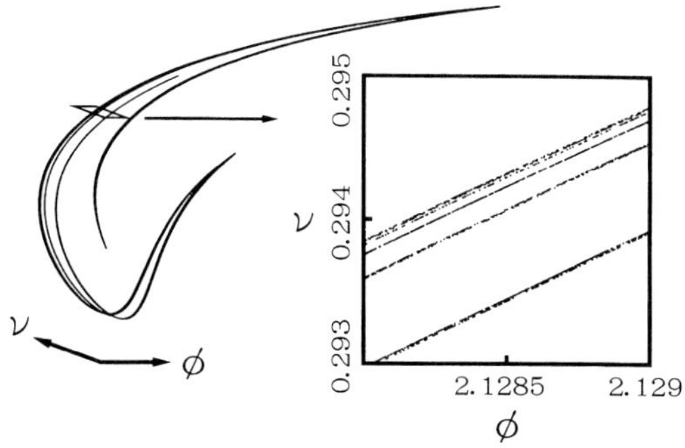

Figure 4.51 Blow-up of the Poincaré surface of section B in Fig. 4.50. A further blow-up of the small squared region is shown on the right.

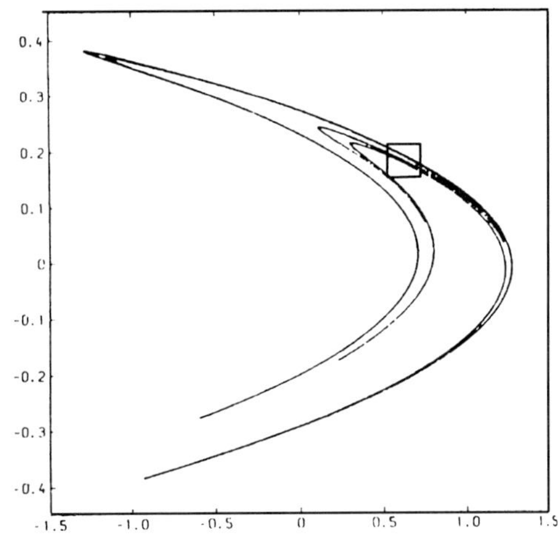

Figure 4.52 Hénon attractor [49, 50].

and the folding at A and D. Section D has almost constant values of the carrier density ν, which means that the impact-ionization avalanche is completed on the upper manifold. Blow-ups of section B are shown in Fig. 4.51. The section is similar to the iterated points of the 2D invertible map by Hénon [49, 50]:

$$x_{n+1} = 1 - ax_n^2 + y_n, \qquad y_{n+1} = bx_n. \tag{4.91}$$

The iterated points and the blow-ups are given in Figs. 4.52 and 4.53 [49, 50], where $a = 1.4$ and $b = 0.3$. Since the Jacobian is b, the area of the set U of the initial conditions contracts by factor b on each application of the mapping (4.41), $T : \mathrm{U} \to \mathrm{U}$. By successive mappings $T^n\mathrm{U}$, the area

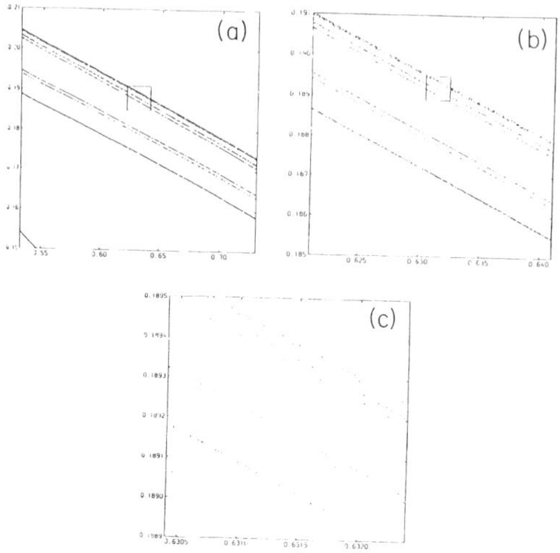

Figure 4.53 Blow-ups of the Hénon attractor [49, 50]. The small squared regions are enlarged in (b) and (c).

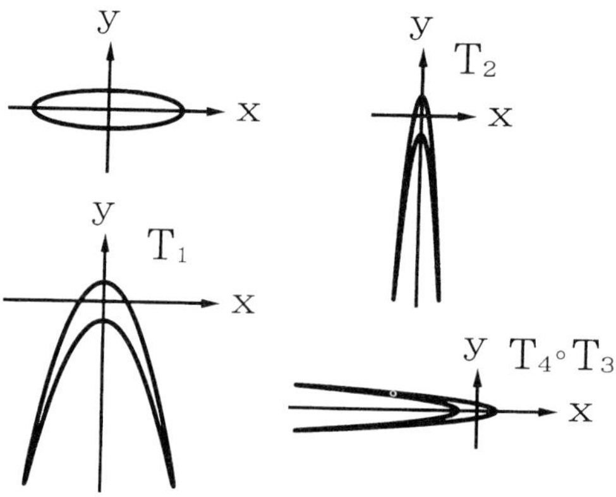

Figure 4.54 Illustration of the transform procedures (4.92) for the Hénon map.

converges to zero, giving rise to the strange attractor $W = \bigcap_{n>0} T^n U$. The Hénon map (4.91) can be expressed as [50]:

$$\begin{aligned}\begin{pmatrix} x_{n+1} \\ y_{n+1} \end{pmatrix} &= \begin{pmatrix} 1 \\ 0 \end{pmatrix} + \begin{pmatrix} 1 & 0 \\ 0 & -1 \end{pmatrix}\begin{pmatrix} 0 & 1 \\ -1 & 0 \end{pmatrix}\begin{pmatrix} b & 0 \\ 0 & 1 \end{pmatrix} \\ &\quad\times \begin{pmatrix} 1 & 0 \\ -ax_n & 1 \end{pmatrix}\begin{pmatrix} x_n \\ y_n \end{pmatrix} \\ &= \begin{pmatrix} 1 \\ 0 \end{pmatrix} + T_4 \circ T_3 \circ T_2 \circ T_1 \begin{pmatrix} x_n \\ y_n \end{pmatrix}, \end{aligned} \tag{4.92}$$

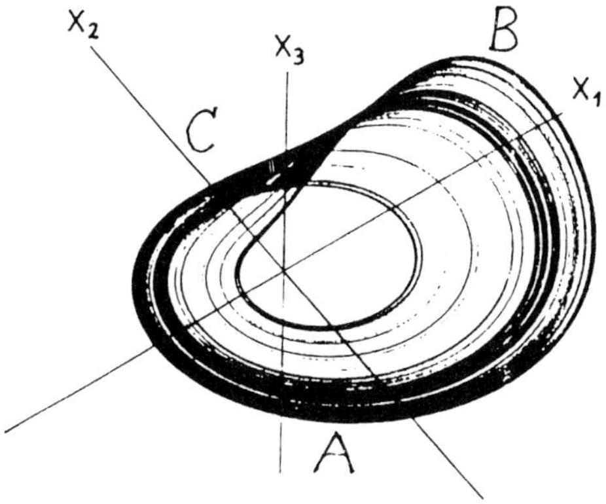

Figure 4.55 Rössler attractor [51]. The sheet of the attractor is stretched at B and folded at C.

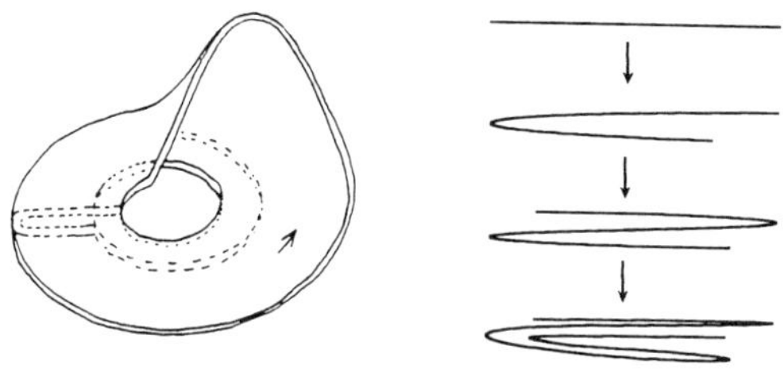

Figure 4.56 Stretching and folding processes of the Rössler attractor. The folding process is shown schematically on the right.

where T_1 is the nonlinear transform $y_{n+1} = -ax_n^2 + y_n$ $(a > 1)$ which stretches and folds the domain in the (x, y)-plane, T_2 the linear transform which contracts interval $[\alpha, \beta]$ of x into $[b\alpha, b\beta]$, where $b < 1$, T_3 the linear transform which rotates the coordinates by $90°(x \mapsto -y,\ y \mapsto x)$, and T_4 the linear transform of reflection $(y \mapsto -y)$. The procedure is shown in Fig. 4.54. The similarity between the differential system (4.88)–(4.90) and the Hénon map (4.91) is very exciting, but it should be mentioned that any kind of nonlinear dynamical system which exhibits chaos more or less shows the universality behavior.

In order to look more intuitively at the self-similar structure, the most famous example is given by the Rössler attractor [51] in Fig. 4.55. Nearby trajectories starting from region A show exponential divergence by their expansive nature, resulting in the stretching of the attractor sheet at region B. From B to C, the attractor sheet is folded into an extremely thin structure. By repeating the process many times, the attractor sheet finally forms a fractal structure. The process is shown schematically in Fig. 4.56. The main part of the chaotic behavior is the sensitive dependence on the initial conditions, which is defined mathematically as follows [15].

4.7.1 Sensitive dependence on initial conditions

For arbitrary $\mathbf{X}_0 \in U$ and the arbitrary neighborhood W of $\mathbf{X}_0$, if there exist $\mathbf{X}'_0 \in W$ and such that $|T^t\mathbf{X}_0 - T^t\mathbf{X}'_0| > \mu$ ($\mu > 0$, $t > 0$), then $T : U \to U$ has sensitive dependence on the initial conditions. The description of sensitive dependence is more accurate than that of the expansive nature stated previously.

In dissipative systems, there is duality of the exponential divergence and convergence of the nearby trajectories, which is shown schematically in Fig. 4.57. If one chooses a disk D_0 as a set of initial points with sufficiently small radius $\delta(\ll 1)$, the disk is transformed into an elliptic disk D by the transitive mapping $T^t D_0 = D$, where the area of the ellipse is smaller than the original disk for the dissipative system. In Fig. 4.57, the disk is stretched in the direction of the y-axis and contracted in the direction of the x-axis. Two nearby points (x_0, y_0, z_0) and (x'_0, y'_0, z'_0) diverge exponentially in the y-direction as $|y - y'| = |y_0 - y'_0|\ \exp(\lambda_y t)\ (\lambda_y > 0)$, converge in the x-direction as $|x - x'| = |x_0 - x'_0|\ \exp(\lambda_x t)\ (\lambda_x < 0)$ and keep the original distance $|z - z'| = |z_0 - z'_0|\ \exp(\lambda_z t)\ (\lambda_z = 0)$. The characteristic exponent $\lambda_1 = \langle\lambda_y\rangle$ is called the largest Lyapunov exponent, $\lambda_2 = \langle\lambda_z\rangle$ the second largest Lyapunov exponent and so on, where $\langle\cdot\rangle$ denotes the time average. The Lyapunov spectrum is defined by the set $(\lambda_1, \lambda_2, \lambda_3)$. Sensitive dependence on initial conditions is now clear if two nearby points are chosen by (x_0, y_0, z_0) and $(x_0, y_0 + \varepsilon, z_0)$ or by (x_0, y_0, z_0) and $(x_0 + \varepsilon, y_0, z_0)$, where $0 < \varepsilon < \delta$.

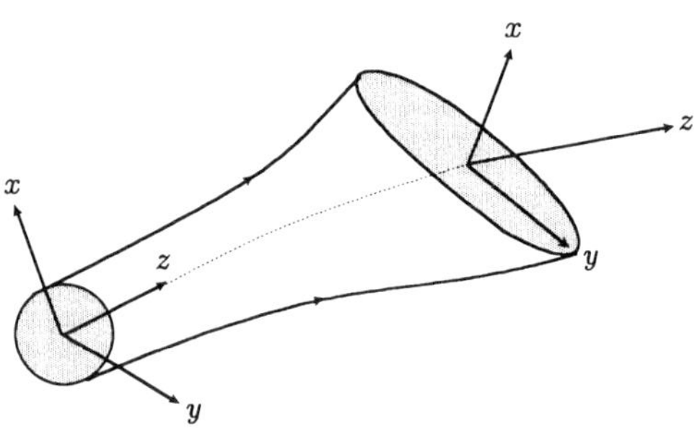

Figure 4.57 Exponential divergence and convergence of the nearby trajectories. Initial points inside a disk diverge in the y-direction after a time t and converge in the x-direction.

For the system of the impact-ionization avalanche, the sensitive dependence on initial conditions is a natural consequence arising from the breakdown of perfect delay, as discussed previously. The damping factor γ_1 in the periodically driven system (4.88)–(4.90) plays a role in determining the contraction rate of the attractors, which is reminiscent of parameter b in the Hénon map.

4.8 Fundamentals of period doubling bifurcation

From the previous section, it was seen that the differential system (4.88)–(4.90) exhibits universal behaviors seen in the difference equations such as the Hénon map, where the fractal structure of the Poincaré surface of section was found to be almost identical to the iterated points of the Hénon map [49, 50]. Moreover, the system of the impact-ionization avalanche shows period doubling bifurcation to chaos, which is a trademark of the logistic map [47, 50]. The Hénon map is identical to the logistic map when $b \to 0; x_{n+1} = 1 - ax_n^2$. A common feature is the existence of the smooth round maximum in the map functions of the form $1 - ax^2$. For nonlinear dynamical systems, in general, there exists a certain hump map similar to the difference equations. In order to see the universality, the fundamentals of the logistic map $f: I \to I$ are described:

$$x_{n+1} = f_\mu(x_n) = 1 - \mu x_n^2, \quad n = 1, 2, \ldots, \tag{4.93}$$

where the map function is equivalent to (1.1). The map function is called unimodal since $\mathrm{d}f_\mu/\mathrm{d}x > 0$ for $x < 0$ and $\mathrm{d}f_\mu/\mathrm{d}x < 0$ for $x > 0$, where $f'_\mu(0) = 0$ and $f_\mu(0) = 1$. The map function $x_{n+1} = \mu x_n(1 - x_n)$ is very popular (see (1.1)), but (4.93) is adopted here. In (4.93), the fixed point is

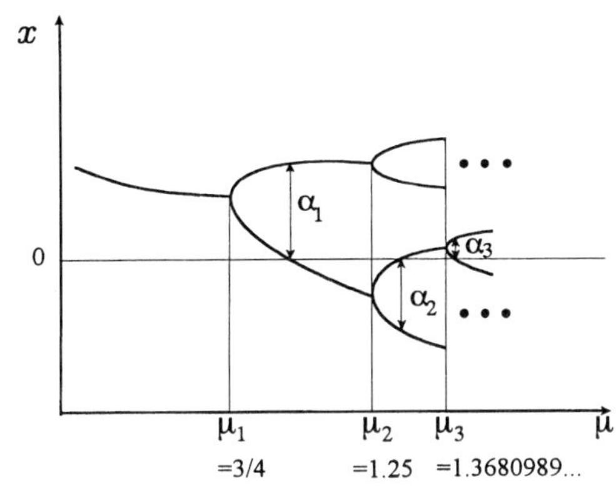

Figure 4.58 Universal scaling in the period doubling bifurcation.

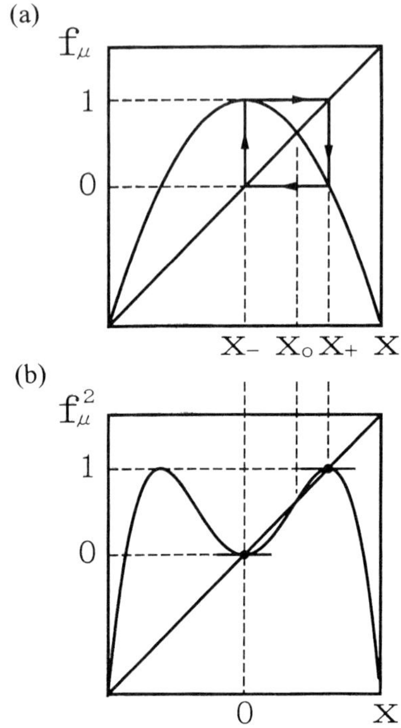

Figure 4.59 (a) Map function f_μ at $\mu_{(1)} = 1.0$, for the superstable cycle of period 2, and (b) the corresponding map function f_μ^2.

given by $x^* = x_0 = (-1 + \sqrt{1 + 4\mu})/2\mu$, by solving $x = f_\mu(x)$. For $0 < \mu < 3/4$, x_0 is stable period 1, namely $x_0 = f_\mu^m(x_0)$, $m = 1, 2, \ldots$, where the criterion is given by $f'_\mu(x_0) > -1$ ($|f'_\mu| < 1$).

The first period doubling bifurcation occurs at $\mu_1 = 3/4$, as shown schematically in Fig. 4.58. When $\mu > \mu_1$, the fixed point becomes unstable, giving birth to the cyclic iteration sequence of period 2: $x_+ x_- x_+ x_- \ldots$ ($x_- = f_\mu(x_+)$, $x_+ = f_\mu(x_-)$).

Taking map function $f_\mu^2(x)$ in Fig. 4.59, then $x_+ = f_\mu^2(x_+)$ and $x_- = f_\mu^2(x_-)$. The criterion for the stability of period 2 is given by the following. Let $x_n = x^* + \varepsilon_n$, then $x^* + \varepsilon_{n+2} = f_\mu^2(x^* + \varepsilon_n)$, where $x^* = x_+$ or x_- and $0 < \varepsilon_n \ll 1$. Expand f_μ^2 into Taylor series $f_\mu^2(x^*) + f_\mu^{2\prime}(x^*)\varepsilon_n$ and find $\varepsilon_{n+2}/\varepsilon_n = f_\mu^{2\prime}(x^*)$. Since $f_\mu^2 = f_\mu \circ f_\mu = f_\mu(f_\mu(x_+ + \varepsilon_n)) = f_\mu(x_-) + f'_\mu(x_-)f'_\mu(x_+)\varepsilon_n$, one obtains:

$$\varepsilon_{n+2}/\varepsilon_n = f_\mu^{2\prime}(x^*) = f'_\mu(x_+)f'_\mu(x_-). \tag{4.94}$$

Period 2 is stable if $|\varepsilon_{n+2}/\varepsilon_n| < 1$, namely

$$|f'_\mu(x_+)f'_\mu(x_-)| < 1. \tag{4.95}$$

Period 2 becomes unstable for $\mu > \mu_2 = 1.25$ and the second bifurcation to period 4 occurs. In general, the stability of period p is discussed similarly by taking $x_n = x^* + \varepsilon_n$, $x_{n+p} = f_\mu^p(x^* + \varepsilon_n)$. From the Taylor expansion of $f_\mu(f_\mu(f_\mu(\ldots f_\mu(x^* + \varepsilon_n))))$, one obtains the criterion of stability by the chain rule [52]:

$$f_\mu^{p\prime} = \left| \prod_{i=1}^{p} f'_\mu(x_i^*) \right| < 1, \tag{4.96}$$

where $p = 1, 2, 4, \ldots, 2^n \ldots$ for the period doubling. With further increases in μ, successive period doubling bifurcations occur: period 4 in the range $\mu_2 < \mu < \mu_3 = 1.3680989\ldots$, period 8 in the range $\mu_3 < \mu < \mu_4 = 1.3940461\ldots$. The accumulation point of the bifurcations is $\mu_\infty = 1.401155\ldots$. There is a universal number (Feigenbaum number) $\delta = 4.66920\ldots$, which is the ultimate value of the ratio of the parameter range $\mu_{n+1} - \mu_n$ of period 2^n to the range $\mu_{n+2} - \mu_{n+1}$ of period 2^{n+1} [47, 52]:

$$\delta = \lim_{n\to\infty} \frac{\mu_{n+1} - \mu_n}{\mu_{n+2} - \mu_{n+1}}. \tag{4.97}$$

The bifurcations occur in a self-similar manner, whereby the pitchfork sizes in the successive bifurcation diagrams can be scaled by the universal numbers δ and α (see Fig. 4.58). The full bifurcation diagram for (4.93) is shown in Fig. 4.60. For the self-similarity among the successive bifurcations, Feigenbaum has given the universal function $g(x)$ [47, 52–54] by using the renormalization group theory [54]. He introduces an operator τ which shifts the bifurcation parameter from some fixed $\mu_{(n)}$ to $\mu_{(n+1)}$ (μ-shifting) and

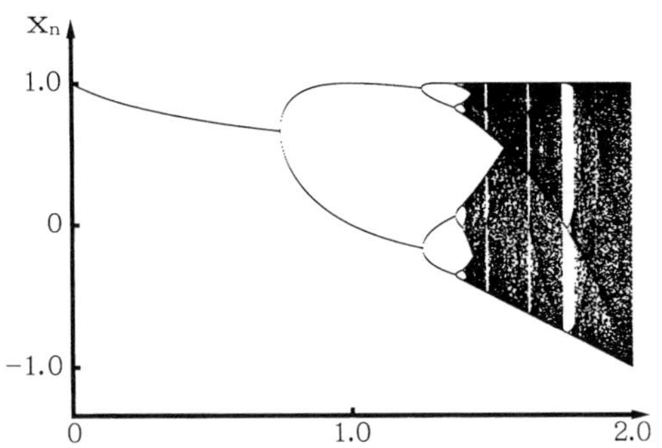

Figure 4.60 Bifurcation diagram in the logistic map: $x_{n+1} = 1 - \mu x_n^2$.

scales the map function by the scaling factor [53]:

$$\begin{aligned}
\mathcal{T} f(\mu_{(n)}, x) &= \alpha f(\mu_{(n+1)}, f(\mu_{(n+1)}, x/\alpha)), \\
\mathcal{T}^2 f(\mu_{(n)}, x) &= \alpha^2 f^2(\mu_{(n+2)}, f^2(\mu_{(n+2)}, x/\alpha^2)), \\
&\dots \\
\mathcal{T}^k f(\mu_{(n)}, x) &= \alpha^k f^{2^k}(\mu_{(n+k)}, x/\alpha^k), \\
&\dots
\end{aligned} \tag{4.98}$$

where $\mu_n < \mu_{(n)} < \mu_{n+1}$. With the limit $k \to \infty$, the universal function is:

$$g(x) = \lim_{k\to\infty} \mathcal{T}^k f(\mu_{(n)}, x). \tag{4.99}$$

From (4.98), the universal function has a property:

$$g(x) = \alpha g(g(x/\alpha)), \tag{4.100}$$

where $g(0) = 1$ and $g'(0) = 0$ are the boundary conditions for the ultimate superstable cycle. In order to see function $g(x)$, draw the map functions $f_\mu^{2^k}$. For the map functions $y = f_\mu(x)$ and $f_\mu^2(x)$, see Fig. 4.59a, b. Shift the parameter value from $\mu_{(1)} = 1$ to $\mu_{(2)} = 1.2943$ for a superstable cycle of period 4 and calculate the map functions $y = f_\mu^2(x)$ and $f_\mu^4(x)$ (Fig. 4.61a, b),

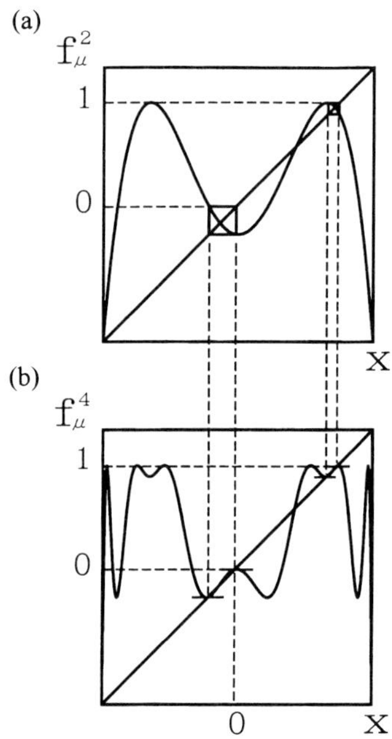

Figure 4.61 Map functions (a) f_μ^2, and (b) f_μ^4 for the superstable cycle of period 4 at $\mu_{(2)} = 1.2943$.

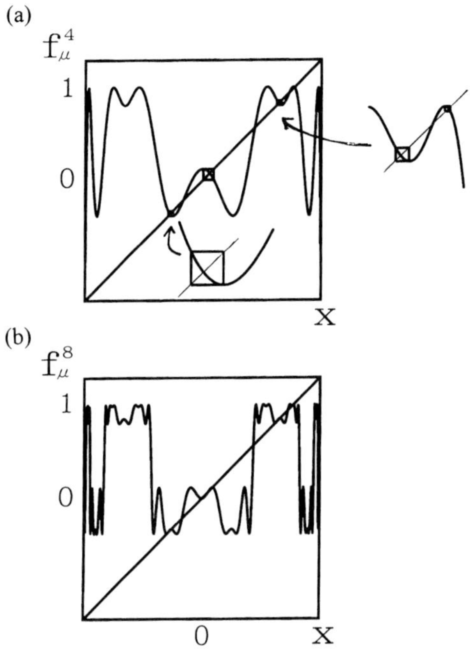

Figure 4.62 Map functions (a) f_μ^4, and (b) f_μ^8 for the superstable cycle of period 8 at $\mu_{(3)} = 1.3810$.

and further shift from $\mu_{(2)}$ to $\mu_{(3)} = 1.3810$ for a superstable cycle of period 8 and calculate $y = f_\mu^4(x)$ and $f_\mu^8(x)$ (Fig. 4.62a, b). In Figs. 4.59, 4.61 and 4.62 local similarity of the map functions is easily found inside squares formed by the superstable cycles. By μ-shifting further and further, the local similarity appears successively, as indicated by the hatched regions in Fig. 4.63. Scaling the functions by $f_\mu^{2^k}(\mu_{(n+1)}, x)$ and repeating the procedures, one obtains the universal functions $g(x)$.

At the accumulation point μ_∞, a set of the iterated points forms a Feigenbaum attractor [55], which is a very special attractor, showing a Cantor-like fractal set but no sensitive dependence on the initial conditions. That is, the distance of two nearby points does not diverge exponentially because the largest Lyapunov exponent λ is zero. The attractor is ergodic but shows no mixing of the iterated points, which is very clear if one considers an infinite number of pitchfork bifurcations in Fig. 4.58.

Invariant probability measure $\rho(x)$ in Fig. 4.64 indicates clearly the singular behavior or no mixing of the iterated points, where the probability density of the iterated points x in a small bin $[x, x + \mathrm{d}x)$ is given by $\rho(x)\mathrm{d}x$. By increasing the strength of the nonlinearity ($\mu > \mu_\infty$), one encounters chaos and many periodic windows, which are described in the next section

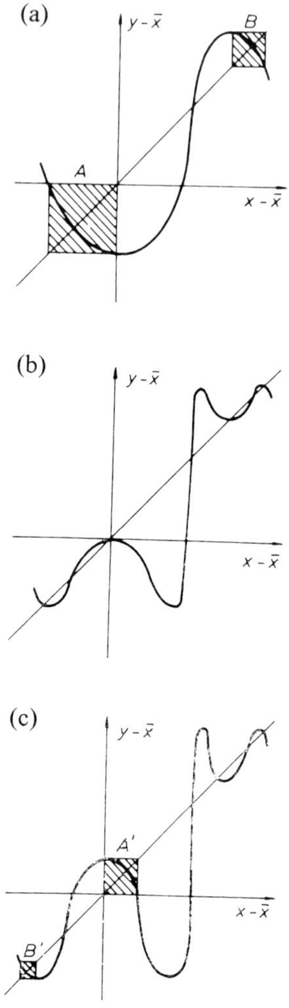

Figure 4.63 Similarity behaviors in the map function. (a) A segment of $f^{(2^{n-1})}(\tilde{\mu}_n, x)$, where $\tilde{\mu}_n$ is the superstable point for $f^{(2^n)}$. (b) A segment of $f^{(2^n)}(\tilde{\mu}_n, x)$. (c) A segment of $f^{(2^n)}(\tilde{\mu}_{n+1}, x)$ to be rescaled and inverted to obtain (a) again [53, p.25].

by taking an example of an impact-ionization avalanche in semiconductors. Since period doubling bifurcation is so popular in a wide variety of dynamical systems, there are numerous reports and books [15, 40, 47–55]. For the universal function and the universal numbers δ, α, the author refers to the original paper by Feigenbaum [54] and the books by Cvitanovic [47], Hao Bai-Lin [53] and Schuster [52]. The editorial books by Cvitanovic [47]

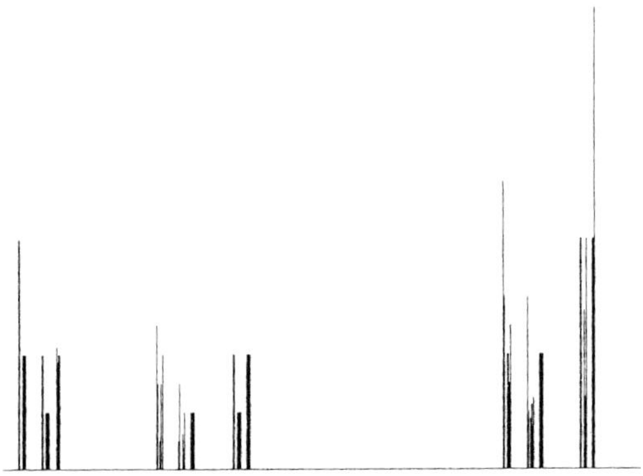

Figure 4.64 The Feigenbaum attractor. Histogram of 50 000 points in 1024 bins. This histogram shows the unique ergodic measure, which is clearly singular [55].

and Hao Bai-Lin [53] include contents of the original papers concerning the experiments and theory of period doubling bifurcation in many dynamical systems.

4.9 Intermittent behavior

One of the routes to chaos is intermittency, which has been observed as a universal route in a large number of experimental systems. By intermittency is meant the interruption of long laminar (periodic or quasi-periodic) phases by relatively short chaotic bursts. In many experiments, the bursts become frequent when a control parameter is varied, and the intermittent state finally merges to a stationary chaos. There are typically three types of intermittency (types I–III). This section, describes briefly the intermittent behaviors, introducing some experimental observations as well as some numerical results in semiconductor systems.

Type I intermittency has been called the Pomeau–Manneville scenario [56–59] and is typical for many low-dimensional dynamical systems. As a pioneering work, Pomeau and Manneville (1979) investigated numerically the type I intermittency in the Lorenz model [57]. In a low-dimensional dynamical system which exhibits logistic map behavior, the intermittency can be expected at an edge of a periodic window embedded in the chaotic regime. Supposing a window of a regular cycle of period q (e.g., $q = 3$), then one expects period doubling bifurcation to chaos ($q \times 2^k, k = 0, 1, 2, \ldots$) at an edge of the window (see Fig. 4.65), and intermittency at another edge of the window with a control parameter at $\mu = \mu_c$. In the logistic map, the intermittency of type I is seen for $\mu < \mu_c = 1 + \sqrt{8}$ at an edge of a period 3 window [56, 58].

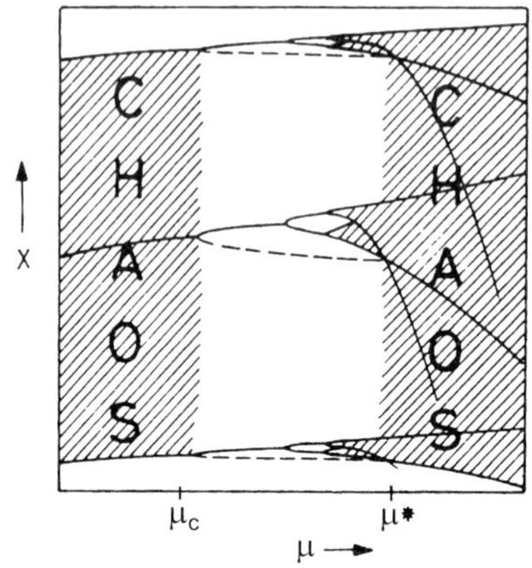

Figure 4.65 Bifurcation diagram of period 3 window for the logistic map. At an edge, $\mu = \mu_c$, type I intermittency occurs [52, 56].

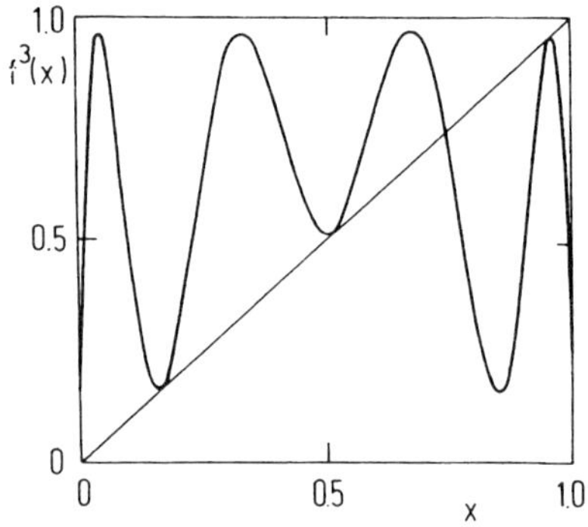

Figure 4.66 The iterative map $f_\mu^3(x)$ at $\mu = \mu_c$ [52, 56].

Period 3 implies $x = f_\mu^3(x)$, where $\mu > \mu_c$. At $\mu = \mu_c$, the iterative map $f_\mu^3(x_c)$ is critical, as given in Fig. 4.66, in the sense that a pair of stable and unstable fixed points x_s and x_u is degenerate ($x_s = x_u = x_c$), that is, the map function f_μ^3 is tangential to the line $y = x$ at x_c near three extrema. For $\mu > \mu_c$, the map function intersects with the line $y = x$ at the fixed points x_s and x_u near each extrema, and thereby one has a stable period 3: $x_s = f_\mu^3(x_s)$. Now, what happens for $\mu < \mu_c$? As expected, there are no fixed points after the coalescence of the stable and unstable fixed points. The time series of the iteration is chaotic, but in the form of intermittency type I.

The waveform of period 3 is interrupted by the chaotic burst. In Fig. 4.67, iterated points x_n are plotted as a function of discrete time n. The chaotic burst becomes more frequent by decreasing the control parameter from $1 + \sqrt{8}$. In order to understand type I intermittency, the map function f_μ^3 is

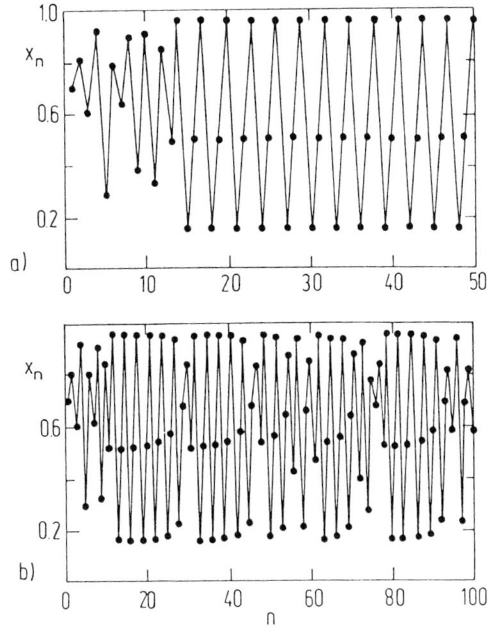

Figure 4.67 Iterates of the logistic map starting from $x = 0.7$ (a) in the stable three-cycle region $\mu_c - \mu = -0.02$; (b) in the intermittent region $\mu_c - \mu = 0.002$ ([58], after Hirsch *et al.*, 1981).

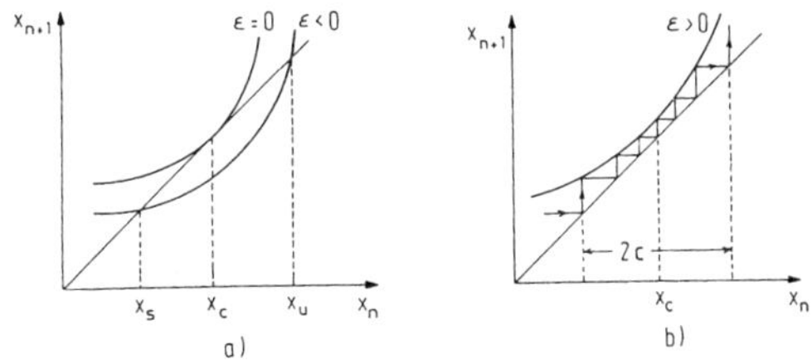

Figure 4.68 Mechanism for type I intermittency. (a) Poincaré map for $\varepsilon = \mu - \mu_c < 0$; (b) Poincaré map for $\varepsilon > 0$ and motion of the trajectory [58].

expanded into series around the points x_c and μ_c [52, 56]:

$$f_\mu^3(x) = x_c + (x - x_c) + a(x - x_c)^2 + b(\mu_c - \mu), \qquad (4.101)$$

where a and b are the derivatives of the Taylor expansion at $x = x_c$ and $\mu = \mu_c$ respectively. With $y = (x - x_c)/b$, $c = a/b > 0$ and $\epsilon = \mu - \mu_c$, the map function (4.101) and thus the iterative map $x_{n+3} = f_\mu^3(x_n)$ in the vicinity of x_c can be transformed into a form of

$$y_{n+3} = y_n + cy_n^2 + \epsilon, \quad n = 1, 2, \ldots, \qquad (4.102)$$

or formally written [52]

$$y_{n+1} = y_n + cy_n^2 + \epsilon. \qquad (4.103)$$

It is noted that $y = y + cy^2 + \epsilon$ has no solution for $\epsilon > 0$, a unique solution for $\epsilon = 0$ and a pair of a saddle and a node for $\epsilon < 0$ (Figs. 46.8a, b).

As shown in Fig. 4.68, the map function is drawn for the three cases, showing the saddle-node bifurcation for $\epsilon < 0$. The iterations in Fig. 4.68b give an explanation for the intermittent behavior. The map function (4.102) in a local region near $y_c = 0$ ($x \sim x_c$) forms a narrow channel in which the iterated points spend much time (forming a laminar phase; see Fig. 4.68b). After escaping the outlet of the channel, the iteration experiences the chaotic burst and shortly it is again reinjected into the inlet of the channel. Type I intermittency occurs via the repeating process. The mean residence time $\langle \ell \rangle$ or mean duration of the laminar region depends on the channel width and thus control parameter ϵ, varying as [52, 57]:

$$\langle \ell \rangle \propto \epsilon^{-1/2}. \qquad (4.104)$$

A famous example of the logistic map behavior was proved in a Bénard experiment by Berge *et al.* [60]. In the semiconductor chaos system, Richter [61] have observed type I intermittency of the spontaneous current oscillation in p-type Ge under the transverse magnetic fields. They have experimentally verified the power law (4.104). Intermittency can also be demonstrated in the present model simulation, where parameters have been chosen at an edge of a period 3 window and the dc bias E_0 slightly smaller than the critical value $E_{0c}(= 15.032249)$.

In Fig. 4.69, the return map ν_{n+3} vs ν_n clearly shows a logistic map behavior, where ν_n is the peak value of the carrier density. The mean time of

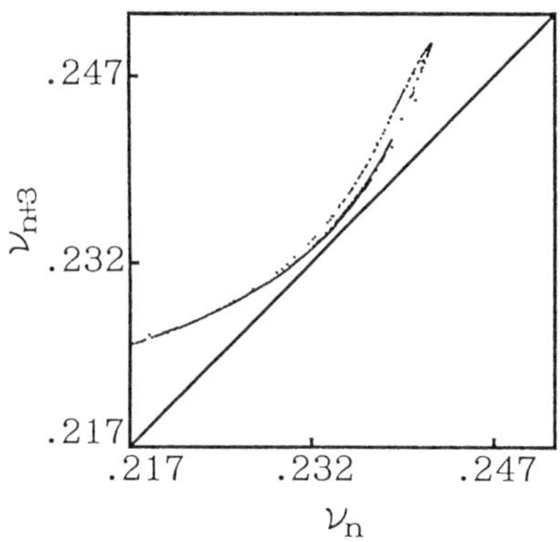

Figure 4.69 Return map $\nu_{(n+3)}$ vs ν_n for intermittency type I, in the approximated impact-ionization model. $E_{ac} = 4.5$, $f_0 = 100/136$ and $E_0 = 15.032249$.

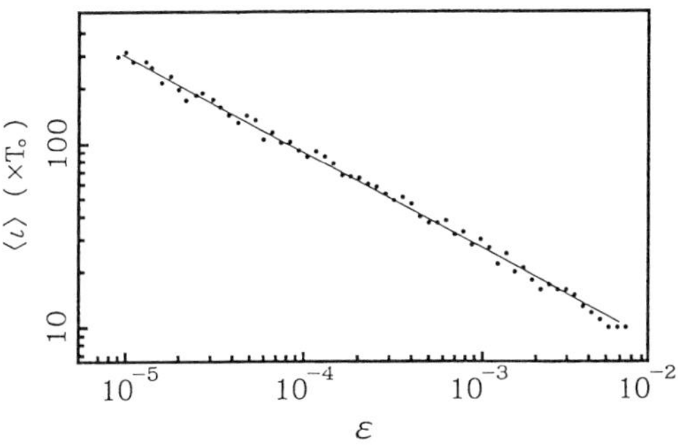

Figure 4.70 Power law of the mean time of the laminar region vs ε, where $\varepsilon = E_{0c} - E_0.E_{ac} = 4.5, f_0 = 100/136$ and $E_0 = 15.032249$.

the laminar region follows exactly the power law $\langle \ell \rangle \propto \epsilon^{-1/2}$ over a wide range of $10^{-5} < \epsilon < 10^{-2}$, where $\epsilon = E_{0c} - E_0$ (see Fig. 4.70).

Type II intermittency has been proposed theoretically by Pomeau and Manneville [57], but has not yet been proved experimentally. Here, the mechanism is briefly supplemented. The return map is assumed to be given by:

$$r_{n+1} = (1+\epsilon)r_n + ur_n^3, \tag{4.105}$$

$$\theta_{n+1} = \theta_n + \Omega, \tag{4.106}$$

where r_n and θ_n are the dynamical variables (radius and angle, respectively, in the sense of torus dynamics), ϵ is a control parameter and $u(> 0)$ and Ω are constants. With $\epsilon < 0$, a unique solution $r_n = 0$ is obtained since the slope of the map (4.105) is smaller than unity at $r = 0$. For $\epsilon > 0$, the torus dynamics (r_n, θ_n) shows a growing spiral oscillation in the phase space (see Table 4.2 after Schuster [52]). The spiral oscillation forms a laminar region but it is sooner or later followed by the chaotic burst. The mean length of the laminar region $\langle \ell \rangle$ is given by a power law $\langle \ell \rangle \propto \epsilon^{-1}$ [52].

Type III intermittency is an intermittency of period 2 oscillation, which appears simultaneously with a period doubling bifurcation [52, 62]. An excellent example has been given experimentally by Dubois *et al.* in an experiment on Bénard convection in a small rectangular cell [52, 62], as shown in Fig. 4.71. Type III intermittency is explained by the return map

$$x_{n+1} = -(1+\epsilon)x_n - ux_n^2. \tag{4.107}$$

Note that a fixed point $x = 0$ is destabilized when $\epsilon > 0$, giving rise to a growing period 2 oscillation (see Table 4.2), where the subharmonic amplitude grows while the fundamental amplitude decreases. When the subharmonic amplitude reaches a maximum, a chaotic burst appears, as found in Fig. 4.71 [52, 62]. The mean time $\langle \ell \rangle$ follows a power law

$$\langle \ell \rangle \propto \epsilon^{-1}. \tag{4.108}$$

In the semiconductor chaos system, type III intermittency has been observed by Richter *et al.* in a p-type Ge under magnetic fields [63, 64].

The temporal current profiles in Fig. 4.72 indicate that the intermittent bursts become more and more frequent by increasing the magnetic field intensity B. Just before each intermittent event, one can recognize the typical time development of the maxima as in (4.107) (see the inset of Fig. 4.72b; see also Table 4.2).

The mean laminar length has been found to scale as $\bar{\tau} \propto (B - B_c)^{-1}$ [63, 64], shown in Fig. 4.73. Peinke *et al.* [63] have also investigated the probability distribution $P(\tau)$ of the laminar length and the integral distribution $N(\tau_0)$, from which they have verified for the first time the type III intermittency in semiconductor systems. Characteristic properties of different types of intermittency are summarized in Table 4.2 see especially the distribution $P(\ell)$] [57].

Another type of intermittency is caused by the boundary crisis as described by Grebogi *et al.* [65–67]. Consider the dynamical behavior in the regime of the Poincaré section or Poincaré map. In the dynamics of the saddle point, the stable manifold W_s is global when it extends to infinity, otherwise it is local. In the global case, a pair of separatrices W_s is a basin boundary which divides the phase space volume into two basins of attraction $\{\mathbf{X}_A\}$ and $\{\mathbf{X}_B\}$ such that the orbit falls into an attractor A (or B) if one chooses an arbitrary initial point in the basin $\{\mathbf{X}_A\}$ (or $\{\mathbf{X}_B\}$). It is also noted that the unstable manifolds W_{uA} and W_{uB} terminate to the attractors A and B, respectively. By shifting a control parameter, it often happens that one of them (or both), say W_{uA}, touches one of the separatrices W_{sA} (homoclinic tangency). By further shifting the control parameter, W_{uA} crosses W_{sA} (homoclinic crossing). The intermittency occurs very often via the boundary crisis by homoclinic crossing. Chaos–chaos transition is an example of the boundary crisis. The most well-known example is given by the Lorenz model [68–70].

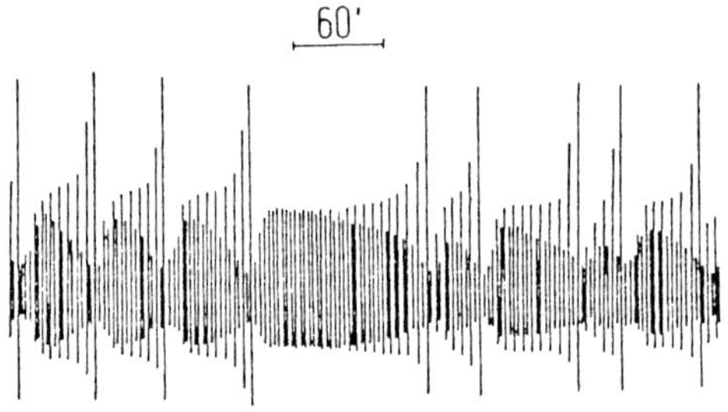

Figure 4.71 Time dependence of the light intensity in the measurements of Bénard convection in a small rectangular cell. The intermittent waveform is characteristic for type III intermittency. The light intensity is roughly proportional to the local horizontal temperature gradient ([52], after Dubois *et al.*, 1983 [62]).

Table 4.2 Characteristic properties of different types of intermittency

Type	Poincaré map	Laminar Signal	Distribution $P(l)$
I	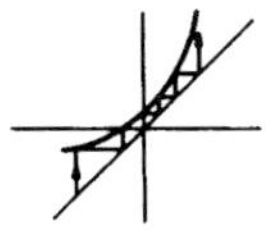	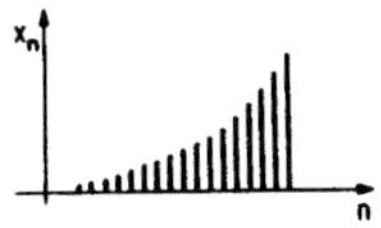	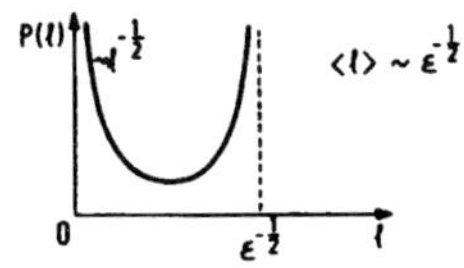
	$x_{n+1} = x_n + x_n^2 + \varepsilon$	increases monotonously	
II	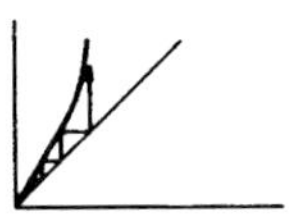		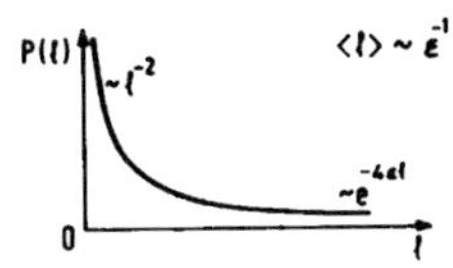
	$r_{n+1} = (1 + \varepsilon) r_n + u r_n^3$ $\theta_{n+1} = \theta_n + \Omega$	spirals	
III	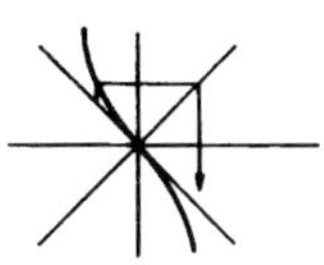	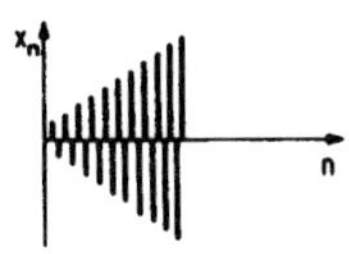	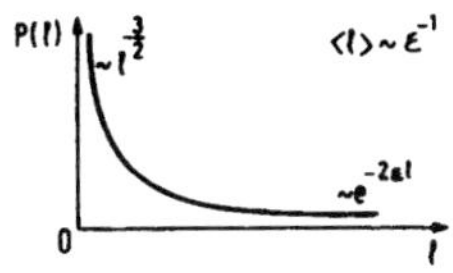
	$x_{n+1} = -(1 + \varepsilon) x_n - u x_n^3$	alternates	

In Fig. 4.74, once W_{uA} touches W_{sA} at a point P_0, it touches the stable orbit W_{sA} an infinite number of times at $P_1, P_2, P_3, \ldots$, where $P_0, P_1, P_2, \ldots$ form homoclinic points. By the transitive map $T: \mathbf{U} \to \mathbf{U}$, a squared region $\mathbf{U}_0$ which contains a homoclinic point P_0 is mapped into a stretched square $\mathbf{U}_1$ which contains P_1, and $\mathbf{U}_1$ to $\mathbf{U}_2$ and so on. When the system is dissipative, the volume $\mathbf{U}_n$ is contracted by the successive mapping, leading to an attractor with zero volume $(W = \bigcap_n T^n \mathbf{U}_n \to 0)$ after an infinite number of transitive mappings [69]. Here, note that the Poincaré map of $\mathbf{U}_0$ is successively stretched and folded in the transverse direction, resulting in a horseshoe-type map near each homoclinic point. The transitive mapping is equivalent to Baker's transformation, which forms a fractal set. Thus, it is found that W_{uA} terminates to a strange attractor A. By the homoclinic

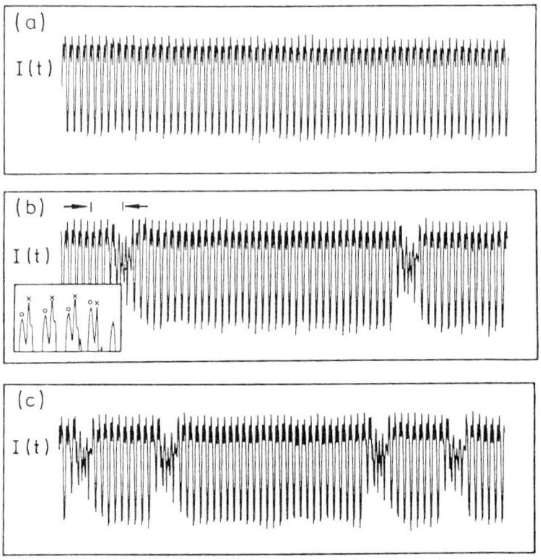

Figure 4.72 Temporal current profiles for distinct (a) periodic and (b), (c) intermittent oscillatory states in p-type Ge at different transverse magnetic fields (a) $B = 1.922$ mT, (b) $B = 1.950$ mT, (c) $B = 1.958$ mT, and the following constant parameters; bias voltage $V_0 = 12.0$ V and temperature $T_L = 1.88$ K (load resistance $R_L = 2.0$ kΩ). The amplitude is in the range of a few μA. The total timespan amounts to 15 ms. The inset in (b) gives a blow-up of the segment of the time trace indicated by the arrows. The development of the maxima $I_n, I_{n+2}, \ldots$ and $I_{n+1}, I_{n+3}, \ldots$ is marked by circles and crosses, respectively [63].

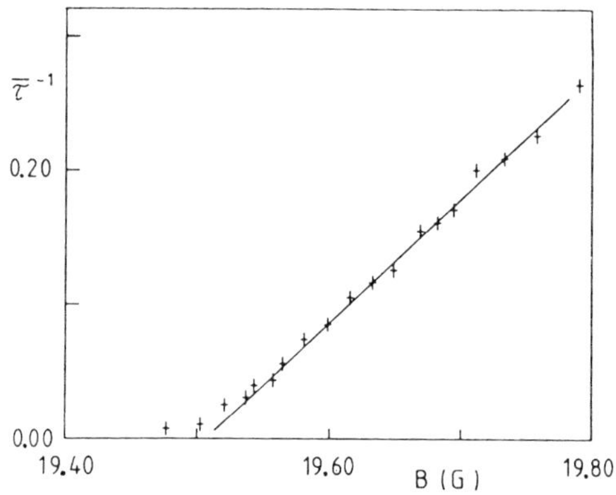

Figure 4.73 Dependence of the inverse of the mean laminar length on the transverse magnetic field (1 G = 0.1 mT) for distinct cases of intermittent oscillatory states as in Fig. 4.72 (same sample and constant parameters). Experimental data are marked by crosses, and their linear scaling by a solid line [63].

crossing, the basin boundary disappears and the chaotic orbit escapes out from the region $\{\mathbf{X}_A\}$ to $\{\mathbf{X}_B\}$ and terminates to an attractor B. If W_{uB} happens to crosses separatrix W_{sB} simultaneously, the attractor B is also

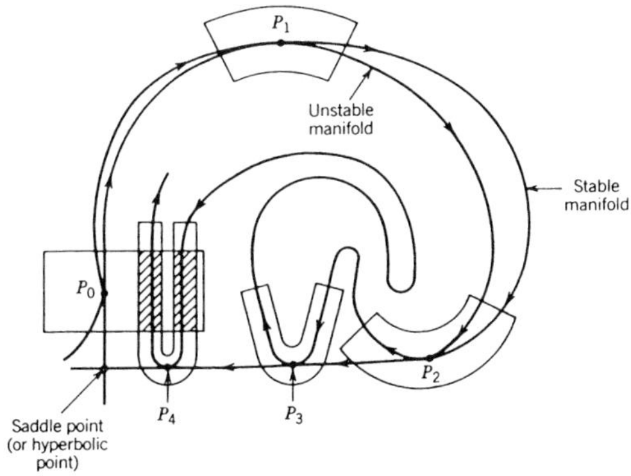

Figure 4.74 Development of a folded horseshoe map for points in the neighborhood of a homoclinic orbit [69].

"strange", for the same reason as discussed above. Then the intermittent switching between A and B appears as a chaos–chaos transition. Normally, one has the mean residence time $\langle t_A \rangle \sim \langle t_B \rangle$ for the chaos–chaos transition [68]. If t_B is very short compared with t_A for any time, the switching behavior is referred to as "reinjective switching". The chaotic time sequence of A is interrupted intermittently by a chaotic burst of B. In any case, the reinjective switching is the crisis-induced intermittency.

4.10 Growth of chaos

At the accumulation point μ_∞ of period 2^∞, the Feigenbaum attractor has a fractal structure but does not exhibit sensitive dependence on the initial conditions. A chaotic attractor grows by increasing a control parameter starting from μ_∞. The bifurcation is known as an inverse bifurcation cascade, which is universal among different dynamical systems. Here, the universal rule is described, which is also expected to be found in the nonlinear carrier transport.

The bifurcation sequence in the logistic map ($f = \mu x(1 - x)$) is given schematically in Fig. 4.75 [52]. In the parameter range $\tilde{\mu}_{n+1} < \mu < \tilde{\mu}_n$, the orbit is 2^n-periodic chaos, which looks like the noisy period 2^n. By taking $f^j_\mu(x)$ ($j = 2^n$), there are 2^n subintervals I_i ($i = 1, 2, \ldots, 2^n$), where for $x \in I_i$ the iterated sequence $f^j(x), f^j(f^j(x)), \ldots, f^j(f^j(f^j(\ldots f^j(x)))), \ldots$ forms a chaotic band in I_i. By shifting the parameter into the range $\tilde{\mu}_n < \mu < \tilde{\mu}_{n-1}$, band merging or band mixing occurs in a pairwise manner, resulting in 2^{n-1} periodic chaos. In Fig. 4.75 the bifurcation points $\tilde{\mu}_j$ scale like $\tilde{\mu}_j - \mu_\infty \sim \delta^{-j}$

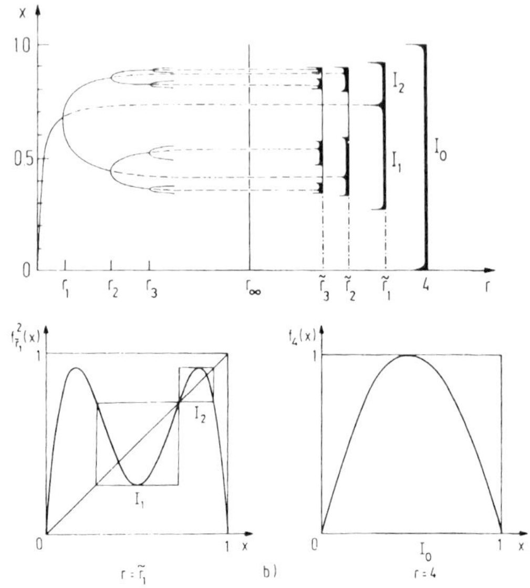

Figure 4.75 Top: Bifurcation diagram for the logistic map $f = rx(1 - x)$. Bottom: Map function f_2 for the onset of band mixing of 2-periodic chaos (left), and the map function f at $r = 4$ for fully developed chaos [52]. The symbol r of the control parameter may be replaced by μ in the text.

[52]. An example of the map function f_μ^2 with $\mu = \tilde{\mu}_1$ is given in Fig. 4.75(bottom, left), where the orbit shows the onset of band mixing of 2-periodic chaos and thus the onset of fully developed chaos. At $\mu = 4$, the fully developed chaos is completed (bottom, right). Growth of chaos is related to the Lyapunov exponents (or Kolmogorov entropy). Since the attractor size increases by increasing μ, the divergence rate of nearby trajectories also increases.

Figure 4.76 shows the global view of the bifurcation diagram and the Lyapunov exponent as a function of μ.

In particular, in Fig. 4.77, the Lyapunov exponent and the topological entropy (smooth solid line) are shown for $\mu > \mu_\infty$ [52]. The exponents exhibit a huge number of discontinuities, which arise due to many periodic windows which appear in the bifurcation diagram. As described in chapter 1, the Lyapunov exponent is given by [52, 55]:

$$\lambda = \int \mathrm{d}x\, \rho(x)\ \ln\ |f'_\mu(x)|, \tag{4.109}$$

$$= \sum_j \int_{x \in I_j} \mathrm{d}x\, \rho_j(x)\ \ln\ |f'_\mu(x)|\ \text{(for } 2^n\text{-periodic chaos)}. \tag{4.110}$$

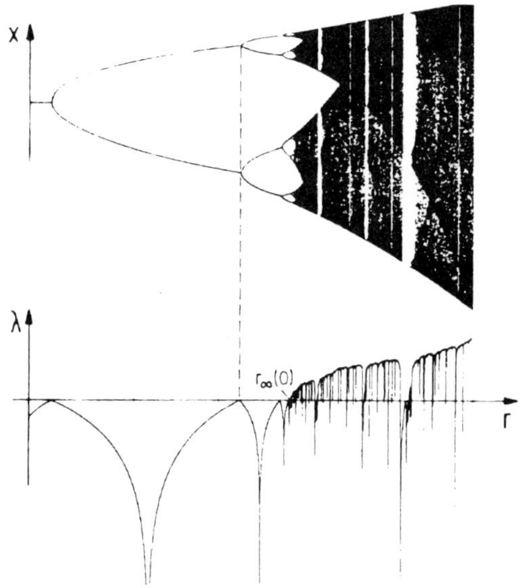

Figure 4.76 Bifurcation diagram and the Lyapunov exponent as a function of r [52]. The symbol r of the control parameter may be replaced by μ in the text.

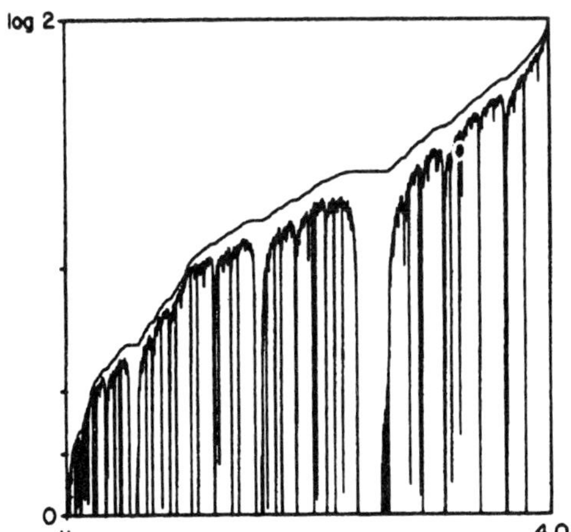

Figure 4.77 Topological entropy (upper curve) and the Lyapunov exponent (lower curve) as a function of μ for the logistic map $f = \mu x(1 - x)$ (graph by J. Crutchfield). Note the discontinuity of the lower curve [55].

An alternative expression is given by:

$$\lambda(x_0) = \lim_{N\to\infty} \lim_{\varepsilon\to 0} \frac{1}{N} \ \ln \left| \frac{f_\mu^N(x_0+\varepsilon) - f_\mu^N(x_0)}{\varepsilon} \right| \tag{4.111}$$

$$= \lim_{N\to\infty} \ln \left| \frac{\mathrm{d}f_\mu^N(x_0)}{\mathrm{d}x_0} \right| \tag{4.112}$$

$$= \lim_{N\to\infty} \ln \prod_{i=0}^{N-1} |f'_\mu(x_i)| \tag{4.113}$$

$$= \lim_{N\to\infty} \frac{1}{N} \sum_{i=0}^{N-1} \ln |f'_\mu(x_i)|, \tag{4.114}$$

where the chain rule is used in (4.111), (4.112). By equating (4.109) to (4.114), the ergodic theorem is typically obtained (see chapter 1). The time average (4.114) of the divergence rate (logarithm of the derivative $|f'_\mu(x)|$) of two nearby trajectories starting from two initial conditions x_0 and $x_0+\varepsilon$ is equal to the spatial average (4.109) with respect to x. In the actual case of the computation, one cannot take the limits with respect to N and ε. Instead, an average is taken as $\lambda = \{\lambda(x_0) + \lambda(x_N) + \lambda(x_{2N}) + \ldots + \lambda(x_{(m-1)N})\}/m$. In the chaotic region, positive λ increases its value by increasing μ, because the contracting region of x ($|f'_\mu| < 1$) becomes narrow ($|x - 1/2| < 1/2\lambda$) and the width of the chaotic bands or the dynamic range increases. The integration (4.110) for the 2^j-periodic chaos is taken over the dynamic range in which the invariant probability measure $\rho_j(x)$ ($j = 0, 1, 2, \ldots, 2^n$) is a smooth function.

An example of the invariant probability measure is shown in Fig. 4.78. Summing the contributions over the 2^n bands, one obtains λ in (4.110). Band merging at $\tilde{\mu}_n$ has no dramatic effect on λ; no critical behaviors are seen before or after $\tilde{\mu}_n$. In Fig. 4.77, the envelope of the structure in λ is very close to the topological entropy. The exponent is ln 2 at $\mu = 4$, which can be obtained by approximating $\rho(x) = 1/\sqrt{x(1-x)}$, while it is zero at $\mu = \mu_\infty$.

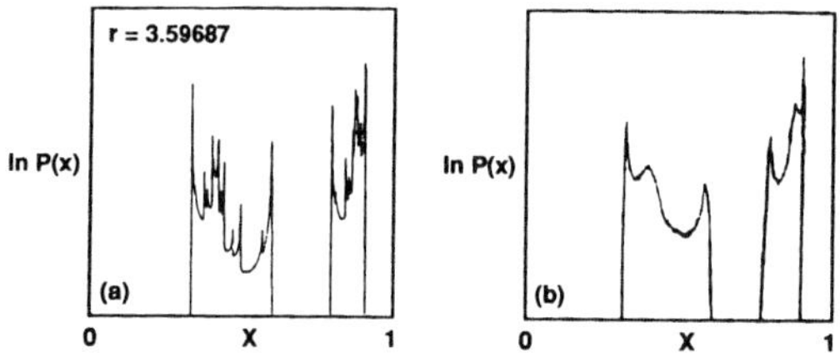

Figure 4.78 (a) Invariant probability measure for the logistic map $f = rx(1-x)$ with $r = 3.59687$. (b) The effect of a small amount of noise on the invariant probability measure [50] (figures taken from Crutchfield *et al.*, 1982).

In the chaotic region such that $\lambda(\mu) > 0$ where $\mu > \mu_\infty$, there has been no complete proof [55] that $\lambda(\mu)$ is continuous in a small subinterval $[\mu, \mu']$. However, it is believed that there are absolutely continuous measures (nonzero Lebesgue measures) of the set $\{\mu'' \epsilon [\mu, \mu'] | \lambda > 0\}$. It is also noticed that the many windows corresponding to attracting periodic orbits of extremely long periods are embedded in the chaotic region. In order to achieve finer and finer structure, we require an unlimited resolution of μ and exhausting time of computation [note: accuracy of λ is attainable only by $t \to \infty$; see (4.114)]. Usually, with limited resolution and time, for economic reasons, the fine structure is smeared out, missing narrow windows of extremely long periods. The situation in the experimental systems becomes worse because the external noise smears the fine structure more strongly.

Briefly, let us supplement the Lyapunov exponents of the periodic orbits during period doubling cascade in Fig. 4.76. For period $j = 2^n$, take $\rho(x) = \frac{1}{j}\delta(x - x_k)$ in (4.109) where δ is the Kronecker delta and $x_k = f^j(x_k)$, $k = 1, 2, \ldots, 2^n$. Then, from (4.109) or (4.114),

$$\lambda = \frac{1}{j}\sum_k \ln |f'_\mu(x_k)| < 0.$$

For a superstable cycle of 2^n at $\mu = \mu_{(n)}$, one of the fixed points is at 1/2 and thus $f'(1/2) = 0$, $\prod_k |f'_\mu(x_k)| = 0$, leading to the limit $\lambda \to -\infty$ with $\varepsilon \to 0$, $N = 2^n$ in (4.111) and (4.113). Examples of the map functions for the superstable cycles are shown in Figs. 4.61, 4.62 and 4.63. With a definite resolution of $\varepsilon \neq 0$ ($|\varepsilon| \ll 1$), however, a negative local minimum is obtained instead of $\lambda \to -\infty$ at $\mu \cong \mu_{(n)}$, where $\mu_n < \mu_{(n)} < \mu_{n+1}$, as shown in Fig. 4.76. At the bifurcation points μ_{n+1} $(n = 0, 1, 2, \ldots)$, $\lambda = 0$ since $|\mathrm{d}f^j_\mu(x_k)/\mathrm{d}x_k| = 1$ ($j = 2^n$, $k = 1, 2, \ldots, 2^n$). Therefore, $\lambda(\mu)$ has cusp structures during the period doubling cascade. With $n \to \infty$, λ approaches asymptotically to zero.

Table 4.3 Bifurcation points and the ratio δ_n in the period-doubling bifurcation, calculated from the one-level impact-ionization model (4.88)–(4.90)

Period	Bifurcation point E_0	Ratio δ_n
$1 \to 2$	16.019250	–
$2 \to 4$	15.800800	–
$4 \to 8$	15.775300	8.567
$8 \to 16$	15.769469	4.373
$16 \to 32$	15.768194	4.573
$32 \to 64$	15.767902	4.366

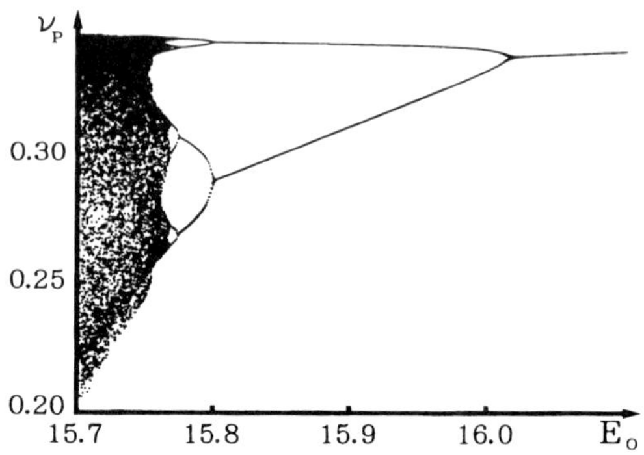

Figure 4.79 Bifurcation diagram in the one-level impact-ionization model (4.88)–(4.90), as a function of the dc bias E_0. $E_{ac} = 4.5$ and $f_0 = 100/136$.

4.11 Semiconductor chaos, U-sequence and Sarkovskii's theorem

This section discusses the fundamental properties of semiconductor chaos by the impact-ionization model, relating to sections 4.8 and 4.10. Usually, in continuous systems which show the nonequilibrium phase transition, the 1D map functions constructed from the systems' equations $\dot{\mathbf{x}} = \mathbf{F}(\mathbf{x}; t)$ are many-to-many (but deterministic) transformations due to the relatively high-dimensional fractal structures of the strange attractors, where the degrees of freedom are normally larger than 3. Among them, some systems [53] exhibit approximately unimodal map functions as obtained in the Rössler equations [50, 51, 53], where the nonlinearity of the systems is not so high. Others show more complicated 1D transformations $T: x_n \to x_{n+1}$. Frequently, in systems with high nonlinearity such as the B-Z reaction [53, 71–73], neuronal dynamics [53, 74] and the impact-ionization avalanche, the 1D transformations are λ-shaped.

For the impact-ionization model, peak values of the carrier density $\nu(t_n)$ are measured, where $n = 1, 2, \ldots;\ t_n - t_{n-1} = T_0(T_0$: driven period). In the postbreakdown regime, as a function of the dc field, one obtains the bifurcation diagram in Fig. 4.79. By decreasing the field E_0 from above, the period doubling cascade is seen, which is very similar to the logistic map behavior in Figs. 4.60 and 4.76.

The parameter values of E_0 for the bifurcation points is summarized in Table 4.3. The asymptotic value of the ratio δ_n is approximately 4.4, which is compared to the Feigenbaum constant $\delta = 4.66292$. Plotting the 1D mapping T: $1 - \nu_n \to 1 - \nu_{n+1}$ of the neutral-donor density $\nu_D = 1 - \nu$ exhibits the λ-shaped transformation, as seen in Fig. 4.80. The result is compared to the 1D transformations in the BZ reaction system [72, 73] and the Onchidium giant neuron [74] in Figs. 4.81 and 4.82.

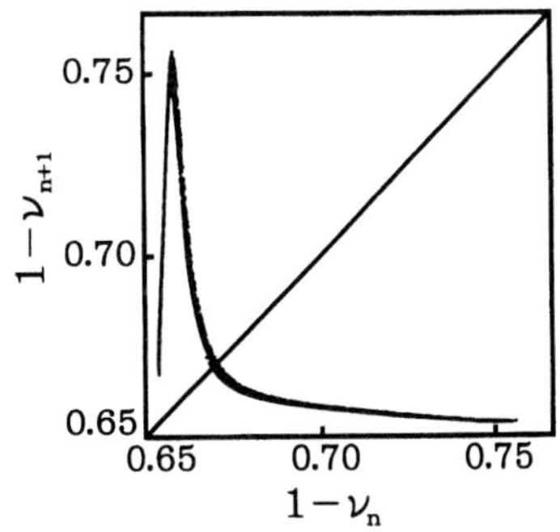

Figure 4.80 One-dimensional mapping $T: 1-\nu_n \rightarrow 1-\nu_{n+1}$, for the one-level impact-ionization model. $E_0 = 15.752$, $E_{ac} = 4.5$ and $f_d = 100/136$. The integration was made up to $i \leq 5.34 \times 10^5$ with the minimum time step of the integration $\Delta t = 0.01$, where $t = i\Delta t (i = 0, 1, 2, \ldots)$.

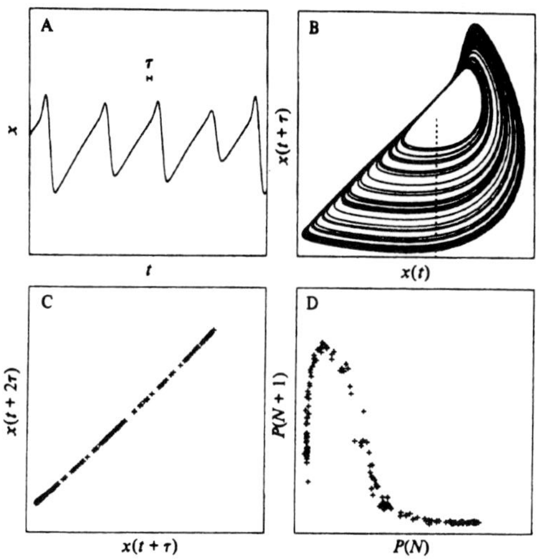

Figure 4.81 Chaotic attractor of the Belousov–Zhabotinskii reaction. The bromide concentration, x, was recorded with a bromide-sensitive electrode. (A) Bromide concentration as a function of time. τ is a time interval selected to construct a three-dimensional phase trajectory $x(t)$, $x(t+\tau)$, $x(t+2\tau)$. (B) Two-dimensional projection of the phase trajectory. The dotted line indicates the position of the cut. (C) Poincaré section of an attractor along the cut. The Poincaré section is very thin, indicating that the flow is confined to an almost two-dimensional surface. (D) Almost one-dimensional map derived from the Poincaré section. Reproduced by Olsen and Degn; see p. 543 in Ref. [53], with permission from Roux, Turner, McCormick & Swinney [72].

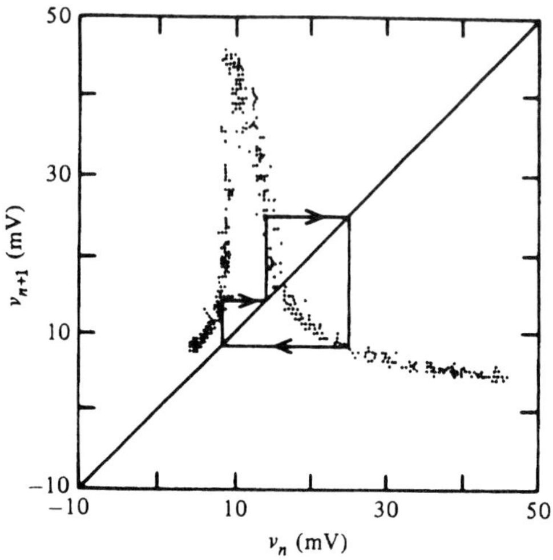

Figure 4.82 Stroboscopic transfer function (one-dimensional mapping) of the chaotic response to periodic current stimulation in the Onchidium giant neuron [74].

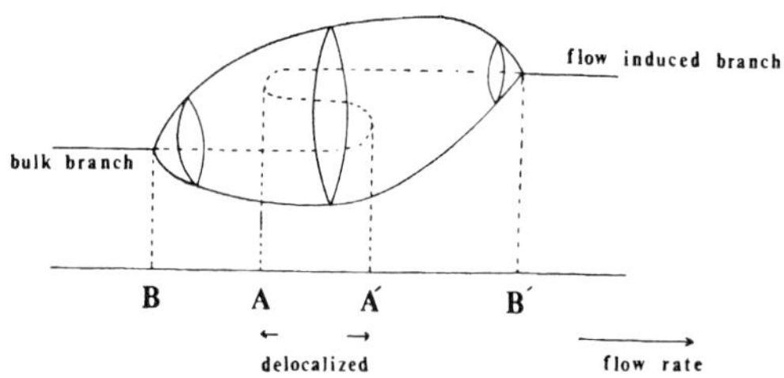

Figure 4.83 Interplay of soft- and hard-mode instabilities (schematic). Phase coordinates of the steady states, represented as a single vertical dimension, are plotted against the control parameter (flow rate) [75].

The interpretation of Hudson's experiment on the B-Z reaction [71] by Tomita and Tsuda [75] is also helpful in explaining semiconductor chaos. For the periodic/chaotic bifurcation sequence observed in the B-Z reaction [71], Tomita and Tsuda [75] have explained the appearance of chaos by the interplay between a soft-mode instability and a pair of hard-mode instabilities, as shown schematically in Fig. 4.83. In the absence of the hard-mode instabilities (B) and (B'), the soft-mode instability means phase transition from bulk branch to flow-induced branch and vice versa, namely bistability, where two basins are separated by an intervening saddle [75]. Under the hard-mode instabilities, all steady states in the bistability become unstable. The limit cycle orbits (solid circle in Fig. 4.83) encircle two branches at least in the middle part of the S-shaped region. Tomita and Tsuda [75] made the interpretation that the chaotic orbits in the periodic/chaotic bifurcation sequence appear as a result of interplay between the upper and lower branches in the S-shaped region via delocalization of the orbits. Based on the experimental data by Hudson *et al.* [71], they have introduced an empirical map function (λ-shaped map function) and argued an essential part of the bifurcation sequence.

In general, the nonlinear chemical reaction systems as argued by Nicolis and Prigogine [76] and neurobiological systems as argued by Zeeman [77] are tightly related to the catastrophe theory [1, 2]. The common features of chaos in Figs. 4.81 and 4.82 suggest that both systems have exhibited nonequilibrium phase transitions with high nonlinearity. In a wider sense, semiconductor chaos can be argued by the common feature of the instabilities in highly nonlinear systems [71–74]. As discussed in section 4.9, limit cycle orbits of the impact ionization avalanche in the postbreakdown regime encircle upper and middle branches (see Fig. 4.47), where the orbits show a sensitive nature in the cliff near the holding field. The interpretation by Tomita and Tsuda [75] is applicable to the semiconductor chaos discussed here.

By taking the 1D mapping $T^2 : 1 - \nu_n \to 1 - \nu_{n+2}$, the symbolic dynamics of the impact-ionization avalanche can be discussed, as has been done in the logistic map [47, 53]. Figure 4.84 shows the result of the T^2 transform for the onset of fully developed chaos, replotted from Fig. 4.80. The chaotic data are almost confined in two regions squared off by U and M, where the symbols U and M means upper and middle branches in the postbreakdown regime and two squares are divided by a saddle point in Fig. 4.84. Just

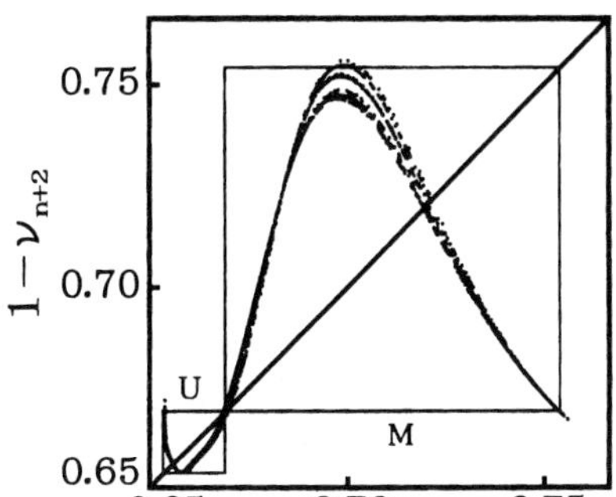

Figure 4.84 One-dimensional mapping $T^2 : 1 - \nu_n \to 1 - \nu_{n+2}$, for the one-level impact-ionization model. $E_0 = 15.752$, $E_{ac} = 4.5$ and $f_d = 100/136$. Values of the parameters are the same as those in Fig. 4.79.

before the onset of fully developed chaos ($E_0 > 15.752$), the chaotic data are completely confined in the two regions. By the T^2 transform, the symbolic sequence is either UUUU... or MMMM..., which is equivalent to the sequence UMUMUM... in the T transform. For the fully developed chaos ($E_0 < 15.752$) as in Fig. 4.84, ergodic mixing of two chaotic bands occurs, followed by the symbolic sequence $\mathrm{U}^k\mathrm{M}^j\mathrm{U}^{k'}\mathrm{M}^{j'}\ldots$ by T^2 and $(\mathrm{UM})^k(\mathrm{MU})^j(\mathrm{UM})^{k'}(\mathrm{MU})^{j'}\ldots$ by T, where $\langle k\rangle$ $(=\langle j\rangle)$ are the mean residence time in units of T_0 that the iterated points by T^2 spend in U or M. The mixing is translated to the intercommunication of the chaotic orbits between the upper and middle branches, as discussed by Tomita and Tsuda [75].

For periodic windows, there is another symbolic dynamics. Metropolis *et al.* [78] have found the U-sequence which holds as a universal rule for different hump maps. For the semiconductor chaos system, the U-sequence is seen only in the limited range of E_0 at the postbreakdown regime. In the logistic map function $f_\mu(x) = \mu x(1-x)$, the symbols "R" is defined when $x > 1/2$ and "L" when $x < 1/2$, where $f^j_\mu(1/2) = 1/2$ and $f'_\mu(1/2) = 0$ for a superstable cycle with period j. Similarly, for other hump maps $g(x)$, R is defined when $x > x_s$ and L when $x < x_s$ where $g^j(x_s) = x_s$ and $g'(x_s) = 0$. In the impact-ionization model, the U-sequence can be speculated numerically, as summarized in Table 4.4.

Taking the symbol "S" for x_S, the repetition sequence of a period 2 is SRSRSR..., and the sequence of period 4 is SRLRSRLR.... In Table 4.4, the symbol S is dropped. By shifting the parameter value E_0 from the superstable points, the repetition sequences of the periodic windows are obtained by replacing S with R or L. For instance, in the period doubling region, the sequence is RRRR... or LRLR... for a period 2 window, and RRLRRRLR... or LRLRLRLR... for a period 4 window. The reader may

Table 4.4 U-sequence found numerically by the one-level impact-ionization model

Impact-ionization model: $E_{ac} = 4.5$, $f_0 = 0.775$		
Period	Symbolic sequence	E_0
2	R	15.733
4	RLR	15.6624
8	RLRRRLR	15.6404
12[a]	RLRRRLRLRLR	15.634473
6[a]	RLRRR	15.625
10[a]	RLRRLRLRR	15.589123
4[b]	–	14.92
5[b]	–	14.75

[a]Narrow periodic windows, coexistence with chaos.
[b]U-sequence unavailable.

try them by using logistic maps. A substitution rule between two windows has been given in detail by Hao and Zheng [79]. In Table 4.3, periods 2, 4 and 8 are the orbits by the period doubling bifurcation. Periods 12, 6 and 10 are found in the narrow windows. Periods 4* and 5* form relatively wide windows, but the U-sequence is unavailable because the 1D transformation deviates from a simple λ-shaped map function as in Fig. 4.80. Metropolis *et al.* [78] have listed numerous U-sequences for many periodic windows, in which the symbolic sequences of Table 4.4 can be found. However, many windows of long periods are too narrow to detect.

As described in section 4.9, one should recognize that the periodic windows are dense in the parameter space. A complete mathematical description for the appearance of windows in a unimodal map $f(x)$ is given by **Sarkovskii's theorem** [50, 52, 80]. This mentions that if $f(x)$ has a solution of period-j cyclic orbit $(x = f^j(x))$ then it must also have solutions of period k for every $k \leftarrow j$ with the sequence, such that:

$$
\begin{aligned}
&1 \leftarrow 2 \leftarrow 4 \leftarrow 8 \leftarrow 16 \ldots 2^n \ldots \leftarrow \\
&\ldots\ 2^n \cdot 9 \leftarrow 2^n \cdot 7 \leftarrow 2^n \cdot 5 \leftarrow 2^n \cdot 3 \ldots \leftarrow \\
&\ldots \\
&\ldots\ 2^2 \cdot 9 \leftarrow 2^2 \cdot 7 \leftarrow 2^2 \cdot 5 \leftarrow 2^2 \cdot 3 \ldots \leftarrow \\
&\ldots\ 2 \cdot 9 \leftarrow 2 \cdot 7 \leftarrow 2 \cdot 5 \leftarrow 2 \cdot 3 \ldots \leftarrow \\
&\ldots\ 9 \leftarrow 7 \leftarrow 5 \leftarrow 3
\end{aligned}
$$

where arrows mean that period j "precedes" period k by $k \leftarrow j$ [50, 52, 80]. Thus, in the theorem, period 3 precedes any other periods, which has been expressed impressively by Li and York, "Period three implies chaos" [81]. However, Sarkovskii's theorem states neither the stability of the periodic orbits nor the range of the parameter values of the windows. In addition to the theorem, there should be many possible windows of a specified period j, as expected from the combination theory.

4.12 Semiconductor chaos and the dissipative structure

So far, focusing on semiconductor chaos, the fundamentals of the nonlinear dynamical theory, the catastrophe theory and the chaos theory have been described. In order to enrich the situation, dynamical behaviors in other dynamical systems have also been introduced. This chapter concludes by supplementing some important physical concepts of the impact-ionization avalanche and chaos in semiconductors from a viewpoint of nonequilibrium thermodynamics.

According to Nicolis and Prigogine [76], when a physical system is biased into the nonequilibrium condition far from the thermal equilibrium state, the system forms a **dissipative structure** or **coherent fluctuation** (spatial pattern). Thom [1] described in his catastrophe theory that the state of the conflict stratum in the cusp catastrophe forms the shock wave and thus the pattern formation. In the nonlinear carrier transport of semiconductors related to the impact-ionization avalanche, the dissipative structure is followed by the filament structure [4], as described in chapter 3. Thermodynamical considerations of the impact-ionization avalanche will give a physical aspect rather than the mathematical part of the chaos theory. In open systems, the total entropy production rate per unit volume is composed of the external entropy flow $\mathrm{d}S_\mathrm{e}/\mathrm{d}t$ and the internal entropy production rate of the system $\mathrm{d}S_\mathrm{i}/\mathrm{d}t$;

$$\mathrm{d}S/\mathrm{d}t = \mathrm{d}S_\mathrm{e}/\mathrm{d}t + \mathrm{d}S_\mathrm{i}/\mathrm{d}t, \tag{4.115}$$

where $S = \int \mathrm{d}V s_\mathrm{V}$ (s_V: the entropy density) and $\mathrm{d}S_\mathrm{i}/\mathrm{d}t > 0$ for the irreversible process. Under the nonequilibrium steady-state condition ($\mathrm{d}S/\mathrm{d}t = 0$),

$$\mathrm{d}S_\mathrm{e}/\mathrm{d}t = -\mathrm{d}S_\mathrm{i}/\mathrm{d}t \leq 0, \tag{4.116}$$

where the equality of the last term holds only under thermal equilibrium, as usual. That is, in order to maintain the nonequilibrium steady state, the positive entropy production rate in the system must be balanced with the negative one flowing into the system from outside. For the dissipative structure of the current filament, the minimum entropy production rate is required:

$$\mathrm{d}_X P = \mathrm{d}_J P \leq 0, \tag{4.117}$$

where $P = \mathrm{d}S/\mathrm{dt}$, X is the generalized force, J is the flow of the particle, and $\mathrm{d}_X, \mathrm{d}_J$ are the variations with respect to X and J. Ridley has discussed the stability of the current filament in the S-shaped J–E characteristic and the domain structure in the N-shaped J–E characteristic based on the minimum entropy production rate. Here, it is shown that the **excess entropy production** $\delta_X P$ [76] describes more precisely the nonequilibrium phase transition and the dynamical behaviors.

The reaction diagram of the impact-ionization process (1.43), (1.44) is rewritten as:

$$\mathrm{D}^0 \underset{B_\mathrm{T}}{\overset{A_\mathrm{T}}{\rightleftharpoons}} \mathrm{D}^+ + \mathrm{e}^-, \tag{4.118}$$

$$\mathrm{e}^- + \mathrm{D}^0 \underset{B_\mathrm{T}}{\overset{A_\mathrm{I}}{\rightleftharpoons}} \mathrm{D}^+ + 2\mathrm{e}^-. \tag{4.119}$$

Using the notations $\overrightarrow{w}$ and $\overleftarrow{w}$ from Nicolis and Prigogine [76], the rate equation for the carrier density n of (4.118) is rewritten as:

$$(\mathrm{d}n/\mathrm{d}t)_1 = \overrightarrow{w}_1 - \overleftarrow{w}/2, \tag{4.120}$$

where $\overrightarrow{w}_1 = A_T N_D^0$ and $\overleftarrow{w} = B_T n\ N_D^+$. Similarly, for (4.119),

$$(\mathrm{d}n/\mathrm{d}t)_2 = \overrightarrow{w}_2 - \overleftarrow{w}/2, \qquad (4.121)$$

where $\overrightarrow{w}_2 = A_I n N_D^0$. For the total reaction system (4.118) and (4.119), one obtains:

$$\begin{aligned} \mathrm{d}n/\mathrm{d}t = (\mathrm{d}n/\mathrm{d}t)_1 + (\mathrm{d}n/\mathrm{d}t)_2 &= \overrightarrow{w}_1 + \overrightarrow{w}_2 - \overleftarrow{w} \\ &= A_I n N_D^0 + A_T N_D^0 - B_T n N_D^+. \end{aligned} \qquad (4.122)$$

The affinity force A_ρ for the ρth reaction ($\rho = 1, 2$) is given by [76]:

$$A_\rho = -\sum_k \mu_k\ \nu_{k\rho}, \qquad (4.123)$$

where μ_k is the chemical potential for the kth component of the particles and $\nu_{k\rho}$ is the stoichiometric coefficient. $\mu_1 = \mu_{1'} = -\varepsilon_D$ is taken for the neutral donors, $\mu_2 = \mu_{2'} = -\varepsilon_D$ for the ionized donors and $\mu_{3'} = \varepsilon'$, $\mu_3 = \varepsilon$ for the electrons, where $\mu_{k'}, \varepsilon'$ are the chemical potentials for the forward process, and μ_k, ε are those for the reverse process in (4.118) and (4.119). The chemical potentials for the neutral and ionized donors cancel each other out in the affinity force (4.123). The chemical potentials ε and ε' for the free electrons are not identical ($\varepsilon \neq \varepsilon'$) under the nonstationary state, where $\varepsilon = \frac{3}{2} k_B T_e$ and $\varepsilon' = \frac{3}{2} k_B T'_e$. From (4.118), (4.119) and (4.123), one obtains

$$A_1 = A_2 = \varepsilon - \varepsilon', \qquad (4.124)$$

where, in (4.123), the stoichiometric coefficients of the neutral donors are $\nu_1 = -1(+1)$ for the forward process (the reverse process) and the values of the ionized donors and the electrons are $\nu_2 = \nu_3 = +1(-1)$ for the forward process (the reverse process). By neglecting the heat flow $\mathbf{J}_q$ and the thermal force $\mathbf{X}_q$ in the entropy production rate given by Ridley [82] and combining it with the formula from Nicolis and Prigogine [76], the production rate of the entropy density s_v is:

$$\frac{\mathrm{d}s_v}{\mathrm{d}t} = -\ \mathrm{div}\ \mathbf{J}_s - \sum_k \mathbf{J}_k \frac{\nabla \mu_k}{T} + \frac{1}{T}[\mathbf{EJ} + \sum_\rho A_\rho w_\rho] \qquad (4.125)$$

$$= -\ \mathrm{div}\ \mathbf{J}_s + s_{iv}, \qquad (4.126)$$

where $\mathbf{J}_s$ is the entropy flow given by

$$\mathbf{J}_s = \mathrm{div}\left\{\frac{1}{T}\sum_k \mu_k \mathbf{J}_k\right\},$$

$\mathbf{J}_k$ is the particle current of the kth component, T is the lattice temperature (fixed), E is the electric field, $\mathbf{J} = \sum e_k \mathbf{J}_k$ is the current density (e_k: the charge of the k-th component) and s_{iv} in (4.126) is the internal production rate of the entropy density. In (4.125), the charge neutrality condition

$\sum_k e_k \rho_k = e(N_D - N_A - n) = 0$ is assumed. In (4.125), the term $\sum_\rho A_\rho w_\rho$ is the expression by Nicolis and Prigogine [76], and the external entropy production rate s_{ev} is formally given by $-$ div $\mathbf{J}_s$. From (4.120), (4.121), (4.123) and (4.124), the last term in (4.125) is rewritten as:

$$\sum_\rho A_\rho w_\rho = (\varepsilon - \varepsilon')(\overrightarrow{w} - \overleftarrow{w}), \tag{4.127}$$

$$= k_B T\{\log(\overrightarrow{w}/\overleftarrow{w})\}(\overrightarrow{w} - \overleftarrow{w}), \tag{4.128}$$

where $\overrightarrow{w} = \overrightarrow{w}_1 + \overrightarrow{w}_2$. The logarithmic term in (4.128) is the definition [76] by

$$A = \varepsilon - \varepsilon' = k_B T \log(\overrightarrow{w} / \overleftarrow{w}). \tag{4.129}$$

Under the steady-state condition, the energy production rate $\sum_\rho A_\rho w_\rho$ by the impact-ionization process is zero, since $\varepsilon' = \varepsilon$ and $\overrightarrow{w} = \overleftarrow{w}$. Then, from the stationary energy balance equation given by

$$\mathbf{EJ} - \frac{\varepsilon - \varepsilon_L}{\tau_\varepsilon} = 0,$$

one obtains

$$\frac{ds_v}{dt} = -\operatorname{div}\mathbf{J}_S + (\mathbf{J}/e)\cdot\frac{\nabla\varepsilon}{T} + \frac{1}{T}\frac{\varepsilon - \varepsilon_L}{\tau_\varepsilon} = 0, \tag{4.130}$$

where the particle current is only by the electrons ($-e\mathbf{J}_e = \mathbf{J}$) and the chemical potential μ_k is replaced by the mean energy ($\varepsilon = \frac{3}{2}k_B T_e$; T_e: the electron temperature). Therefore, under the nonequilibrium steady-state condition, the homogeneous part of the internal production rate ($\mathbf{EJ} + \sum_\rho A_\rho w_\rho$) of the energy density is equal to the rate of energy loss to the lattice.

Now, let us discuss the excess entropy production. The excess entropy production [76] due to the homogeneous part of the entropy production rate in (4.125) and (4.126) is given by $\delta_x P = \int dV[\delta\mathbf{J}\cdot\delta\mathbf{E} + \delta_n A\ \delta_n w]/T$, where the term $\delta_n A\ \delta_n w$ is the product of the variation in the reaction speed w and the variation in the affinity force A. Then,

$$\delta_x P = \int dV\left[\frac{\delta\mathbf{J}\ \delta\mathbf{E}}{T} + \frac{k_B}{\overrightarrow{w}}\left(\frac{\partial\overrightarrow{w}}{\partial n} - \frac{\partial\overleftarrow{w}}{\partial n}\right)^2 (\delta n)^2\right]. \tag{4.131}$$

The stability condition of the system is given by [76]:

$$\begin{aligned} &\delta_X P < 0 : \text{unstable}, \\ &\delta_X P > 0 : \text{asymptotically stable}, \\ &\delta_X P = 0 : \text{critically stable}. \end{aligned} \tag{4.132}$$

For the voltage-controlled bias condition ($\delta E = 0$), the critical state $\delta_X P = 0$ in (4.131) corresponds to the transition points between the stable and the unstable states at the breakdown field E_b and the holding field E_h, since $\partial\overrightarrow{w}/\partial n = \partial\overleftarrow{w}/\partial n$, and $\delta E = 0$. Actually, it is well known that the switching occurs either at E_b or at E_h in the S-shaped J–E curve. For the bias condition of an inclined load line ($\delta E \neq 0$), $\delta_X P < 0$ may occur just after the transition points, since $\delta J > 0$, $\delta E < 0$ at E_b, and $\delta J < 0$, $\delta E > 0$ at E_h. From the thermodynamical picture in (4.125), the pattern formation of the current filament will proceed via the energy flow terms (the first and second terms in (4.125)). However, the stable pattern formation in the static regime is allowed only by the breakdown of the charge neutrality condition (see Chapter 3), as stated by Schöll [4], whereby the transverse drift current counterbalances the diffusion current.

For the dynamical regime, the entropy production rate and the excess entropy production oscillate around the value of the unstable equilibrium state. In the numerical simulation using the approximated impact-ionization model, the density of entropy production rate $(\mathbf{JE} + \sum_\rho A_\rho w_\rho)/T$ is replaced by:

$$s_i = \frac{\mathbf{JE} + Aw}{T} = K[\nu \mathbf{E}^2 + (\overrightarrow{w} - \overleftarrow{w}) \log(\overrightarrow{w} / \overleftarrow{w})] \tag{4.133}$$

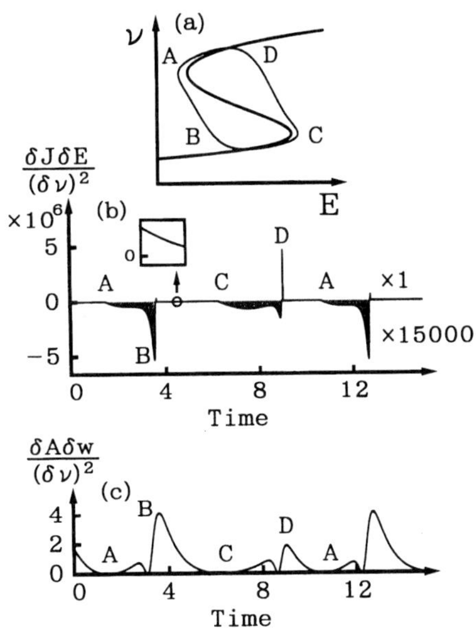

Figure 4.85 (a) (ν, E) phase portrait of self-sustained relaxation oscillation in the approximated one-level impact-ionization model, where $E_0 = 14.8$ and $c = 0.78$ in (4.80), (4.81). (b) Contribution to the density of the excess entropy production by $\delta J \delta E/(\delta\nu)^2$, where $J = \nu E$, $\delta J = (E + \nu\partial E/\partial\nu)(\delta\nu)$, and $\delta E = (\partial E/\partial\nu)(\delta\nu)$ in (4.134). Note that the negative part (gray region) was magnified by a factor of 15 000. (c) Contribution to the density of the excess entropy by $\delta A \delta w/(\delta\nu)^2$.

and the density of the excess entropy production by

$$\delta_x p = K\left[\left(E + \nu\frac{\partial E}{\partial \nu}\right)\left(\frac{\partial E}{\partial \nu}\right) + \frac{1}{\overrightarrow{w}}\left(\frac{\partial w}{\partial \nu}\right)^2\right](\delta\nu)^2, \tag{4.134}$$

where $w = \overrightarrow{w} - \overleftarrow{w}$, and the normalization constant K is given by $K = e(N_D - N_A)/(\mu_0 T_L)$: the zero-field electron mobility. In (4.133), (4.134) note that all of the variables $\nu, E, \overrightarrow{w}$ and $\overleftarrow{w}$ are scaled quantities.

In (4.134), the numerical calculation can be done replacing $\partial E/\partial \nu$ with $\dot{E}/\dot{\nu}$. Figure 4.85b, c shows the evolution of the density of the excess entropy production for the case of self-sustained relaxation oscillation, where the parameters are $c = 0.78, E_0 = 14.8$ and $E_{ac} = 0$ in (4.80)–(4.82). During one cycle of the oscillation $A \rightarrow B \rightarrow C \rightarrow D \rightarrow A$; see the $\nu - E$ phase portrait in Fig. 4.85a), the excess entropy production by $\delta\mathbf{E}\ \delta\mathbf{J}$ becomes negative just after each transition (the gray regions in Fig. 4.85b). At transition points A and C, the excess entropy production by Aw is critically zero in Fig. 4.85c.

In Fig. 4.86b, c, evolution of the entropy production rates **JE** and Aw are shown for driven chaos, where $c = 0.78$, $E_0 = 15.7$, $E_{ac} = 4.5$ and $f_0 = 0.735$ (the parameters are the same as in Figs. 4.80 and 4.84). For reference,

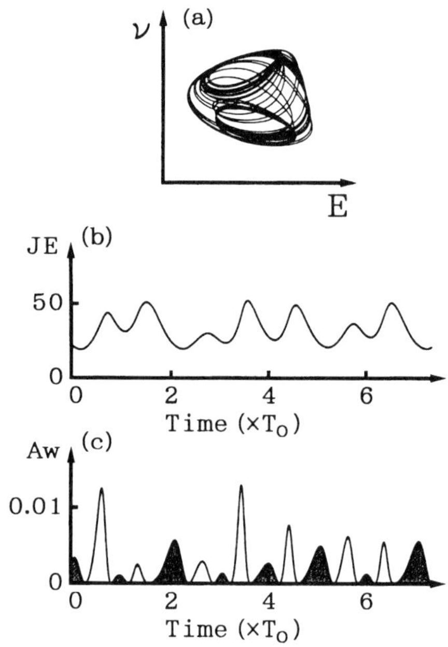

Figure 4.86 (a) (ν, E) phase portrait of chaos under the ac drive in the approximated impact-ionization model, where $E_0 = 15.7$, $E_{ac} = 4.5$ and $f_d = 100/136$. (b) JE, and (c) Aw as a function of time in units of the driving period $(T_0 = 1/f_d)$.

the $\nu - E$ phase portrait is given in Fig. 4.86a. First, note that $\mathbf{JE} \gg Aw > 0$, which means that the energy production rate per unit volume due to the catalytic process (4.118), (4.119) is much less than **JE** supplied by the voltage source (i.e., ac + dc bias). The gray regions in Fig. 4.86c indicate the recombination part in the catalytic g-r process such that $A < 0$, $w < 0$ and $Aw > 0$ ($\dot{\nu} < 0$ in Fig. 4.86a).

Figure 4.87b, c shows the excess entropy production by JE and Aw respectively. For reference, the J–E phase portrait is also shown in Fig. 4.87a. In Fig. 4.87b, the negative part of the excess entropy production is

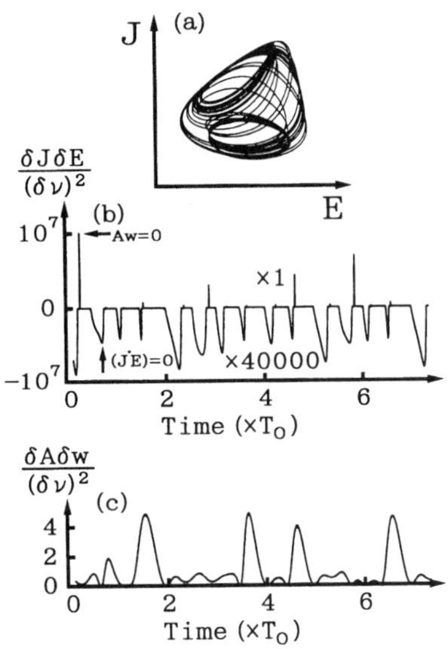

Figure 4.87 (a) (J, E) phase portrait of chaos under the ac drive in the approximated impact ionization model, where $E_0 = 15.7$, $E_{ac} = 4.5$ and $f_d = 100/136$. (b) Contribution to the density of the excess entropy production by $\delta J \delta E/(\delta \nu)^2$. (c) Contribution to the density of the excess entropy by $\delta A \delta w/(\delta \nu)^2$.

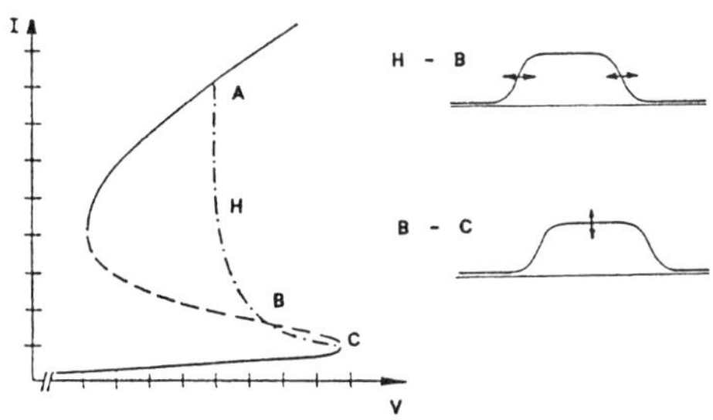

Figure 4.88 Left: Stationary current–voltage characteristic including uniform states (full line: stable; broken line: unstable) and filamentary states (dot–dashed). Right: Breathing mode (H-B) and a bulk-dominated oscillation mode (B-C) [83]

magnified by a factor of 4×10^4. Again, we note that $\delta\mathbf{J}\ \delta\mathbf{E} \gg \delta A\ \delta w$ and therefore $\delta\mathbf{J} \cdot \delta\mathbf{E} + \delta A \delta w \sim \delta\mathbf{J} \cdot \delta\mathbf{E}$. The positive sharp spikes in Fig. 4.87b correspond to $Aw = 0$ and thus $\vec{w} = \overleftarrow{w}$ in Fig. 4.86c. The negative minima of $\delta\mathbf{J} \cdot \delta\mathbf{E}$ correspond to the local minima and the local maxima of **JE** (i.e., $\mathrm{d}(\mathbf{JE})/\mathrm{d}t = 0$) in Fig. 4.86b. The negative part of the excess entropy production $\delta\mathbf{J} \cdot \delta\mathbf{E}$ is highly important: it may be a source of **spatiotemporal turbulence** or the chaotic dissipative structure. If the carrier transport is ohmic ($\mathbf{J} \propto \mathbf{E}$), the excess entropy production is positive and no dissipation structures appear. In order for $\delta\mathbf{J} \cdot \delta\mathbf{E} < 0$, $(\mathrm{d}J/\mathrm{d}t)\ (\mathrm{d}E/\mathrm{d}t) < 0$ is required (see Fig. 4.86a). The dynamical behaviors in an S-shaped J–E characteristic may reveal the electronic turbulence of the current filament in the chaotic regime.

Concerning the spatiotemporal behaviors of the current filament, Schöll [83] has classified formally two types of spontaneous current instability: (a) breathing mode (H-B) and (b) bulk-dominated mode (B-C), as shown in Fig. 4.88. In the breathing mode, the size of the current filament (kink–antikink pair) oscillates periodically or in a chaotic manner. The breathing mode is relevant in the postbreakdown regime, while the bulk-dominated mode is the nucleation dynamics of the current filament in the prebreakdown regime. Although Schöll's model is restricted to spontaneous current oscillations, the physical picture may also be applied to the periodically driven instability in n-GaAs [19, 32, 33]. Under dc bias, no spontaneous current oscillations occur in n-GaAs. In the postbreakdown regime, a stationary filamentary current flows. By applying a dc + ac bias, chaotic behaviors and bifurcations are observed in the postbreakdown regime. By considering an experimental fact observed under the dc bias, that the size of the filament increases monotonically by increasing the dc bias, it is very natural to conjecture that the periodic or chaotic behaviors in the postbreakdown regime are due to the enforced breathing mode [83]. Studies on spatiotemporal behaviors have just begun [24]. Further discussion of this problem is presented in chapter 6.

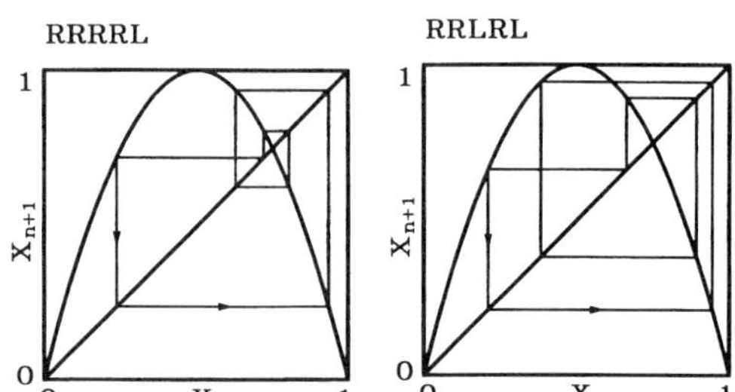

Figure 4.89 Unstable periodic orbits constructed by the time reversal $x = f^{-1}(y)$ in the logistic map, where $\mu = 4$.

This chapter will be summarized from a global viewpoint. The semiconductor chaos system is a typical example which is characterized by the nonlinear dynamical theory, the chaos theory, the catastrophe theory and the nonequilibrium thermodynamical theory. By the chaos theory, the deterministic nature of the chaotic current instability is found in terms of Lyapunov exponents [50, 52, 53], Kolmogorov entropy [50, 52, 53] and fractal dimensions [44, 50, 52, 53]. Since semiconductor chaos arises from the nonequilibrium phase transition, the nonequilibrium thermodynamical nature could be a true origin of the chaotic behaviors. Then, if this is so, are there any common features between the chaos theory and the thermodynamical theory?

The answer may be quite difficult. However, one of the features is the loss of memory in dissipative structure. In chapters 1 and 4, chaos was shown to possess many unstable periodic orbits, i.e., chaos retains the bifurcation history. In the logistic map $y = f_\mu(x)$ with $\mu = 4$, one can visualize any kind of unstable periodic orbit by the time reversal $x = f_\mu^{-1}(y)$ if the rule is settled artificially, e.g., $R^2L^3R^2L^3\ldots$ (see Fig. 4.89).

In that sense, the dynamical structure has not lost the memory completely. However, without control of chaos, one cannot realize any kind of periodic orbit in the chaotic regime. Thus, a trademark of chaos is the loss of memory and information [52]. In the chaos theory, the Lyapunov exponent or the Kolmogorov entropy is a good indicator of the dynamical structure.

In the theory of nonequilibrium thermodynamics, the excess entropy production is a good indicator of the dissipative structure. By the phase transition, the system loses the old phase and falls into a new phase of the dissipative structure, where the excess entropy production is critically zero or negative at the transition point. Once the system gains the new phase, the system loses the memory of the previous phase. In the static regime of bistability, the relation between the criterion of structural stability in the catastrophe theory [1, 2] and the criterion of excess entropy production is apparent. In the dynamical regime, the dissipative structure is violated, followed by successive memory loss during the periodic or chaotic evolution of the structure. However, the latter is not fully understood in semiconductor systems.

References

[1] R. Thom, *Structural Stability and Morphogenesis – An Outline of a General Theory of Models* (Benjamin/Cumings, London, 1975).

[2] T. Poston and N. Stewart, *Catastrophe Theory and its Applications* (Pitman, London, 1978).

[3] J.M.T. Thompson and G.W. Hunt, *A General Theory of Elastic Stability* (Wiley & Sons, New York, 1973).

[4] E. Schöll, *Nonequilibrium Phase Transitions in Semiconductors – Self-Organization Induced by Generation and Recombination Processes* (Springer, Berlin, 1987).
[5] M.V. Berry, J. Phys. **A8** (1975) 566.
[6] J. Grasman, *Asymptotic Methods for Relaxation Oscillations and Applications,* Appl. Math. Sciences, Vol. **63** (Springer, Verlag, Berlin, 1987).
[7] H. Haken, Z. Phys. **B20** (1975) 413.
[8] H. Haken, *Synergetics – An Introduction* (Springer, Berlin, 1978).
[9] A.E.R. Woodcock and T.A. Poston, *A Geometrical Study on the Elementary Catastrophes,* Lecture Notes in Mathematics 373 (Springer, Berlin, 1974).
[10] Th. Bröcker and L. Lander, *Differentiable Germs and Catastrophes,* London Mathematical Society Lecture Notes 17 (Cambridge University Press, London, 1975).
[11] K. Aoki, Y. Kawase, K. Yamamoto and N. Mugibayashi, J. Phys. Soc. Jpn **59** (1990) 20.
[12] K. Aoki and T. Kondo, Phys. Lett. **154A** (1991) 281.
[13] K. Aoki, Solid State Commun. **77** (1991) 87.
[14] K. Aoki, Phys. Lett. **152A** (1991) 485.
[15] R.L. Devaney, *An Introduction to Chaotic Dynamical Systems,* 2nd ed. (Benjamin/Cumings, London, 1989).
[16] Van der Pol, Phil. Mag. **2** (1926) 978.
[17] K.F. Bonhoeffer, J. Gen. Physiol. **32** (1948) 69.
[18] R. FitzHugh, Bull. Math. Biophys. **17** (1955) 257.
[19] K. Aoki and K. Yamamoto, Appl. Phys. **A48** (1989) 111.
[20] J. Argemi and B. Rossetto, J. Math. Biol. **17** (1983) 67.
[21] V.I. Arnold, *Catastrophe Theory,* 2nd ed. (Springer, Verlag, Berlin, 1986); for the seven elementary catastrophe see I.N. Stewart, New Scientist **68** (1975) 447.
[22] M.W. Hirsch and S. Smale, *Differential Equations, Dynamical Systems, and Linear Algebra* (Academic Press, New York, 1974); for an application of catastrophe theory to the Duffing's equation see P.J. Holmes and D.A. Rand, J. Sound Vibr. **44** (1976) 237.
[23] E.C. Zeeman, Scient. Am **234** (1976) 65.
[24] F.-J. Niedernostheide ed., *Nonlinear Dynamics and Pattern Formation in Semiconductors and Devices* (Springer, Berlin, 1995).
[25] J.M.T. Thompson and H.B. Stewart, *Nonlinear Dynamics and Chaos* (Wiley and Sons, Chichester, 1986); see also T. Fang and E.H. Dowell, Int. J. Non-Linear Mech. **22** (1987) 401.
[26] Y. Ueda, J. Stat. Phys. **20** (1979) 181; see also Y. Ueda, Int. J. Non-Linear Mech. **20** (1985) 481.
[27] J. Guckenheimer and P.J. Holmes, *Nonlinear Oscillations, Dynamical Systems and Vector Fields* (Springer, Berlin, 1982).
[28] B. Van der Pol and J. Van der Mark, Nature **120** (1927) 63; M. Levi, Mem. Am. Math. Soc. **32**, No. 244 (1981) 1.
[29] C. Grebogi, E. Otto and J.A. Yorke, Phys. Rev. Lett. **50** (1983) 935.
[30] C. Grebogi, E. Otto and J.A. Yorke, Phys. Rev. Lett. **56** (1986) 1011.
[31] W.H. Press, B.P. Flannery, S.A. Teukolsky and W.T. Vettering, *Numerical Recipes in C – The Art of Scientific Computing* (Cambridge University Press, Cambridge, 1988).
[32] K. Aoki, K. Yamamoto and N. Mugibayashi, J. Phys. Soc. Jpn **57** (1988) 26.

[33] K. Aoki, K. Yamamoto and N. Mugibayashi, Proc. 19th Int. Conf. on Phys. Semicond., ed. W. Zawadzki (Institute of Physics, Polish Academy of Science, Warsaw, 1988) p. 1415.
[34] J. Peinke, U. Rau, W. Clauss, R. Richter and J. Parisi, Europhys. Lett. **9** (1989) 743.
[35] B. Röhricht, B. Wessely, J. Peinke, A. Mühlback, J. Parisi and R.P. Huebener, Physica **134B** (1985) 293.
[36] K. Aoki, O. Ikezawa, N. Mugibayashi and K. Yamamoto, Physica **134B** (1985) 288.
[37] G.N. Maracas, W. Porod, D.A. Johnson, D.K. Ferry and H. Goronkin, Physica **134B** (1985) 276.
[38] S.B. Bumeliène, Yu.K. Pozhela, K.A. Pyragas and A.V. Tamaševičius, Proc. 18th Int. Conf. on Phys. Semicond., ed. O. Engstrom (World Scientific, Singapore, 1987) p. 1563.
[39] J. Parisi, J. Peinke, B. Röhricht and K.M. Mayer, Z. Naturforsch. **42a** (1987) 329; J. Peinke, J. Parisi, B. Röhricht, B. Wessely and K.M. Mayer, Z. Naturforsch. **42a** (1987) 841.
[40] J.-P. Eckmann, Rev. Mod. Phys. **53** (1981) 643.
[41] X.N. Song, D.G. Seiler and M.R. Loloee, Appl. Phys. **A48** (1989) 137; D.G. Seiler, C.L. Littler, R.J. Justice and P.W. Milonni, Phys. Lett. **108A** (1985) 462.
[42] G.-X. Li and F.C. Moon, Nonlinear Dynam. **1** (1990) 209.
[43] B.B. Mandelbrot, *Fractals, Form, Chance, and Dimensions* (W.H. Freeman, San Francisco, 1977); *The Fractal Geometry of Nature* (W.H. Freeman, San Francisco, 1982).
[44] J.D. Farmer, E. Ott and J.A. Yorke, Physica **7D** (1983) 153.
[45] K. Aoki, K. Yamamoto and N. Mugibayashi, Proc. 18th Int. Conf. on Phys. Semicond., ed. O. Engstrom (World Scientific, Singapore, 1987) p. 1543; K. Aoki, N. Mugibayashi and K. Yamamoto, Physica Scripta **T14** (1986) 76.
[46] K. Aoki and N. Mugibayashi, Appl. Phys. **A48** (1989) 161.
[47] P. Cvitanovic and P. Predrag, *Universality in Chaos* (Heyden, Philadelphia, 1984).
[48] R.H. Abraham and C.D. Shaw, *Dynamics – The Geometry of Behavior*; Parts 1–4 (Aerial Press, Santa Cruz, 1983–1985).
[49] M. Hénon, Commun. Math. Phys. **50** (1976) 69.
[50] F.C. Moon, *Chaotic and Fractal Dynamics – An Introduction for Applied Scientists and Engineers* (John Wiley & Sons, New York, 1992).
[51] O.E. Rössler, Z. Naturforsch. **38a** (1983) 788; Fig. 4.56 was taken from J. Peinke, J. Parisi, O.E. Rössler and R. Stoop, *Encounter with Chaos* (Springer, Berlin, 1992) p. 112.
[52] H.G. Schuster, *Deterministic Chaos,* 2nd ed. (VHC Publishers, Weinheim, 1988).
[53] Hao Bai-Lin, *Chaos II–An Introduction and Reprints Volume* (World Scientific, Singapore, 1990).
[54] M.J. Feigenbaum, J. Stat. Phys. **19** (1978) 25.
[55] J.-P. Eckmann and D. Ruelle, Rev. Mod. Phys. **57** (1985) 617.
[56] H.G. Winful, Y.C. Chen and J.M. Liu, Appl. Phys. Lett. **48** (1986) 616; see also H.G. Schuster, *Deterministic Chaos–An Introduction,* 2nd ed. (Physik-Verlag, Weinheim, 1988).
[57] Y. Pomeau and P. Manneville, Commun. Math. Phys. **74** (1980) 189.
[58] J.E. Hirsch, B.A. Huberman and D.J. Scalapino, Phys. Rev. **A25** (1982) 519.

[59] Hao Bai-Lin, *Chaos* (World Scientific, Singapore, 1984) and references therein.

[60] P. Berge, M. Dubois, P. Manneville and Y. Pomeau, J. Phys. (Paris) Lett. **41** (1980) L-344.

[61] R. Richter, Z. Naturforsch **46a** (1991) 1012; Diploma Thesis (Tübingen, 1989).

[62] M. Dubois, M.A. Rubio and P. Berge, Phys. Rev. Lett. **51** (1983) 1446.

[63] J. Peinke, J. Parisi, O.E. Rössler and R. Stoop, *Encounter with Chaos* (Springer, Berlin, 1992).

[64] R. Richter, J. Peinke, W. Clauss, U. Rau and J. Parisi, Europhys. Lett. **9** (1989) 743.

[65] C. Grebogi, S.W. Macdonald, E. Otto and J.A. York, Phys. Lett. **A99** (1983) 415.

[66] C. Grebogi, E. Otto and J.A. York, Physica **7D** (1983) 181.

[67] C. Grebogi, E. Otto and J.A. York, Science **238** (1987) 632.

[68] Y. Aizawa and T. Uezu, Prog. Theor. Phys. **68** (1982) 1864.

[69] F.C. Moon, *Chaotic and Fractal Dynamics* (John Wiley & Sons, New York, 1992).

[70] E.N. Lorenz, J. Atmos. Sci. **20** (1963) 130.

[71] J.L. Hudson, M. Hart and D. Marinko, J. Chem. Phys. **71** (1979) 1601; see also J.L. Hudson and J.C. Mankin, J. Chem. Phys. **74** (1981) 6171.

[72] J.C. Roux, J.S. Turner, W.D. McCormick and H.L. Swinney, in *Non-Linear Problems: Present and Future,* ed. A.R. Bishop, D.K. Campbell and B. Nicolaenko (North-Holland, Amsterdam, 1982) pp. 409–422; see also J.C. Roux, R.H. Simoyi and H.L. Swinney, Physica **8D** (1983) 257.

[73] P. Richetti, J.C. Roux, F. Argoul and A. Arneodo, J. Chem. Phys. **86** (1987) 3339.

[74] C. Hayashi, S. Ishizuka and K. Hirakawa, Phys. Lett. **98A** (1983) 474.

[75] K. Tomita and I. Tsuda, Prog. Theor. Phys. **64** (1980) 1138.

[76] G. Nicolis and I. Prigogine, *Self-Organization in Nonequilibrium Systems – From Dissipative Structures to Order through Fluctuations* (John Wiley & Sons, New York, 1977).

[77] E.C. Zeeman, *Brain Modeling – In Structural Stability, the Theory of Catastrophes, and Applications in the Sciences.* Lecture Notes in Mathematics 525, ed. P.J. Hilton (Springer, Berlin, 1976).

[78] N. Metropolis, N.L. Stein and P.R. Stein, J. Combinatorial Theory **15** (1973) 25.

[79] B.-L. Hao and W.-M. Zheng, J. Mod. Phys. **B3** (1989) 235.

[80] A.N. Sarkovskii, Ukranian Math. J. **16** (1964) 61.

[81] T.Y. Li and J.A. Yorke, Am. Math. Monthly **82** (1975) 985.

[82] B.K. Ridley, Proc. Phys. Soc. **82** (1963) 954.

[83] E. Schöll, in *Nonlinear Dynamics in Solids,* ed. H. Thomas (Springer, Berlin, 1992) p. 46.

Chapter 5

Observation of chaos and experimental methods

5.1 Encounter with chaos

As described in Chapter 2, research on the nonlinear carrier transport and current instabilities in semiconductor physics progressed significantly during the 1960s and 1970s. Especially in the 1960s, during the early years of the semiconductor physics, a huge number of papers reported on the impact-ionization mechanism [1, 2], plasma instability [3], helical instability under longitudinal magnetic fields [4–7] and Gunn oscillation [8, 9]. The enormous efforts of the researchers can be seen in the books by Bonch-Bruevich [10] and Pozhela [3]. The 1960s were the pioneering age of semiconductor physics and modern electronics technology, following the development of the transistor by Bardeen and Brattain in 1948 [11]. Researchers tried to find a new paradigm in solid-state electronics, which led to a breakthrough towards the brilliant age of the electronics revolution. There were investigations applying the high-nonlinearity phenomena of the carrier transport such as the Gunn effect [8] and also the tunneling effect by Esaki [12].

Studies on plasma instability under longitudinal magnetic fields [4–7] and the impact-ionization avalanche [1, 2] fell into the same research field. For instance, Larrabee and Steele [13] proposed a device, the "**oscillistor**", which is a kind of oscillator, using plasma instability in n-Ge under a longitudinal magnetic field. McWhorter and Rediker [14] proposed a device the "**cryosar**" [low-temperature (*cryo*-) *s*witching by *a*valanche and *r*ecombination], which is a switching element (memory device) using the impact-ionization avalanche of shallow neutral impurities and the negative differential resistance (monostable cryosar in In-doped p-type germanium, bistable cryosar in In-doped p-type germanium compensated with Sb). Melngailis and Rediker [15] proposed the "**madistor**" a magnetic-controlled semiconductor plasma device using InSb. Madistors are active elements (four-terminal amplifier, flip-flops, stepping resistors and variable gain transistors), as shown in Fig. 5.1. For device fabrication, the following conditions were required:

- high-speed operations
- clear-cut switching operation of "on" and "off"
- low dissipation power
- no significant spiky reaction just after the on-state switching and low noise
- room-temperature operation.

By forming an array of N cryosars in a matrix on a wafer, McWhorter and Rediker [14] proposed a new device which can even be applied to digital computers. The proposals by McWhorter and Rediker [14] and Melngailis and Rediker [15] seemed to succeed, but failed mainly because of their low-temperature operations.

The case of the oscillistor presented by Larrabee and steele [13] was more difficult. They described in their paper: "The variety of frequencies and waveforms of the self-generated plasma oscillations are observable in a single specimen. Although complicated waveforms seem to be the most common, waveforms approaching sinusoidal are also observed". From a review of many articles published during the 1960s, it was found that many researchers were urgently seeking simple sinusoidal oscillation with high frequency, independent of the sample characters (the device length, surface treatment of the sample, nature of the contacts, etc.). However, these researchers actually experienced irregular current oscillations as an unexpected noisy phenomenon. Holonyak and Bevacqua [16] observed self-generated oscillation in the S-shaped *I–V* characteristic of an alloyed double-injection p-i-n diode (Fig. 5.2). The current oscillations in the postbreakdown regime (the positive resistance region) of Fig. 5.2 look more irregular than simply sinusoidal.

More interesting are the fully irregular oscillations observed under a magnetic field [16]. Similar results were observed in p-type germanium by

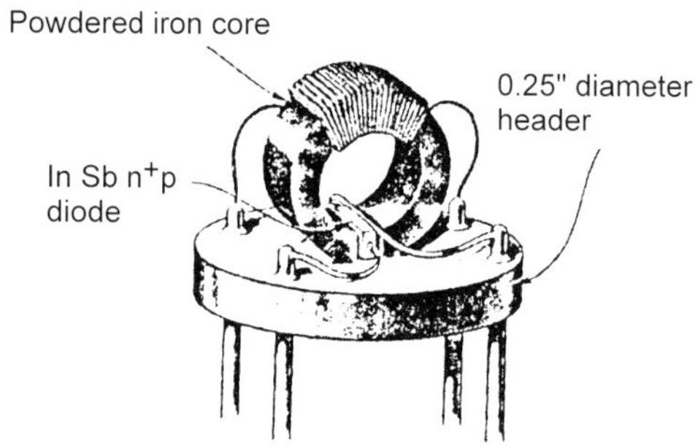

Figure 5.1 Sketch of an experimental diode madistor, showing the diode mounted in the air gap of ferromagnetic toroid [15].

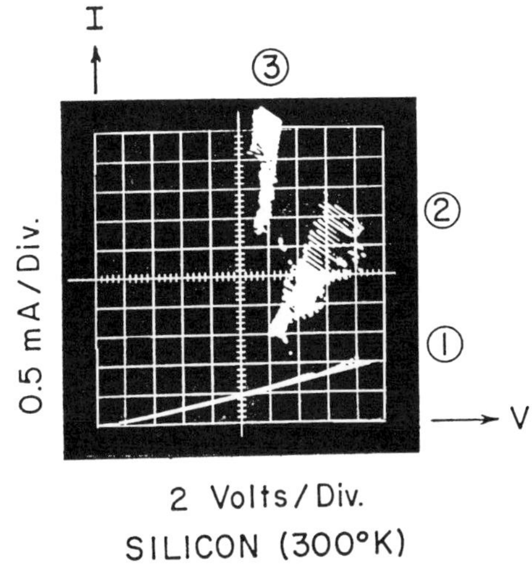

Figure 5.2 I–V characteristics of alloyed (Al-B, Au-Sb) double-injection p-i-n diode fabricated on 9.1 Ω cm n-type Si which was compensated to 104 cm by Co diffused at 1150°C. Positive-resistance-region oscillations evident in regions 1, 2, and 3 [16].

Cardona and Ruppel [17], as pointed out by Peinke [18]. They may have looked like "chaos" and confused with the noisy phenomenon, but chaos theory and the term "chaos" had not yet been proposed during that period. In spite of the intensive research in the field of plasma instability and the impact-ionization avalanche, the carrier transport theory had not yet explained satisfactorily the wide diversity of observed current instabilities.

The first encounter with chaos was reported in 1981–1982 by Aoki *et al.* [19–21]. The discovery was quite accidental. In order to obtain some information about exciton–electron collision and also impact ionization of donor-bound excitons in high purity n-GaAs at 4.2 K [22–29], photoconductivity spectra [20, 29] were measured at 4.2 K by a standard method under dc bias, using a light chopper (200 Hz) and a lock-in amplifier, where the carrier concentration was $n \sim 2 \times 10^{14}/\text{cm}^3$ and the dimension of the sample surface was typically $3 \times 2\ \text{mm}^2$ [19–21, 28, 29]. A 100 W halogen lamp and a grating monochromator were used for the measurements of photoconductivity spectra. The electric signal of the light chopper was fed into the lock-in amplifier as a reference signal. The dc bias was settled near the breakdown voltage for the impact-ionization avalanche. By illuminating a monochromatic light at the wavelengths of the donor-bound exciton and neutral donor, and also at the excited states, the lock-in amplifier became inactive owing to a failure in phase-locking the photocurrent signals. It was soon clarified that the failure was caused by the self-generated spikes of the photocurrent due to the impact-ionization avalanche, where the spikes were generated in a stochastic manner [19, 20, 29] independently of the chopping frequency. Therefore, the measurements were shifted to the different methods using:

(a) the dc bias without light chopping [19, 20, 28, 29]
(b) the pulse voltage (the repetition of 38 kHz) without [21, 30, 31] (or with [32]) light chopping.

With method (a), the photocurrent was recorded directly on the recorder charts (see Fig. 5.3). Figure 5.3 shows the photocurrent spectra [19, 20, 28, 29] observed by method (a) under various photoexcitation densities. The selective generations of spike currents at the wavelength of donor-bound excitons (D^+, X) and (D°, X) and excited donor state are shown in Fig. 5.3. Aoki *et al.* [19, 20, 28] also paid attention to the waveforms and the Fourier power spectra of the spike currents, and found small amplitudes of the current oscillations superimposed on each spike current. Waveforms of the spike currents observed by He–Ne laser excitation [28] are shown in Fig. 5.4 (a, b), where the photoexcitation density is (a) $J_p = 3.2\ \mathrm{mW/cm^2}$ and (b) $J_p = 80\ \mathrm{mW/cm^2}$. In Fig. 5.4a, oscillatory structure is clearly seen on the spike current, while the waveform in Fig. 5.4b is quite turbulent. A similar

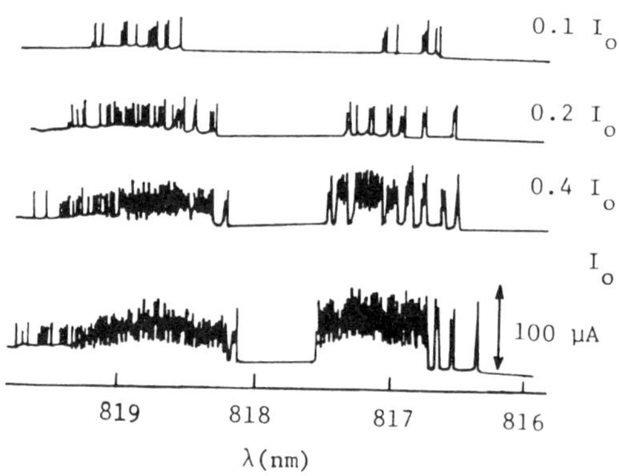

Figure 5.3 Spectra of the photocurrent noise on the recorder at 4.3 V/cm, as a function of excitation wavelength. The maximum photoexcitation density is $I_0 \sim 2\ \mu\mathrm{W/cm^2}$ [19].

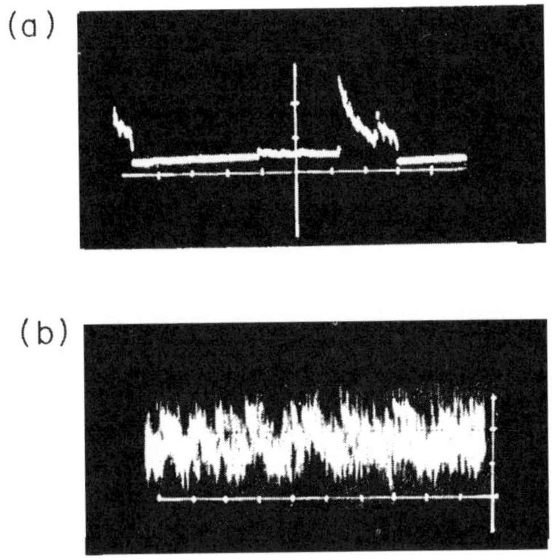

Figure 5.4 Typical waveforms of the current noise observed with photoexcitation density of (a) 3.2 mW/cm^2 and (b) 80 mW/cm^2. The electric field is fixed at 4.5 V/cm. The horizontal scale is 5 ms/div in (a) and 0.5 ms/div in (b). The vertical scale is 67 μA/div in (a) and 17 μA/div in (b). With the photoexcitation on density of 3.2 mW/cm^2 in (a), the oscillatory structure superimposed on each shot noise can be seen [28].

oscillatory structure has been measured for the photocurrent spikes in Fig. 5.3 [19]. The Fourier power spectra observed by the resonant photoexcitation at the (D°, X) line ($J_p = 2\ \mu$W; $\lambda = 818.5$ nm; see the bottom part of Fig. 5.3) have shown two incommensurate frequencies f_1 and f_2 as shown in Fig. 5.5 [19]. "Incommensurate" means that the ratio f_1/f_2 is irrational. By increasing the light intensity, broad band spectra of the power have been observed [19, 20, 28]. The peculiar behaviors have been concluded as being due to chaos [19]. In Ref. [19], the terminology "turbulence of hot carrier plasma" was used.

By method (b), Aoki *et al.* [21, 30–32] have observed chaotic behaviors including period doubling bifurcation at extremely low frequencies (0.1 to a few Hz) and analyzed them using a model simulation [32]. They termed the oscillatory behaviors "firing wave instability" (see section 5.2). Aoki *et al.* have studied further periodically driven chaos (section 5.5), crossover instability (sections 4.1, 7.4) and many other chaotic behaviors (discussed in this chapter).

In 1983, Teitsworth *et al.* [33] reported self-generated nonlinear oscillations in a photoconductor (ultrapure p-type Ge) under far-infrared (FIR) irradiation at 4.2 K. The current instability is induced essentially by the impact ionization of shallow acceptors [33]. The ohmic contacts (injecting p$^+$ layers) have been formed by the B$^+$ ion implantation. They have obtained excellent results of period doubling bifurcation, as shown in

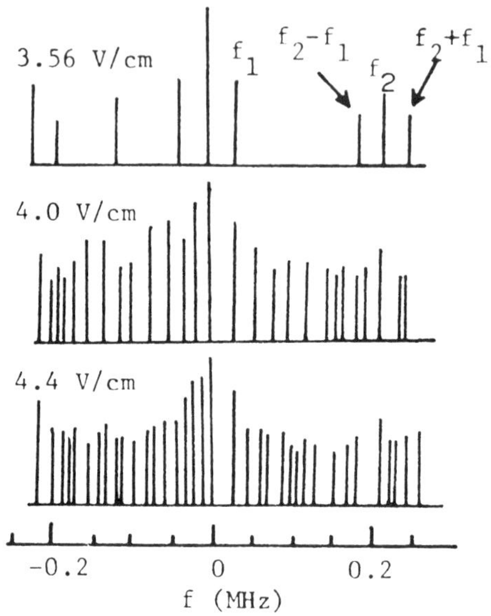

Figure 5.5 Power spectra observed by resonant photoexcitation [19].

Figs. 5.6 and 5.7. In Fig. 5.6, the phase portrait $I(t)$ vs -d$I(t)/dt$ shows the bifurcation from period 1 to period 2, period 4 and chaos by increasing applied bias V. The bifurcation diagram in Fig. 5.7 has been obtained by plotting the measured current minima I_{min} on the vertical axis as a function of V, where the applied bias V has been slowly varied.

Teitsworth and Westervelt [34, 35], Teitsworth [36] and Westervelt and Teitsworth [37, 38] have also investigated periodically driven chaos in the photoconductive p-type Ge ($N_A \sim 3 \times 10^{10}/\text{cm}^3$) under constant FIR irradiation. For the experimental results in the prebreakdown regime [35–38], the authors have presented model simulations [34–36] in order to interpret the physical origin of intrinsic chaos in semiconductors. Primarily, the model is given for the modulated FIR irradiation under dc bias and then developed to the driven system [34]:

$$\dot{p} = \gamma(a - a_0) + p\kappa(a - a_0) - pra^*, \tag{5.1}$$

$$\epsilon\dot{E} = J_{\text{ext}} - epv, \tag{5.2}$$

$$p = a^* - d, \tag{5.3}$$

$$J_{ext} = J_0 + \Delta J \sin(2\pi f_{\text{dr}} t). \tag{5.4}$$

In (5.1), p is the hole concentration, a the total acceptor concentration, a^* the time-dependent ionized acceptor concentration, γ the generation rate,

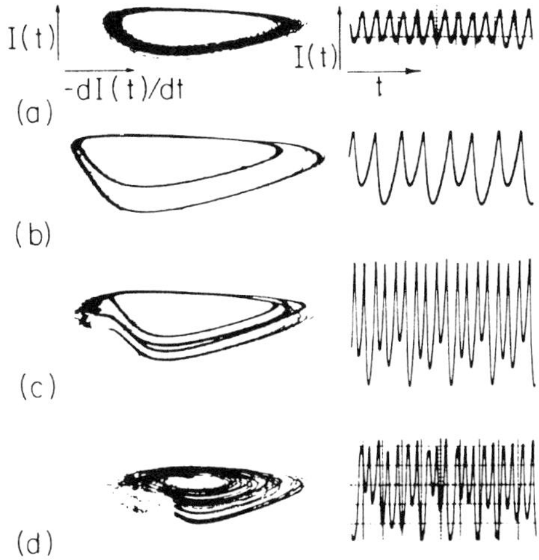

Figure 5.6 Oscillatory current $I(t)$ and phase portrait, $I(t)$ vs $\mathrm{d}I(t)/\mathrm{d}t$, for Ge sample at $T = 4.2$ K with increasing applied voltage V from (a) to (d); $I(t)$ scale/div: (a) 1 nA, 2 ms; (b) 5 nA, 2 ms; (c) 5 nA, 5 ms; (d) 10 nA, 5 ms. Sample geometry $0.4 \times 1 \times 8$ mm^3, two B-implanted p^+ contacts 1.0×1.8 mm^2 at opposite ends of one 1×8 mm^2 face [33].

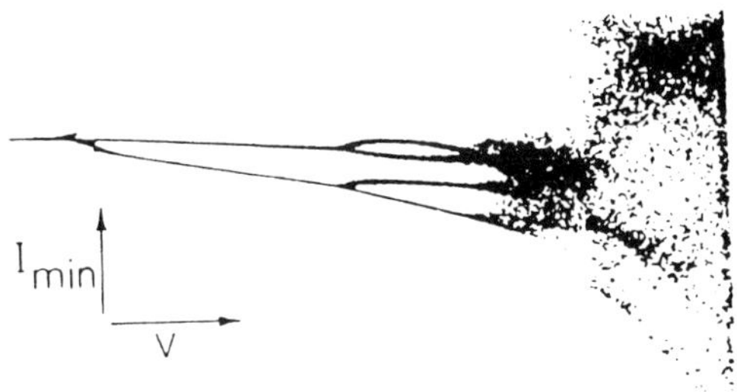

Figure 5.7 Bifurcation diagram of current minima $I_{\min}$ vs applied voltage V for Ge $T = 4.2$ K, for range $V = 11.1$ to 18.4 V; breakdown of the entire sample occurred at $V_{\mathrm{b}} = 20$ V [33].

and $\kappa(E)$ and $r(E)$ denote the field-dependent impact ionization and capture rates, respectively. Equation (5.2) is Ampère's law $J_{\mathrm{ext}} = \dot{D} + \sigma E$, where J_{ext} is the external current density, $\dot{D} = \epsilon\, \dot{E}$ is the displacement current density, and $\sigma E = epv = ep\, \mu_{\mathrm{p}} E$ ($v = \mu_{\mathrm{p}} E$: drift velocity). The authors have assumed the charge neutrality condition (5.3), where d is the compensating donor concentration. In (5.4), J_0 is the dc current and the square-wave FIR irradiation has been assumed to induce the sinusoidal photocurrent $\Delta J\, \sin(2\pi f_{\mathrm{dr}} t)$. In Fig. 5.8, a numerical result of phase diagram for the system response (period doubling sequence) is shown at different drive

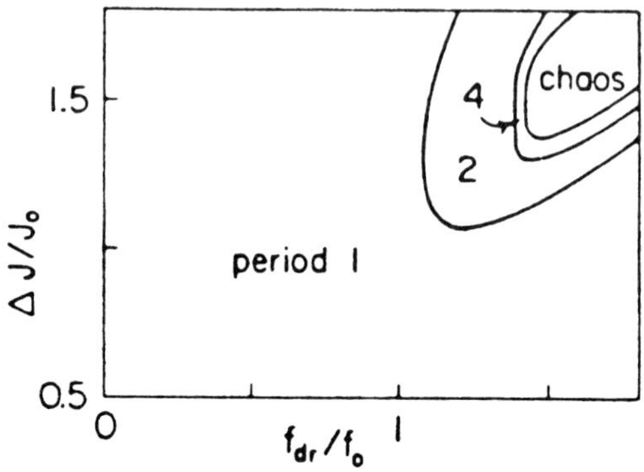

Figure 5.8 Phase diagram for the system response at different drive strengths J/J_0 and frequencies f_{dr}/f_0. The region marked 1 denotes simple periodic response, 2 denotes period 2 response, and 4 denotes period 4 and higher response; the chaotic region is also indicated [34].

strengths $\Delta J/J_0$ and frequencies f_{dr}/f_0 [34], where f_0 is the characteristic frequency of the damped oscillatory response ($f_0 = 6422$ Hz). Later, they showed that the model equations (5.1)–(5.4) are identical to the periodically driven case by the sinusoidal drive $V_0 + V_{\mathrm{ac}} \sin(2\pi f_{\mathrm{dr}} t)$ and constant FIR irradiation [36, 37]. A measured phase diagram in the prebreakdown regime (Fig. 5.9) has been compared to the numerical phase diagram in Fig. 5.8 [36], resulting in good qualitative agreement. In the experiments, the characteristic frequency f_0 has been estimated to be 80 Hz [36], while it is 6000 Hz in the numerical simulations [34]. The difference in time scale has been interpreted as being due to the effect of space charge trapped near the positive contact [36].

In the postbreakdown regime, Westervelt *et al.* [38] and Gwinn and Westervelt [39, 40] tested the universal properties derived from the chaos theory [41, 42]. For the photoconductive p-type Ge, the I–V characteristic is N-shaped in the postbreakdown regime due to the electric field dependence of the impact-ionization rate $\kappa(E)$ [38], whereby the spontaneous oscillation occurs at the resonant frequency $f_0 \cong 6$ kHz. They have attributed the oscillation to a spatial (domain-type) instability of the impact-ionization avalanche [38, 43, 44]. By adding the sinusoidal drive $V_{\mathrm{ac}} \sin(2\pi f_{\mathrm{dr}} t)$, the photocurrent system exhibits circle map behavior [41, 42]. Frequency locking, quasi-periodic states and transition to chaos have occurred in a universal manner, depending on the driving amplitude V_{ac} and the normalized frequency f_0/f_{dr}. The system behavior has been proved to follow the circle map [41, 42];

$$\theta_{n+1} = f_n(\theta_n) = \theta_n + \Omega + g(\theta_n) \quad \mathrm{mod}\ 1, \tag{5.5}$$

where θ_n is the angle of the circle at discrete time n, $g(\theta_n)$ is a periodic

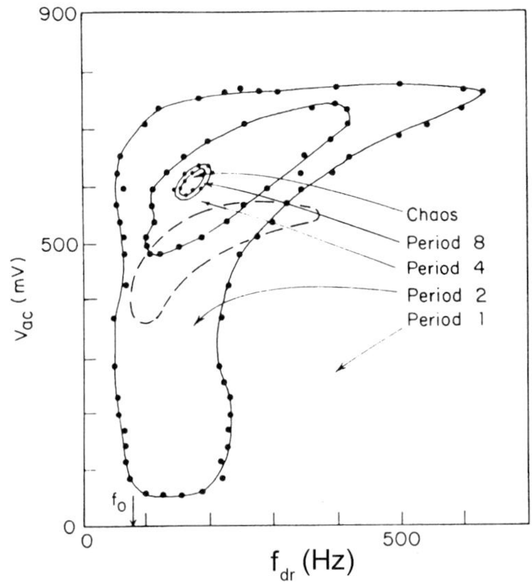

Figure 5.9 Phase diagram: qualitative type of current response vs drive amplitude (V_{ac}) and frequency (f_{dr}) for dc bias voltages $V_{dc} = 0.94$ V. The small-signal resonant frequency $f_0 = 80$ Hz without ac drive is marked by an arrow [34].

function of θ, say $g(\theta) = -(K/2\pi)\sin(2\pi\theta)$, and Ω corresponds to the ratio f_0/f_{dr}. In the semiconductor system, the amplitude K is related to the driving amplitude V_{ac}. Just above the critical amplitude of $V_{ac} = 153$ mV (the value approximately at the golden mean), the authors [37, 38] have shown from the normalized power spectrum, the onset of chaos which originates from the overlap of the Arnold tongues [40–42]. Relating to the transition from quasi-periodicity to chaos, Gwinn and Westervelt tested the universality of the $f(\alpha)$ spectrum [40]. These problems will be introduced later in this chapter. For the moment, readers who are unfamiliar with chaos theory may refer to the terminology stated above.

Regarding the dissipation structure of the current filament measured by the low-temperature scanning electron microscope (LTSEM) [45–49], Peinke *et al.* [50] investigated the hierarchical complexity of the current instabilities caused by the impact-ionization avalanche in p-type Ge, where the carrier concentration was approximately $p = 2 \times 10^{14}/\text{cm}^3$ and the lattice temperatures were 4.2 K and 1.78 K. They investigated complex behaviors mainly under transverse magnetic fields. For the summarized contents, refers to the book by Peinke *et al.* [50]. The main parts of the observations are transition from chaos to hyperchaos [50, 51], quasi-periodic oscillations [50, 52], test of the circle map behavior [50, 52], blue sky catastrophe [50, 53] and intermittency [50, 54]. Among the observations, the authors have discussed the physical origin of quasi-periodic oscillations in the experimental set-up of Fig. 5.10. For the quasi-periodic oscillatory state of the total current $I(t)$ and the Fourier power spectrum, the authors measured the potential drops $V_1(t)$ and $V_3(t)$ as seen in Fig. 5.11, where $V_1(t)$ and $V_3(t)$ oscillate sinusoidally with the fundamental frequencies f_1 and f_2 respectively. The Fourier power spectrum of $I(t)$ has been satisfactorily explained by the combination of the power spectra of $V_1(t)$ and $V_3(t)$. From the results, it was concluded that the quasi-periodic oscillatory state arises owing to nonlinear coupling between two regions of the oscillation centers. In the experiments of p-Ge, the basic idea obtained from their elaborations of the LTSEM measurements [45–50] is as follows. For the spatially extended dissipative structures of the current filaments, the instabilities and the nonlinear dynamical structures (behaviors) show self-organized hierarchical complexity due to nonlinear coupling between (or among) different parts of the oscillation centers. The quasi-periodic state is one of the fundamental properties of complex behaviors [49, 50].

The appearance of hyperchaos under the transverse magnetic field observed by Peinke *et al.* [50] provided another viewpoint on hierarchical complexity. Provided that the irregular but macroscopic fluctuations in the carrier transport of semiconductors are deterministic in a sense of usual chaos, the fractal dimension of the fluctuation should be less than three, and normally greater than unity (e.g., $d_f = 1.6$). Then, the number of system equations to describe the dynamical structure should be at least three in the autonomous system. However, if the fractal dimension happens to be larger than three [50], the number of system equations may be four or more, at least greater than three. Of course, it could be confirmed that the homogeneous system equations with $d_f = 3$ (or $d_f = 4$) give a good approximation to the spatiotemporal behaviors of the system. Elimination of the space variables may be a good approximation owing to self-organization of the system (e.g., the whole region is divided into two oscillation centers) in the sense of synergetics and the slaving principle presented by Haken [55], in

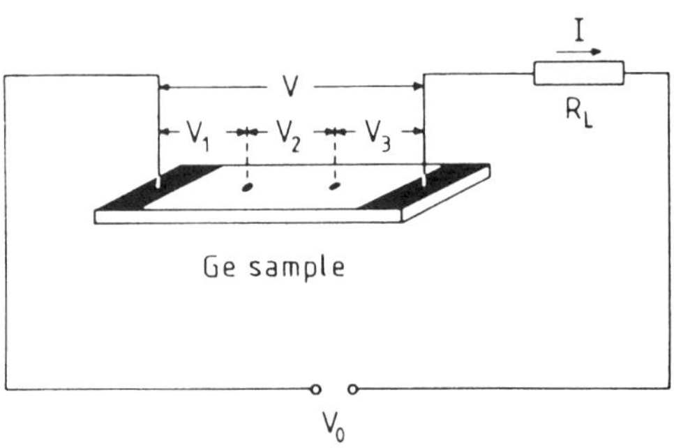

Figure 5.10 Scheme of the experimental arrangement applied to a Ge sample [50].

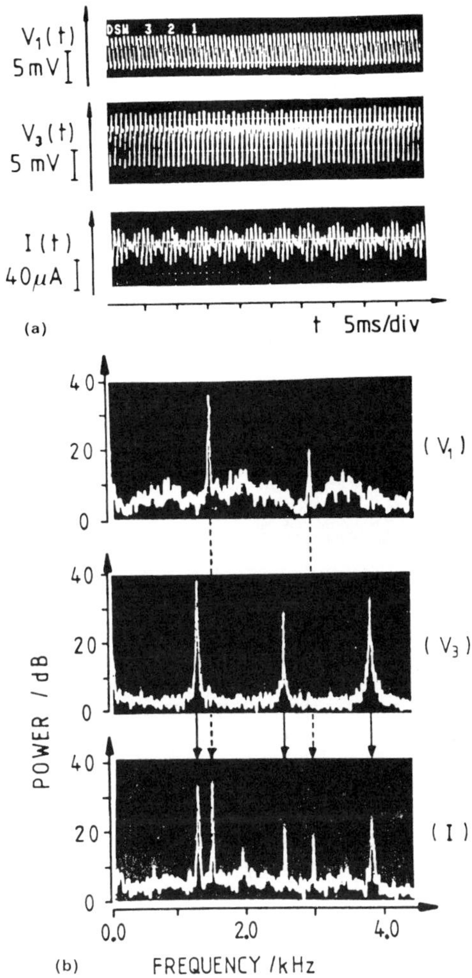

Figure 5.11 (a) Temporal profiles and (b) power spectra for the case of the quasi-periodic oscillatory state obtained from the partial voltage drops $V_1(t)$ and $V_3(t)$ and the current signal $I(t)$ according to the experimental set-up sketched in Fig. 5.10 [50].

spite of a huge number of spatial degrees of freedom. Peinke *et al.* [50] have evaluated the fractal dimension for hyperchaos to be $D_1 = 3.6$, where D_1 is the information dimension. For the observation of hyperchaos [50, 51] in Fig. 5.12e, the authors have suggested that one of the additional degrees of freedom may originate from the lattice heat (phonon) propagation between separated parts of the oscillation centers in Fig. 5.10. Figure 5.12d, e appears to show evidence for the transition from ordinary chaos to hyperchaos.

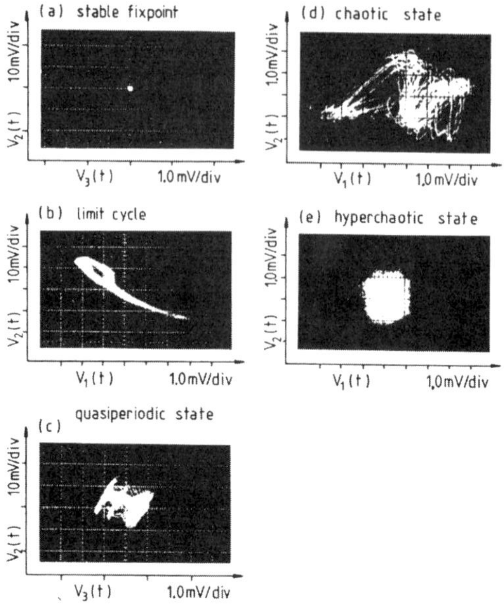

Figure 5.12 Phase portraits for distinct (a) nonoscillatory, (b) periodic, (c) quasi-periodic, (d) chaotic, and (e) hyperchaotic oscillatory states obtained from the experimental set-up in Fig. 5.10 by plotting pairs of partial voltage drops V_i ($i = 1, 2, 3$) against one another at different sets of parameter values: (a) bias voltage $V_0 = 2.220$ V, and transverse magnetic field $B = -0.4$ mT; (b) $V_0 = 2.184$ V, $B = 0$ mT; (c) $V_0 = 2.220$ V, $B = 0.21$ mT; (d) $V_0 = 2.145$ V, $B = 3.15$ mT; (e) $V_0 = 2.145$ V, and $B = 4.65$ mT. The lattice temperature $T = 4.2$ K, and the load resistance $R_L = 100\ \Omega$ [50].

Brandl *et al.* [56–58] have investigated current filaments and the self-generated nonlinear oscillation in n-GaAs at 4.2 K or 1.8 K under transverse magnetic fields, where the carrier concentration was in the order of $n = 10^{14}/\text{cm}^3$ and the electron mobility $\mu = 5 \times 10^4\ \text{cm}^2/\text{Vs}$ at 77 K. They have developed both the spatial measurements of the current filament by scanning optical microscopy and the dynamical behaviors of the current and the Hall voltage [58]. Their main result on the dynamical behaviors is the finding of Hopf bifurcation, seen as a function of the dc bias under a fixed magnetic field intensity (Fig. 5.13). In the negative differential resistance (NDR) region of the I–V curve, the spontaneous periodic oscillation occurs under zero magnetic field, with the fundamental frequency typically $f_1 = 1$ MHz (rising up to 5 MHz with increasing average current) [58]. Under a fixed magnetic field intensity ($B = 100$ mT), the current bifurcates from the stationary state to the periodic state and further to the quasi-periodic state and chaos by increasing the dc bias, as seen in Fig. 5.13. As well as the quasi-periodicity, the Farey-tree type phase locking ($f_2/f_1 = 1/2,\ 1/3,\ 1/4,\ 1/6$) has also been observed. By combination with the

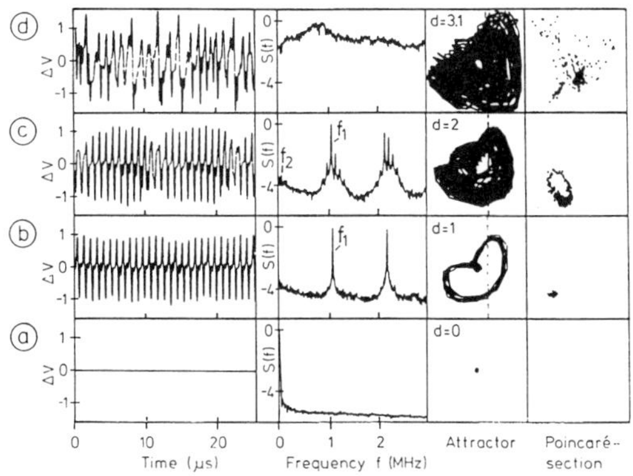

Figure 5.13 Ruelle–Takens–Newhouse scenario obtained at $B = 100$ mT with a load resistor $R_L = 200$ k Ω. From the left to right are shown: the voltage across the sample as a function of time and the corresponding power spectra, phase portraits of the attractor where d denotes the dimension, and the Poincaré sections which were determined in a plane normal to the broken lines of the phase portraits. The bias voltages are (a) 1.500 V, (b) 5.102 V, (c) 5.213 V, and (d) 5.709 V [58].

scanning optical microscope, they have observed self-generated oscillations at the filament borders. From the measurements of the time-dependent Hall voltage $V_H(t)$, the authors explained the Hall-field instability using the two-level impact-ionization model by Schöll [59–61].

Among a large number of observations of chaos, many other authors have also investigated the current instabilities caused by the impact-ionization avalanche of shallow impurities under magnetic fields, and further studied the NDR instabilities due to impact ionization of deep-trap levels. Many of them have been reported in the Proceedings of the International Conference on Hot Carriers in Semiconductors [62–64] and in a special issue of Applied Physics (Vol. A48: The Nonlinear and Chaotic Carrier Transport Phenomena in Semiconductors, edited by Abe [64]). Here, because of limited space, the contents are listed in brief, although many other authors have also presented the original and pioneering works. Seiler *et al.* [64–66] investigated period doubling bifurcation and quasi-periodic behavior under transverse magnetic fields (~ 1T) in undoped n-InSb samples ($n \cong 1 \times 10^{14}/\text{cm}^3$) at 4.2 K. The I–V curve does not show any negative differential resistance, but exhibits magnetic freeze-out effects and the impact ionization of the shallow impurities [64]. Fujii *et al.* [64] observed similar behavior in high-purity n-InSb under magnetic fields (up to 1 T, ⟨111⟩ direction of the sample). Yamada *et al.* [64, 67] reported chaotic conductivity oscillation and period doubling bifurcation in n-type Si under pulsed high magnetic fields (18 T, transverse and longitudinal).

Other groups have investigated different categories of the impact-ionization avalanche via deep-trap levels [62–64, 68–71], and separately

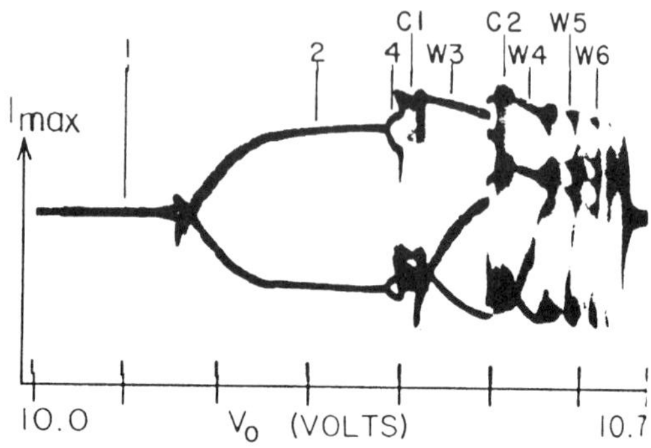

Figure 5.14 Bifurcation diagram of local current maxima, I_{max}, vs applied voltage V_0, for Ge sample. Labels (1), (2), (4), (C1) and (W3) refer to periods 1, 2, 4, chaos and window of period 3, respectively. Labels (W4), (W5) and (W6) refer to windows of periods 4, 5 and 6, respectively [77].

chaotic behaviors in p-i-n diodes [72–74] and in GaAs MESFET [75]. Maracas *et al.* [62, 68, 69] showed low-frequency oscillations (LFO) and chaos caused by the impact-ionization avalanche of deep-level EL2 centers in semi-insulating GaAs. The oscillation frequency ranges from a few Hz to several kHz depending on the electric fields (a few V/cm to 2.4 kV/cm), lattice temperature and light illumination. Knap *et al.* also investigated the LFO and chaos related to the EL2 centers in semi-insulating GaAs [70], and in cooperation Rau *et al.* [71] reconstructed the spatiotemporal behavior of the LFO using a scanning electron microscope. Chaotic behaviors of hot electron plasma in Ni-compensated Ge have been reported by Bumelière *et al.* [63] and analyzed by Pozhela *et al.* [64, 76]. An important study was conducted by Symanczyk *et al.* [74], who measured the spatiotemporal pattern of period 2 oscillation in the period doubling regime in the Si p-i-n diode. Chaos in a device of GaAs MESFET was studied by Yano *et al.* [75], who showed side-gate instability due to the impact ionization of deep-trap levels in the i-layer between a drain and a side-gate.

Finally, note the plasma (two-carrier) instability reported by Held *et al.* [77–80]. The electron-hole (e-h) plasma ($n_e \sim n_h \sim 10^{13}/\text{cm}^3$) is created at $T_L = 77$ K by double injection or laser pumping in a rod of n-type Ge of size $1 \times 1 \times 10$ mm^3, with the donor concentration $N_D \sim 10^{12}/\text{cm}^3$. Under a longitudinal magnetic field (1–15 kG), the e-h plasma instability shows transition to a turbulent state or chaos with increasing dc bias V_0. They obtained period doubling [77, 78], Hopf bifurcation [78], periodic/chaotic bifurcation sequence [77] similar to Hudson's experiment in the chemical reaction [81] and circle map behavior [80, 82]. The period doubling and periodic/chaotic bifurcation sequence are shown in Fig. 5.14. It is quite impressive that the e-h instability is helical or screwed instability [3], and the experiment was the first observation of chaos in the oscillistor effect [13].

5.2 Firing wave instability

Firing wave instability occurs in a peculiar manner. The observations and modeling, are described in this section.

The experimental set-up is shown in Fig. 5.15 [21]. All measurements were taken at 4.2 K. The samples used were undoped n-GaAs, epitaxially grown on Cr-doped semi-insulating substrate, where the carrier concentration was $n \sim 2 \times 10^{14}/\mathrm{cm}^3$ at 300 K and the epitaxial layer was about 12 μm. Typical dimensions of the sample surfaces were 5–3 mm in length and 3–2 mm in width. Spacing between the planar-type ohmic contacts was $1.2 \sim 4$ mm. In order to illuminate the sample surface with a monochromatic light around a band gap energy of GaAs, a 100 W halogen lamp and a grating monochromator were used. The photoexcitation density was approximately 2μW or more. The pulse voltage with a frequency of 38 kHz and the width of 20 μs was applied between the contacts by a pulse generator. The load resistance was 7–200 Ω. The photocurrent noise induced by the impact-ionization avalanche of neutral shallow donors was detected and coarse-grained by a lock-in amplifier (PAR M126) with the time constant of 1 ms, and the output signals were recorded on recorder charts [21, 30, 31] or stored by a microcomputer system in the later experiments [32, 64].

The pulse voltage has been settled in the prebreakdown regime such that no avalanche breakdown occurs under the dark condition [21, 30, 31, 64]. By photoexcitation, the impact-ionization avalanche does not occur on every pulse voltage but occurs rarely in a stochastic manner, as shown in Fig. 5.16. An example of the output signals $S(t)$ which are coarse-grained by the lock-in amplifier is shown in Fig. 5.17. The peculiarity of the experimental results is indicated by the fact that the deterministic oscillatory behaviors appear when measured by the lock-in amplifier in spite of the seemingly stochastic avalanche breakdown. The temporal behavior of firing density of the impact-ionization avalanche per unit time shows the stationary state, periodic and chaotic states, depending on the control

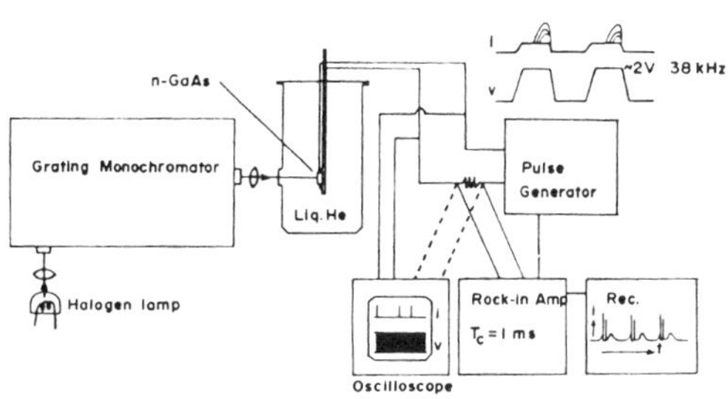

Figure 5.15 Experimental set-up for the measurement of the electrical avalanche breakdown in n-GaAs [21].

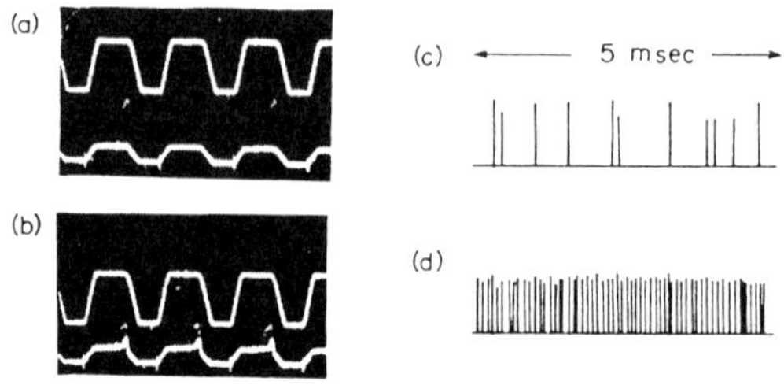

Figure 5.16 Typical waveforms of the electrical avalanche breakdown observed at the electric field of 4.4 V/cm and with the photoexcitation at 812 nm. The time-scale in (a) and (b) is 10 μs/div. The avalanche breakdown in the longer time-scale is shown schematically in (c) and (d). The periodic oscillation occurs in the process of (a)–(b)–(a)–(b)... or (c)–(d)–(c)–(d)... in the figure [21].

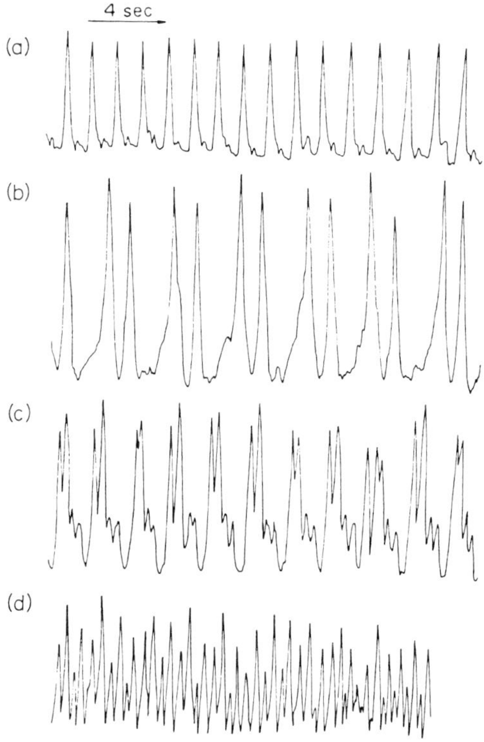

Figure 5.17 Continuous bifurcations of the oscillations observed at 812 nm, as a function of the applied electric fields: (a) 3.8 V/cm, (b) 3.25 V/cm, (c) 3.1 V/cm, and (d) 3.0 V/cm. The frequency of the pulse voltage is 38 kHz. The rise- and decay-times of the pulse voltage are 10 and 5 μs, respectively [21].

parameters. Figure 5.17 show the waveforms of the oscillation as a function of the pulsed electric field. The oscillation frequencies are much lower (a few Hz) than those of the pulse voltage and the inverse time constant of the lock-in amplifier. Some bifurcations have been observed [21]. Period doubling is seen in Fig. 5.17a, b and chaotic motion in Fig. 5.17c, d. In particular, the waveform in Fig. 5.17d reveals quite irregular chaotic motion but possesses a fundamental frequency ($f_1 \sim 1.7$ Hz).

Phenomenologically, the oscillatory behaviors can be described by the **firing density wave** (FDW) of the impact-ionization avalanche [21, 30–32]. The instability is termed **firing wave instability**. In Fig. 5.16a, b, photographs of the pulse voltage and the corresponding pulse current (bottom) are shown, where the photocurrent by the impact-ionization avalanche is seen at the end of the plateau of pulses. By looking with a macroscopic time scale (~ 5 ms), the time series of the stochastic avalanche breakdown looks like the sketch in Fig. 5.16c, d. Periodic oscillation, for instance, occurs by the repeated sequence (a)–(b)–(a)–(b) ... or (c)–(d)–(d)–(d) .... More precisely, one can look at simultaneous changes of the firing density S per unit time as well as the current amplitudes during one cycle of the periodic oscillation in Fig. 5.18a–e, where Fig. 5.18f is the coarse-grained signal S by the lock-in amplifier [30]. In Fig. 5.18a–e, the filamentary currents (denoted by f.c.) have several characteristic amplitudes, which means that the firings among several current filaments are incorporated. Although the FW instability is a kind of self-organization among multicurrent filaments, understanding of the physical origin for the novel phenomenon seemed to be difficult. For chaotic behavior, the FDW behaves in a complicated manner.

Another remarkable point is that the oscillatory state depends sensitively on the strength of the applied field, the waveform of the pulse, and the wavelength and intensity of the photoillumination. An example is given in Fig. 5.19. Figure 5.19 shows spectra of the photocurrent noise (PCN) signal as a function of wavelength [31]. The PCN signal is coarse-grained by the lock-in amplifier, as described above. The scanning speed of the wavelength was 40 nm/min, as indicated by the arrow (from right to left). Spectrum (a) was obtained by the pulsed electric field of $E = 4.4$ V/cm with the pulse shape of the rise time $\tau_r = 5.5$ μs and the decay $\tau_d = 10$ μs The oscillatory behaviors are seen selectively above band-gap energy E_g and at the resonant wavelength of free exciton. In the range of 812 $\sim$ 816 nm, four-peak or five-peak erratic oscillations are clearly seen. By shortening the rise time of the pulse to 1 μs, while keeping $E = 4.4$ V/cm, one obtains the spectrum (b), where the oscillations are seen with relatively high frequency (~ 3 Hz) at the wavelengths of donor-bound excitons (D^+, X), (D°, X), free exciton and excited donor states. No oscillations are seen above the band-gap energy in spectrum (b). Selective excitation in the spectrum of Fig. 5.19b is closely related to the selective excitation of the filamentary currents in Fig. 5.3 [19, 20, 29].

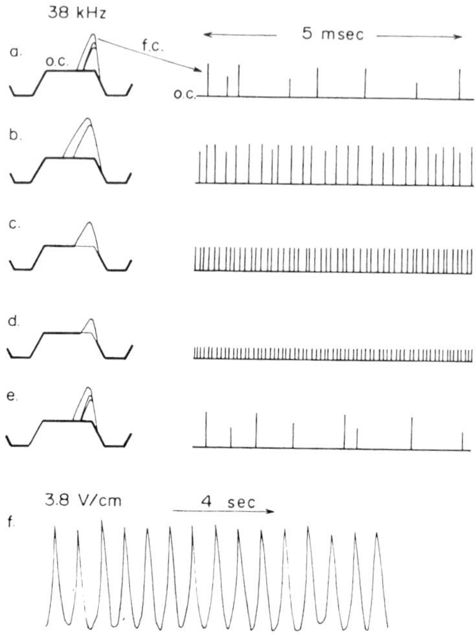

Figure 5.18 Firing waveforms of the filamentary current during one period of a simple oscillation (a–e), and the output signals detected by a lock-in amplifier (f). The rise- and decay-times are both 3 μs. o.c. denotes the background level of the ohmic current [30].

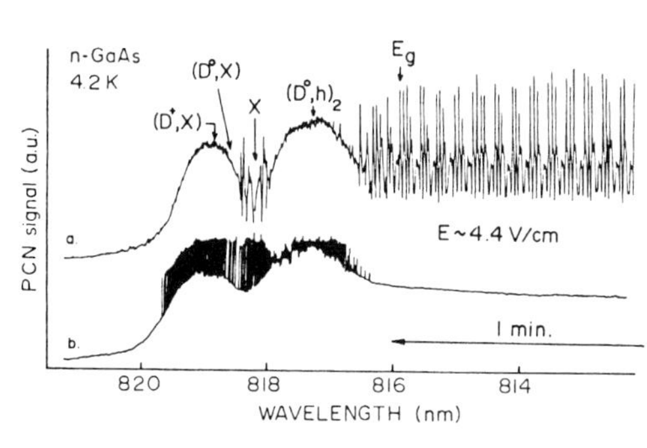

Figure 5.19 PNC spectra obtained by the pulse voltage with the frequency of 38 kHz. The electric field is 4.4 V/cm. Spectrum (a) was observed by the rise- and decay-times of 5.5 and 10 μs, respectively. Spectrum (b) was obtained by the shortened rise-time of 1 μs. Peak positions of free and bound excitons were determined by the photoluminescence spectrum of the same sample [31].

Control parameters in the system are wavelength λ, light intensity, electric field intensity E, rise time τ_r and decay time τ_d. The oscillatory states have mainly been analyzed by fixing the wavelength $\lambda = 812$ nm [21, 30, 31]. With so many control parameters, however, systematic investigations of the bifurcation schemes are impossible. In particular, failure in the complete adjustment of light intensity at a constant level in the routine experiments seriously affects the investigations. Among the observations, many peculiar oscillations with long periods have been observed [31]. A striking feature of the long periodicity is shown in Fig. 5.20, where $E = 3.63$ V/cm, $\tau_r = 2$ μs and $\tau_d = 5$ μs [31]. The multipeak oscillation has a long period of $\sim$ 18s and contains many characteristic oscillating structures. Extremely small variation in the pulsed electric field from $E = 3.63$ to 3.66 V/cm, results in the quite different but simpler waveform in Fig. 5.21. The return map of the coarse-grained signal S_{n+1} vs S_n in Fig. 5.21 clearly shows that the erratic oscillation is deterministic chaos [31], where S_n $(n = 1, 2, 3, \ldots)$ is the peak value of the oscillation.

The difficulty in understanding the dynamical behaviors in the FW instability comes from an experimental fact of the seemingly stochastic property of the impact-ionization avalanche in the micro-time-scale ($\lesssim$ 1 ms) and also comes from another fact of the deterministic nature which appears on measuring the coarse-grained signal in the macro-time-scale ($\gg$ 1 ms). How can it be proved that the present semiconductor system behaves truly in a deterministic manner? Do any other systems behave like this? In order to answer these questions, Aoki and Yamamoto [30] and Aoki *et al.* [31, 83–85] have taken an analogy with neurodynamics [86–88], especially with a neural network model based on the McCulloch–Pitts formal neurons given by Amari [87, 88].

In the system of a neuron, a chemical reaction in the synapse induced by external stimuli raises the synaptic potential. If the synaptic potential

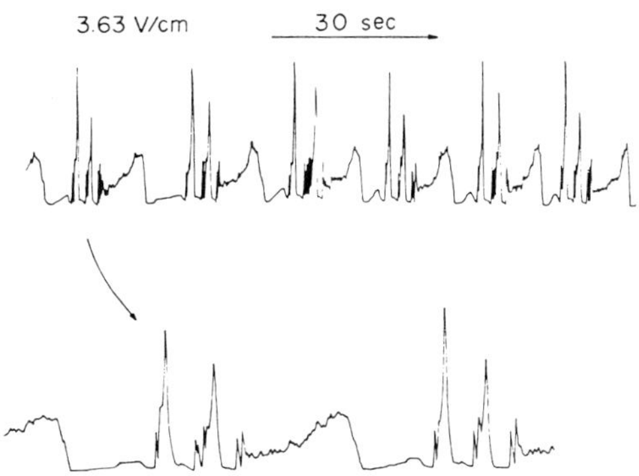

Figure 5.20 Example of a long period observed at 3.63 V/cm [30].

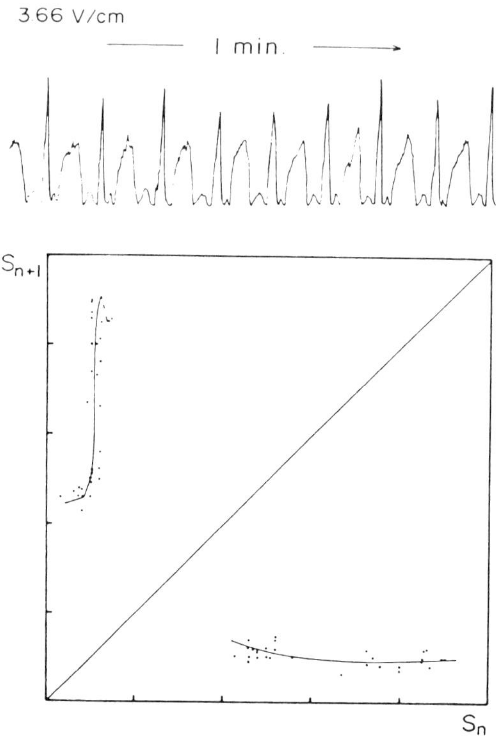

Figure 5.21 Example of chaotic motion and the Poincaré return map observed at 3.66 V/cm [30].

exceeds the threshold voltage, depolarization of Na^+ and K^+ ions occurs, resulting in the propagation of a solitary wave of depolarization along the axon. The behavior is similar to the high-field domain in the Gunn diode. There exists some refractory period such that successive firing of the neuron is inhibited during some period after the latest firing (see Fig. 5.22 [89, 90]). A neuron is a switching element which is characterized by the none-quilibrium phase transition, called the "**all-or-none transition**" [90]. In the neural network model by Amari [87, 88], a huge number of neurons is connected in a random manner. One of the state variables is given by the firing activity of the system S, defined by $S = N^{-1} \sum X_j$, where X_j is the state symbol 0, 1 in the jth element (1: fire; 0: rest) and the spatial summation is taken over the N-connected neurons of the system. The second state variable is given by the threshold H Then, S and H represent the macrostate of the neural network [87, 88]. In order to study some possibilities of intellectual information processing in the random neural networks, Amari introduced a couple of macrostate equations which describe coupling between or among

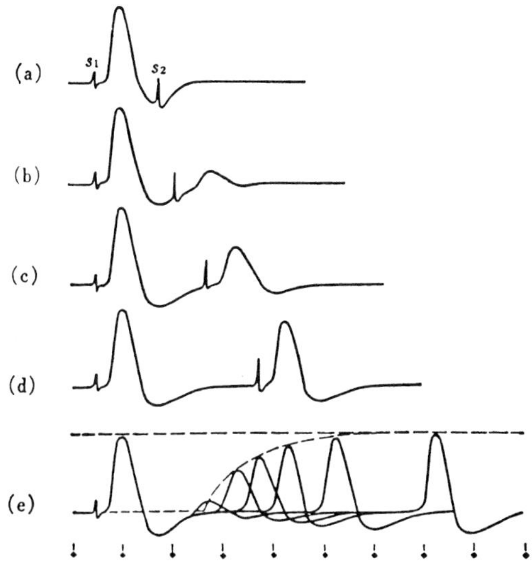

Figure 5.22 Refractory period in the response of a nerve axon. In (a), there exists an absolute refractory period; i.e., no firing is seen by the second stimulus (s_2) just after the first firing by the first stimulus (s_1). In (b)–(d), firing occurs by delaying the timing of the second stimulus. Note that the amplitude of the second firing by s_2 is smaller than the first one by s_1. (e) Summary of the results: the arrows indicate the timings of the second stimulus [89].

subsystems of two or more neural networks [87, 88]. Stationary state, bistability, oscillatory state and complex irregular oscillations have been shown.

How can an analogy be made between neurodynamics and the firing wave instability in semiconductors? Aoki and Yamamoto [30] have discussed this as follows. In Fig. 5.23, a current filament could be treated as a switching element which has a trigger zone (or possibly many trigger zones) under the photoillumination. The filamentary channel corresponds to the axon in neuron since it works as an opened conducting channel once formed. Clearly, the switching behavior is regarded as an all-or-none transition. The photon flux corresponds to the external stimuli. By the photon flux, current filamentation may be initiated locally from the trigger zone(s) [30]. By photoexcitation, the breakdown field at the trigger zone is lowered E_b^* from the original value E_b under the dark condition, as shown in Fig. 5.24. Assume $E^* + \xi(t)$ is the local field at the trigger zone and let $h = E_\mathrm{b} - E^*$ be the threshold ($h > 0$), where $\xi(t)$ is the random fluctuating force. Firing of the current filament occurs whenever $\xi(t)$ exceeds the threshold h. By analogy with the neuron [30, 83–85], the input–output function is formally

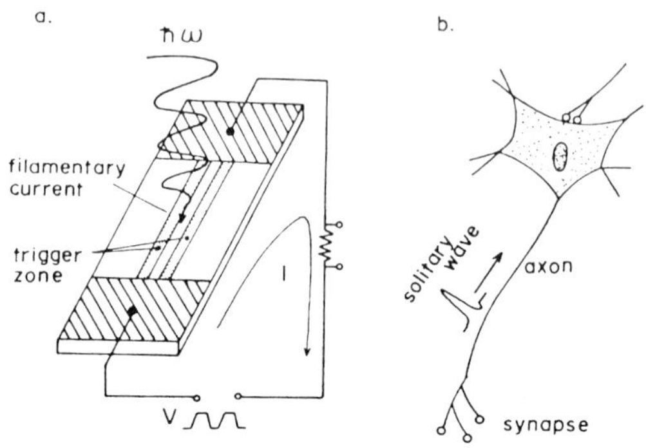

Figure 5.23 Schematic drawing of (a) current filaments in n-GaAs, and (b) solitary wave propagation in a nerve axon. Erratum: the arrow in the nerve axon must be reversed.

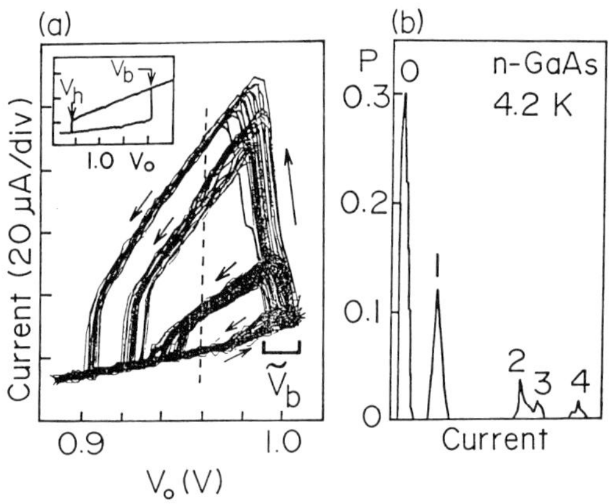

Figure 5.24 (a) I–V characteristics during the firing of the multicurrent filaments in n-GaAs under photoexcitation, at 4.2 K. The firings occur with several current amplitudes. The inset denotes the well-defined I–V characteristic under the dark condition. (b) Probability of the firings for the multicurrent filaments as a function of the current amplitude at the broken line in the left figure.

obtained

$$X = \Psi(\xi(t) - h), \tag{5.6}$$

or

$$X = \begin{cases} 1 & \xi(t) \geq h \\ 0 & \xi(t) < h. \end{cases} \tag{5.7}$$

If the discrete time t_n $(n = 0, 1, 2, \ldots)$ is introduced, where $t_n - t_{n-1} = T_0$ is the period of the applied pulse voltage in Fig. 5.18, the output $\{X_n\}$ is given by the random symbolic sequence 010010 . . ..

The ensemble average $\bar{X} = N^{-1} \sum X_n$ over a macro-time-scale gives the firing activity S of the current filament. There may be two kinds of memory effects which are responsible for the oscillatory behavior of $S(t)$. In the neural network system [87, 88], spatial summation of the firing acts as an activator effect, and the refractory period as an inhibitory effect. The activator effect lowers the threshold, while the inhibitor effect raises it. In semiconductor systems, Aoki and Yamamoto [30] considered joule heating as a possible physical origin for the activator effect. However, the present author (Aoki) has paid careful attention to this: joule heating is inconclusive since the lattice heating effect is considered to be negligibly small in the low-field regime of the impact-ionization avalanche.

The inhibitor effect has been found experimentally [30], as seen in Fig. 5.25. By applying double pulses, the firing behavior on the second pulse is measured, where the time interval between two pulse voltages is denoted by τ in Fig. 5.25. The repetition period of the double pulses was $T =$ 180 μs ($f = 5.6$ kHz) in Fig. 5.25a–c. Figure 5.25a shows that the firing

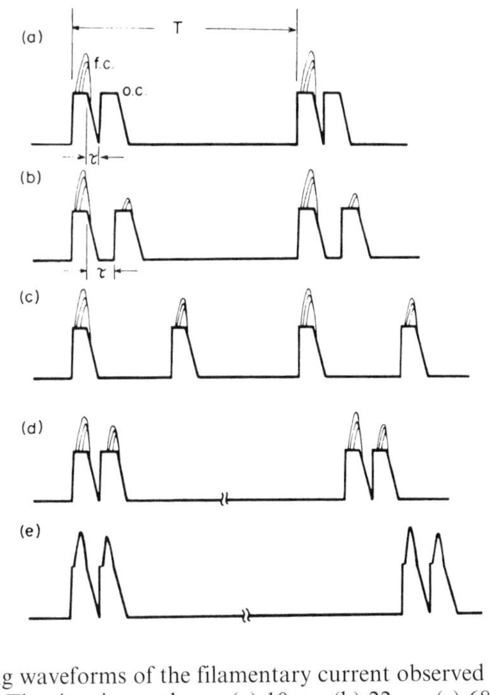

Figure 5.25 Firing waveforms of the filamentary current observed under the double-pulse conditions. The time intervals are (a) 10 μs, (b) 22 μs, (c) 68 μs, (d) 10 μs and (e) 10 μs. The periods T are (a–c) 180 μs, (d) 333 μs, and (e) 1 ms [30].

occurred stochastically on the first pulse, while the firing on the second pulse was completely inhibited by the previous current flow (even by the ohmic current), mainly due to the latest pulse, where the time interval τ was short enough ($\tau = 10\ \mu$s). By increasing τ (Fig. 5.25b, c), the stochastic firing can be seen also on the second pulse, but the amplitudes of the filamentary current are less than those on the first pulse, where $T = 180\ \mu$s. The observation indicates clearly the refractory effect. Figure 5.25d, e shows the firing waveforms under long periods with a short time interval ($\tau = 10\ \mu$s), where $T = 333\ \mu$s in Fig. 5.25d and $T = 1$ ms in Fig. 5.25e. By increasing the period, the firing density on the double pulse increases (e.g., in Fig. 5.25d), and finally the firing can be observed on every pulse when $T \gtrsim 1$ ms ($f \lesssim 1$ kHz, see Fig. 5.25e). From these results, both the ohmic and filamentary currents act as inhibitors to keep the higher threshold level h in the trigger zone(s). That is, the memory of the previous firing events tends to inhibit the successive firing.

Suppose that the threshold h is given by the Langevin-type equation [91]:

$$\frac{\mathrm{d}h}{\mathrm{d}t} = F(t) - \gamma(h - h_0), \tag{5.8}$$

where γ is the damping constant, h_0 is the threshold level under the steady state ($\mathrm{d}h/\mathrm{d}t = 0$) without any perturbation ($F = 0$), and $F(t)$ is assumed to be proportional to the ohmic current density $J_0(t)$ and the filamentary current density $J_f(t)$. From (5.8), one has

$$h(t) = \int_{-\infty}^{t} F(\tau)\exp[\gamma(\tau - t)]\mathrm{d}\tau + h_0. \tag{5.9}$$

The response (5.9) is a non-Markov process if $\gamma \ll f$ and almost a Markov process if $\gamma \gtrsim f$, where f is the frequency of the pulse voltage in Fig. 5.18. The present system follows the non-Markov process since $\gamma \ll f$. Note that (5.8) is equivalent to the dielectric relaxation equation under the dark condition with $\gamma_{\text{dark}} \gg f = 38$ kHz, where $h = E_\mathrm{b} - E$, $h_0 = E_\mathrm{b} - E_0$ and $\mathrm{d}h/\mathrm{d}t = -\mathrm{d}E/\mathrm{d}t$. For the linear response with $F(\tau) = A\ \delta(\tau)$, $h(t)$ is given by $A\ \exp(-\gamma t) + h_0$, where $A > 0$. The refractory period is approximately of the order of γ^{-1} ($\sim$1 ms in the experiments; see Fig. 5.25e). A random sequence

$$F(\tau) = A\sum_{k=0} \delta(\tau - t_{-k}),$$

where $t > t_0 > t_{-1} > \ldots$ (t: present time), has the linear response

$$h(t) = A\sum_{k=0} \exp[-\gamma(t - t_{-k})] + h_0.$$

Since the ohmic current is a periodic function of time, the effect due to the ohmic current can be included in the background level h_0 of (5.9) as a definite increment Δh ($h'_0 = h_0 + \Delta h$). Now the input–output function (5.6) is replaced by:

$$X(t) = \Psi(\xi(t) - h(t)). \tag{5.10}$$

Namely, the inhibitor effect is given by the slowly varying threshold $h(t)$ in the non-Markov process (5.8) and the linear response (5.9). Unfortunately, the physical origin of the inhibitory effect has only been speculated upon [30]. Carrier injection at a contact (negative contact for electrons) may cause the recombination (discharging [30]) of the charge carriers at the surface states. Suppose that the space charge region near the contact deviates slightly from ideal ohmic property, then the gradient of the potential $E = -\text{grad}\ \phi$ may be affected by the g-r process of the surface states, possibly with a long lifetime (Fig. 5.26). By carrier injection due to the impact-ionization avalanche, the recombination of the injected carriers at the surface states modulates the electric field, which may possibly lower the local field at the trigger zones and thus increase the threshold level h during a long period after firing.

The quantitative description of the FDW model is given as follows [30, 83–85]. In the neural network model [87,88], connected neurons fire stochastically whenever the synaptic potential exceeds the threshold level. Suppose a specified neuron (jth threshold element) receives the inputs $\sum w_{ji}X_i$ ($i = 1, 2, \dots, n$) from n-connected neurons, where w_{ji} is the synaptic weight of the jth element ($j = 1, 2, \dots, \mathrm{N}$), The output X_j is given by:

$$X_j = \Psi\left(\sum w_{ji}X_i - h_j\right). \tag{5.11}$$

The macrostate of a random net is characterized by the firing activity as

$$S = N^{-1}\sum X_j, \qquad 0 \leq S \leq 1. \tag{5.12}$$

In a closely connected random net, w_{ji} and h_j are stochastically independent and assumed to be normally distributed. By introducing another macrostate

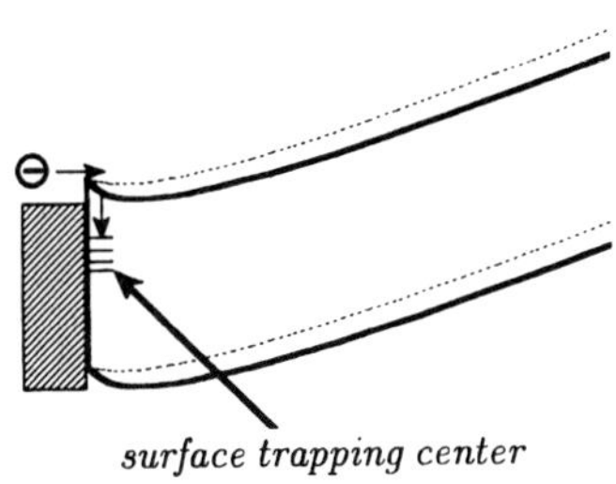

Figure 5.26 Band bending of the semiconductor surface at the interface between metal and semiconductor.

variable H (threshold), Amari [87, 88] deduced the macrostate equation:

$$S_{n+1} = \Phi(W\,S_n - H) \quad (n = 0,\ 1,\ 2, \ldots), \tag{5.13}$$

$$W \sim w/\sigma, \qquad H \sim h/\sigma, \tag{5.14}$$

and

$$\Phi(y) = \int_{-\infty}^{y} (2\pi)^{-1/2} \exp\left(-x^2/2\right) \mathrm{d}x, \tag{5.15}$$

where w and h are the expectation values of the synaptic weights and the threshold, respectively, and σ^2 is the variance given by $\sigma^2 = N\,S\sigma_w^2 + \sigma_h^2$. The error function (5.15) is a monotonically increasing function (sigmoid function) which satisfies $\Phi(+\infty) = 1$, $\Phi(0) = 1/2$ and $\phi(-\infty) = 0$. The difference equation is the most fundamental case among the system equations [87, 88].

In semiconductor systems, temporal summation is considered instead of spatial summation $\sum w_{ji} X_i$ taken in the neural network model. Let $\{X(t_i),\ X(t_i + T), \ldots, X(t_i + (m-1)T\}$ be a state vector which describes the random sequence of the symbols 010010..., where the time interval $I_i = (t_i,\ t_i + (m-1)T]$ is a measure of the macro-time-scale with $m \gg 1$, and T is the period of the pulse voltage. Temporal summation in the time interval I_{i+1} is assumed to be given by the linear combination $(\varepsilon/m)\sum_j X(t_i + jT)$, where ε is the temporal weight which corresponds to the synaptic weight w in (5.14) [87, 88]. As found experimentally [21, 30–32], the macro-time-scale is in the order of or greater than the time constant of the lock-in amplifier (>1 ms) and much greater than the period T ($= 26.3\ \mu$s). Here, it is postulated that the firing activity in I_{i+1} is affected by the latest firing activity in I_i. The input-output function is rewritten as:

$$X(t_{i+1} + jT) = \Psi\left\{\xi(t_{i+1} + jT) + (\varepsilon/m)\sum_j X(t_i + j'T) - h_{i+1}\right\}. \tag{5.16}$$

where the random fluctuating force ξ is considered to be homogeneous and rapidly fluctuating in any sectional interval. The threshold $h(t)$ in (5.10) varies microscopically according to the linear response after each firing, but it is slowly varying and replaced artificially by a macroscopic value h_{i+1}, which is assumed to have a constant value in the time interval I_{i+1}, as discussed later. The firing activity can be defined as:

$$S_i = \frac{1}{m}\sum_{j=0}^{m-1} X(t_i + jT). \tag{5.17}$$

If and only if ξ and ε are statistically independent and normally distributed in any time interval, (5.16) can be converted to the macrostate equation:

$$S_{i+1} = \Phi(\Lambda + \tilde{\varepsilon}_1 S_i - H_{i+1}), \tag{5.18}$$

where $\Lambda = \langle \xi \rangle / \sigma$, $\tilde{\varepsilon}_1 = \langle \varepsilon \rangle / \sigma$ and formally $H_{i+1} = h_{i+1}/\sigma$ (σ^2: variance of the system; $\sigma^2 = S\sigma_\varepsilon^2 + \sigma_\xi^2$, see (5.14), and Φ is the same as (5.15)). Equation (5.18) is formally converted to the nonlinear differential equation:

$$\dot{S} = -S + \tilde{\Phi}(\tilde{\varepsilon}_1 S + \Lambda - H). \tag{5.19}$$

Current flow in the filamentary channel raises the threshold due to the refractory period. In order to introduce the macrostate equation of H, consider the Langevin-type equation (5.8) for h. As described before, the contribution of the ohmic current is omitted from $F(t)$ since it is homogeneous on every pulse voltage and its contribution is included in h_0. Therefore:

$$F(t) = Cj_f X(t), \tag{5.20}$$

where j_f is the filamentary current density, and C is a constant. The solution of (5.8) is given by:

$$h(t) = Cj_f \int_{-\infty}^{t} X(\tau) \exp[\gamma(\tau - t)] \mathrm{d}\tau + h_0. \tag{5.21}$$

The integral part of (5.21) is the weighted mean of the past firing. That is, the microscopic threshold $h(t)$ at time t is affected by the previous firing during $t - \gamma^{-1} < \tau < t$. Since the time constant $\gamma^{-1}(\sim 1$ ms) is in the order of the macro-time-scale and $h(t)$ responds slowly to $X(\tau)$, $h(t)$ is approximated by defining the macroscopic threshold h_{i+1}, which is a constant value in I_{i+1} $(t \in I_{i+1})$ such that h_{i+1} is uniquely determined by the previous firing S_i. Corresponding to (5.21), $h_{i+1} \sim \varepsilon_2 S_i + h_0$, where $\varepsilon_2 \sim mCj_f$. By this approximation with $H_{i+1} = h_{i+1}/\sigma$ and formally converting to the differential equation, one obtains:

$$\dot{H} = \tilde{\varepsilon}_2 S - \gamma(H - H_0), \tag{5.22}$$

where $\tilde{\varepsilon}_2$ is the temporal weight. The ODEs (5.19) and (5.22) describe the FW instability and the dynamical behaviors of a filamentary current.

Let us investigate coupling between two subsystems (two filamentary currents), each subsystem having the firing activity S_i and the threshold

H_i ($i = 1, 2$). The proposed model is given by [83–85]:

$$\dot{S}_1 = -S_1 + \Phi(\varepsilon_1 S_1 - H_1), \tag{5.23}$$

$$\dot{H}_1 = -\gamma_1 H_1 + \varepsilon_2 S_1 + \varepsilon_3 H_2, \tag{5.24}$$

$$\dot{S}_2 = -\gamma_2 S_2 + \Phi(\varepsilon_4 S_2 - H_2), \tag{5.25}$$

$$\dot{H}_2 = -\gamma_3 (H_2 - H_2^0) + \varepsilon_5 S_2, \tag{5.26}$$

$$\Phi(y) = \frac{1}{2}(1 + \tanh(y)), \tag{5.27}$$

where ε_i are the temporal weights (the tilde is dropped for simplicity), γ_i are the damping rates, and H_2^0 is the background threshold level. Averaged fluctuating force Λ_i and background level H_1^0 are dropped without loss of generality. The function $\Phi(y)$ in (5.27) is an approximation form of the error function (5.15), replaced by a sigmoid function, where $\Phi(+\infty) = 1$, $\Phi(0) = 1/2$ and $\Phi(-\infty) = 0$. Subsystem 2 with the macrostate (S_2, H_2) in (5.25), (5.26) is adjusted to work as a driving oscillator which is isolated from subsystem 1 (filamentary channel 1). That is, the parameters are selected so that subsystem 2 oscillates spontaneously. Subsystem 1 with the macrostate (S_1, H_1) is driven periodically by subsystem 2 via the term $\varepsilon_3 H_2$ in (5.24). The parameter values are specified as $\gamma_i = \gamma_2 = 1.5$, $\gamma_3 = 0.5$, $\varepsilon_1 = 6.0$, $\varepsilon_2 = 4.0$, $\varepsilon_3 = 1.5$ and $\varepsilon_4 = 5.0$, while ε_5 is varied as a control parameter. H_2^0 is adjusted such that the equilibrium point ($\bar{S}_2$, $\bar{H}_2$) is independent of ε_5 and fixed at $(1/(2\gamma_2),\ \varepsilon_4/(2\gamma_2))$. That is, $H_2^0 = \varepsilon_4/(2\gamma_2) - \varepsilon_5/(2\gamma_2\varepsilon_3)$ is automatically adjusted as a function of ε_5 when ε_5 is varied. Under these conditions, there are three equilibrium points of ($\bar{S}_1$, $\bar{H}_1$), two of which are stable if $\varepsilon_3 = 0$ is put in (5.24) and the third is an unstable saddle point ($\bar{S}_1 = 0.5$, $\bar{H}_1 = 0.5\ \varepsilon_1$). Subsystem 1 shows bistability in the absence of a driving force from subsystem 2 (when $\varepsilon_3 = 0$).). Equation (5.23) forms a double-well potential, which is driven periodically by the term $\varepsilon_3 H_2$ in (5.24). The model simulation was made by the Runge–Kutta–Gill method. The minimum time step was $\Delta t = 0.01$ [83–85].

The frequency f_s of the spontaneous oscillation in subsystem 2 ranges from 0.0538 to 0.0964 when ε_5 is varied from 1.0 to 1.6. As a function of ε_5, the bifurcation diagram is shown in Fig. 5.27, and the corresponding Lyapunov exponent in Fig. 5.28 [85]. The period doubling cascade can be seen from both sides of the diagram. By decreasing ε_5 from 1.6, for instance, period 2 appears below 1.52, period 4 below 1.322, and period 8 is located below 1.3155. The accumulation point of period 2^∞ locates at around $\varepsilon_5 \sim 1.3149$. Below 1.3149, the Lyapunov exponent is positive, followed by an inverse cascade or band merging of periodic chaos, where 2^k chaotic bands merge into 2^{k-1} chaotic bands [85]. The Fourier power spectrum of

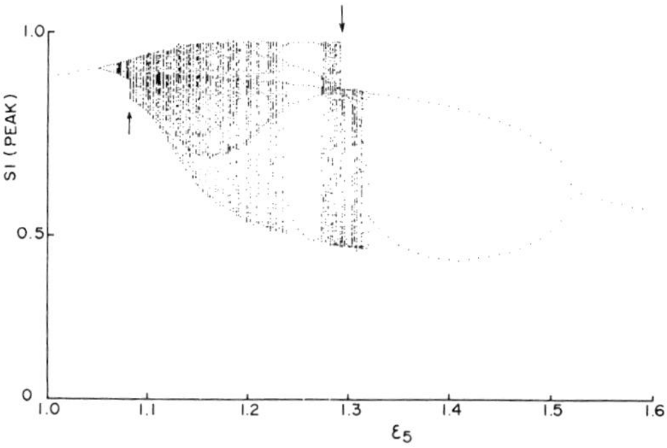

Figure 5.27 Bifurcation diagram of the FDW, $S_1(t)$ as a function of ε_5. Arrows show the chaos–chaos transition [85].

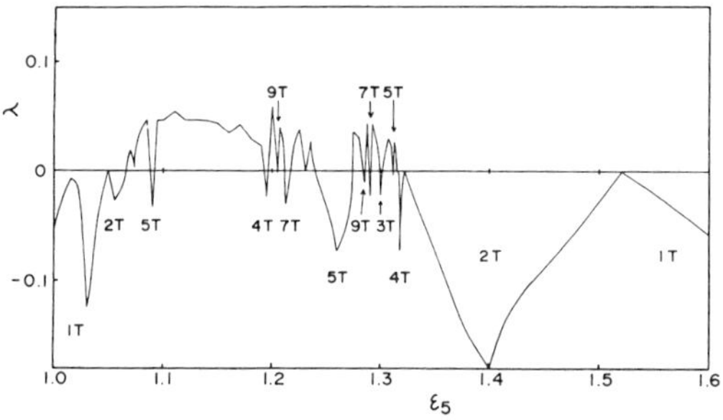

Figure 5.28 Lyapunov characteristic exponents as a function of ε_5 [85].

2^{k-1} periodic chaos looks like a 2^{k-1} periodic limit cycle, and the lowest frequency component $2^{-k}f_s$ loses its sharpness. The universal situation was recognized with $k \leq 4$ in the system [85].

With a further decrease in ε_5 many periodic windows appear, as seen clearly in Figs 5.27 and 5.28. In the present system, a basin boundary exists for both limit cycle orbits and chaos [85]. The basin boundary appears by driving the double-well potential of subsystem 1. After Aizawa and Uezu [92], the asymmetric limit cycles are denoted as $\pi^{\pm}$ (as), where π^{+} (as) and π^{-} (as) form a symmetric pair. Similarly, chaotic attractors are denoted by χ (s) and $\chi^{\pm}$ (as). When one of the local minima in the double-well potential is driven periodically in subsystem 1, an attractor labeled as + is obtained, while in the other case the attractor is labeled −. For instance, the appearance of χ^{+} (as) or χ^{-} (as) depends on the initial conditions due to the basin boundary. For instance, Fig. 5.29 shows a χ^{+} (as) orbit at $\varepsilon_0 = 1.29272$, where there is also a χ^{-}(as) orbit in a pairwise manner with different initial conditions. By decreasing ε_5, chaos–chaos transition occurs at $\varepsilon_5 = 1.2925$, as indicated by an arrow in Fig. 5.27. The transition regime

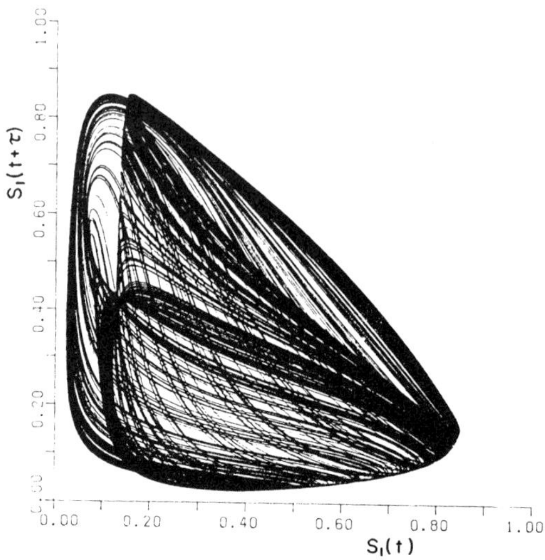

Figure 5.29 Two-dimensional phase portrait of the asymmetric orbits $\chi^+(\mathrm{as})$ at $\varepsilon =$ 1.29272 [85].

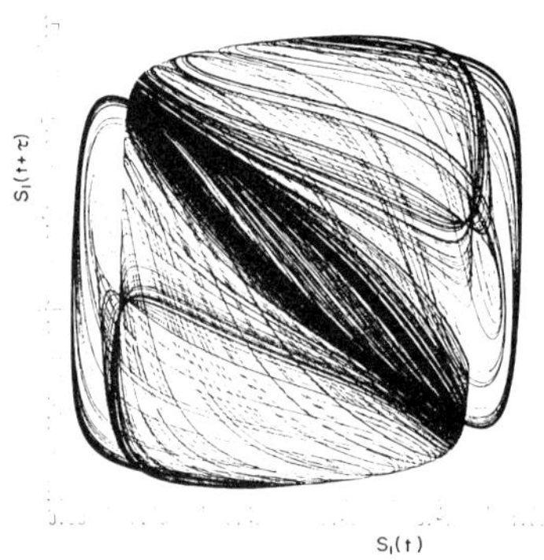

Figure 5.30 Two-dimensional phase portrait of the symmetric orbits $\chi(\mathrm{s})$ at $\varepsilon_5 =$ 1.2925 [85].

is $\chi^{\pm}(\mathrm{as}) \to \chi(\mathrm{s})$, where two basins of $\chi^+(\mathrm{as})$ and $\chi^-(\mathrm{as})$ coalesce into a symmetric basin of $\chi(\mathrm{s})$ [92], as shown in Fig. 5.30. The chaos-chaos transition is described by a random but deterministic switching between two orbits $\chi^+(\mathrm{as})$ and $\chi^-(\mathrm{as})$ as seen in the waveform of Fig. 5.31. As a result, the Fourier power spectrum exhibits $1/f^{\alpha}$ behavior with $\alpha \sim 2$ in Fig. 5.31. In Fig. 5.31, the broken line is the background noise level in the spectrum of $\chi^+(\mathrm{as})$ orbit at $\varepsilon_5 = 1.29272$. With $\varepsilon_5 < 1.292$, the intermittent behavior becomes more frequent and the spectra flatten ($\alpha < 2$). In their pioneering work, Aizawa and Uezu [92] investigated the chaos–chaos transition in the forced Lorenz model and analyzed the intermittent behavior by means of symbolic dynamics. The chaos–chaos transition in Figs. 5.30 and 5.31 is explained well by Aizawa and Uezu's interpretation [92]. In the model simulation (5.23)–(5.27), intermittency of period 4 is seen at an edge of the window ($\varepsilon_5 = 1.19697$), where the transition is $\pi^+(\mathrm{as}) \to \pi^-(\mathrm{as}) \to \pi^+(\mathrm{as}) \to \ldots$. For details, see Ref. [85]. A quasi-periodic route to chaos (Hopf bifurcation) has been also predicted for the different parameters [84].

Returning to the experimental observations, some aspects are now supplemented. As stated previously, it is difficult to observe clear-cut bifurcation phenomena because of the extremely sensitive nature of the system. Later, the experimental method was shifted by adding a light chopper, as shown by the experimental set-up in Fig. 5.32. Figure 5.33a–c

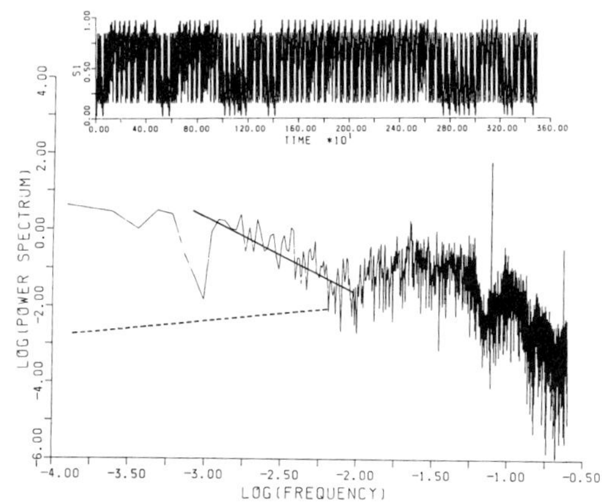

Figure 5.31 The waveform and the Fourier power spectrum of the chaos–chaos transition at $\varepsilon_5 = 1.2925$. The solid straight line shows the $1/f^2$ noise [85].

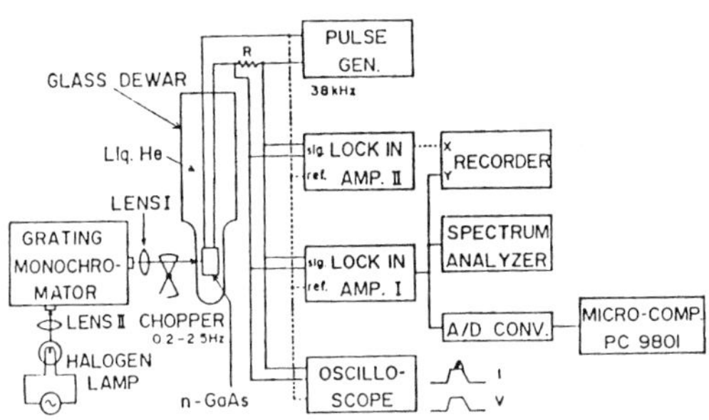

Figure 5.32 Apparatus for the observation of the firing wave instability in high-purity n-GaAs [93].

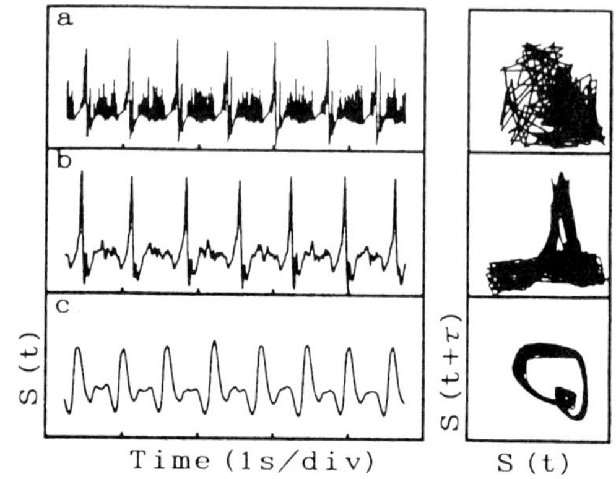

Figure 5.33 Waveforms $S(t)$ and the phase portraits in the FWI regime. (a) $T = 1$ ms, (b) $T = 10$ ms, and (c) $T = 100$ ms [94].

shows the coarse-grained signals $S(t)$ and correspondingly the phase portraits measured in n-GaAs with different time constants τ of the lock-in amplifier [94], where a pulse voltage of 0.952 V with repetition frequency 8 kHz was applied between two ohmic contacts of the sample, and a monochromatic light ($\lambda = 812$ nm) was chopped with a frequency of 1.42 Hz [94]. Output signals were stored on a microcomputer system (Fig. 5.32). The coarse-graining procedure is clearly seen by increasing τ in Fig. 5.33. In particular, with $\tau = 1$ ms (Fig. 5.33a), the phase portrait is irregular and somehow stochastic rather than deterministic. The waveform exhibits microsize instability. With increasing τ, rapid fluctuations are coarse-grained and only a periodic oscillation is detected with $\tau = 100$ ms [94] (Fig. 5.33c). The Lyapunov spectrum is $(0, -, -, -)$ for Fig. 5.33c, while it is $(+, -, -, -)$ for Fig. 5.33a, b (the largest Lyapunov exponent is non-negligibly positive), where the Lyapunov spectra have been evaluated by a fixed condition of the algorithm [94]. The spectrum $(+, -, -, -)$ cannot be understood from the standard chaos theory, but it has been attributed to the coexistence of different sizes of instabilities in space and time [94]. By coarse-graining with $\tau = 100$ ms, only a large size instability is detected. The spatiotemporal behavior of the FW instability measured by the LTSEM method will be discussed in section 5.3.

As a function of chopping frequency f_d, period doubling bifurcation [32, 64] is observed (Fig. 5.34), where $E = 6.3$ V/cm. A period 1 limit cycle is observed with $f_d \lesssim 1$ Hz (Fig. 5.34a), period 2 with $f_d \sim$1.0–1.3 Hz (Fig. 5.34b) and chaos with $f_d \gtrsim 1.3$ Hz (Fig. 5.34c). Unfortunately, periods 4 and 8 have not been resolved. Figure 5.34d shows a period 3 limit cycle observed with an electric field slightly less than 6.3 V/cm and $f_d = 1.225$ Hz.

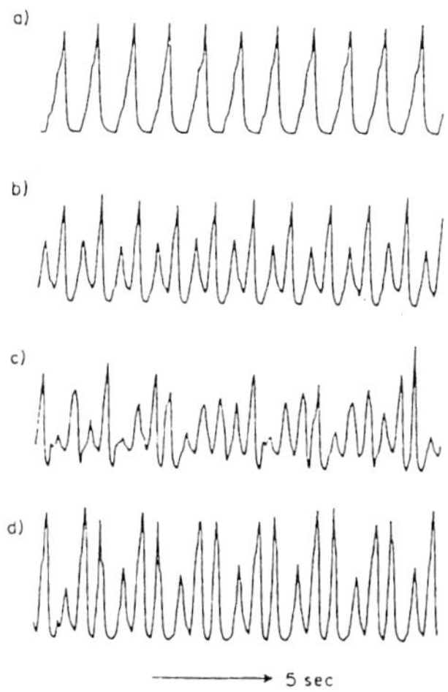

Figure 5.34 Typical example of period doubling bifurcation observed at $E = 6.3$ V/cm. The chopping frequency f_d is: (a) 0.675 Hz, (b) 1.25 Hz, (c) 1.475 Hz, and (d) obtained with slightly lowered electric field and $f_d = 1.225$ Hz. [93].

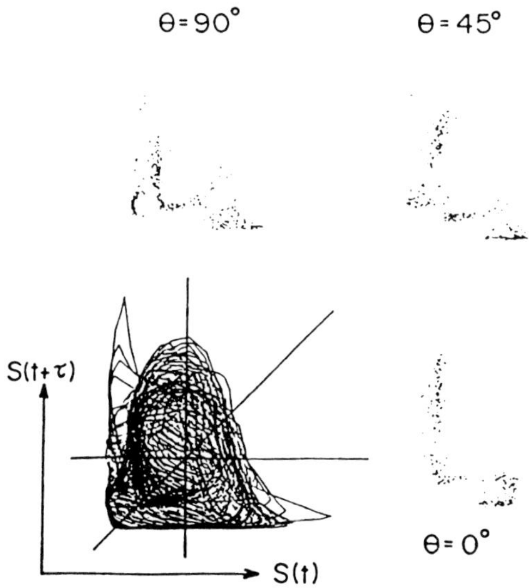

Figure 5.35 The 2D phase portrait $S(t)$ vs $S(t+\tau)$, and the Poincaré sections of the extended 3D phase portrait, cut by the solid lines at various angles [93].

Figure 5.35 shows the 2D phase portrait of chaos in Fig. 5.34c ($S(t+\tau)$ vs $S(t)$) and the Poincaré surface of section obtained by slicing the extended 3D phase portrait at various angles [93].

To conclude, the FW instability and chaos in GaAs appear as a complex hierarchical situation of the impact-ionization avalanche. The complexity arises from the experimental fact that the firing process of the filamentary current is basically stochastic on a micro-time-scale, but the coarse-grained signals are deterministic on a macro-time-scale. Poor control of the bifurcation phenomena in the experiments strongly suggests that many oscillatory states are densely distributed in a narrow range of control parameters (electric field intensity, photoexcitation density, rise time and decay time, and chopping frequency). Fluctuations in the control parameters make it harder to measure the bifurcation processes more in detail. The oscillatory behaviors have been found to reveal deterministic chaos [82] and are well characterized by the proposed model described above [83–85]. The hierarchical complexity (degree of complexity) is related to the spatially extended instability, which will be discussed in section 5.3. Clear evidence for the deterministic nature will be discussed again in section 6.1 by introducing an experimental method to analyze the fractal dimension.

5.3 Electron-beam-induced instability

In this section, the hierarchical complexity of the FW instability in n-GaAs is discussed. Relating to the previous section, the electron-beam-induced FW instability measured by Rau *et al.* [95] and Aoki *et al.* [96] is described. The present discussions are based mainly on the interpretation by Rau *et al.* [95]. As mentioned in Chapter 3, various probing techniques [18, 45–54, 56–58, 97–99] such as the LTSEM method [18, 45–54] have provided precise informations about the spatial pattern images of the stationary filamentary currents formed during the low-temperature impact-ionization avalanche in bulk semiconductors [18, 45–54, 56–58], in $p^+ - n^+ - p - n^-$ [97] and p-i-n diodes (at 300 K) [73, 74, 98]. Nevertheless, under a certain bias condition of a GaAs sample settled in the NDR region, probing by beam scans (electron beam [95, 96] and laser beam [58, 99]) induces current instability, which may prevent precise measurement of the stationary pattern images of the current filaments. Since the unstable behaviors occurring during the measurements [58, 95, 96, 99] could be due to an extremely sensitive nature and thus an essential character of the nonequilibrium phase transition in the spatially extended impact-ionization avalanche, one can also investigate some essential parts of the spatiotemporal pattern formation [95]. In particular, some important features of FW instability such as described in section 5.2 can be expected by measuring the electron-beam-induced instability.

The experimental set-up of the LTSEM method is shown schematically in Fig. 5.36. The sample was a 12-μm-thick n-GaAs on a semi-insulating substrate, which is the same one as used for the experiments on FW instability in section 5.2 (donor concentration $2 \times 10^{14}/\text{cm}^3$, donor level

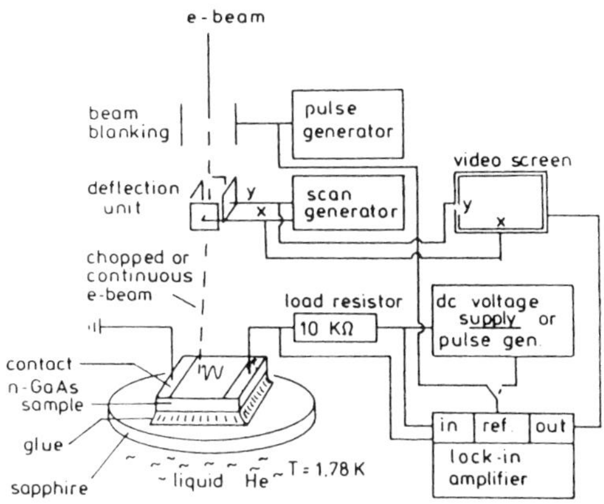

Figure 5.36 Schematic of the experimental set-up for the LTSEM [95].

about 6 meV). The surface area dimension was about 1.3 mm width and 1.2 mm length between two planar Sn contacts. The sample was glued onto a sapphire disk which was mounted in a scanning microscope equipped with a liquid helium stage (Fig. 5.36). The bath temperature was kept at 1.78 K. In Fig. 5.36, the focus of the chopped electron beam was scanned over the surface of the sample to detect the spatially inhomogeneous transport structure [96]. The electron beam current was about 10 pA, supplied by a gun voltage of 26 kV, resulting in a beam power of about 0.3 μW. The chopping frequency of the beam was kept at 5 kHz for measurement by the phase-sensitive technique [95, 96]. For measurement of the FW instability, the continuous electron beam was also scanned without chopping. The sample current was detected via the voltage drop at the load resistor (10 k Ω). The response signal was measured phase-sensitively with respect to the chopping frequency of the electron beam (by a lock-in amplifier) and recorded by a video screen as a function of the beam coordinates (Fig. 5.36).

Figure 5.37 shows the *I–V* characteristic of the n-GaAs sample mounted in the scanning electron microscope. A low conduction branch from zero up to a sample voltage of 0.56 V is followed by a high conducting branch, which is due to the onset of avalanche breakdown of neutral donors and the formation of current filaments. When the electron beam is scanned rapidly over the sample surface, the *I–V* characteristic does not show any difference from the nonirradiated case except at the hysteresis region (hatched area in Fig. 5.37). In the hatched area, the electron beam destabilizes the hysteresis, followed by current instability between the low and the high conducting states of the *I–V* curve. Note that the spontaneous current oscillations do not emerge in the absence of the electron beam. When the GaAs sample is biased to a high conducting state (upper branch) of the *I–V* curve, a spatial image of the current filament can be measured typically by the phase-

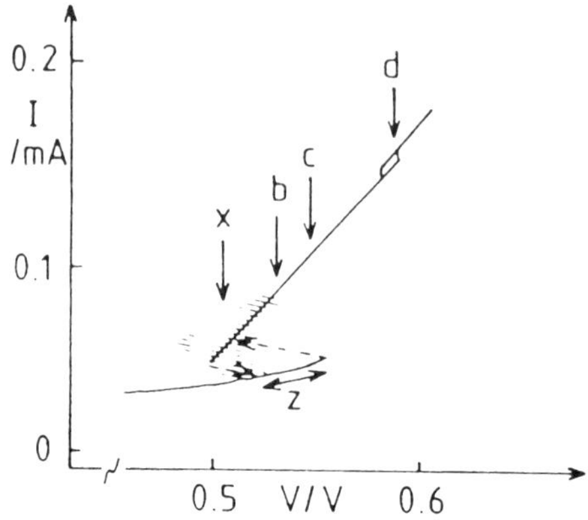

Figure 5.37 *I–V* characteristic of the n-GaAs sample mounted in the scanning electron microscope. The hatched area indicates the region of electron-beam-induced instability. Points b–d refer to the spatial images displayed in Fig. 5.38b–d, respectively, the point *x* to those in Fig. 5.39a, b, and the arrow at the bottom of the characteristic (*z*) to the ac voltage applied to the sample during the experiments of firing instability [95].

sensitive measurement technique [95, 96]. If the sample is biased at an unstable region (hatched area), the pattern image no longer gives precise information about the stationary structure. Instead, one can expect some important features of the spatially extended dynamical behaviors of the current filament using several measurement techniques.

Before discussing the spatial and temporal structure, let us recap on the imaging method of the stable filament structures by using the phase-sensitive measuring technique [95, 96]. The chopped electron beam is scanned over the sample surface of the geometry, as sketched in Fig. 5.38a. The direction of the linescan is from left to right in the transverse direction of the sample, where the scanning velocity is about 0.2 mm/s [95]. The pattern images at three different bias conditions denoted by arrows b, c and d are shown correspondingly in Fig. 5.38b–d. The beam-induced excess current is detected by a lock-in amplifier (Fig. 5.36), where the reference signal (5 kHz) is fed from the pulse generator used for the beam blanking. In Fig. 5.38, a filament is indicated by an increased response signal at both of its facing borders. The pattern images in Fig. 5.38 are well understood, since a considerable change in conductivity is expected only if the electron spot touches the boundaries of the current filament. The transverse profile of the current filament is in the form of a kink–antikink pair, which is stabilized via the counterbalance between the transverse drift and the diffusion currents at the borders [59]. When the electron spot touches either one of the filament boundaries, it disturbs the counterbalance and induces locally the impact-

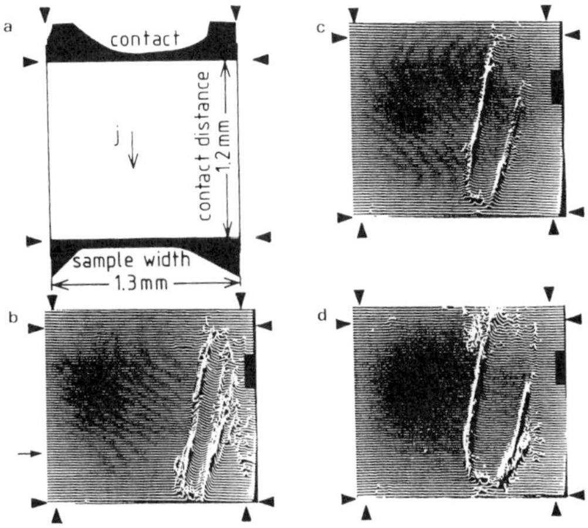

Figure 5.38 (a) Sketch of the sample geometry. Contact distance and sample width are indicated. The current flow extends in the vertical direction. (b)–(d) Two-dimensional images (y-modulation) of the transport structure in the n-GaAs sample at different sample voltages (referred to points b–d in Fig. 5.37). The sample voltages are (b) $V = 0.524$ V, (c) 0.545 V, and (d) 0.584 V. The results are obtained using the phase-sensitive measurement technique. The edges of the sample and of the contact bars are marked by black triangles in each picture. The scanning direction was from the left to the right, and the scanning velocity was about 0.2 mm/s. The vertical position of the linescan shown in Fig. 5.40a is indicated by an arrow in (b) [95].

ionization avalanche, resulting in generation of the excess current and then a considerable increase in conductivity. Inside the current filament, the electron beam does not induce any excess current since a large part of the impurities is already impact-ionized. If the electron beam is located far from the filamentary region, it does not cause any change in conductivity, since the ohmic region is electrically insulated owing to the low-temperature freeze-out of the impurities and thus stable against small local disturbance [95]. Therefore, the pattern images in Fig. 5.38b–d are well explained by a growing filament as a function of the sample bias.

Now, we discuss the electron-beam-induced instability occurring in the hysteretic region denoted by x in Fig. 5.37. The phase-sensitive measuring technique alone no longer gives realistic imaging of the filament structure because of the oscillatory behavior. In order to understand the spatiotemporal behaviors, careful investigations are required, using several measuring techniques. The best way is to use an appropriate technique to localize the instability (e.g., resonant imaging technique [45, 46]). However, as stated below, the present system, despite having a simple structure of a static current filament (Fig. 5.38), exhibits complex behaviors of electron-beam-induced instability. Full understanding of the spatiotemporal structure is complicated. Figure 5.39a shows the two-dimensional spatial image of the filament structure in the instability region (point x in Fig. 5.37) measured by the phase-sensitive technique, where the electron beam chopped with a frequency of 5 kHz was scanned over. Apparently, the structure is never explained by a simple structure of the stationary filament such as measured in Fig. 5.38, but is composed of some oscillatory structure. Detection of the response signals with an ac voltmeter, produces a spatial image as shown in Fig. 5.39b, where the continuous electron beam was scanned over. With the ac voltmeter measurement, only the sensitive regions of the structure showing local instability are detected, since only the time-dependent signals with relatively large amplitudes are integrated owing to the slow response of the voltmeter. Therefore, Fig. 5.39b indicates the oscillatory spatial structure rather than the stationary structure. The structure is measured approximately along the stable filament borders in Fig. 5.38b. A small side-

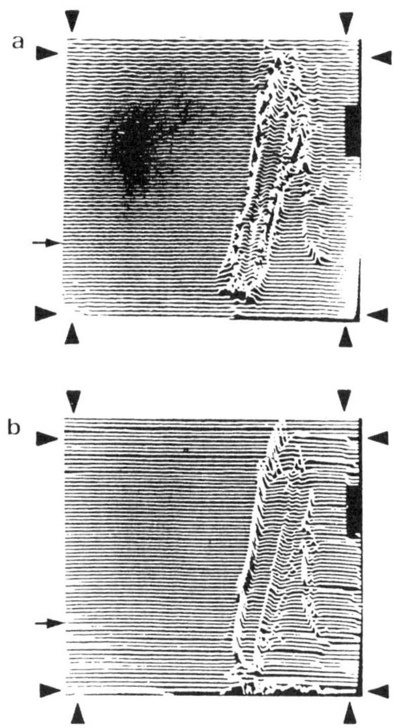

Figure 5.39 (a) Two-dimensional spatial image of the filament structure in the instability region obtained by the phase-sensitive technique (bias voltage $V_0 = 2.12$ V). (b) Two-dimensional spatial images of the ac voltmeter measurement, indicating the sensitive regions of the structure where the electron-beam-induced instability arises ($V_0 = 2.12$ V) [95].

branch of the oscillatory filament structure has also been imaged in the lower right part of the sample in Fig. 5.39a, b [95, 96]. The difference between two measurements in Fig. 5.39a, b can be discussed by looking at the transverse profiles in Fig. 5.40b, c, respectively, where the profiles have been measured by linescans at a specified vertical position, indicated by an arrow in Fig. 5.39a, b. For reference, the transverse profile of the filament (note: bias point b in Fig. 5.37; see Fig. 5.38b) measured by the phase-sensitive technique is shown in Fig. 5.40a. The response signal in Fig. 5.40a indicates clearly a well-defined double-peak structure reflecting two facing borders of the stable current filament. In the instability regime of Fig. 5.40b the signal measured by the phase-sensitive technique consists of several peaks and noisy regions. Moreover, the structure depends on the scanning direction of the electron beam. Whenever the electron beam enters the filament region, a significant peak of the response signal (dashed curves in Fig. 5.40b) could be measured, where the arrows in Fig. 5.40b mark the scanning directions. In Fig. 5.40c, the profile measured by the ac voltmeter does not show any strong dependence on the scanning direction.

The behaviors in Fig. 5.40 can be understood by measuring directly the specific time signals of the sample currents obtained by focusing the chopped electron beam onto a specified position. When the chopped electron beam is focused onto the boundary in Fig. 5.40a (point 1), the

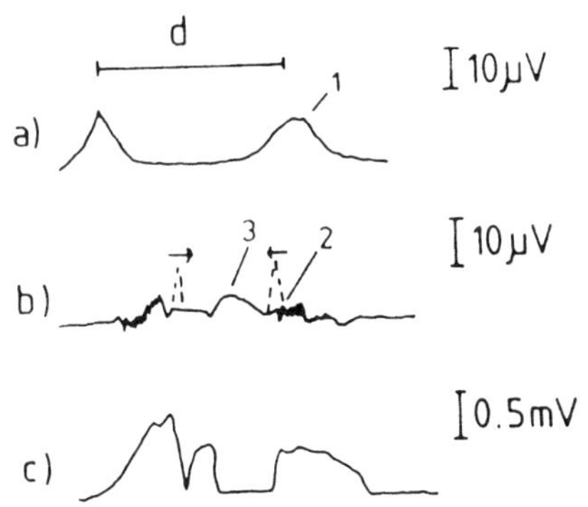

Figure 5.40 Linescans of the spatial signal taken perpendicular to the current flow in the center of the sample. (a) and (b) were obtained by the phase-sensitive method at voltages corresponding to Figs. 5.38c and 5.39a, respectively. In (a) the filament cross-section is displayed as the distance d (140 μm) between the two peaks of the linescan. Points 1 and 2 indicate the location of the electron beam during the measurements of the time signals displayed in Fig. 5.41a and b, respectively. (c) Linescan taken from the ac voltmeter measurement corresponding to Fig. 5.39(b). These signals were almost independent of the scanning direction, except for the two peaks in (b) (indicated by dashed curves with arrows giving the scanning direction). The scanning velocity was about 0.02 mm/s in (a) and 0.01 mm/s in (b, c) [95].

modulated current shows an ordinary time signal in Fig. 5.41a, where the on/off state of the chopped electron beam is marked by black/white bars in the upper trace of Fig. 5.41a, respectively. The time signal is strongly correlated with the modulation of the electron beam. Thus, the current filament is in the stable regime to the perturbation. Remember that the profile in Fig. 5.40a is obtained by averaging the modulated currents using a lock-in amplifier. In the unstable regime, focusing the electron beam onto point 2 in Fig. 5.40b, a spontaneous current oscillation with large amplitude and low frequency (~35 Hz) is measured (Fig. 5.41b). The self-generated oscillation induced by the electron beam is uncorrelated with the beam modulation (~5 kHz). The ordinary response to the chopped electron beam (~5 kHz) can be recognized only in part as superimposed on the self-generated oscillation (see inset of Fig. 5.41b). Note that the self-generated oscillation is observed in the same way when using a continuous electron

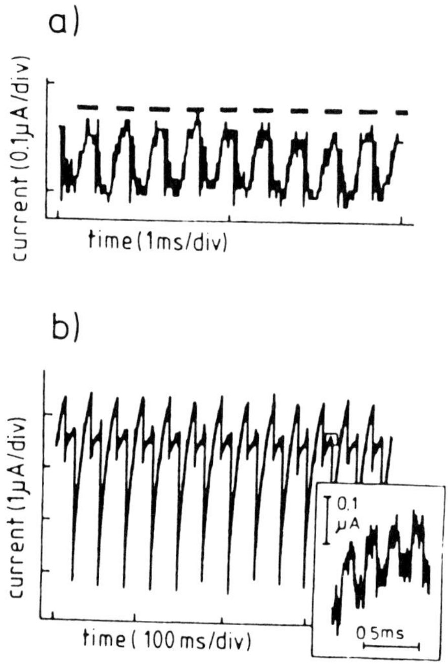

Figure 5.41 (a) Time signal obtained in the case when the chopped electron beam is focused on the boundary of a current filament (point 1 in Fig. 5.40a). An identical signal could be obtained at one of the inner peaks of the linescan in Fig. 5.40b (indicated by point 3). (b) Time signal in the case of electron-beam-induced instability (points 2 in Fig. 5.40b). The inset shows the direct response to the electron beam, as can be found in parts of the total signal marked by the square box [95].

beam and that the oscillation does not emerge without the electron beam. Since the amplitude of the spontaneous oscillation is relatively large ($\sim 4\ \mu A$), it is suggested that the oscillation arises by a repeating cycle of g-r processes of the current filament.

The ac voltmeter measurement in Fig. 5.40c is an adapted method to localize this instability. As in Fig. 5.39b, the continuous electron beam was scanned over the transverse direction across the sample width in Fig. 5.40c. The self-generated oscillations are responsible for the structure of Fig. 5.40c. By comparing Fig. 5.40b with Fig. 5.40c, a remarkable point is observed, as follows. In Fig. 5.40b, the peak structure at the marked position of point 3 is referred to the ordinary modulated signal as expected in Fig. 5.41a and not referred to the self-generated oscillation. Apparently, the structure is absent in Fig. 5.40c. From this fact, it is found that a stable current filament having nearly the minimal diameter can be destabilized by an electron beam focused onto a location close to its boundaries.

Firing wave instability induced by electron beam excitation is discussed in the following paragraphs. As shown later, in the regime of the FW instability the electron beam is no longer a simple tool to detect the transport structure in the sample, but an important control parameter which determines the nonlinear dynamical behaviors of the system as a function of its coordinates of the sample surface. In the experiment, an ac bias with the frequency of $f = 40$ kHz is superimposed on a dc bias in such away that the firing of the filament does not occur in the absence of the electron beam. Using a continuous electron beam under the bias condition stated above, one can observe electron-beam-induced FW instability, the behaviors of which are similar to those described in section 5.2. The I–V phase portrait during the FW instability is shown in Fig. 5.42. Apparently, the stochastic firing in Fig. 5.42 occurs via different paths in the phase portrait with several characteristic amplitudes of the filamentary currents. The recurrent firing forms the density wave, which is similar to the observation in Fig. 5.18 of the previous section.

The FW instability is coarse-grained by using the lock-in amplifier with a reference signal of 40 kHz. Choosing a dynamical range of some ms, the amplifier integrates the firing events over the micro-time-scale giving an output of the course-grained signal in the macro-time-scale. Figure 5.43a indicates a nearly periodic oscillation with the frequency of about 2.7 Hz [96], where the electron beam has been focused on a fixed position inside the filamentary channel. Again, it is emphasized that no firing events occur in the absence of the electron beam. Frequencies (some Hz) of the FW instability are much lower than the driving frequency (40 kHz) and even lower than the frequencies of the spontaneous oscillations (30–100 Hz) such as discussed in Fig. 5.41b. By scanning the electron beam over the whole sample surface, complex behavior of the spatiotemporal dynamical structure [95, 96] is obtained as given in Fig. 5.43b and the blow-up in Fig. 5.44. The

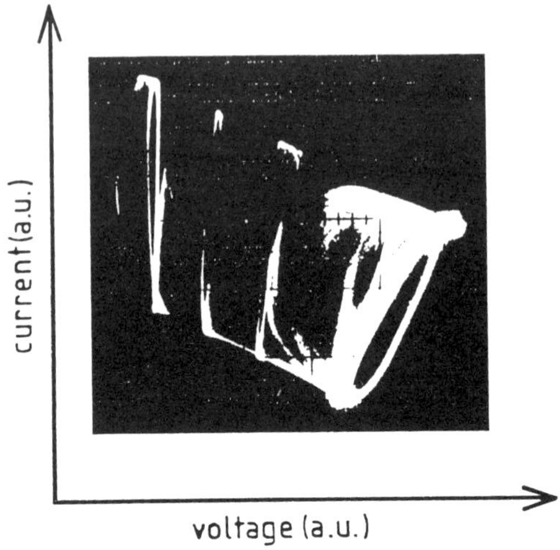

Figure 5.42 Phase portrait in the case of firing wave instability, where the continuous electron beam is applied to a distinct fixed location on the sample surface [95].

seemingly turbulent image in Figs. 5.43b and 5.44 indicates that the oscillating states depend sensitively on the coordinates of the beam spot. The image pattern is quite irregular but distributed approximately along the filament structure in Figs. 5.38 and 5.39. In Fig. 5.43b, the linescan of the electron beam is not simply a tool for imaging, but its location is apparently an important control parameter for the space-dependent dynamical behaviors. The oscillation frequency and amplitudes vary in an extremely sensitive manner depending on the local position of the electron beam.

Comparing the present measurements with the observations of the FW instability in section 5.2, there are several new findings. Apparently, the measurements in Figs. 5.43b and 5.44 deny the single trigger zone but show a great number of trigger zones along or inside the filamentary channel. In addition, a single filament with a certain branching structure (Figs. 5.38b, 5.39) is measured instead of two filaments. A great difference is that the sample surface is uniformly irradiated by photon flux in section 5.2, while the beam spot can be specified at an arbitrary trigger point on the sample surface in the present experiment. Competition modes (different sizes of the instability) may occur since many trigger points are excited by uniform photoirradiation. This explains well the coexistence of different sizes of the instabilities observed in Fig. 5.33. The two filamentary channels have been assumed in the FW model in section 5.2. The assumption seems to be simplified. Under uniform photoexcitation, there may be some cooperative oscillating centers (regions) inside the filamentary channel to generate chaos

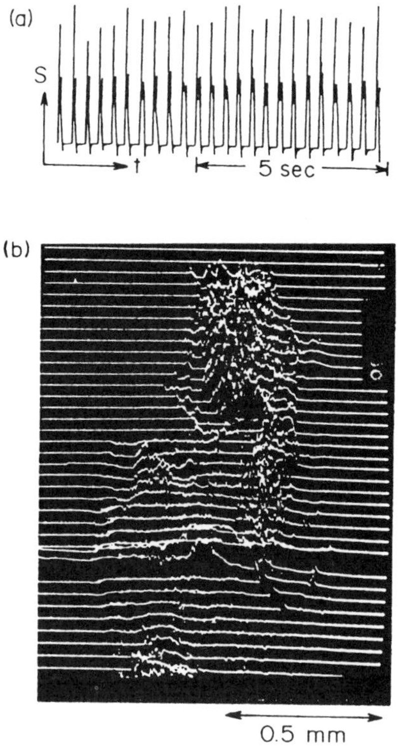

Figure 5.43 (a) Waveform of the coarse-grained signal from a lock-in amplifier observed in beam-induced instability: (b) turbulent image observed in firing wave instability [96].

via the mutual coupling of the FW. In the FW model (5.23)–(5.27), the assumption of two filaments should be replaced by two oscillating centers inside a filamentary channel with the firing activities S_1 and S_2 as a realistic modification, without revision of the ODEs in (5.23)–(5.27). The proposed model is considered to be the simplest case to explain the real system of FW instability.

5.4 Periodically driven chaos

Among various types of current instability concerning the impact-ionization avalanche [50, 59] (see also this chapter), the periodically driven chaos system introduced in Chapter 4 exhibits the most fundamental dynamical behaviors of the NDR instability and even provides a prototype of the nonlinear response in the nonequilibrium phase transition. This section will

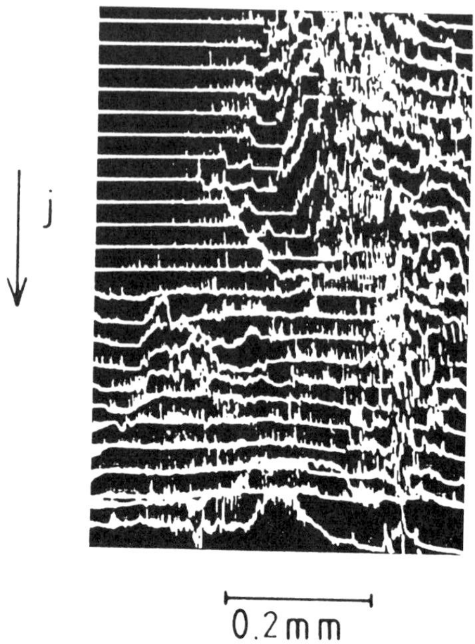

Figure 5.44 Blow-up of the spatiotemporal dynamical structure in Fig. 5.43 [95].

describe mainly the experimental aspects of periodically driven chaos, which will be discussed and evaluated by model simulations.

Binding energies of the ground states of shallow donors or acceptors in semiconductors are relatively small, ranging from a few meV to 40 meV. Especially in n-GaAs, the binding energies are 5.8 $\sim$ 6.1 meV depending on the impurity atoms [100]. Therefore, the freeze-out effect of the impurities is limited only at low temperatures near 4.2 K. As a result of the freeze-out effect, semiconductor materials behave like insulators, showing low conductivity. By applying relatively low electric fields (a few V/cm [59] for shallow impurities), one can observe the hysteretic I–V curves formed by the impact-ionization avalanche of the neutral impurities. The nonequilibrium phase transition is described as a first-order phase transition from low- to high-conductivity states. The general form of the J–E characteristic is written as:

$$J = en(E)\mu(E)E, \tag{5.28}$$

$$\mu(E) = e\tau(E)/m^*, \tag{5.29}$$

where e is the electron charge, $\mu(E)$ the carrier mobility, m^* the effective mass of free carriers, and $\tau(E)$ the momentum relaxation time of the carriers due to phonon scatterings and ionized impurity scattering. The nonlinearity arises from $n(E)$ and $\mu(E)$. In the approximated model, a constant electron mobility was assumed in Chapter 4. Remember that the rate equation of the electron density n is approximately expressed by the derivative of a potential function $V(n, E; T_L)$:

$$\dot{n} = -\gamma \frac{\partial V(n, E; T_L)}{\partial n}, \tag{5.30}$$

where γ is the damping constant and T_L is the lattice temperature. In the approximated model, the potential function has been given by (3.85). In the experiments, a dc + ac bias of the form $E_0 + E_{ac} \sin(2\pi f_0 t)$ is applied, where f_0 is the driving frequency. The electric field E responds to the external driving force as:

$$\dot{E} = -\gamma_d (E - E_0 - E_{ac} \sin(2\pi f_0 t) + cJ(n, E)). \tag{5.31}$$

As expected from the model simulations in sections 4.7 and 4.11 a rich structure of the dynamical behaviors and chaos is observed when the driving frequency satisfies the relation $f_0 \sim 1/\gamma_d$.

The experimental set-up is shown in Fig. 5.45. A typical example of the I–V curve at 4.2 K is shown in Fig. 5.46, where the broken line indicates the expected load line, V_{dc} is the sample voltage and V_0 is the applied dc voltage [101]. The hysteretic behavior caused by the impact-ionization avalanche is quite stable in Fig. 5.46, showing no oscillatory instability. The currents and voltages are monitored by a digital storagescope with a sampling frequency of 40 MHz. The experimental data are accumulated into the microcomputer system in order to analyze statistical quantities such as the correlation dimensions and $f(\alpha)$ spectra. A dc+ac bias of the form of $V_0 + V_{ac} \sin(2\pi f_0 t)$ is applied to the ohmic contacts, where $f_0 \gtrsim 0.9$ MHz.

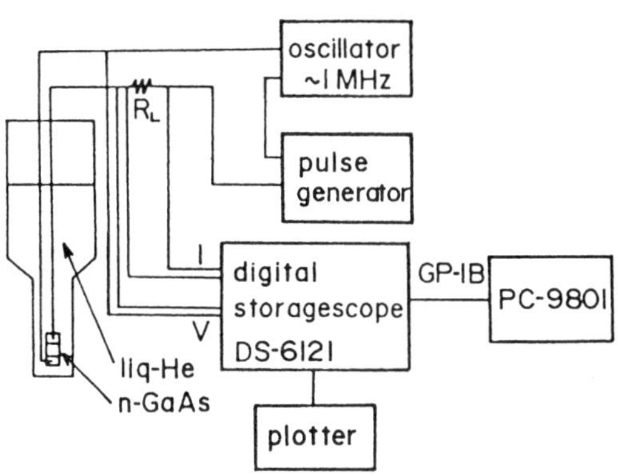

Figure 5.45 Experimental set-up for the observation of chaos in the periodically driven current filament. A pulse generator can also be used as a dc voltage supplier.

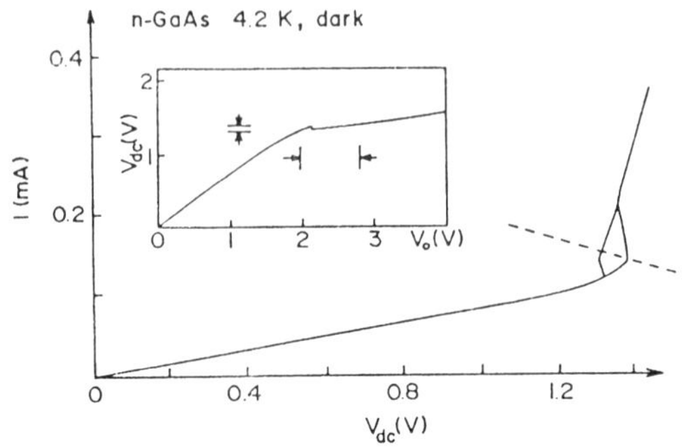

Figure 5.46 The S-shaped *I–V* curve observed in high-purity n-GaAs at 4.2 K under the dark condition, with the static regime. The inset shows the relation between V_0 and V_{dc}, where V_0 is the applied voltage and V_{dc} is the sample voltage [101].

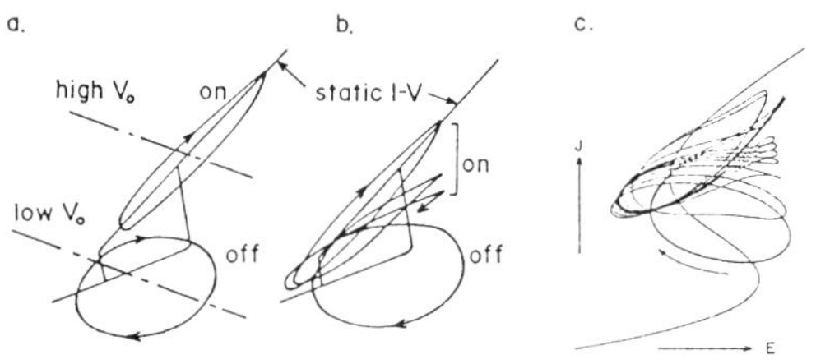

Figure 5.47 (a, b) Schematic presentation of Lissajous' figure, (c) numerical results of the strange attractor [101].

Chaotic behavior of the current oscillation appears approximately above 500 kHz. Typically, the bifurcation routes to chaos can be clearly observed above 900 kHz [101]. In order to observe chaos, one must choose the amplitudes of the ac voltage to be larger than the hysteresis width ($2V_{ac} >$ hysteresis width $\sim$0.06 V). Before proceeding to the experimental details, the nonlinear response and chaos will be described qualitatively. Figure 5.47 shows the schematic presentation of the Lissajous' figures observed in the experiment with $f_0 \sim 1$ MHz and $V_{ac} = 0.1$–0.2V, where the static *I–V* curve is superimposed on the Lissajous' figures in Fig. 5.47a, b. The clockwise looping of an ellipse at low V_0 and high V_0 in Fig. 5.47a is caused by a large displacement current. The looping at low V_0 indicates simply phase locking of ohmic and displacement currents in the lower branch of the S-shaped *I–V* curve (i.e., off-state), while it means an enforced breathing mode of a current filament in the upper branch (on-state). With an intermediate value between low and high V_0, chaotic nonlinear response can be observed due to intercommunication between the lower and higher branches, as shown schematically in Fig. 5.47b. The behavior is well explained by the approximated model of the impact ionization avalanche as described in Chapter 4. The essential part is the breakdown of perfect delay when the hysteresis is driven periodically. Figure 5.47c shows a drawing of the chaotic attractor calculated by the approximated model, where the displacement current is taken into account by the form of $J = \nu E + \alpha \dot{E}$ ($\alpha = 27$).

Typical examples of a photograph of the Lissajous' figure, the waveform of the chaotic oscillation and the Fourier power spectrum are shown in Fig. 5.48. The current waveform is irregular and the corresponding Fourier power spectrum shows continuous noise background [101]. In the current waveform, the large peaks are attributed to the formation of the current filament, which fires randomly but in a deterministic manner. In the Fourier power spectrum, the background level increases by 30 dB in comparison with the phase-locked states. In the early days of nonlinear science, just before the brilliant days of chaos theory, the accepted theory predicted that a gradual increase in the number of independent frequencies would lead to a turbulent state in the fluid when a control parameter is varied [102] (Landau–Hopf picture). In the semiconductor experiments as well as in other experimental systems, the Landau–Hopf picture can be denied since a continuous spectrum of chaos can be obtained by definite bifurcation

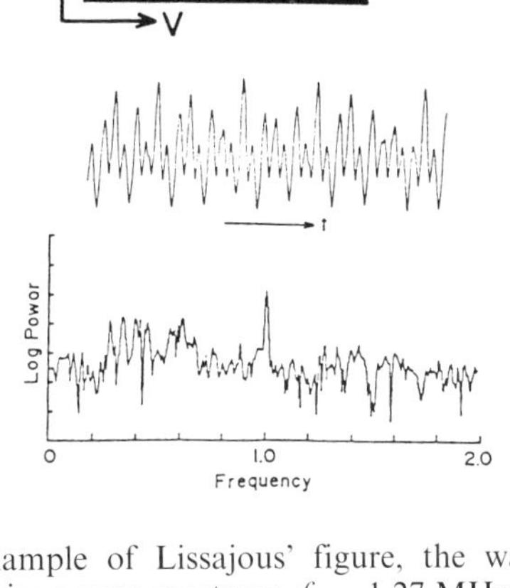

Figure 5.48 Typical example of Lissajous' figure, the waveform of the current oscillation and the Fourier power spectrum. $f_0 = 1.27$ MHz, $V_0 = 2.52$ V and $V_{ac} = 0.15$ V [93,101].

regimes such as period doubling bifurcation, not by adding successively independent frequencies. The photograph of the phase portrait in Fig. 5.48 is explained satisfactorily by the model simulation in Fig. 5.47c. Just before the onset of chaos, one never observes a Fourier power spectrum composed of many sharp lines as expected from the Landau–Hopf scenario.

As stated later, the bifurcation diagrams differ with different values of the load resistance R_L Here, the results obtained by relatively small resistance ($R_L = 200\ \Omega$ and $1\ k\Omega$) are discussed. The control parameters are V_0, V_{ac} and f_0. With fixed V_{ac}, typical examples of the phase diagram are shown in Fig. 5.49, where $V_{ac} = 0.14$ V, $R_L = 200\ \Omega$ and two control parameters V_0 and f_0 are varied successively. In the phase diagram, the numbers n ($n = 1, 2, 3, 4$) denote the period n limit cycle and the symbol ch implies chaos. Period 1 at low V_0 and high V_0 indicates a simple limit cycle, as explained in Fig. 5.47a. When V_0 is increased from below at a fixed value of f_0, the period 1 limit cycle bifurcates abruptly to a chaotic phase [93, 101]. The long-term behavior of the current oscillation looks like intermittency; the laminar phase of period 1 is suddenly interrupted by a burst of chaotic phase. In the phase diagram of Fig. 5.49, the intermittency is observed in a relatively narrow region ($\Delta V_0 \sim 0.06$ V) along the phase boundary near $V_0 = 2.28$ V. With a slight increase in V_0 near the phase boundary, bursts are frequently observed and the average duration of the chaotic phase becomes longer than that of the laminar phase of period 1. With a further increase in V_0, the laminar phase finally disappears and stationary chaos appears. There exists a wide region of chaotic sea near the center of the phase diagram. In the chaotic sea, there is an island of period 3 (see the photograph of the phase portrait in the inset figure of Fig. 5.49) and peninsula of period 2 and period 4). By decreasing V_0 from 2.9 V, period 1 bifurcates to period 2 via unusual

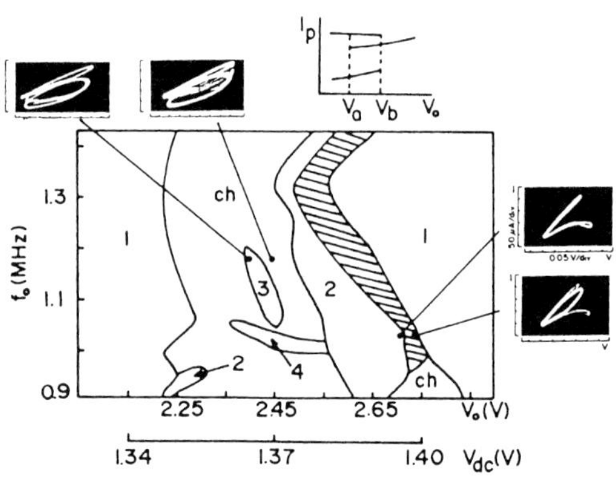

Figure 5.49 Phase diagram observed with $V_{ac} = 0.14$ V and $R_L = 200\ \Omega$. Two pairs of Lissajous' figures are shown [93,101].

period doubling bifurcation. In the hatched region of Fig. 5.49, random switching between period 1 and period 2 occurs: the period doubling bifurcation is not smooth, but rather indicates a hysteretic behavior of the bifurcation in a certain range of $V_a < V_0 < V_b$ as shown schematically in the inset of Fig. 5.49. Under the influence of small external noise, the hysteretic bifurcation or subcritical bifurcation may induce an extrinsic crisis if two limit cycles of period 1 and period 2 are very close each other [93, 101]. The photograph of the phase portrait in the hatched region of Fig. 5.49 indicates the occurrence of extrinsic crisis, which is compared to the photograph for a period 2 limit cycle. In the long-term behavior of the extrinsic crisis, the Fourier power spectra show Lorentzian shapes. At the center of the hatched region, the cut-off frequency is typically about 2 kHz, which means that the average recurrence time is approximately 0.5 ms for both limit cycles. The origin of the external noise was later attributed [103, 104] to radiowaves from broadcasting stations, which will be discussed in section 7.2.

With $V_{ac} = 0.21$ V and $R_L = 1k\Omega$, an alternating periodic–chaotic bifurcation sequence can be observed, as shown in Fig. 5.50. The intermittent behavior at low V_0 and the extrinsic crisis are very similar to Fig. 5.49. In addition, five islands of period $n+1$ (denoted P_1^n ($n = 2$–6) for period $n+1$ limit cycle [103, 104]) are embedded in parallel within the chaotic sea, where P_1^n means one spark (current filamentation) per $(n+1)$-drive. That is, the orbit winds n times around the lower branch of the I–V curve and jumps up once to the upper branch (see the photograph for period 3 in Fig. 5.49). The periodic sequence is physically meaningful since the spark rate $1/(n+1)$ is a decreasing function by decreasing V_0. The observed periodic–chaotic bifurcation sequence is similar to those in the Belousov–Zhabotinsky

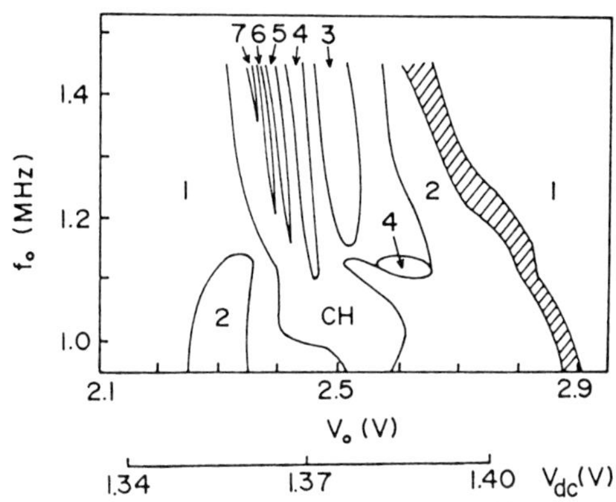

Figure 5.50 Phase diagram observed with $V_{ac} = 0.21$ V and $R_L = 1$ kΩ [93, 101].

(B-Z) reaction [81, 105–107] and also to the bifurcation phenomenon observed in the chaotic behavior of electron-hole plasma in n-Ge [77]. The argument by Tomita and Tsuda [105] seems to be applicable to the present system, as discussed in section 4.11.

Many periodic islands, as in Fig. 5.50, are observed with relatively large values of V_{ac} and with a large load resistance ($R_L \sim 1$ kΩ). However, it is noted also that the bifurcation diagram differs from sample to sample in a sensitive manner, depending especially on the contact formations. Among many samples of n-GaAs, several samples exhibit many periodic islands, while others show simply a few islands in the phase diagram.

The alternating periodic–chaotic bifurcation sequence can also be seen as a function of V_{ac}. Figure 5.51 shows the current waveforms, the Fourier power spectra and the phase portraits for various values of the ac amplitudes V_{ac}. In the figure, the phase portraits are drawn by an embedding method, by plotting $I(t+\tau)$ vs $I(t)$ in the 2D phase space. By decreasing V_{ac} from $V_{ac} = 82$ mV, period 1 (Fig. 5.51a) bifurcates to period 2 (Fig. 5.51c) via an extrinsic crisis (Fig. 5.51b). Note that the phase portrait in Fig. 5.51b is composed of two components of period 1 and period 2 (see

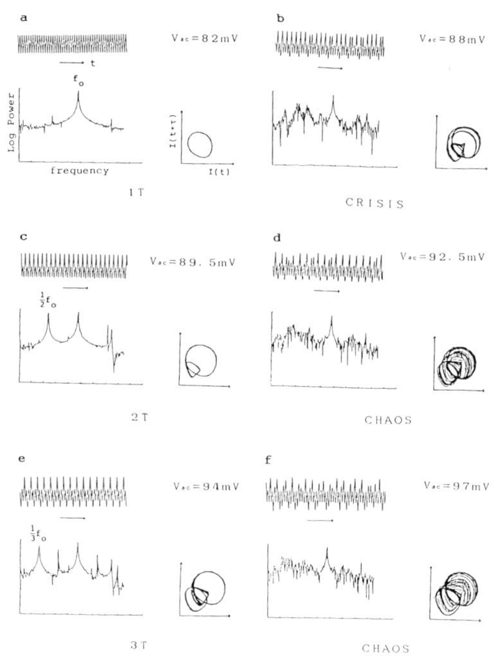

Figure 5.51 Alternating periodic–chaotic bifurcation sequence (a–l) as a function of V_{ac} where $f_0 = 1.4$ MHz, $V_{dc} = 1.54$ V, and $R_L = 500\Omega$. Note that V_{dc} is the sample voltage [93]. Continued overleaf.

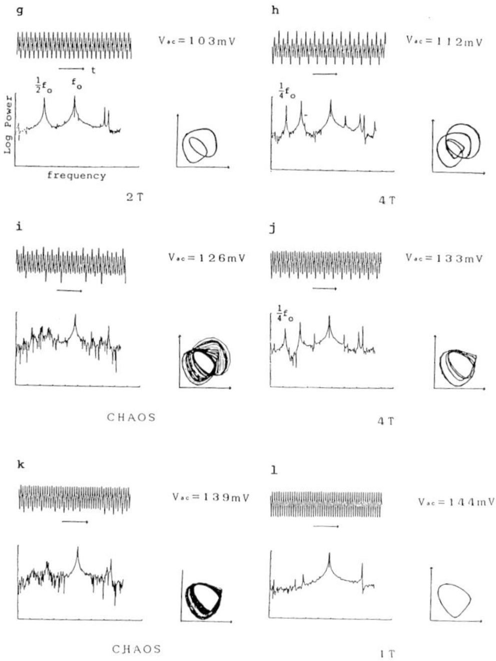

Figure 5.51 Continued

Fig. 5.51a, c). With a further decrease in V_{ac}, an alternating periodic–chaotic bifurcation sequence appears, as shown in Fig. 5.51c–j. The bifurcation scheme is similar to Fig. 5.50. The Fourier power spectra of periods 2, 3 and 4 exhibit sharp subharmonic frequency components, while they are board-band spectra for chaotic states. Between Fig. 5.51g and h, a smooth period doubling from period 2 to period 4 is observed. The bifurcation scheme as a function of V_{ac} is classified phenomenologically into three parts as (i) the postbreakdown regime in Fig. 5.51a, b, (ii) the interplay between two branches (Fig. 5.51c–i), and (iii) the prebreakdown regime (Fig. 5.51j–l). The bifurcation in the prebreakdown regime is caused by a weak nonlinearity near the lower branch.

Qualitatively, bifurcation phenomena such as the alternating periodic–chaotic bifurcation sequence can be explained using mathematical models. Assume that the switching phenomena of the current filament are modeled mathematically by 1D mappings. First, consider the following map function [93]:

$$
\begin{aligned}
X_{n+1} &= TX_n \quad 0 \le X_n < 1/T, \\
X_{n+1} &= A + S(X_n - 1/T) \quad \text{otherwise,}
\end{aligned}
\tag{5.32}
$$

where T is a control parameter, while other parameters are held constant

($A = 0.3$, $S = 0.12$). The iterative equation (5.32) (see Fig. 5.52a) is considered to represent behaviors of a switching element, where an interval $[0, 1/T]$ corresponds to, say, the lower branch and (1/T, 1] to the upper branch. Note that the map function is linear in each interval. The iterated points give periodic cycles with period m for any given T. The bifurcation sequence as a function of T exhibits the devil's staircase (Farey sequence) [108, 109]. In Fig. 5.52b, the rational number $\rho = 1/m$ is given as a function of T. The rational number means the firing rate, namely that the iterated points take $m - 1$ round trips in the lower branch and one firing in the upper branch. The staircase in Fig. 5.52b includes a finer and finer structure of self-similarity. For instance, between the rational number 1/5 and 1/6, there is a locking state 2/11 (period 11) which is given by the sum of the numerators divided by the sum of denominators. Between 1/5 and 2/11, there is 3/16, and so on. The Farey sequence is then given as:

$$1/6 < \ldots < 4/23 < \ldots < 3/17 < \ldots < 5/28 < \ldots < 2/11 < \ldots < 5/27$$
$$< \ldots < 3/16 < \ldots < 4/21 < \ldots < 1/5, \tag{5.33}$$

where the sequence continues infinitely in a self-similar manner to form the devil's staircase. The Farey sequence has been observed in a Si p-i-n diode by Coon *et al.* [110], and in the optical bistability of CdS by Wegener and Klingshirn [109]. In order to explain the alternating periodic–chaotic sequence, nonlinearity is introduced in the map function [101]:

$$X_{n+1} = (X_n - B)^2 + A \quad 0 \le X_n \le C,$$
$$X_{n+1} = A + (0.23 + 0.41A)(X_n - C^{1/2}) \quad \text{otherwise}, \tag{5.34}$$

where A is a control parameter, $B = C - \sqrt{1 - A}$, and $C = 0.8$. The Farey sequence is now destroyed by the introduced nonlinearity. Instead, the

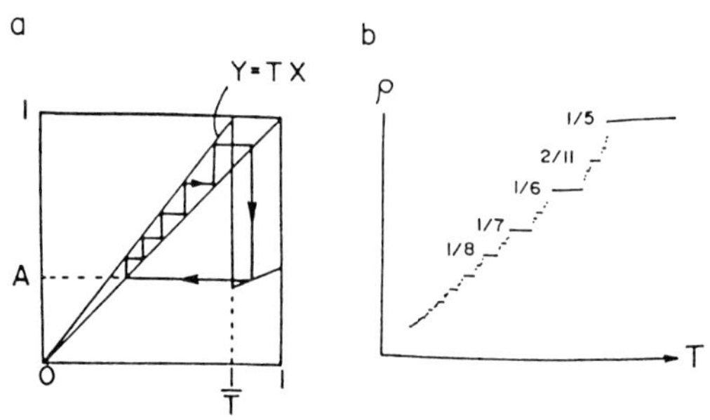

Figure 5.52 A folder 1D linear map, and the rational number as a function of T [93].

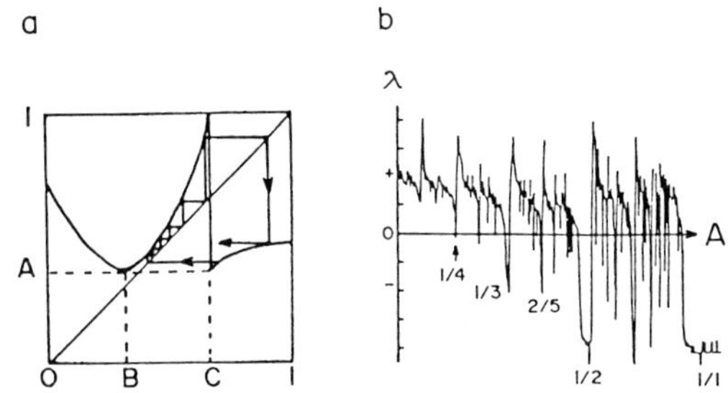

Figure 5.53 A folder 1D nonlinear map, and the Lyapunov exponent as a function of A [93].

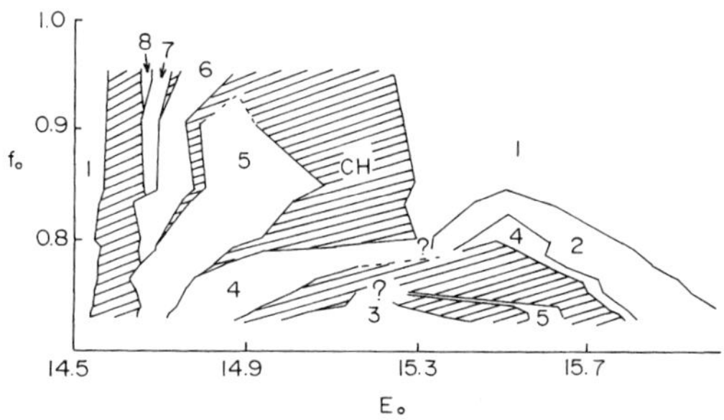

Figure 5.54 Phase diagram obtained by computer simulation, where the numbers indicate the frequency locking, and CH denotes the chaotic phase [93].

alternating periodic–chaotic bifurcation sequence is found as a function of A. In the map function (5.34) or Fig. 5.53a, an interval $[0, C]$ is defined as the lower branch and $(C, 1]$ as the upper branch. The bifurcation sequence is investigated by calculating the Lyapunov exponent as a function of A, as given in Fig. 5.53b, where rational numbers are given for several periodic cycles. The positive Lyapunov exponent means chaos, while the negative one means periodic cycle. With increasing A, the bifurcation sequence of period 4 → chaos → period 3 → chaos → ... → period 2 → ... appears. Between two periods (e.g., period 3 and period 2), subsequent periods are embedded in the chaotic region (e.g., period 5). The bifurcation is somehow similar to the alternating periodic–chaotic bifurcation sequence observed in the experiment of Fig. 5.50, where the control parameter A is considered as a dc bias.

However, this simple mathematical model does not explain the realistic physical situation of the impact-ionization avalanche. In order to see realistic features, the bifurcation scheme described by the approximated model (4.88)–(4.90) of the impact-ionization avalanche will be studied. Here, the ODEs are rewritten [111]:

$$\dot{\nu} = -\gamma_1[\nu^3 + (\kappa_1 + \kappa_3 - 1)\nu^2 - (\kappa_1 - \kappa_2 - \kappa_3)\nu - \kappa_2], \tag{5.35}$$

$$\dot{E} = -\gamma_2[E - (E_0 + E_{ac}\sin(2\pi f_0 t) + c_1\nu E + \phi], \tag{5.36}$$

$$\dot{\phi} = -\gamma_3\phi + c_2\nu E, \tag{5.37}$$

$$\kappa_1 = 1/4\exp(-6/E). \tag{5.38}$$

With $\gamma_1 = 250$, $\gamma_2 = 1$, $\gamma_3 = 0.2$, $\kappa_2 = 0.005$, $\kappa_3 = 0.25$, $c_1 = 0.78$ and $c_2 = 0.15$, the phase diagram is investigated by varying f_0 and E_0 simultaneously as given in Fig. 5.54. Many periodic islands are embedded in the chaotic sea, as indicated by numbers, where the detailed bifurcation schemes, such as period doubling and intermittency which are found near the phase boundaries of the islands and also very narrow islands in the chaotic sea, have been omitted in Fig. 5.54. The phase diagram is similar to the experimental result in Fig. 5.50, except for a smooth period doubling bifurcation in the postbreakdown regime in Fig. 5.54. At the onset of chaos in the prebreakdown regime (l.h.s. of Fig. 5.54, $E_0 \sim 14.55$), intermittent behavior is found. Tominaga and Mori [112] have also investigated the model equations (5.35)–(5.38), where they have shown in detail the Hopf bifurcation route to chaos and crisis-induced intermittency in the prebreakdown regime. A typical example of the bifurcation map is shown as a function of E_0 in Fig. 5.55, where $f_0 = 100/136$ [111, 112]. Note in Fig. 5.55(a) that relatively wide windows of period 5, period 4 and period 3 appear in order as E_0 is increased. Figure 5.55b is a magnification of Fig. 5.55a, in which Hopf bifurcation occurs with $E_0 > 14.5508$ and crisis-induced intermittency with $E_0 > E_0^L = 14.55501\ldots$ [112] (see section 5.6).

As already mentioned, period doubling bifurcation is a universal scenario (Feigenbaum scenario) of a bifurcation route to chaos which has been shown in a variety of nonlinear dynamical systems [113–116]. The scenario predicts that if a period 1 limit cycle bifurcates to period 2 generating a subharmonic component $f_0/2$ when a control parameter is varied, then a period 4 limit cycle with the subharmonic component $f_0/4$ is likely to

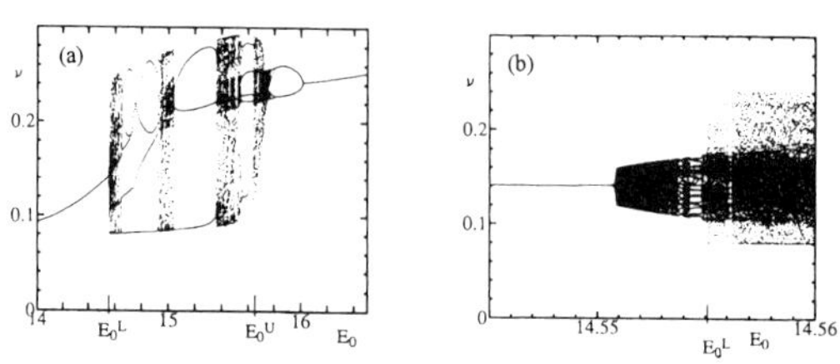

Figure 5.55 Bifurcation diagram with $f_0 = 100/136$ (computer stimulation) [112].

happen with further change in the control parameter, and so on. In order for the dynamical system to reach the edge of chaos, it is not necessary to exhibit an infinite number of independent (incommensurate) frequencies. For the sake of nonlinearity, spatially extended physical systems such as semiconductor chaos systems very often show cooperative behaviors. "Cooperative" means coherent motion of the particles (electrons). Therefore, in spite of the spatial degrees of freedom, the phase space dimension is considerably reduced and the system behaves as if it is a low-dimensional system. The cooperative phenomenon in the present semiconductor system appears as a dissipative structure or current filamentation, which occurs as a thermodynamical bifurcation induced by the impact-ionization avalanche. The collective motion of impact-ionized electrons behaves coherently inside the filamentary channel, although microscopically individual electrons experience various types of scattering. By reduction of the dimensionality, truncated homogeneous models such as in (5.35)–(5.38) may be good approximations to explain experimental observations. Suppose that a dynamical system exhibits a dynamical structure with N independent frequencies $f_1, f_2, \ldots, f_N$ and that the dynamical structure is approximated by linear combination of the N local oscillators. Then the degrees of freedom of the system may be $d_f \geqq 3N$. With $N \to \infty$ ($\mathrm{d}_f \to \infty$), the system might behave in a random manner. However, the physical reality of the nonlinear dynamical systems is macroscopic and cooperative, originating from the collective motion and from the reduction of the dimensionality. For the impact-ionization avalanche, the coherent size could be in the order of or larger than the Debye length $L_\mathrm{D} = \sqrt{D\tau}$, where transverse diffusion of the impact-ionized carriers plays an important role in the coherency.

Thus, the Landau–Hopf picture is not required, but the Feigenbaum scenario is supported as a natural consequence of the nonlinearity world in which the systems' dynamical structure bifurcates to more and more complicated forms simply by a single frequency f_0 followed by the subharmonic frequency components $f_0/2, f_0/4, \ldots$. In the phase space, the trajectory of period 2^k lies on a Möbius band in a pathological manner (Chapter 4). As a function of the control parameter E_0, the universal constant for period doubling bifurcation has been numerically shown using the ODEs (5.35)–(5.38):

$$\frac{E_0(k) - E_0(k-1)}{E_0(k+1) - E_0(k)} \sim 4.4, \tag{5.39}$$

where $k = 3 \sim 5$.

A typical example of the observed period doubling bifurcation is shown in Fig. 5.56, where $V_0 = 1.4V$, $f_0 = 1.42$ MHz and the amplitude of the ac bias E_ac is varied as a control parameter [117]. In Fig. 5.56, the peak values of the currents are plotted at each value of E_ac, where the numbers denote period, C is chaos and 2/4 is the coexisting region of periods 2 and 4. Period

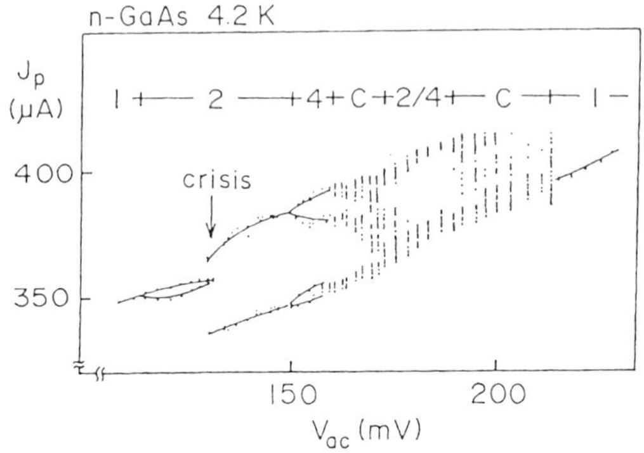

Figure 5.56 Bifurcation diagram for the period doubling cascade measures in n-GaAs as a function of V_{ac}, where $V_0 = 1.4$ V, $f_d = 1.42$ MHz, and $T_L = 4.2$ K. The chaotic region is denoted by C, and 2/4 denotes the coexisting region of periods 2 and 4 [117].

doubling occurs by increasing V_{ac}; period 1 bifurcates to period 2 at $V_{ac} =$ 113 mV, period 2 bifurcates to period 4 at $V_{ac} = 150$ mV, and period 4 bifurcates to chaos at $V_{ac} = 158$ mV. In the experiments, there is a resolution limit to obtain fine period doubling (up to period 4 at most) due to the influence of external noise. The extrinsic crisis between two different attractors of period 2 caused by the hysteretic bifurcation is seen at $V_{ac} =$ 129 mV, which is very similar to the crisis between period 1 and period 2 seen in Fig. 5.50. Fixing V_{ac} at 129 mV, intermittent switching between two kinds of period 2 limit cycle is seen.

In Fig. 5.57, time series of the currents and the Fourier power spectra are shown for period doubling observed under different conditions $V_0 = 1.038$ V and $f_0 = 2.2$ MHz. By increasing V_{ac} from 76.8 mV, period 1 is observed at 78 mV, period 2 at 96 mV, period 4 at 111 mV and chaos above 112 mV. The bifurcation is similar to those in Fig. 5.56. Correspondingly, the Fourier power spectra exhibit the subharmonic sequence. Weak chaos at $V_{ac} = 112$ mV looks like a period 8 limit cycle, as seen from the power spectrum, where subharmonic components $f_0/2, f_0/4$ and $f_0/8$ are clearly seen, but it is a periodic chaos of period 4, as judged from the phase portrait. The current waveform at $V_{ac} = 116$ mV is periodic chaos of period 2, and consequently the subharmonic component of $f_0/8$ disappears and the component $f_0/4$ broadens. Therefore, the bifurcation observed above 111 mV is identified as the inverse bifurcation cascade (see section 4.10). The accumulation point for the period doubling regime (i.e., edge of chaos) may locate between 111 and 112 mV. From the logistic map, it has been shown by Feigenbaum [82, 118, 119] that there is a universal scaling in the subharmonic spectra peaks of the Fourier transform, namely that the subharmonic peaks for period 2^k

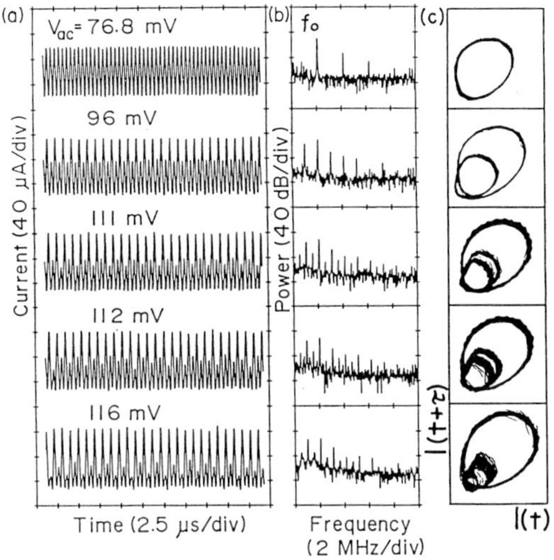

Figure 5.57 Period doubling cascade observed as a function of V_{ac}. (a) Current waveforms, and (b) the Fourier power spectra.

are lower by 8.2 dB below the previous subharmonic peaks for period 2^{k-1}. This scaling property has been confirmed in many experiments [120, 121]. In Fig. 5.58, the universal scaling from an experiment on Rayleigh–Bénard thermal convection obtained by Libchaber and Maurer [120] is compared with the result in Fig. 5.57 (periodic chaos at $V_{ac} = 112$ mV).

In order to see the dynamical structure and fractal sets of chaos, it is necessary to analyze the experimental data series using various techniques in the chaos theory [82, 118]. As seen in Chapter 4, the model simulation predicts that the strange attractors are very thin in the phase space owing to strong dissipation. The stereographic view of the phase portrait is given in Fig. 5.59, where the Cartesian coordinates x, y and z correspond to $I(t), I(t+\tau)$ and $I(t+2\tau)$, respectively (I: current; τ: time lag) and the trajectory is given by the long-term behavior of the vector point $\mathbf{X}(t)$:

$$\mathbf{X}(t) = \{I(t),\ I(t+\tau),\ I(t+2\tau)\} \tag{5.40}$$

The fractal sets can be seen by taking a Poincaré surface of section. For the periodic orbits, the closed trajectory intersects the surface of section once (or at most several times if the period is 2 or more). The dimension of the point(s) is zero on the surface of section. For the quasi-periodic motion, the intersections form a circle, giving dimension 1 on it. If the orbit is chaotic, the intersections form a fractal set, thereby the fractal nature of chaos can be recognized.

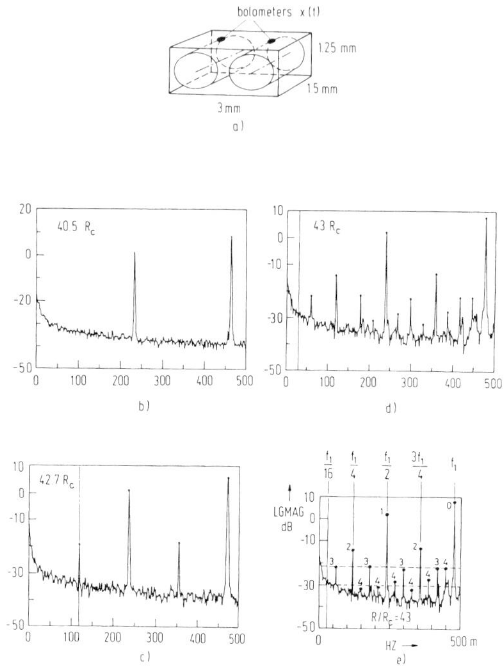

Figure 5.58 (a) Bénard cell with only a two-roll convection pattern of liquid helium, basic frequency $f = 0.5\ s^{-1}$. (b–d) Power spectrum of the temperature $x(t)$ with increasing Rayleigh number. (e) The heights of the nth subharmonics are compared with Feigenbaum's theory (horizontal lines) [122], after Libchaber and Maurer [120].

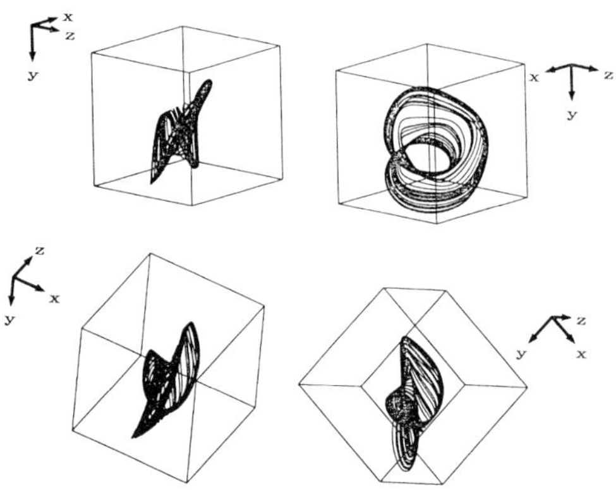

Figure 5.59 Zoological garden of chaos in the 3D phase space, drawn by the one-level impact ionization model [94].

For fully developed chaos, a typical example of the phase portrait and the Poincaré surface section [93, 94] is shown in Fig. 5.60a, where the 2D phase portrait is given simply as a projection of the 3D phase portrait onto the xy-plane. The Poincaré surface of section is given by cutting the center of the 3D phase portrait (see Fig. 5.59) on a plane parallel to the z axis at an angle θ relative to the x axis. The experimental result is excellently simulated in Fig. 5.60b by the approximated model (5.35)–(5.38). The Poincaré surface of section in Fig. 5.60 predicts that the fractal dimension of the intersections is close to but larger than one, and thus the extended dimension in the 3D phase space is slightly greater than 2. In the experiments, fine fractals are destroyed at the microscopic level as a result of the noise floor. In the model simulation, a finer and finer fractal structure can be found by blowing up the Poincaré surface of section (Fig. 5.61, where $\theta = 0$) [93, 94]. The successive blow-up in Fig. 5.61 exhibits many fine divisions, such as Cassini's division of Saturn. A similar structure is seen in the Hénon map and many other dynamical systems [123, 124]. Gaps among the rings could be constructed as a fractal hierarchy. Correspondingly, it can be concluded that the strange attractor of fully developed chaos is composed of an infinite number of belts when it is embedded in the 3D phase space (belt chaos).

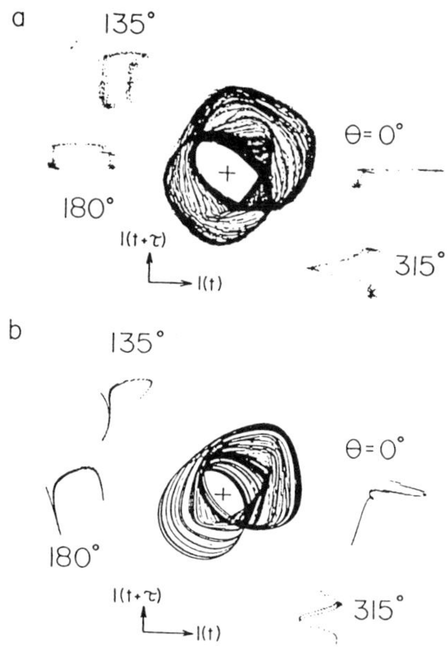

Figure 5.60 (a) The observed 2D phase portrait of chaos and the Poincaré section of the 3D extended phase portrait, (b) the computer simulation by the approximated impact ionization model (5.35)–(5.38) [93, 94].

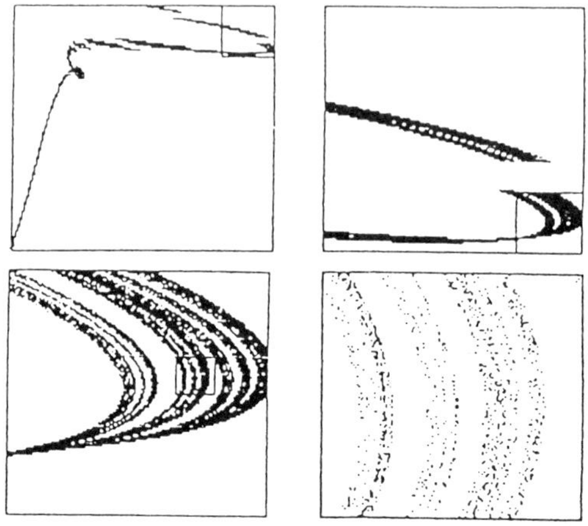

Figure 5.61 Successive blow-ups of the Poincaré section [93, 94].

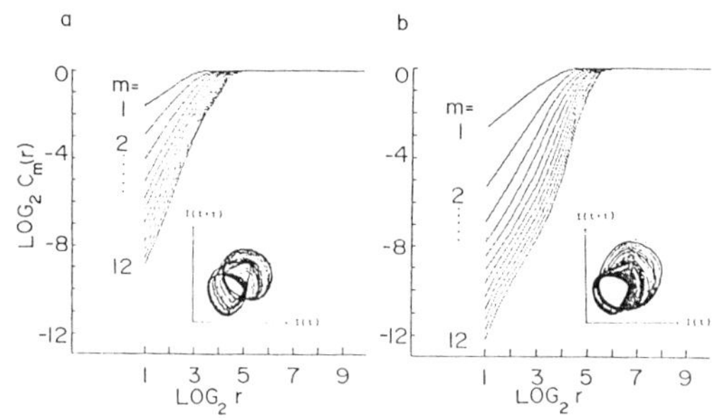

Figure 5.62 The correlation integral $C_m(r)$ as a function of r, for two kinds of strange attractor, where $N = 1000$ [93].

Therefore, how large is the fractal dimension? By using the Grassberger–Procaccia algorithm (see section 6.1), the correlation integral is calculated by:

$$C_m(r) = \lim_{N\to\infty} (1/N^2) \sum_{i}^{N} \sum_{j}^{N} H(r - \|\mathbf{X}_i - \mathbf{X}_j\|). \tag{5.41}$$

In Fig. 5.62, two examples of the correlation integral are shown as a function of r for fully developed chaos, where $N = 1000$. The inset figures show the corresponding 2D phase portraits. The slope $\tilde{\nu}$ of the correlation integral $C_m(r) \propto r^{\tilde{\nu}}$ is plotted as a function of the embedded dimension m in Fig. 5.63, where the open circles are evaluated from Fig. 5.62a and the solid circles from Fig. 5.62b. The saturated values give the correlation dimension, which are $\tilde{\nu} = D_2 = 2.19 \pm 0.1$ for Fig. 5.62a and $\tilde{\nu} = D_2 = 1.93 \pm 0.07$ for Fig. 5.62b.

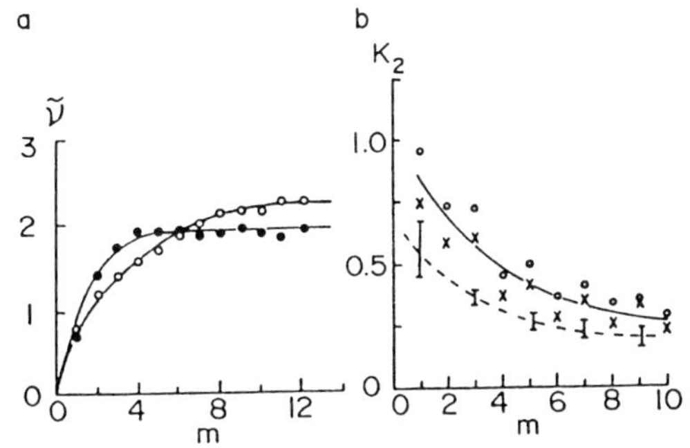

Figure 5.63 (a) The critical exponents as a function of m (open circles for Fig. 5.62a, solid circles for Fig. 5.62b), and (b) the K_2 entropy as a function of m. Open circles and crosses were evaluated for the attractor in Fig. 5.62a, with small values of r. Error bars are for the attractor in Fig. 5.62b [93].

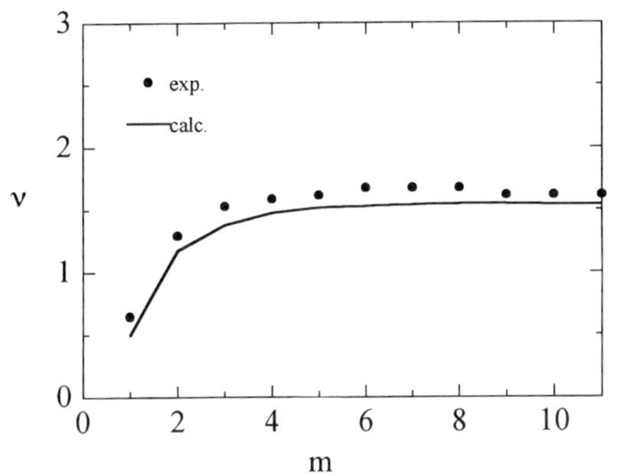

Figure 5.64 The slope of the correlation integral $C_m(r)$ for a strange attractor of Fig. 5.57 (V_{ac} = 112 mV). The solid line indicates the result by the model simulation.

At an accumulation point of period doubling, the attracting set in the semiconductor chaos system resembles the Feigenbaum attractor of the logistic map (see section 6.1). A quality test of the dimensions near the accumulation point may be an important factor to show how the experimental system exhibits a universal behavior. In Fig. 5.64, the slope of the correlation integral $C_m(r)$ for a strange attractor of Fig. 5.57 is compared with the result by the model simulation (solid line). The correlation dimension is obtained as $D_2 = 1.6 \pm 0.006$, the value of which is equivalent to $D_2 = 0.6 \pm 0.006$ by subtracting a trivial dimension. The value is somehow larger than the correlation dimension of $D_2 = 0.5 \pm 0.005$ for the logistic map [125]. The correlation dimension increases by shifting the parameter value (V_{ac} or V_0) beyond the accumulation point.

5.5 Route to crisis-induced intermittency

In the semiconductor chaos system, the reinjective switching can be observed frequently in the prebreakdown regime (see periodically driven chaos, section 5.4). The intermittent behaviors are described by bursts of current filamentation. It seems that the reinjective switching is a universal phenomenon in the semiconductor chaos system when the S-shaped $I-V$ curve is driven periodically in the prebreakdown regime. Then, how does the reinjective switching occur? In the approximated model A of impact-ionization avalanche, the reinjective switching can be reproduced via the Hopf bifurcation route to chaos (Ruelle–Takens–Newhouse scenario) [111, 112]. Although the bifurcation scheme is highly complicated in the parameter space (E_0, f_0) [111, 112], the route to reinjective switching reveals a common feature at the boundary of the phase diagram in Fig. 5.54. In Fig. 5.55b, the bifurcation diagram in the prebreakdown regime indicates clearly an explosion of the attractor size at the phase boundary [112], where $E_{ac} = 4.5$ and $f_0 = 100/136$ (see also the bifurcation diagram in [111]). Just before the explosion, the attractor shows Hopf bifurcation via the Ruelle–Takens–Newhouse scenario; period 1 bifurcates to a torus and the torus bifurcates to chaos via collapse of the toric surface when E_0 is increased. By increasing E_0 further, the chaotic attractor touches a separatrix and the reinjective switching occurs.

As described by Tominaga and Mori [112], there exist two types of Hopf bifurcation (subcritical and supercritical Hopf bifurcations), both of which are followed by crisis-induced intermittency (reinjective switching). They described the bifurcations by normal form analysis, as follows. For the numerical solutions of (5.35)–(5.38) one can construct the Poincaré map by stroboscopic vector point $\mathbf{X}_n$:

$$X_{n+1} \equiv \begin{pmatrix} \nu_{n+1} \\ E_{n+1} \\ \phi_{n+1} \end{pmatrix} = F(X_n) \equiv \begin{pmatrix} f_\nu(\nu_n,\ E_n,\ \phi_n) \\ f_E(\nu_n,\ E_n,\ \phi_n) \\ f_\phi(\nu_n,\ E_n,\ \phi_n) \end{pmatrix}, \qquad (5.42)$$

where f_ν, f_E and f_ϕ are nonlinear functions of ν_n, E_n and ϕ_n taken at discrete time $t = n/f_0$ $(n = 0, 1, 2, \ldots)$. In (5.42) a limit cycle of period Q means $\mathbf{X}_i^* = F^Q(\mathbf{X}_i^*) = \mathbf{X}_{i+Q}^*(i = 1, 2, \ldots, Q)$. When the absolute values of the eigenvalues of the Jacobi matrix $DF^Q(\mathbf{X}_i^*)$ are less than one, the periodic points are stable against a small perturbation. There are three types of bifurcations whereby this limit cycle loses its stability; namely, when the eigenvalues are 1 or -1 or a complex conjugate on the unit circle. For the complex conjugate pair on the unit circle, the bifurcation is called the Hopf bifurcation, leading to a closed orbit on the Poincaré section which is the section of torus [112]. Now, according to the center manifold theorem, (5.42) can be reduced to a two-dimensional map $f_\mu(x) : \mathbf{R}^2 \to \mathbf{R}^2$ with a smooth family of fixed points $x(\mu)$ on the surface spanned by the eigenvectors of complex eigenvalues using coordinate transformations [126], where μ is the control parameter of the system and it is assumed that μ_0 is the value of the Hopf bifurcates to an invariant circle with $\mu > \mu_o$. Then, the map function $f_\mu(x)$ can be transformed into the normal form in the polar coordinate $(r, 2\pi\theta)$ as:

$$f_\mu(x) - hf_\mu h^{-1}\begin{pmatrix} r \\ \theta \end{pmatrix} = \begin{pmatrix} r\{1 + d(\mu - \mu_0) + a_1 r^2 + a_2 r^4\} + O(r^7) \\ \theta + c_0 + b_1 r^2 + b_2 r^4 + O(r^6) \end{pmatrix}, \quad (5.43)$$

where $d > 0$ and $(r, 2\pi\theta) = (0, 0)$ is the fixed point. For the map, it is noted that the signs of the coefficients a_i $(i = 1, 2)$ and $\epsilon = d(\mu - \mu_0)$ determine the direction and stability of the bifurcating periodic orbits, while c_0 and b_i $(i = 1, 2)$ give asymptotic information on the rotation number [112].

For $a_1 < 0$ there is a small torus around the limit cycle, just after the bifurcation. By increasing μ, it is found from (5.43) that the radius of torus r^* on the two-dimensional space in the normal form increases as $r^* \propto \sqrt{\epsilon}$. In the semiconductor system, $\epsilon \propto (E_0 - E_0^H)/E_0^H$, where E_0^H represents the parameter value at the Hopf bifurcation point [112]. In this case, the orbit spirals out from the neighborhood of the destabilized fixed point but it is suppressed within a region owing to the nonlinear term $a_1 r^3$, forming a steady-state small torus just after the bifurcation. With this regime $(a_1 < 0)$ one has the normal Hopf bifurcation, which is called the supercritical Hopf bifurcation.

For $a_1 > 0$, negative values of higher-order coefficients than a_1 are required for the Hopf bifurcation. For instance, with $a_2 < 0$, there can exist a torus with radius $r^* = \sqrt{a_1/|a_2|}$ even just after the bifurcation point, $\epsilon = +0$, as shown schematically in Fig. 5.65 [112]. Even if $a_2 > 0$, the higher-order negative-value coefficient $(a_4 < 0)$ may guarantee the Hopf bifurcation. Such a bifurcation as shown in Fig. 5.65 is called the subcritical Hopf bifurcation.

In Fig. 5.66, a part of the phase diagram of the model system is given in the control parameter space (E_0, f_0) [112], where the solid curve indicates the onset point E_0^H of the Hopf bifurcation and the dashed line represents the onset point E_0^L of the crisis-induced intermittency. Figure 5.66 shows that the torus region is quite narrow before the emergence of the crisis-induced intermittency. In Fig. 5.67, the coefficient a_1 was estimated as a function of the driving frequency f_0. At $f_0 = f_0^c \simeq 0.82$, the sign of a_1 changes from positive to negative so that at that point the nature of the Hopf bifurcation changes from subcritical $(f_0 < f_0^c)$ to supercritical $(f_0 > f_0^c)$.

Accordingly, Tominaga and Mori [112] have selected two values for $f_0, f_0 = 1.0$ and $f_0 = 100/136 \simeq 0.753$, and investigated the torus behavior and the subsequent intermittency in detail. Such behaviors can also be reproduced in the model simulations (5.35)–(5.38), and the reinjective

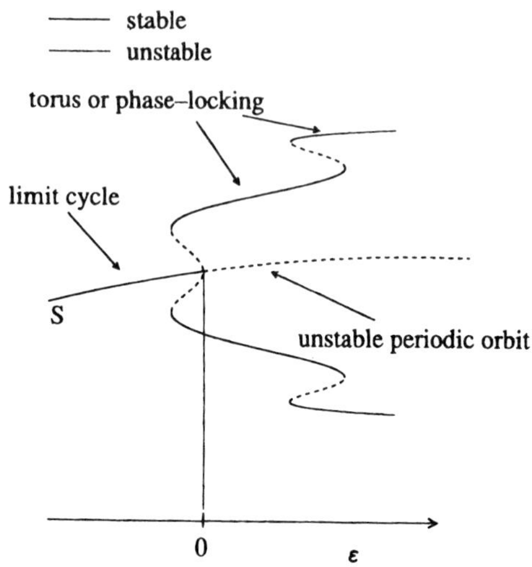

Figure 5.65 Schematic diagram for the subcritical Hopf bifurcation caused by higher order effects in the normal form, where the solid lines denote stable orbits and dashed lines unstable orbits, and $E_0 - E_0^{\mathrm{Hopf}}$. This diagram leads to a bifurcation from a limit cycle to a torus with a finite amplitude, where two attractors exist just before the bifurcation [112].

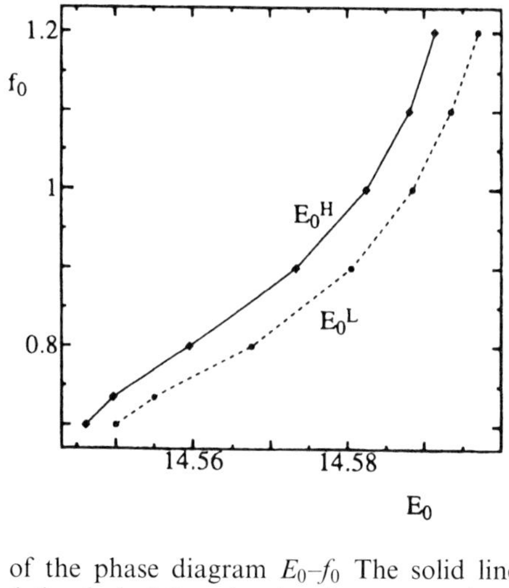

Figure 5.66 A part of the phase diagram E_0–f_0 The solid line denotes the Hopf bifurcation point and the dashed line denotes the onset point of the intermittency [112].

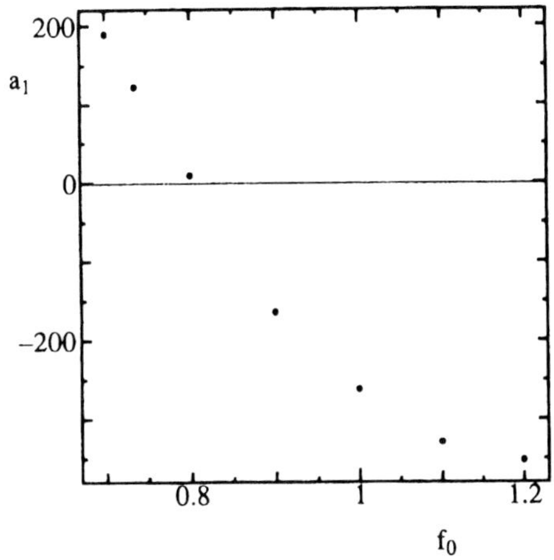

Figure 5.67 Plot of the third-order coefficient a_1 vs f_0 [112].

switching is argued by combining it with the detailed results given by Tominaga and Mori [112]. The parameters in (5.35)–(5.38) are chosen as $\gamma_1 = 250,\ \gamma_2 = 1,\ \gamma_3 = 0.2,\ \kappa_2 = 0.0055,\ \kappa_3 = 0.25,\ c_1 = 0.78,\ c_2 = 0.15$ and $E_{ac} = 4.5$ [111, 112]. First, the driving frequency is fixed at $f_0 = 100/136$, in the regime of the subcritical Hopf bifurcation, where the bifurcation point is $E_0 = E_0^H = 14.54966340\ldots$. Just below the bifurcation point, there exists a stable fixed point of period 4, which is due to the fact that a torus caused by a negative coefficient of order more than five in the normal form (5.43) is phase-locked accidentally [112]. When E_0 exceeds E_0^H slightly, the fixed point seems to continue, but this is due to the extremely slow transient. The period 4 (p-4) orbit is already stable for $E_0 < E_0^H$ and becomes unstable for $E_0 > E_0^{p4} \sim 14.5506$. For $E_0 > E_0^{p4}$, a stable torus which may already exist when the p-4 periodic orbit is stable, becomes stable (see the schematic diagram in Fig. 5.65). Figure 5.68 shows the time series of the carrier density $\nu_t(t = 1, \ldots, 2000)$ and correspondingly the Poincaré section in the $\nu - \phi$ plane after the subcritical Hopf bifurcation, where (a) $E_0 = 14.553$, (b) $E_0 = 14.555$, and (c) $E_0 = 14.556$ [112]. The torus in Fig. 5.68a is already distorted, but it is still smooth. As E_0 is increased, the torus bifurcates to the chaotic wrinkled torus (the fractal torus) as in Fig. 5.68b, and finally explosion of the fractal torus results in crisis-induced intermittency (Fig. 5.68c; see also [111]).

In the parameter regime of quasi-periodicity ($E_0^H < E_0 < 14.555$), several periodic windows appear due to the phase-locking of torus. Tominaga and Mori [112] have examined the construction of the devil's staircase, as given by the rotation number ρ in Fig. 5.69. Although an infinite number of periodic windows may be embedded in the quasi-periodic sea, only a few

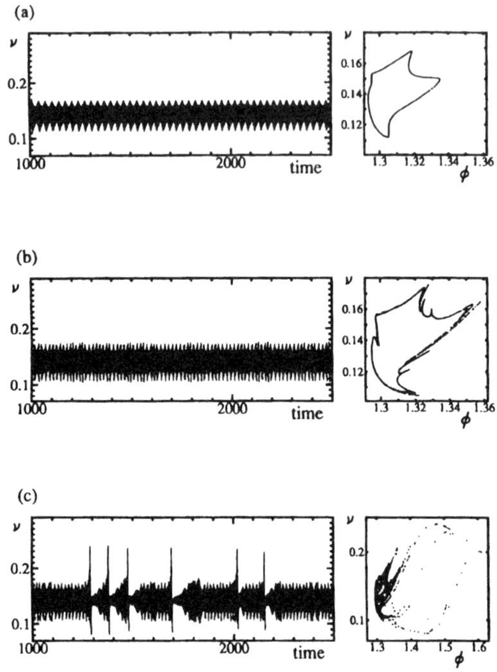

Figure 5.68 Time series of ν_t with attractors after the subcritical Hopf bifurcation for $f_0 = 100/136$. (a) $E_0 = 14.553$ (distorted torus). (b) $E_0 = 14.555$ (chaotic wrinkled torus), (c) $E_0 = 14.556$ (crisis-induced intermittent chaos) [112].

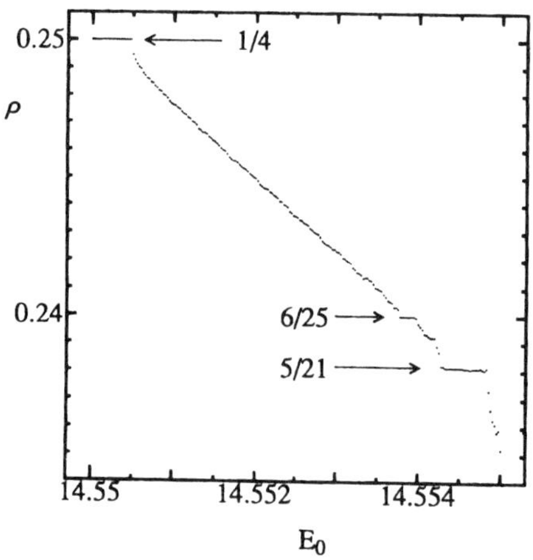

Figure 5.69 Plot of the rotation number ρ vs E_0, leading to a devil's staircase. The steps for the phase locking regions with $\rho = 5/21$ and $6/25$ are easily observed [112].

periodicities are indicated because of the extremely narrow windows in Fig. 5.69. They found the windows of period 171, 96, 46 and 42 [112]. Figure 5.69 was drawn by finding the rotation angle $2\pi\theta$ along the Poinearé section of the torus on the ν–ϕ plane and then taking the rotation number $\rho = \lim_{n\to\infty} \{f^n(\theta_0) - \theta_0\}/n$ without taking mod 1, where one can construct a circle map $\theta_{n+1} = f(\theta_n)$ and θ_0 is an arbitrary initial point (see section 5.4). The behavior in Fig. 5.69 may be extended more globally by taking the phase diagram of phase-locked states in the parameter space of (f_0, E_0), whereby many Arnold tongues may be embedded in the quasi-periodic sea (section 6.4). The fractal torus in Fig. 5.68 is actually caused by a strange attractor due to chaos. Clear evidence of chaos in Fig. 5.68b is shown by taking the return maps $\theta_{n+1} = f(\theta_n)$ in Fig. 5.70; that is, the map is invertible for the

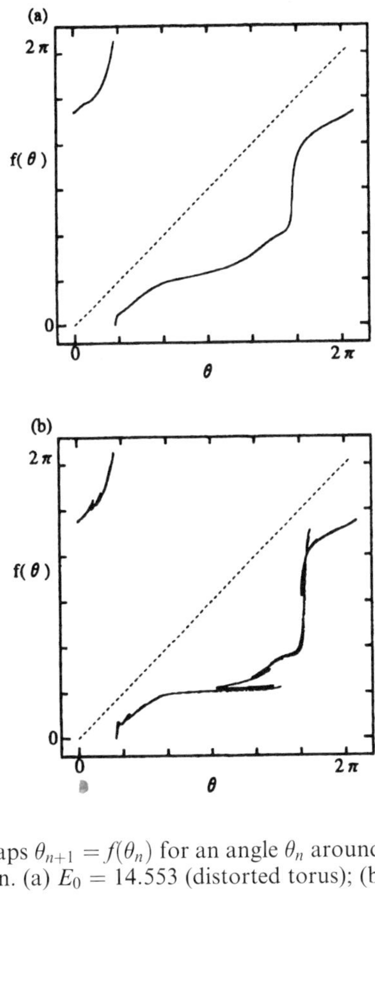

Figure 5.70 Return maps $\theta_{n+1} = f(\theta_n)$ for an angle θ_n around an unstable fixed point on the Poincaré section. (a) $E_0 = 14.553$ (distorted torus); (b) $E_0 = 14.555$ (wrinkled torus) [112].

non-fractal torus corresponding to Fig. 5.68a, while it becomes noninvertible for the chaotic wrinkled torus in Fig. 5.68b.

By taking $f_0 = 1.0$, the bifurcation scheme becomes a supercritical Hopf bifurcation [112]. By increasing dc bias E_0 from $E_0^H = 14.58244539914\ldots$, a small torus grows up continuously from a fixed point, where the toric surface is quite smooth without being distorted until $E_0 = 14.587$ (see Fig. 5 in Ref. [112]. The size of torus is proportional to $\sqrt{\epsilon}$ ($\epsilon \propto E_0 - E_0^H$), as expected from the normal form (5.43). During the course of the bifurcation, an infinite number of phase-locking windows appears [112]. As shown by the Poincaré sections in Fig. 5.71a–f the torus doubling occurs successively by increasing E_0 (Fig. 5.71a–c), but it stops after the fourfold torus in Fig. 5.71c. After the fourfold torus, a cylinder torus (chaos) terminates (Fig. 5.71d) and finally reinjective switching via a homoclinic crossing is observed (Fig. 5.71e, f). In more detail, there is a fourfold fractal torus with a twofold inhomogeneous cylinder between the fourfold torus (Fig. 5.71c) and a fully developed cylinder (Fig. 5.71d), as given in Fig. 5.72a–d, drawn by Tominaga and Mori [112], where the applied electric fields are (a) $E_0 = 14.588225$, (b) $E_0 = 14.588231$, (c) $E_0 = 14.588235$ and (d) $E_0 = 14.5887$. Figure 5.73a–g shows the magnification of the attractors at a certain part selected arbitrarily, where the electric fields are (a) $E_0 = 14.588225$, (c) $E_0 = 14.588231$, (e) $E_0 = 14.588233$, (f) $E_0 = 14.58835$ and (g) $E_0 = 14.58836$. Here, Fig. 5.73b is a further blow-up of Fig. 5.72a, d is a blow-up of Fig. 5.72c. These figures indicate how a smooth torus surface becomes wavy, and then fractal and a further twofold cylinder [112]. Apparently, the surface oscillates in Fig. 5.73a. Figure 5.73c, corresponding to Fig. 5.72b, shows clearly the fractal surface, which is no

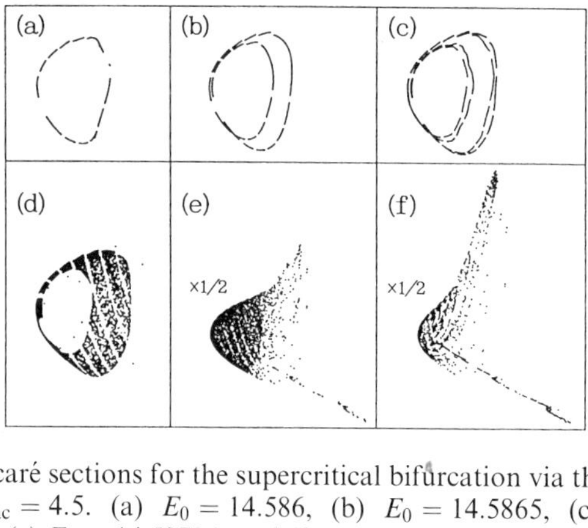

Figure 5.71 Poincaré sections for the supercritical bifurcation via the torus doubling. $f_0 = 1.0$, and $E_{ac} = 4.5$. (a) $E_0 = 14.586$, (b) $E_0 = 14.5865$, (c) $E_0 = 14.58675$, (d) $E_0 = 14.5869$, (e) $E_0 = 14.58706$, and (f) $F_0 = 14.59$. Values of E_0 were found to be slightly lower than those in [112] by Tominaga and Mori (the reason was not clear). Drawn by K. Aoki.

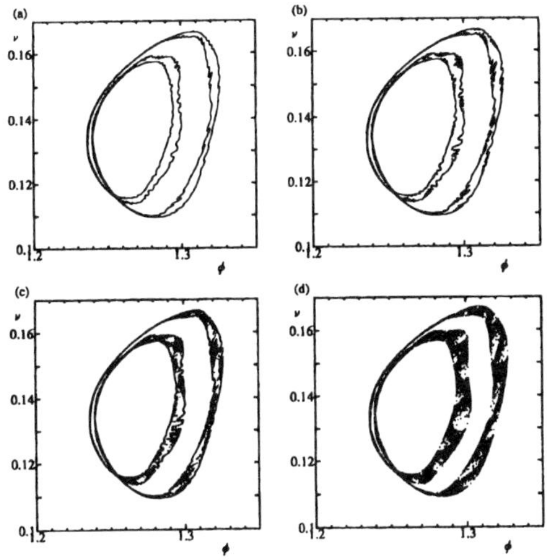

Figure 5.72 Attractors with $f_0 = 1.0$ for (a) $E_0 = 14.588225$ (fourfold distorted torus), (b) $E_0 = 14.588231$ (fourfold fractal torus), (c) $E_0 = 14.588235$ (twofold inhomogeneous cylinder), and (d) $E_0 = 14.58827$ (twofold inhomogeneous cylinder). (c) and (d) are chaotic attractors [112].

longer a torus in the usual sense. Indeed, the fractal torus in Figs. 5.72b and 5.73c could be critical between torus and chaos. Thus, the cylinder in Fig. 5.73f, g is a direct result of the bifurcation of torus oscillation into chaos via fractalization.

Such torus oscillation and its bifurcation have been observed in a coupled map by Kaneko [127] and in experiments on a two resistively coupled p-n junction circuit [128]. Tominaga and Mori [112] have given a possible explanation for the torus oscillation and its fractalization in Fig. 5.74. They explain that the torus oscillation resembles the unstable manifold of the saddle-node points, which exists just at the saddle-node bifurcation for collapse of a phase-locking of long period [126, 129] (for details see Ref. [112]). The inhomogeneous cylinder in Fig. 5.72c, d becomes wider, losing the oscillation structure when E_0 is increased, and finally the inhomogeneous cylinder becomes a homogeneous one (Fig. 5.71d), which is a fully developed homogeneous cylinder. Further increase in E_0 induces reinjective switching via crisis. Figure 5.75 summarizes such behavior in the regime of supercritical Hopf bifurcation [112].

5.6 Mechanism of crisis-induced intermittency

In Fig. 5.71e, f, the toric chaos is interrupted by bursts of current filamentation, where reinjective switching causes a structural change in the

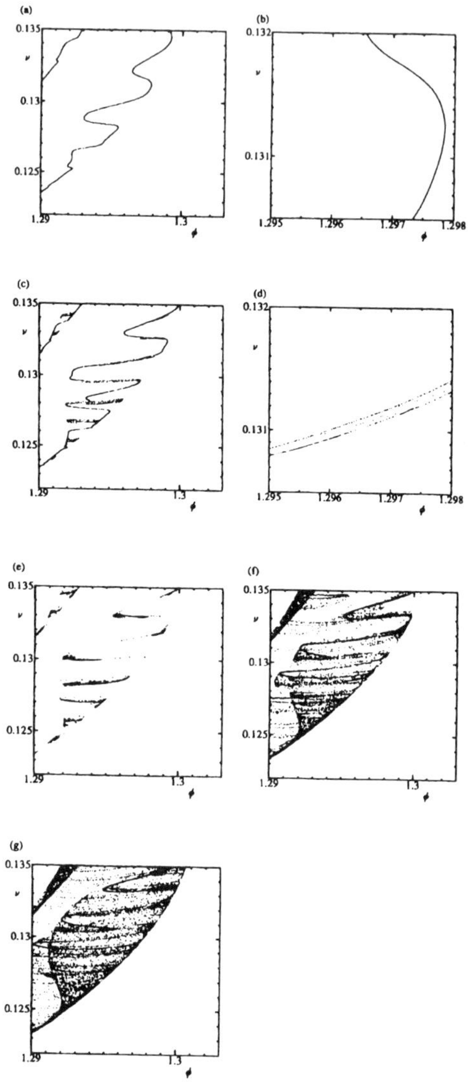

Figure 5.73 Magnification of a certain part of the attractor with $f_0 = 1.0$ for (a) $E_0 = 14.588225$ (distorted torus), (c) $E_0 = 14.588231$ (fractal torus), (e) $E_0 = 14.588233$ (wrinkled torus), (f) $E_0 = 14.588235$ (inhomogeneous cylinder), and (g) $E_0 = 14.588235$ (inhomogeneous cylinder). (b) and (d) are higher magnifications of (a) and (c), respectively. The existence of a fractal structure can be observed clearly in (d), but in (b) there is no fractal structure [112].

attractor. In order to look at the details, the evolution of the peak current is plotted in Fig. 5.76. In Fig. 5.76a–c, the orbit is reinjected near a ghost of period 1 unstable orbit after each burst, and the number of bursts increases

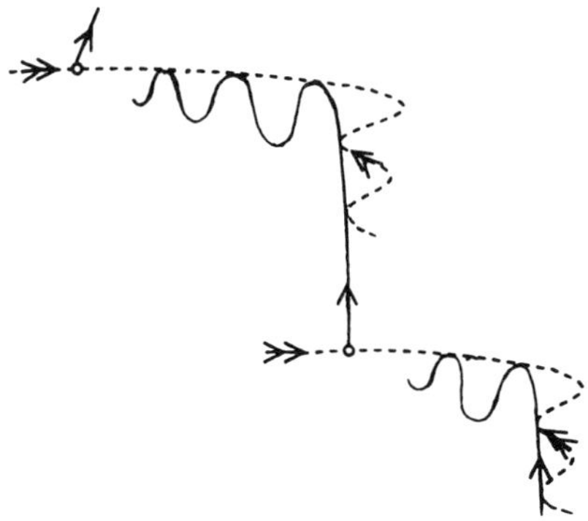

Figure 5.74 Invariant manifold of the saddle-node points of period $Q(\gg 1)$, where the $^\circ$ symbols denote the saddle-node points, the solid lines indicate the unstable manifold and the dashed lines the stable manifold. The tangent points accumulate at the saddle-node points. The figure only indicates a type of possible phase space structures between nearly two saddle-node points, since it may be a candidate for the damped oscillation of torus [112].

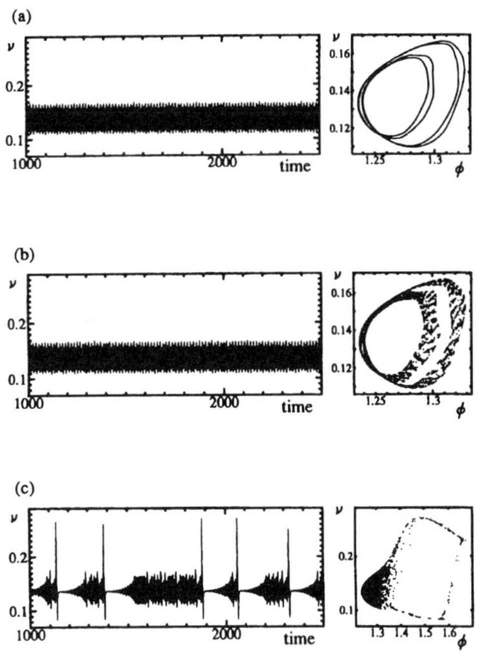

Figure 5.75 Time series ν_t with attractors after the supercritical Hopf bifurcation with $f_0 = 1.0$. (a) $E_0 = 14.5882$, (b) $E_0 = 14.5883$, and (c) $E_0 = 14.5888$ [112].

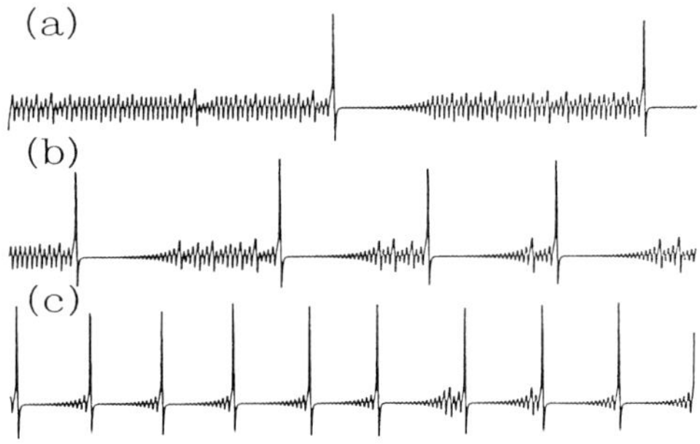

Figure 5.76 Evolution of the peak current for the crisis-induced intermittency. $f_0 = 1.0$ and $E_{ac} = 4.5$. (a) $E_0 = 14.58706$, (b) $E_0 = 14.58707$, and (c) $E_0 = 14.589$. Drawn by K. Aoki.

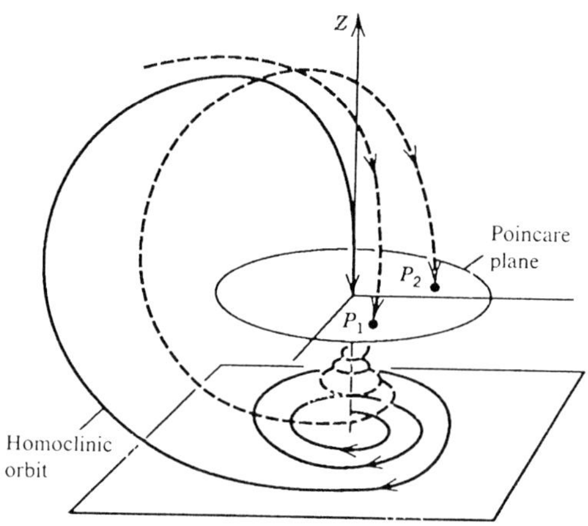

Figure 5.77 Homoclinic orbit in the 3D phase space generated by a saddle-focus fixed point [118].

with increasing E_0, where (a) $E_0 = 14.58706$, (b) $E_0 = 14.58708$ and (c) $E_0 = 14.589$. Here, it is noted that the selected values of E_0 for the dynamics in Figs. 5.71 and 5.76 and in Ref. [111] are slightly smaller than these in Ref. [112], presumably owing to different numerical algorithms and different resolutions. The behavior in Fig. 5.76 is reminiscent of Shil'nikov chaos [130–132] observed in the autonomous systems (laser system by Arecchi *et al.* [130] and Swetitis and Buoncristiani [131], and chemical reaction system by Argoul *et al.* [132]. The Shil'nikov theorem suggests that the existence of a homoclinic orbit in an autonomous system implies chaotic motion in 3D phase space. The schematic in Fig. 5.77, drawn by Moon [118], explains very well that the homoclinic orbit is injected near a saddle focus and spirals outwards. The burst mechanism of the present system in Fig. 5.76 may be interpreted as an extended version of the Shil'nikov chaos in the nonautonomous system, where the existence of a saddle focus is recognized from the dynamics of the peak currents in Fig. 5.76. More precisely, one can show the saddle focus by extending the 2D Poincaré section (ν, ϕ) into a 3D Poincaré section (ν, ϕ, t_n) by adding an axis of discrete time t_n. Let $t_n (n = 0, 1, 2, \ldots)$ be the time when the orbit passes through a plane of the Poincaré section every period $(t_n - t_{n-1} \sim T_0)$. Then, by taking appropriate modulo such that $t_n = 0$ after each burst, one obtains the 3D Poincaré section in Fig. 5.78. The figure clearly indicates the Shil'nikov chaos or the spiral chaos [118]. The saddle focus (unstable fixed point) in the Poincaré section is a section of a ghost of the limit cycle in the present regime. As seen in Figs. 5.75c and 5.76b, the time series reveals three types of subsequent motion, i.e., laminar motions, chaotic motions and burst motions. The orbits are reinjected into the neighborhood of the unstable periodic orbit S just after the bursts and spiral out gradually with growing amplitudes, leading to regular laminar motions (see Figs. 5.75 and 5.76), and then exhibit irregular chaotic motions within the regime of cylinder torus until the subsequent burst occurs. Tominaga and Mori [112] have shown clearly an unstable periodic orbit (saddle S) by a Poincaré section (Fig. 5.79), where, starting from the initial points which lie on a line segment along the burst branch, the iterated points are plotted until they approach asymptotically to the saddle S. Noted especially in Fig. 5.79 that the reinjections into the neighborhood of S take place along the neighborhood of the stable manifold. Figure 5.80 shows a blow-up, which shows that the reinjections from the bursts are restricted very near S, leading to the suppression of the variety of the laminar lifetime before entering into the chaotic phase [112] (see the transients just after the burst between bursts in Figs 5.75c and 5.76b, c).

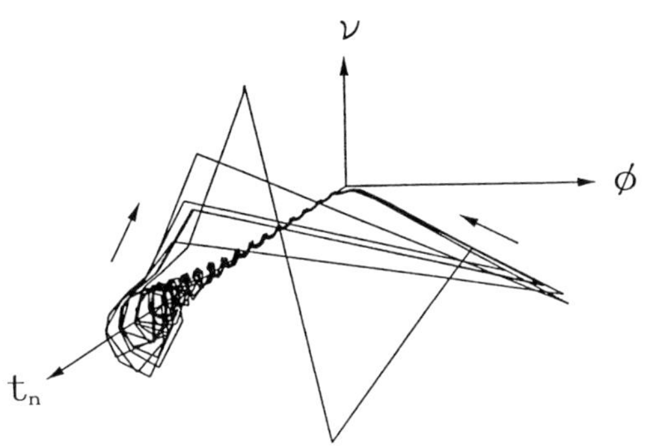

Figure 5.78 3D Poincaré section (ν, ϕ, t_n), where $n = 0, 1, 2, \ldots$.

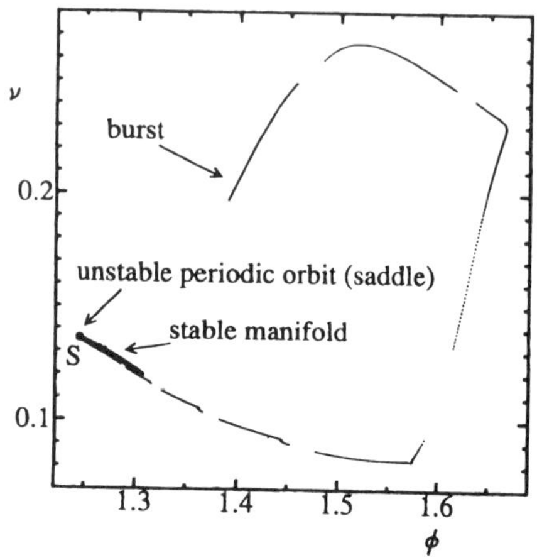

Figure 5.79 Poincaré section which shows a burst branch, an unstable periodic orbit S which lost its stability through the Hopf bifurcation, and its stable manifold, where 150 initial values are taken along a line segment tangent to the burst branch in order to draw the burst branch. Orbits are reinjected into the neighborhood of S along the stable manifold after bursts [112].

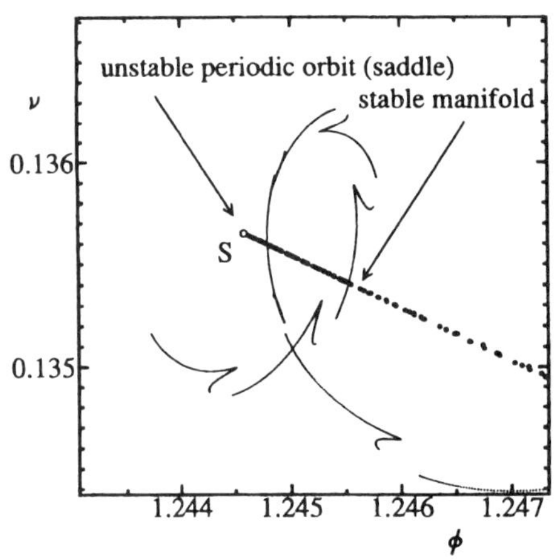

Figure 5.80 Magnification of Fig. 5.79 around S. Orbits on a line segment along the burst branch are reinjected near S and then spiral out on a surface spanned by the unstable manifold of S [112].

As pointed out by Tominaga and Mori [112], the reinjection switching is different in the long-term behavior from type II intermittency. Although the time series (Fig. 5.75) are similar to those of type II intermittency, their statistical properties are quite different, as follows. In type II intermittency, the probability distribution $P(\tau)$ of the residence time τ (laminar lifetime) in the phase of torus is given by $P(\tau) \sim \epsilon^2 \exp(4\epsilon\tau)/(\exp(4\epsilon\tau) - 1)^2$, so that $P(\tau) \sim \tau^{-2}$ for $\epsilon\tau \ll 1, P(\tau) \sim \exp(-4\epsilon\tau)$ for $\epsilon\tau \gtrsim 1$. The mean residence time follows as $\langle\tau\rangle \sim \epsilon^{-1}$ where $\epsilon = |E_0 - E_0^{\mathrm{Hopf}}|$ with E_0^{Hopf} being the Hopf bifurcation point [112, 122]. In the case of crisis-induced intermittency, the probability $P(\tau)$ is proportional to the phase space volume Γ of the chaotic attractor (toric chaos) just before the crisis, so that the lifetime τ of a chaotic orbit staying within the chaotic region Γ must be considered [112]. Then, the probability distribution for many chaotic orbits obeys the exponential decay [133, 134]:

$$P(\tau) = \langle\tau\rangle^{-1} \exp[-\tau/\langle\tau\rangle] \tag{5.44}$$

with

$$\langle\tau\rangle \propto \epsilon^{-\gamma}, \tag{5.45}$$

where $\epsilon = |E_0 - E_0^L|$, with E_0^L the bifurcation point of the crisis in the present system (see Fig. 5.55), and γ is an exponent. In the case of $f_0 = 1.0$, the probability distribution and the mean lifetime are shown in Figs. 5.81 and 5.82, respectively. The laminar lifetime of the transients just before entering to the chaotic phase (see Fig. 5.75c) is about 140–160 round trips for the

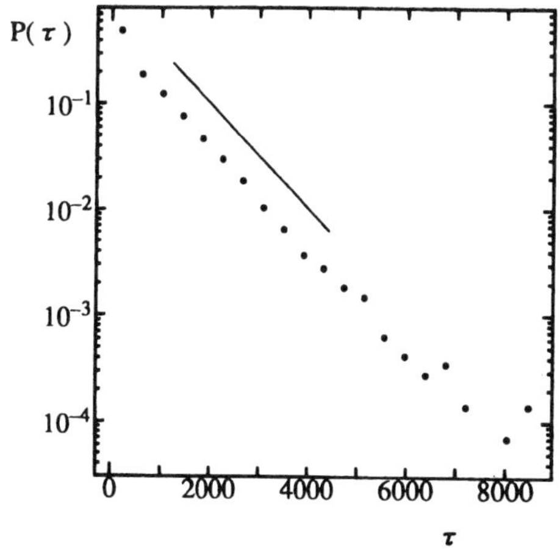

Figure 5.81 Distribution $P(\tau)$ of the lifetime τ for staying in the chaotic region Γ with $f_0 = 1.0$ and $E_0 = 14.5885$. The mean lifetime $\langle\tau\rangle \simeq 860$ can be obtained from the slope 0.00116 by using (5.44) [112].

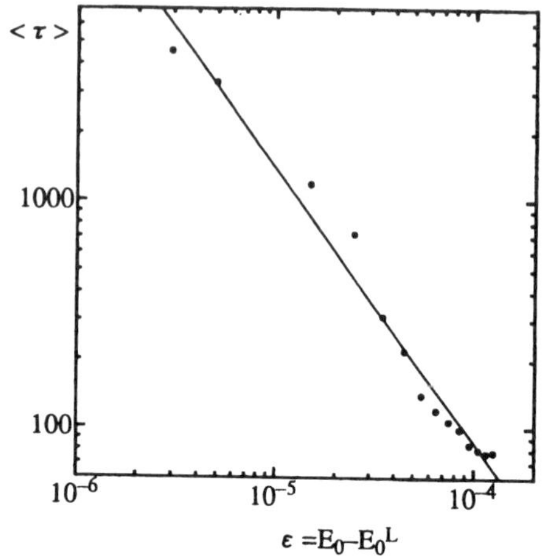

Figure 5.82 Log–log plot of the mean lifetime $\langle\tau\rangle$ for staying in the chaotic region Γ vs $\varepsilon = E_0 - E_0^{\mathrm{L}}$. The solid line with slope $\gamma \simeq 1.2$ is obtained from the least squares method fitting [112].

range $\epsilon = |E_0 - E_0^L| = 10^{-6} \sim 10^{-4} (E_0^L = 14.588475063\ldots)$, giving a sharp probability distribution $P(\tau)$ peaking at $\tau = 150$–160 [112]. Such a sharp distribution is understood by considering again that the orbits just after the bursts are reinjected very close to the saddle point without any exceptions and then exhibit a characteristic transient before entering the chaotic phase. The subsequent chaotic phase exhibits the scaling behavior of the lifetime (5.44), as given in Fig. 5.81, where $E_0 = 14.5885$ and $\langle\tau\rangle \simeq 860$, estimated from the slope 0.00116 using (5.44). The total lifetime of the laminar phase plus chaotic phase between bursts follows distribution (5.44). Figure 5.82 shows the log–log plot of the mean lifetime $\langle\tau\rangle$ as a function of ϵ, which is apparently ruled by the power law (5.45) with $\gamma \simeq 1.2$.

An example in the experiments is given in Fig. 5.83, where $V_{\mathrm{ac}} = 101.2$ mV and $f_0 = 1.91$ MHz. By increasing the dc bias, period 1 bifurcates to the torus of period 2 (Fig. 5.83c). With $V_0 > 2.668$ V (Fig. 5.83d–f), reinjective switching occurs, where the bursts become frequent by increasing the dc bias. The waveforms are similar to Fig. 5.75, but with an exception that the laminar phase is not a simple form of a torus or toric chaos in Fig. 5.83 (note: torus of period 2 in Fig. 5.83c, toric chaos of period 2 in Fig. 5.83d, e). For this point, it is conjectured that period 2 is generated by the period doubling bifurcation in the prebreakdown regime, and period 2 bifurcates to quasi-periodic motion. Since the bias region is quite narrow and the noise floor makes it difficult to observe the bifurcation, the period doubling point was missed. Thus, the reinjective switching in Fig. 5.83 is essentially explained by the same mechanism as shown by the mode simulations [111,

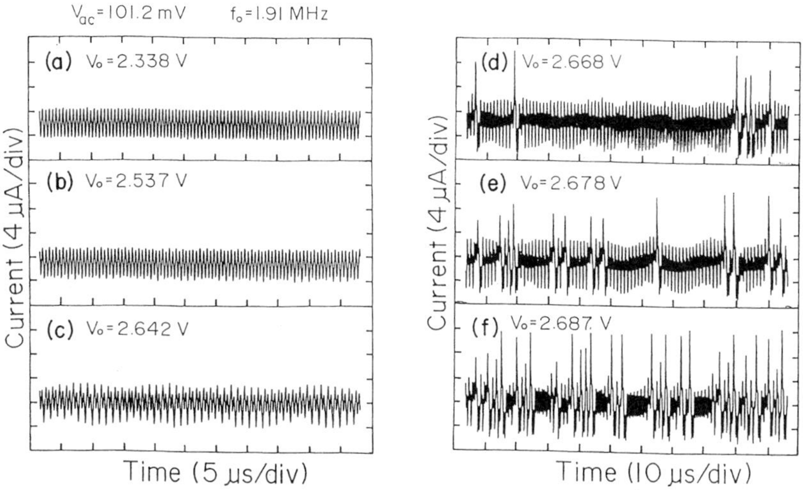

Figure 5.83 Intermittent bursts after Hopf bifurcation in n-GaAs under the periodically driven condition. $V_{ac} = 101.2\text{mV}$, $f_0 = 1.91$ MHz, and $T_L = 4.2$ K. (a) $V_0 = 2.338$ V, (b) $V_0 = 2.537$ V, (c) $V_0 = 2.642$ V, (d) $V_0 = 2.668$ V, (e) $V_0 = 2.678$ V, and (f) $V_0 = 2.687$ V. Measurements by K. Aoki.

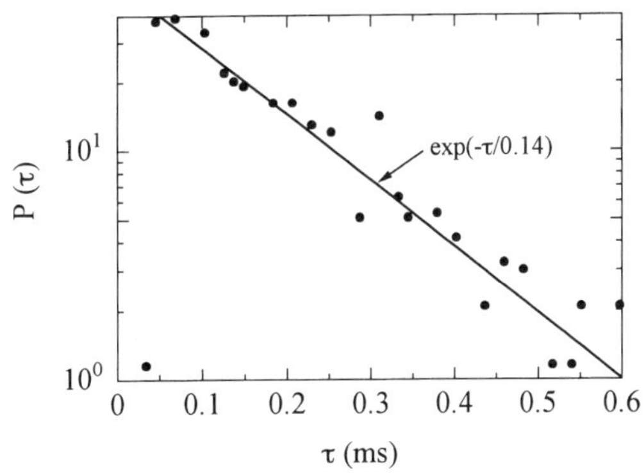

Figure 5.84 Probability distribution $P(\tau)$ of the residence time between bursts at the dc bias $V_0 = 2.66$ V.

112]. Figure 5.84 shows the probability distribution $P(\tau)$ of the residence time between bursts at the dc bias $V_0 = 2.66$ V. Although the experimental data exhibit large fluctuations, the reinjective switching seems to be ruled by (5.44), as shown by a solid line, where it is estimated that $\langle \tau \rangle \simeq 0.14$ ms, which means 267 round trips with $f_0 = 1.91$ MHz.

References

[1] N. Sclar and E. Burstein, J. Phys. Chem. Solids **2** (1957) 1.
[2] S.H. Koenig and G.R. Gunter-Mohr, J. Phys. Chem. Solids **2** (1957) 268; S.H. Koenig, Phys. Rev. **110** (1958) 986.
[3] J. Pozhela, *Plasma and Current Instabilities in Semiconductors* (Pergamon, Oxford, 1981).
[4] M. Glicksman and M.C. Steel, Phys. Rev. Lett. **2** (1959) 461.
[5] M. Glicksman, Phys. Rev. **124** (1961) 1655; C.E. Hurwitz and A.L. McWhorter, Phys. Rev. Lett. **10** (1963) 20.
[6] F. Okamoto, T. Koike and S. Tosima, J. Phys. Soc. Jpn **17** (1962) 804; T. Misawa and T. Yamada, Jpn. J. Appl. Phys. **2** (1963) 19.
[7] J. Bok, Proc. 1966 Int. Conf. on Phys. Semicond. (Suppl. J. Phys. Soc. Jpn 21, 1966) p. 685.
[8] J.B. Gunn, Solid State Commun. **1** (1963) 88; IBM J. Res. Develop. **8** (1964) 141.
[9] C. Hilsum, Proc. IEEE **52** (1964) 1736.
[10] V.L. Bonch-Bruevich, I.P. Zvyagin and A.G. Mironov, *Domain Electrical Instabilities in Semiconductors* (Consultants Bureau, New York, 1975).
[11] J. Bardeen and W.H. Brattain, Phys. Rev. **74** (1948) 230.
[12] L. Esaki, Phys. Rev. **109** (1958) 603.
[13] R.D. Larrabee and M.C. Steele, J. Appl. Phys. **31** (1960) 1519.
[14] A.L. McWhorter and R.H. Rediker, Proc. IRE **47** (1959) 1207.
[15] I. Melngailis and R.H. Rediker, Proc. IRE (1962) 2428.
[16] N. Holonyak and S.F. Bevacqua, Appl. Phys. Lett. **2** (1963) 71.
[17] M. Cardona and W. Ruppel, J. Appl. Phys. **31** (1967) 1826.
[18] J. Peinke, in *Nonlinear Dynamics in Solids*, ed. H. Thomas (Springer, Berlin, 1992) pp. 51–67.
[19] K. Aoki, T. Kobayashi and K. Yamamoto, Proc. 3rd Int. Conf. on Hot Carriers in Semicond. (Suppl. J. de Phys. No. 10, 1981) pp. c7-51.
[20] K. Aoki, K. Miyamae, T. Kobayashi and K. Yamamoto, Proc. 6th Int. Conf. on Noise in Physical Systems, eds. P.H.E. Meijer, R.D. Mountain and R.J. Soulen, Jr (National Bureau of Standards Special Publication 614, Washington DC, 1981) 93.
[21] K. Aoki, T. Kobayashi and K. Yamamoto, J. Phys. Soc. Jpn **51** (1982) 2373.
[22] K. Aoki, Y. Okuyama, T. Kobayashi, K. Yamamoto and S. Namba, Solid State Commun. **25** (1978) 717.
[23] K. Aoki, T. Kobayashi, K. Yamamoto and S. Namba, Proc. 14th Int. Conf. on Phys. Semicond. (Edinburgh, 1978) 319.
[24] K. Aoki, T. Kinugasa and K. Yamamoto, Phys. Lett. **72A** (1979) 63.
[25] K. Aoki, Y. Okuyama, T. Kobayashi and K. Yamamoto, J. Phys. C: Solid State Phys. **13** (1980) L29.
[26] K. Aoki, T. Kobayashi and K. Yamamoto, Phys. Lett. **79** (1980) 445.
[27] K. Aoki, K. Miyamae, T. Kobayashi and K. Yamamoto, J. Phys. Soc. Jpn **51** (1982) 3920.
[28] K. Aoki, K. Miyamae, T. Kobayashi and K. Yamamoto, Jpn. J. Appl. Phys. **19** (1980) L657;
[29] K. Aoki, K. Miyamae, T. Kobayashi and K. Yamamoto, J. Phys. Soc. Jpn **50** (1981) 357.
[30] K. Aoki and K. Yamamoto, Phys. Lett. **98A** (1983) 72.
[31] K. Aoki, K. Miyamae, T. Kobayashi and K. Yamamoto, Physica **117/118 B+C** (1983) 570.

[32] K. Aoki, O. Ikezawa, N. Mugibayashi and K. Yamamoto, Physica **134B** (1985) 288.
[33] S.W. Teitsworth, R.M. Westervelt and E.E. Haller, Phys. Rev. Lett. **51** (1983) 825.
[34] S.W. Teitsworth and R.M. Westervelt, Phys. Rev. Lett. **53** (1984) 2587.
[35] S.W. Teitsworth and R.M. Westervelt, Physica **23D** (1986) 181.
[36] S.W. Teitsworth, PhD, Thesis, Harvard University (1986).
[37] R.M. Westervelt and S.W. Teitsworth, Physica **23D** (1986) 187.
[38] R.M. Westervelt, S.W. Teitsworth and E.G. Gwinn, Phys. Scr. **T14** (1986) 65.
[39] E.G. Gwinn and R.M. Westervelt, Phys. Rev. Lett. **57** (1986) 1060.
[40] E.G. Gwinn and R.M. Westervelt, Phys. Rev. Lett. **59** (1987) 157.
[41] J. Belair and L. Glass, Physica **16D** (1985) 143.
[42] T. Bohr, P. Bak and M.H. Jensen, Phys. Rev. **A30** (1984) 1970.
[43] A.M. Kahn, D.J. Mar and R.M. Westervelt, Solid-State Electron. **32** (1989) 1143.
[44] A.M. Kahn, D.J. Mar and R.M. Westervelt, Proc. 1st Experimental Chaos Conference (World Scientific, 1992) p. 355.
[45] K.M. Mayer, J. Peinke, B. Röhricht, J. Parsi and R.P. Huebener, Phys. Scr. **T19** (1987) 505.
[46] K.M. Mayer, R. Gross, J. Parisi, J. Peinke and R.P. Huebener, Solid State Commun. **63** (1987) 55; K.M. Mayer, J. Parisi, J. Peinke and R.P. Huebener, Physica **32D** (1988) 306.
[47] U. Rau, J. Peinke, J. Parisi and R.P. Huebener, Z. Phys. **B71** (1988) 305.
[48] K.M. Mayer, J. Parisi, J. Peinke and R.P. Huebener, Z. Phys. B–Condensed Matter **71** (1988) 171.
[49] U. Rau, K.M. Mayer, J. Parisi, J. Peinke, W. Clauss and R.P. Huebener, Solid State Electron. **32** (1989) 1365.
[50] J. Peinke, J. Parisi, O.E. Rössler and R. Stoop, *Encounter with Chaos* (Springer, Berlin 1992).
[51] R. Stoop, J. Peinke, J. Parisi, B. Röhricht and R.P. Huebener, Physica **35D** (1989) 425; see also O.E. Rössler, Z. Naturforsch. **38a** (1983) 788.
[52] J. Parisi, J. Peinke, B. Röhricht and K.M. Mayer, Z. Naturforsch. **42a** (1987) 329; J. Peinke, J. Parisi, B. Röhricht, B. Wessely and K.M. Mayer, Z. Naturforsch. **42a** (1987) 841.
[53] J. Peinke, U. Rau, W. Clauss, R. Richter and J. Parisi, Europhys. Lett. **9** (1989) 743.
[54] R. Richter, J. Peinke, W. Clauss, U. Rau and J. Parisi, Europhys. Lett. **14** (1991) 1.
[55] H. Haken, *Synergetics – An Introduction,* 2nd ed. (Springer, Berlin, 1978)
[56] A. Brandl, T. Geisel and W. Prettl, Europhys. Lett. **3** (1987) 401.
[57] A. Brandl, M. Völcker and W. Prettl, Appl. Phys. Lett. **55** (1989) 238; Solid State Commun. **72** (1989) 847.
[58] A. Brandl, W. Prettl, Festkörperprobleme **30** (1990) 371.
[59] E. Schöll, *Nonequilibrium Phase Transitions in Semiconductors* (Springer, Berlin, 1987).
[60] E. Schöll, Proc. 4th Int. Conf. on Hot Carriers in Semiconductors, Physica **134B** (1985) 271.
[61] E. Schöll, Phys. Rev. **B34** (1986) 1395.
[62] G.N. Maracas, W. Porod, D.A. Johnson, D.K. Ferry and H. Goronkin, Physica **134B** (1985) 276.
[63] S.B. Bumelìene, Yu. K. Pozhela, K.A. Pyragas and A.V. Tamaševičius, Physica **134B** (1985) 293.

[64] Y. Abe ed., *Nonlinear and Chaotic Transport Phenomena in Semiconductors*, Appl. Phys. **A48**, pp. 93-191 (1989)
[65] D.G. Seiler, C.L. Littler, R.J. Justice and P.W. Milonni, Phys. Lett. **108A** (1985) 462.
[66] D.G. Seiler, C.L. Littler, R.J. Justice and P.W. Milonni, Proc. 17th Int. Conf. Phys. Semicond., eds. J.D. Chadi and W.A. Harrison (San Francisco, 1984) 1385.
[67] K. Yamada, N. Miura and C. Hamaguchi, Solid-State Electron. **31** (1988) 809.
[68] G.N. Maracas, D.A. Johnson and H. Goronkin, Appl. Phys. Lett. **46** (1985) 305.
[69] G.N. Maracas, D.A. Johnson, R.A. Puechner, J.L. Edwards, S. Myhajlenko, H. Goronkin and R. Tsu, Solid-State Electron. **32** (1989) 1887.
[70] W. Knap, M. Jezewski, J. Lusakowski and W. Kuszko, Solid-State Electron. **31** (1988) 805.
[71] U. Rau, J. Peinke, K. Karpierz, J. Lusakowski and W. Knap, Phys. Lett. **152A** (1991) 356.
[72] R. Symanczyk, in *Nonlinear Dynamics and Pattern Formation in Semiconductors and Devices*, ed. F.-J. Niedernostheide (Springer, Berlin, 1995) pp. 200–219.
[73] D. Jäger and R. Symanczyk, in *Nonlinear Dynamics in Solids*, ed. H. Thomas (Springer, Berlin, 1992) pp. 68–87.
[74] R. Symanczyk, S. Knigge and D. Jäger, Proc. of a NATO Advanced Research Workshop on Negative Differential Resistance and Instabilities in 2-D Semiconductors, eds. N. Balkan, B.K. Ridley and A.J. Vickers (Plenum, London, 1993) NATO ASI Series B: Physics Vol. 307, p. 269.
[75] H. Yano, N. Goto and Y. Ohno, Sremicond. Sci. Technol. **7** (1992) B491.
[76] J. Pozhela, A. Tamaševičius and J. Ulbikas, Solid-State Electron. **31** (1988) 805.
[77] G.A. Held, C. Jeffries and E.E. Haller, Phys. Lett. **52** (1984) 1037.
[78] G.A. Held, C. Jeffries and E.E. Haller, Proc. 17th Int. Conf. Phys. Semicond., eds. J.D. Chadi and W.A. Harrison (San Francisco, 1984) 1289.
[79] G.A. Held and C. Jeffries, Phys. Rev. Lett. **55** (1985) 887; G.A. Held and C. Jeffries, in *Dimensions and Entropies in Chaotic Systems*, ed. G. Mayer-Kress (Springer, Berlin, 1986) pp. 158–170.
[80] G.A. Held and C. Jeffries, Phys. Rev. Lett. **56** (1986) 1183.
[81] J.L. Hudson, M. Hart and P. Marinko, J. Chem. Phys. **71** (1979) 1601.
[82] H.G. Winful, Y.C. Chen and J.M. Liu, Appl. Phys. Lett. **48** (1986) 616; see also H.G. Schuster, *Deterministic Chaos – An Introduction*, 2nd ed. (Physik-Verlag, Weinheim, 1988) [122].
[83] K. Aoki, O. Ikezawa and K. Yamamoto, Phys. Lett. **98A** (1983) 217.
[84] K. Aoki, O. Ikezawa and K. Yamamoto, J. Phys. Soc. Jpn **53** (1984) 5.
[85] K. Aoki, O. Ikezawa and K. Yamamoto, Phys. Lett. **106A** (1984) 343.
[86] J.J. Hopfield, Proc. Natl Acad. Sci. **79** (1982) 2554; J.J. Hopfield and D.W. Tank, Biol. Cybern. **52** (1985) 141.
[87] S. Amari, IEE Trans. **SMC-2** (1972) 643.
[88] S. Amari, Kybernetik **14** (1974) 201.
[89] B. Katz, *Nerve, Muscle, and Synapse* (McGraw-Hill, New York, 1966).
[90] J.S. Griffith, *Mathematical Neurobiology* (Academic Press, New York, 1971).
[91] H. Haken, *Synergetics – An Introduction Nonequilibrium Phase Transitions and Self-Organization in Physics, Chemistry and Biology*, 2nd ed. (Springer, Berlin, 1978).
[92] Y. Aizawa and T. Uezu, Prog. Theor. Phys. **68** (1982) 1864.
[93] K. Aoki and K. Yamamoto, Appl. Phys. **A48** (1989) 111.

[94] K. Aoki, Proc. of a NATO Advanced Research Workshop on Negative Differential Resistance and Instabilities in 2-D Semiconductors, eds. N. Balkan, B.K. Ridley and A.J. Vickers (Plenum, London, 1993) NATO ASI Series B: Physics Vol. 307, pp. 393–407.
[95] U. Rau, K. Aoki, J. Peinke, J. Parisi, W. Clauss and R.P. Huebener, Z. Phys. B–Condensed Matter **81** (1990) 53.
[96] K. Aoki, U. Rau, J. Peinke, J. Parisi and R.P. Huebener, J. Phys. Soc. Jpn **59** (1990) 420.
[97] F.-J. Niedernostheide, in *Nonlinear Dynamics and Pattern Formation in Semiconductors and Devices,* ed. F.-J. Niedernostheide (Springer, Berlin, 1992) pp 168–199, and references therein.
[98] D. Jäger, H. Baumann and R. Symanczyk, Phys. Lett. **117A** (1986) 141; R. Symanczyk, E. Pieper and D. Jäger, Phys. Lett. **143A** (1990) 337; R. Symanczyk, S. Gaelings and D. Jäger, Phys. Lett. **160A** (1991) 397.
[99] V. Nov́ak and W. Prettl, in *Nonlinear Dynamics and Pattern Formation in Semiconductors and Devices,* ed. F.-J. Niedernostheide (Springer, Berlin, 1992) pp. 144–167.
[100] E.W. Williams and H.B. Bebb, *Semiconductors and Semimetals,* eds. R.K. Willardson and A.C. Beer (Academic Press, New York, 1972).
[101] K. Aoki, K. Yamamoto and N. Mugibayashi, J. Phys. Soc. Jpn **57** (1988) 26.
[102] L.D. Landau, Akad. Nauk Dokl. **44** (1944) 339.
[103] K. Aoki and K. Yamamoto, Proc. 21st Int. Conf. Phys. Semicond. (Beijing, 1992) eds. P. Jiang and H. Zheng (World Scientific, Singapore, 1992) 261.
[104] K. Aoki, A. Matsuura and Y. Takahashi, Semicond. Sci. Technol. **9** (1994) 607.
[105] K. Tomita and I. Tsuda, Prog. Theor. Phys. **64** (1980) 1138.
[106] J.S. Turner, J.C. Roux, W.D. McCormick and H.L. Swinney, Phys. Lett. **85A** (1981) 9.
[107] A.S. Pikovsky, Phys. Lett. **85A** (1981) 13; A.S. Pikovsky and M.I. Rabinovich, Physica **2D** (1981) 8.
[108] T. Bohr, P. Bak and M.H. Jensen, Phys. Rev. **A30** (1984) 1960; P. Bak and R. Bruinsma, Phys. Rev. Lett. **49** (1982) 429; J.A. Glazier and A. Libchaber, IEEE Trans. Circ. Syst. **35** (1988) 790.
[109] M. Wegener and C. Klingshirn, Phys. Rev. **A35** (1987) 1740.
[110] D.D. Coon, S.N. Ma and A.G.V. Perera, Phys. Rev. Lett. **58** (1987) 1139.
[111] K. Aoki, N. Mugibayashi, Appl. Phys. **A48** (1989) 161.
[112] H. Tominaga and H. Mori, Prog. Theor. Phys. **91** (1994) 1081.
[113] A. Arneodo, P. Coullet, C. Tresser, A. Libchaber, J. Maurer and D. d'Humieres, Physica **6D** (1983) 385.
[114] W. Lauterborn and E. Cramer, Phys. Rev. Lett. **47** (1981) 1445.
[115] J. Testa, J. Perez and C. Jeffries, Phys. Rev. Lett. **48** (1982) 714.
[116] R.V. Buskirk and C. Jeffries, Phys. Rev. **A31** (1985) 3332.
[117] K. Aoki, Semicond. Sci. Technol. **7** (1992) B474.
[118] F.C. Moon, *Chaotic and Fractal Dynamics* (John Wiley & Sons, New York, 1992).
[119] M.J. Feigenbaum, Los Alamos Sci. (Summer) (1980) 4-27.
[120] A. Libchaber and J. Maurer, J. Phys. (Paris) Coll. **41** (1980) C3-51.
[121] P.S. Linsay, Phys. Rev. Lett. **47** (1981) 1349.
[122] H.G. Schuster, *Deterministic Chaos,* 2nd ed. (Physik-Verlag, Weinheim, 1988).
[123] M. Hénon, Commun. Math. Phys. **50** (1976) 69.
[124] e.g., P. Bryant and C. Jeffries, Phys. Rev. Lett. **53** (1984) 250.
[125] J.D. Farmer, Physica **4D** (1982) 366.

[126] P. Berge, Y. Pomeau and C. Vidal, *Order within Chaos* (John Wiley & Sons, New York, 1984).
[127] K. Kaneko, Prog. Theor. Phys. **68** (1982) 669; **69** (1983) 403; **71** (1984) 282; **72** (1984) 202.
[128] C.D. Jeffries, Physica Scripta **T9** (1985) 11.
[129] D.G. Aronson, M.A. Chory, G.R. Hall and R.D. McGehee, Commun. Math. Phys. **83** (1982) 303.
[130] F.T. Arecchi, R. Meucci and W. Gadomski, Phys. Rev. Lett. **58** (1987) 2205.
[131] J.J. Swetitis and A.M. Buoncristiani, Phys. Rev. **38A** (1988) 5430.
[132] F. Argoul, A. Arneodo, P. Richetti, Phys. Lett. **A120,** 269 (1987).
[133] C. Grebogi, S.W. Mcdonald, E. Ott and J.A. Yorke, Phys. Lett. **99A** (1983) 415.
[134] G. Grebogi, E. Otto, F. Romeiras and J.A. Yorke, Phys. Rev. **A36** (1987) 5356.

Chapter 6

Characterizing chaos

This chapter describes the analytical methods used to characterize the chaotic behaviors observed in semiconductors. Self-similarity or fractal structure is one of the most important concepts in chaos theory. Fractal dimensions and fractal sets are introduced in section 6.1. In order to characterize the sensitive dependence of the chaotic orbits on the initial conditions, analysis using the Lyapunov spectrum is given in section 6.2. The global structure of the fractal sets can be found by the $f(\alpha)$ spectrum in section 6.3. Circle-map behaviors are described in section 6.4.

6.1 Fractal dimensions and fractal sets

In this section, the important concepts of fractal sets and fractal dimensions are introduced, which have become a powerful guideline to find the deterministic nature of nonlinear dynamical systems. In particular, some experimental methods to analyze the fractal dimensions from experimental time signals are introduced. Using these methods one can assign a relevant dimension to the deterministic chaos, as well as speculate on the degrees of freedom of the order parameters which predominate the nonlinear chaotic dynamics in a spatially extended instability (such as the firing wave (FW) instability) of semiconductor transport. No definite definition of fractal dimension is given, but the fractal dimension is used as the general terminology of the fractional dimensions defined by various computational methods [1, 2]. Very often the fractal dimension represents the Hausdorff–Besicovitch dimension or the capacity dimension, which are introduced below [1].

In order to understand the concept of fractal dimensions in the sense of the fractal geometry introduced by Mandelbrot [3], several well-known examples are presented. The terminology "fractal" was given by Mandelbrot [3]. In the Euclidean space $\mathbf{R}^E$ (E: phase space dimension; Euclidean dimension), any geometric shape has its topological dimension D_T (an integer), and the Hausdorff–Besicovitch dimension $D_{\rm H}$ (or simply Hausdorff dimension), where $0 \leq D_{\rm H} \leq E$ and $0 \leq D_{\rm T} \leq E$. The fractal is a geometrical set in the Euclidean space such that the Hausdorff–Besicovitch dimension satisfies

$$D_{\rm H} \geq D_{\rm T}. \tag{6.1}$$

The topological dimension must have an integer which satisfies (6.1). The equality in (6.1) holds typically for the ordinary Euclidean geometry (non-fractal smooth curve, surface and volume). The Hausdorff–Besicovitch dimension is defined as follows.

6.1.1 Hausdorff–Besicovitch dimension

The Hausdorff–Besicovitch dimension is defined and obtained by taking the most efficient covering of the fractal set (fractal curve, surface and volume). It is a critical dimension d such that measure M_d of a geometrical set varies critically from 0 to ∞ (at $D = D_{\rm H}$) when D is varied as a parameter [1, 2]:

$$M_{\rm D} = \alpha(D)N(\epsilon)\epsilon^D \rightarrow \begin{cases} 0, & D > D_{\rm H} \\ \infty, & D < D_{\rm H}, \end{cases} \tag{6.2}$$

where $\alpha(D)$ is the geometrical factor. In Fig. 6.1, in order to measure a finite surface (e.g., a square sheet with the dimension L^2) by the most efficient covering with a cube of size ϵ, then the measure M_D is the total volume $N(\epsilon)\epsilon^D$ covered by $N(\epsilon)$ cubes, where $\alpha(D) = 1$. Since $N(\epsilon) = (L/\epsilon)^2$, $M_d = L^2\epsilon^{D-2}$. By taking the limit $\epsilon \rightarrow 0, M_D \rightarrow 0$ for $D > 2$ and $M_D \rightarrow \infty$ for $D < 2$. Then, $D_H = 2$ for the dimension of smooth surface. Many fractal sets normally have non-integer values of the Hausdorff–Besicovitch dimensions. Some exceptions are seen in special cases among Brownian fractals [3] (e.g., the Brownian motion $D_{\rm H} = 2, D_{\rm T} = 1$).

The following description is more general [1]. When the geometric structure of a set S embedded in the Euclidean space is covered by a countable number of spheres (or circles) with diameters $d_1, d_2, \ldots, d_k$ which

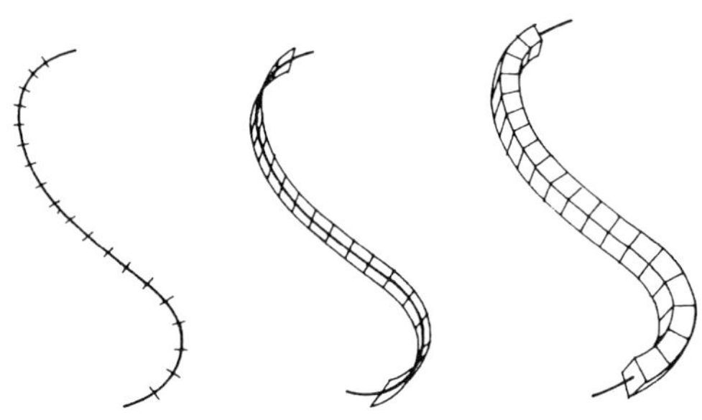

Figure 6.1 The most efficient covering of a curve [2].

are less than a definite value ϵ, the D-dimensional Hausdorff measure [1] is defined by:

$$M_D(S) = \lim_{\epsilon \to 0} \inf_{d_k < \epsilon} \sum_k d_k^D. \tag{6.3}$$

Thus, set S is covered most effectively by spheres with different sizes (see Fig. 6.2). When D is varied as a parameter in (6.3), the Hausdorff measure shows a transition from zero to infinity at $D = D_H$. The critical value D_H is called the Hausdorff dimension. Taking an example of a Koch curve in Fig. 6.2, the Hausdorff measure for $D = 1$ gives the dimension of length $d_1 + d_2 + \ldots + d_k$. With the limit $\epsilon \to 0$ ($k \to \infty$), the measure diverges to infinity, since the fractal structure continues down to the micro-size and the total length goes to infinity. Here, fine coverage over the fractal means a reduction in the coarse-graining degree. For $D = 2$, the measure $d_1^2 + d_2^2 + \ldots + d_k^2$ has the dimension of area, which converges to zero by taking the limit since the fractal is composed of the self-similar line structure. More precisely, $M_D(S) \to 0$ for $D > \log 4/\log 3$ and $M_D(\mathrm{S}) \to \infty$ for $D < \log 4/\log 3$ and a finite value at $D_H = \log 4/\log 3$ (see below).

Fractal geometry is mathematically pathological (e.g., nondifferentiable everywhere). The universal feature is the self-similar geometry; fractal everywhere from the macroscopic down to the microscopic structures. An alternative expression of the fractal dimension is given by the similarity dimension D_S [1, 2]:

$$N(\epsilon) = 1/\epsilon(N)^{D_s},$$

i.e.,

$$D_S = -\ln\ N(\epsilon)/\ln\ \epsilon(N), \tag{6.4}$$

where $N(\epsilon)$ is the total number necessary to cover the fractal set by a small segment (a small square or box) and $\epsilon(N)$ is the scaling factor [2]. For the

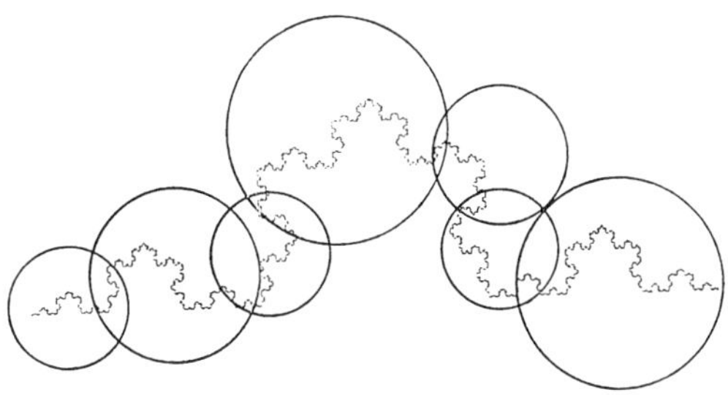

Figure 6.2 Covering the Koch curve effectively with circles of different sizes [1]. For the generator see Fig. 6.4.

self-similar fractals note especially that the similarity dimension is equivalent to the Hausdorff–Besicovitch dimension ($D_S = D_H$).

An introductory example is given by the Cantor set in Fig. 6.3. The Cantor set can be formed by removing the central third of the interval [0,1], and successively removing the central third of each remaining line segment. After the k steps of the removal process, the remaining total length is $(2/3)^k$, where the total number of the line segments is $N = 2^k$ and the length of each line segment $\epsilon = 3^{-k}$. From (6.4), one has:

$$2^k = (3^{-k})^{-D}. \tag{6.5}$$

By taking the limit $k \to \infty$, one has:

$$D_H = \frac{\log 2}{\log 3} = 0.6309\ldots. \tag{6.6}$$

By the limit $k \to \infty$, the set (Cantor dust [3]) is discontinuous everywhere with the topological dimension $D_T = 0$ (note: each line segment covers Cantor dust with the Lebesgue measure zero). After Mandelbrot [3], the initial interval [0,1] is termed the "initiator" and the removing process the "generator". Here, the reader is reminded of the similarity between the Cantor set and a set of the iterated points at the onset of chaos (Feigenbaum attractor; see the bifurcation maps in Fig. 4.76). The band merging process (inverse cascade of period doubling) is reminiscent of the construction process in Fig. 6.3. Apparently, the self-similar structure can be found everywhere.

A second example is given by the Koch curve as given in Fig. 6.2. The initiator is the unit interval [0,1] and the generator is given in Fig. 6.4. In order to obtain the Koch curve, the unit interval is first divided into three segments and the central segment replaced by two line segments of length 1/3. The first step produces the generator with four sides, each of length 1/3 in Fig. 6.4, where the two line segments of the central part form two sides of a regular triangle. Thus, the total length of the generator is 4/3. The same

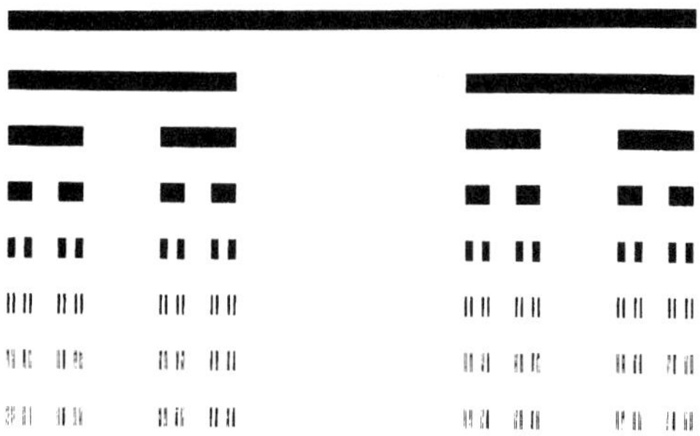

Figure 6.3 Construction of the Cantor set [3].

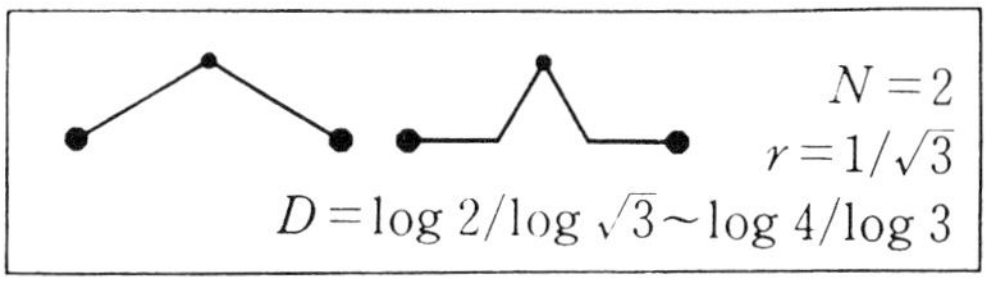

Figure 6.4 Generator for the Koch curve [3].

Figure 6.5 Construction of the Sierpinski triangle [3].

process is repeated on each line segment of the generator, giving four self-similar structures with total length 16/9 (16 line segments, each of length 1/9). The process is repeated by k steps, giving the total length $4^k/3^k$. The number of line segments $N(\epsilon) = 4^k$ scales as $(3^{-k})^{-D}$. With $k \to \infty$ ($\epsilon \to 0$), the Hausdorff–Besicovitch dimension is given by:

$$D_{\mathrm{H}} = \frac{\log 4}{\log 3} = 1.26185\ldots.$$

Figure 6.5 shows the procedure forming the Sierpinski triangle [3]. At the first step (generator), an original triangle (initiator) is divided into four regular triangles and the central triangle erased. Repeating the process for each remaining triangle produces the Sierpinski triangle. The Hausdorff–Besicovitch dimension is $D_{\mathrm{H}} = \log 3/\log 2 = 1.5849$, since $N(\epsilon) = 3^k, \epsilon = 2^{-k}$ after k steps, where ϵ is the side length of the divided triangles. An example of the Sierpinski carpet is given in Fig. 6.6. A square is divided into nine equal squares and the central part is erased. The process is repeated for the remaining part of the squares and so on. Noting $N(\epsilon) = 8^k$ and $\epsilon = 3^{-k}$, then $D_{\mathrm{H}} = \log 8/\log 3 = 1.8928$ [3].

Figure 6.6 Construction of the Sierpinski carpet [3].

The examples given above are very fundamental and form a small part of an enormous number of fractal sets [3]. More sophisticated fractals are shown by the Julia set and the Mandelbrot set [3, 4], which are given by the boundary of the basin of attraction solving the iterative map: $z_{n+1} = f_c(z_n)$ in the complex plane for the complex parameter values c (z: complex variable). An example of a Julia set is given in Fig. 6.7. The pathological geometry of the fractals seems to be nothing but mathematical productions. However, fractals are found everywhere in nature: river fractals (e.g., the Amazon) and various types of geographical fractals, woods, lightning patterns [6], earthquakes, $1/f$ noise [7], human body (lung, blood tubes, etc.), aggregations (metallic leaf, dendrite patterns) [8], percolations and metal clusters [9] and, of course, the dynamical structure of chaos. Therefore, fractals are not restricted to the mathematical world. In many cases, the fractals are patterns that have grown from generalized bifurcations [10], e.g., canalizations for geographical reasons (branching the rivers or land shape by erosion) or canalizations for biological reasons (branching of blood vessels, etc.). The dimensions of fractals in nature are apparently less than dimension 3 in our realistic Euclidean space. Many fractals such as the patterns of a tree can be reproduced by computer simulations [1,3]. A typical example of fractals in nature is the coast line formed by erosion. In Fig. 6.8, one can measure the total length of the coast line on a map from point A to point B using a scale (or a divider) of size ϵ. The total number of measurements $N(\epsilon)$ to cover the coast line depends on the size ϵ and increases on using a smaller scale with the power law:

$$N(\epsilon) \propto \epsilon^{-D}. \tag{6.8}$$

The length of the coast line is given by $L(\epsilon) = \epsilon\, N(\epsilon) \sim \epsilon^{1-D}$. By plotting

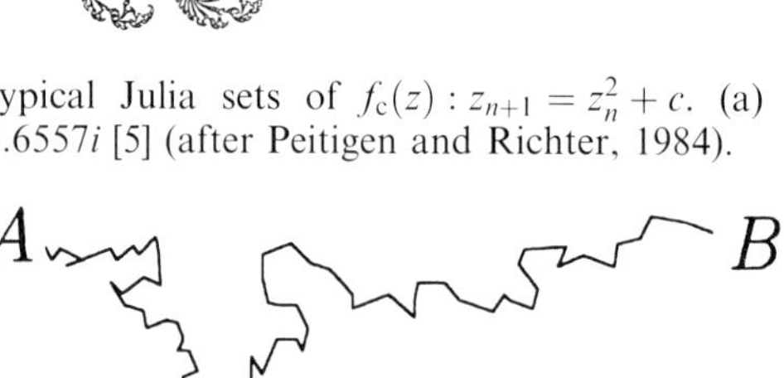

Figure 6.7 Two typical Julia sets of $f_c(z) : z_{n+1} = z_n^2 + c$. (a) $c = 0.32 + 0.043i$, (b) $c = -0.194 + 0.6557i$ [5] (after Peitigen and Richter, 1984).

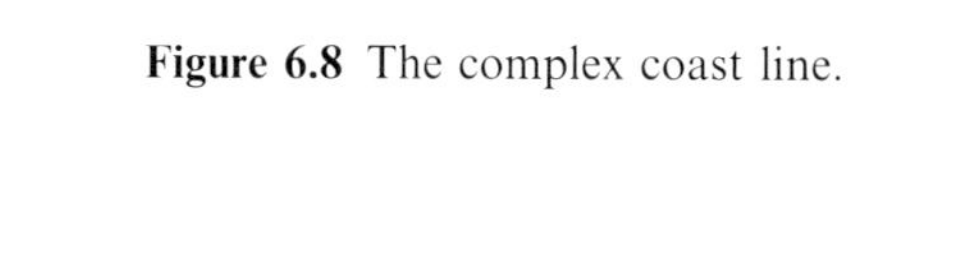

Figure 6.8 The complex coast line.

$\log L(\epsilon)$ vs $\log \epsilon$, one can obtain the slope $1 - D$ and thus the Hausdorff dimension D_H [3]. The length $L(\epsilon)$ increases with finer and finer resolution of ϵ since the coast line shows a fractal structure, which can be confirmed by blow-ups of the map. Depending on the degree of complexity of the coast lines, the fractal dimension may take the values $1 < D_H < 2$.

Fractals may be classified into two categories: real space fractals (sea lines, rivers, etc.), which do not involve temporal behaviors, and the fractals of dynamical structures ($1/f$ noise, chaos, etc.). The purpose of this section is to describe the fractals and the dimensions of the dynamical structures and to apply the fractal theory to chaos in semiconductors. The experiments measure output signals $I(t)$ of chaos. If the system equations are unknown, how can the time series $I(t)$ be recognized as an output of the deterministic chaos? The answer can be obtained by measuring the fractal dimensions from $I(t)$. There are many analytical methods and relevant definitions of the dimensions, as stated below.

6.1.2 Embedding method

Suppose that the behaviors of a dynamical system can be approximately described by n independent variables $\{Y_i(t)\}$, $i = 1, 2, \ldots, n$. Very often, the degree of freedom is unknown and one can measure only a few variables among Y_is. However, even with the measurements of one physical quantity, say current $I(t)$, one can construct an attractor embedded in the m-dimensional phase space [11,12]. Let

$$\mathbf{X}(t_i) = \{I(t_i),\ I(t_i + \tau),\ I(t_i + 2\tau), \ldots,\ I(t_i + (m-1)\tau)\} \qquad (6.9)$$

be a vector point in the phase space, where $i = 0, 1, 2, \ldots, \Delta t = t_{i+1} - t_i$ is the sampling time of the measurements and τ is the time delay which is taken arbitrarily but fixed at some value. Typically, the time delay is fixed at a value between $T/4$ and $3T/4$, where T is the fundamental period of the oscillation. By doing so, the set $\{\mathbf{X}(t_i)\}, i = 0, 1, 2 \ldots$, forms an attractor in the m-dimensional phase space. Many fractal dimensions [1,2] are related to and helpful in the analysis of dynamical behaviors. Many of them are introduced here (see also Moon [13]).

6.1.3 Capacity dimension

The capacity dimension was introduced by Kolmogorov. A finite set (or attracting set) embedded in the m-dimensional Euclidean space (or m-dimensional phase space) is covered by a minimal number $n(\epsilon)$ of m-dimensional spheres with the diameter ϵ. Then, the capacity dimension is defined by:

$$D_C = \lim_{\epsilon \to 0} \frac{\log n(\epsilon)}{\log(1/\epsilon)}. \tag{6.10}$$

Similar to the Hausdorff measure, the D-dimensional measure is approximately proportional to $N(\epsilon) \sim \epsilon^D$, which is zero for $D > D_C$ and infinity for $D < D_C$. Very often the capacity dimension coincides with the Hausdorff–Besicovitch dimension (note: $D_C > D_H$).

6.1.4 Information dimension

In the following section, the problems are restricted to the fractals of dynamical structures. For an attractor $\{\mathbf{X}(t_i)\}$ $(i = 0, 1, 2, \ldots)$ constructed by the embedding method (6.9), the information dimension can be obtained by the following procedure. The m-dimensional phase space can be divided into cells, i.e., cubes of size ϵ. Then the attracting set $\{X\}$ which forms a phase portrait can be covered by M cells. Among a large number of vector points N (a long enough trajectory), the number of points N_i which enter each of the M cells is counted. Then, the probability $P_i(\epsilon)$ of finding the vector points in a specified cell is given by:

$$P_i(\epsilon) = \frac{N_i}{N}, \qquad \sum_{i=1}^{M} P_i(\epsilon) = 1, \tag{6.11}$$

where the sum is taken over the M cells which cover the attractor. The information entropy is given by:

$$I(\epsilon) = -\sum_i P_i(\epsilon) \log P_i(\epsilon). \tag{6.12}$$

By varying ϵ, $I(\epsilon)$ behaves as

$$I(\epsilon) \propto -D_I \log \epsilon, \tag{6.13}$$

where ϵ takes small values. Then, the information dimension D_I is defined by:

$$D_I = \lim_{\epsilon \to 0} \frac{I(\epsilon)}{\log(1/\epsilon)}. \tag{6.14}$$

The information dimension satisfies the relation

$$D_I \le D_C. \tag{6.15}$$

The simplest case is given if the probabilities P_i are equal for all cells. Summing over the M cells, the information in (6.12) is given by $I = \log M$, from which $D_I = D_C$. For the Cantor set, note that $D_I = D_C$ exactly. Normally, $D_I < D_C$ for any dynamical system since chaos in the deterministic systems has the invariant probability measure. For the logistic map $x_{n+1} = \mu x_n(1 - x_n)$, the Feigenbaum attractor at the accumulation point of the 2^∞ period ($\mu_\infty = 3.5699345\ldots$) has a Cantorian structure with the information dimension $D_{\text{I}} = 0.5170976$ [14], the value of which has been given analytically by

$$D_{\text{I}} = \lim_{k\to\infty} \frac{\ln\ 2}{\ln\ \alpha + 2^{-(k+1)} \sum_{i=1} g'(\xi_{2i-1})}, \tag{6.16}$$

where $g(x)$ is the universal function or the Feigenbaum function $g(g(x)) = -g(\alpha x)/\alpha, \alpha = 2.5029$ and the sum is taken over $i = 1 \sim 2^k$. For the fully developed chaos at $\mu = 4$, the dimensions are found to be $D_{\text{C}} = D_{\text{I}} = 1$ using the invariant probability measure $\rho(x) = 1/\pi[x(1-x)]^{-1/2}$ [14], the value of which is consistent with the intuitive consideration that the dimensions could be unity since the iterated points are dense everywhere in the interval [0,1].

In the continuous dynamical systems with the degrees of freedom $d_f = 3$ or more, such as in the semiconductor chaos system, no analytical methods are available to obtain the fractal dimensions. One must use the box-counting algorithm for the capacity dimension (6.10) or calculate the information entropy (6.12) for the information dimension. In general, for D_{C} in (6.12) one must count the number of sufficiently small boxes or hyperspheres which cover effectively the strange attractor embedded in the m-dimensional phase space with $m = 3$ or more. However, this is a tedious task (the same thing happens for D_{I}). With $m = 2$, the problem seems to be simpler, as shown in Fig. 6.9. With $m = 1$, the problem becomes quite simple.

As an exercise for $m = 1$, one can easily find out the dimensions D_{C} and D_{I} in the approximated model of the impact-ionization avalanche (4.88)–(4.90). Figure 6.10 shows the invariant probability measure $\rho(\nu_{\text{p}})$ for the strange attractor obtained just at the onset of chaos ($E_0 = 15.767854$), where ν_{p} is the peak value of the carrier density. A blow-up of Fig. 6.10 a shows a self-similar structure (see Fig. 6.10b). The invariant measure is obtained by dividing the whole interval (the dynamic range) of ν_{p} into M bins and calculating the probabilities $P_j = N_j/N$ ($j = 1, 2, \ldots, M$), where N is the total number of the data points and N_j the number of points located in the jth bin. The general form of $\rho(\nu_{\text{p}})$ is given by:

$$\rho(\nu_{\text{p}}) = \lim_{N\to\infty} \frac{1}{N} \sum_{i=1}^{N} (\nu_{\text{p}}(iT) - \nu_{\text{p}}), \tag{6.17}$$

where $\nu_{\text{p}}(iT)$ is the peak value of $\nu(t)$ at time $t = iT$ ($i = 1, 2, \ldots, N$ and T: period). In Fig. 6.11, the information (6.12) is plotted as a function of $\log_2 \epsilon$, from which the information dimension is obtained as $D_{\text{I}} = 0.548 \pm 0.12$, where $N = 57753$, $M = 2^{14}$ and $\epsilon = (0.14746)2^{k-14}$ ($k =$

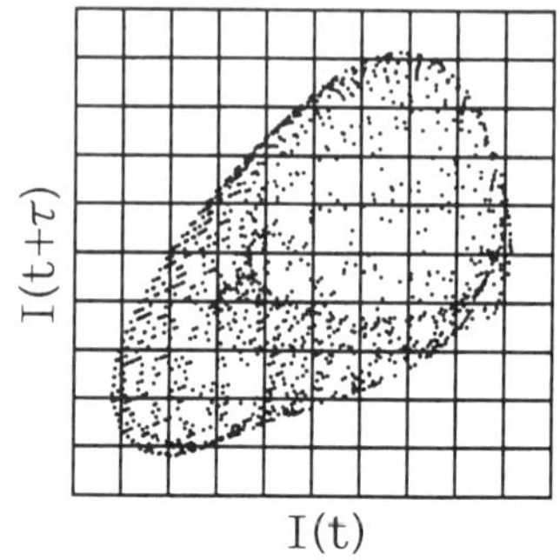

Figure 6.9 Box counting method in the 2D phase space (schematic). First, the strange attractor is projected onto the 2D phase space. Then, the phase space is divided into cells (squares) of side ϵ. The minimal number $n(\epsilon)$ of cells which cover the attracting set is counted. With $\epsilon \to 0$, we find the capacity dimension D_c according to (6.10).

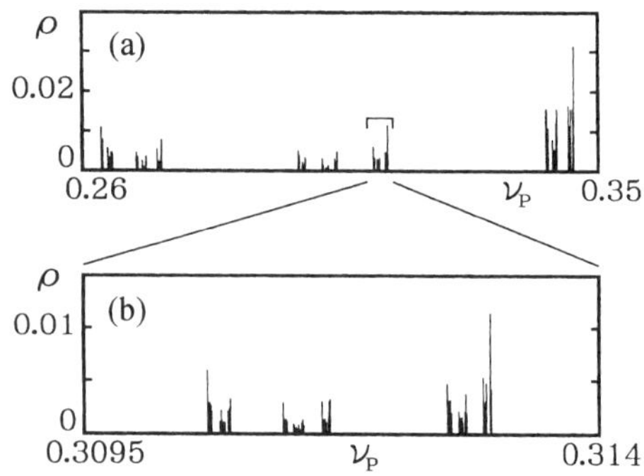

Figure 6.10 Invariant probability measure for the strange attractor in the approximated model of the impact-ionization avalanche (4.88), (4.89) $E_{ac} = 4.5, f_d = 100/136$. The dc bias was taken at the onset of chaos ($E_0 = 15.767854$).

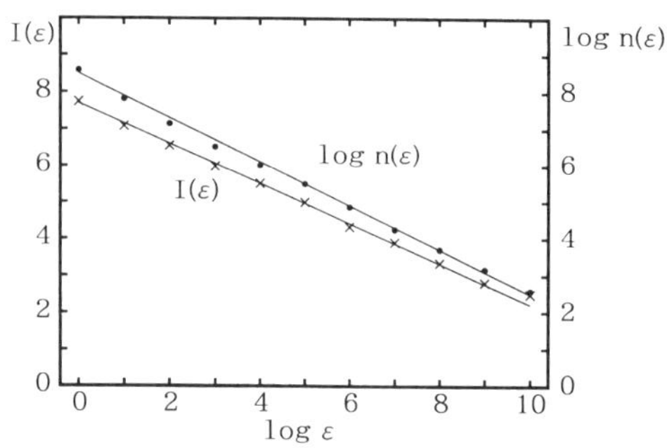

Figure 6.11 The information $I(\epsilon)$ and the number of bins $n(\epsilon)$ as a function of $\log_2 \epsilon$. For simplicity the values of the abscissa are given by $k(= \log_2 2^k,\ k = 0, 1, 2, \ldots.)$ and are proportional to $\log_2 \epsilon$.

$0, 1, 2, \ldots, 14$). For reference, the number of bins $n(\epsilon)$ necessary for covering the attracting set in Fig. 6.10 is also plotted as a function of ϵ in the logarithmic scale, showing the power law $n(\epsilon) = \epsilon^{-D}$. The capacity dimension is found to be $D_C = 0.6 \pm 0.09$. Note that $D_C > D_I$.

6.1.5 Lyapunov dimension

Here, the Lyapunov dimension is simply described since the Lyapunov spectrum has not yet been introduced. Assuming a strange attractor is embedded in the m-dimensional phase space, then the dynamical structure of the whole orbits can be characterized by the Lyapunov exponents λ_1 $(i = 1, 2, \ldots, m)$. A set of the exponents $(\lambda_1, \lambda_2, \ldots, \lambda_m)$ is called the Lyapunov spectrum. In the 3D phase space $(m = 3)$, $\lambda_1 > 0$, $\lambda_2 = 0$ and $\lambda_3 < 0$ for low-dimensional chaos. Then the Lyapunov dimension D_L or the Kaplan–Yorke dimension (D_{KY}) [14–17], defined by Kaplan and Yorke [15] and discussed by Farmer [16] and others [17], is given by:

$$D_L = 2 + \frac{\lambda_1}{|\lambda_3|}. \tag{6.18}$$

For instance, Rössler chaos has $\lambda_1 = 0.13, \lambda_2 = 0$ and $\lambda_3 = -14.1$, from which $D_L = D_{KY} = 2.0092$. is estimated. The value indicates that the Rössler attractor is an almost 2D structure of a very thin folded sheet (see Fig. 4.55). In general, with the m-dimensional phase space, the Lyapunov dimension is given by:

$$D_L = D_{KY} = j + \frac{\sum_i^j \lambda_i}{|\lambda_{j+1}|}, \tag{6.19}$$

where $j + 1$ is the smallest index which makes the sum $\sum \lambda_i$ negative [17]. The Lyapunov spectrum and the Lyapunov dimension, are discussed in more detail in section 6.2 Kaplan and Yorke [15] suggested that D_L is a lower bound of the capacity dimension $(D_L \leq D_C)$.

6.1.6 Correlation dimension

The most relevant dimension in this book is the correlation dimension (Grassberger dimension) D_G introduced by Grassberger and Procaccia [14] and Hentschel and Procaccia [18]. Its greatest advantage is that the lower bound of the fractal dimension can be estimated from a single-variable time series, e.g., the current $I(t)$, without measuring the time series of other variables. Recall the embedding method (6.9). From an experimental time series $I(t)$ of chaos, the strange attractor can be embedded into the m-dimensional phase space by constructing the m-dimensional vector points $\{\mathbf{X}_i(t)\}$ $(i = 1, 2, \ldots, N$; see (6.9)). Then the strange attractor can be characterized by a set of N vector points. Suppose a hypersphere with small radius r is centered on a specified vector point $\mathbf{X}_i$, and count the number of the vector points included inside the hypersphere. The distances between pairs of vector points, say $d_{ij} = \|\mathbf{X}_i - \mathbf{X}_j\|$ are calculated for all j, where the distance may be the Euclidean norm or another relevant norm. Then the number can be obtained by comparing d_{ij} with r. A correlation function $C(r)$ is calculated by the definition [13,14,18]:

$$C_m^i(r) = \lim_{N\to\infty} \frac{1}{N} \text{ (number of } X_j \text{ such that } d_{ij} < r), \tag{6.20}$$

$$C_m(r) = \lim_{N\to\infty} \frac{1}{N^2} \begin{pmatrix} \text{number of ordered pairs } (i,j) \\ \text{such that the distance } d_{ij} < r \end{pmatrix}, \tag{6.21}$$

$$= \lim_{N\to\infty} \frac{1}{N^2} \sum_i^N \sum_j^N H(r - \|X_i - X_j\|), \tag{6.22}$$

where $H(x)$ is the Heaviside function satisfying $H = 1$ if $x \geq 0$ and $H = 0$ if $x < 0$. Note that the sum is taken among all pairs (i,j) of the vector points. The correlation integral depends on the radius r with a certain scaling behavior:

$$\lim_{N\to\infty} \lim_{r\to 0} C_m(r) = r^D. \tag{6.23}$$

Thus, the correlation dimension D_G is defined by:

$$D_G = \lim_{N\to\infty} \lim_{r\to\infty} \frac{\log\ C_m(r)}{\log\ r}. \tag{6.24}$$

By increasing the embedded dimension m (namely the phase space dimension), the dimension D_G approaches a saturated value which is truly the dimension meaningful as a fractal dimension. For the correlation dimension D_G, note that many authors use different notations, such as ν [14]. The correlation dimension is a kind of information dimension [1], but the algorithm is more convenient than (6.14). The meaning of (6.23) is very clear. For a fixed point $\mathbf{X}_0$ embedded in the 3D phase space, $C_3(r) = 1$ from (6.22) and $D_G = 0$ from (6.23), in accordance with the intuition that the point is zero-dimensional. A limit cycle orbit forms a closed trajectory and thus a curvilinear set. Then, the correlation integral $C_3(r)$ increases linearly with r, giving the power law $C_3(r) \propto r$ and $D_G = 1$. For a torus (quasi-periodic orbit), the vector points $\mathbf{X}_i$ distribute uniformly on the toric surface, giving $C_3(r) \propto r^2$ and $D_G = 2$. A strangely attracting set (chaos) has a non-integer dimension. At the onset of chaos reached after the period doubling cascade, the value is between 1 and 2 (e.g., $D_G = 1.6$). By the torus (Hopf) bifurcation, the chaos may exhibit a correlation dimension between 2 and 3. For chaos, the correlation dimension for $m > 3$ increases slightly more than the value for $m = 3$ and normally saturates at a definite value.

The advantage of this method is that one can judge the deterministic nature of a dynamical system from the experiments without any knowledge about the system equations. That is, it is easily judged from the relatively small size of the experimental data ($N \approx 1000$ or more) whether the irregular oscillation is chaos or simply stochastic noise. A typical example of the Rayleigh–Bénard experiments measured by Malraison *et al.* [19, 20] is shown in Fig. 6.12. In Fig. 6.12a, the correlation functions are obtained as a function of r for different values of the embedded dimension m. From the slope of each curve in Fig. 6.12a, the measured dimension is given as a function of m in Fig. 6.12b. The value saturates with $m \gtrsim 4$, giving a fractal

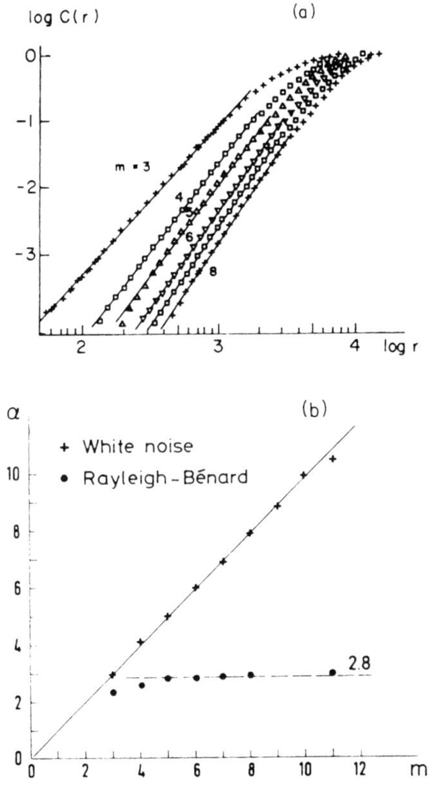

Figure 6.12 Experimental results from Malraison *et al.* and Atten *et al.* (1982); see also Dubois (1982). (a) The plots show $\log C$ vs $\log r$ for different values of the embedding dimension, for the Rayleigh–Bénard experiment. (b) The measured dimension as a function of the embedding dimension m, both for the Rayleigh–Bénard experiment and for numerical white noise. Note that α becomes nearly constant (but not quite) at $m = 3$. The α for white noise is nearly equal to m (but not quite) [19, 20].

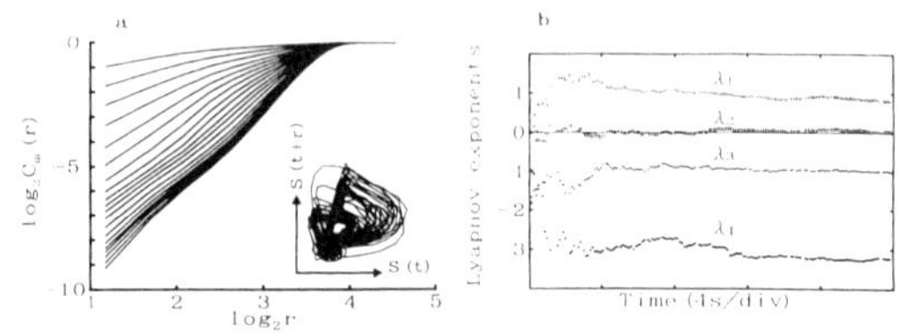

Figure 6.13 (a) Correlation integral $C_m(r)$ vs r in logarithmic scale: (b) Lyapunov spectrum [21].

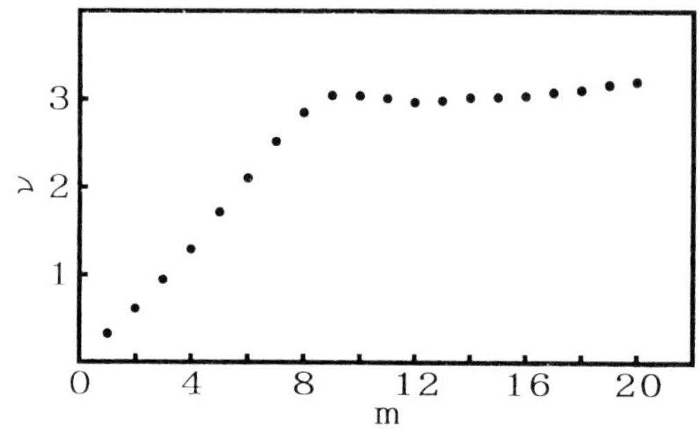

Figure 6.14 Slope of the correlation integral in Fig. 6.13 as a function of m.

dimension of 2.8. There, a clear-cut difference between the deterministic chaos and the stochastic noise (white noise, marked + in Fig. 6.12b) is shown. Since the white noise exhibits no attracting set, the uncorrelated vector points are uniform and dense in a finite region of the m-dimensional phase space, resulting in no saturation of the dimension for $m > 3$ $(D_G = m)$.

The present aim is to know the deterministic nature and the fractal dimension in experiments on semiconductor chaos (FW instability and others). An example of FW instability is given in Fig. 6.13a. The slope of each curve saturates with $m > 8$, as shown in Fig. 6.14. Thus, the correlation dimension is $D_G = 3.06 \pm 0.1$ for $m > 8$. From the measured Lyapunov spectrum $\{\lambda_i\}$ for $m = 4$ in Fig. 6.13b, the Lyapunov dimension is estimated to be $D_L = 2 + \lambda_1/|\lambda_3| \sim 2.9$. Though the fractal dimension is found to be relatively high in the experiments, the deterministic nature of the FW instability is strongly confirmed. The correlation dimensions for periodically driven chaos and crossover instability will be discussed in section 7.4.

6.1.7 Relation among the dimensions

By introducing the generalized dimension D_q, the relation among the various dimensions becomes apparent. In order to cover an attracting set by the total number $n(\epsilon)$ of hypercubes of each side ϵ in the m-dimensional phase space, the attracting set is assumed to be composed of N vector points. Then the probability p_i is estimated of finding the vector points which enter the specified box labeled $i; p_i = \lim_{N \to \infty} N_i/N$. The generalized dimension is introduced by [1,18]:

$$D_q = \lim_{\epsilon \to 0} \frac{1}{q-1} \frac{\log\left(\sum^{n(\epsilon)} p_{\rm i}^q\right)}{\log \epsilon}, \tag{6.25}$$

where $-\infty < q < +\infty$. For $q = 0$, D_0 is the capacity dimension (6.10) since $\sum_i p_i^0 = n(\epsilon)$. By putting $q = 1, D_1$ is identical to information dimension (6.14). For $q = 2, D_2$ is the same as the correlation dimension (6.24) by:

$$D_2 = \lim_{\epsilon \to 0} \frac{\log\left(\sum^{n(\epsilon)} p_i^2\right)}{\log\ \epsilon}, \tag{6.26}$$

where $\sum p_i^2 = \sum p_i C_m^i(r) = \langle C_m^i(r) \rangle = C_m(r)$. For $q = 2, 3, \ldots, n$ generalized dimensions are associated with correlation integrals of triplets, quadruplets ... and n-tuplets of points on the attractor [18]. The generalized dimensions D_q will be described in more detail concerning the f(α) spectrum in section 6.3. It has been proved that $D_q > D_{q'}$ for any $q' > q > 0$ [18]. Therefore, an important relation is:

$$D_{\rm C} \geq D_{\rm I} \geq D_{\rm G}, \tag{6.27}$$

where equalities hold for the homogeneous set of uniformly distributed vector points in the Euclidean space.

6.2 Lyapunov spectrum

Very often, one is required to show clear evidence of deterministic chaos from an experimental time series. Lyapunov exponents are very useful quantities to characterize the chaotic behaviors. The Lyapunov exponents are the averaged rate of the exponential divergence (positive exponents) or convergence (negative exponents) of nearby orbits in phase space. In d-dimensional phase space, a strange attractor forms a complicated fractal structure due to stretching (divergent) and folding (convergent) properties of the nearby orbits, where there is a pair of averaged exponents $\lambda_1, \lambda, \ldots$, and λ_d (some are positive and others are zero and negative; normally $\lambda_1 > 0$, $\lambda_2 = 0$, $\lambda_3, \lambda_4, \ldots, \lambda_d < 0$ such that $\sum_i \lambda_i < 0$ in the dissipative systems). The spectrum $(\lambda_1, \lambda_2, \ldots, \lambda_d)$ of the Lyapunov exponents (sometimes simply called the Lyapunov spectrum) provides important information on the deterministic nature of observed irregular signals, estimations of the Lyapunov dimension and Kolmogorov entropy. Following the pioneering work by Benettin *et al.* [22] and Shimada and Nagashima [23], Wolf *et al.*

[24] and Sano and Sawada [25] have further developed the algorithms so that the Lyapunov spectrum can be obtained from the experimental time series of only one variable in any experimental system. Here, an algorithm by Sano and Sawada [25] is introduced and applied to the semiconductor chaos system.

From an experimental time series of some physical quantity $x(t)$, one can embed an attractor $\{x(t)\}$ in d-dimensional phase space using the time-delay method [11,12]:

$$\mathbf{x}(t) = \{x(t),\ x(t+t_{\rm d}), \ldots,\ x(t+(d-1)t_{\rm d})\}, \tag{6.28}$$

where $t_{\rm d}$ is the time delay and t is the time measured by a discrete time interval (or sampling time) Δt, namely, $t = (j-1)\Delta t$ $(j = 1, 2, \ldots)$. Denoting $\mathbf{x}(t)$ by $\mathbf{x}_j$, a set $\{\mathbf{x}_1,\ \mathbf{x}_2, \ldots\}$ forms an attractor in the phase space. Assume that the system behavior in an experimental system is characterized by some certain but unknown ordinary differential equations:

$$\dot{\mathbf{x}} = F(\mathbf{x}). \tag{6.29}$$

Once the system equations (6.29) are known, the observed attractor can be reproduced exactly by computer simulation (but it is almost impossible to deduce from the experimental time series). Here, consider the divergence rate between nearby orbits $\mathbf{x}$ and $\mathbf{x}'(\boldsymbol{\xi} = \mathbf{x} - \mathbf{x}')$ by linearizing (6.29) as [25]:

$$\dot{\xi} = \mathbf{T}(x(t)) \cdot \boldsymbol{\xi}, \tag{6.30}$$

where, $\mathbf{T} = (\partial F/\partial x)$ is the Jocobian matrix of F and $\boldsymbol{\xi}$ is a tangent vector in a tangent space. By integrating formally (6.30), one obtains:

$$\boldsymbol{\xi}(t) = \mathbf{A}^t \boldsymbol{\xi}(0), \tag{6.31}$$

where $\mathbf{A}^t$ is the linear operator (transitive map) which maps a tangent vector $\boldsymbol{\xi}(0)$ to $\xi(t)$. The mean exponential rate of divergence of λ is defined by:

$$\lambda(\mathbf{x}(0),\ \xi(0)) = \lim_{t\to\infty} \frac{1}{t} \frac{\|\boldsymbol{\xi}(t)\|}{\|\boldsymbol{\xi}(0)\|}, \tag{6.32}$$

where $\| \ldots \|$ denotes a norm with respect to some Riemannian metric. More precisely, one has the d Lyapunov exponents $\{\lambda_1,\ \lambda_2, \ldots,\ \lambda_d\}$ with respect to the orthogonal coordinates by a set of d-dimensional basis $\{\mathbf{e}_i\}$ $(i = 1, 2, \ldots, d)$, which are constructed along a fiducial trajectory $\mathbf{x}(t)$.

Each element $\lambda_i(\mathbf{x}(0))$ of the Lyapunov spectrum can be written formally by:

$$\lambda_i(\mathbf{x}(0)) = \lambda(\mathbf{x}(0), e_i), \tag{6.33}$$

where $i = 1, 2, \ldots, d$. The exponents can be ordered with their magnitudes $\lambda_1 > \lambda_2 > \ldots > \lambda_d$. With $d = 3$, one obtains the Lyapunov spectrum $(\lambda_1,\ \lambda_2,\ \lambda_3)$. The signs of the exponents are $(-,\ -,\ -)$ for a fixed point,

(0, −, −) for a limit cycle, (0, 0, −) for a two-torus and (+, 0, −) for a strange attractor [5].

The basis $\{\mathbf{e}_i\}$ are anchored to the fiducial trajectory and thus the orthogonal coordinates of $\{\mathbf{e}_i\}$ move and must rotate along the trajectory in a complex screwed manner, as shown in Fig. 6.15. The difficulty in tracing the orthogonal coordinate system is overcome by the Gram–Schmidt procedure in the comoving vector frame [22–25]:

$$\begin{aligned}
\mathbf{e}_{1'} &= \frac{\mathbf{e}_1}{\|\mathbf{e}_1\|}, \\
\mathbf{e}_{2'} &= \frac{\mathbf{e}_2 - \langle \mathbf{e}_2 \cdot \mathbf{e}_{1'} \rangle \, \mathbf{e}_{1'}}{\|\mathbf{e}_2 - \langle \mathbf{e}_2 \cdot \mathbf{e}_{1'} \rangle \, \mathbf{e}_{1'}\|}, \\
&\dots \\
\mathbf{e}_{d'} &= \frac{\mathbf{e}_d - \langle \mathbf{e}_d \cdot \mathbf{e}'_{d-1} \rangle \, \mathbf{e}'_{d-1} - \dots \langle \mathbf{e}_d \cdot \mathbf{e}'_1 \rangle \, \mathbf{e}'_1}{\|\mathbf{e}_d - \langle \mathbf{e}_d \cdot \mathbf{e}'_{d-1} \rangle \, \mathbf{e}'_{d-1} - \dots \langle \mathbf{e}_d \cdot \mathbf{e}'_1 \rangle \, \mathbf{e}'_1\|},
\end{aligned} \tag{6.34}$$

where $\langle \cdot \rangle$ denotes the inner vector product.

In order to carry out the algorithm presented by Sano and Sawada [25], one must calculate (6.33) in a realistic manner, as follows. Let $\mathbf{x}_j$ be a vector point on the fiducial trajectory at time $t = (j-1)\,\Delta t$, and $\mathbf{x}_{k_i}$ $(i = 1, 2, \dots)$ be any vector points among a set of data points $\{\mathbf{x}_j\}$. Find all of the orbital points $\mathbf{x}_{k_i}$ $(i = 1, 2, \dots, N)$ that fall inside a small hypersphere of radius ϵ centered at $\mathbf{x}_j$. This results in a set of the displacement vector $\{\mathbf{y}^i\}$ between $\mathbf{x}_{k_i}$ and $\mathbf{x}_j$:

$$\{\mathbf{y}^i\} = \{\mathbf{x}_{k_i} - \mathbf{x}_j | \, \|\mathbf{x}_{k_i} - \mathbf{x}_j\| \le \epsilon\}, \tag{6.35}$$

where $\| \quad \|$ denotes the Euclidean norm. After the evolution of a time interval $\tau = m\,\Delta t$, the orbit point will proceed to $\mathbf{x}_{j+m}$ and neighboring

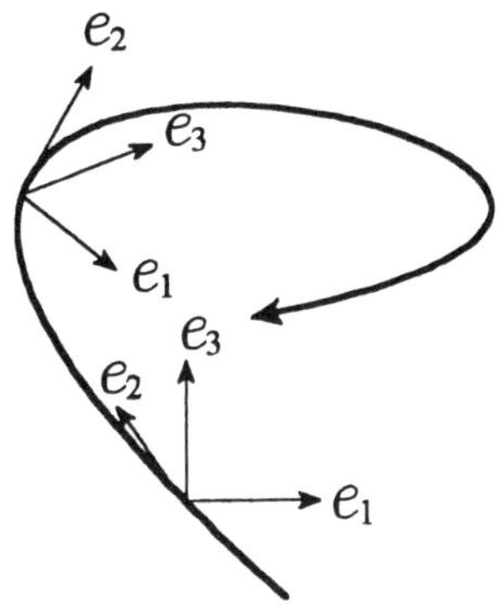

Figure 6.15 Successive construction of the basis $\{e_i\}$ anchored to the fiducial trajectory.

points $\{\mathbf{x}_{k_i}\}$ to $\{x_{k_i+m}\}$ [25]. Correspondingly, as shown schematically in Fig. 6.15, a set $\{\mathbf{y}^i\}$ of the displacement vectors in (6.30) is mapped to a set $\{\mathbf{z}^i\}$ by:

$$\{\mathbf{z}^i\} = \{\mathbf{x}_{k_i+m} - \mathbf{x}_{j+m}|\ \|\mathbf{x}_{k_i} - \mathbf{x}_j\| \leq \epsilon\}. \tag{6.36}$$

The map $\mathbf{A}^t$ in (6.31) is approximated by the linearized flow map (matrix) $\mathbf{A}_j$ as:

$$\mathbf{z}^i = \mathbf{A}_j \mathbf{y}^i. \tag{6.37}$$

After Sano and Sawada [25], the matrix $\mathbf{A}_j$ can be obtained by the optimal estimation using the data sets $\{\mathbf{z}^i\}$ and $\{\mathbf{y}^i\}$. By the least-square-error algorithm, the average of the squared error norm between $\{\mathbf{z}^i\}$ and $\mathbf{A}_j\mathbf{y}^i$ is minimized as follows:

$$\min_{\mathbf{A}_j} S = \min_{\mathbf{A}_j} \frac{1}{N} \sum_{i=1}^{N} \|\mathbf{z}^i - \mathbf{A}_j \mathbf{y}^i\|^2. \tag{6.38}$$

By taking variations of S with respect to the (k, l) component a_{kl} of the $d \times d$ matrix $\mathbf{A}_j$ to be zero, $\partial S/\partial a_{kl} = 0$, the problem of the least-square-error algorithm can be easily converted to the following expression:

$$\mathbf{A}_j\mathbf{V} = \mathbf{C}, \quad (\mathbf{V})_{k\ell} = \frac{1}{N}\sum_{i=1}^{N} y^{ik} y^{i\ell},$$
$$(\mathbf{C})_{k\ell} = \frac{1}{N}\sum_{i=1}^{N} z^{ik} y^{i\ell}, \tag{6.39}$$

where $\mathbf{V}$ and $\mathbf{C}$ are $d \times d$ matrices, and y^{ik} and z^{ik} are the k components of vectors $\mathbf{y}^i$ and $\mathbf{z}^i$, respectively [25]. The matrices $\mathbf{V}$ and $\mathbf{C}$ are constructed from the experimental data, and $\mathbf{A}_j$ is obtained by $\mathbf{A}_j = \mathbf{C}\ \mathbf{V}^{-1}$. If the total number of vector points which fall inside the ϵ-ball centered at $\mathbf{x}_j$ is larger than the embedded dimension, a solution is found for $a_{k\ell}$ without any degeneracy. Finally, the Lyapunov exponents can be computed by:

$$\lambda_i = \lim_{n\to\infty} \frac{1}{n\tau} \sum_{j=1}^{n} \ln \|\mathbf{A}_j \mathbf{e}_i^j\|, \tag{6.40}$$

where $i = 1, 2, \ldots, d$.

For the actual computation, some technical aspects on the computing program are provided. The programs may be composed of five subprograms: (a) reading data, (b) vector construction of $\mathbf{y}$ and $\mathbf{z}$, (c) matrices $\mathbf{V}$ and $\mathbf{C}$, (d) matrix $\mathbf{A}_j$ (by the Gauss–Jordan algorithm), (e) the Gram–Schmidt procedure, and (f) Lyapunov exponents and a set of basis vectors $\{\mathbf{e}_i^j\}$ $(i = 1, 2, \ldots, d)$. Procedures (e) and (f) are now explained. For the set of basis vectors $\{\mathbf{e}_1^j, \mathbf{e}_2^j, \ldots, \mathbf{e}_d^j\}$ of the tangent space at $\mathbf{x}_j$, one puts:

$$\mathbf{e}_i^j = \{e_{1i},\ e_{2i}, \ldots e_{ii}, \ldots e_{di}\}, \qquad i = 1, 2, \ldots, d, \tag{6.41}$$

where $\|\mathbf{e}_i^j\| = 1$. Each of the basis elements is newly constructed at $\mathbf{x}_j$ by a comoving vector frame, using the Gram–Schmidt procedure (6.34). Note that (6.41) has been decomposed to m components $e_{i\ell}$ ($\ell = 1, 2, \ldots, d$). Suppose that $\mathbf{A}_j$ and $\{\mathbf{e}_i^j\}$ at $\mathbf{x}_j$ have already been computed. By operating with the matrix $\mathbf{A}_j$ on $\{\mathbf{e}_i^j\}$, then

$$\{\tilde{\mathbf{e}}_i^{j+m}\} = \mathbf{A}_j\{\mathbf{e}_i^i\}, \tag{6.42}$$

after the evolution of τ. By the Gram–Schmidt procedure (6.34), one can first reconstruct $\{\mathbf{e}_i\}$ without normalizing by the Euclidean norms, but maintaining mutual orthogonality, as follows:

$$\begin{aligned}
\tilde{\tilde{\mathbf{e}}}_1 &= \tilde{\mathbf{e}}_1, \\
\tilde{\tilde{\mathbf{e}}}_2 &= \tilde{\mathbf{e}}_2 - \langle \tilde{\mathbf{e}}_2 \cdot \tilde{\tilde{\mathbf{e}}}_1 \rangle \tilde{\tilde{\mathbf{e}}}_1, \\
&\cdots \\
\tilde{\tilde{\mathbf{e}}}_i &= \tilde{\mathbf{e}}_i - \sum_{m=1}^{i-1} \langle \tilde{\mathbf{e}}_i \cdot \tilde{\tilde{\mathbf{e}}}_m \rangle \tilde{\tilde{\mathbf{e}}}_m, \\
&\cdots
\end{aligned} \tag{6.43}$$

Taking the Euclidean norm $\|\tilde{\tilde{\mathbf{e}}}_i\|$, one puts:

$$\lambda_i = (1/n\tau) \sum_{j=1}^{n} \ln\ \|\tilde{\tilde{\mathbf{e}}}_i^{j+m}\|, \tag{6.44}$$

where $i = 1, 2, \ldots, d$. Note that the above equation is equivalent to formula (6.40). After calculating $\ln\ \|\tilde{\tilde{\mathbf{e}}}_i^{j+m}\|$, (6.43) is normalized (6.43) and a new orthonormal set of basis vectors $\{\mathbf{e}_i\}$ is obtained at the orbital point $\mathbf{x}_{j+m}$: $\mathbf{x}_{j+m}$:

$$\mathbf{e}_i^{j+m} = \tilde{\tilde{\mathbf{e}}}_i / \|\tilde{\tilde{\mathbf{e}}}_i\|, \quad i = 1, 2, \ldots, d, \tag{6.45}$$

where $\mathbf{e}_i \cdot \mathbf{e}_i = 1$ and $\mathbf{e}_i \cdot \mathbf{e}_j = 0$ for any $i \neq j$. Orbital points are found which locate inside the ϵ-ball centered at $\mathbf{x}_{j+m}$. By repeating the whole process after each evolution of τ, using five subprograms and summing up the exponents (6.44), one can reach the asymptotic values of (6.40).

The algorithm is of great advantage since it can deal with any experimental data. If ordinary differential equations of the related model system are available, the exponents can even be deduced from the computed data series. An example of the Lyapunov spectrum of the Lorenz system has been given by Sano and Sawada [25] in Fig. 6.16. It is shown that the convergence to the asymptotic values is relatively fast. From the Lyapunov exponents, the Lyapunov (or Kaplan–Yorke) dimension D_{KY} and the

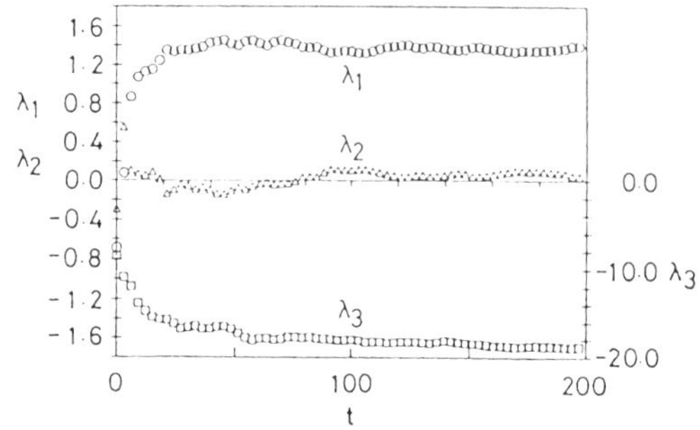

Figure 6.16 Stability of the Lyapunov spectrum of the Lorenz system for a special choice of parameters [25].

Kolmogorov entropy K are easily estimated (see 6.19)) by:

$$D_{\mathrm{KY}} = j + \frac{\sum_{i=1}^{j} \lambda_i}{\|\lambda_{j+1}\|}, \qquad K \sim \lambda_1. \tag{6.46}$$

In Fig. 6.16, $\lambda_1 = 1.37 \pm 0.08$, $\lambda_2 = -0.02 \pm 0.09$ and $\lambda_3 = -15.2 \pm 2.1$, thereby $D_{\mathrm{KY}} = 2.09$ and $K = 1.37$ [25]. For the computation of the exponents, one can choose arbitrarily the delay time t_{d} and the evolution time τ. In order to solve (6.39) from a small amount of experimental data with the total number of orbital points $N_{\mathrm{tot}} \sim 10^3 - 10^4$, it is recommended to settle the embedding dimension as a relatively small value ($d \leq 4$), otherwise a large amount of data points and a large d [25] are required. It is not necessary to search all of the points contained in the ϵ-ball, since this is often time consuming, but an upper limit is set to the number N of points included [25]. In addition, the number of points inside the ϵ-ball may not satisfy $N > d$ (embedded dimension). In that case, the solution for $\mathbf{A}_j$ is never found and the point $\mathbf{x}_j$ is skipped; going to the next point $\mathbf{x}_{j+n}$.

In the semiconductor chaos system, some examples are given in p-type Ge by Peinke *et al.* [26] and in n-type GaAs by Aoki [21] and Aoki *et al.* [27]. Peinke *et al.* [26] analyzed two types of chaotic oscillatory states A and B (Fig. 6.17). They evaluated $\lambda_1 = 0.095 \pm 0.005$, $\lambda_2 = 0.003 \pm 0.005$ and $\lambda_3 = -0.72 \pm 0.02$ for state A, and $\lambda_1 = 0.159 \pm 0.005$, $\lambda_2 = 0.076 \pm 0.005$, $\lambda_3 = -0.021 \pm 0.005$ and $\lambda_4 = -0.77 \pm 0.03$ for state B. From the numerical evaluations using several different embedding dimensions up to 11, they remarked that state A in Fig. 6.17a is ordinary chaos possessing one positive exponent λ_1 and state B is hyperchaos possessing two positive exponents λ_1 and λ_2. From the exponents, the Kaplan–York dimension was estimated to be $D_{\mathrm{KY}} = 2.1 \pm 0.2$ for state A and $D_{\mathrm{KY}} = 3.3 \pm 0.1$ for state B [26]. The values seem to be quite reasonable since the ordinary chaos (low-dimensional chaos) could be modeled by three independent variables

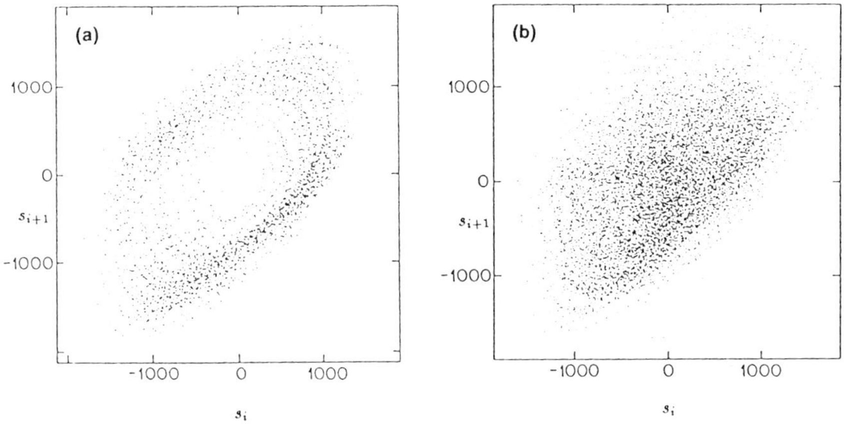

Figure 6.17 Time series of (a) the chaotic and (b) the hyperchaotic state of the p-Ge semiconductor. Points with coordinates (s_{i+1}, s_i) are plotted for parts of two characteristic time series [26].

(degrees of freedom 3) in the autonomous case and thus, topologically, $1 < D_{KY} < 3$ (note: the fractal dimension of toric chaos is close to but larger than the dimension 2.0 of torus; the fractal dimension of the fully developed ordinary chaos may take its value between 2.0 and 3.0). Hyperchaos is rather exceptional among many dynamical systems, and little is known about the transition from ordinary chaos to hyperchaos. Since there are two principal axes of divergence ($\lambda_1 > 0, \lambda_2 > 0$), the topology of the strange attractor is hardly imagined. It is interesting to note that Rössler-hyperchaos is obtained numerically by modifying the Rössler equation [28] and adding an independent variable (degrees of freedom 4) [28]. In semiconductor systems, the current instabilities are often spatially extended, where the degrees of freedom are, in principle, infinite. In particular, in experiments on the impact-ionization avalanche, Peinke *et al.* [26, 29] have shown complex dynamical behaviors of filamentary currents. From measurements of the Lyapunov spectrum, they suggested that it is capable of distinguishing the hierarchical order inherent to chaos.

In the periodically driven chaos system using n-type GaAs, the Lyapunov spectrum indicates that the system behavior is relatively low-dimensional in a wide range of experimental parameters. In some limited range, however, the Kaplan–Yorke dimension is estimated to be as large as 3.0. For the chaotic current oscillations in Fig. 6.18, where $V_0 = 3.252$, $V_{ac} = 185$ mV and $f_0 = 1.85$ MHz, the initial evolution of the Lyapunov spectrum is typically obtained (Fig. 6.19). The exponents are $\lambda_1 = 1.1 \pm 0.1$, $\lambda_2 = 0.05 \pm 0.15$, $\lambda_3 = -1.2 \pm 0.1$ and $\lambda_4 = -5.35 \pm 0.35$, from which the Kaplan–Yorke dimension is estimated to be $D_{KY} = 2.93 \pm 0.16$. This value is

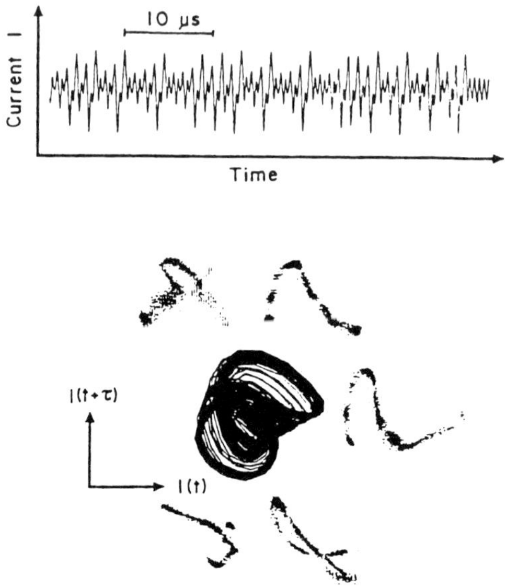

Figure 6.18 Chaotic current oscillation and the Poincaré section, where $V_0 = 3.252$ V, $V_{ac} = 185$ mV and $f_0 = 1.85$ MHz.

unexpectedly large, while a smaller value lying close to 2.0 may be expected from the Poincaré section in Fig. 6.18. Under the influence of external noise, the Kaplan–Yorke dimension might be overestimated for the following reason, as explained by Sano and Sawada [25]: the possibility of measuring the negative exponents depends on the signal-to-noise ratio of the experimental data. If contraction of the phase-space volume is so strong that the thickness of the attractor becomes smaller than the resolution of observation, one can never have information concerning the direction of contraction. In the worst case, experimental noise may give spurious zero exponents for these directions, because, on average, the perturbation caused by noise is neither contracting nor expanding. Using their algorithm, Sano and Sawada [25] have obtained for the Rössler equations $\lambda_1 = 0.073 \pm 0.004$, $\lambda_2 = 0.0003 \pm 0.002$ and $\lambda_3 = -0.7 \pm 0.3 (D_{DK} = 2.1)$, while the numerical values (for a computing program see, e.g., a paper by Wolf *et al.* [24]) are $\lambda_1 = 0.069 \pm 0.003$, $\lambda_2 = -0.0002 \pm 0.0002$ and $\lambda_3 = -4.987 \pm 0.002$ $(D_{KY} = 2.01)$. With the underestimated value of $\lambda_3 = -4.987$, the dimension is overestimated as $D_{KY} = 2.1$, which is larger than the reliable value 2.01. However, as a rough estimate, the algorithm has given reliable values whenever $\lambda_1/|\lambda_3| << 1$ (note: $D_{KY} \sim 2.0$ anyway). When $\lambda_1/|\lambda_3| \sim 1$, the Kaplan–Yorke dimension may take its value between 2.0 and 3.0, depending on λ_3 (e.g., $D_{KY} = 3.0$ with $\lambda_1/|\lambda_3| = 1.0; D_{KY} = 2.5$ with

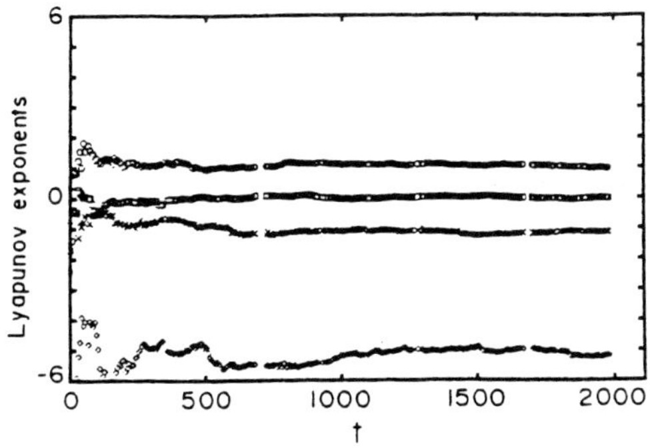

Figure 6.19 Initial evolution of the Lyapunov spectrum for the chaotic oscillation in Fig. 6.18. $\lambda_1, \lambda_2, \lambda_3$, and λ_4 from above.

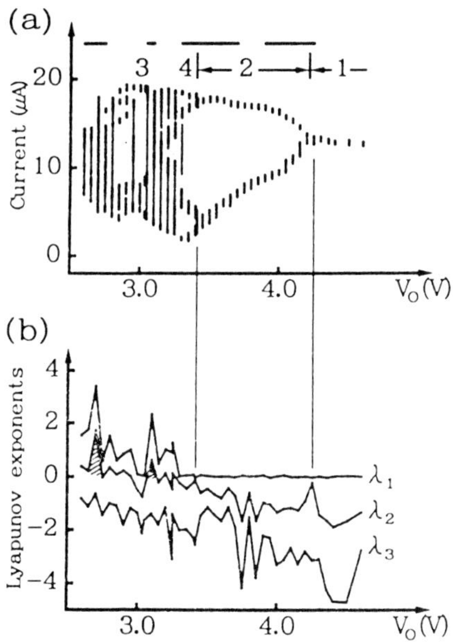

Figure 6.20 (a) Observed bifurcation diagram and (b) Lyapunov spectrum as a function of V_0, $V_{ac} = 170$ mV and $f_0 = 2.33$ MHz. Numbers indicated show periods. Horizontal bars indicate regions in which radio signals are clearly detected. In (b), λ_4 is omitted [27].

$\lambda_1/|\lambda_3| = 0.5$. Therefore, in the case in Fig. 6.19, the Kaplan–Yorke dimension may have been overestimated as a result of external noise. In Fig. 6.20, the bifurcation diagram and the corresponding Lyapunov spectrum are shown as a function of the dc bias V_0, where $V_{ac} = 170$ mV and $f_0 = 2.33$ MHz [27]. The bifurcation diagram was obtained by measuring the dynamic ranges of the currents at a fixed voltage on the $I - V$ phase portrait for each V_0. The embedding dimension was $d = 4$, but λ_4 has been omitted in Fig. 6.20. By decreasing V_0, period doubling bifurcation to chaos is typically observed, where period 8 has not been resolved in Fig. 6.20a. In the periodic regime, the largest Lyapunov exponent is zero, while $\lambda_2, \lambda_3 < 0$ ($D_{KY} = 1.0$). At the bifurcation points, as indicated by the vertical lines in Fig. 6.20, λ_2 becomes critically zero. Near $V_0 = 3.1$ V, there is a narrow window of period 3, where $\lambda_1 = 0$. In the chaotic regime, typical values are $\lambda_1 > 0, \lambda_2 \sim 0$ and $\lambda_3 < 0$, i.e., $(+, 0, -)$, but with some exceptions, $\lambda_2 > 0$ near $V_0 \sim 2.7$ V and 3.15 V (hatched region). As seen from Fig. 6.20b, the Kaplan–Yorke dimension fluctuates between 2.0 and 3.0 depending on V_0. In the hatched regions, the dimension may be higher than 3.0. The horizontal bars in Fig. 6.20a indicate the regions where the external noise effect (deterministic noise amplification [21, 27, 30]; see section 7.2) is remarkable.

Details of the noise effect are given in section 7.2. Some numerical aspects expected from the impact-ionization model are discussed in brief. By adding the external noise $\xi(t)$ to the electric field in the model simulation, the bifurcation diagram becomes fuzzy (Fig. 6.21) [27]. The inset of Fig. 6.21 shows a noisy period 4 near the bifurcation point at $E_0 = 15.8015$. In the bifurcation diagram, the effect of noise is remarkable in the lower part of the pitchfork, especially at the bifurcation point of period 4. That is, when the orbit is reinjected in the NDR region, it shows local instability and thus sensitive dependence on the external noise (see the inset figure). The bifurcation diagram in Fig. 6.20a is quite similar to Fig. 6.21 in the region of period doubling, especially noting that the noise effect is significant in the lower part of the pitchfork in Fig. 6.20a.

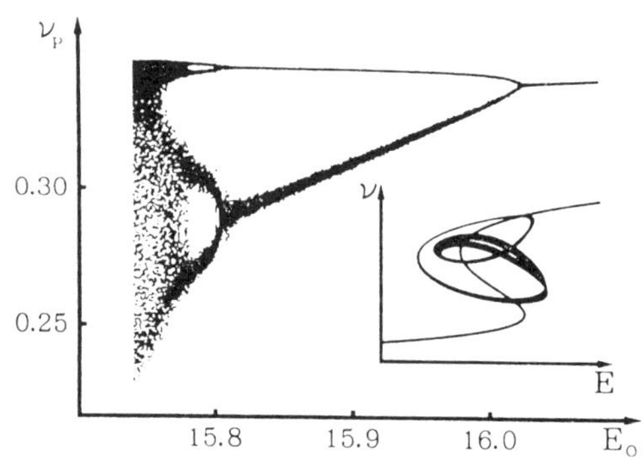

Figure 6.21 Fuzzy bifurcation diagram of period doubling under the external noise $\xi(t)$ in the model simulation. Parameter values are $E_{ac} = 4.5$, $f_0 = 100/136$. $\langle \xi \rangle = 0$ and $\langle \xi^2 \rangle = 0.00021$ [27].

6.3 $f(\alpha)$ spectrum

From the Lyapunov spectrum, one can recognize the deterministic nature of chaos and its low dimensionality (found from the Lyapunov dimension). In terms of the fractal dimensions (generalized dimensions D_q in section 6.1), one learns the degree of strangeness of chaos. More precisely, the local fractal dimension distributes over a strange attractor with different values depending on the local region of the attractor. The local dimension may be called singularity [5, 31]. Namely, the strange attractor has a multifractal nature [5, 26, 31–36]. This section describes the multifractal spectrum or $f(\alpha)$ spectrum introduced by Halsey *et al.* [31] and Jensen *et al.* [36] in order to look a global view of multifractal attractors. Characterizations by the $f(\alpha)$ spectrum provide strong evidence of the universality which appears at the onset of chaos (2^∞ cycle of period doubling, onset of chaos via the Ruelle–Takens–Newhouse scenario, onset of chaos in the circle maps with golden-mean winding number, etc.).

Recall the definition of the generalized dimension D_q in

$$D_q = \lim_{\ell \to 0} \frac{1}{q-1} \frac{\ln \left(\sum^{n(\ell)} p_i^q\right)}{\ln \ \ell}, \tag{6.47}$$

where ϵ in (6.25) has been replaced by ℓ. The dimension D_q is obtained by calculating the probability p_i that the orbital points locate inside a box i of side l and by taking the qth moment of p_i. The phase space may be partioned into $n(\ell)$ boxes (see Fig. 6.9) and the number of orbital points N_i included in box i is counted. Then, the probability p_i is given by:

$$p_i = \lim_{N \to \infty} N_i/N, \tag{6.48}$$

where N is the total number of orbital points in the phase space. The probability may increase by increasing ℓ with a scaling law:

$$p_i = \ell^{\alpha_i}, \tag{6.49}$$

where the exponent α_i gives a kind of dimension (local dimension near box i) following the definition in section 6.1. Suppose that the values of the local dimension α_i distribute over the whole attractor with a smooth function $\rho(\alpha)$. Let $n(\alpha, \ell)$ be the number of boxes of side ℓ with the local dimension $\alpha \epsilon [\alpha, \alpha + \mathrm{d}\alpha]$. Denoting the fractal dimension $f(\alpha)$ (fractal dimension of a set of the boxes which are specified by the local dimension, and labeled by the singularity α rather than by index i) [5, 31, 36] and recalling the scaling law $n \propto \ell^{-D}$ in section 6.1, then

$$n(\alpha, \ell) = \mathrm{d}\alpha \rho(\alpha) \ell^{-f(\alpha)}, \tag{6.50}$$

where one must multiply $d\alpha$ and the probability measure $\rho(\alpha)$. Choosing the boxes labeled by α (the total number is given by (6.50)), the probability p_α of each box is given by ℓ^{α} instead of (6.49)). By taking the qth moment and multiplying (6.50) the probability $p(\alpha, \ell)$ is given by [5, 31, 32]:

$$p_q(\alpha, \ell) = d\alpha\rho(\alpha)\ell^{-f(\alpha)}\ell^{q\alpha}. \tag{6.51}$$

By taking a limit $\ell \to 0$ and integrating over α (over the whole attractor), one has:

$$\chi(q) = \int \mathrm{d}\alpha\rho(\alpha)\ell^{-\mathrm{f}(\alpha)}\ell^{q\alpha}. \tag{6.52}$$

Note that $\chi(q)$ is equivalent to the sum of the qth moment p_i^q:

$$\chi(q) = \sum_i p_i^q. \tag{6.53}$$

For $q = 0$, it is clear that $p_q(\alpha, \ell) = n(\alpha, \ell)$, and $\chi(q) = n(\ell)$. The generalized dimension (6.47) is therefore rewritten as:

$$D_q = \lim_{\ell \to 0} \frac{1}{q-1} \frac{\ln\ \chi(q)}{\ln\ \ell}. \tag{6.54}$$

For small ℓ, $\chi(q)$ in (6.52) is dominated by the smallest value of $q\alpha - f(\alpha)$ if $\rho(\alpha) \neq 0$ [31]. The extreme conditions are given by [5, 31, 36]:

$$\frac{\mathrm{d}}{\mathrm{d}\alpha}[q\alpha - f(\alpha)] = 0, \tag{6.55}$$

$$\frac{\mathrm{d}^2}{\mathrm{d}\alpha^2}[q\alpha - f(\alpha)] < 0. \tag{6.56}$$

Since α depends on q [31], the smallest value of α which satisfies (6.55) and (6.56) is denoted by $\alpha(q)$. From (6.52), (6.54) [31, 32]:

$$D_q = \frac{1}{q-1}[q\alpha(q) - f(\alpha(q))]. \tag{6.57}$$

By the Legendre transform of $(q-1)D_q$, $\alpha(q)$ is given by:

$$\alpha(q) = \frac{\mathrm{d}}{\mathrm{d}q}\left[(q-1)D_q\right], \tag{6.58}$$

and from (6.57),

$$f(\alpha) = \alpha(q)q - (q-1)D_q. \tag{6.59}$$

By calculating D_q from a set of experimental data, one can obtain $\alpha(q)$ from (6.58) and $f(\alpha(q))$ from (6.59). By the $f(\alpha(q))$ spectrum, it is found that points in the boxes with a singularity $\alpha(q)$ lie on a support of the set of the fractal dimension $f(\alpha)$. It is apparent from (6.59) that

$$f(\alpha(0)) = D_0 \text{ (capacity dimension for } q = 0\text{)},$$
$$\text{and } f(\alpha(1)) = \alpha(1) = D_1 \text{ (information dimension for } q = 1\text{)},$$

where $f'(\alpha(0)) = 0$ and $f'(\alpha(1)) = 1$.

The $f(\alpha)$ spectrum can be obtained analytically for some simple fractal geometry such as Cantor set [31]. For the inhomogeneous (two-scale) Cantorian set (Fig. 6.22), one puts the probability p_1 and p_2 ($p_1 > p_2$) for the line segment ℓ_1 and $\ell_2(\ell_1 < \ell_2)$ at the first stage of the initiator. After n steps, one of the segments has a length $\ell_i = \ell_1^m \ell_2^{n-m}$ and the probability $p_i = p_1^m p_2^{n-m}$. Since $p_i = \ell^{\alpha_i}$ (see (6.49)), one obtains

$$p_1^m p_2^{n-m} = (\ell_1^m \ell_2^{n-m})^{\alpha}, \tag{6.60}$$

and thereby the singularity

$$\alpha = \alpha(m, n) = \frac{(m/n) \ln p_1 + (1 - m/n) \ln p_2}{(m/n) \ln \ell_1 + (1 - m/n) \ln \ell_2}, \tag{6.61}$$

where $0 < m < n$. The total number of segments is 2^n, among which ${}_nC_m$ segments have the same singularity (6.61). It is apparent that the singularity distributes over a certain range depending on n and m. From the original assumption of the scaling (6.50),

$${}_nC_m \sim (\ell_1^m \ell_2^{n-m})^{-f(\alpha)} \tag{6.62}$$

For $n >> 1$, one can apply the Stirling formula and obtain

$$f(\alpha) = \frac{(m/n) \ln (m/n) + (1 - m/n) \ln (1 - m/n)}{(m/n) \ln \ell_1 + (1 - m/n) \ln \ell_2}. \tag{6.63}$$

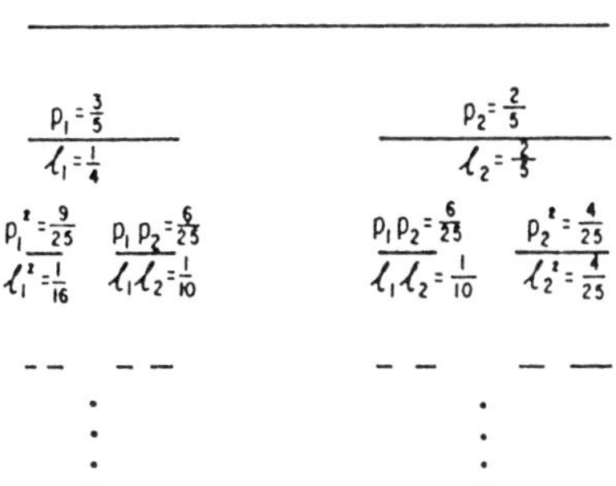

Figure 6.22 A Cantor-set construction with two rescalings $\ell_1 = 0.25$ and $\ell_2 = 0.4$ and respective measure rescalings $p_1 = 0.6$ and $p_2 = 0.4$. The division of the set continues self-similarly [31].

Thus, from (6.61), (6.63), one can plot $f(\alpha)$ vs α as a function of $b = m/n$, as given in Fig. 6.23, where $\ell_1 = 0.25$, $\ell_2 = 0.4$ and $p_1 = 0.6, p_2 = 0.4$ [31].

The $f(\alpha)$ spectrum can also be solved starting from the partition function [31] which is defined by:

$$\Gamma(q, \tau, \ell) = \sum_i \frac{p_i^q}{\ell_i^\tau}, \tag{6.64}$$

where $\tau = (q-1)D_q$, and ℓ is the maximum length of the segments $(\ell_i \leq \ell)$. By putting $p_i = p_1^m p_2^{n-m}$ and $\ell_i = \ell_1^m \ell_2^{n-m}$, one obtains [31]

$$\Gamma(q, \tau, \ell) = \sum_m {}_nC_m p_1^m p_2^{n-m} (\ell_1^m \ell_2^{n-m})^{-\tau} = 1. \tag{6.65}$$

After Halsey *et al.* [31], the final results are given by:

$$\tau = \frac{\ln\ (n/m - 1) + q\ \ln\ (p_1/p_2)}{\ln\ (\ell_1/\ell_2)}, \tag{6.66}$$

$$\ln(n/m)\ln(\ell_1/\ell_2) - \ln(n/m - 1)\ \ln \ell_1 = q(\ln p_1\ \ln \ell_2 - \ln p_2\ \ln \ell_1). \tag{6.67}$$

For any given q, a value of n/m is obtained from (6.67) and thus D_q is obtained from (6.66); the D_q curve is plotted in Fig. 6.24. Using (6.58) and (6.59), the $f(\alpha)$ spectrum is drawn in Fig. 6.23. The densest region of the segments in Fig. 6.22 is located on the left-hand edge $(m = n)$, where the (probability) measure is most concentrated, while the most rarefied region is on the right-hand edge $(m = 0)$. The minimum value of the singularity $\alpha_{\min}$

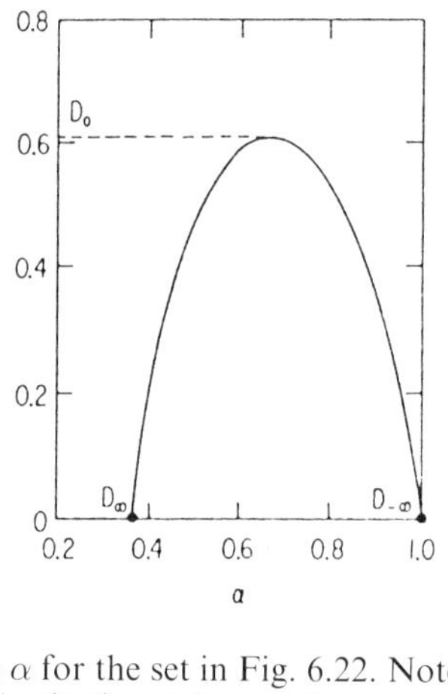

Figure 6.23 The plot of f vs α for the set in Fig. 6.22. Note that $f = 0$ corresponds to α values $D_{-\infty} = \ln(0.4)/\ln\ (0.4) = 1.0$ and $D_{+\infty} = \ln\ (0.6)/\ln\ (0.25) = 0.3684$ [31].

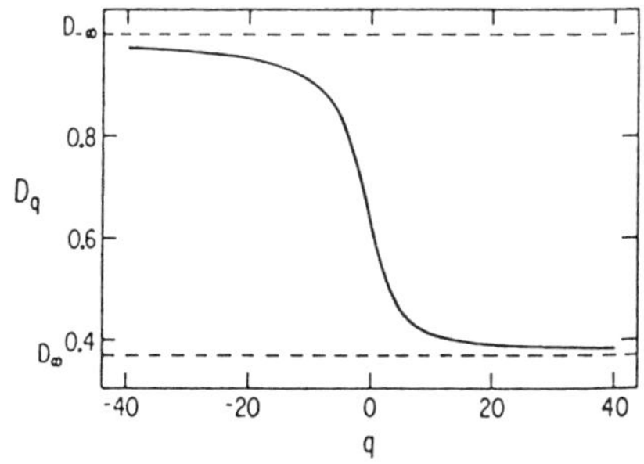

Figure 6.24 D_q plotted vs q for the two-scale Cantor set of Fig. 6.22 [31].

corresponding to the dense region is given by $\alpha_{\min} = D_\infty = \ln\, p_1 / \ln \ell_1 = 0.3684$, while $\alpha_{\max} = D_{-\infty} = \ln\, p_2 / \ln\, \ell_2 = 1$ for the rarefied region (see (6.61); also take a limit $q \to \pm\infty$ in (6.59). Apparently, $f(\alpha(\pm\infty)) = 0$ from (6.63).

An example of a two-scale Cantor set provides a helpful guideline to interpret the multifractal spectrum in experimental systems. Imagine that the Cantor set with a limit $n \to \infty$ is equivalent to a set of iterated points in the map systems or a Poincaré section of the orbits in differential dynamical systems. The Cantor dust is disjointed and singular everywhere, each point having the probability $\lim_{n\to\infty} {}_mC_n p_1^m p_2^{n-m}$. At the onset of chaos after successive period doubling, the set of the iterated points in the logistic map or in the semiconductor model forms a Cantor-like structure, where the iterated points are singular everywhere, as found from the invariant probability measure obtained in the semiconductor model (see Fig. 6.10). In Fig. 6.25, the $f(\alpha)$ spectrum is shown for the 2^{11} cycle element, which is very close to the Feigenbaum attractor in the logistic map [31]. The values D_∞ and $D_{-\infty}$ are found by using the Feigenbaum constant $\alpha_{\rm PD}$ (simply denoted by α in chapter 4) as [31]:

$$D_{-\infty} = \frac{\ln\ 2}{\ln\ \alpha_{\rm PD}} = 0.7551\ldots, \tag{6.68}$$

$$D_\infty = \frac{\ln\ 2}{\ln\ \alpha_{\rm PD}^2} = 0.37775\ldots. \tag{6.69}$$

It is expected that at the onset of chaos after the period doubling cascade many dynamical systems may exhibit the $f(\alpha)$ spectrum which is almost identical to that in Fig. 6.25.

From the experimental data, one can calculate the $f(\alpha)$ spectrum as follows. By the time-delay method, a set of the vector points $\{\mathbf{x}_i\}$ $(i = 1, 2, \ldots, N\}$ is first constructed in the m-dimensional phase space. Here, an algorithm for the correlation $C_i(\ell)$ integral by Grassberger and Procaccia

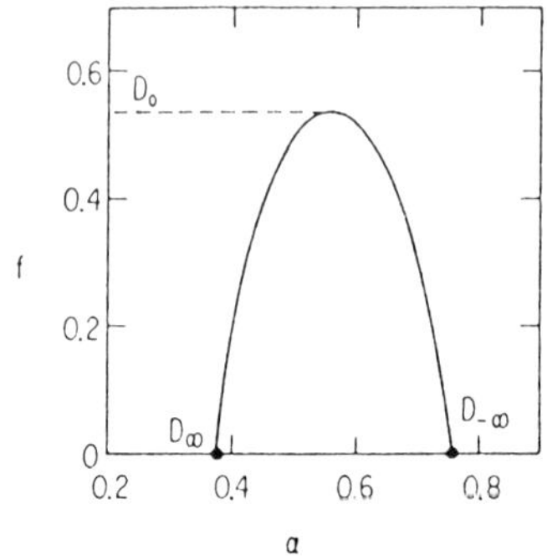

Figure 6.25 The function $f(\alpha)$ for the period doubling attractor of the 2^{11} cycle in a logistic map [31].

[14] and Hentschel and Procaccia [18] is adopted. Picking M points randomly, the correlation integral around each point is taken as:

$$C_i(\ell) = \frac{1}{N}\sum_{j=1}^{N} H(\ell - \|\mathbf{x}_i - \mathbf{x}_j\|), \tag{6.70}$$

where $i = 1, 2, \ldots, M$, ℓ is the radius of a hypersphere centered at $\mathbf{x}_i$, $\|\mathbf{x}_i - \mathbf{x}_j\|$ is the Euclidean norm and H is the Heaviside function (see (6.20)). The above method is similar to the algorithm to find the point-wise (correlation) dimension [14]. Since $C_i(\ell)$ is the probability $p_i(\ell)$ that the vector points enter inside the hypersphere, the partition function $\Gamma(q, \ell) = \Gamma_q$ is given by:

$$\Gamma(q, \ell) = \Gamma_q = \sum_i p_i^q = \sum_i p_i(\ell) C_i^{q-1}(\ell) = \langle C_i^{q-1}(\ell) \rangle, \tag{6.71}$$

where $\langle\cdot\rangle$ is an average taken over M vector points. Then the generalized dimension (6.54) is obtained by:

$$(q-1)D_q = \frac{\ln \langle C_i^{q-1}(\ell)\rangle}{\ln \ell}; \tag{6.72}$$

note also the power law $\Gamma(q, \ell) \propto \ell^{(q-1)D_q}$. One can measure a slope $(q-1)D_{\text{q}}$ from the graph of $\ln\langle C_1^{q-1}(\ell)\rangle$ vs $\ln\,\ell$. By the Legendre transform (6.58) and (6.59), one obtains the $f(\alpha)$ spectrum. In the experiments, the external noise influences strongly the spectrum for $q < 0$ [35], because the rarefied region of the vector points is more rarefied by the random kicks of external noise. When the hypersphere includes only a few vector points, one cannot obtain the spectrum satisfactorily (for $q < 0$ and small p_i, p_i^q may lead to a considerable error). By abandoning such a hypersphere, the whole spectrum may be recovered.

Figure 6.26 shows the results of the $f(\alpha)$ spectrum for the period doubling cascade in an experiment of driven chaos in n-GaAs (see Fig. 5.57 [37]), where the experimental parameters were $V_0 = 1.038$ V, $f_0 = 2.2$ MHz, $V_{ac} = 112$ mV (in Fig. 6.26a), and $V_{ac} = 114$ mV (Fig. 6.26b). The total number of vector points was $N = 10^4$, the sampling number $M = 10^3$, the embedding dimension $m = 4$, and the time delay $\tau = 0.25$ μs. The capacity dimension is estimated to be 1.6 for both spectra. The spectrum in Fig. 6.26b shows some broadening, which is very consistent with the fact that the waveform and the Fourier power spectrum for $V_{ac} = 116$ mV in Fig. 5.57 indicate some growth of chaos (probably in a period 2 chaotic band state). Large errors are found on the right-hand-side of the spectra, where q takes large negative values. Some numerical results from the model of the impact-ionization avalanche are shown in Fig. 6.27, where the parameters are $E_{ac} = 4.5$, $f_0 = 0.735$, $E_0 = 15.767852$ (onset of chaos; Fig. 6.27a) and $E_0 = 15.752$ (fully developed chaos; Fig. 6.27b). Other parameters are $N = 8 \times 10^4$, $M = 3 \times 10^3$, $m = 4$ and $\tau = 0.257/f_0$. The capacity dimension is $D_0 = 1.56$ in Fig. 6.27a and $D_0 = 1.99$ in Fig. 6.27b. By decreasing E_0 from its value at

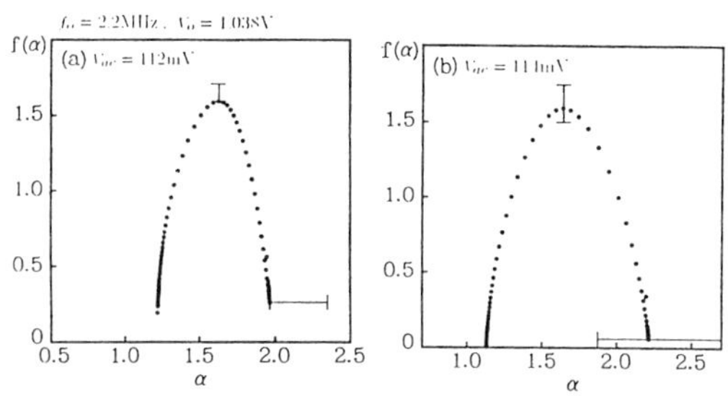

Figure 6.26 $f(\alpha)$ spectrum for the period doubling cascade in an experiment of driven chaos in n-GaAs. Experimental parameters are: (a) $V_0 = 1.038$ V, $f_0 = 2.2$ MHz, $V_{ac} = 112$ mV, and (b) $V_0 = 1.038$ V, $f_0 = 2.2$ MHz, $V_{ac} = 114$ mV.

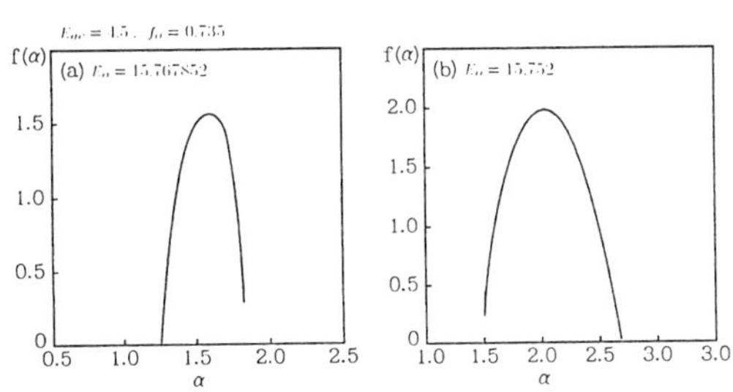

Figure 6.27 $f(\alpha)$ spectrum for (a) the onset of chaos and (b) fully developed chaos, simulated in the approximated model of an impact-ionization avalanche. $E_{ac} = 4.5$, $f_0 = 0.735$. dc bias: (a) $E_0 = 15.767852$, and (b) $E_0 = 15.752$. Calculated by K. Aoki.

the onset of chaos, the chaotic attractor grows via the band merging and correspondingly the $f(\alpha)$ spectrum broadens and shifts towards larger values of singularity (Fig. 6.27a). The main interest here is, however, the $f(\alpha)$ spectrum at the onset of chaos. Figure 6.28 gives a comparison between the experiment (dots) and the model simulation (solid line). One can subtract a trivial dimension and look at the boxed region in Fig. 6.28, which is replotted in Fig. 6.29. For reference, the broken line in Fig. 6.29 indicates the $f(\alpha)$ spectrum obtained in the logistic map. Although the numerical result by the model simulation (solid line) shows a broader spectrum with $D_\infty = 0.35$ and $D_{-\infty} = 0.775$ (cf. (6.68), (6.69)), it may be concluded that $f(\alpha)$ spectrum of semiconductor chaos exhibits a universality (the spectrum is almost identical to that of the Feigenbaum attractor). The experimental

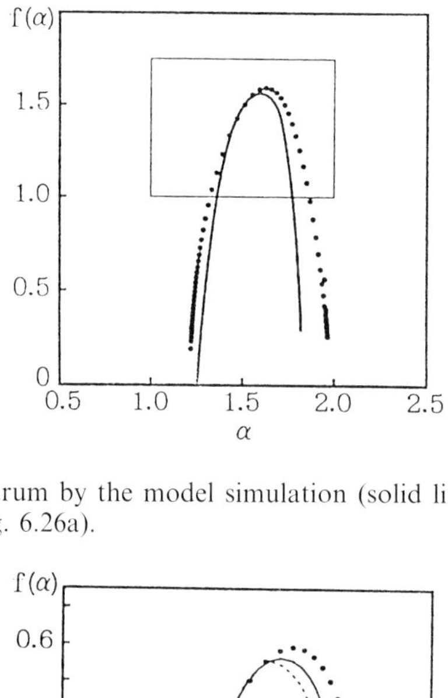

Figure 6.28 $f(\alpha)$ spectrum by the model simulation (solid line, Fig. 6.27a) and the experiments (dots, Fig. 6.26a).

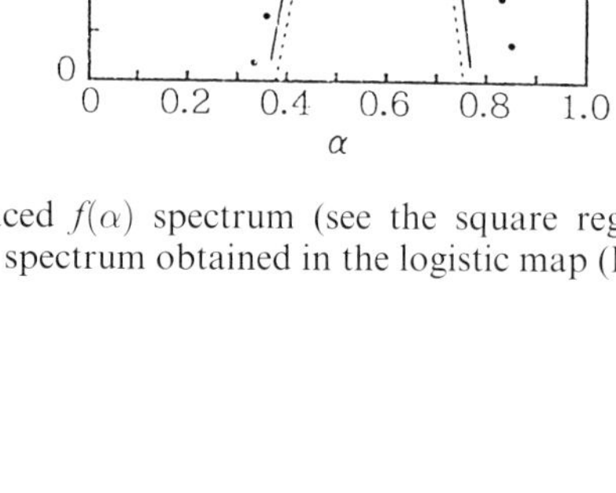

Figure 6.29 The reduced $f(\alpha)$ spectrum (see the square region in Fig. 6.28. The broken line is the $f(\alpha)$ spectrum obtained in the logistic map (Feigenbaum attractor). Drawn by K. Aoki.

$f(\alpha)$ curve (dots) is slightly broader than the numerical result, indicating some noise effect and also difficulty in adjusting the experimental parameters precisely at the onset of chaos. The $f(\alpha)$ spectrum observed under a longitudinal magnetic field will be discussed in section 7.4, where quality tests of the spectrum will be argued in detail.

Another example of universality is found at the onset of the transition from quasi-periodicity to chaos. Many dynamical systems show the transition via circle-map behavior with the golden-mean winding number W^*, where the bifurcation scheme is far more complicated than the simple Ruelle–Takens–Newhouse scenario. A famous example of the $f(\alpha)$ spectrum was presented by Jensen *et al.* [36] in an experiment on forced Rayleigh–Bénard convection, in which they showed $\alpha_{\text{max}} = D_{-\infty} = \ln w^*/\ln \alpha_{\text{GM}}^{-1} = 1.8980\ldots$, $\alpha_{\text{min}} = D_{\infty} = \ln w^*/\ln \alpha_{\text{GM}}^{-3} = 0.6326\ldots$, and $D_0 = 1.0$, where $w^* = (\sqrt{5}-1)/2$, and $\alpha_{\text{GM}} = 1.2885\ldots$ is the universal local scale factor [37]. In semiconductor chaos systems, Gwinn and Westervelt have shown universality by using a p-type Ge sample with the acceptor concentration $N_{\text{A}} = 1 \times 10^{11}/\text{cm}^3$ [32, 33]. The $I-V$ characteristic shows an abrupt rise in current due to impact ionization of shallow acceptor levels. In the postbreakdown regime $E_{\text{dc}} > 5$ V/cm, the $I-V$ characteristic exhibits a voltage-controlled (N-shaped) negative differential region [32]. Similar to the Gunn oscillation, they have observed spontaneous oscillation ($f_0 > 8020$ Hz) caused by the traveling of planar electric field domains in the transverse direction [32, 33, 38] (note: there is longitudinal traveling of the electric field domain between ohmic contacts for the Gunn oscillation). In order to drive the sample to the transition from quasi-periodicity to chaos, they superimposed sinusoidal and dc voltages of amplitudes $V_{\text{ac}}(= 153$ mV) and V_{dc} ($= 3.05$ V), respectively, and adjusted the driving frequency so that the condition of golden-mean winding number could be reached. The measured Poincaré section in Fig. 6.30 indicates fractal folding of the toric surface. Using a ridge-finding algorithm [33, 34] they located a high-probability path around the Poincaré section and counted the number of points $N_s(\ell)$ along the path, where the full length of the Poincaré section in Fig. 6.30 is normalized to be unity and the section is divided into $1/\ell$ bins (thus N_s is the number of points included in a bin specified by an integer s which indicates the location of the bin along the path of the Poincaré section: $s = 1, 2, \ldots$). Since $p_s(\ell) = N_s(\ell)/N_{\text{tot}}$, where N_{tot} is the total number of points, the partition function is given by $\Gamma(q, \ell) = \langle p_s(\ell)^{q-1} \rangle$. Once $\alpha(q)$ is evaluated from the slope of the log–log plot of $\Gamma(q, \ell)$ vs ℓ, the $f(\alpha)$ spectrum is obtained as shown in Fig. 6.31. In Fig. 6.31, the experimental result agrees with the universal $f(\alpha)$ curve predicted from the circle-map theory (dots).

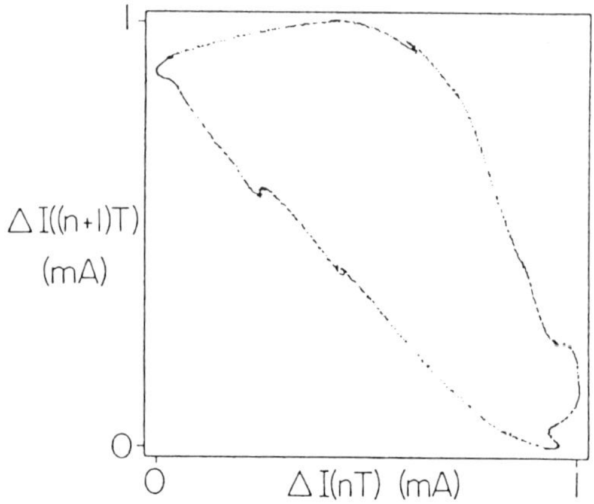

Figure 6.30 Experimental Poincaré section reconstructed from the current time series $I(t) = I_0 + \Delta I(t)$ at the transition from quasi-periodicity to chaos with golden-mean winding number, where $I_0 \simeq 5$ mA; n and T are the number and period of drive cycles [32].

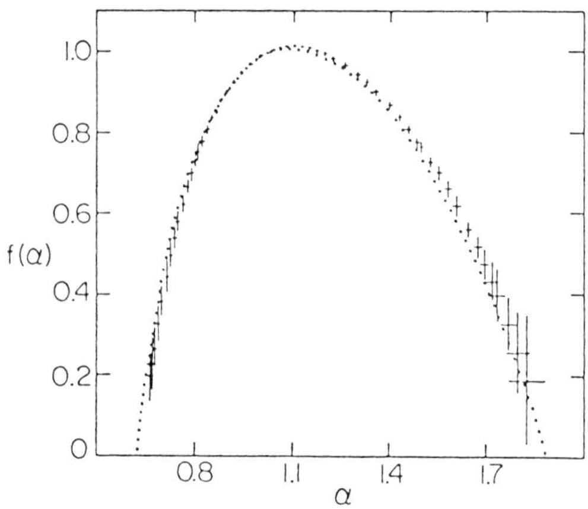

Figure 6.31 The fractal dimension $f(\alpha)$ of the subset of the experimental attractor with scaling index α, shown with error bars indicating the standard deviation of the mean of three data sets. The universal circle-map result is shown as a dotted curve for comparison [32].

6.4 Circle-map behavior

One of the universal properties of nonlinear dynamical systems is the transition from quasi-periodicity to chaos which can be viewed globally in the whole parameter space, say, $\{K, \Omega\}$, where the bifurcation scheme is found to possess a much more complex hierarchical structure than simply a meaning of Hopf bifurcation to chaos via "break-up of a torus" [5, 39, 40]. The global phase diagram shows the Farey tree structure of the mode-locking states (Arnold tongues), overlap of Arnold tongues and locally critical emergence of chaos via the transition from a rational to an irrational winding number w^* (golden mean) [41], where the technical terms will be clarified below. In nonlinear dynamical systems, many systems have exhibited such a criticality (onset of the break-up of the torus), for instance, in experiments on fluid dynamics [42, 43] chemical systems [44], solid-state systems [26, 32, 34, 45, 46] and biological systems [47]. The observed critical behaviors are well characterized by a model of a sine circle map (or simply a circle map) [5, 48–52]. This section introduces circle-map behavior as a universal feature in nonlinear dynamical systems and relates it to observations in semiconductor systems [26, 32, 34, 45].

In many semiconductor systems, self-generated spontaneous current oscillation can be observed approximately in a form of $I_0 \sin(2\pi f_0 t)$ with the natural unperturbed frequency f_0 [26]. When the autonomous system is driven periodically by an external driving force $K_e \sin(2\pi f_d t)$, the simplest nonlinear response of the system may be either quasi-periodic or phase-locked periodic states with the perturbed response frequency f'_0, where the ratio f'_0/f_d may be either irrational (for the former) or rational (for the latter). The current $I(t)$ can be reconstructed as a phase space trajectory by the embedding method. Suppose that the phase space trajectory in 3D phase space can be represented by a simple torus given schematically in Fig. 6.32. The trajectory is a helical motion on the torus with the small diameter corresponding to a rotation angle $\theta_1 = f_0 t$ and the large diameter corresponding to $\theta_2 = f_d t$, as shown in Fig. 6.32. If a slice is taken through the torus at each driving period nT_d ($n = 0, 1, 2 \ldots; T_d = 1/f_d$), say, at a fixed

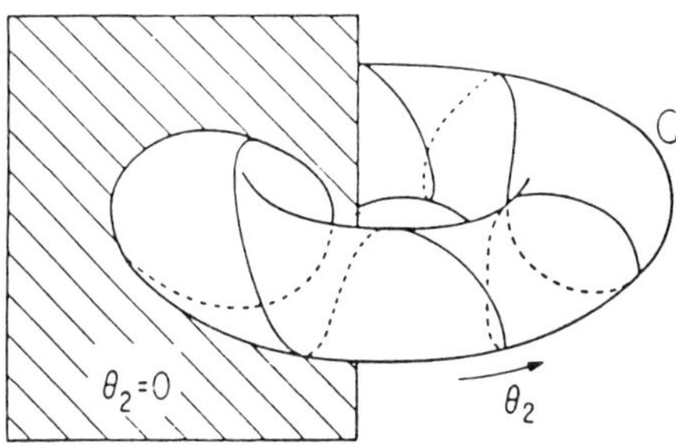

Figure 6.32 Schematic diagram of the two-frequency torus and a Poincaré section. The helical line on the torus traces out the system's trajectory in phase space. Angular coordinates θ_1 and θ_2 are indicated. The vertical plane indicates the stroboscopy at times $i\tau$. The Poincaré is the intersection between the plane and the system trajectory [43].

value of θ_2 $(=0)$; vertical plane in Fig. 6.32, then the system (four-dimensional including time t) is reduced to a 1D map system $\theta_{n+1} = f(\theta_n)$, where the angle θ_1 is measured stroboscopically at each driving period in Fig. 6.32 (replace it with θ_n). For a uniform rotation, the map function is given by the rotation map: $f(\theta) = \theta + \Omega$, where $\Omega = f_0/f_\mathrm{d}$ [5, 43, 53, 54]. For a nonuniform rotation (in the nonlinear dynamical systems), the map function is a nonlinear function of $\theta : f(\theta) = \theta + \Omega + g(\theta)$, where the most famous map function is the sine circle map [5, 43, 53, 54], as discussed later. Thus, the problem of solving the dynamical behavior is reduced to investigating the winding number given by [5, 48–52, 53]:

$$w = \lim_{n\to\infty} \frac{\theta_n - \theta_0}{n}. \tag{6.73}$$

The simplest consideration is the linear response approximation, which corresponds to the trivial uniform rotation and is given by a linear combination:

$$I(t) = I_0 \sin(2\pi f_0 t) + I'_0 \sin(2\pi f_\mathrm{d} t). \tag{6.74}$$

Then, for the beat signal one has the frequency components f_0 and f_d in the Fourier power spectrum. The amplitude of the beat signal is obtained from the cosine formula:

$$\sqrt{I_0^2 + I_0'^2 + 2I_0 I'_0 \cos\{2\pi(f_\mathrm{d} - f_0)t\}}. \tag{6.75}$$

The beat occurs with a period $1/|f_\mathrm{d} - f_0|$. From (6.74), the stroboscopic record of the current at the discrete time $t_n = nT_\mathrm{d} = n/f_\mathrm{d}$ $(n = 0, 1, \ldots)$ is given by:

$$I(t_n) = I_0 \sin(2\pi n(f_0/f_\mathrm{d})) = I_0 \sin(2\pi\theta_n). \tag{6.76}$$

Therefore the linear response approximation gives a uniform rotation via $\theta_n = (f_0/f_\mathrm{d})n$. Obviously, $w = f_0/f_\mathrm{d} = \Omega$ from (6.73), and the rotation map is $\theta_{n+1} = \theta_n + \Omega$. If a positive integer q exists such that $I(t_n) = I(t_{n+q})$ but $I(t_n) \neq I(t_{n+q'})$ for any $q' < q$, then the beat signal is a q-cycle, where $\Omega q =$ "integer" $= p$ and then $\Omega = f_0/f_\mathrm{d} = p/q$ (rational winding number; 1/2, 3/5, etc.). In Fig. 6.32, the winding number p/q means that the orbit completes the q-cycle by taking q circuits around the big circle with respect to θ_2 and simultaneously p rotations around the small circle with respect to θ_1. The beat signal is quasi-periodic if $I(t_n) \neq I(t_{n+q})$ for any integer q in (6.76), then $\Omega(f_0/f_\mathrm{d})$ is an irrational winding number. For $\Omega \in [0, 1]$, there is a countable and infinite number of rational numbers, but the set $\{\Omega\}$ is measure-zero. The trivial case here is no more interesting simply because of the linear-algebraic world.

In the actual dynamical system, the response is highly nonlinear, where the system response (6.73) depends strongly on both the driving amplitude (or the nonlinearity strength) K_e and the ratio $\Omega = f_0/f_\mathrm{d}$. Taking a 2D

Poincaré surface of section, the reduced dynamics of the current may be well characterized by an invertible analytical map of the annulus [48–50]:

$$r_{n+1} - 1 = \lambda(r_n - 1) + Kg(\theta_n). \tag{6.77}$$

$$\theta_{n+1} = \theta_n + \Omega + r_{n+1} - 1, \tag{6.78}$$

where K_e is replaced by the reduced variable K, (r, θ) are polar coordinates (here, ϕ [49, 50] is replaced by θ), r_n is the radius of the circle in the Poincaré section, θ_n is the rotation angle, $g(\theta_n)$ is the sinusoidal function of θ_n, and λ is the Jacobian ($0 \le \lambda < 1$). When $K = 0$, $r_n = 1$ is an attracting invariant circle, and θ_n shows a trivial uniform rotation. When $\lambda = 1$, the map is the standard area-preserving map of Chirikov [55]. If $\lambda = 0$ (infinite contraction), (6.77) and (6.78) can be reduced to an analytical map of the circle:

$$\theta_{n+1} = \theta_n + \Omega + Kg(\theta_n) \quad \text{mod } 1. \tag{6.79}$$

Imagine that the global bifurcation of the nonlinear response (6.79) can be investigated by varying the amplitude K and the driving frequency f_d simultaneously. The perturbed response frequency $f_0{}'$ depends on both K and f_d (say, $f'_0(K, f_d)$ or $f'_0(K, \Omega)$), giving the winding number $w = f'_0(K, \Omega)/f_d \neq \Omega$. Limiting discussion to the nonchaotic regime of the parameter space $\{K, \Omega\}$, a set $\{\Omega | f'_0/f_d = p/q\} = \{\Omega_{p/q}\}$ of a q-cycle for a fixed K is no longer measure-zero owing to mode locking $f'_0/f_d = p/q$ ($\Omega_{p/q} \in [\Omega_{\min}, \Omega_{\max}]; \Delta\Omega_{p/q} = \Omega_{\max} - \Omega_{\min}$), where the response frequency $f'_0 \to f_0$ and the width $\Delta\Omega_{p/q} \to 0$ when $K \to 0$.

A detailed model of the circle map is given in Refs [5, 48–52]:

$$\theta_{n+1} = f(\theta_n) = \theta_n + \Omega - \frac{K}{2\pi} \sin(2\pi\theta_n) \quad \text{mod } 1. \tag{6.80}$$

In analogy with the Reynold's number, K (or K_e in semiconductor systems [32]) provides the nonlinearity strength. For $K = 0$, the problem reduces to the uniform rotation in the linear case as described above. For $K > 0$, the winding number w is calculated in (6.73). As shown in Fig. 6.33, the map function $f(\theta)$ is invertible for $K < 0 < 1$ ($f^{-1}(\theta)$ is differentiable and $f(\theta)$: diffeomorphism [5]) for $f(\theta)$ is still differentiable but $f^{-1}(\theta)$ becomes nondifferentiable for the critical value $K = 1$; and finally $f(\theta)$ becomes noninvertible for $K > 1$. As the nonlinearity strength K is increased, the width of each locked interval p/q increases superlinearly, forming a cusp-shaped region, the Arnold tongue, which is embedded in the sea of quasi-periodicity (Fig. 6.34) [5]. Since there is a countable and infinite number of rational numbers at $K = 0$, then an infinite number of Arnold tongues is embedded in the sea of quasi-periodicity, which form a fractal set in the $K - \Omega$ space, as in the Cantor set [3, 5, 13]. Just above a critical value $K = 1$, chaos emerges critically at a localized value of Ω owing to overlap of the Arnold tongues [5, 43, 48–52]. In Fig. 6.34, the solid lines for $K > 1$

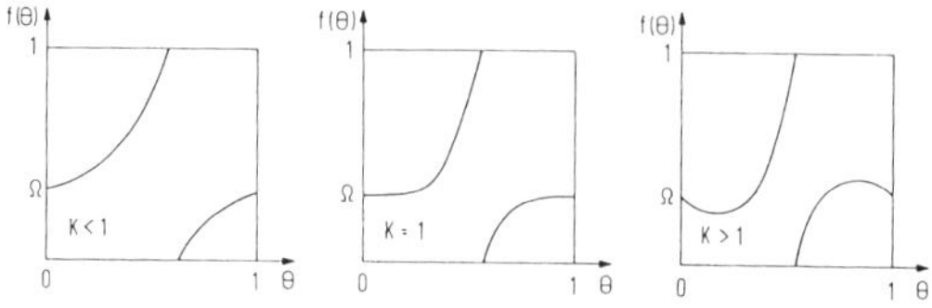

Figure 6.33 Variation in the map $f(\theta)$ with the parameter K. Note that for $K > 1$, the map becomes noninvertible [5].

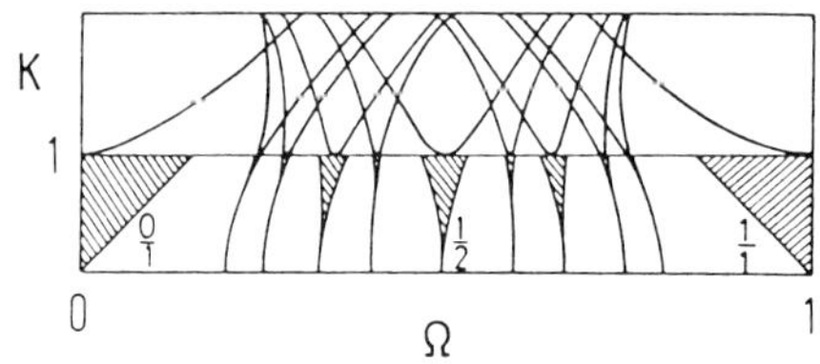

Figure 6.34 Schematic phase diagram of the circle map $K < 1$: within the Arnold tongues (hatched) the winding number w is rational, and there is mode locking. $K = 1$; the Arnold tongues have moved together, and the remaining nonmode-locked "holes" form a Cantor set. $K > 1$: chaos becomes possible, but coexists with order. The lines correspond to the parameter values for superstable, that is nonchaotic cycles which are associated with the mode-locking regions [5].

correspond to the parameter values for superstable cycles (the mode-locking states with the minimum Lyapunov exponents). For $K > 1$, chaos coexists with the mode-locking states exhibiting hysteretic behaviors [32, 43]. Experimentally, varying the driving frequency f_d at a fixed value K_e is equivalent to varying Ω at a fixed value K in (6.80). By increasing f_d (decreasing Ω) from outside the Arnold tongue, the response frequency f'_0 shows a transient with a gradual shift (namely, f'_0 shifts gradually, keeping its irrational number and its quasi-periodicity) until f'_0 "pulls in" to the ratio $f'_0/f_d = p/q$ inside the Arnold tongue [43].

The winding number is a function of Ω, say, $w(\Omega)$. For $0 < K < 1, w(\Omega)$ forms the devil's staircase, as shown in Fig. 6.35. The staircase with rational winding numbers shows, pathologically, a self-similar structure, whereby successive blow-ups of the local regions the self-similar staircase is found again and again. The devil's staircase in Fig. 6.35 is tightly related to the Farey tree given in Fig. 6.36. The natural order of the rational numbers is as follows. Between 1/2 and 1/1, there is a rational number 2/3 with the order $1/2 < 2/3 < 1/1$, and subsequently between 1/2 and 2/3 one has 3/5 with the order $1/2 < 3/5 < 2/3$, etc. In general, there is a rational winding number $(p + p')/(q + q')$ between p/q and p'/q' in Fig. 6.35 and 6.36. It is noted that the width of the Arnold tongue for the locked state p/q becomes smaller with larger values of q. Surprisingly, at the critical line $K = 1$, the Devil's staircase is complete [5, 43, 56] because the width of the quasi-

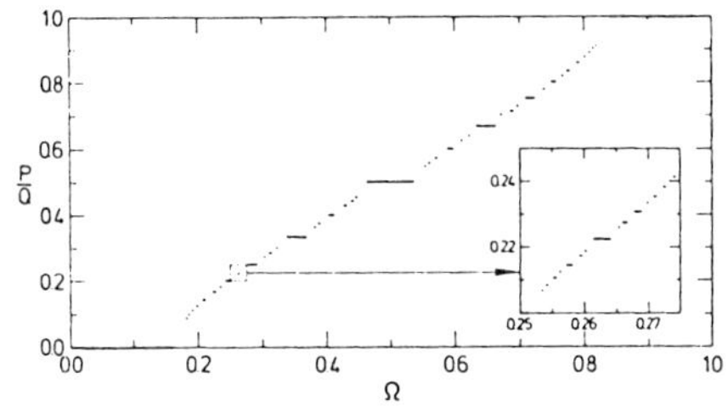

Figure 6.35 Winding number w vs Ω for the circle map at $K = 1$. Steps with $\Delta(P/Q) > 0.0015$ are shown. The inset shows intervals with $\Delta > 0.00015$. Note the self-similar nature of the staircase under magnification [43,56].

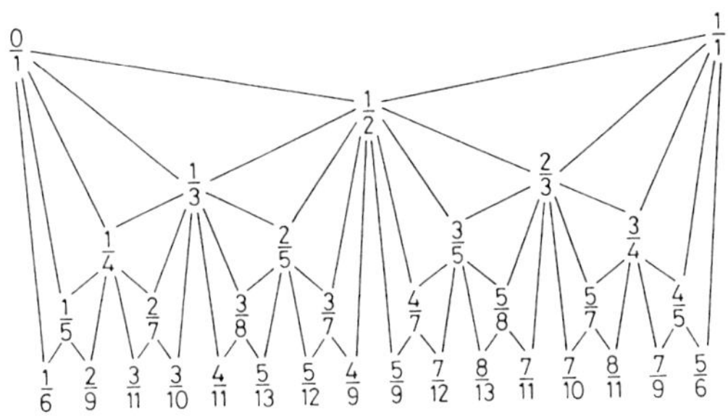

Figure 6.36 The Farey tree orders all rationals in [0,1] with increasing denominators according to the rule that the largest rational between p/q and p'/q' is $(p + p')/(q + q')$ [5], after Cvitanovic and Söderberg, 1985).

periodic region becomes zero and then $\sum \Delta\Omega_{p/q} = 1$, while $\sum \Delta\Omega_{p/q} < 1$ for $K < 1$ because of the nonzero width of the quasi-periodic region. The situation at $K = 1$ is not visualized in Fig. 6.37 since one cannot draw a complete set of Arnold tongues by taking the limit $q \to \infty$ and $p \to \infty$. The ultimate situation can only be imagined from a schematic figure (Fig. 6.37) drawn by Ostlund *et al.* [49]. As seen later, at $K = 1$, an overlap of the Arnold tongues begins at the golden mean $w^* = \sigma_G = (\sqrt{5} - 1)/2$. However, it is also noted that the overlap of two Arnold tongues with simpler rational numbers such as 1/2 and 3/5 occurs at larger values, $K > 1$. The critical line $K = 1$ is special in the sense that one encounters signaling of the first onset of chaos at a special control parameter Ω_∞, which gives the special winding number $w^* = \sigma_G$ when K approaches to unity from below, where $\Omega_{p/q}$ was replaced by Ω_i for $p/q = F_i F_{i+1}$ (see below) and the limit $i \to \infty$ was taken. The circle-map behavior, the transition from quasi-periodicity to chaos, at the critical point $K = 1$, has been well predicted by the theoretical approach using renormalization-group methods [48–52].

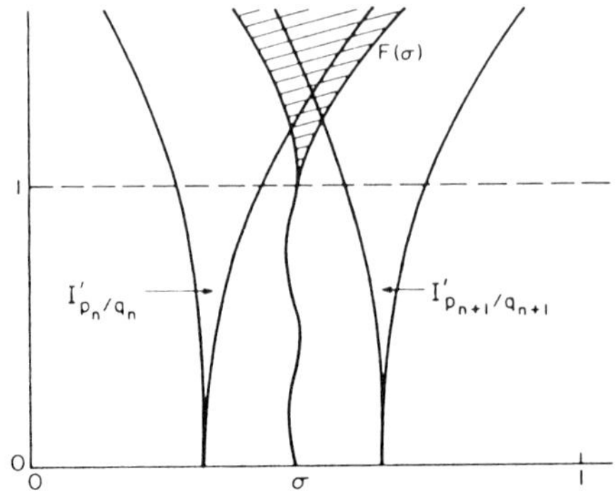

Figure 6.37 Conjectured structure of $F(\sigma_G)$ and its approximation by the tongues $I'_{pn/qn}$ [49].

Any generalized mean $\sigma \in [0, 1]$ of irrational numbers can be obtained theoretically by taking a continued fraction form [5, 17, 43, 48–52]:

$$\sigma = \cfrac{1}{n_1 + \cfrac{1}{n_2 + \cfrac{1}{n_3 + \dots}}} = \langle n_1, n_2, n_3, \dots \rangle \tag{6.81}$$

where n_i $(i = 1, 2, \dots)$ are positive integers. The truncated form of σ is rational and given by

$$\sigma_{i+1} = \sigma_i + \sigma_{i-1} = F_i/F_{i+1}, \tag{6.82}$$

where F_i is the ith Fibonacci number defined by:

$$F_{i+1} = F_i + F_{i-1} \; (i > 1); \quad F_0 = 0 \text{ and } F_1 = 1. \tag{6.83}$$

One can trace the specified root to the golden mean in the Farey tree in Fig. 6.36 as: $\sigma_2 = 1/1 \to \sigma_3 = 1/2 \to \sigma_4 = 2/3 \to \sigma_5 = 3/5 \to \sigma_6 = 5/8 \to \dots$. (6.82) is rewritten as:

$$\sigma_{i+1} = \frac{F_i}{F_i + F_{i-1}} = \frac{1}{1 + \sigma_i}. \tag{6.84}$$

By taking a limit $i \to \infty$, we have $\lim_{i\to\infty} \sigma_{i+1} = \lim_{i\to\infty} \sigma_i = w^* = \sigma_G$, then from (6.84) $w^*(1 + w^*) = 1$, which gives $\sigma_G = (\sqrt{5} - 1)/2$. The silver mean σ_S is given by $\sigma_S = \langle 2, 2, 2, \dots \rangle$ in (6.81). One can also trace to the silver mean in the Farey tree as: $\sigma_0 = 1/2 \to \sigma_2 = 2/5 \to \sigma_4 = 5/12 \to \sigma_6 = 12/29 \to \sigma_8 = 29/70 \to \sigma_{10} = 70/169 \to \dots$, where the sequence is ruled by $\sigma_i = 1/(2 + \sigma_{i-2})$ $(i > 2;$ even). Taking the limit $i \to \infty$, then $\sigma_S = \sqrt{2} - 1$. The golden mean is the most irrational, and the silver mean slightly less irrational [43]. It is commonly recognized that the golden mean is the easiest place to observe quasi-periodicity in the experiments because of its distance from rational approximants (note: other winding numbers are more likely to lock to low-denominator tongues) [43]. For any case, the generalized means (6.81) are locally cusp points (end points) of quasi-periodic sea at $K = 1$.

The quasi-periodicity can be well characterized by the Fourier power spectrum, and the discrete line spectrum is given by:

$$f_{nm} = nf'_0 \pm mf_{\mathrm{d}}, \tag{6.85}$$

where n and m are arbitrary integers ($n, m \in Z$). If $w = f'_0/f_{\mathrm{d}}$ is any irrational number, then $f(\theta)$ is topologically equivalent to a simple rotation by w. In what follows, the frequency is normalized by the driving frequency f_{d}, and f'_0 replaced by ω and σ, respectively. (6.85) is transformed formally to:

$$\omega_{nm} = n\sigma \pm m. \tag{6.86}$$

What is the most significant difference between the normal quasi-periodicity for $K < 1$ and critical quasi-periodicity at $K = 1$ with Ω_∞? The Fourier power spectrum at the critical point ($K = 1, \Omega_\infty$) exhibits various universal scaling behaviors related to the property of the golden mean (i.e., $\sigma_{\mathrm{G}}^2 = 1 - \sigma_{\mathrm{G}}$). The Fourier transform of the rotation angle $\theta(t_j)$ on an attracting invariant circle can be investigated numerically by taking [5, 50]:

$$u(t_j) = \theta(t_j) - t_j, \tag{6.87}$$

where the discrete time t_j is given by

$$t_j = j, \quad j = 0, 1, 2, \ldots. \tag{6.88}$$

For the critical quasi-periodicity, the fundamental response frequency is given by $\omega_{10} = \sigma_{\mathrm{G}}$ from (6.86) ($n = 1, m = 0$), where the successive iterations (6.80) never complete their orbit on the attracting invariant circle since σ_{G} means the ∞-cycle ($F_i \to \infty$ with $i \to \infty$). For an actual problem of finding the universal scaling structure in the power spectrum, the critical quasi-periodicity is approximated by the F_i-cycle and then the fundamental response frequency by $\omega_{10} = \sigma_i \sim \sigma_{\mathrm{G}}$ (almost quasi-periodicity if $i \gg 1$), where $\sigma_i = \langle 1, 1, \ldots, 1 \rangle = F_{i-1}/F_i$ given by a truncated form of continued fraction (6.81) with $i \gg 1$. The problem is then reduced to find the Fourier transform from a finite time sequence $\{u^k(t_j)\}$ of the F_i-cycle, where $t_j = 0, 1, \ldots, F_i - 1$, and $u(t_j)$ is replaced by $u^k(t_j)$. For the F_i-cycle the following properties are added:

$$\theta^k(t_{j+1}) = f(\theta^k(t_j)), \tag{6.89}$$

and

$$\theta^k(0) = 0, \tag{6.90}$$

$$\theta^k(t_j + 1) = \theta(t_j) + 1, \tag{6.91}$$

where $k = F_i$. From (6.87), (6.89)–(6.91) $u^k(0) = u^k(F_i - 1) = 0$. The power spectrum is given by [43, 49, 50]:

$$f^k(\omega) = \frac{1}{F_i}\sum_{j=0}^{F_i-1} u^k(t_j)\exp(2\pi i\omega t_j). \tag{6.92}$$

The function $u^k(t_j)$ numerically calculated by Shenker [50] is shown in Fig. 6.38, where $K = 0.5$, $\sigma_i = 1597/2587$ ($i = 17$ for F_i) in (a), and $K = 1, \sigma_{17}$ in (b). The normal quasi-periodicity for $K = 0.5$ shows a smooth periodic function in Fig. 6.38a, while for $K = 1$ the time sequence becomes bumpy in Fig. 6.38b, which signals the first transition from quasi-periodicity to chaos (namely critical quasi-periodicity). Figure 6.39 shows the power spectrum $|f(\omega)/\omega|^2$ in the log–log plot [49, 50]. The spectrum is formally written as:

$$|f(\omega)/\omega|^2 = \sum a_{nm}\delta(\omega - \omega_{nm}). \tag{6.93}$$

The following universal behaviors are included in the spectrum [50–52].

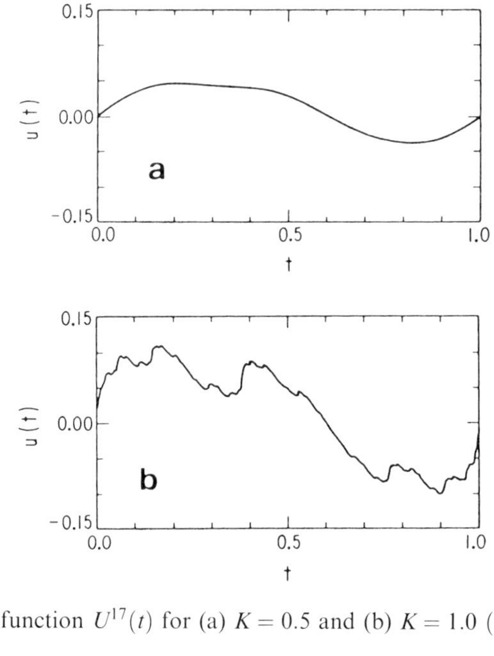

Figure 6.38 The function $U^{17}(t)$ for (a) $K = 0.5$ and (b) $K = 1.0$ ($\sigma_{17} = 1597/2584$) [50].

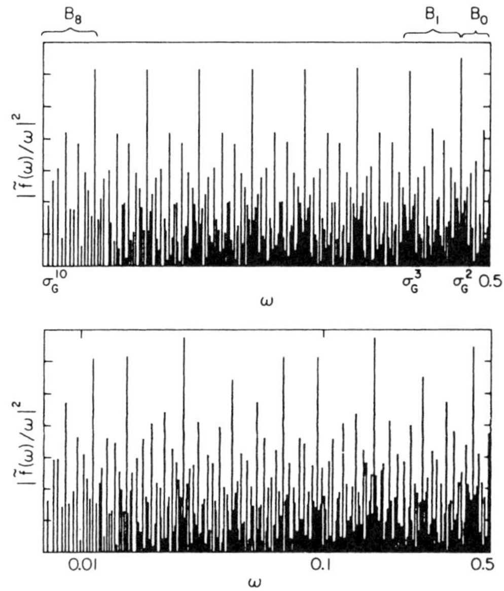

Figure 6.39 Top: power spectrum of a time series (6.93) of the sine map with $K = 1$ and $\rho = \sigma_G$ on a log-log plot. A normalization factor of ω^2 has been divided out of the power. The lines in each band B_j are in one-to-one correspondence and the associated complex amplitudes become universal for $j \to \infty$. The principal peaks fall at $\omega = \sigma_G^i$. Bottom: power spectrum of the sine map at $K = 1$ as in the top except that the winding number has a random bounded continued fraction, $\langle$ 2, 2, 1, 1, 2, 1, 1, 2, 1, 1, 1, 1, 2, 2, 2, ...$\rangle$ based on the binary expansion of π. The low-frequency spectrum is universal but not self-similar [49].

(1) In each frequency interval $B_i = (\sigma_G^{i+2}, \sigma_G^{i+1}], i = 1, 2, \ldots,$ there is a universal structure, where the spectrum B_i is identical to the spectrum in B_{i+1} in the log–log plot.
(2) The largest peaks (main peaks) locate at $\omega = \sigma_G^{i+1}$ $(i > 1)$ with the same strength $a_{10} = a_{11} = \ldots = a_{nm} = \ldots$
(3) In order of decreasing amplitudes of a_{nm}, the next six peaks in each band B_i correspond to the pairs of the Fibonacci numbers (Q_1, Q_2) = (2, 2), (1, 3), (3, 3), (1, 4), (2, 5) and (4, 4), and are positioned at $\omega_{rs} = |r\sigma_G - s|\,|\sigma_G|^i$, where $r = Q_1$ and $r + s = Q_2$ [49].

The positions of the main peaks and the successive generations of subpeaks are found as follows. If $\omega_i = n\,\sigma_G - m \in B_i$, then $\omega_{i+1} = -(n+m)\,\sigma_G + n \in B_{i+1}$ since $\omega_{i+1} = n\,\sigma_G^2 - m\,\sigma_G = \sigma_G\,\omega_i$ (note: $\sigma_G^2 = 1 - \sigma_G$). Similarly, $\omega_{i+2} = (2n+m)\,\sigma_G - (n+m) = \sigma_G^2\,\omega_i \in B_{i+2}$, and so on. Starting from the fundamental response frequency $\omega_0 = \sigma_G(n = 1, m = 0$ in (6.86)), then $\omega_1 = |\sigma_G - 1| = 1 - \sigma_G = \sigma_G^2$, $\omega_2 = |2\sigma_G - 1| = \sigma_G - \sigma_G^2 = \sigma_G^3, \ldots,$ $\omega_{i-1} =$

$|F_i\sigma_G - F_{i-1}| = \sigma_G^i$ where

$$\omega_{i-1} = |(F_{i-1} + F_{i-2})\sigma_G - F_{i-1}|$$
$$= \sigma_G|F_{i-1}\sigma_G - F_{i-2}| = \sigma_G\omega_{i-2}. \tag{6.94}$$

Then the main peaks (the first generation) are positioned at $\omega_{i-1} = \sigma_G^i$ ($i > 1$). Successive (second, third, etc.) generations of the subpeaks in B_i can be confirmed by taking linear combinations of the frequencies of the main peaks between σ_G^{i+3} and σ_G^{i+2}:

$$\omega_{rs}^i = |r\sigma_G^{i+3} - s\sigma_G^{i+2}|,$$
$$= |r\sigma_G - s|\ \sigma_G^{i+2} \in B_i, \tag{6.95}$$

where $r = Q_1$ and $r + s = Q_2$. Corresponding to the pairs $(Q_1, Q_2) = (2, 2)$ and $(2, 3)$, $(r, s) = (2, 0)$ and $(1, 3)$, one finds $\omega_{20} = 2\sigma_G^{i+3} \in B_i, \omega_{12} = (2 - \sigma_G)\sigma_G^{i+2} \in B_i$, where $\omega_{20} < \omega_{12}$. Similarly, corresponding to $(Q_1, Q_2) = (3, 3)$, $(1, 4)$, $(2, 5)$ and $(4, 4)$, one has $(r, s) = (3, 0)$, $(1, 3)$, $(2, 3)$ and $(4, 0)$ and then $\omega_{30} < \omega_{13} < \omega_{23} < \omega_{40}$, where $\omega_{30} \in B_i$ and $\omega_{13}, \omega_{23}, \omega_{40} \in B_{i-1}$. However, because of the repeated structure of the peaks for any subsets B_i ($\omega_{rs}^{(i)} = \sigma_G\omega_{rs}^{(i-1)}$), the successive six generations of the peaks in a specified subset B_i appear at ω_{rs} in the order:

$$\omega_{23} < \omega_{30} < \omega_{20} < \omega_{12} < \omega_{13} < \omega_{40}. \tag{6.96}$$

The power amplitudes a_{rs} in $|f(\omega)/\omega|^2$ are ordered corresponding to their magnitudes as [35]:

$$a_{20} > a_{12} > a_{30} > a_{13} > a_{23} > a_{40}. \tag{6.97}$$

In accord with the order of the power amplitudes (6.97), the second-generation peaks are positioned at ω_{20}, the third-generation peaks at ω_{12}, etc. (for details see Table 6.1). From the linear combinations (6.95) with any arbitrary integers $r, s \in Z$ such that $\omega_{rs} \in [0, 0.5]$, one may find the universal structure $\{\omega_{rs}\} \in B_i$ which is composed of an infinite number of subpeaks. Since $\log(\omega_{rs}^{(i-1)}) - \log(\omega_{rs}^{(i)}) = \log(\sigma_G)$ for any two adjacent subsets, the

Table 6.1 Number of sub-generation peaks ω_{rs} and the frequencies

Number of subpeaks	ω_{rs} in B_i	Frequencies
6	ω_{23}	$(3 - 2\sigma_G)\sigma_G^{i+3}$
4	ω_{30}	$3\sigma_G^{i+4}$
2	ω_{20}	$2\sigma_G^{i+3}$
3	ω_{12}	$(2 - \sigma_G)\sigma_G^{i+2}$
5	ω_{13}	$(3 - \sigma_G)\sigma_G^{i+2}$
7	ω_{40}	$4\sigma_G^{i+4}$

universal structure appears identical in all of the subsets. From Fig. 6.39, the reader may easily see the ordering of the line spectrum. For the circle-map behavior, theoretical treatment by renormalization-group methods has shown a complete analogy with the period doubling route to chaos. Readers who are interested in renormalization-group methods are refered to a book by Schuster [5] and articles by Feigenbaum *et al.* [48], Ostlund *et al.* [49] and Rand *et al.* [51].

Excellent experimental measurements have been taken in a driven Rayleigh–Bénard system using mercury [42, 43], in electrical conductivity in barium sodium niobate (BSN) crystals under a humidified atmosphere [46], and in the periodically driven chaos systems of semiconductors [26, 32, 34]. In an experiment of forced Rayleigh–Bérnard convection of mercury in a small-aspect-ratio cell ($0.7 \times 0.7 \times 1.4$ cm^3), Stavans *et al.* [42] have obtained, for circle-map behavior, universal structure and universal scaling constants at the golden mean winding number. The rescaled experimental Fourier spectrum in Fig. 6.40 [42] may be helpful to understand the ordering of the successive generation peaks as labeled by numbers. In the electrical measurements of the BSN crystal by Martin and Martienssen [46], a simple spontaneous current oscillation (observed under a constant external current) has been modulated periodically by an ac current. They have shown excellent results on the circle-map behavior and Devil's staircase near the transition to chaos.

In the semiconductor systems, Held and Jeffries [34, 57] have investigated quasi-periodic transition to chaos in an electron-hole plasma instability. The

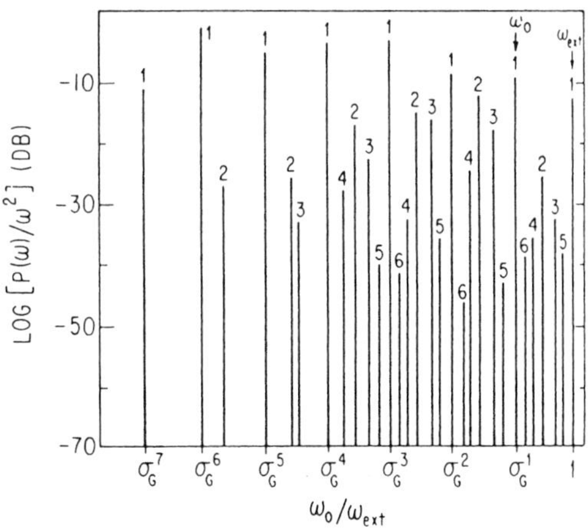

Figure 6.40 Scaled spectrum for the frequency ratio $\omega_0/\omega_{\text{ext}} = \sigma_G \pm 2 \times 10^{-4}$ and pulse amplitude $A = 17.4$ mA (nearest approximation to the critical line), where ω_0 is the renormalized oscillatory instability frequency and ω_{ext} the external pulse frequency. $\text{Log}[P(\omega)/\omega^2]$ vs $\log \omega$. Each peak is labeled by an integer showing to which generation it belongs [42].

experiments were performed on a $1 \times 1 \times 10$ mm^3 sample cut from a large single crystal of ultra-high-purity n-type Ge with a net donor concentration $N_D \sim 3.7 \times 10^{12}/\text{cm}^3$. A Li-diffused n$^+$ contact for electron injection and a B-implanted contact for hole injection were formed on opposite 1×1 mm^2 ends. Under a longitudinal magnetic field, spontaneous current oscillation occurs at 77 K due to a traveling screw-shaped helical density wave of electron-hole plasma [34, 57, 58], where the angle between the dc field (E_0) and the magnetic field (B_0) has been settled typically within $\theta = 0 \pm 3°$. With a dc bias $V_0 = 15.02$ V and $B_0 = 9.32$ kG, the plasma instability shows a simple periodic oscillation at a fundamental frequency $f_0 = 235.47$ kHz (some higher harmonics have also been recognized). The simple periodic oscillation is then perturbed by an ac voltage $V_1 \sin(2\pi f_1 t)$. At the critical line of $V_1 = 274$ mV, Held and Jeffries [34] measured the length S [42, 43] of the interval between two locked band parents around the irrational number σ_G, where $S = |\Omega_i - \Omega_{i-1}|$ and Ω_i is the control parameter corresponding to a locked state with the rational winding number σ_i in (6.82). This interval includes the locked daughter at Ω_{i+1}. Then, measuring the lengths S_1 and S_2, where $S_1 = |\Omega_{i+1} - \Omega_{i-1}|$ and $S_2 = |\Omega_{i+1} - \Omega_i|$, they estimated the fractal dimension D of the Cantor-type set of the successive locked states around σ_G, by

$$(S_1/S)^D + (S_2/S)^D \approx 1. \tag{6.98}$$

The estimated value $D = 0.90 \pm 0.03$ is found to be close to the predicted value 0.868 ... [52, 60], where the dimension in (6.98) is one of the universal scaling factors for the local universality [48–52, 55, 60] at the golden mean. Held and Jeffries also measured another universal constant δ, which is characterized by:

$$\delta = \lim_{i\to\infty} \frac{\Delta\Omega_i - \Delta\Omega_{i+1}}{\Delta\Omega_{i+1} - \Delta\Omega_{i+2}}, \tag{6.99}$$

where $\Delta\Omega_i$, $\Delta\Omega_{i+1}$ and $\Delta\Omega_{i+2}$ are the widths of the control parameter for the locking states σ_i, σ_{i+1} and σ_{i+2}, respectively. By measuring the widths for the locking states 3/5, 5/8 and 8/13, they obtained the value $\delta = 2.7 \pm 0.5$, in agreement with the theoretical value of 2.83... expected for $w = \sigma_G$. The Fourier spectrum for the electron-hole plasma instability is shown in Fig. 6.41, where the bias condition has been settled at almost the critical point. The first-generation peaks labeled by 1 are found to be nearly constant in height, but other generation peaks (subpeaks labeled 2–4) deviate from the predicted theory (see Fig. 6.39). The deviation indicates a difficulty in adjusting the experimental bias condition exactly on the critical point.

Gwinn and Westervelt [32] presented excellent measurements in a high-purity p-Ge with shallow acceptor concentration $p = 10^{11}/\text{cm}^3$. The sample dimension is $4 \times 8 \times 8$ mm^3. B-implanted p^+ layers were formed as ohmic contacts on each of the broad 8×8 mm^2 surfaces to produce a spatially

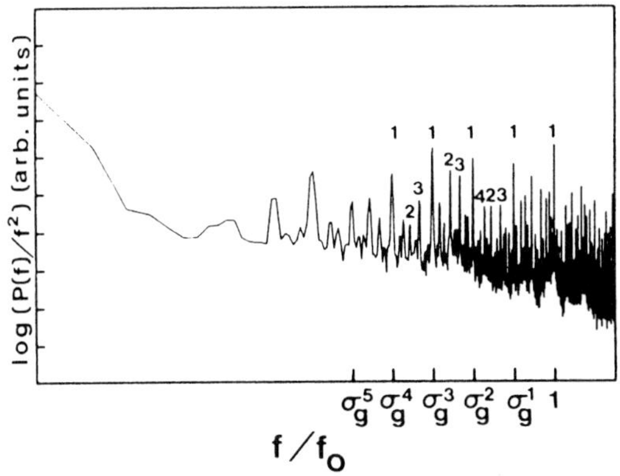

Figure 6.41 Power spectrum of total current $I(t)$ at onset of chaos, with $\log f$ (frequency) plotted on the horizontal axis and $\log\,[P(f)/f^2]$ plotted on the vertical axis, where $P(f)$ is the spectral power density: f_0 is normalized to unity [34].

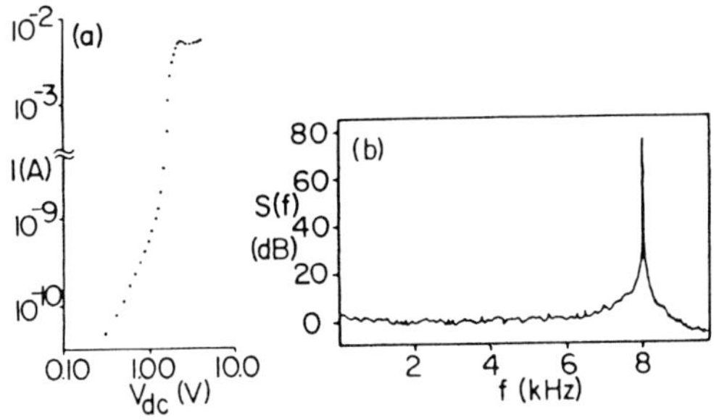

Figure 6.42 (a) Time-average dc I–V characteristic. (b) Current power spectrum of the spontaneous oscillation [32].

uniform electric field. At liquid helium temperature (4.144 K), the time-average dc I–V characteristic in Fig. 6.42 shows a superlinear curve due to the impact-ionization avalanche of the shallow acceptors. At a threshold applied electric field (5.5 V/cm), free holes gain sufficient energy (> 10 meV) from the applied bias to ionize neutral acceptors, producing an abrupt increase of roughly five orders of magnitude in the current in Fig. 6.42a. The most significant result in the p-type Ge is the N-shaped negative differential resistance (NDR) caused in the postbreakdown regime by the field dependence [32] of the impact-ionization rate. By the NDR, a simple spontaneous current oscillation occurs owing to planar high-field domains which sweep across the sample when the applied dc bias V_{dc} is fixed [32]. At $V_{dc} = 3.05$ V, the Fourier spectrum of the current oscillation in Fig. 6.42b is very clean at the fundamental frequency $f_0 = 8021$ Hz with no subharmonic or spurious components over a dynamic range that exceeds 70 dB. Drift in the frequency of the spontaneous oscillation has been measured to be 1 Hz per 20 min. Such clear and high stability is the main reason for the successful results obtained for circle-map behaviors. Circle-map behavior was investigated by applying an external ac drive $K_e \sin(2\pi f_d t)$. In the parameter space (K_e, f_d), the frequency-locking diagram (Arnold tongues) was measured, as shown in Fig. 6.43, where the dashed line indicates the position corresponding to σ_G. Adjusting the frequency ratio on σ_G, Gwinn and Westervelt have observed the current power spectra and Poincaré sections of Fig. 6.44, where the Poincaré sections are plotted stroboscopically at constant drive phase and the drive amplitude is settled at three values of K_e: well below ($K_e = 138$ mV), just above ($K_e = 153$ mV) and well above ($K_e = 168$ mV) the critical amplitude [32]. Well below the critical point, the Poincaré section in Fig. 6.44a shows a smooth curve, indicating no folding of the toric surface. Just above the transition from quasi-periodicity to chaos, the attractor in Fig. 6.44b clearly indicates wrinkling of the toric surface, signaling the break-up of the invariant two-torus, and correspondingly a larger number of mixing frequencies appears in the Fourier power spectrum. Finally, well above the transition point, the Poincaré section in Fig. 6.44c shows a strange attractor (chaos), whereby the current power spectrum shows a large increase in the broad-band noise level as shown. Then, by replotting the power spectrum $S(f)/f^2$ [32] in the log–log scale for the nearly critical state of Fig. 6.44b, the universal structure in Fig. 6.45 is seen, which is similar to the theoretical one as predicted in Fig. 6.39. Agreement between the experiment and the theory is excellent. The first-generation peaks are almost constant over a wide frequency range.

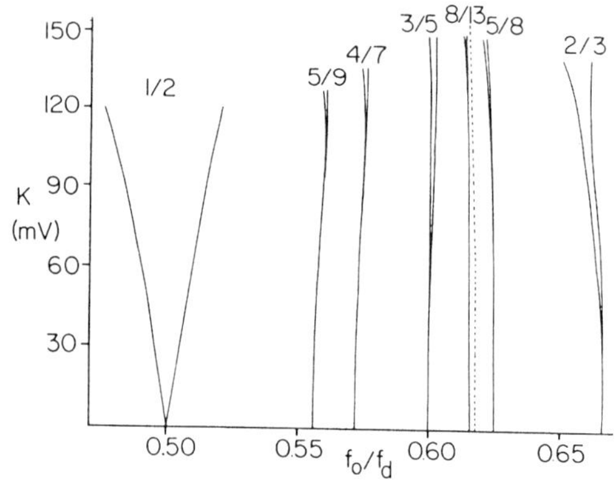

Figure 6.43 Experimental frequency-locking diagram for drive amplitude K_e and frequency f_d normalized to the spontaneous oscillation frequency f_0 [32].

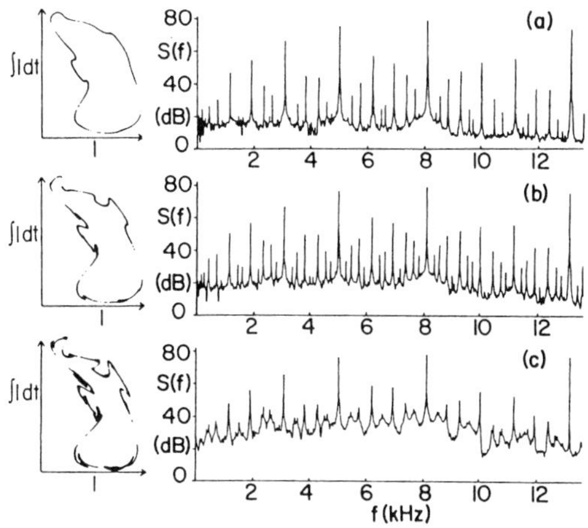

Figure 6.44 Experimental Poincaré sections and corresponding current power spectra. (a) $K_e = 138$ mV, below the critical amplitude K_{crit}. (b) $K_e = 153$ mV, just above K_{crit}. (c) $K_e = 168$ mV, well above K_{crit} [32].

In the frequency interval $[\sigma_G, 1]$, 25 generation peaks are visible. In particular, in the subsets B_0–B_2, the spectrum shows the characteristic ordering and peak heights for many-generation peaks [32].

6.5 Fractal basin boundary

Many nonlinear dynamical systems may have more than one possible time-asymptotic final state, say, attractors A, B, ..., depending on the initial conditions. Consequently, basins of attractions $\{\mathbf{X}_A\}, \{\mathbf{X}_B\}, \ldots$ occur, namely the sets of initial conditions from which the transient phase space orbits are attracted and reach asymptotically to the final states of A, B, ..., respectively. The basin boundaries which divide the phase space into the basins of attractions may be either smooth or fractal [62–65]. For fractal basin boundaries, uncertainty in initial conditions leads to uncertainty in the final state (final state sensitivity [62–65]). The section describes such behavior in semiconductor chaos systems. Under the influence of external noise, there may be intermittent switching behavior between two oscillatory states owing to the final state sensitivity.

In order to characterize the fractal basin boundaries, Grebogi *et al.* [62–64] have introduced the fraction of the uncertain phase space volume f. In Fig. 6.46, the basin boundary Σ divides the phase space into basins of attractions for the two attractors A and B [62]. If the initial conditions are

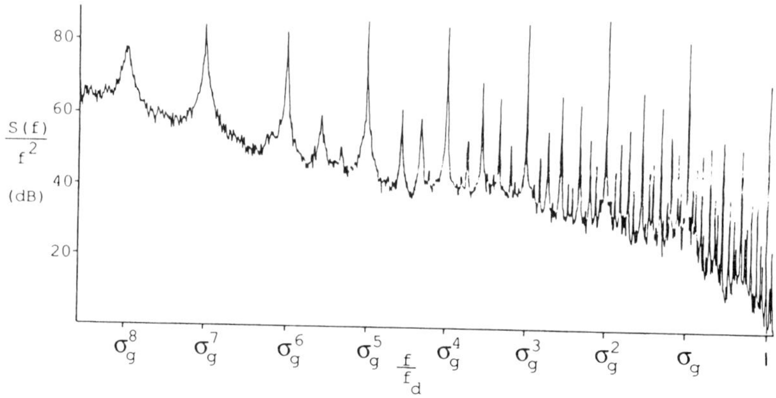

Figure 6.45 Log-log plot of the normalized power spectrum $S(f)/f^2$ for drive at the critical amplitude $K_c = 153$ mV and the golden mean σ_G [32].

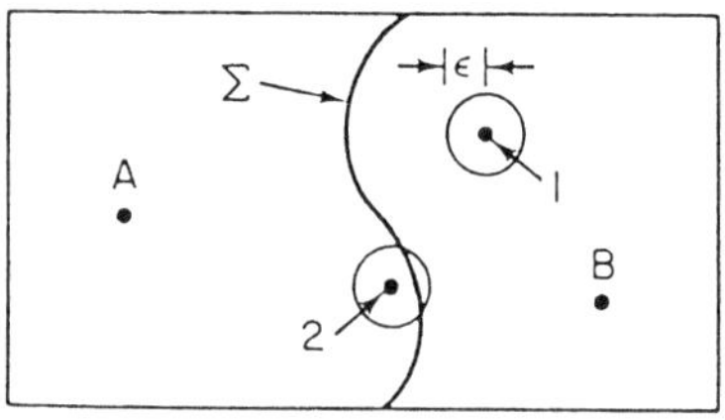

Figure 6.46 Region of phase space divided by the basin boundary into basins of attraction for the two attractors A and B. 1 and 2 represent two initial conditions with uncertainty [61].

uncertain by an amount ϵ, then, for those initial conditions within the boundary (initial condition 2 in Fig. 6.46), one cannot say a priori to which attractor the orbit eventually tends, while there is no doubt that the initial condition 1 reveals no uncertainty in the final state. For the smooth basin boundary in Fig. 6.46, one apparently has $f \sim \epsilon$ [62]. Many nonlinear dynamical systems often exhibit fractal basin boundaries [66–68]. In these cases, one may have

$$f \sim \epsilon^{\alpha} \tag{6.100}$$

with the uncertainty exponent $\alpha < 1$ [62–64]. Correspondingly, the dimension of the fractal basin boundary is related to by:

$$d = D - \alpha, \tag{6.101}$$

where D is the dimension of the phase space (or the parameter space) [62, 64]. Since the total number of D-dimensional cubes of side ϵ needed to cover the basin boundary is given by $N(\epsilon) \sim \epsilon^{-d}$ (d: capacity dimension) and the volume of one of these D-dimensional cubes is ϵ^D, the uncertain volume of the phase space is apparently of the order of $\epsilon^D N(\epsilon) \sim \epsilon^{D-d} = \epsilon^\alpha = f$. Thus, d is the capacity dimension. The fractal basin boundaries with index $\alpha < 1$ possess a self-similar structure like Cantor sets, followed by a fine-scale structure in the basins of attraction on a smaller and smaller scale. What effects of a fractal basin boundary are expected in the experimental systems under ambient noise? If attractors A and B are locally close to the fractal basin boundary, even a small amount of ambient noise power can kick randomly and disturb an intrinsic orbital trajectory of, say, attractor A, forcing the orbit to switch from attractor A to B. As a result, noise-induced intermittency can be observed. Iansiti *et al.* [67] observed intermittent switching behavior in a Josephson junction and explained it by numerical simulation. Relating to the experimental results, they studied numerically the current equation of the resistively shunted junction (RSJ) model:

$$\ddot{\phi} + \frac{1}{\sqrt{\beta}}\dot{\phi} + \sin\phi = i_{\rm ac} + i_{\rm L}\sin\left(\frac{\omega_{\rm L}}{\omega_{\rm p}}t\right) + i_{\rm N}, \tag{6.102}$$

where $i_{\rm dc}, i_{\rm L}$, and $i_{\rm N}$ are the dc, the laser and the noise currents, respectively, all quantities normalized to the critical current of the junction, and ϕ is the phase difference across the junction given by the Josephson equation $\dot{\phi} = 4\pi{\rm eV}/\hbar$ [67]. The ac drive is made using an optically pumped far-infrared laser with $f_{\rm L} = \omega_{\rm L}/2\pi = 419$ GHz. In the absence of noise current ($i_{\rm N} = 0$ in (6.102)) they obtained the fractal basin boundary in the parameter space $(\phi, \dot{\phi})$ of the initial values in Fig. 6.47, where the black region is the basin of the zeroth-step solution (period 1 limit cycle, white circle), while the white region is the basin of the 2/3-step solution (period 3 limit cycle, black circles). By adding noise current $i_{\rm N}$ (thermal noise, $T > 5$ K), the noise pushes the orbit out into the basin boundary, leading to noise-induced intermittency [67]. The simulant results explain well the experimental $I - V$ curve at the noisy transition region between two flat regions of the zeroth and 2/3-steps, where low-frequency $1/f$ noise (10 kHz–100 Hz) has been clearly shown.

In semiconductor chaos systems, little has been known about the fractal basin boundaries. In a model of a periodically driven chaos system, a fractal basin boundary can be seen in the prebreakdown regime [68]. The bifurcation diagram in Fig. 6.48 indicates the existence of a fractal basin boundary, where $f_0 = 10/15, E_{\rm ac} = 4.5$ and the dc bias E_0 is increased from 14.5225. For the bifurcation diagram, time integration has been done up to $i = 30000$, where $t = i\,\Delta{\rm t}$ ($\Delta t = 0.01$), and the peak values of the carrier density $\nu_{\rm p}$ have been plotted for $10\ 000 \leq i \leq 30\ 000$, at each dc bias E_0. The final values (E, ν, ϕ) at E_0 have been adopted as the initial values at the next step of the dc bias $E_0 + \Delta E_0$. By increasing E_0 from 14.5225, two solutions of period 1 limit cycle (attractor A) and chaos (attractor B) appear randomly in an unpredictable manner (region I in Fig. 6.48), which suggests

Figure 6.47 Basins of attraction for $i_L = 1$ and $i_{dc} = 0.15$. The white region is the basin of the $\frac{2}{3}-$ step solution, corresponding to the three-point Poincaré section indicated by the centers of the black circles. The black region is the basin of the zeroth-step solution, with a one-point Poincaré section (white circle). No noise is included [67].

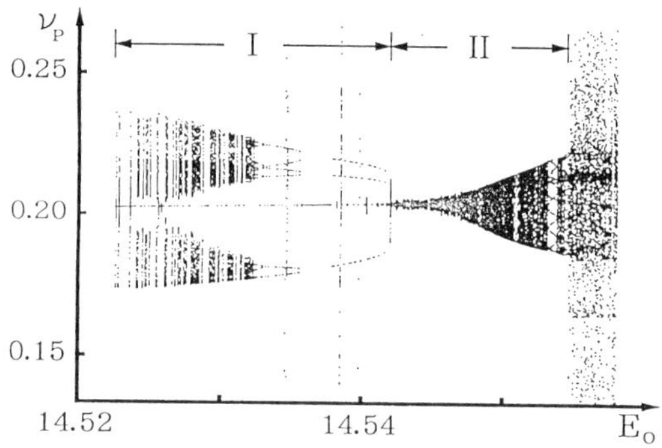

Figure 6.48 Bifurcation diagram as a function of E_0, showing the existence of a fractal basin boundary. Simulated by the approximated model of the impact-ionization avalanche. $E_{ac} = 4.5$ and $f_0 = 10/15$ [68].

the fractal basin boundary between the two basins of attraction. In region I, the chaotic phase is apparently a consequence of period doubling bifurcation of period 3 [68]. For $E_0 > 14.542$, period 1 bifurcates to torus and toric chaos, and the toric chaos is further destabilized via the reinjective switching above 14.556, as mentioned in section 5.6. In Fig. 6.49, the basins of attraction are shown by black and white regions, where $E_0 = 14.53$, 300×300 grids of initial conditions $(E(0), \nu(0))$ are used (where $\phi(0) = 1.322693$) is fixed), and the black and white regions indicate basins of period 1 and chaos, respectively. By blowing up a small squared region, the fractal structure in Fig. 6.50a, b is found simultaneously. The self-similar structure at a different portion is also given in Fig. 6.51a, b, where a white circle near the center in Fig. 6.51b indicates the Poincaré surface of section for period 1, where the point attractor locates at $(E, \nu) = (11.280063, 0.202946)$. As seen in Fig. 6.51b, there may be a different attractor (chaos) with initial con-

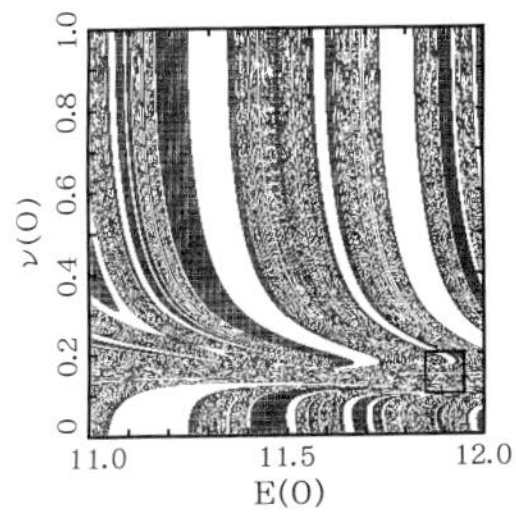

Figure 6.49 Basins of attraction by black and white regions, calculated for the approximated model of the impact-ionization avalanche. $E_{ac} = 4.5$, $f_0 = 10/15$ and $E_0 = 14.530$. Black: period 1; white: chaos. 300×300 grids of the initial conditions $(E(0),\ \nu(0))$ were plotted. Note that $\phi(0) = 1.322693$.

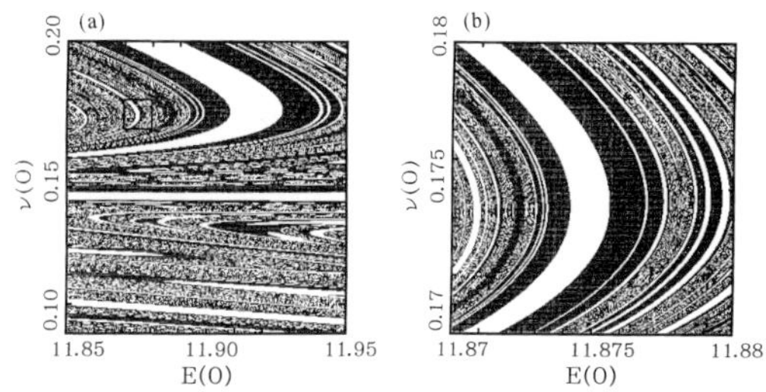

Figure 6.50 Blow-up of a small squared region of the basins of attraction. (a) Magnification of the square in Fig. 6.49 and (b) magnification of a square in (a). Black: period 1; white: chaos.

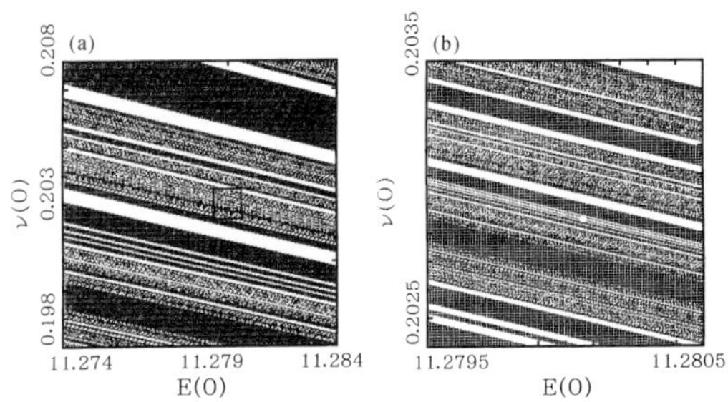

Figure 6.51 Self-similar structure of the basins of attraction at different portions of the phase space $(E(0),\ \nu(0))$. The parameters are the same as those in Fig. 6.49. Black: period 1; white: chaos.

ditions which deviate slightly from the point attractor by an amount of 10^{-6}. Actually, two oscillatory solutions are obtained in the $\nu - E$ characteristics in Fig. 6.52.

Let us find the dimension d of the fractal basin boundary in Fig. 6.51b. To do this, consider an initial condition $(E(0), \nu(0))$ and ask, what is the effect of a small change ϵ in the initial condition? The initial conditions

$(E(0),\ \nu(0))$, $(E(0),\ \nu(0)+\epsilon)$, and $(E(0),\ \nu(0)-\epsilon)$ are iterated until they approach one of the attractors [13, 62–64]. If one or both of the perturbed initial conditions do not approach the same attractor as the unperturbed initial condition $(E(0),\ \nu(0))$, then $(E(0),\ \nu(0))$ is said to be uncertain. In Fig. 6.51b 300 randomly chosen initial conditions are used in the range of $\nu(0); [0.2025, 0.2035]$, where $E(0) = 11.2800630$. By counting the total number n of the initial conditions that are uncertain, a fraction of the uncertainty f is obtained as $\bar{f} = n/300$, which gives approximately the fraction of uncertain phase space volume f originally given in (6.100) and hence $\bar{f} \propto e^{D-d}$. The scaling of $\bar{f}$ with ϵ is given in Fig. 6.53, from which $\alpha \sim 0.37 \pm 0.09$. Since $D = 1$ for the present case, the fractal dimension is found to be $d \sim 0.63 \pm 0.09$. It is interesting to note that $d_c = \ln 2/\ln 3 \sim 0.631$ for a Cantor set. By extending this argument to $D = 2$, it may be estimated that the dimension d of the fractal basin boundary in Fig. 6.51 is 1.63 ± 0.09. Obviously, the structure in Fig. 6.51 is bands of the Cantor set. As mentioned by Grebogi *et al.* [62–64], even apparently small uncertainty in the initial conditions yields substantial fractions of the phase space which are uncertain as to which final state is eventually attained. In the present

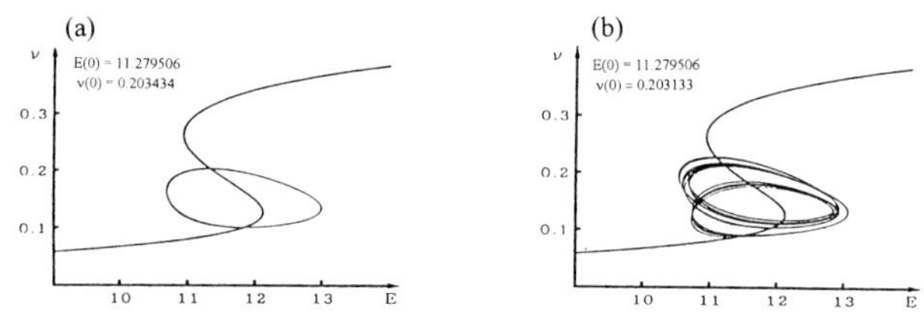

Figure 6.52 Two oscillatory states obtained by slightly different initial conditions. Left: $E(0) = 11.279506$ and $\nu(0) = 0.203134$; right: $E(0) = 11.279506$ and $\nu(0) = 0.203133$.

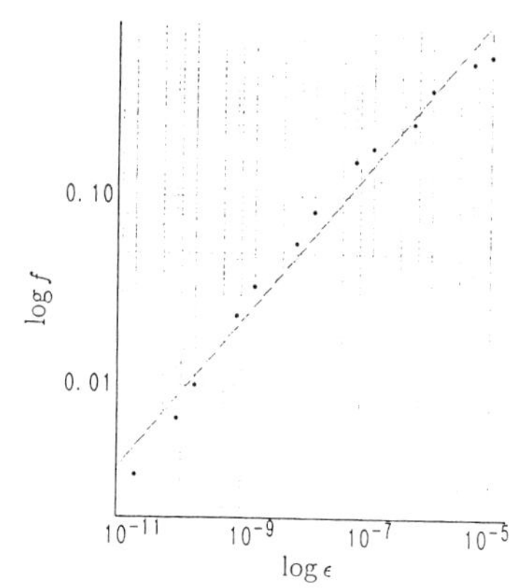

Figure 6.53 $\bar{f}$ vs ϵ in the log–log scale.

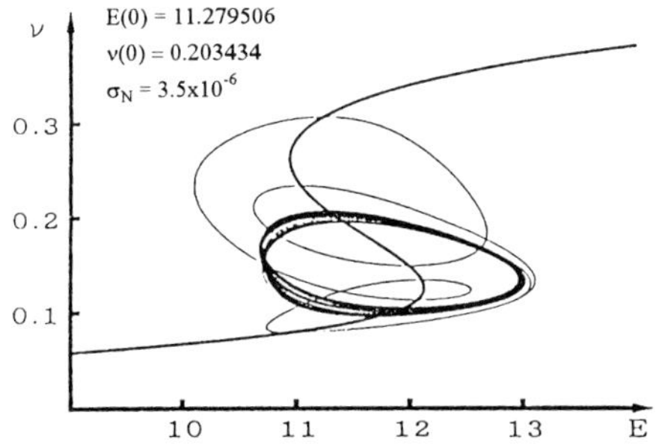

Figure 6.54 Noise-induced intermittency in the model simulation. The parameters are $E_{ac} = 4.5$, $f_0 = 10/15$ and $E_0 = 14.530$. The initial conditions are $E(0) = 11.279506$ and $\nu(0) = 0.203134$. Noise was added ($\sigma_N = 3.5 \times 10^{-6}$).

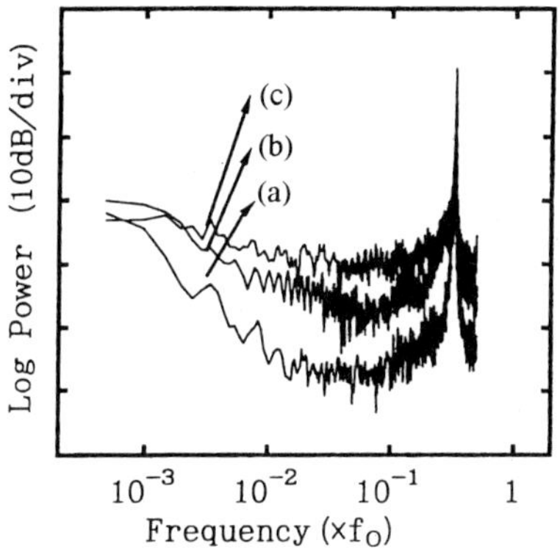

Figure 6.55 Fourier power spectrum for the noise-induced intermittency with different noise level. Simulated by the approximated model of the impact-ionization avalache. (a) $\sigma_N = 2.07 \times 10^{-6}$, (b) $\sigma_N = 3.50 \times 10^{-6}$ and (c) $\sigma_N = 5.31 \times 10^{-6}$.

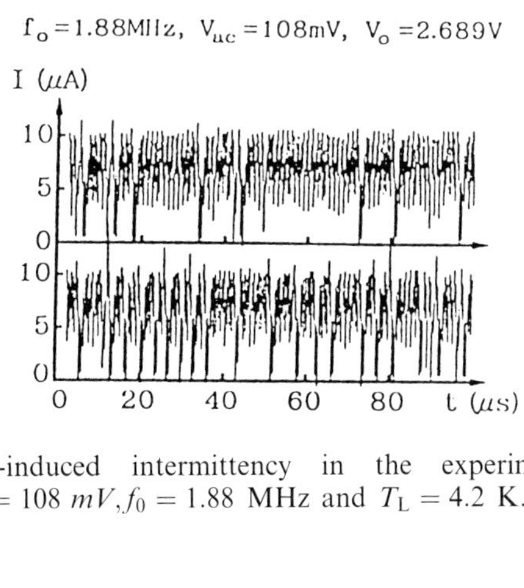

Figure 6.56 Noise-induced intermittency in the experiments of n-GaAs. $V_0 = 2.689$ V, $V_{ac} = 108\ mV$, $f_0 = 1.88$ MHz and $T_L = 4.2$ K. Experiments by K. Aoki.

case, the numerical results in Fig. 6.53 are: 54.7% of the initial conditions are uncertain for $\epsilon = 1.0 \times 10^{-5}$; $\bar{f} = 5.67\%$ for $\epsilon = 5.0 \times 10^{-9}$; $\bar{f} = 0.33\%$ for $\epsilon = 1.0 \times 10^{-11}$. This means that an extraordinarily high accuracy of the initial conditions is required for the reliable prediction of the eventual final state [62–64].

Such a final state sensitivity induces noise-induced intermittency in semiconductor chaos systems. In particular the intermittency may be quite significant even under imperceptive ambient noise if the attractors locate close to the basin boundary. For instance, by adding a small random fluctuating force $\xi(t)$ in a system equation ($\xi(t)$ is added in the rate equation for ν), the noise-induced intermittency is simulated as in Fig. 6.54, where $\langle\xi(t)\rangle = 0, \sigma_{\mathrm{N}} = 3.5 \times 10^{-6}$. Consequently, the Fourier power spectra show $1/f$ noise in Fig. 6.55 where the exponent decreases by increasing the noise power level, especially noting $1/f$ noise for $\sigma_{\mathrm{N}} = 2.07 \times 10^{-6}$. In the experiments, two different types of intermittency are observed in the prebreakdown regime (see section 5.4); one is due to the reinjective switching (or boundary crisis) as given in section 5.5 (Fig. 5.83), and the other to the noise-induced switching between the periodic attractor and chaos. The former exhibits spike currents due to current filamentation, while the latter is the intermittent behavior between two small-size attractors without showing the spike currents. A typical example of the current waveform is given in Fig. 6.56. The intermittent behavior is best understood by the fractal basin boundary.

References

[1] H. Takayasu, *Fractals* (Asakura, Tokyo, 1986) (in Japanese).
[2] J. Feder, *Fractals* (Plenum Press, New York, 1988).
[3] B.B. Mandelbrot, *The Fractal Geometry of Nature* (W.H. Freeman and Company, New York, 1983).
[4] H.O. Peitgen and P.H. Richter, *Beauty of Fractals* (Springer, Berlin, 1986).
[5] H.G. Schuster, *Deterministic Chaos*, 2nd ed. (Physik-Verlag, Weinheim, 1988).
[6] A.A. Few, J. Geophys. Res. **75** (1970) 7517; L. Niemeyer, L. Pietronero and H.J. Wiesmann, Phys. Rev. Lett. **52** (1984) 1033.
[7] e.g., see Proc. Int. Conf. on Noise in Physical Systems and 1/f Fluctuations, eds. T. Musha, S. Sato and M. Yamamoto (Ohmusha, Tokyo, 1991).
[8] S.R. Forrest and T.A. Witten Jr, J. Phys. **A12** (1979) L109.
[9] R.F. Voss, *Lecture Note in Math*, Vol. **1035** (Springer, Berlin, 1983) p. 153.
[10] R. Thom, *Structural Stability and Morphogenesis* (Benjamin/Cummings, London, 1975).
[11] F. Takens, *Lecture Notes in Math.* Vol. **898**, eds. D.A. Rand and L.S. Young (Springer, Berlin, 1981) pp. 366–381; N.H. Packard, J.P. Crutchfield, J.D. Farmer and R.S. Shaw, Phys. Rev. Lett. **45** (1980) 712.
[12] H.D.I. Abarbanel, *Analysis of Observed Chaotic Data* (Springer, Berlin, 1996).
[13] F.C. Moon, *Chaotic and Fractal Dynamics* (John Wiley & Sons, New York, 1992).

[14] P. Grassberger and I. Procaccia, Physica **9D** (1983) 189.
[15] J.L. Kaplan and J.A. Yorke, *Functional Differential Equations and Approximations of Fixed Points*, eds. by H.-O. Peitgen and H.-O. Walther, *Lecture Notes in Math*, Vol. **730** (Springer, Berlin, 1979) p. 204.
[16] J.D. Farmer, Physica **4D** (1982) 366.
[17] R. Conte and M. Dubois, in *Chao's II*, ed. by H. Bai-Lin (World Scientific, Singapore, 1990) 441.
[18] H.G.E. Hentschel and I. Procaccia, Physica **8D** (1983) 435.
[19] B. Malraison, P. Atten, P. Berge and M. Dubois, J. Phys. (Paris) Lett. **44** (1983) L-897.
[20] J.-P. Eckmann and D. Ruelle, Rev. Mod. Phys. **57** (1985) 617.
[21] K. Aoki, Proc. of a NATO Advanced Research Workshop on Negative Differential Resistance and Instabilities in 2-D Semiconductors, eds. N. Balkan, B.K. Ridley and A.J. Vickers (Plenum, London, 1993) NATO ASI Series B: Physics Vol. 307, pp. 393–407.
[22] G. Benettin, L. Galgani and J.M. Strelcyn, Phys. Rev. **A14** (1976) 2338.
[23] I. Shimada and T. Nagashima, Prog. Theor. Phys. **61** (1979) 1605.
[24] A. Wolf, J.B. Swift, H.L. Swinney and J.A. Vasano, Physica **D16** (1985) 285.
[25] M. Sano and Y. Sawada, Phys. Rev. Lett. **55** (1985) 1082.
[26] J. Peinke, J. Parisi, O.E. Rössler and R. Stoop, *Encounter with Chaos* (Springer, Berlin, 1992).
[27] K. Aoki, A. Matsuura and Y. Takahashi, Semicond. Sci. Technol. **9** (1994) 607.
[28] O.E. Rössier: Phys. Lett. **57A** (1976) 397; Z. Naturforsch **31a** (1976) 259.
[29] U. Rau, K.M. Mayer, J. Parisi, J. Peinke, W. Clauss and R.P. Huebener, Solid State Electron. **32** (1989) 1365.
[30] K. Aoki and K. Yamamoto, Proc. 21st Int. Conf. Phys. Semicond. (Beijing, 1992) eds. by P. Jiang and H. Zheng (World Scientific, Singapore, 1992) 261.
[31] T.C. Halsey, M.H. Jensen, L.P. Kadanoff, I. Procaccia and B. Shraiman, Phys. Rev **A32** (1986) 1141.
[32] E.G. Gwinn and R.M. Westervelt, Phys. Rev. Lett. **57** (1986) 1060.
[33] E.G. Gwinn and R.M. Westervelt, Phys. Rev. Lett. **59** (1987) 157.
[34] G.A. Held and C. Jeffries, Phys. Rev. Lett. **56** (1986) 1183.
[35] R.R. Prasad, C. Meneveau and K.R. Sreenivasan, Phys. Rev. Lett. **61** (1988) 74.
[36] M.H. Jensen, L.P. Kadanoff and A. Libchaber, I. Procaccia and J. Stavans, Phys. Rev. Lett. **55** (1985) 2798.
[37] K. Aoki, Semicond. Sci. Technol. **7** (1992) B474.
[38] A.M. Kahn, D.J. Mar and R.M. Westervelt, Phys. Rev. **B43** (1991) 9740.
[39] M. Dubois, P. Berge and V. Croquette, J. Phys. (Paris) Lett. **43** (1982) L295.
[40] A. Brandstater, J. Swift, H.L. Swinney, A. Wolf, D. Farmer, E. Jen and P. Crutchfield, Phys. Rev. Lett. **51** (1983) 1442.
[41] J.L. Hudson, M. Hart and P. Marinko, J. Chem. Phys. **71** (1979) 1601.
[42] J. Stavans, F. Heslot and A. Libchaber, Phys. Rev. Lett. **55** (1985) 596.
[43] J.A. Glazier and A. Libchaber, IEEE Trans. Circ. Syst. **35** (1988) 790.
[44] H.L. Swinney and J. Maselko, Phys. Rev. Lett. **55** (1985) 2366.
[45] G.A. Held and C. Jeffries, Phys. Rev. Lett. **55** (1985) 887; G.A. Held and C. Jeffries, in *Dimensions and Entropies in Chaotic Systems*, ed. G. Mayer-Kress (Springer, Berlin, 1986) pp. 158–170.
[46] S. Martin and W. Martienssen, Phys. Rev. Lett. **56** (1986) 1522; in *Dimensions and Entropies in Chaotic Systems*, ed. G. Mayer-Kress (Springer, Berlin, 1986) pp. 191–197; for the forced charge density waves, see S.E. Brown, G. Mozurkewich and G. Gruner, Phys. Rev. Lett. **52** (1984) 2277.

[47] M.R. Guevara, L. Glass and A. Shrier, Science **214** (1981) 1350.
[48] M.J. Feigenbaum, L.P. Kadanoff and S.J. Shenker, Physica **5D** (1982) 370.
[49] S. Ostlund, D. Rand, J. Sethna and E.D. Siggia, Physica **8D** (1983) 303.
[50] S.J. Shenker, Physica **5D** (1982) 405.
[51] D. Rand, S. Ostlund, J. Sethna and E.D. Siggia, Phys. Rev. Lett. **49** (1982) 132.
[52] M.H. Jensen, P. Bak and T. Bohr, Phys. Rev. **A30** (1984) 1960.
[53] Hao Bai-Lin, *Chaos* (World Scientific, Singapore, 1984) and references therein.
[54] Hao Bai-Lin, *Chaos II* (World Scientific, Singapore, 1990).
[55] B.V. Chirikov, Phys. Rep. **52** (1979) 264.
[56] T. Bohr, P. Bak and M.H. Jensen, Phys. Rev. **A30** (1984) 1960.
[57] G.A. Held, C. Jeffries and E.E. Haller, Phys. Lett. **52** (1984) 1037.
[58] G.A. Held, C. Jeffries and E.E. Haller, Proc. 17th Int. Conf. Phys. Semicond. eds. J.D. Chadi and W.A. Harrison (San Francisco, 1984) 1289.
[59] P. Cvitanovic, M.H. Jensen, L.P. Kadanoff and I. Procaccia, Phys. Rev. Lett. **55** (1985) 343.
[60] R.M. Westervelt and S.W. Teitsworth, J. Appl. Phys. **57** (1985) 5457.
[61] C. Grebogi, S.W. Mcdonald, E. Ott and J.A. Yorke, Phys. Lett. **99A** (1983) 415.
[62] C. Grebogi, E. Ott and J.A. Yorke, Physica **7D** (1983) 181.
[63] C. Grebogi, E. Ott and J.A. Yorke, Science **238** (1987) 632.
[64] G. Grebogi, E. Otto, F. Romeiras and J.A. Yorke, Phys. Rev. **A36** (1987) 5356.
[65] P. Berge and M. Dubois, Phys. Lett. **93A** (1983) 365.
[66] J.A. Yorke and E.D. Yorke, J. Stat. Phys. **21** (1979) 263.
[67] M. Iansiti, Qing Hu, R.M. Westervelt and M. Tinkham, Phys. Rev. Lett. **55** (1985) 746.
[68] K. Aoki and N. Mugibayashi, Appl. Phys. **A48** (1989) 161.

Chapter 7

Novel phenomena in semiconductors

7.1 Dissipation energy and its fluctuation

In chaos theory, little attention has been paid to energy dissipation and its fluctuation [1] in the dynamical systems approach, while there are important theorems such as Nyquist theorem and Onsager's theorem in the traditional statistical mechanics [2–4]. Rather, the main part of chaos theory covers the universal features of bifurcations, structure of strange attractor and various kinds of universality in deterministic chaos in the dissipative dynamical systems [5–8]. It might seem curious that the dissipation and fluctuation problems do not enter the dynamical laws of chaos theory. However, it may be very natural as only the dynamics of macroscopic physical quantities are treated and one can only find the deterministic nature of the dynamical systems from chaos theory. In nonlinear dynamical systems, there apparently exists time's arrow or the broken time-reversal symmetry which is tightly related to the cooperative phenomenon and the energy dissipation among a huge number of molecules [9]. So, more precisely, it may be true to say that spatially extended dynamical systems with infinite degrees of freedom are also ruled by energy dissipation and fluctuation. If the dissipation rate (or energy) is known or defined adequately in a nonlinear dynamical system, one can argue chaotic fluctuation from the viewpoint of the dissipation/fluctuation problem. Here the semiconductor chaos system is presented from a new approach given by Mori *et al.* [1]. In semiconductors the dissipation energy is well known, so that the argument seems to be more realistic than many other dynamical systems. The approach [1] is an extended version of a statistical–mechanical formalism which characterizes globally the chaotic fluctuation by the coarse-grained divergence rate of nearby orbits [10, 11]. The q-phase transition found by the statistical–mechanical formalism is also supplemented here.

Let us start with an example of the driven pendulum given by Mori *et al.* [1]:

$$\ddot{x} + \gamma\dot{x} + \omega_0^2 \sin x = a \cos(\omega t), \tag{7.1}$$

where $x(t)$ is the pendulum angle, γ (> 0) is the friction constant, ω_0 the natural frequency for small oscillations, and a and ω are the amplitude and frequency of a periodic driving force. The dissipation rate is given by $W_f(t) = \gamma\dot{x}^2(t)$, which is the work done by the system against friction per unit time. With the driving period $T = 2\pi/\omega$, one can take the Poincaré section plot $\mathbf{X}_i = \{x(t_i), \dot{x}(t_i)\}$ at time $t_i = iT$ $(i = 0, 1, 2, \ldots)$. Then, the two-dimensional (2D) dynamics is given by a Poincaré map $\mathbf{X}_{i+1} = F(\mathbf{X}_i)$. The long-term average of the dissipation rate $\bar{W}_f(t)$ can be written as: $\bar{W}_f = \lim_{n\to\infty} U_n(\mathbf{X}_0)$, with

$$U_n(X_0) = \frac{1}{n}\sum_{i=0}^{n-1} u(\mathbf{X}_i), \tag{7.2}$$

$$u(\mathbf{X}_i) = \frac{1}{T}\int_{iT}^{(i+1)T} \mathrm{d}t, \qquad W_f(t) = \frac{\gamma}{T}\int_{iT}^{(i+1)T} \dot{x}(t)\,\mathrm{d}x(t). \tag{7.3}$$

For a limit cycle of period mT, one has $\bar{W}_f = U_m = \gamma S/mT$, where $S \equiv \oint \dot{x}(t)\,\mathrm{d}x(t)$ is the area enclosed by the limit cycle in the $(x, \dot{x})$ phase space [1]. Thus, the T-average $u(X_i)$ reduces to the action integral in the phase space [1]. For chaos, the average of the dissipation rate over n periods $U_n(\mathbf{X}_{jn})$, $j = 0, 1, 2, \ldots$, in (7.2) fluctuates from interval to interval, having a probability density $P(U; n)$ for $U_n(\mathbf{X}_{jn})$ (or simply U_n) to take a value around U for given n:

$$P(U;n) = \langle\delta(U_n - U)\rangle = \exp[-n\psi(U)]P(U_\infty; n) \tag{7.4}$$

for large n, where $\psi(U)$ is a concave function of U with minimum $\psi(U_\infty) = 0$ at $U = U_\infty = \bar{W}_f$ [1]. From the discussions by Mori *et al.* [1], the spectrum $\psi(U)$ describes the statistical structure of the fluctuations of U_n with the power law:

$$\psi(U) = \begin{cases} \hat{U}^2 & \text{for } |\hat{U}| < \hat{U}^*, \\ \hat{U}^{1/\nu_\pm} \quad (1/2 < \nu_\pm < 1) & \text{for } |\hat{U}| > \hat{U}^*, \end{cases} \tag{7.5}$$

where $\hat{U} = U - U_\infty$. By investigating the spectrum $\Psi(U)$, the dissipation rate can be related to the dynamical structure of chaos. Mori *et al.* [1] further extended the problem to 1D map (the logistic map) and 2D map (the Poincaré map for the kicked rotor) by defining the truncated dissipation rate (TDR):

$$u^T(\mathbf{X}_i) = \dot{x}^2(t_{i+1}) = (x_{i+1} - x_i)^2/\tau^2, \tag{7.6}$$

where $\tau = (1 - b)/\gamma b$, and $b = \exp(-\gamma T)$. Hence, the average $U_n^T(X_0)$ over

n periods is:

$$U_n^T(X_0) = (1/n) \sum_{i=0}^{n-1} u^T(\mathbf{X}_i), \tag{7.7}$$

whereby the spectrum $\psi_T(U)$ has been argued [1]. Also, in continuous dynamical systems such as the driven pendulum (7.1), one can argue the TDR and $\psi_T(U)$ by taking the Poincaré map $\mathbf{X}_{i+1} = F(\mathbf{X}_i)$ and compare with the spectrum $\psi(U)$. In Fig. 7.1, the spectrum $\psi_T(U)$ (TDR) and the Poincaré section plot are shown for the band-merging crisis, as a function of the driving force amplitude a [1]. Before the crisis (Fig. 7.1a), the attractor consists of three bands. After the crisis, the three bands merge into one band, spreading over a chaotic repeller which lies among the three bands (Fig. 7.1d). Just before the crisis, $\psi_T(U)$ is approximately quadratic with $\nu_\pm \sim 1/2$ in accordance with the central limit theorem [4, 12]. From the results by Mori *et al.* [1] and the results in the semiconductor chaos system (see later), one can say that the probability density $P(U; n)$ for the stationary chaos is simply Gaussian or composed of several Gaussian distributions. Here, a reminder is given of the central limit theorem [4, 12]. Let X_j $(j \geq 1)$ be a sequence of independent and identically distributed random variables, where the mean $\langle X_j \rangle = m$ and the variance $\langle (X_j - m)^2 \rangle = \sigma^2$. Then the sum $S_n = X_1 + X_2 + \ldots + X_n$ $(n > 1)$ is again a random variable, where the mean $\langle S_n \rangle = nm$ and the variance $\sigma^2(S_n) = n\sigma^2$. By converting to a new variable

$$Y_n = \frac{S_n - nm}{n}, \tag{7.8}$$

one apparently has $\langle Y_n \rangle = 0$ and $\sigma^2(Y_n) = 1$. Then the central limit theorem

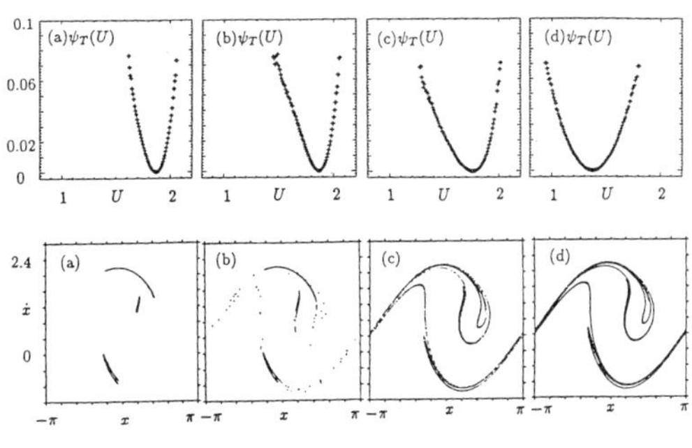

Figure 7.1 Spectrum $\psi_T(U)$ of $U_n^T(X_0)$ and the Poincaré section plot for the driven pendulum (7.1) with $\gamma = 0.258$, $\omega_0 = 1$, $\omega = 0.65$ near the band merging crisis $a_\omega = 0.728384\ldots$, where $n = 100$, and $N = 5 \times 10^5$. (a) Just before the crisis, $a = a_\omega - 0.8 \times 10^{-6}$. (b) Just after the crisis, $a = a_\omega + 1 \times 10^{-6}$, where $\psi_T(U)$ has a linear slope $S_U \simeq -0.2$ on the left-handside of U_∞^T [1]. (c) $a = a_\omega + 0.0004$, where the variance is about three times larger than that at (a). (d) $a = a\omega + 0.02$ [1].

mentions the probability distribution of Y_n in the limit $n \to \infty$. The probability $P(x < Y_n < x + dx)$ of Y_n taking a value around x can be approximated as:

$$\lim_{n\to\infty} P(x < Y_n < x + dx) = \frac{1}{\sqrt{2\pi}} \exp(-x^2/2)\, dx. \tag{7.9}$$

The central limit theorem can be applied to all cases where there are no correlations among different events but in which the outcome of the different events piles up via a sum (S_n) [12]. In deterministic chaos systems, the Poincaré map X_{i+1} vs X_i has a strong correlation, but the correlation among $S_{n,0}$, $S_{n,1}$, $S_{n,2}, \ldots$ is lost considerably by taking a temporal sum

$$S_{n,j} = \frac{1}{n}\sum_{i=0}^{n-1} X_{i+jn}$$

where $j = 0, 1, \ldots$. Therefore, one may expect a Gaussian probability density $P(U; n)$ for a large n, as seen in Fig. 7.1a, d, for stationary chaos. Returning to the problem in Fig. 7.1, the central limit theorem breaks down just after the crisis, as shown in Fig. 7.1b, where the spectrum $\psi_T(U)$ has a linear slope $S_U \sim -0.2$ in a regime $U_1^* < U < U_2^*$ ($< U^T$), implying $\nu \simeq 1$ there (see (7.5)) [1]). As described later, the linear slope indicates a discontinuous phase transition in the thermodynamical formalism [10, 11, 13, 14], which is caused by the collision of the chaotic attractor with an unstable periodic orbit (unstable period 3) at the crisis point. By increasing the driving force amplitude further, the linear slope bends down as shown in Fig. 7.1c, and finally in Fig. 7.1, the spectrum $\psi_T(U)$ becomes quadratic again [1]. In Fig. 7.1, the mean rate U^T decreases with increasing a, while the mean value of the coarse-grained local expansion rate or the Lyapunov exponent Λ^∞ increases [1, 11, 13]. This indicates the decrease in the mean spatial extent of the attractor for one iterate. Thus, energy dissipation decreases as chaos develops [1]. Can a similar relation be found in semiconductor chaos systems? The aim is to find the relation at the crisis point for the reinjective switching (see section 5.6). In the following, the ac-mode dissipation power density is denoted $U(t) = \tilde{J}\tilde{E}$ in the model simulation (or the ac-mode dissipation power $U(t) = \tilde{I}\tilde{V}$ in the experiments), where the current density $\tilde{J} = J - \langle J \rangle$ and the electric field $\tilde{E} = E - \langle E \rangle$ with the means $\langle \tilde{J} \rangle = 0$ and $\langle \tilde{E} \rangle = 0$ (or the current $\tilde{I} = I - \langle I \rangle$, the sample voltage $\tilde{V} = V - \langle V \rangle$ in the experiment). The system is a series circuit, composed of the semiconductor sample, the load resistor, the dc voltage source E_0 and the ac voltage source $E_{ac} \sin(2\pi f_0 t)$. The averaged power density of the chaotic fluctuation in the model simulation is given by:

$$U_n = -c\langle \tilde{J}^2 \rangle_n + \langle \tilde{J} E_{ac} \sin(2\pi f_0 t) \rangle_n, \tag{7.10}$$

or

$$U_n = \frac{1}{nT} \int_{njT}^{n(j+1)T} U(t)\, \mathrm{d}t, \quad (j = 0, 1, 2, \ldots), \tag{7.11}$$

where c is the slope of the load line, and $\langle \ldots \rangle_n$ denotes the average over n periods ($T = 1/f_0$). In the r.h.s. of (7.10), the first term is the energy dissipation rate by the load resistor and the second term is the energy supply rate by the ac voltage source. Equation (7.10) is equivalent to the average U_n:

$$U_n = -R_{\mathrm{L}} \langle \tilde{I}^2 \rangle_n \; + \; \langle \tilde{I} V_{\mathrm{ac}} \; \sin{(2\pi f_0 t)} \rangle_n, \tag{7.12}$$

where R_{L} is the load resistance. Here, in addition $\langle J \rangle \langle E \rangle = -c \langle J \rangle^2 + \langle J \rangle E_0$, or $\langle I \rangle \langle V \rangle \; = -R_{\mathrm{L}} \langle I \rangle^2 + \langle I \rangle V_0$. Using the approximated model (5.35)–(5.38) of the impact-ionization avalanche, the energy dissipation rate is investigated at the crisis point in the postbreakdown regime. Figure 7.2 shows the results of the probability density $P(U; n)$ in (7.4) and the spectrum $\psi(U)$, where $n = 11$, and (a) just before crisis: $E_0 = 14.555$, $E_{\mathrm{ac}} = 4.5, f_0 = 0.735$, and (b) just after crisis (reinjective switching): $E_0 = 14.56$, $Eac = 4.5$, $f_0 = 0.735$. For the stationary chaos in Fig. 7.2a, the probability density is not a simple Gaussian shape but is composed of several Gaussian distributions, which may be caused by a relatively small value of n (for $n \approx 1$, one can see many sharp structures similar to the invariant probability measure). Correspondingly, the spectrum $\psi(U)$ shows some deviation from a simple quadratic shape. Just after the crisis (Fig. 7.2b), the probability density shows a tail at the negative part of U, giving rise to a linear slope in $\psi(U)$ in the range $-0.025 < U < -0.05$. The negative values of U_n indicate that the semiconductor system supplies the ac-mode power due to negative differential resistance, which means that the dissipation power by the load

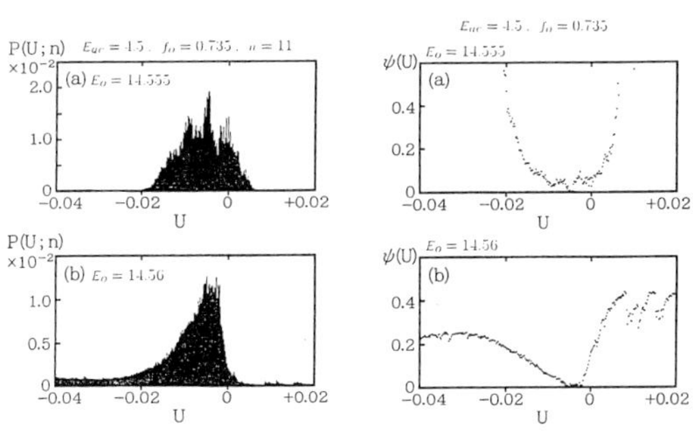

Figure 7.2 The probability density $P(U; n)$ and the spectrum $\psi(U)$ calculated by the approximate model of the impact-ionization avalanche, where $n = 11$. (a) Just before the crisis: $E_0 = 14.555$, $E_{\mathrm{ac}} = 4.5$, $f_0 = 0.735$, and (b) just after the crisis; $E_0 = 14.56$, $E_{\mathrm{ac}} = 4.5$, $f_0 = 0.735$.

resistance is larger than the power supplied by the ac voltage source (see (7.12)). The experimental results in Fig. 7.3 provide clear evidence for the universal behavior [1]. In Fig. 7.3, the spectrum $\psi(U)$ just before the crisis (broken line in Fig. 7.3a and the blow-up in Fig. 7.3b) agrees with a quadratic curve, satisfying the central limit theorem, where $U_\infty = 0.145\ \mu$W. Just after the crisis, the linear slope (solid straight line) can be seen in the range $0.134\ \mu\text{W} < U < 0.142\ \mu$W and U_∞ shifts towards a lower value, taking $U_\infty = 0.143\ \mu$W (Fig 7.3b). However, more significant behavior in the semiconductor chaos system is the appearance of a broad band peaking at $U \sim 0.115\ \mu$W (Fig. 7.3a). In the model simulation a flat band region was seen in Fig. 7.2b. This differs from typical results of the driven pendulum (Fig. 7.1) and the logistic map presented by Mori *et al.* [1] who found no such additional broad band. A possible reason for the large difference is the following. For the logistic map, the transition [1] is due to the interior crisis. The result for the logistic map after Mori *et al.* [1] is shown in Fig. 7.4. In the logistic map $x_{i+1} = f(x_i) = a - x_i^2$, the attractor forms three narrow bands (namely period 3 chaotic band) just before the crisis [1]. Just after the merging crisis, the iterated points distribute over the

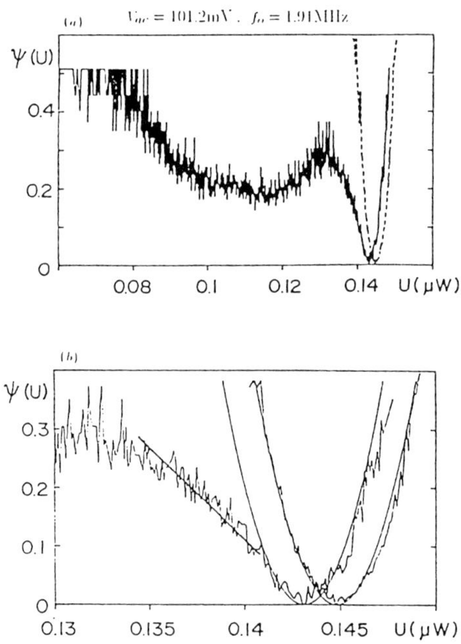

Figure 7.3 The spectrum $\psi(U)$ measured for the crisis in the periodically driven chaos experiments in n-GaAs. $V_{ac} = 101.2$ mV, $f_0 = 1.91$ MHz. (a) The solid line indicates the spectrum just after the crisis, and the broken line just before the crisis. (b) Magnification of the spectrum (U) in (a); the spectrum before the crisis agrees with the solid quadratic curve.

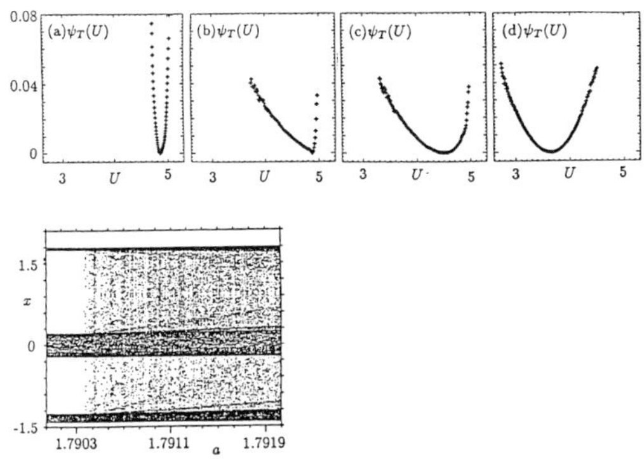

Figure 7.4 Spectrum $\psi_T(U)$ of $U_n^T(X_0)$ and the bifurcation diagram for the logistic map near the band merging crisis $a_\omega = 1.79032\ldots$, where $N = 1.3 \times 10^6$. (a) Just before the crisis, $a = a_\omega - 3 \times 10^{-5}$, where $n = 100$ (b) Just after the crisis, $a = a_\omega + 8 \times 10^{-5}$, where $n = 201$, and $\psi_T(U)$ has a linear slope $S_U \simeq -0.04$ on the left-handside of U_∞^T (c) $a = a_\omega + 0.001$, where $n = 180$, and the variance is about 40 times larger than that at (a). (d) $a = a_\omega + 0.03$, where $n = 150$ [1].

whole range, but never spread beyond the outermost boundaries, as seen in the bifurcation diagram in Fig. 7.4, which is a characteristic of the interior crisis. Therefore, the spatial extent of the attractor becomes smaller when it is averaged, contributing to the tail (linear slope) of $\psi_T(U)$ (which may be found from the truncated dissipation rate $(x_{i+1} - x_i)^2/\tau^2$ in (7.6) and from U_n^T in (7.7)). For the driven pendulum, the velocity of the pendulum distributes over the whole band-merged attractor taking effectively smaller values on average after the crisis (Fig. 7.1), which results in a similar effect such as in the logistic map. In the semiconductor chaos system, the transition (reinjective switching) is the **boundary crisis** or the **exterior crisis** [15–17], as described insections 5.6 and 6.5, where two effects are expected on looking, for instance, at the Poincaré section plots for the reinjective switching in Fig. 5.78. Just after the crisis, the iterated points are reinjected into the unstable saddle and fill the inner part of the Poincaré section, and thus the average spatial extent of the attractor becomes smaller, which gives rise to a linear slope in $\psi(U)$. At the same time, the iterated points spread beyond the old boundary by the crisis, meaning that the chaotic orbit experiences switching from a small size attractor of the low conductivity state to a high conductivity state. The high conductivity state formed by the impact-ionization avalanche produces the lowest energy dissipation (and it is negative in the ac-mode power density), revealing a broad band in $P(U;n)$. Thus, besides the effect seen in the driven pendulum and the logistic map, there is also a remarkable effect in the dissipation energy induced by the boundary crisis.

For the autonomous case, where $E_{ac} = 0$, $U_n = -R_L\langle\tilde{I}^2\rangle_n$ from (7.12). Thus, it is more obvious that the dissipation power of chaotic fluctuation is always negative. It can be said that the semiconductor chaos system behaves as an active element in which the spontaneous chaotic fluctuations generate power. This significant behavior is caused by the appearance of the negative differential resistivity in the S-shaped I–V characteristic in the dynamical regime, whereby the orbital $(I(t),\ V(t))$ produces negative dissipation power when referenced by the offset levels $\langle I \rangle$ and $\langle V \rangle$. By taking a limit $n \to \infty$, the average $U_\infty = -R_\mathrm{L}\,\sigma_I$, where σ_I is the variance of the current fluctuation. The relation between the dissipation energy rate and the fluctuation at the crisis point is a universal property of chaos which can be applied to both the periodically driven systems and the autonomous systems (spontaneous generation of chaos under magnetic fields [18–20]). Note especially that the central limit theorem holds for steady-state chaos, while it is broken at the crisis point. When viewed from the fluctuation theory [12, 21], the finding may be important compared with the microscopic fluctuation theory, such as in the Markov and non-Markov processes [12, 22, 23] and the related Langevin equation [12, 22]. The intermittency just after the crisis point corresponds to the non-Markov process with the long time tail in the autocorrelation function.

Finally, the statistical–mechanical formalism which describes the q-phase transition [10, 11, 13, 14] is given. Nonlinear dynamical systems exhibit complex local structures along the strange attractors, ruled by the dynamical law $F(\mathbf{X})$. The local divergence rate of nearby orbits along an orbit $\{\mathbf{X}_t\}$ $(t = 0, 1, 2, \ldots)$ generated by the map $\mathbf{X}_{t+1} = F(\mathbf{X}_t)$ is given by:

$$\lambda_1(X_t) = \log |DF(\mathbf{X}_t) \cdot \mathbf{U}_1(\mathbf{X}_t)|, \tag{7.13}$$

where $DF(\mathbf{X})$ is the Jacobian matrix and $\mathbf{U}_1(\mathbf{X})$ is the unit vector tangential to an unstable manifold of $F(\mathbf{X})$ at $\mathbf{X}$. Since nearby orbits are exponentially stretched along an unstable manifold and contracted along a stable manifold, chaotic orbits near a specified orbit $\{\mathbf{X}_t\}$ lie on the closure of the unstable manifold [11]. Therefore, a vector $\delta\mathbf{X}_t$ from $\mathbf{X}_t$ to a nearby orbit at time t lies on the extremely thin chaotic band and is thus tangential to the unstable manifold. Then the unit vector in (7.13) is given by $\mathbf{U}_1(\mathbf{X}_t) = \delta\mathbf{X}_t / \ |\delta\mathbf{X}_t|$. Since the linearized response at a vector point $\mathbf{X}_t$ is given by $\delta\mathbf{X}_{t+1} = DF(\mathbf{X}_t) \cdot \delta\mathbf{X}_t$, (7.13) shows the local divergence rate of nearby orbits. The local divergence rate $\lambda_1(\mathbf{X}_t)$ characterizes the local structure of the vector flows around $\mathbf{X}_t$. Along the whole strange attractor, the time-average of the local expansion rate fluctuates, taking its value from a negative to a positive one. Then, one must find the global spectrum of the time-averaged local expansion rate along the whole strange attractor. For continuous dynamical systems such as the Duffing equation or experimental systems, one should take the Poincaré section.

In order to describe the global spectrum by the statistical–mechanical formalism, the time average of $\lambda_1(\mathbf{X}_t)$ is introduced by:

$$\Lambda_n(\mathbf{X}_0) = (1/n) \sum_{t=0}^{n-1} \lambda_1(\mathbf{X}_t) \tag{7.14}$$

with a finite but large n, which is called the coarse-grained divergence rate of nearby orbits. By taking the limit $n \to \infty$, one has a Lyapunov number,

$$\Lambda^\infty = \lim_{n\to\infty} (1/n) \sum_{t=0}^{n-1} \lambda_1(\mathbf{X}_t) \tag{7.15}$$

which is equal to the largest Lyapunov number λ_1.For a given but larger n, in (14), $\Lambda_n(\mathbf{X}_0)$ fluctuates around Λ^∞, taking its value in the range $[\Lambda_{\mathrm{min}}, \Lambda_{\mathrm{max}}]$, where $\Lambda_{\mathrm{max}} > 0$. Thus, the values of $\Lambda_n(\mathbf{X}_0)$for different $\mathbf{X}_0$ distribute randomly in that range. The probability $P(\Lambda; n)$ of $\Lambda_n(\mathbf{X}_0)$ taking a value in the interval $[\Lambda, \Lambda + \mathrm{d}\Lambda]$ is given as [10, 11, 13, 14]:

$$P(\Lambda; n) \equiv \langle \delta(\Lambda_n - \Lambda) \rangle, \tag{7.16}$$

where $\delta(g)$ is the δ-function of g and $\langle \ldots \rangle$ denotes the long-term average

$$\langle G(\mathbf{X}_0) \rangle = \lim_{N\to\infty} (1/N) \sum_{t=0}^{N-1} G(\mathbf{X}_t). \tag{7.17}$$

Here, the δ-function (7.16) is the collection of the coarse-grained divergence rate taking its value Λ at many local regions of the whole attractor, which is possible by taking many initial conditions successively by (7.17). Then $P(\Lambda; n)$ takes the asymptotic scaling form [10, 11, 13, 14]:

$$P(\Lambda; n) = \exp\{-n\psi(\Lambda)\} P(\Lambda^\infty; n) \tag{7.18}$$

for $n \to \infty$, where $\psi(\Lambda) \geq \psi(\Lambda^\infty) = 0$. For (7.18), $\psi(\Lambda)$ can be approximated by [13]:

$$\psi(\Lambda) = -(1/n) \ln [P(\Lambda; n)/P(\Lambda^\infty; n)]. \tag{7.19}$$

In order to discuss large fluctuations Λ, the partition function is introduced through thermodynamic formalism [10, 11, 13, 14, 24–26] by:

$$Z_n(q) = \int \mathrm{d}\Lambda P(\Lambda; n) \exp[-n(q-1)\Lambda], \quad (-\infty < q < \infty) \tag{7.20}$$

and

$$\Phi_n(q) = -(1/n) \ln [Z_n(q)], \tag{7.21}$$

$$\Lambda_n(q) = \mathrm{d}\Phi_n(q)/\mathrm{d}q, \tag{7.22}$$

$$\sigma_n(q) = -\mathrm{d}\Lambda_n/\mathrm{d}q > 0. \tag{7.23}$$

Functions (7.19)–(7.23) are called the dynamical structure functions [12, 13]. Using (7.16), one has:

$$\Lambda_n(q) = \langle\Lambda\rangle_{q,n} = \frac{\int \mathrm{d}\Lambda P(\Lambda; n) \mathrm{e}^{-n(q-1)}\Lambda}{\int \mathrm{d}\Lambda P(\Lambda; n) \mathrm{e}^{-n(q-1)}}, \tag{7.24}$$

$$\sigma_n(q) = n\langle\{\Lambda - \Lambda_n(q)\}^2\rangle_{q,n}, \tag{7.25}$$

where $\Lambda_n(1) = \Lambda^\infty$, and $\langle\ldots\rangle_{q,n}$ is the q-weighted average. From (7.22), $\Phi_n(q) = \int_1^q \mathrm{d}q\Lambda_n(q)$, which is called the q-potential of $\Lambda_n(\mathbf{X}_0)$. For $q \to \infty$, the largest integrand in (7.24) is $P(\Lambda_{\min}; n) \exp[-nq\Lambda_{\min}]$, leading to $\Lambda_n(\infty) = \Lambda_{\min}$. Since $\mathrm{d}\Lambda_n(q)/\mathrm{d}q < 0$ from (7.23) and (7.25), the q-average $\Lambda_n(q)$ is a decreasing function of q, where $\Lambda_n(-\infty) = \Lambda_{\max}$ and $\Lambda_n(\infty) = \Lambda_{\min}$. Inserting (7.18) into (7.20) and taking the largest integrand for $n \to \infty$, the variational principle is given by:

$$\Phi_\infty(q) = \min_\Lambda \{\psi(\Lambda) + (q-1)\Lambda\}. \tag{7.26}$$

Sine $\psi(\Lambda)$ is differentiable, $\Lambda(q)$ is obtained by solving $\psi'(\Lambda) = 1 - q$ [11]. The coarse-grained divergence rate $\Lambda_n(X)$ fluctuates about the largest Lyapunov number $\Lambda^\infty = \lambda_1 > 0$ along the chaotic attractor. The positive Lyapunov number means mixing of chaotic orbits (ergodic), and thereby it means loss of memory. If the fluctuation of Λ around Λ^∞ is small, the spectrum $\psi(\Lambda)$ behaves as the Gaussian spectrum $\psi(\Lambda) \propto (\Lambda - \Lambda^\infty)^2$ in accordance with the central limit theorem. A typical example has been given [11] for the Lozi map [27]. When the strange attractor experiences homoclinic tangency or heteroclinic tangency, the coarse-grained divergence rate fluctuates largely in the range $[\Lambda_{\min}, \Lambda_{\max}]$, where $\Lambda_{\min} < 0$ and $\Lambda_{\max} > 0$. The present interests lie in such a large fluctuation. Mori *et al.* [11] have given an example (Fig. 7.5), by the Hénon map [6, 17, 28–30]:

$$\begin{bmatrix} x_{n+1} \\ y_{n+1} \end{bmatrix} = F(\mathbf{X}_n) \begin{bmatrix} 1 - ax_n^2 + by_n \\ x_n \end{bmatrix}, \quad (n = 0, 1, 2, \ldots) \tag{7.27}$$

The map function (7.27) [11, 17] is equivalent to, but a slightly different form of (4.91) [6, 28, 29], where the Jacobian is $J = -b$ for both cases. Figure 7.5a shows a two-band chaotic attractor obtained for $a = 1.10357$, $b = 0.3$ as denoted by A, where a stable manifold is shown by W^s and an unstable fixed point $\mathbf{X}^* = F(\mathbf{X}^*)$ is also indicated. With increasing a, the attractor touches tangentially the stable manifold (homoclinic tangency), resulting in continuous band merging (one-hand attractor), where the chaotic attractor lies on the unstable manifold. Figure 7.5c is the well-known chaotic attractor obtained for $a = 1.4, b = 0.3$ [6, 17, 28–30], where $\hat{W}^s$'s are parts of the stable manifold of another unstable fixed point $\mathbf{S} = F(\mathbf{S})$ [11]. As a is increased further, the chaotic attractor touches the stable manifold at $a = a_\mathrm{d} = 1.42692$ (Fig. 7.5d). Since W^s's are the basin

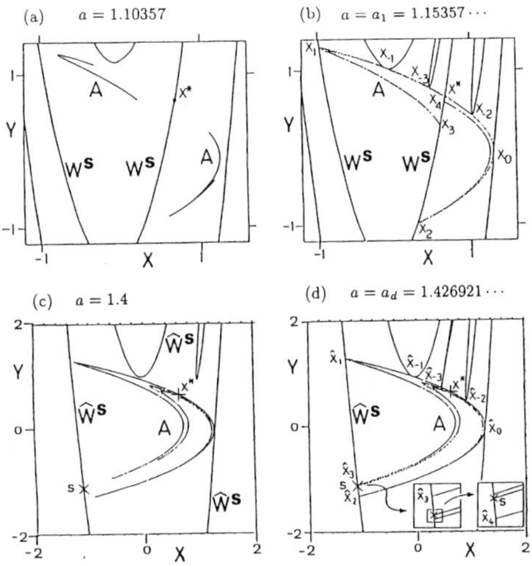

Figure 7.5 Chaotic attractor of the Hénon map (7.27) with $b = 0.3$: (a) two-band attractor ($a = 1.10357$); (b) continuous band merging ($a = a_1 = 1.15357\ldots$); (c) Hénon attractor ($a = 1.4$; (d) attractor destruction ($a = a_\mathrm{d} = 1.426921\ldots$), where A indicates chaotic attractors, W^s's the stable manifold of X^*, and $\hat{W}^s$'s the stable manifold of S [11].

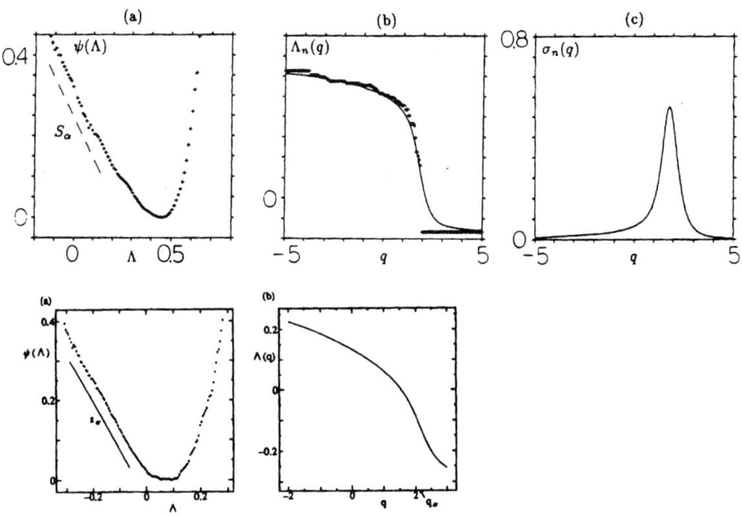

Figure 7.6 Spectrum $\psi(\Lambda)$, q-average $\Lambda_n(q)$ and q-variance $\sigma_n(q)$ for the Hénon attractor shown in Fig. 7.5 (c), where $n = 25$, $N = 5 \times 10^7$, $\Lambda^\infty \simeq 0.432$ and $s_\alpha \simeq -1.0$ [11].

boundary, the chaotic attractor for $a > a_d$ with $b = 0.3$ exhibits a boundary crisis via the heteroclinic tangency [15–17]. The chaotic orbit on the attractor escapes to its outside at the intersection points so that the attractor is destroyed for $a > a_d$ (for a details see Ref. [11]). Corresponding to Fig. 7.5, the dynamical structure functions are shown in Figs. 7.6 and 7.7 [11]. For the famous one-band Hénon attractor in Fig. 7.5, spectrum $\psi(\Lambda)$ of Fig. 7.6a exhibits a negative linear slope $s_\alpha = -1$, which means a large deviation from the Gaussian spectrum by taking its negative values, where the minimum of $\psi(\Lambda)$ locates at $\Lambda_\infty \sim 0.432$, the value of which is the largest Lyapunov number of the attractor. The linearity of $\psi(\Lambda)$ (Fig. 7.6a) which extends to the negative part of Λ means that the chaotic orbits include the unstable fixed point $\mathbf{X}^*(\mathbf{X}^* = F(\mathbf{X}^*))$. In Fig. 7.6b, the crosses, + represent the largest term approximation $\Lambda_n(q)$ obtained from $\psi(\Lambda) + (q-1)\Lambda =$ min, and exhibit a discontinuous transition at $q = q_\alpha = 2.0$ from a positive to a negative value [11], where $s_\alpha = 1 - q_\alpha$. This is called the q-phase transition [10, 11, 13, 14]. In Fig. 7.6(c), the q-variance $\sigma_n(q)$ has a sharp peak at $q = q_\alpha$ which must diverge in the limit $n \to \infty$ [11]. In the semiconductor model system (5.35)–(5.38) (one-level impact-ionization model A′), Tominaga and Mori [31] have shown the q-phase transition for toric chaos (see Fig. 7.6). Just after the homoclinic tangency ($a = a_1 = 1.15357$ in Fig. 7.5b, a positive linear slope $s_\beta \sim 1.80$ also appear (Fig. 7.7a). The homoclinic tangency gives rise to infinite number of singular tangency points $\mathbf{X}_0, \mathbf{X}_1, \ldots, \mathbf{X}_j, \ldots$ in Fig. 7.5b, where $\mathbf{X}_{j+1} = F(\mathbf{X}_j)$ and $\mathbf{X}^* = \lim_{j\to\infty} F^j(\mathbf{X}_0)$, and then such local structures of the tangency may be reflected in the spectrum $\psi(\Lambda)$. The q-phase transition in Fig. 7.7 can be best understood by a schematic figure of Fig. 7.8 [11]. Since the local regions of the tangency points show significant stretching of the chaotic attractor along the unstable manifold (on which the chaotic attractor lies) and since a tangency structure is mapped infinitely approaching an unstable fixed point $\mathbf{X}^*$ (Fig. 7.5b), such tangency structures may give the maximum divergence

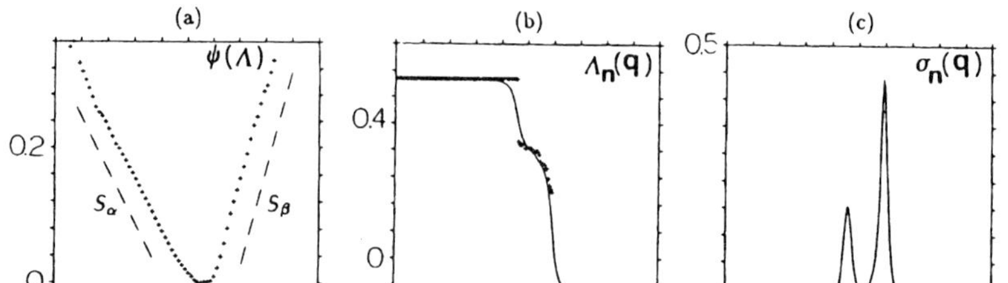

Figure 7.7 Dynamic structure functions for the continuous band merging in the Hénon map shown in Fig. 7.5 (b), where $n = 40$, $N = 10^8$, $\Lambda^\infty \simeq 0.315$, $\Lambda_{\max} = \Lambda_\infty(X^*) \simeq 0.550$, $s_\alpha \simeq -1.0$ and $s_\beta \simeq 1.80$ [11].

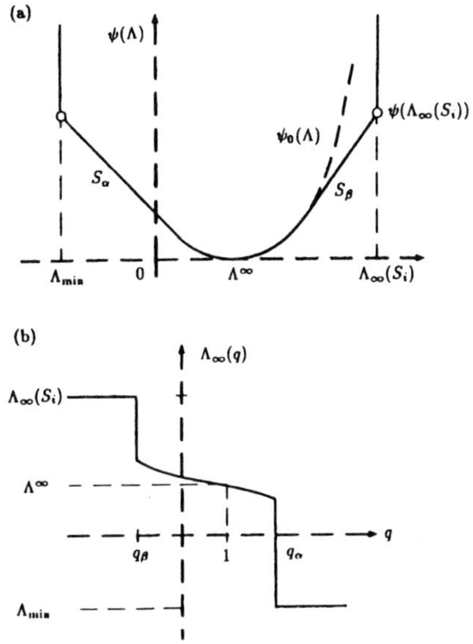

Figure 7.8 (a) Collision with a periodic orbit $\{S_i\}$ and the tangent s_β from $\psi(\Lambda_\infty(S_i))$ to $\psi_0(\Lambda)$ (b) The q_β-phase $q < q_\beta$, the normal phase $q_\beta < q < q_\alpha$ and the q_α-phase $q > q_\alpha$ [11].

rate, say, $\Lambda_\infty(S_i)$ in Fig. 7.8. Then, after the collision with the stable manifold, the value $\psi(\Lambda)$ lies outside the spectrum $\psi_0(\Lambda)$ which is one before the homoclinic tangency (Fig. 7.8). For detailed discussion of the positive linear slope s_β, see Ref. [11]. Far beyond the homoclinic tangency $(a > a_1)$, more positive Λ prevails along the attractor, and then the linearity in the positive region may be destroyed as in Fig. 7.6a. For the heteroclinic tangency in Fig. 7.5d, the spectrum $\psi(\Lambda)$ also exhibits the q-phase transition for the same reason for Fig. 7.7a. For the chaotic repeller (in the pendulum; see Fig. 7.1c, and in the Duffing equation), the q-phase transition also appear [11, 13]. The dynamical structure functions (7.19)–(7.23) were found by analogy with statistical thermodynamics. The parameter $q - 1$ corresponds to the inverse temperature or the magnetic field H for ferromagnets. Mori *et al.* [11] made the analogy, regarding $1 - q$, Λ, $\sigma_\infty(q)$ as the magnetic field H, the magnetization m per spin, and the susceptibility $\chi(H)$. The q-phase transition corresponds to the ferromagnetic phase transition of $m(H)$ below the Curie temperature [11].

7.2 Deterministic noise amplification

This section discusses a novel phenomenon related to the effect of external noise in semiconductor chaos systems. In experiments, external noise has a strong influence on dynamical systems in a way such that the fine structures of the deterministic properties are distracted, resulting in fuzzy bifurcation (section 6.2) and fuzzy fractal structures of the strange attractors. This means that nonlinear systems respond sensitively to the external noise due to local instability, which is an intrinsic nature of chaos systems (e.g., recall the exponential divergence of nearby trajectories or the sensitive dependence on initial conditions). In other words, nonlinear dynamical systems can amplify imperceptive ambient noise. Relating to the noise effects, Deissler and Farmer [32] have presented a new aspect theoretically using the term "deterministic noise amplifiers". In experiments on the impact-ionization avalanche, deterministic noise amplification occurs owing to extremely sensitive behavior of a current filament to an external perturbation or noise.

Let us start with the qualitative descriptions of ideas on deterministic noise amplifiers [32]. In nonlinear dynamical systems, the deterministic trajectory in the phase space possesses an essential character of local instabilities. Remember that the Lyapunov exponents (6.40) are the ensemble average of the local divergence rates, say, $\lambda_i(t)$ [32]; $\lambda_i = \langle \lambda_i(t) \rangle$. The local instabilities mean a positive local divergence rate $\lambda_i(t) > 0$. Therefore, a phase space point in a region of local instability can be strongly influenced by ambient noise. That is, local instabilities cause small fluctuations to be temporarily amplified, even though the system behavior (deterministic motion) is quite stable when averaged over the long term [32]. Whether the system behavior is periodic or chaotic, the noise amplifications occur when the local divergence rate is locally positive. As described by Deissler and Farmer [32], the positive divergence rate implies noise amplification and the spontaneous production of new macroscopic information on small fluctuations. However, the existence of deterministic chaos is not necessary for the amplification to occur. Even in the periodic regime, departures from a deterministic trajectory may be dramatic, showing unpredictable, chaotic-looking behavior. In the chaotic regime, the amplified signals are superimposed on the deterministic fluctuation (chaos), so that one cannot separate the signals from the superimposed fluctuations. For the above cases, the system behavior shows sensitive dependence on noise (note: sensitive dependence on both noise and initial conditions in the chaotic regime). Since a deterministic trajectory is asymptotically stable, ongoing fluctuations are required rather than small perturbations of initial conditions. One must pay careful attention to the term "deterministic amplifier", in order not to confuse it with "deterministic chaos". Deissler and Farmer [32] have given numerically several examples of deterministic amplification and discussed it quantitatively in terms of the covariance matrix and local divergence rate. By adding ongoing external random perturbations (Gaussian white noise) to the model systems, for instance to the Rössler equations, a limit cycle orbit begins to reveal chaotic-looking behavior, as shown in Fig. 7.9. Deissler and Farmer have given possible situations where noise amplification can be observed significantly, as described below [32].

- **Critical phenomena or bifurcations**. At a bifurcation parameter value a dynamical system becomes neutral–stable, which means that orbits are "unrestrained". Consequently, in the presence of noise a point in phase space can execute a random walk and thus cause a large fluctuation.
- **Noise-induced deterministic chaos**. At parameter values where regular behavior occurs, but where there is deterministic chaos nearby, the addition of external noise can cause a transition to chaos. In this case, the noise effectively shifts the bifurcation parameter value so that the transition to chaos occurs earlier.
- **Metastability**. Fluctuations whose amplitude exceed a certain critical size can induce transitions between the basins of two attractors, or between different wells of a potential. In order to make a transition between two regions of stability it is necessary to pass through a region of instability (e.g., the basin boundary). This type of the noise amplification is quite different from other types in that the instability is not on the attractor but occurs across the basin boundary. Therefore, the exponential noise amplification is a rare event and only occurs for perturbations exceeding a certain critical size. The noise amplification might be observed in the form of an intermittent burst (see section 6.5).
- **Convective instabilities**. In a spatially extended system, a localized perturbation can be carried along by the mean flow so that it grows only in a moving frame of reference, eventually damping at any given stationary point. In the limit of zero noise level (deterministic limit) there is no motion. If there are ongoing perturbations, the resulting behavior downstream in the stationary frame of reference looks very much like deterministic chaos. Examples of three model systems have

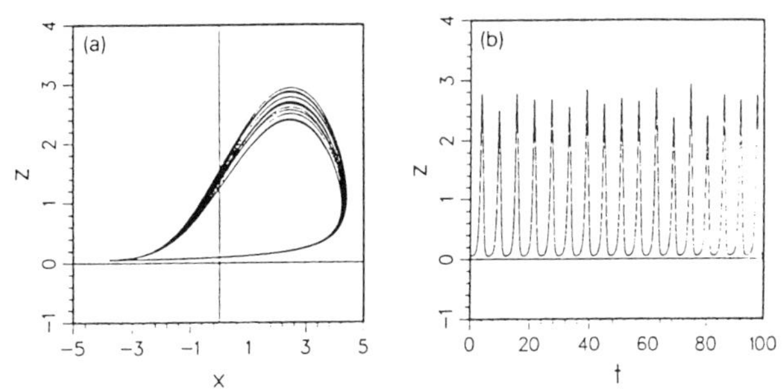

Figure 7.9 Typical trajectory of z on the Rössler attractor as a function of x and t, but with a Gaussian white noise of standard deviation 10^{-3} added to x, y and z at each time step $\Delta t = 0.01$ [32].

been given in a one-dimensional (1D) lattice of quadratic maps with asymmetric coupling (Deissler [33]), the 1D Ginzburg–Landau equation (Deissler [21]) and the Kuramoto–Sivashinsky equation (Brand and Deissler [34]).

This section discusses the deterministic noise amplification observed in a periodically driven chaos system of n-type GaAs [35–37]. Figure 7.10 shows a schematic of the experimental set-up, where $R_L = 10\ \text{k}\Omega$, $R'_L = 500\ \Omega$, r is the resistance of the circuit wire, and i_N is the noise current. Experiments have been performed at 4.2 K using n-type GaAs ($n \sim 2 \times 10^{14}/\text{cm}^3$) grown epitaxially on a semi-insulating substrate [36]. In this system, astonishingly, noise coming from radiowaves is amplified by the chaos element of n-GaAs (denoted by CH in Fig. 7.10). Voltages $e(t)$ and $V^*(t)$ are measured by a digital storagescope (sampling time of 40 MHz), and the time series can be stored in a microcomputer system. Total current I and sample voltage are evaluated by $I = e(t)/(R_L + R'_L)$ and $V = V^*(t) - e(t)$. The spatial pattern of a filamentary current for the sample used in Fig. 7.10 is of a plane filament [38]. In the postbreakdown regime, the filamentary current may experience enforced breathing under the dc + ac bias. During the breathing, the filamentary current is very sensitive to external noise owing to the nonlinear character of the I–V characteristic and the local instability of the orbit. As Deissler and Farmer [32] have suggested, chaos is not necessarily required for deterministic noise amplification. Periodic orbits also exhibit local instability against noise and thus are able to produce spontaneous macroscopic information via noise amplification. The deterministic noise amplification is especially remarkable near the bifurcation points of the period doubling sequence in the postbreakdown regime [35–37]. Near the bifurcation point at $V_0 = 4.425$ V, the phase portrait in Fig. 7.11(a) becomes fuzzy between periods 1 and 2 where the fuzzy bifurcation has been found to be due to the deterministic noise amplification by detecting radiowaves in

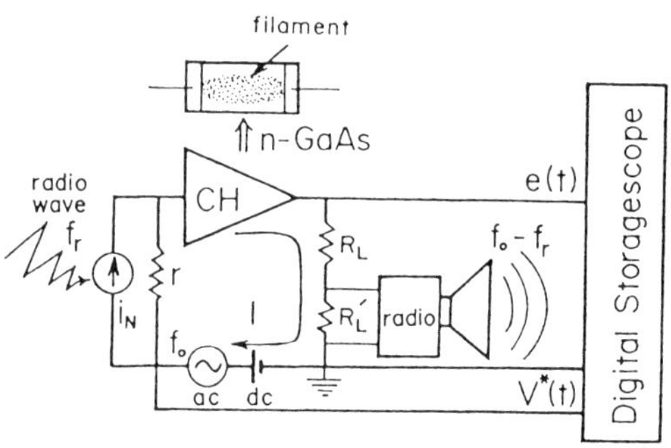

Figure 7.10 Noise equivalent circuit for the amplification of the radio signal by the current filament in n-GaAs [36].

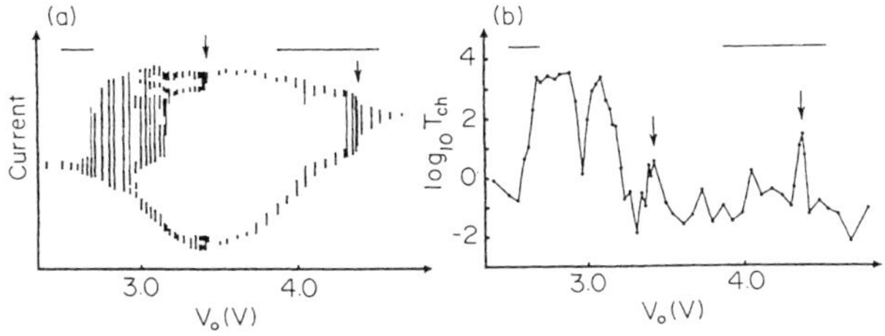

Figure 7.11 (a) Bifurcation diagram; (b) chaos temperature in n-GaAs at $T_L = 4.2$ *K*. $V_{ac} = 190$ mV, $f_0 = 1.83$ MHz. The bars indicate the bias region in which the radio signals were amplified [36].

Fig. 7.10. As shown later, the long-term behavior of the current looks like a torus of period 2. In Fig. 7.10, by feeding the total current into the antenna terminal of a radio, one can clearly hear the radio signal from broadcasting station(s) with frequency f_r, but by tuning the frequency to $f_0 - f_r$. The tuned frequency can be varied by varying f_0. The S-shaped I–V curve and the deterministic noise amplification disappear when the lattice temperature is raised (by raising the sample holder above the sea level of the liquid helium). By applying a magnetic field, the radio signal becomes noisy. These results indicate the extremely sensitive nature of the filamentary current to imperceptive ambient noise. In order to quantify the noise amplification, chaos temperature T_{ch} is defined by an extended version of the Nyquist theorem as follows [36]. Let

$$\tilde{I}(t_i) = I(t_i) - \langle I(t_i)\rangle, \tag{7.28}$$

$$\tilde{V}(t_i) = V(t_i) - \langle V(t_i)\rangle, \tag{7.29}$$

where $i = 0, 1, 2, \ldots,$ $\Delta t = t_i - t_{i-1}$ is the sampling time and $\langle I(t_i)\rangle$ and $\langle V(t_i)\rangle$ are the averaged values over N periods ($N \to \infty$). Then the averaged values $\langle\tilde{I}(t_i)\rangle$ and $\langle\tilde{V}(t_i)\rangle$ are calculated successively over m periods, which are denoted by I_m^k and V_m^k at the kth interval ($k = 0, 1, 2, \ldots, k_{\max}$; $k_{\max} \sim$ N/m; each interval has the data length of m/f_0):

$$I_m^k = \langle\tilde{I}(t_i)\rangle, \qquad V_m^k = \langle\tilde{V}(t_i)\rangle. \tag{7.30}$$

During the long-term sequence, I_m and V_m fluctuate from interval to interval and hence have variances σ_{I_m} and σ_{V_m}. By applying the Nyquist theorem, the chaos temperature T_{ch} is defined as:

$$\sigma_{I_m} = 4k_B T_{ch} m^{-2\alpha}\ \tilde{Y} f_0, \tag{7.31}$$

where $\tilde{Y}$ is the effective admittance given by $\tilde{Y} = a\bar{I}\bar{V}/\sigma_{V_m}$($a$: proportionality constant; $\bar{I}$, $\bar{V}$: dc components). Since the variances follow the scaling law $\sigma_{I_m} \propto m^{-\alpha}$, $\sigma_{V_m} \propto m^{-\alpha}$ ($\alpha \sim 3/2$ in the experiments) for deterministic chaos and noise amplification, T_{ch} does not depend on the choice of m

values if m is relatively large. For a periodic oscillation of period j, one must choose a specified m value such that m is equal to j times an integer. For the period doubling bifurcation, for instance, $m = 8$ and $a = 16$. Definition (7.31) is similar to the formula of noise temperature. For the deterministic noise amplification in the periodic regime, the temperature T_{ch} is not given in the frame of deterministic chaos but is an effective temperature which reflects the degree of local instability. In the chaotic regime, T_{ch} is best expressed as "chaos temperature" since the temperature reflects mainly the attractor size of chaos and partly the degree of local instability under the ambient noise.

In Fig. 7.11, the bifurcation diagram and the chaos temperature are shown as a function of dc bias, where $V_{ac} = 190$ mV and $f_0 = 1.83$ MHz. By decreasing V_0, period 1 in the postbreakdown regime bifurcates to period 2 at $V_0 \sim 4.384$ V, period 2 to period 4 at $V_0 \sim 3.415$ V, and finally chaos sets in for $V_0 \lesssim 3.154$ V (Fig. 7.11a). In Fig. 7.11b note that T_{ch} is plotted on a logarithmic scale. The horizontal bars mark the regions in which the amplified radio signals were picked up clearly. At the bifurcation points (see arrows), one can recognize large fluctuations which are due to deterministic noise amplification. Accordingly, in Fig. 7.11b, T_{ch} increases sharply near the bifurcation points (see arrows). The chaos temperature ranges widely from 0.01 to 5000 K throughout the entire region. In the chaotic regime, where T_{ch} is as high as 10^2–10^3 K, the radio signal mixed with the original chaos sound could be picked up. In the periodic regime, depending on the driving frequency f_0 ($f_0 = 1.83$ MHz in Fig. 7.11), the radio signal with frequency f_r could be picked up either by the torus modulation with $f_0 - f_r$ or by the mode-pulling $f'_r \to f_0$ when $f_r \sim f_0$. For the chaotic regime, it is also possible to pick up the radio signal in the range of wide band frequencies near $f_0/2$ or $f_0/4$. Figure 7.12 shows the waveform of torus modulation of period 2 and the phase portrait near the bifurcation point (note that the experimental parameters are different from those in Fig. 7.11), where $T_{ch} \sim 300$ K. Near the bifurcation points, especially near the bifurcation point between periods 1 and 2, the noise amplification factor β has been evaluated to be 10^3–10^5 [36], the values of which, depend on the control parameters V_{ac} and f_0 (β is $\sim 2 \times 10^3$ at $V_0 = 4.3$ V in Fig. 7.11). From the experimental results, it is found that T_{ch} is a good indicator for the

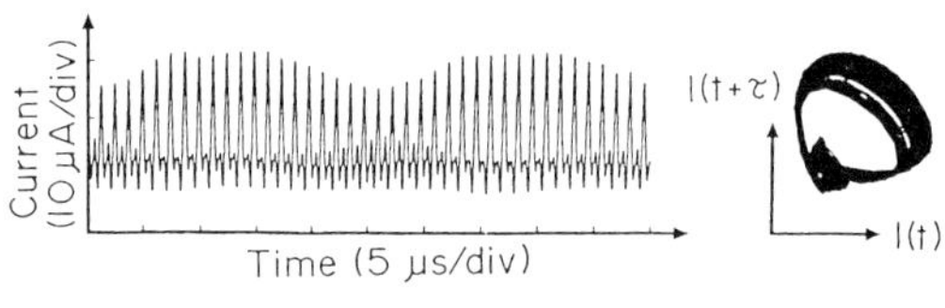

Figure 7.12 Waveform of torus modulation of period 2 and the phase portrait near the bifurcation point.

extremely sensitive dependence on noise (i.e., deterministic noise amplification) and also for the attractor size (for deterministic chaos). In the chaotic regime, high chaos temperature is a consequence of nonlinear mixing of deterministic chaos and amplified noise, so that one cannot separate the amplified information from the total signal.

The Lyapunov spectrum corresponding to Fig. 7.11 is shown as a function of V_0 in Fig. 7.13, where the spectrum has been evaluated with the embedding dimension four, and λ_3, λ_4 have been omitted from Fig. 7.13. In the experiments shown in Fig. 7.10, the circuit wire acts as an antenna, and thereby imperceptive noise current from radiowaves is effectively injected into the semiconductor system. The filamentary current responds sensitively to external perturbation under the oscillatory mode since the trajectory experiences the local instability region (NDR region) every period. For the oscillatory mode, instantaneous dissipative power is locally negative at the local instability region, suggesting that the filamentary current behaves like an active element in the ac-mode operation and thus amplifies the noise current. In the postbreakdown regime, the total current is given by [38]:

$$I(t) = \frac{V(t)}{\ell}\{\sigma_0(S_0 - S(t)) + \sigma S(t)\}, \tag{7.32}$$

where ℓ is the contacts spacing, $\sigma(\sigma_0)$ is the conductivity inside (outside) the filamentary current, and $S(t)$ and $S_0(t)$ are the cross-section of the filamentary current and the remaining part of the ohmic current, respectively. Near the bifurcation point, where the trajectory is neutral–stable under the noise-free condition, it is conjectured that the enforced breathing mode of $S(t)$ is sensitively coupled to amplify the noise current from the radiowave. In Fig. 7.12, the Poincaré section indicates chaotic-looking behavior.

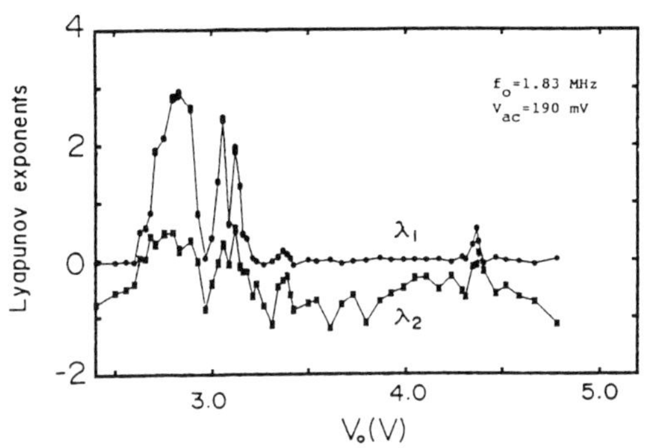

Figure 7.13 The Lyapunov spectrum corresponding to Fig. 7.11.

7.3 Hall-field instability

Section 5.1 briefly introduced the experimental observations of complex chaotic behaviors under dc bias with the applications of a parallel or perpendicular static magnetic field [18–20, 39–56]. A large part of the observations was associated with the experiments of low-temperature impact-ionization avalanche of shallow impurities under the transverse magnetic field in p-Ge [18, 46], n-InSb [52, 55, 56] and n-GaAs [20, 47–49, 53]. In a wide sense the chaotic current instability under magnetic fields may be termed Hall-field instability. Some have been related to positive differential conductivity (PDC) [52, 56] and many others to NDC [18–20, 43–49], where several observations [18, 46–49, 53] have been explained by a model of the dynamic Hall effect coupled with the impact-ionization avalanche [57]. In this section, the theoretical investigations of the dynamic Hall effect by Hüpper and Schöll [57], Hüpper *et al.* [58] and Schöll *et al.* [59] are introduced, in relation to some experimental results in p-Ge [46] and n-InSb [52, 56].The dynamic Hall-effect model [57–59] applies the classical magnetotransport theory [60] to the nonlinear dynamic regime. The basic idea is to consider both the applied drift field, E_x, and the induced Hall field, E_z, as dynamical variables whose time dependence is governed by dielectric relaxation, while the magnetic induction, B, is considered as an external control parameter [57–59]. For the dynamics of the electric field it follows from Maxwell's equations that an externally applied current density $\mathbf{J}_0$ is composed of the conduction current density $\mathbf{J}$ and the displacement current density $\epsilon\dot{\mathbf{E}}$:

$$\mathbf{J}_0 = \mathbf{J} + \epsilon\,\dot{\mathbf{E}}, \tag{7.33}$$

where ϵ is the dielectric constant. With the sample configuration as given in Fig. 7.14, a magnetic field B applied perpendicularly to the external current density gives rise to a Hall field which leads to a distortion in the transverse

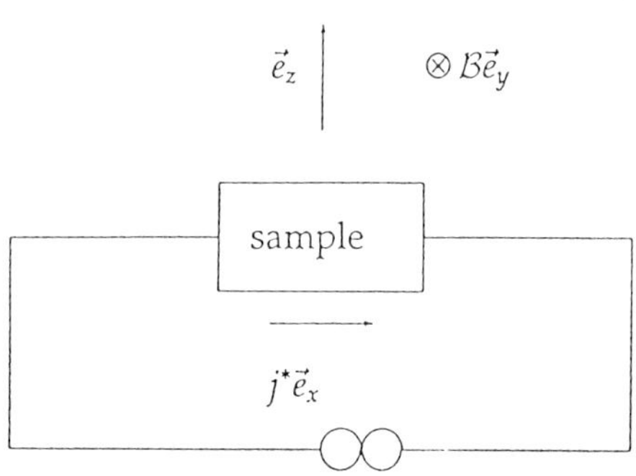

Figure 7.14 Orientation of the sample. Symbols e_x, e_y and e_z are replaced by E_x, E_y and E_z, in the text

carrier density profile. In the simplest model, for moderate magnetic fields, one can neglect these nonuniformities [57–59]. Then the current density takes the form

$$\mathbf{J} = en\ \mu_B \mathbf{E}\ +\ qn\mu\ \mu_B \mathbf{E} \times \mathbf{B}\ +\ en\ \mu^2\ \mu_B\ \mathbf{B}(\mathbf{E} \cdot \mathbf{B}), \tag{7.34}$$

where $e > 0$ is the elementary charge, n is the carrier density, $\mu = e\tau_m/m^*$ is the zero field mobility of the carriers, $q = \mp e$ for electron and holes, respectively, τ_m is the mean momentum relaxation time for zero magnetic field, and m^* the effective mass of the carriers. Owing to the magnetoresistance [57–60] the carrier mobility μ_B decreases as the magnetic field intensity increases:

$$\mu_B = \frac{\mu}{1 + \mu^2\ B^2}. \tag{7.35}$$

Here a semiconductor sample is considered of length L and cross-section A, with a load resistor, R, and a dc bias V_0 [57–59]. By combining (7.33)–(7.35) and the circuitry condition, the dielectric relaxation equation is obtained [57–59]:

$$\dot{E}_x = J_0 - (en\ \mu_B + \sigma_\mathrm{L})E_x\ +\ qn\ \mu\ \mu_B\ BE_z, \tag{7.36}$$

$$\dot{E}_z = -en\ \mu_B E_z - qn\ \mu\ \mu_B\ BE_x, \tag{7.37}$$

where $J_0 = V_0/RA$ and $\sigma_\mathrm{L} = L/RA$. The multilevel impact-ionization model describes the dynamics of the free (n) and bound (n_i, $i = 1, 2, \ldots, M$) carriers governed by the respective nonlinear generation–recombination (g-r) rates ϕ, $\phi_1, \ldots,\ \phi_M$:

$$\dot{n} = \phi(n,\ n_1, \ldots, n_M, E) \text{ and } \dot{n}_i = \phi_i(n,\ n_1, \ldots, n_M, E), \tag{7.38}$$

with $E = (E_x^2\ +\ E_z^2)^{1/2}$. The free electrons (or free holes) are supplied by the impact ionization of shallow donors in n-type (or p-type) semiconductors. Note that one of the variables n, $n_1, \ldots,$ n_M can be eliminated by assumption of the local charge-neutrality condition $\rho^{\mathrm{tot}} = 0$ [61]. The realistic model for (7.38) includes the ground states of several impurity levels, $i = 1$ (7.38) or the ground state and an excited state ($i = 2$) [57–59, 61]. In Schöll's model [61], in which an equation of the energy relaxation rate has been neglected, the multilevel impact-ionization processes ($i = 2$) are essential to exhibit the NDR or the S-shaped J–E characteristic. The model simulations present two examples of the dynamical Hall effect for (1) the PDR (by a one-level impact-ionization process; $i = 1$ in (7.38)), and (2) the NDR (by two-level impact-ionization processes; $i = 2$).

7.3.1 Simulations with positive differential conductivity

The dynamical system for the one-level impact-ionization process was originally given by Hüpper and Schöll [57]:

$$\begin{aligned}
\dot{n} &= X^s(N_D{}^* - n) - T^s n(n + N_A) + Xn(N_D{}^* - n), \\
\epsilon \dot{E}_x &= J_0 - en\, \mu_B E_x + qn\, \mu\, \mu_B B E_z, \\
\epsilon \dot{E}_z &= -qn\, \mu\, \mu_B B\, E_x - en\, \mu_B E_z,
\end{aligned} \tag{7.39}$$

where X is the impact-ionization rate, X^s the photo- and thermal ionization rate, T^s the capture rate, $N_D{}^* = N_D - N_A$ the effective donor concentration for n-type material, and N_A the compensating acceptor concentration (for p-type material N_A and N_D should be interchanged). In (7.39) an additional term $\sigma_L = 0$. The steady state of (7.39) describes the classical Hall effect: $E_z^* = -\mu B E_x^* q/e$ and $J_0 = en^* \mu\, E_x^*$, where the steady-state values are indicated by *. Corresponding to p-Ge at 4.2 K, the material parameters are $N_A^* = 10^{14}$ cm $^{-3}$, $N_D = 10^{11}$ cm^{-3}, $X^s = 10^{-4}$s, $\mu = 10^6$ cm ^{2}/Vs, and the effective dielectric relaxation time $\tau_d = 10^{-13}$ s. X is modeled as a threshold-like function:

$$X(E) = X_0 \Sigma^{1/2} \Big/ \left[(1 + \Sigma) \left(1 + \exp\left(\frac{3}{2} \frac{\varepsilon_{th} - \varepsilon'}{\varepsilon - \varepsilon'} \right) \right) \right], \tag{7.40}$$

where ε is the mean kinetic energy of the free carriers given by $\varepsilon = \varepsilon_0 + 2\varepsilon'$ with $\varepsilon_0 = (3/2)k_B T_L$ and $\varepsilon' = (1/2)m^* \mu\, \mu_B E^2$, ε_{th} is the acceptor energy and $\Sigma = \varepsilon'/\varepsilon_{th}$. The capture rate is also field dependent via $T^s = T_0^s(\varepsilon_0/\varepsilon)^{3/2}$. A linear stability analysis of (7.39) shows that for constant mobility and carrier density, self-generated oscillation cannot occur in the absence of the magnetic field for any values of the applied electric field [57–59]. However, under the magnetic field, a Hopf bifurcation is possible if the necessary condition

$$\tau_d \left| \frac{\partial \dot{n}}{\partial n} \right| < \frac{n^*}{N_D} \tag{7.41}$$

is satisfied. Details of the linear stability analysis are given in Ref. [58]. Note especially that the Hopf bifurcation is possible if $\sigma_{diff} > -\sigma_L$ [61]. This means positive differential conductivity $\sigma_{diff} > 0$ for current control ($\sigma_L = 0$). Figure 7.15 shows the regime of the oscillatory instability in the parameter space $(B,\ E_x^*)$; the steady-state value E_x^* is regarded as a control parameter. In Fig. 7.15, there is a minimum threshold magnetic field for the Hopf bifurcation to occur [57]. Fixing the magnetic field at $B = 100$ mT, period doubling bifurcation to chaos is seen as a function of E_x^*, as found in the bifurcation diagram of Fig. 7.16. A typical example of the chaotic behavior is shown in Fig. 7.17. The Poincaré section and return map in Fig. 7.17 suggest that the correlation dimension may take a value close to 2.

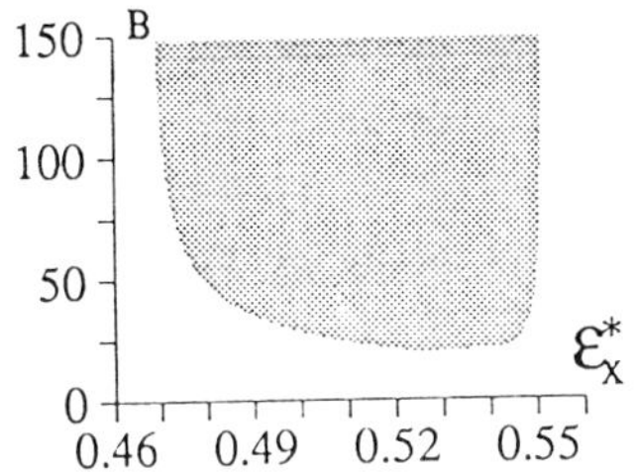

Figure 7.15 Phase diagram of the control parameter plane (B in mT, E_x^* in units of $E_0 = 6$ V/cm). The dashed area indicates the unstable regime. The material parameter are: $N_D^* = 10^{14}$ cm^{-3}, $N_A = 10^{11}$ cm^{-3}, $X^S = 10^{-4}$ s, $\tau_M = 10^{-13}$ s, $\mu = 10^6$ cm^2/V s. X is modeled as a threshold-like function, and T^S as a decreasing function of the electric field [57].

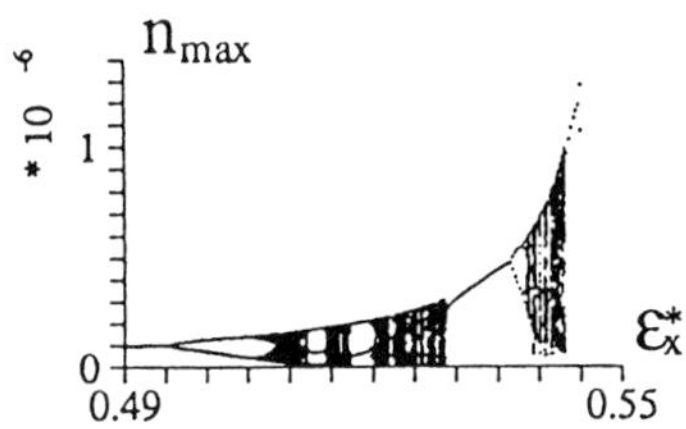

Figure 7.16 Bifurcation diagram of the maxima of the carrier density n plotted vs the control parameter $E_x^* = j_0/(en^*\mu)$ for $B = 100$ mT [57].

Hüpper *et al.* [58] estimated the Kaplan–Yorke dimension to be 2.1–2.5 for fully developed chaos. Further, Hüpper and Schöll [57], Hüpper *et al.* [58] and Schöll *et al.* [59] have demonstrated type I intermittency, showing the scaling law of the mean laminar length $\langle \ell \rangle \propto (B - B_c)^{-1/2}$ and also the scaling law of the variance $\langle \Delta \ell^2 \rangle \propto (B - B_c)^{-1}$. The model simulations with the PDR regime [57–59] have shown some correspondence with the experiments performed in p-Ge [46] and n-InSb [52, 56].

7.3.2 Simulations with negative differential conductivity

The model simulations have been based on Schöll's model [61], which require multilevel impact ionization processes for the SNDC to occur. The g-r kinetics in (7.39) is replaced by (7.38):

$$\frac{\partial n}{\partial t} = X_1^s n_2 - T_1^s n(n + N_t) + X_1 n\, n_1 + X_1^* n\, n_2, \tag{7.42}$$

$$\frac{\partial n_1}{\partial t} = T^* n_2 - X^* n_1 - X_1 n\, n_1, \tag{7.43}$$

where n is the concentration of free carriers, n_1 and n_2 are the concentration of carriers bound at the ground state and first-excited state of the impurity, respectively, $n_2 = N_A - n - n_1$ from the requirement of the local charge neutrality, and X_1^s, T_1^s, X_1, X^*, T^* denote the appropriate g-r coefficients [61]. The impact-ionization rates of the ground state and the first-excited state are given by $X_1 = X_1^0 \exp(-0.016\alpha)$, $X_1 = X_1^{*0} \exp(-4 \times 10^{-3}\alpha)$, where $\alpha = (\mu_B/\mu_d)^{1/2}(E_1/E)$ [59]. With the material parameters appropriate to n-GaAs (see Table 7.1 [58]), two different oscillation mechanisms are shown: (i) oscillations arising for B = 0 on the NDC branch [61]; and (ii) oscillations due to the dynamic Hall effect, arising for B larger than a threshold value on the low-conductivity PDC branch (cf. Fig. 7.15), and also extending to the middle branch and the high conductivity PDC branch with increasing B. Figure 7.18 shows the regime of oscillatory instability for case (ii), where the inset shows the static SNDC characteristic which is independent of B [58]. From the model simulations, a large difference in the dynamical behaviors between PDC and NDC has been predicted [58]: in the SNDC regime the Hall field E_z does not play an active part in the oscillation cycle, but follows with some inertia E_x. Under the transverse field, one can expect experimentally a rich structure of the dynamical behaviors, in which the dynamical Hall effect may be coupled with the breathing mode of a filamentary current [62] or coupled with the spontaneous current oscillation arising as a result of the SNDC characteristic. For the latter, roughly speaking, the longitudinal and transverse fields (E_x, E_z) form a driving oscillator and perturb periodically the SNDC manifold, and

Table 7.1 Material parameters for the two-level model; for case (ii) only the differences with respect (i) are shown [58]

n-GaAs	Case (i)	Case (ii)
T_1^s	10^{-6} cm^3 s^{-1}	
T^*	10^6 s^{-1}	
X_1^s	10^4 s^{-1}	5×10^5 s^{-1}
X^*	10^4 s^{-1}	4.7×10^5 s^{-1}
X_1^0	5×10^{-8} cm^3 s^{-1}	
X_1^{*0}	10^{-6} cm^3 s^{-1}	
E_1	256 V/cm	
μ_d	$\mu_0 \tan^{-1}(3000E/E_1)/3000E/E_1$	μ_0
μ_0	500 cm^2/Vs	10^5 cm^2/Vs
μ_H	μ_d	
N_D	1.3×10^{15} cm^{-3}	
N_A	3×10^{-14} cm^{-3}	
N_L	4 K	

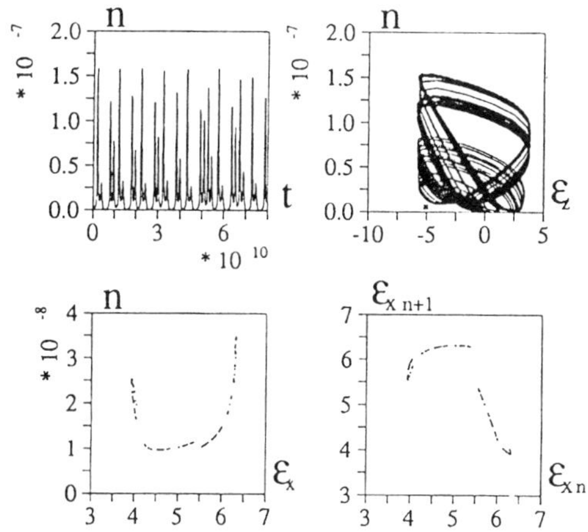

Figure 7.17 Chaos for $E_x^* = 0.508E_0$. Top: time series and phase portrait; bottom: Poincaré section and return map reconstructed from the maxima E_{x_n} of E_x. The figure shows the folding of the map which is necessary for chaos (n in units N_D^*, t in τ_M, E in E_0). B = 100 mT [57].

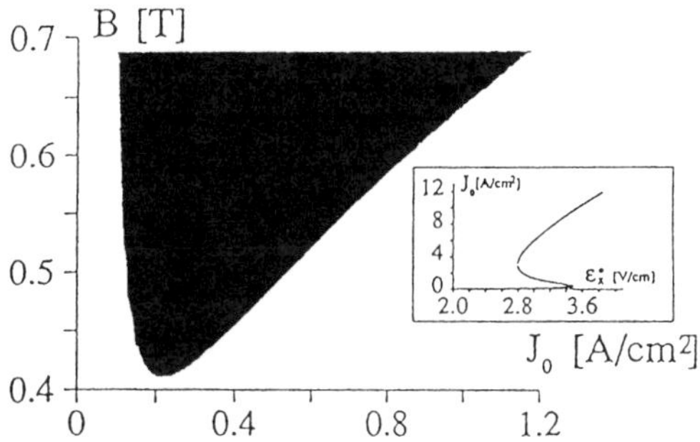

Figure 7.18 Regime of oscillatory instability in the (J_0, B) control plane for n-GaAs. The inset shows the corresponding static current density vs electric field characteristic [58].

the effect may be expected to be similar to that in periodically driven chaos systems, as given in section 5.4.

7.4 Cross-over current instability

Among the current instabilities and chaos observed under magnetic fields [18, 19, 39–56, 63–70], several experiments performed in n-GaAs under a longitudinal magnetic field [66–70] have revealed a novel phenomenon of carrier dynamics when viewed from the nonequilibrium phase transition in semiconductors. The observed instability, termed "cross-over current instability", includes a new result concerning first-order phase transition in the carrier transport theory [60], and it can be even extended to a mathematical model as a generalized cross-over instability [69] (see butterfly catastrophe and a mathematical model in section 4.1). In this section, the cross-over current instability will be characterized theoretically in terms of a realistic physical model. The latter part of this section includes the experimental observations and their analysis.

The observed cross-over instability is described phenomenologically as follows. A typical example is given in Fig. 7.19. Under a longitudinal magnetic field $B(=B_x)$, the holding voltage V_h of the hysteretic I–V curve formed by current filamentation shifts towards a higher voltage as a result of the magnetoresistance effect of the current filament, while the breakdown voltage $V_b(V_b > V_h)$ is almost fixed (more precisely, it shifts slightly towards a lower voltage) [66–68]. Above a critical magnetic field intensity B_c, V_h crosses over V_{b} ($V_{\mathrm{b}} < V_{\mathrm{h}}$), resulting in spontaneous current oscillations with natural frequencies in the order of 10^5 Hz. As has been observed experimentally by low-temperature scanning microscopy [18, 19, 38–40, 71] and the laser scanning method [47–49], spatial patterns of the filament boundaries are often found to be concave. A schematic of the sample geometry is shown in Fig. 7.20. The cross-over current instability is concerned with the concave spatial pattern of the filament combined with the impact-ionization avalanche. From experimental findings in n-GaAs [38, 71], it has been confirmed that the filament boundaries are concave, especially near the contacts. Under a relatively low longitudinal magnetic fields ohmic current does not suffer any magnetoresistance effect in the prebreakdown regime since $I \parallel B$, while, once a current filament is formed, a Lorentz force may act locally on the filament boundaries since the current

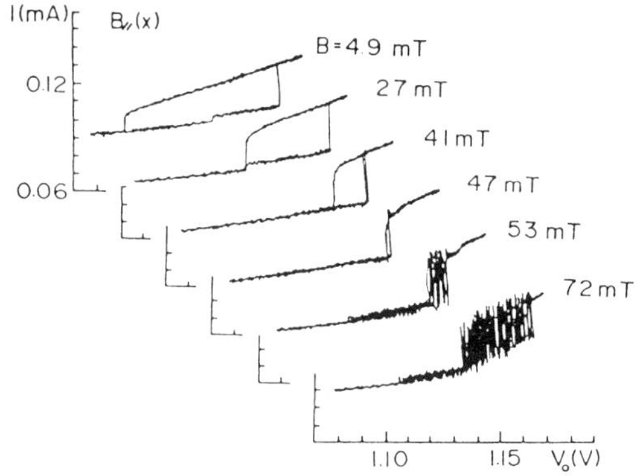

Figure 7.19 Cross-over current instability under longitudinal magnetic fields in n-GaAs. $T_{\mathrm{L}} = 4.2$ K [66].

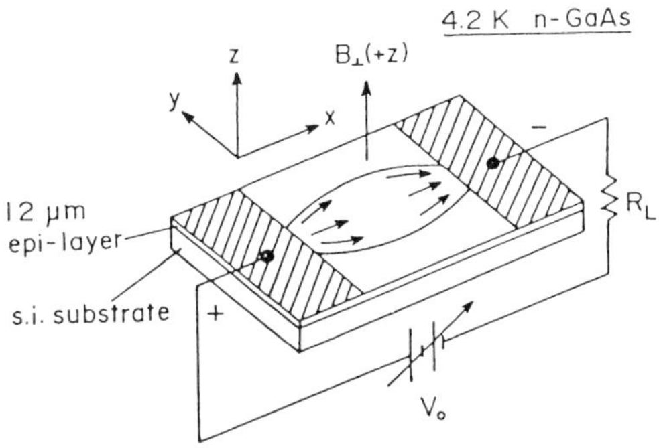

Figure 7.20 Sample geometry. A plane current filament is drawn schematically [72].

flows are not completely parallel to B, especially near the contacts (see Fig. 7.20). As a result, inhomogeneous magnetoresistance induced at the filament boundaries can destabilize the filament with an applied bias V between V_b and V_h, where $V_b < V_h$ for $B > B_c$. After annihilation of the filament, it can be soon regenerated by the impact-ionization avalanche since $V > V_b$. Therefore, the cross-over current instability can be described by a repeating cycle of a generation–annihilation process of a current filament [67–70, 72, 73].

For a realistic physical model of the cross-over current instability, the effective magnetoresistance factor h and its steady-state value h_0 are introduced in the formula of the carrier mobility μ_B [70, 72, 73]:

$$\mu_B = \mu h \ (h = h_0 \text{ under the steady state}), \tag{7.44}$$

$$h_0 = 1 - \frac{1}{2}\, T_M(\mu_0 B)^2 \, \{1 + \tanh[\alpha(n - n_c)]\}, \tag{7.45}$$

where μ is the carrier mobility in zero-magnetic field, T_M is the magnetoresistance scattering factor [70, 72, 73], μ_0 is the zero-field mobility, n is the carrier density, n_c is the critical density considered necessary for the current filamentation, and α (> 0) is the inhomogeneity factor of the current filament. For an ideal case of the filamentary current completely parallel to the longitudinal magnetic field ($I \parallel B$), one puts $h_0 = 1$ and then no electromagnetic force acts on the filament. With $\alpha \to \infty$ and $h_0 = 1 - T_M(\mu_0 B)^2$, the problem simply reduces to the transverse magnetoresistance in the ohmic regime under the weak magnetic field [60]. A threshold-like function $\{1 + \tanh[\alpha(n - n_c)]\}/2$ was introduced in (7.45) such that $h_0 \sim 1$ for the ohmic current ($n \ll n_c$) and $h_0 \sim 1 - T_M(\mu_0 B)^2$ for the filamentary current ($n \gg n_c$).From the transverse magnetoresistance theory [60], the ohmic current density is expressed as $J_x = \sigma_B E_x$ and the relative change in the resistivity $\Delta\rho/\rho_B$ is given by:

$$\Delta\rho/\rho_B = T_M(e\langle\tau_m\rangle B_z/m^*)^2 = T_M(\mu_0 B_z)^2, \tag{7.46}$$

where ρ_B is the resistivity under the transverse magnetic field B_z, $\Delta\rho$ is the increment of the resistivity $\rho_B - \rho_0$, e is the elementary charge, m^* is the effective mass, and $\langle\tau_m\rangle$ is the mean momentum relaxation time. The magnetoresistance scattering coefficient T_M is defined by [60]

$$\begin{aligned} T_{\mathrm{M}} &= (\langle\tau_{\mathrm{m}}^3\rangle/\langle\tau_{\mathrm{m}}\rangle - \langle\tau_{\mathrm{m}}^2\rangle^2)/\langle\tau_{\mathrm{m}}\rangle^4 \\ &= \frac{9\pi}{16}\,\frac{(3r+3/2)!\,(r+3/2)! - [(2r+3/2)!]^2}{[(r+3/2)!]^4}. \end{aligned} \tag{7.47}$$

T_M takes the value of 0.38 for the acoustic deformation potential scattering ($r = -1/2$ in (7.47)), and 2.15 for the ionized impurity scattering ($r = 3/2$) [60], then the value depends strongly on the scattering mechanisms. Although the cross-over instability has been measured with a longitudinal configuration in Fig. 7.20, an essential part of the electromagnetic force comes from the transverse component as discussed above. Then (7.45) is an approximated form applicable to the cross-over instability.

The magnetoresistance factor h is treated as an order parameter. Corresponding to the generation–annihilation dynamics of the filamentary current, a simple relaxation-type equation is introduced for h:

$$\dot{h} = -\gamma(h - h_0), \tag{7.48}$$

where γ is a damping constant. In doped semiconductors, the recombination time $1/\gamma$ is considered theoretically [61] to be much longer than the dielectric relaxation time τ_r of the transverse field and the effective relaxation time τ_d of the longitudinal field ($1/\gamma \gg \tau_r \sim \tau_d$; "lifetime semiconductor" [61], but it will be comparable to or longer than τ_r and τ_d in high-purity semiconductors at low temperatures ($1/\gamma \gtrsim \tau_r \sim \tau_d$; "relaxation semiconductor" [61]). Schöll has argued the g-r instability of the Goldstone mode and the unstable mode based on linear stability analysis [61]; the dielectric-relaxation mode in high-purity semiconductors may couple with the g-r instability (the unstable mode), and may lead to oscillatory unstable modes (breathing mode, i.e., periodic expansion and shrinking of a cylindrical current filament [61, 62]). In the present case (a plane current filament under a longitudinal magnetic field), Schöll's argument [61] cannot simply be applied, but one may reasonably impose the condition $1/\gamma \gtrsim \tau_r \sim \tau_d$. Therefore, the recombination time may be an important factor which determines the oscillation frequency.

The cross-over instability can be modeled by coupling the inhomogeneous magnetoresistance effect in (7.44), (7.45) and (7.48) with the one-level impact-ionization model B in section 3.2 for n-type semiconductors. Then, ODEs for the mean carrier energy ε, the effective magnetoresistance factor

h, the carrier density n and the electric field E are given by [70, 72, 73]:

$$\dot{\varepsilon} = -(\varepsilon - \varepsilon_L)/\tau(\varepsilon) + e\mu_B E^2, \tag{7.49}$$

$$\dot{h} = -\gamma(h - h_0), \tag{7.50}$$

$$\dot{n} = n(N_D - N_A - n)A_I + (N_D - N_A - n)A_T - n(n + N_A)B_T, \tag{7.51}$$

$$\dot{E} = -\gamma_d(E - E_0 + c\ ne\ \mu_B E), \tag{7.52}$$

and

$$A_I = C\sqrt{\varepsilon_i/k_B T_e}\ (1 + \sqrt{2m^*\varepsilon_i/3}\ \mu_B E/k_B T_e)\exp(-\varepsilon_i/k_B T_e), \tag{7.53}$$

where the local charge neutrality condition is assumed, ε_L is the thermal energy ($\varepsilon_L = \frac{3}{2}k_B T_L$; T_L: lattice temperature), $\varepsilon = \frac{3}{2}k_B T_e$ (T_e:electron temperature), $\tau(\varepsilon) \propto \varepsilon^{-1/2}$ is the energy relaxation time in (3.88), A_I is the impact-ionization coefficient, A_T the thermal ionization coefficient, B_T is the thermal recombination coefficient, and N_D and N_A are the donor and acceptor concentrations, respectively. One can apply (7.49)–(7.53) to p-type semiconductors by replacing N_D with N_A. An important aspect of cross-over instability is as follows. The mean energy gain of impact-ionized carriers by an electric field E is given by $e\ \mu_B E^2$. Under the magnetic field, the holding field E_h shifts towards a higher field by the relation $E'_h = \sqrt{\mu/\mu_B}\ E_h$, where $e\ \mu_B E'^2_h = e\ \mu E_h$; E_h is the holding field under zero magnetic field. In contrast, the breakdown field E_b does not shift because under the longitudinal configuration, $n \ll n_c$ and thus $\mu_B = \mu$. The cross-over instability occurs when the condition $E_h > E_b$ meets $B > B_c$. The carrier mobility under zero magnetic field is supplemented by:

$$\mu = \frac{a\varepsilon^{3/2}}{(N_I + n)\ g(\beta_{BH})} \tag{7.54}$$

$$g(\beta_{BH}) = \ln(1 + \beta_{BH}^2) - \beta_{BH}^2/(1 + \beta_{BH}^2). \tag{7.55}$$

It is again stressed that during the impact-ionization process the carrier mobility and the electron temperature play an essential role for the occurrence of the first-order phase transition and thus the SNDC [70, 72, 73] (see section 3.2). By applying a longitudinal magnetic field, a new phase transition regime appears in the J–E characteristics by the inhomogeneous magnetoresistance effect, as seen below.

For the simulations the following dimensionless parameters are used:

$$\tilde{t} = t/\tau_r, \quad \nu = n/(N_D - N_A), \quad \nu_I = N_I/(N_D - N_A),$$

$$\tilde{\varepsilon} = \varepsilon/\varepsilon_L \quad (\tilde{T}_e = T_e/T_L), \quad \tilde{\mu}_B = \mu_B/\mu_0, \quad \tilde{E} = (\sqrt{\mu_{ac}^0 \mu_0}/u_\ell)E,$$

$$\kappa_1 = A_I/\bar{A}_I, \quad \kappa_2 = A_T/(N_D - N_A)\bar{A}_I, \quad \kappa_3 = B_T/\bar{A}_I,$$

where μ_{ac}^{0} is the zero field electron mobility due to acoustic deformation potential scattering [60, 70], u_{ℓ} is the sound velocity, τ_d the dielectric relaxation time of the longitudinal electric field ($\tau_d = 1/\gamma_d$ in (7.52)), and $\bar{A}_I$ is the impact-ionization coefficient at zero electric field. Reasonable values of the parameters for n-GaAs [70, 72, 73] are: $m^* = 0.067\ m_0$, $T_L = 4.2$ K, $\varepsilon_i = 6.8$ meV, $T_M = 2.1$ for the ionized impurity scattering [60], $\alpha = 8.44 \times 10^{-13}\mathrm{cm}^3$, $n_c = 9.47 \times 10^{12}/\mathrm{cm}^3 (\nu_c = 0.2)$, $\mu_0 = 1 \times 10^4\ \mathrm{cm}^2/\mathrm{Vs}$, $\mu_{ac}^0 = 1.57 \times 10^5\ \mathrm{cm}^2/\mathrm{Vs}$, $1/\gamma = \tau = 2\ \mu\mathrm{s}$, $\tau_d = 1\ \mu\mathrm{s}$, $A_I = 4.22 \times 10^{-7}\ \mathrm{cm}^3/\mathrm{s}$, $A_T = 10^6/\mathrm{s}\ (\kappa_2 = 0.05)$, $B_T = 1.097 \times 10^{-6}\ \mathrm{cm}^3/\mathrm{s}\ (\kappa_3 = 2.6)$ and $N_D = 1 \times 10^4$ ($N_D/N_A = 1.9$). Figure 7.21 shows the static J–E characteristics in the parameter space (B, $\tilde{E}$, J) under steady-state conditions ($\dot{\varepsilon} = \dot{n} = 0$, $h = h_0$, $E = E_0$ in (7.49)–(7.52)) and with voltage-controlled conditions ($c = 0$ in (7.52)), where $J = \nu\mu_B E$. By applying a longitudinal magnetic field, the $J - E$ curves exhibit doubly S-shaped characteristics which result directly from the threshold-like function h_0. The plateau (or pocket) of the manifold is seen in Fig. 7.21, which is formed around $\nu_c (= 0.2)$ in the negative differential resistivity region (i.e., $\partial \mathrm{J}/\partial\tilde{E} < 0$) of the unperturbed J–$\tilde{E}$ curve with $B = 0$. The manifold in Fig. 7.21 is similar to the manifold of butterfly catastrophe in section 4.1. The characteristics indicate a novel phenomenon and a new nonequilibrium phase transition in semiconductor carrier transport, although the behavior appears as an effect of inhomogeneous magnetoresistance modeled by (7.44) and (7.45). For several magnetic field intensities, the characteristics of the carrier density vs the electric field are shown in Fig. 7.22, where unity of the normalized field $\tilde{E}$corresponds to $E = 12.6$ V/cm. For the doubly S-shaped $\nu - \tilde{E}$ curve in Fig. 7.22a, the middle branch (MB) is found, from the linear stability analysis (as discussed later), to be unstable for the chosen parameter $1/\gamma = 2\ \mu$s, so that phase transitions occur from the lower branch (LB) to the upper branch (UB) at the breakdown field $\tilde{E}_b$, or from UB to LB at the holding field $\tilde{E}_h$, as shown by arrows. By increasing the magnetic field intensity, the holding field shifts towards higher fields, while the breakdown field is fixed since no magnetoresistance is induced for $\nu < \nu_c$. As a result, the actual hysteresis width $\tilde{E}_b - \tilde{E}_h$ decreases with increasing B and finally the critical behavior emerges at $B = 0.143$ T, where

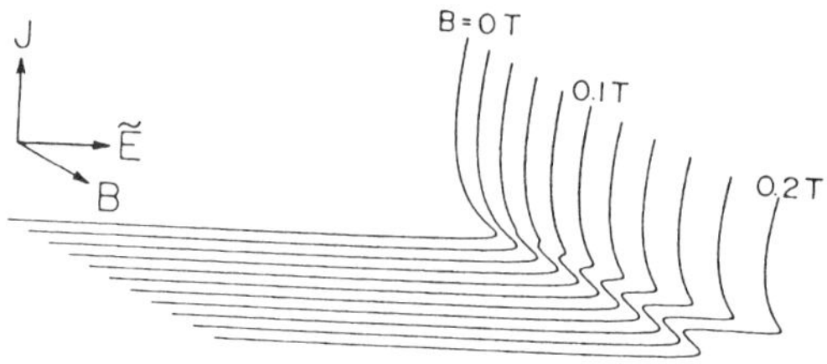

Figure 7.21 Steady-state J–$\tilde{E}$ characteristics as a function of B [73].

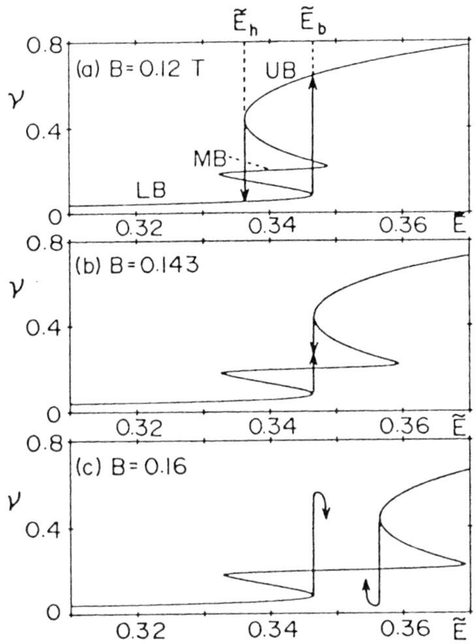

Figure 7.22 The $\nu - \tilde{E}$ characteristics for (a) $B = 0.12$ T, (b) $B = 0.143$ T, and (c) $B = 0.16$ T [73].

$\tilde{E}_b = \tilde{E}_h$ in Fig. 7.22b. With $B > 0.143$ T, the holding field crosses over the breakdown field ($\tilde{E}_h > \tilde{E}_b$ in Fig. 7.22c), whereby spontaneous oscillations are expected with the voltage-controlled bias condition $\tilde{E}_b < \tilde{E} < \tilde{E}_h$.

Steady-state solutions of ν, $\tilde{T}_e$ and $\tilde{\mu}_B$ are shown as a function of $\tilde{E}$ in Fig. 7.23a–c, respectively, where $B = 0.2$ T. At an electric field $\tilde{E} = 0.37$, about 80% of the neutral donors are impactionized ($\nu \sim 0.8$) in Fig. 7.23a, and correspondingly $\tilde{T}_e \sim 1.2$ and $\tilde{\mu}_B \sim 4$. The remarkable increase in electron mobility after the phase transition is apparently caused by the screening effect of the ionized impurity scattering in (7.54). Hall mobility has been observed in high-purity n-GaAs at 4.2 K [74]. In Fig. 7.23c, a negative differential mobility (NDM) is seen in the range $0.33 \lesssim \tilde{E} \lesssim 0.4$. Surprisingly, the NDM does not play an essential role in oscillatory cross-over instability, for the following reason deduced from the linear stability analysis. With an applied electric field $\tilde{E}_0$ in the range $\tilde{E}_b < \tilde{E} < \tilde{E}_h$, there exists a unique steady-state solution $(\tilde{T}_e^*, h^*, \nu^*, \tilde{E}_0)$ for the voltage-controlled condition ($c = 0$), where $\nu^* \sim \nu_c$ (see Fig. 7.23a). Since the energy relaxation time $\tau(\varepsilon)$ is very short compared with the recombination time $1/\gamma$, (7.49) can be safely eliminated adiabatically for the purpose of linear stability analysis. By putting $\dot{\varepsilon} = 0$ and linearizing (7.50), (7.51) around the

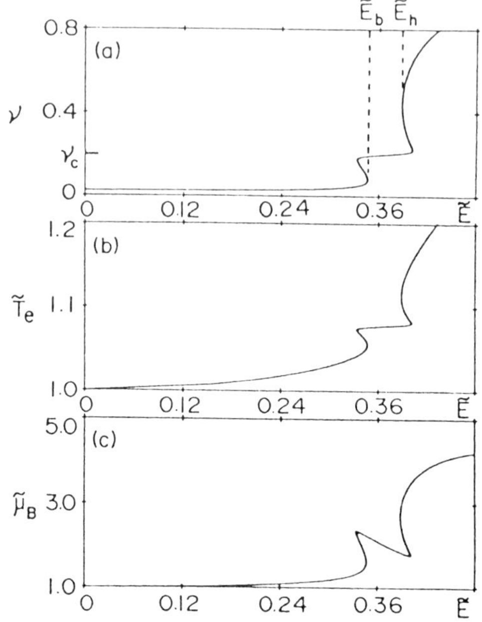

Figure 7.23 Characteristics of (a) ν vs $\tilde{E}$, (b) $\tilde{T}_e$ vs $\tilde{E}$, and (c) ν_B vs $\tilde{E}$. $B = 0.2$ T [73].

steady-state solution (ν^*, h^*), where $\xi_\nu = \nu - \nu^*$; $\xi_h = h - h^*$, then:

$$\dot{\xi}_\nu = a\xi_\nu + b\xi_h, \qquad \dot{\xi}_h = c\xi_\nu + d\xi_h, \tag{7.56}$$

where

$$\begin{aligned} a &= \delta\{(1 - 2\nu^*)\kappa_1^* - \kappa_2^* - [2\nu^* + N_A/(N_D - N_A)] \\ &\quad + \nu^*(1 - \nu^*)(\partial\kappa_1/\partial\nu)^*\} \\ &\quad \sim \delta\nu^*(1 - \nu^*)(\partial\kappa_1/\partial\nu)^*, \\ \delta &= \tau_d(N_D - N_A)\bar{A}_I, \quad b = \delta\nu^*(1 - \nu^*)(\partial\kappa_1/\partial h)^*, \\ c &= (\gamma/\gamma_d)(\partial h_0/\partial\nu)^*, \quad d = -\gamma/\gamma_d = -\tau_d/\tau, \end{aligned}$$

where the asterisks denote the equilibrium values. From the linear stability analysis around the equilibrium point $(\tilde{T}_e^*, h^*, \nu^*, \tilde{E}_0)$, the necessary (but not sufficient) condition for the oscillatory behavior is found as:

$$1/\tau_c = (N_D - N_A)\bar{A}_I\nu^*(1 - \nu^*)(\partial\kappa_1/\partial\nu)^* > \gamma. \tag{7.57}$$

$1/\tau_c$ is interpreted as the linearized impact-ionization rate per carrier at around ν^*. The inequality in (7.57) describes that the recombination time $\tau(1/\gamma)$ of the inhomogeneous magnetoresistance effect should be longer than the linearized mean collision time τ_c. Note that the value $\tau = 2\ \mu$s satisfies the inequality (7.57). If $1/\tau_c < \gamma$, however, the MB may be stable.

By scanning the dc bias $\tilde{E}_0$ quite slowly, corresponding to the experimental observation in Fig. 7.19, the $\tilde{J} - \tilde{E}$ characteristics for various magnetic fields in Fig. 7.24 are obtained. By increasing the magnetic field intensity, only the holding field $\tilde{E}_h$ moves towards higher fields, and eventually the hysteresis width becomes narrow. Above the critical field ($B_c \sim 0.15$), spontaneous cross-over instability occurs. Then the simulant results show qualitatively good agreement with the experimental observations [66–70]. The model can even be extended to the mathematical model of generalized cross-over instability (see section 4.1), and thereby similar results have been obtained in Fig. 4.17 [69].

By fixing the magnetic field intensity, e.g., $B = 0.2$ T, and specifying the load-line condition as $\tilde{c} = ce(N_{\rm D} - N_{\rm A})\mu_0 = 0.14$, one can investigate the dynamical behavior and the bifurcation phenomena. Figure 7.25 shows an example of the bifurcation map as a function of E_0, where the peak values $\nu_{\rm p}$ have been plotted on the ordinate. Note that the values of the applied bias $\tilde{E}_0$ in Fig. 7.25 are larger than the values in the plateau region (the MB region) of Fig. 7.22 owing to the inclined load line (i.e., $\tilde{E}_0 = \tilde{E}^* + c\tilde{J}^*$). In Fig. 7.25a, a period doubling bifurcation is seen in the range $0.437 < \tilde{E}_0 < 0.4393$. With $\tilde{E}_0 > 0.4403$, an alternating periodic–chaotic bifurcation sequence appears, where the periodic states are denoted by P_n^1 $(n = 1, 2, \ldots)$. For the bifurcation sequence, several examples of the current waveforms and the $\tilde{J}$–$\tilde{E}$ phase portraits are shown in Fig. 7.26. As

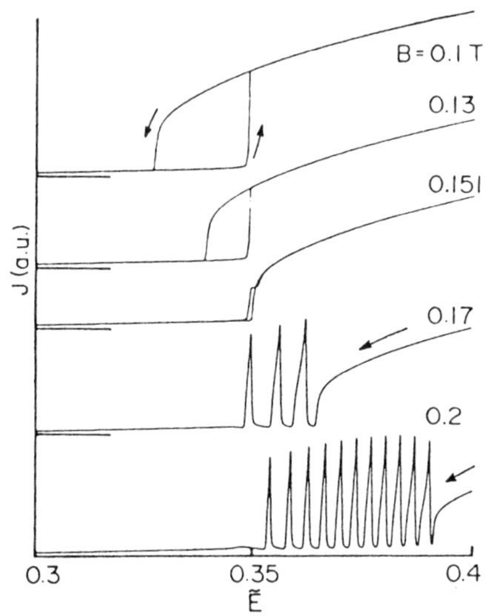

Figure 7.24 $J - \tilde{E}$ characteristics as a function of the magnetic field intensities. Curves for $B < 0.151$ T(or $B > 0.17$ T) were obtained by the iterative method (or by Runge–Kutta–Gill) [69].

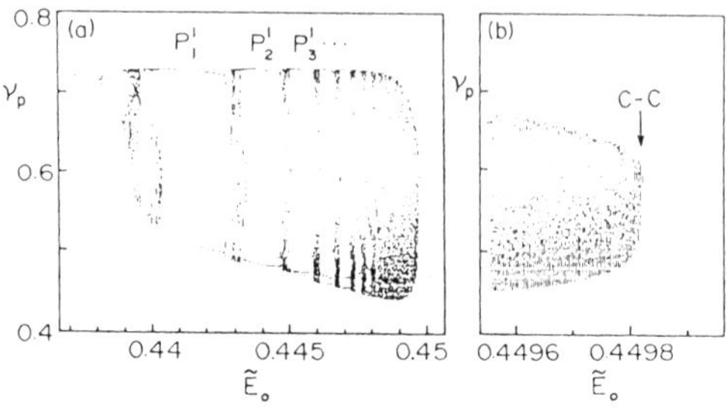

Figure 7.25 Bifurcation map as a function of $\tilde{E}_0$. Peak values ν_p of the carrier density are plotted on the ordinate. The magnified bifurcation map is shown in (b). $B = 0.2$ T [73].

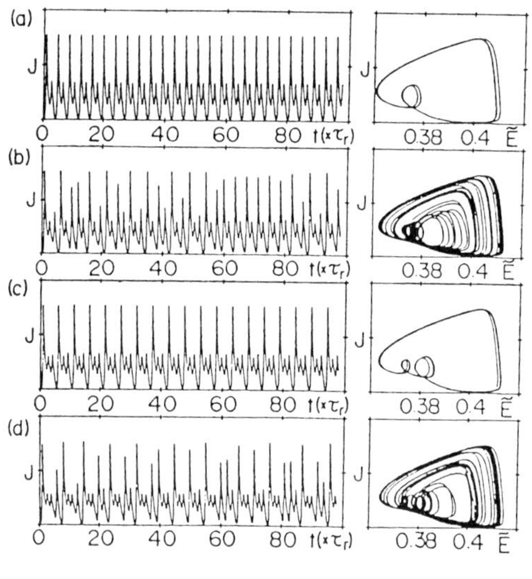

Figure 7.26 Current waveforms and $J - \tilde{E}$ phase portraits. (a) $\tilde{E}_0 = 0.4415$, (b) $\tilde{E}_0 = 0.44325$, (c) $\tilde{E}_0 = 0.444$, and (d) $\tilde{E}_0 = 0.445$. $B = 0.2$ T [73].

seen in Fig. 7.26a, c, the periodic states P_n^1 are characterized by the n times small loops (spikes) and a large loop (large spike current). Since the period is added by the bifurcation $P_n^1 \rightarrow P_{n+1}^1$, the bifurcation is also termed period adding bifurcation. Such a bifurcation sequence has been observed for the electron-hole plasma instability in n-Ge at 77 K under a longitudinal magnetic field [64] and also for the periodically driven chaos system with $B = 0$ in GaAs [66, 75]. For the present case, small loops may correspond to the breathing mode of a small-size filament around the MB and a large loop to relaxation-type oscillation of a large-size filament. Chaos–chaos transition can be seen with $\tilde{E}_0$ in the blow-up of the bifurcation map in Fig. 7.25b (denoted by C–C), where the C–C transition occurs via random switching between the small-size and large-size strange attractors. The small-size attractor ($\nu_p \sim 0.6$) at $\tilde{E}_0 = 0.449815$ results from another period doubling bifurcation (small pitchfork near $\tilde{E}_0 = 0.4498$ in Fig. 7.25b when $\tilde{E}_0$ is decreased from its above value. With $\tilde{E}_0 \sim 0.449814$, the attractor experiences the boundary crisis critically by coalescence with the large-size attractor, showing an f^{-1}-type noise spectrum. The correlation dimension of the small-size attractor has been estimated to be $D_2 \sim 1.6$, while it is approximately 2.0 ($1.8 \lesssim D_2 \lesssim 2.2$) for the large-size attractor [73]. In the model simulation, the dynamics of the electron temperature and the mobility are slaves of the slow variable $\tilde{E}$, where $\tilde{T}_e$ evolves almost in phase with $\tilde{\mu}_B$, as evidenced from the $\tilde{T}_e$ vs $\tilde{\mu}_B$ phase portrait [72], as shown in Fig. 7.27.

The experiments have been performed at 4.2 K using the same high-purity n-GaAs as used in the periodically driven experiments [75–77], with the typical surface dimension of 1 mm × 3 mm [66–69]. A longitudinal electric field is applied along the x-direction between planar-type ohmic contacts in Fig. 7.20, where the distance between the ohmic contacts is typically 1.5 mm. All of the measurements are taken under dark conditions in order to avoid a photo-induced impact-ionization avalanche. Angles between the magnetic

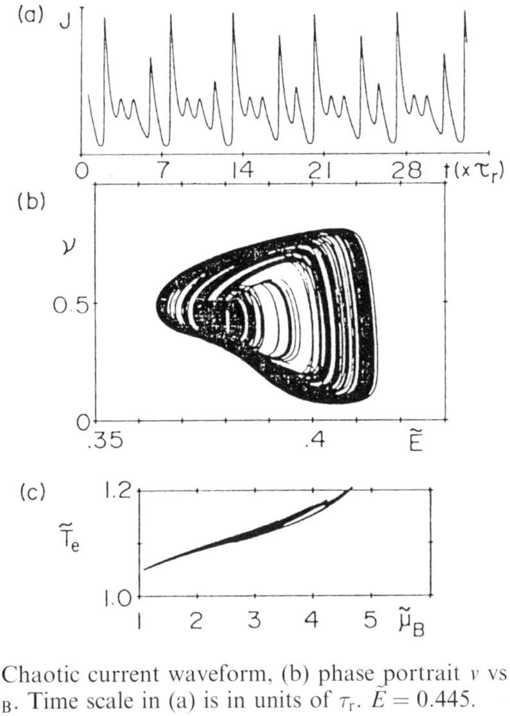

Figure 7.27 (a) Chaotic current waveform, (b) phase portrait ν vs $\tilde{E}$, and (c) phase portrait $\tilde{T}_e$ vs $\tilde{\mu}_B$. Time scale in (a) is in units of τ_r. $\tilde{E} = 0.445$.

field and current can be settled arbitrarily by rotating the sample holder. The I–V curves can be observed by a digital storagescope with a minimum sampling time of 0.025 μs, and the experimental data are analyzed by a microcomputer system. With relatively small load resistance ($R_L = 500\Omega$), a hysteretic I–V curve can be observed with the breakdown voltage of ~ 1.2 V and hysteresis width of ~ 0.1 V. It has been shown experimentally by the LTSEM that a large-scale plane current filament is formed on the upper branch of the hysteresis [38, 71], where the width of the plane current filament ranges from 200 to 700 μm depending on the applied bias. The shape of the filament is shown schematically in Fig. 7.20. Knowledge about the plane filament is helpful in interpreting the longitudinal and transverse magnetic field effects [68], as discussed later.

Many samples show simple cross-over instability behavior [67, 68], as shown in Fig. 7.19, which is explained very well by the model simulation above, while several samples show more complicated instability behavior [69]. By rotating the sample holder by 180°, the instability can also be seen, but the breakdown field differs slightly. During the cross-over instability chaotic behaviors are encountered [66–70]. As a function of applied bias V_0, Hopf bifurcation [67, 69, 70] as well as period doubling bifurcation [70, 78] have been observed. Figure 7.27a shows a typical example of chaotic current oscillation and the 2D phase portrait. The natural frequency f_0 vs V_0 is plotted in Fig. 7.27b. Here, note that a large value of load resistance ($R_L =$ 10 kΩ has been used for measurement of the dynamical behaviors, so that the applied bias is three times larger than those in the I–V curve measurements in Fig. 7.19. The natural frequency decreases from 400 to 70 kHz as V_0 is increased from 3.0 to 3.9 V. This behavior is explained by the V_0-dependent width of the current filament [38, 71]. Since the width of the plane filament increases as V_0 is increased, it is naturally deduced that the relaxation (recombination) time of the filament becomes longer for the larger size with high V_0.

Hopf bifurcation can be observed as a function of V_0, as shown in Fig. 7.28, where $B = 105$ mT and the applied bias is in the prebreakdown regime. In Fig. 7.28a, a simple form of cross-over current oscillation is seen with a natural frequency $f_0 = 356$ kHz. With a slight increase in V_0, the oscillation bifurcates to a torus, followed by an incommensurate frequency $f_1 = 53.6$ kHz, where $f_0 = 429$ kHz [67]; with a further increase in V_0, the torus collapses to fully developed chaos (Fig. 7.28d). With $V_0 > 2.770$ V, the chaotic oscillation exhibits intermittent bursts in Fig. 7.28e, caused by a boundary crisis between the small-size and large-size current amplitudes [67, 70]. The observed bifurcation map for the Hopf bifurcation is given in Fig. 7.29, where the intermittency regime is labeled "I". The deterministic nature of fully developed chaos in Fig. 7.28d has been clarified by the global spectrum or $f(\alpha)$ spectrum in Fig. 7.30, where the total number of data points is $N = 5000$. From several examples, the correlation dimension has

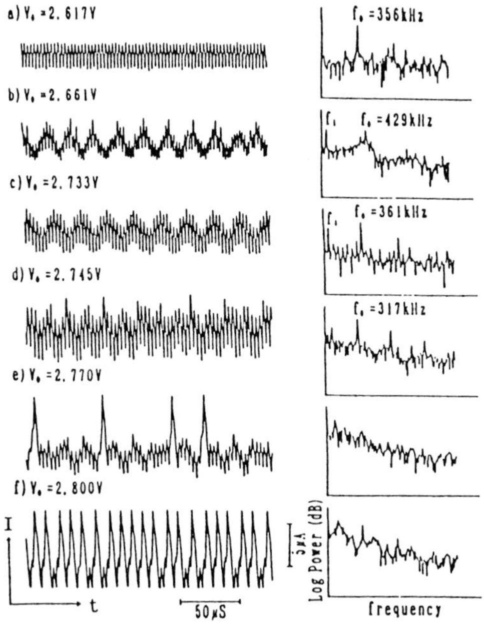

Figure 7.28 Time series of current oscillations and the Fourier power spectra for the Hopf bifurcation route to chaos [67].

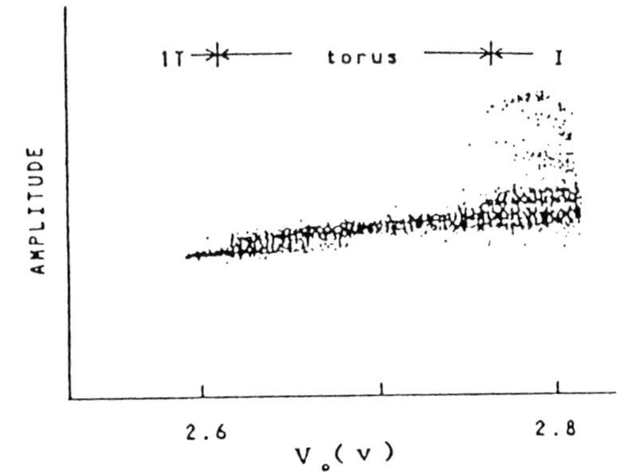

Figure 7.29 Bifurcation map for the Hopf bifurcation, observed as a function of V_0. Measurements by K. Aoki.

been found from the $f(\alpha)$ spectrum to range from 2.7 to 3.3 [69, 70]. Therefore, the degrees of freedom (d_f) required to explain the cross-over instability should be four or more. Then, the truncated model (7.49)–(7.52) may be said to be fairly reasonable with the minimum degrees of freedom $d_f = 4$ to explain the experimental results, although an essential part of

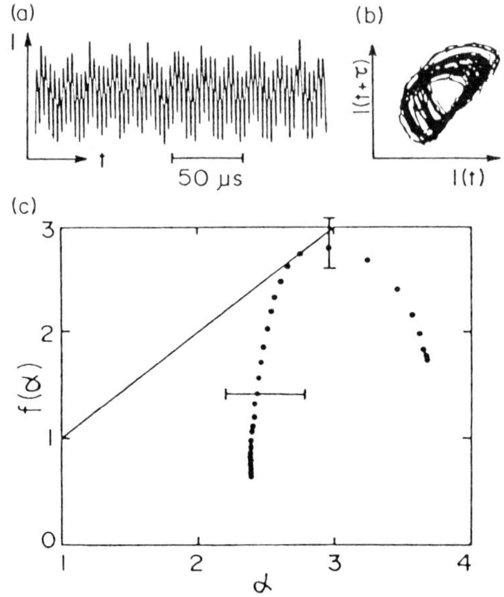

Figure 7.30 (a) Chaotic current waveform; (b) phase portrait $I(t)$ vs $I(t+\tau)$ in n-GaAs, where $\tau = 1$ μs, and (c) spectrum $f(\alpha)$. The applied bias is 2.745 V. $B = 105$ mT. $R_L = 10$ kΩ, $T_L = 4.2$ K [69].

spatial inhomogeneity is truncated and replaced by the magnetoresistance factor h.

The noise effect is also pronounced for the cross-over current instability [70, 78]. One can test crucially the underlying multifractal nature of chaos under the noise floor. Here the multifractal nature for chaos observed just after the period doubling and the further succeeded boundary crisis will be investigated. The noise floor effect to the $f(\alpha)$ spectrum is also demonstrated in the model simulation by adding an external noise $\xi(t)$. Period doubling bifurcation can be seen in the postbreakdown regime. Figure 7.31 shows the time sequences of the current oscillations (left column) and the corresponding Fourier power spectra (right column) observed as a function of B, where $V_0 = 2.339$ V. The multifractal nature of chaos is analyzed for the time series in Fig. 7.31e, g, which resulted from period doubling bifurcation [78]. Owing to the noise floor, period 4 has not been clearly resolved within an extremely narrow range of the control parameter (69.9 mT $< B <$ 79.8 mT) for the periodic regime. With $B >$ 84.6 mT the Fourier power spectrum shows broadening of the subharmonic frequency component $f_0/2$ (marked by an asterisk in Fig. 7.31d) as well as an increase in the background noise level (an increase by approximately 20 dB in Fig. 7.31h, j).

The Poincaré section analyses of the phase portraits for Fig. 7.31e, g are shown in Fig. 7.32a, b, respectively. The results indicate that the period-2-

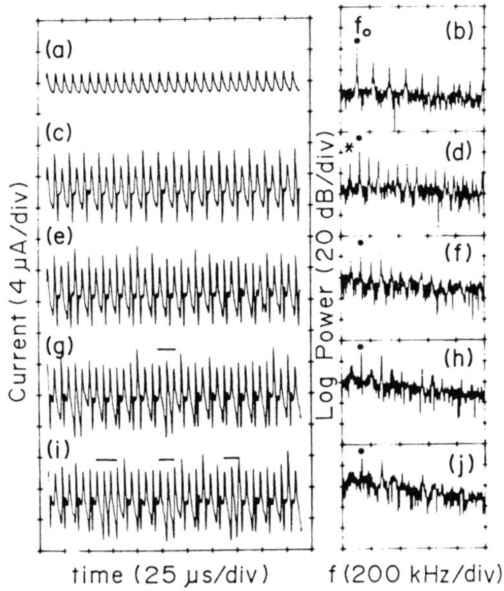

Figure 7.31 Series of the current waveforms (left column) and the Fourier power spectrum (right column) as a function of B in n-GaAs, where $B = 59$ mT in (a, b), $B = 69.9$ mT in (c, d), $B = 79.8$ mT in (e, f), $B = 84.6$ mT in (g, h), and $B = 85.3$ mT in (i, j). The natural frequencies f_0 are (b) 113 kHz, (d) 131 kHz, (f) 139 kHz and (j) 141 kHz. $R_L = 10$ kΩ, $T_L = 4.2$ K. The applied bias $V_0 = 2.339$ V [78].

like chaotic attractor in Fig. 7.31e undergoes a boundary crisis [15–17] by increasing the magnetic field intensity from 79.8 to 84.6 mT (Fig. 7.31g) or 85.3 mT (Fig. 7.31i), where the intermittent switching to a new phase (a new oscillatory state) is indicated by bars in the current time sequences in Fig. 7.31g, i. In Fig. 7.32b, possible separatrices (stable orbits of unstable manifold) are indicated by a dashed line. It is considered that the boundary crisis in Fig. 7.32b occurs by the homoclinic tangency. The singularity spectra $f(\alpha)$ are evaluated by following the procedure of Mino *et al.* [79] in an experiment of parallel-pumped spin-wave instability. The procedure is identical to the procedure used in (6.71) and (6.72). The strange attractors are constructed using the time lagging method. In order to obtain the generalized dimension D_q, the partition function $\Gamma(q, \ell) = (1/n) \sum_i P_i(\ell)^{q-1}$ is calculated, where $P_i(\ell)$ is the probability which is obtained by counting the number of data points within the ith hypersphere of radius ℓ ($i = 1, 2, \ldots, n$) and dividing it by the total number N of data points of the attractor. The total number of data points is $N = 6000$ for the attractor in Fig. 7.32a and $N = 30\,000$ for the attractor in Fig. 7.32b. The hyperspheres are chosen at random, with the total number of $n = 1000$ for the attractor of

Fig. 7.32a and $n = 3000$ for the attractor of Fig. 7.32b. The D_q curve is obtained from the scaling law $\Gamma(q, \ell) \propto \ell^{(q-1)D_q}$ and finally the $f(\alpha)$ spectrum is calculated from the Legendre transform (6.58) and (6.59). As discussed in section 6.3, local singularities of different strengths are distributed on the interwoven sets of the dimensionality $f(\alpha)$, where densely (sparsely) distributed fractalities are related to $\Gamma(q, \ell)$ or $P_i(\ell)^q - 1$ for large positive (negative) q values. When the external noise is added in the system, it affects seriously the r.h.s. of $f(\alpha)$. Very often the r.h.s. cannot be obtained [80] because sparsely distributed regions of the attractor become sparser due to the external noise, and the scaling law for $q < 0$ is broken by the most rarefied hyperspheres which include only a few data points inside the small radius ℓ. By filtering (rejecting) such rarefied hyperspheres, one can recover the r.h.s. of $f(\alpha)$, where the scaling in $\Gamma(q, \ell)$ can be taken typically for $q > -6$ and with a precision approximately up to the scale $\ell \sim 2^{-4}$ [78]. The numerical procedure is termed a filtering method.

Figure 7.33a, b shows the $f(\alpha)$ spectra obtained for the period-2-like chaotic attractor ($B = 79.8$ mT, Fig. 7.31e) and for the extended attractor after the crisis ($B = 84.6$ mT, Fig. 7.31g), where the solid circles are the experimental results obtained by the filtering method and the solid lines are those of the model simulation [78]. The capacity dimensions are $D_0 = 1.69 \pm 0.1$ in Fig. 7.33a, and $D_0 = 2.24 \pm 0.1$ in Fig. 7.33b. With further increase in the magnetic field intensity, the $f(\alpha)$ suggests growth to the higher dimensional attractor (for instance $D_0 \sim 2.49$ with $B = 115.3$ mT). Apparently, the experimental results in Fig. 7.33a, b have exhibited much broader spectra than the noise-free model simulation.

In order to investigate the noise effect, random fluctuating force $\xi(t)$ is added to the dielectric relaxation equation (7.52) of the cross-over instability

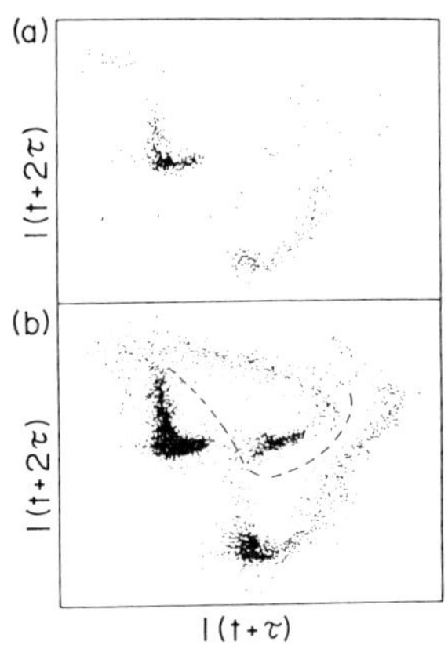

Figure 7.32 Poincaré sections of the 3D phase portraits: (a) $B = 79.8$ mT and (b) $B = 84.6$ mT. The time lag is chosen to be $\tau = 1.5$ μs [78].

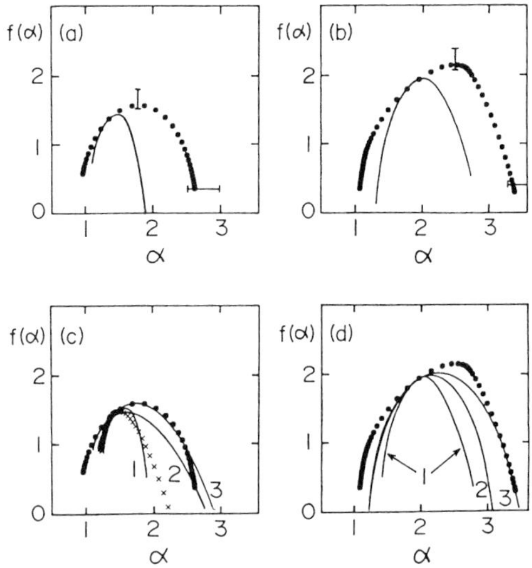

Figure 7.33 Experimental $f(\alpha)$ spectra (solid circles): (a) $B = 79.8$ mT and (b) $B = 84.6$ mT. Solid lines are from the model simulations: (a) $B = 154.921$ mT and (b) $B = 155$ mT. Corresponding to (a) and (b) broadening effects of the simulated f (α) spectra are shown in (c) and (d), respectively, as a function of noise level: (1) $\sigma = 2.886 \times 10^{-5}$, (2) $\sigma = 8.66 \times 10^{-5}$, and $\sigma = 1.442 \times 10^{-4}$. Crosses in (c) indicate the $f(\alpha)$ spectrum typically obtained for periodically driven chaos [78].

model (7.49)–(7.53) [78]. Using the same parameter values as those used in [70, 72, 73] (listed above), one can show the period doubling cascade in the postbreakdown regime and the further succeeding boundary crisis as a function of B [78]. The noise-free $f(\alpha)$ spectra in Fig. 7.33a, b (solid lines; $\xi(t) = 0$)) are calculated for the period-2-like attractor at $B = 0.154921$ T and the fully extended attractor at $B = 0.155$ T. The boundary crisis (chaos–chaos transition) appears with $B \sim 0.145494$ [78]. The solid line in Fig. 7.33a is relatively similar to (but slightly broader than) the $f(\alpha)$ spectrum of the Feigenbaum attractor in Fig. 6.25 [70,78], by subtracting a trivial dimensionality. Namely, $D_0 = 0.52, D_\infty = 0.25, D_{-\infty} = 0.8$ in Fig. 7.33a, while $D_0 = 0.538$ [81], $D_\infty = 0.37775\ldots, D_{-\infty} = 0.75551\ldots$ [82] for the Feigenbaum attractor. By adding the external noise to (7.52), the $f(\alpha)$ spectra are numerically obtained as the solid lines in Fig. 7.33c, d, which are labeled 1–3 for three different noise levels: (1) $\sigma = 2.886 \times 10^{-5}$, (2) $\sigma = 8.66 \times 10^{-5}$, and (3) $\sigma = 1.442 \times 10^{-4}$, where $\langle\xi(t)\rangle = 0$, $\sigma^2 = \langle\xi^2(t)\rangle$, and the filtering method has been used to calculate $f(\alpha)$ spectra. By increasing the noise level, the singularity spectrum broadens, especially towards the higher values of α, while the capacity dimension D_0 increases only by a small amount, as seen in Fig. 7.33c, d. The experimental results (solid circles) are best fitted to the simulant results with $\sigma = 1.442 \times 10^{-4}$

(curve 3). For reference, crosses in Fig. 7.33c show the $f(\alpha)$ spectrum obtained in the experiments of periodically driven chaos [70, 77] for the onset of chaos which has resulted from the period doubling cascade (bifurcated from period 4). The spectrum (see also Fig. 6.26) is much narrower than that observed in the cross-over instability (present case) and fairly close to the spectrum of the Feigenbaum attractor [82]. Since both of the results (crosses and solid circles in Fig. 7.33c) have been obtained with the same quality of minimum noise levels in the experimental systems, the external noise is found to influence the multifractal nature of the strange attractors in different manners depending on the types of spatially extended instabilities of the current filament. The hierarchical complexity has been given in [35, 77]. A possible interpretation may be that the attractor is well controlled by the periodic driving force in the driven chaos system, while the freely oscillating autonomous system receives the noise effect strongly by random kicks of the fluctuating force (in particular the oscillation phase may be remarkably influenced by the random kicks for spatially extended autonomous systems).

So far, the experimental cross-over instability under a longitudinal magnetic field has been shown to be explained satisfactorily by the model simulation. Some additional results are now given for the effects of a transverse magnetic field [66, 68, 70]. In the epitaxial GaAs sample, self-generated oscillation has not been observed under transverse magnetic fields [66–68, 70]. An exceptional case is seen in Ref. [66]. Therefore, the Hall-field instability (the dynamic Hall effect) proposed by Hüpper and Schöll [57], Hüpper *et al.* [58] and Schöll *et al.* [59] may not be universal, and the experimental observations [66–70] indicate that the current instabilities may strongly depend on the boundary conditions such as sample geometry (thin epitaxial layer for the present case) and type of current filament (plane [66–70] or cylindrical [61]). Limiting the discussion to the present case [66–70], the effects of the transverse magnetic field will be discussed from a viewpoint of the semiconductor physics. With the sample geometry of Fig. 7.20, the transverse magnetic fields can be applied in the possible directions $B_\perp(+y), B_\perp(-y), B_\perp(+z)$ and $B_\perp(-z)$, which are perpendicular to the x-direction of the applied electric field and thus to ohmic current I_x. Under the transverse field, except for $B_\perp(+y)$, hysteresis of the I–V characteristic shifts towards higher voltages as a function of B due to the transverse magnetoresistance effect. Qualitatively, the carrier mobility μ and the mean energy gain $e\mu E_x^2$ are reduced by the transverse magnetic field via magnetoresistance scattering, whereby it requires higher breakdown voltage for the heated carriers to cause an impurity avalanche breakdown. Two distinct behaviors have been observed for $B_\perp(-y)$ in Fig. 7.34(a) and for $B_\perp(+y)$ in Fig. 7.34b. With $B_\perp(-y)$, the hysteresis shifts significantly towards higher voltages as a function of the magnetic field intensity, while it shifts towards lower voltages with $B_\perp(+y)$. With the $B_\perp(+z)$ and $B_\perp(-z)$

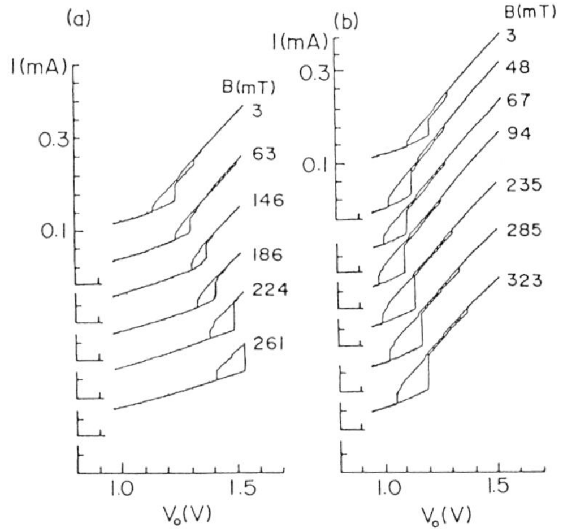

Figure 7.34 *I–V* characteristics under transverse magnetic fields in n-GaAs: (a) $B_\perp(-y)$; (b) $B_\perp(+y)$ [68, 70].

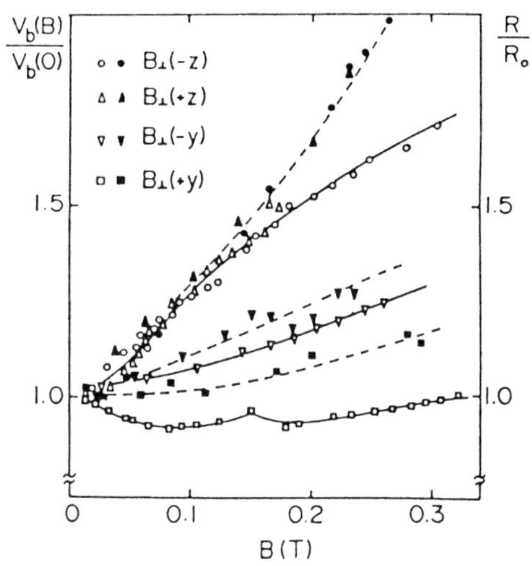

Figure 7.35 Normalized breakdown voltage (open symbols) and differential resistance R/R_0 (solid symbols) as function of the transverse magnetic field intensity. The differential resistances were measured on the upper branch just above V_b. $V_b(0) = 1.26$ V, $R_0 = 1$ kΩ [68, 70].

configurations, which are typically the Hall-field configuration in Fig. 7.20, the shifts are much more significant than those in Fig. 7.34a [66–70].

Figure 7.35 summarizes the shifts of the normalized breakdown voltage $V_b(B)/V_b(0)$ as well as the normalized differential resistance R/R_0 as a function of B for various transverse configurations, using open symbols (○,

△, ▽, ❑) for $V_b(B)$ and solid symbols (●, ▲, ▼, ■) for R. $V_b(0)$ and R_0 are the breakdown voltage and the differential resistance at zero magnetic field intensity, where $V_b(0) = 1.26$ V and $R_0 = 1$ kΩ (note: the load resistance $R_L = 500\Omega$). The differential resistance R_0 has been measured on the upper branch of the hysteresis just above V_b. The large shifts of the hysteresis are seen for $B_\perp(\pm z)$, followed by considerable increases in the differential resistance $V_b(B)/V_b(0) \sim 2.0$ at $B = 0.26$ T. Although some difference is seen in the I–V curves when the direction of the magnetic field is inverted [66], Fig. 7.35 indicates clearly that the inversion from $B_\perp(+z)$ to $B_\perp(-z)$ does not make any difference to the normalized quantities $V_b(B)/V_b(0)$ and R/R_0 [68, 70]. Therefore, the Lorentz force acts on the filament wall in a manner symmetrical to the inversion of the magnetic field under the Hall-field configuration.

For the $B_\perp(+y)$ and $B_\perp(-y)$ configurations the behaviors are quite different (Figs 7.34 and 7.35). In Fig 7.20, the Lorentz force tends to press the plane current filament towards the interface layer between the epitaxial layer and the semi-insulating substrate for the $B_\perp(-y)$ configuration or towards the broad crystal surface for the $B_\perp(+y)$ configuration. Since the sheet-like plane current filament is confined within the epitaxial layer, where the thickness of the plane filament is supposed to be as large as the epitaxial layer, it is considered that the transverse magnetoresistance may be less significant that one can expect under $B_\perp(\pm z)$. Although probably not significant through the restriction of the macroscopic charge neutrality condition, the current path for the $B_\perp(\pm z)$ configurations may be more or less curved due to the Lorentz force as in the gaseous discharge, since the filament walls are free from the sample boundaries in the y-direction (cf. Fig. 7.20). Thus, the significant magnetoresistance effect in Fig. 7.35 for the Hall-field configuration is not so surprising. The large difference between $B_\perp(-y)$ and $B_\perp(\pm z)$ in Fig. 7.35 has been attributed to a kind of Corbino magnetoresistance effect [68], which depends strongly on the sample geometry [60]. The magnetoresistance effect for the $B_\perp(+y)$ configuration may be more complicated [68], as the Lorentz force may press the plane filament to the crystal surface. The increase in differential resistance is relatively small (■ in Fig. 7.35) and the breakdown voltage shifts towards lower voltages (❑). Such anomalous behavior is possibly related to the surface bending effect, as discussed below. The effects of surface band bending have been investigated by preparing a test Schottky-gate point contact between the planar-type ohmic contacts near the central position of the broad crystal surface [68]. By applying the reverse bias of –25 V to –35 V at the Schottky gate, the hysteretic I–V curve shows considerable shifts towards higher breakdown voltage, by approximately 0.02–0.06 V under the $B_\perp(+y)$ configuration. The result indicates clearly that the depletion layer extending near the point contact on the crystal surface prevents current filamentation. Without the Schottky-gate contact, it is likely, although not necessarily so, that electrons pressed onto the depleted layer by the Lorentz force become hot and facilitate current filamentation at lower breakdown voltages.

For any cases of transverse magnetic fields, the *I–V* characteristics keep the hysteretic behaviors ($V_b > V_h$), showing neither cross-over instability nor the dynamic Hall-effect predicted by Hüpper *et al.* [58].

7.5 Cross-talk coupling

In semiconductor systems, self-organized hierarchical complexity appears by the mutual coupling between or among different oscillating centers in the nonlinear carrier transport caused by the low-temperature impact-ionization avalanche [18,19]. Peinke *et al.* have shown that a quasi-periodic current oscillation can be generated by two independent oscillating centers, with incommensurate frequencies f_1 and f_2, at two localized parts of a broad p-type Ge sample, and that more complex hyperchaotic states are observed when the relevant control parameters (applied voltage, transverse magnetic field intensity) are varied [18, 19]. These findings have been highly important in resolving the complex behaviors of solid-state turbulence in semiconductors from a viewpoint of synergetics [12], although many problems remain unresolved. This section investigates mutual coupling between or among multicurrent filaments from a different approach, by applying the cross-over instability described in the previous section. The experimental results give some fundamental properties on (i) the sensitive nature of the filamentary current to the external perturbation, and (ii) enforced breathing dynamics.

The mutual coupling between two filamentary currents can be investigated by forming two subsystems I and II in a GaAs sample, with the circuitry condition given in Fig. 7.36. Using high-purity n-type GaAs with

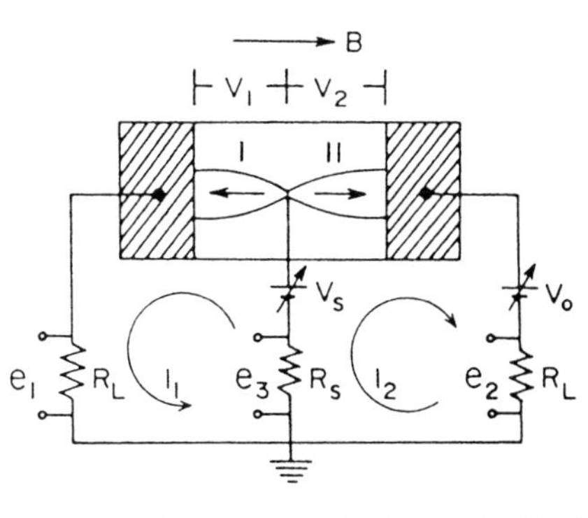

Figure 7.36 Sample geometry of a three-terminal circuit [35, 83, 84].

typical surface area dimensions of ~ 1.7 mm $\times$ 4 mm, the planar-type ohmic contacts are prepared and a Schottky-type point contact (~ 0.18 mm $\varnothing$ Cu-wire) is pressed directly on the broad crystal surface near the center position (Fig. 7.36). By applying a forward voltage V_s at the point contact, cross-talk coupling of the current flows between two subsystems I and II is possible, where above a threshold value the forward bias V_s causes the impact-ionization avalanche in each subsystem (see below). A longitudinal magnetic field is applied, with intensity less than 0.4 T. For measurement of the I–V curves, a triangle pulse voltage superimposed on a dc bias V_0 (or V_s) is applied, where the repetition frequency f_d ($\sim$ 1 kHz) of the triangle pulse must be chosen such that $f_d \ll f_0$ ($f_0 \sim 10^5$ Hz is the natural frequency of the self-generated oscillating current by the cross-over instability [66–70, 72, 73]). The currents I_1 and I_2 are estimated by the voltage drops across the load resistors R_L and R_s, where $R_L = R_s = 470\ \Omega$ and the sample voltages are denoted by V_1 and V_2 in Fig. 7.36. In the following, V_s is termed the "source voltage" and V_0 the "balancer voltage".

By applying the source bias V_s, I_1 vs V_s (or I_2 vs V_s) shows hysteretic behaviors due to impurity avalanche breakdown, and thereby filamentary currents I and/or II flow in the opposite directions. Depending on the position of the Schottky point contact, the breakdown voltage V_1^b for the current filamentation in subsystem I differs from the breakdown voltage V_2^b in subsystem II. For instance, when the point contact is located near the center but closer to the ohmic contact of subsystem II, V_1^b is larger than V_2^b ($V_1^b \sim 2.1$ V, $V_2^b \sim 1.6$ V [35, 83, 84]. With increasing V_s, the impact-ionization avalanche occurs first in subsystem II with $V_s \gtrsim 1.6$ V and subsequently in subsystem I with $V_s \gtrsim 2.1$ V. Then, for $V_s \gtrsim 2.1$ V, both filamentary currents I and II flow in Fig. 7.36, as expected. In such a case, the balancer voltage V_0 is useful in investigating the cross-talk coupling since one can control the breakdown voltages V_1^b and V_2^b freely [83, 84]. Under the longitudinal magnetic field, filamentary currents I and/or II exhibit cross-over instability, which can be described by a repeating cycle of generation–annihilation processes of the current filament(s), as explained in section 7.4. By adequately adjusting the control parameters V_s, V_0 and B, any types of cross-talk coupling between subsystems I and II can be reached: (i) filament I (or II) oscillates spontaneously and it modulates on-state filament II (or I), leading to the enforced breathing mode of filament II (or I): (ii) filament I (or II) oscillates spontaneously while filament II (or I) is in the off-state; and (iii) both filaments oscillate spontaneously.

Before describing the details, the basic I–V curves under zero magnetic field are shown in Fig. 7.37 [83]. Figure 7.37a, b shows a series of I_1 vs V_0 curves and $(-I_3)$ vs V_0 curves, respectively, where the total current injected from the Schottky point contact $I_3 = I_1 + I_2$ (see Fig. 7.36), I_1 and I_3 are evaluated by the voltage drops e_1 and e_3 across R_L and R_s, respectively, and the parameter V_s is kept constant during each bias scan of V_0. With $V_s =$

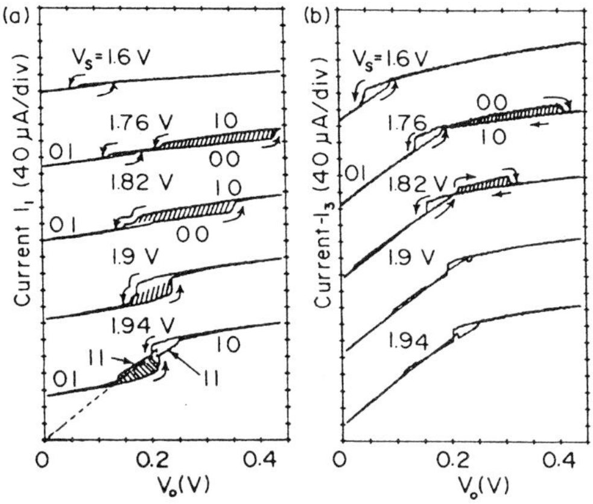

Figure 7.37 (a) I_1–V_0 characteristics; (b) $(-I_3) - V_0$ characteristics. Source voltage V_s was varied as a parameter [83].

1.6 V (top view), only a hysteresis for the filamentary current II is seen (note that filament I is in the off-state). The hysteresis for filament II is reversed in both the I_1 vs V_0 and $(-I_3)$ vs V_0 curves, which is easily understood by considering Kirchhoff's law in Fig. 7.36:

$$V_1 = V_s - e_1 + e_3, \tag{7.58}$$

$$V_2 = V_s - V_0 - e_2 + e_3, \tag{7.59}$$

where $e_1 = R_L I_1$, $e_2 = R_L I_2$ and $e_3 = -R_s I_3 = -R_s(I_1 + I_2)$. In subcircuit II, the I_2 vs V_2 characteristic should be given by a standard S-shaped curve, with the breakdown voltage V_2^b and the holding voltage $V_2^h(V_2^h < V_2^b)$. Correspondingly, one has V_0^b and V_0^h with the reversed relation $V_0^h > V_0^b$ from (7.59), where V_s is fixed, say, $V_s = 1.6$ V. Hence, the standard hysteresis expected in the I_2 vs V_2 curve can be reversed when it is measured by the I_1 vs V_0 or $(-I_3)$ vs V_0 curves. Another remarkable fact is that the hysteresis height is very small in the top view of Fig. 7.37a compared with that in Fig. 7.37b. Since filament I is in the off-state for $V_s = 1.6$ V, the resistance of subsystem I is relatively high, while filament II is in the on-state. Therefore, only a small fraction of filamentary current II leaks into subsystem I and the cross-talk current is small. With increasing V_s, the hysteresis shifts towards higher voltages of V_0, as seen in Fig. 7.37, which is also explained by Kirchhoff's law (7.59), mainly via the term $V_s - V_0$. With further increase in V_s, a hysteresis of filament I appears for $V_s > 1.76$ and moves towards lower values of V_0 (as indicated by hatched regions). For the hysteresis, there is a breakdown voltage V_1^b and a holding voltage V_1^h with the normal relation $V_1^b > V_1^h$. The voltage V_1 is an implicit function of the balancer voltage V_0 via voltage drop e_3 in (7.58) and (7.59). By this relation, one has V_0^b and $V_0^h (V_0^h < V_0^b)$ corresponding to V_1^b and V_1^h; then the hatched hysteresis is not reversed in Fig. 7.37a. The most remarkable finding in Fig. 7.37a is that the reversed hysteresis of filament II becomes large in height once filament I switches on. In this case, the resistance of subsystem I is very small, so that approximately half of the filamentary current II leaks into subsystem I according to the ratio $R_L/R_s = 1$. Then the hysteretic behavior of filament II is observed remarkably in the I_1 vs V_0 curve. Here the on/off states are denoted by the symbols 00, 10, 01 and 11 (01: filament I is in the off-state and filament II is in the on-state, etc.).

In order to show the sensitive nature of the filamentary current, the mutual differential resistance r_1 of subsystem I is defined by [83]:

$$r_1/R_s = -\frac{\partial I_3/\partial V_0 + \partial I_1/\partial V_0}{\partial I_1/\partial V_0}. \tag{7.60}$$

Definition (7.60) is interpreted as follows. Assume that a perturbation ΔV_0 of the balancer voltage produces an excess current ΔI_2 in subsystem II which branches away with ΔI_1 and ΔI_3, as shown by the equivalent circuit in the schematic figure of Fig. 7.38. From Kirchhoff's law $(r_1 + R_L)\Delta I_1 = R_s \Delta I_3$, one has:

$$r_1/R_s = \frac{\Delta I_3 - (R_L/R_s)\ \Delta I_1}{\Delta I_1}. \tag{7.61}$$

The relation (7.61) is equivalent to (7.60) by putting $R_L = R_s$ and $\partial I_3/\partial V_0 = -\Delta I_3/\Delta V_0$. The mutual differential resistance gives a useful physical quantity for the sensitive nature of the filamentary current. If $\Delta I_1 = \Delta I_3$ for $R_L = R_s$, then $r_1 = 0$ and 100% current injection into subsystem I is possible by a perturbation ΔI_2 from subsystem II. Figure 7.38 shows the mutual differential resistance as a function of V_0 obtained by

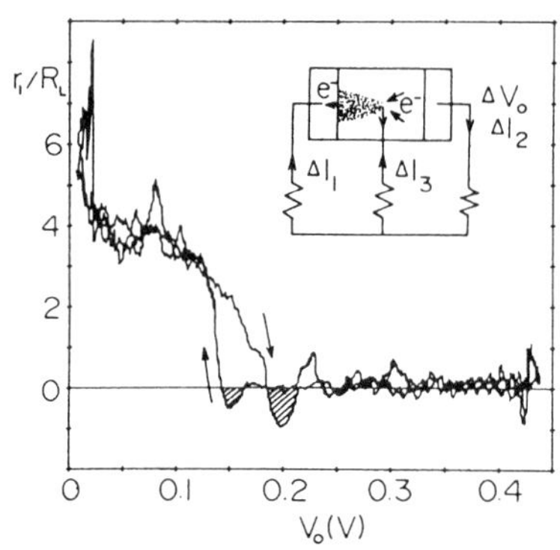

Figure 7.38 Mutual differential resistance r_1 normalized by R_L as a function of V_0 [83].

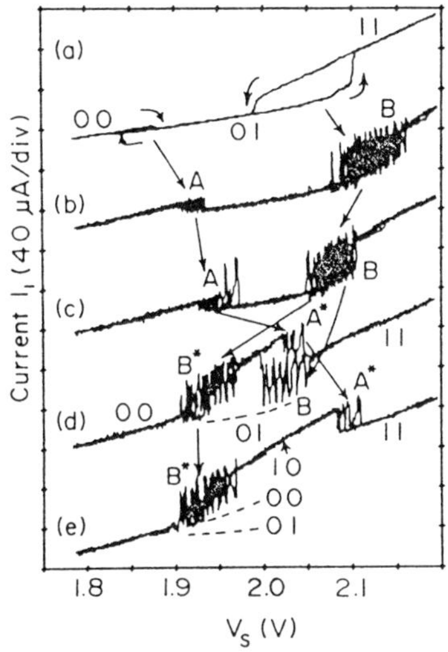

Figure 7.39 I_1–V_s characteristics. B = 0 T in (a), and B = 0.136 T in (b)–(e). Balancer voltage was varied as a parameter: (a) 0.075 V, (b) 0.09 V, (c) 0.13 V, (d) 0.163 V, and (e) 0.188 V [83].

measuring the slopes of the I_1 vs V_0 curve and the $(-I_3)$ vs V_0 curve for Fig. 7.37a, $V_s = 1.94$ V. Apparently, $r_1 \sim 0$ after filament I switches on. It is noted in Fig. 7.37 that the I_1 vs V_0 curve and the $(-I_3)$ vs V_0 curve are almost identical once filament I switches on (the dashed line in Fig. 7.37a shows part of the $(-I_3)$ vs V_0 curve). The hatched regions in Fig. 7.38 indicate the negative mutual differential resistance caused by the switching behavior at the transition points of hysteresis I (the hatched hysteresis in Fig. 7.37a). When filament I is in the off-state, subsystem I behaves as a normal resistor with $r_1 > 0$. One can also define r_2 for subsystem II. The above finding is helpful in interpreting the dynamical behaviors as discussed below.

The cross-over current instability can be observed by applying a longitudinal magnetic field. In Fig. 7.39, a series of $I_1 - V_s$ characteristics is shown, where $B = 0.136$ T. For reference, the current–voltage characteristic under the zero magnetic field is shown in Fig. 7.39a. In Fig. 7.39, note that the source voltage V_s is bias scanned instead of the balancer voltage V_0, and that V_0 has been varied as a parameter: (a) $V_0 = 0.075$ V, (b) $V_0 = 0.09$ V, (c) $V_0 = 0.13$ V, (d) $V_0 = 0.163$ V, and (e) $V_0 = 0.188$ V [83]. By the upward bias scan of V_s in Fig. 7.39a, filamentary current II flows first with $V_s \gtrsim 1.88$ V (indicated by 01 mode) and then filamentary current I with $V_s \gtrsim 2.1$ V (11 mode), followed by the hysteretic behavior by the downward bias scan. On applying the magnetic field, the cross-over

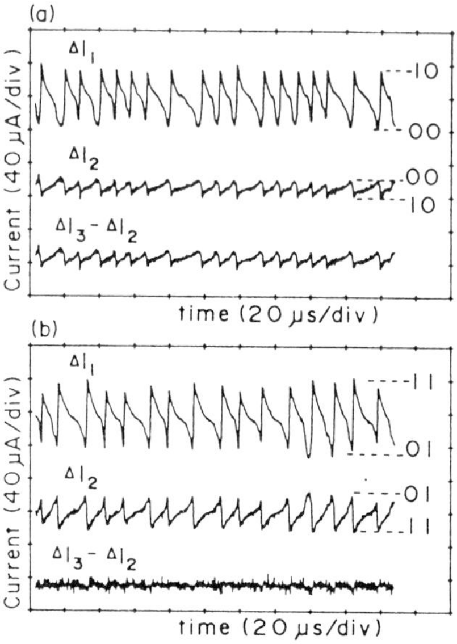

Figure 7.40 (a) B^*-mode oscillation and (b) B-mode oscillation. The experimental parameters are: (a) $B = 0.185$ T, $V_s = 1.962$ V, $V_0 = 0.31$ V, and (b) $B = 0.185$ T, $V_s = 2.127$ V, $V_0 = -0.125$ V [83].

instability of the hysteresis occurs with $B \gtrsim 0.052$ T for filament I and with $B \gtrsim 0.11$ T for filament II. In Fig. 7.39, the spontaneous current oscillation (SO) of filament II is denoted by mode A (SO between 00 and 01 states), while the spontaneous oscillation of filament I is denoted by mode B (SO between 01 and 11 states). By increasing the balancer voltage, mode A shifts towards higher voltages of V_s, while mode B shifts towards the lower voltages. With $V_0 \gtrsim 0.163$ V, mode A switches abruptly to mode A* (SO between 10 and 11 states) and the mode B to mode B* (SO between 00 and 01 states). The critical coexistence of modes B and B* can be seen with $V_0 \approx 0.163$ V in Fig. 7.39d. The current amplitude of the mode A (filament II) measured by I_1 is relatively small in Fig. 7.39b, c, since the filament I is in the off-state ($r_1 > 0$), but it becomes considerably large for the mode A* in Fig. 7.39d, e in the presence of filament I ($r_1 \sim 0$).

By measuring the spontaneous oscillation of mode A* (or mode B), one can investigate the nonlinear response of filamentary current I (or II) enforced by the SO of filament II (or I). For instance, Fig. 7.40a shows the current oscillations ($\Delta I_1, \Delta I_2$ and $\Delta I_3 - \Delta I_2$) in mode B*, and Fig. 7.40b the current oscillations in mode B, where the currents ΔI_1, ΔI_2 have been measured by the ac mode of the digital storagescope (i.e., $\Delta I_1 = I_1 - a$, $\Delta I_2 = I_2 - b; a, b$: offset levels). The rapid rise and the slow relaxation in the waveform of the spike current ΔI_1 clearly indicate self-generated oscillation of filament I (not of filament II), while the signal has been reversed when measured by ΔI_2. Thus, the SO current in subsystem I leaks to subsystem II with the ratio of the amplitudes $\Delta I_1/\Delta I_2 \sim -3$ for mode B* (Fig. 7.40a) and $\Delta I_1/\Delta I_2 \sim -2$ for mode B (Fig. 7.40b). For mode B, it is especially noted that the mutual differential resistance of filament II should be $r_2 \sim 0$ in Fig. 7.40b since $\Delta I_3 - \Delta I_2 \sim 0$ ($\Delta I_1 \sim \Delta I_3 + \Delta I_2$; suppose the reversed relation in the equivalent circuit in Fig. 7.38. This implies a physical situation of the cross-talk coupling in which filament I is oscillating by a repeating cycle of the generation–annihilation process (cross-over instability), while filament II keeps the on-state during the oscillatory behavior in subsystem I. During the oscillation of subsystem I, efficient current injection (or ejection) into (or from) the filamentary channel II occurs. This means a periodic expanding and shrinking (i.e., breathing mode) of filament II enforced by the SO in subsystem I.

More precisely, one can discuss the nonlinear response and the breathing dynamics by measuring the instantaneous power P_j ($j = 1, 2$):

$$P_j = (\Delta I_j - \langle \Delta I_j \rangle)(\Delta V_j - \langle \Delta V_j \rangle), \tag{7.62}$$

where $\Delta V_j = R_L(\Delta I_3 - \Delta I_j)$ and $\langle \cdot \rangle$ denotes the time averages. In Fig. 7.41, the instantaneous powers are shown for mode B* (Fig. 7.41a) and mode B

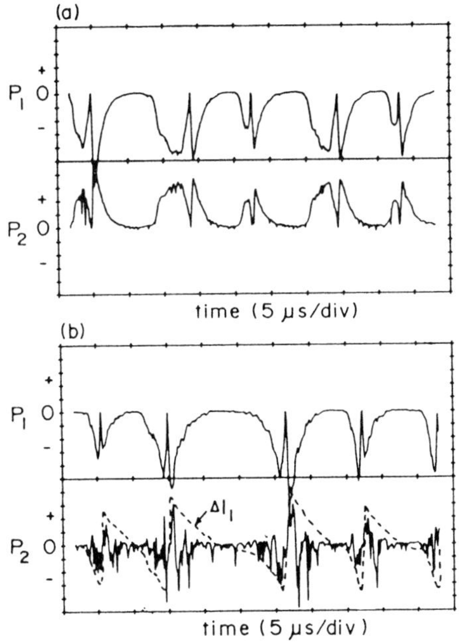

Figure 7.41 Instantaneous electric powers P_1 and P_2 for (a) B^*-mode oscillation, and (b) B-mode oscillation. The experimental parameters are the same as those in Fig. 7.40 The instantaneous electric powers are drawn in units of $0.02\mu W/div$ [83].

(Fig. 7.41b). Corresponding to Fig. 7.40a, b, it is obvious that P_1 is always negative since filament I is self-oscillating and generating power (i.e., is an active element) with NDM [72,73]. The positive power P_2 in Fig. 7.41a is also obvious because the current injection into subsystem II is dissipative (resistive) in the absence of filament II. In contrast, in Fig. 7.41b, P_2 is negative for a large part of the time series. For reference, the dashed line indicates the current waveform ΔI_1 in Fig. 7.41b. P_2 shows the noisy fluctuation, which is locally positive corresponding to the sharp rise in ΔI_1, and locally negative during the slow current relaxation. Now, the breathing dynamics can be approximately described by:

$$\Delta I_2 - \langle \Delta I_2 \rangle \approx \frac{\Delta S(t)(\sigma - \sigma_0)}{\ell} V_2, \tag{7.63}$$

$$\Delta V_2 = R_L(\Delta I_3 - \Delta I_2) = R_L(1/m(t) - 2)\Delta I_2, \tag{7.64}$$

where $\sigma(\sigma_0)$ are the conductivity inside (outside) the filamentary channel $(\sigma \gg \sigma_0)$, ℓ is the distance between the Schottky point contact and the ohmic contact, $\Delta S(t) = S(t) - \langle S(t) \rangle$ is an increment of the cross-sectional area [83] of filament II $(\langle \Delta S(t) \rangle = 0)$, and $m(t)$ is the current injection efficiency from subsystem I defined by $\Delta I_2 = -m(t)\Delta I_1$. In (7.63), the breathing dynamics are characterized by a repeating cycle of the expanding $(\Delta \dot{S}(t) > 0)$ and the shrinking $(\Delta \dot{S}(t) < 0)$ of filament II. The noisy fluctuation of P_2 around zero in Fig. 7.41b may be caused by the fluctuation of the current injection efficiency around 100% efficiency, $m = 1/2$, since the instantaneous power is given by:

$$P_2 \sim R_L(1/m(t) - 2)(\Delta I_2)^2. \tag{7.65}$$

Note that $P_2 < 0$ (> 0) for $m(t) > 1/2$ $(< 1/2)$. The positive spikes of P_2 in Fig. 7.41b may be attributed to the inertia of the nonlinear response [62] to the rapid current ejection by a sharp rise in ΔI_1 (i.e., inertia to the rapid shrinking of filament II with $m(t) < 1/2$). The negative part in P_2 indicates that filament II behaves as a quasi-active element for the slow current injection (expansion of the filament occurs with $m(t)$ slightly larger than 1/2).

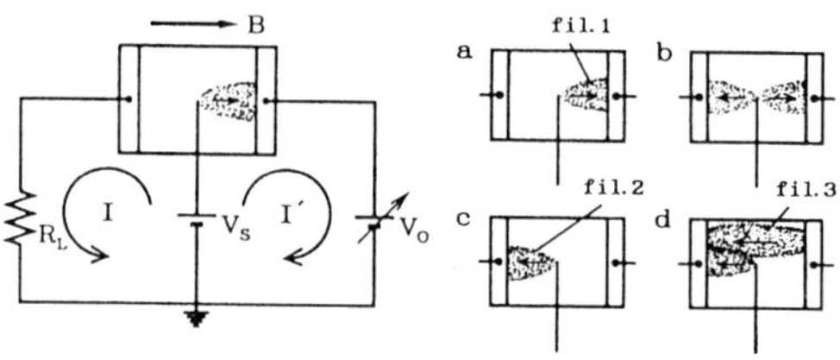

Figure 7.42 Three-terminal circuit. (a)–(d) correspond to the a–d branches of the $I - V_0$ curve in Fig. 7.43 [35].

With a circuit in Fig. 7.42, one may observe a remarkable effect of cross-over instability and the related cross-talk coupling in the set of I–V_0 curves of Fig. 7.43. In Fig. 7.43, three hystereses are observed for $B = 3.6$ mT, where the branches denoted by a–d correspond to the filamentary currents shown schematically in Fig. 7.42a–d. Note that the middle hysteresis is reversed. The branch d observed at $V_0 > 1.1$ V is caused by the filamentary current (fil. 3 in Fig. 7.42d) which flows between ohmic contacts [38], where V_0 is no longer the balancer voltage for the large values but works as a supply voltage. Filament 3 is found to be possible by considering the experimental fact that the typical breakdown voltage is 1.2 V for ohmic contacts with contact spacing of 1.5 mm. On increasing the magnetic field intensity, only filament 2 (see Fig. 7.42b) typically shows cross-over instability with $B > 96$ mT, as seen in Fig. 7.43. The branch d (filament 3) exhibits current oscillation with a small amplitude as soon as the cross-over instability of filament 2 sets in. The oscillatory region spreads into a wide range of V_0 for $B > 222$ mT. The oscillatory behaviors of the currents observed at a fixed bias of V_0 show complex behaviors of cross-talk coupling between filaments 1 and 2 or between the filaments 2 and 3. However, the correlation dimensions of the erratic (chaotic) current oscillations are found to be considerably high (4–5) and no clear bifurcations have been found. Therefore, the problem of complex behaviors has not been well resolved. The experimental findings in this section are rather more fundamental, i.e., zero mutual differential resistance, evidence of an enforced breathing mode by cross-talk coupling, and clear evidence of a filamentary current working as an active element. By taking V_0 at a critical value, symmetry-breaking switching between two oscillatory modes of subsystems I and II can be observed, whereby $1/f^{\alpha}$ noise ($\alpha \sim 0.5$–0.75) is generated [84].

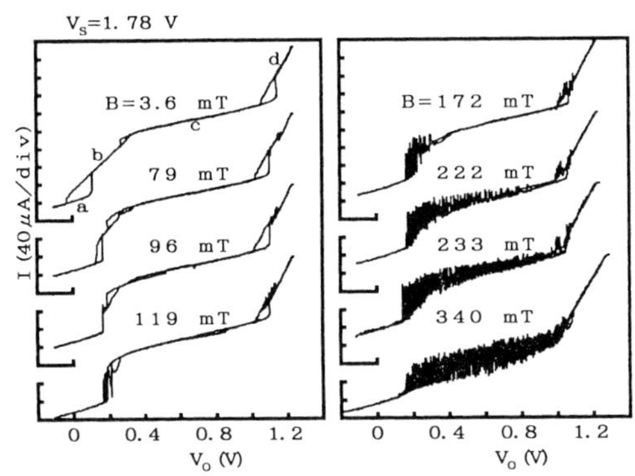

Figure 7.43 $I - V_0$ characteristics under the various magnetic field intensities [35].

References

[1] H. Mori, H. Okamoto and H. Tominaga, Prog. Theor. Phys. **85** (1991) 1143.
[2] H. Nyquist, Phys. Rev. **32** (1928) 110.
[3] L. Onsager, Phys. Rev. **37** (1931) 405; Phys. Rev. **38** (1931) 226.
[4] C. Kittel, *Elementary Statistical Physics* (John Wiley & Sons, New York, 1958).
[5] H.G. Winful, Y.C. Chen and J.M. Liu, Appl. Phys. Lett. **48** (1986) 616; see also H.G. Schuster, *Deterministic Chaos – An Introduction,* 2nd ed. (Physik-Verlag, Weinheim, 1988).
[6] F.C. Moon, *Chaotic and Fractal Dynamics* (John Wiley & Sons, New York, 1992).
[7] Hao Bai-Lin, *Chaos* (World Scientific, Singapore, 1984) and references therein.
[8] Hao Bai-Lin, *Chaos II* (World Scientific, Singapore, 1990).
[9] J. Brigges and F.D. Peat, *Turbulent Mirror,* perennial library edition (Harper & Row, New York, 1990).
[10] T. Horita, H. Hata, H. Mori, T. Morita, K. Tomita, S. Kuroki and H. Okamoto, Prog. Theor. Phys. **80** (1988) 793.
[11] H. Mori, H. Hata, T. Horita and T. Kobayashi, Prog. Theor. Phys. Suppl. **99** (1989) 1; see also K. Tomita, H. Hata, T. Horita, H. Mori, T. Morita, H. Okamoto and H. Tominaga, Prog. Theor. Phys. **81** (1989) 1124.
[12] H. Haken, *Synergetics – An Inttroduction, Nonequilibrium Phase Transitions and Self-organization in Physics, Chemistry and Biology*, 2nd edn (Springer, Berlins, 1987).
[13] K. Tomita, H. Hata, T. Horita, H. Mori, T. Morita, H. Okamoto and H. Tominaga, Prog. Theor. Phys. **81** (1989) 1124.
[14] T. Murayama, H. Tominaga, H. Mori, H. Hata and T. Horita, Prog. Theor. Phys. **83** (1990) 649.
[15] C. Grebogi, S.W. Mcdonald, E. Ott and J.A. Yorke, Phys. Lett. **99A** (1983) 415.
[16] C. Grebogi, E. Ott and J.A. Yorke, Physica. **7D** (1983) 181.
[17] C. Grebogi, E. Ott and J.A. Yorke, Science **238** (1987) 632.
[18] U. Rau, K.M. Mayer, J. Parisi, J. Peinke, W. Clauss and R.P. Huebener, Solid State Electron. **32** (1989) 1365.
[19] J. Peinke, J. Parisi, O.E. Rössler and R. Stoop, *Encounter with Chaos* (Springer, Berlin, 1992).
[20] A. Brandl, W. Kröninger, W. Prettl and G. Obermair, Phys. Rev. Lett. **64** (1990) 212.
[21] R.J. Deissler, J. Stat. Phys. **40** (1985) 371; Physica **D25** (1987) 233; J. Stat. Phys. **54** (1989) 1459.
[22] M. Lax, Rev. Mod. Phys. **32** (1960) 25; **38** (1966) 358, 541.
[23] H. Haken, Rev. Mod. Phys. **47** (1975) 67.
[24] J.-P. Eckmann and I. Procaccia, Phys. Rev. **A34** (1986) 659.
[25] H. Fujisaka and M. Inoue, Prog. Theor. Phys. **77** (1987) 1334; M. Sano, S. Sato and Y. Sawada, Prog. Theor. Phys. **76** (1986) 945.
[26] P. Grassberger, R. Badii and A. Politi, J. Stat. Phys. **51** (1988) 135.
[27] R. Rozi, J. Phys. **39** (1978) 9.
[28] J.-P. Eckmann and D. Ruelle, Rev. Mod. Phys. **57** (1985) 617.
[29] M. Hénon, Commun. Math. Phys. **50** (1976) 69.
[30] J.P. Crutchfield, J.D. Farmer, N.H. Packard and R.S. Shaw, Sci. Amer. **254** (1986) 46.
[31] H. Tominaga and H. Mori, Prog. Theor. Phys. **91** (1994) 1081.
[32] R.J. Deissler and J.D. Farmer, Physica **D55** (1991) 155.

[33] R.J. Deissler, Phys. Lett. **100A** (1984) 451.

[34] H.R. Brand and R.J. Deissler, Phys. Rev. **A39** (1989) 462.

[35] K. Aoki, Proceedings of a NATO Advanced Research Workshop on Negative Differential Resistance and Instabilities in 2-D Semiconductors, eds. N. Balkan, B.K. Ridley and A.J. Vickers (Plenum, London, 1993) NATO ASI Series B: Physics Vol. 307, pp. 393–407.

[36] K. Aoki and K. Yamamoto, Proc. 21st Int. Conf. Phys. Semicond. (Beijing, 1992) eds. P. Jiang and H. Zheng (World Scientific, Singapore, 1992) 261.

[37] K. Aoki, A. Matsuura and Y. Takahashi, Semicond. Sci. Technol. **9** (1994) 607.

[38] K. Aoki, U. Rau, J. Peinke, J. Parisi and R.P. Huebener, J. Phys. Soc. Jpn **59** (1990) 420.

[39] K.M. Mayer, J. Peinke, B. Röhricht, J. Parsi and R.P. Huebener, Phys. Scr. **T19** (1987) 505.

[40] K.M. Mayer, R. Gross, J. Parisi, J. Peinke and R.P. Huebener, Solid State Commun. **63** (1987) 55; K.M. Mayer, J. Parisi, J. Peinke and R.P. Huebener, Physica **32D** (1988) 306.

[41] U. Rau, J. Peinke, J. Parisi and R.P. Huebener, Z. Phys. **B71** (1988) 305.

[42] K.M. Mayer, J. Parisi, J. Peinke and R.P. Huebener, Z. Phys. B – Condensed Matter **71** (1988) 171.

[43] R. Stoop, J. Peinke, J. Parisi, B. Röhricht and R.P. Huebener, Physica **35D** (1989) 425; see also O.E. Rössler, Z. Naturforsch. **38a** (1983) 788.

[44] J. Parisi, J. Peinke, B. Röhricht and K.M. Mayer, Z. Naturforsch. **42a** (1987) 329; J. Peinke, J. Parisi, B. Röhricht, B. Wessely and K.M. Mayer, Z. Naturforsch. **42a** (1987) 841.

[45] J. Peinke, U. Rau, W. Clauss, R. Richter and J. Parisi, Europhys. Lett. **9** (1989) 743.

[46] R. Richter, J. Peinke, W. Clauss, U. Rau and J. Parisi, Europhys. Lett. **14** (1991) 1.

[47] A. Brandl, T. Geisel and W. Prettl, Europhys. Lett. **3** (1987) 401.

[48] A. Brandl, M. Völcker and W. Prettl, Appl. Phys. Lett. **55** (1989) 238; Solid State Commun. **72** (1989) 847.

[49] A. Brandl and W. Prettl, Festkörperprobleme **30** (1990) 371.

[50] G.A. Held and C. Jeffries, Phys. Rev. Lett. **55** (1985) 887; G.A. Held and C. Jeffries, in *Dimensions and Entropies in Chaotic Systems,* ed. G. Mayer-Kress (Springer, Berlin, 1986) pp. 158–170.

[51] G.A. Held and C. Jeffries, Phys. Rev. Lett. **56** (1986) 1183.

[52] X.N. Song, D.G. Seiler and M.R. Loloee, Appl Phys. **A48** (1989) 137.

[53] J. Spangler, A. Brandl and W. Prettl, Appl Phys. **A48** (1989) 143.

[54] K. Yamada, N. Miura and C. Hamaguchi, Appl. Phys. **A48** (1989) 149.

[55] K. Fujii, T. Ohyama and E. Otsuka, Appl. Phys. **A48** (1989) 189.

[56] D.G. Seiler, C.L. Littler, R.J. Justice and P.W. Milonni, Phys. Lett. **A108** (1985) 462.

[57] G. Hüpper and E. Schöll, *Proc. 20th Int. Conf. on Physics of Semiconductors,* eds. E. Anastassakis and J.D. Joannopoulos (World Scientific, Singapore, 1990) p. 2463.

[58] G. Hüpper, E. Schöll and A. Rein, Mod. Phys. Lett. **B6** (1992) 1001.

[59] E. Schöll, G. Hüpper and A. Rein, Semicond. Sci. Technol. **7** (1992) B480.

[60] K. Seeger, *Semiconductor Physics – An Introduction,* (3rd ed.) (Springer, Berlin, 1985), p. 160.

[61] E. Schöll, *Nonequilibrium Phase Transitions in Semiconductors* (Springer, Berlin, 1987)

[62] E. Schöll, Proc. 19th Int. Conf. On Physics of Semiconductors, ed. W. Zawadzki (Institute of Physics, Polish Academy of Sciences, Warsaw, 1988) 1407.
[63] K. Yamada, N. Miura and C. Hamaguchi, Solid-State Electron. **31** (1988) 809.
[64] G.A. Held, C. Jeffries and E.E. Haller, Phys. Lett. **52** (1984) 1037.
[65] G.A. Held, C. Jeffries and E.E. Haller, *Proc. 17th Int. Conf. Phys. Semicond* eds. J.D. Chadi and W.A. Harrison (San Francisco, 1984) 1289.
[66] K. Aoki, Y. Kawase, K. Yamamoto and N. Mugibayashi, J. Phys. Soc. Jpn **59** (1990) 20.
[67] K. Aoki, *Proc. 20th Int. Conf. on Physics in Semiconductors,* eds. E.M. Anastasakis and J.D. Joannopoulos (World Scientific, Singapore, 1990) 2578.
[68] K. Aoki, T. Kondo and T. Watanabe, Solid State Commun. **77** (1991) 91.
[69] K. Aoki and T. Kondo, Phys. Lett. **154A** (1991) 281.
[70] K. Aoki, T. Kondo and Y. Takahashi, Memoirs of the Faculty of Engineering, Kobe University, No. 38 (1991) 81
[71] U. Rau, K. Aoki, J. Peinke, J. Parisi, W. Clauss and R.P. Huebener, Z. Phys. B–Condensed Matter **81** (1990) 53.
[72] K. Aoki, Solid State Commun. **77** (1991) 87.
[73] K. Aoki, Phys. Lett. **152A** (1991) 485.
[74] R.S. Crandall, J. Phys. Chem. Solids **31** (1970) 2069; Phys. Rev. **B1** (1970) 730.
[75] K. Aoki, K. Yamamoto and N. Mugibayashi, J. Phys. Soc. Jpn **57** (1988) 26.
[76] K. Aoki and K. Yamamoto, Appl. Phys. **A48** (1989) 111.
[77] K. Aoki, Semicond. Sci. Technol. **7** (1992) B474.
[78] K. Aoki and T. Kondo, Phys. Rev. **B45** (1992) 3830.
[79] M. Mino, H. Yamazaki and K. Nakamura, Phys. Rev. **B40** (1989) 5279.
[80] R.R. Prasad, C. Meneveau and K.R. Sreenivasan, Phys. Rev. Lett. **61** (1988) 74.
[81] H.G.E. Hentschel and I. Procaccia, Physica **8D** (1983) 435.
[82] T.C. Halsey, M.H. Jensen, L.P. Kadanoff, I. Procaccia and B. Shraiman, Phys. Rev **A32** (1986) 1141.
[83] K. Aoki, T. Kondo and Y. Takahashi, Solid State Commun. **81** (1992) 711.
[84] K. Aoki, Proc. Int. Conf. on Noise in Physical Systems and 1/f Fluctuations, eds. T. Musha, S. Sato and M. Yamamoto (Ohmsha, 1991) 191.

Chapter 8

Towards a second new paradigm of semiconductor chaos

8.1 Towards an understanding of fully developed electronic turbulence

In Chapters 5–7 chaotic behaviors have been described, mainly associated with nonlinear carrier transport via the impact-ionization avalanche in various bulk-semiconductor materials. Sections 8.1 and 8.2 of this final chapter, will introduce up-to-date problems related to nonlinear carrier transport and chaos, focusing especially on the spatial and spatiotemporal behaviors of current filaments. In section 8.3, chaos in device structures is briefly covered.

Among a huge number of nonlinear dynamical systems in biology, chemistry, physics and engineering, many systems exhibit universal behaviors of bifurcation and chaos when control parameters are adequately and precisely varied. Such a **universality** is quite surprising because each dynamical system is almost uncorrelated with others except for the nonlinearity character. In spatially extended systems, it is well recognized that pattern formations and spatiotemporal behaviors are highly complex and in many cases different from each other. The precise differences among many dynamical systems may be termed **individuality**. The target pattern in the Belousov–Zhabotinsky (BZ) reaction system and the roll pattern in Rayleigh–Bérnard convection are different in their physical origin. In fluid dynamical systems, pattern formations are quite different, depending on the types of convection instability and their boundary conditions. Fractals in nature provide more examples, which are various growth patterns (trees, rivers, snow crystals, dendrite patterns of bacteria). However, it is also true that similarity in pattern formations can be seen among different physical systems. Roll patterns in liquid crystal (Williams domain) [1] and Rayleigh–Bérnard convection [2] are quite similar, where the rich pattern structures in the liquid crystal can be essentially described by electrohydrodynamic instability, the fundamental equations for which are highly complex. The similarity may be related to a strong analogy between the two systems; the motional force (buoyancy), the viscosity of the fluid, the bifurcation parameter (temperature difference or Rayleigh number) and the Prandtl number in the Rayleigh–Bérnard convection correspond to the electrostatic force, the elasticity, the voltage (V^2) and the driving frequency (by $(f/f_c)^{-1}$) in the liquid crystal. The spiral patterns in the BZ reaction system resemble the wave pattern of chemotactic activity in dense cell layers of slime mold. Such a variety of pattern formations and spatiotemporal behaviors is not so easy to understand. Although the universal rule of low-dimensional chaos has observed among many dynamical systems and the chaos theory has been now fully established and has prevailed, these systems also exhibit many aspects of unresolved spatiotemporal complexities. The goal may be the full understanding of spatiotemporal complex behaviors. Historically, extensive studies on spatiotemporal behaviors (pattern dynamics) have been done in the fluid system [3], the BZ system [4] and liquid crystal [1]. In semiconductors, the most prominent examples of spatial patterns are current filaments and high-field domains. Stable filaments and domains have been well known for several decades from experimental and theoretical investigations [5, 6]. However, only recently has attention been paid to the field of pattern dynamics, mainly due to the difficulty in visualizing time-resolved spatial patterns. Many more problems remain unanswered than resolved.

8.1.1 Numerical aspects of pattern dynamics in bulk semiconductors

Progress in numerical studies by a variety of models for the spatial and spatiotemporal dynamical behaviors of filamentary current flow caused by low-temperature impurity breakdown has been made by Schöll and collaborators, as introduced below. The present aim is to study how the current filaments in semiconductors behave under various experimental conditions. The spatial structure of a current filament is a dissipative structure deviated far from the thermal equilibrium, which is caused by an abrupt transition from a low conduction state (ohmic conduction state) to a high conduction state by the impact-ionization avalanche, followed by a spatially localized high-conduction channel of the current flow.

Since the filamentary current circulates through the external circuit, dynamical information on the integrated value over the spatially extended current density (down flow) is fed back nonlinearly and quasi-instantaneously. So, the system is a globally coupled one as well as being a locally diffusive-coupling system via the diffusive reaction of carrier generation and recombination, as in coupled map lattices [7], from which one can observe a wide diversity of complex dynamical behaviors such as hierarchical complexity of the current instability (Chapter 5). A particular feature which distinguishes semiconductors from other self-organized nonequilibrium systems, such as chemical reactions, is the ubiquitous presence of space charge [8]. The nonlinear feedback between charge carrier distribution and electric potential distribution and carrier diffusion are highly important factors in pattern formation. Such nonlinear feedback inside a semiconductor may be termed **local coupling**. A great wealth of complex dynamical phenomena is a direct result of **global coupling** and local coupling.

In bulk semiconductors, depending on the types of ohmic contacts (planar-type contacts [9–12], triangular contacts [13, 14] and point contacts [15, 16]), formation of the current filament and the dynamical behaviors may be different. With planar-type ohmic contacts, self-organization of a current filament [12, 13] and multifilaments [17, 18] is possible, since the field distribution is homogeneous in the transverse direction. With point contacts, only one current filament may be formed [15, 16] owing to the dipole-like field distribution. Several types of pattern dynamics have been confirmed from experiments and numerical simulations, as follows.

8.1.1.1 Breathing

The breathing mode has been proposed by Schöll [19, 20] and Kunz and Schöll [21]. The model of the breathing filament is described by a couple of truncated basic rate equations for the filament radius R_0, the position of the peak transverse electric field R_E and the longitudinal electric field $E_{\|}$. It has been shown that the breathing filament is possible at electric fields in the S-shaped negative differential conductivity (SNDC) region below the upper branch of an inhomogeneous current–voltage (I–V) curve of filamentary state (dash–dotted line H-B in Fig. 8.1) only when a large load resistance is used, where the oscillatory instability is basically confirmed due to the negative differential conductivity with the large load resistance settled in the SNDC region. An essential part of the breathing mode is reflected by the small-amplitude oscillation of the total current (Fig. 8.2b), which is based on the following two features [19, 21]:

(i) an instability of the filament radius $R_0(t)$, such that R_0 tends to increase further if R_0 is above a critical radius $R_0^*(E_{\|})$. Thus, R_0^* is analogous to the critical droplet radius in equilibrium or nonequilibrium phase transitions.
(ii) A restoring force, which is provided by dielectric relaxation of the longitudinal electric field $E_{\|}$. It is essentially controlled by the average carrier density $\bar{n}(R_0)$ which forces $E_{\|}$ to decrease with increasing R_0.

The phase portrait R_0 vs $E_{\|}$ is shown for n-GaAs in Fig. 8.2a, where the increase in the filament radius R_0 is a direct result of feature (i) given in the rate equation of R_0, while the decrease is caused by the dielectric relaxation of the longitudinal field $E_{\|}$ which is regulated by feedback via $\bar{n}(R_0)$. Such a

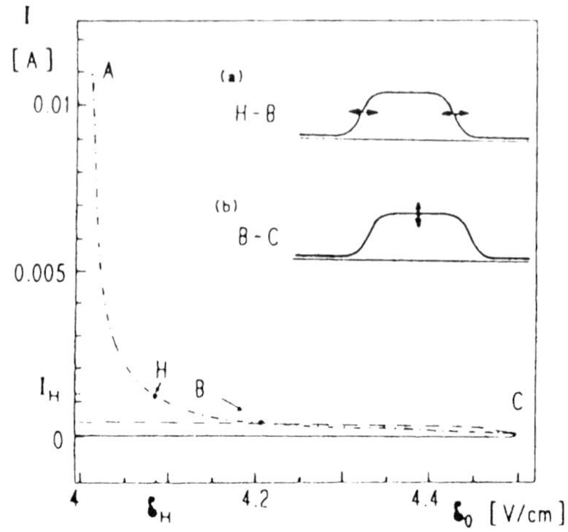

Figure 8.1 Regimes of stable stationary filaments (A-H), breathing current filaments (H-B) and bulk-dominated oscillations (B-C), drawn in the part of the S-shaped current-field $(I_0 - E_0)$ characteristically just below the postbreakdown regime, calculated for p-Ge. Full line: stable homogeneous; broken line: unstable humogeneous (estimated); dash-dotted line: filamentary. The Hopf bifurcation point E_H is marked. The breakdown field (C) is assumed to be arbitrary as $E_0 \sim 4.5$ V/cm. The inset shows (a) the breathing mode and (b) a bulk-dominated oscillation mode (schematically), corresponding to regimes H-B and B-C, respectively [19].

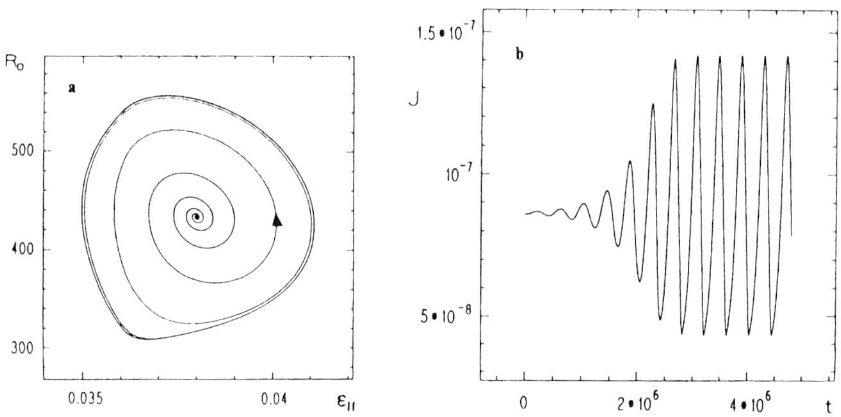

Figure 8.2 Phase portraits and time series of a limit cycle corresponding to a breathing current filament for p-Ge at $E_0 = 0.038$. (a) Filament radius R_0 in units of L_D vs longitudinal field E in units of $k_B T_\mathrm{L}/(-qL_\mathrm{D})$. (b) R_0 vs R_E of peak transverse field in units of L_D. (c) Current density J in units of $e/(L_\mathrm{D}^2 \tau_\mathrm{M})$ vs time in units of τ_M [19].

breathing mode has been observed in p-type Ge [22] and n-type GaAs [15]. For details see Refs [19–21].

8.1.1.2 *Traveling or locking filaments*

For planar-type ohmic contacts in bulk semiconductors [23] and for diode structures such as the p^+–n^+–p–n^- diode [24], the filament may slide freely in the transverse direction owing to the uniform field distribution. The locking filament and traveling filament appear in the same oscillatory regime of the p^+–n^+–p–n^- diode [25], as shown by the numerically calculated spatiotemporal evolution in Fig. 8.3 [26]. As a function of the driving voltage V_s, a static localized filament (Fig. 8.3a) bifurcates to a locking filament (Fig. 8.3b) and further bifurcates to a traveling filament (Fig. 8.3c). The traveling filaments and the locking filaments are like propagation and reflection of a soliton bounded within walls, without changing the shape of the kink–antikink pair. The locking filament oscillates bounded within the center region, while the traveling filament propagates and is reflected at the boundaries [26]. The details will be described in section 8.1.2.

8.1.1.3 *Spatiotemporal spiking*

This type of spiking is quite typical when the load line is settled in the S-shaped region of the *I–V* curve. Spontaneous current oscillations with large amplitudes are typically observed as a result of interplay between lower and upper branches of the S-shaped *I–V* curve. The spiking has been observed and explained by a repetitive cycling of ignition and extinction of a current filament by Brandl *et al.* [16]. For theoretical aspects see by Wacker and Schöll [27] and Brandl and Prettl [28]. The spatiotemporal spiking is not

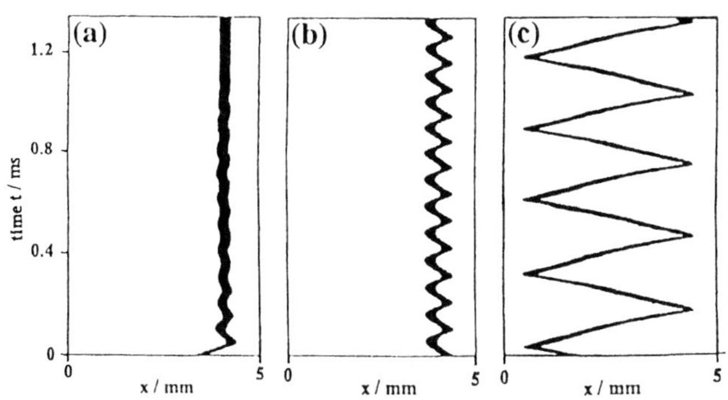

Figure 8.3 Numerically calculated spatiotemporal evolution of $j_c(x, t)$: (a) transient of a disturbed filament with $V_s = 32.85$ V, (b) rocking filament with $V_s = 32.90$ V, (c) traveling filament with $V_s = 33.05$ V [26].

restricted to the spontaneous case, but it is quite a popular oscillatory regime which can be seen easily under the periodically driven case in the SNDC region [9, 29].

Studies on pattern dynamics are widespread but can be classified into two parts: (i) spatiotemporal patterns under various experimental conditions, and (ii) nucleation process of the current filament. Here, the numerical investigations which have been done by Schöll and collaborators are introduced.

8.1.2 Dynamical Hall-field instability in crossed electric and magnetic fields

Hüpper and Schöll [30] and Hüpper *et al.* [14] have investigated numerically the dynamical Hall-field instability of a current filament observed in p-type Ge [23, 31], n-GaAs [32] and n-InSb [33] under crossed electric and magnetic fields. Hüpper *et al.* extended their original homogeneous model [30] into the spatiotemporal problem [14] and explained an asymmetric breathing of the filament walls observed in n-GaAs [16]. Their model is based on Schöll's model (multilevel impact-ionization model) [20]. In their model they strongly support that SNDC is explained only when the excited state(s) of the impurity participates effectively in the generation–recombination (g-r) processes of the carriers. The numerical study has been based on semiclassical transport theory combined with Monte Carlo (MC) simulations [34] of the g-r processes, and is able to account for the spatiotemporal degrees of freedom of nonlinear filamentary conduction and instabilities in crossed electric and magnetic fields. In the classical regime (with low magnetic field intensity) they neglect quantum effects leading to the splitting of the conduction band into Landau levels.

In the presence of a magnetic field the current density $\mathbf{j} = qn\mathbf{v}$ is given by [14, 30]:

$$\mathbf{j} = en\,\mu_B\mathbf{F}/q - qn\,\mu\,\mu_B(\mathbf{B}\times\mathbf{F}/q) + n\,\mu^2\,\mu_B\mathbf{B}(\mathbf{B}\cdot\mathbf{F}), \tag{8.1}$$

where n is the carrier density, v the mean group velocity, $q = \pm e$ is the charge, $\mu = e\,\tau_{\mathrm{m}}/m^*$ (mobility for $B = 0$), $\mu_B = \mu/(1+\mu^2B^2)$, $\mathbf{F} = q\mathbf{E} - eDn^{-1}\mu_B^{-1}\nabla_r n$ (E: electric field), and D is the diffusion constant [14].

The dynamical variables (independent variables) are n, $n_i (i = 1, 2 \ldots, M)$, where n_i is the carrier density at the excited state labeled by i of the impurity, $\mathbf{E} = (E_x, E_z)$. Both the applied drift field E_x and the induced Hall field E_z are treated dynamically, reflecting their finite dielectric relaxation time [14]. The spatial degrees of freedom are considered only in the transverse direction (the z-direction) since the Hall-field effect and the dominant spatial inhomogeneity in the SNDC regime occur perpendicular to the drift field in the form of current filaments, but later the model simulations have been extended to the two-dimensional (2D) case [35]. Assuming a sufficiently long sample with ideal planar contacts connected in series with a voltage source

U_0 and a load resistor R and in parallel with a capacitance C, a set of spatiotemporal dynamic equations is composed of the continuity equations

$$\frac{\partial n}{\partial t} + \frac{\partial (n v_z)}{\partial z} = \phi(n, \underline{n}_t, E), \qquad \frac{\partial n_t}{\partial t} = \underline{\phi}_t(n, \underline{n}_t, E), \tag{8.2}$$

and the dielectric relaxation equations

$$c_{\mathrm{r}} \frac{\partial E_x}{\partial t} = j_0 - (\mu_B \langle n \rangle + \sigma_\ell) E_x - \mu_B \, \mu B \left[\frac{\Delta n}{W} - \langle n E_z \rangle \right], \tag{8.3}$$

$$\frac{\partial E_z}{\partial t} = - n v_z, \tag{8.4}$$

supplemented by Maxwell's equations

$$\partial E_x / \partial z = 0, \qquad \partial E_z / \partial z = \rho = n + \sum_i^M n_i - 1, \tag{8.5}$$

where $c_{\mathrm{r}} = 1 + LC/(\epsilon A)$, $A = bWL_{\mathrm{D}}^2$ is the lateral cross-section of the sample, L is the length in the x-direction, ϵ is the dielectric constant, $j_0 = U_0/(RAe\mu_\ell N_{\mathrm{A}}^* E_0)$, $\sigma_\ell = \tau_{\mathrm{M}} L L_{\mathrm{D}}/(RA\epsilon)$, $v_z = \mu_B[E_z - (1/n)\partial n/\partial z + \mu B E_x]$ is the mean velocity in the z-direction and ρ is the local charge density. In (8.2) ϕ, $\underline{\phi}_t$ are the g-r rates, where $\underline{\phi}_t = (\phi_1, \ldots, \phi_{\mathrm{M}})$. In (8.3) the angular brackets denote the spatial mean value $\int_0^W \mathrm{d}z/W$, where W is the width of the sample ($W \ll L$), and $\Delta n = n(W) - n(0)$ [14]. These equations hold for holes ($q = e$) in p-type semiconductors, but can be easily converted to the spatiotemporal dynamical equations for electrons in n-type semiconductors by inverting the sign of E_z. For the computer simulations all quantities μ, t, z, E and n, $\underline{n}_t$ are given in dimensionless units, scaled by the low field mobility μ_ℓ, the dielectric relaxation time $\tau_{\mathrm{M}} = \epsilon/(e\,\mu_\ell N_{\mathrm{A}}^*)$, the Debye length $L_{\mathrm{D}} = [k_B T_{\mathrm{L}} \epsilon/(e^2 N_{\mathrm{A}}^*)]^{1/2}$, the thermal field $E_0 = k_B T_{\mathrm{L}}/(e L_{\mathrm{D}})$ and the effective doping concentration $N_{\mathrm{A}}^* = N_{\mathrm{A}} - N_{\mathrm{D}}$, respectively.

To obtain a physical idea of the transverse motion of a current filament under the crossed electric and magnetic fields, Hüpper *et al.* considered a carrier density fluctuation around the homogeneous state with and without a magnetic field [14]. From the arguments of the linear stability analysis of (8.2)–(8.5), it has been shown that the fluctuation around the homogeneous state at the turning point of the SNDC grows as a result of the g-r instability for both cases $B = 0$ and $B \neq 0$ and moves with velocity v in the direction of the Lorentz force for $B \neq 0$ [14]. A fluctuation of the transverse electric field will be built up according to (8.4), whereby the drift and diffusion currents will tend to cancel each other out, i.e., $n\,\delta E_z = \partial \delta n/\partial z$. In the case $B = 0$ the fluctuation leads to a symmetric change of the absolute value of E_z, since the homogeneous steady state $E_z^* = 0$ from (8.4). This local increase in the electric field results in a symmetric increase in the fluctuation due to the g-r processes. For the steady-state distribution of the transverse electric field and the carrier density of a current filament see Ref. [20]. In the case $B \neq 0$ the fluctuation δE_z will reduce the absolute value of E_z on one side, while it will increase on the opposite site. Owing to the g-r processes, including the impact-ionization process, the side of the fluctuation with the lower electric field will decrease and the opposite side will increase. As a result, the fluctuation will move in the direction of the Lorentz force [14]. In particular Hüpper *et al.* stressed that only the shape of the fluctuation moves transversally, while the carriers themselves remain at their transverse positions, performing free to bound transitions and vice versa [14]. In contrast to the previous view of traveling filaments being due to individual carriers moving in the Lorentz force [23], this interpretation may be obvious if one considers the charge neutrality condition inside and outside the current filament. It is only the case at the filamentary boundary that the charge neutrality is broken. Hüpper *et al.* [14] suggested that the situation is analogous to the motion of water molecules in a water wave.

A numerical simulation has been performed for the system (8.2)–(8.5), where the g-r parameters have been adequate for p-type Ge at 4.2 K with $c_r = 1$ and $\sigma_\ell = 0$ (current control) [14]. The g-r rates in (8.2) are $\phi = X_1^s n_2 - T_1^s n(1 + c - n_1 - n_2) + X_1 n n_1 + X_1^* n n_2$, $\phi_1 = T^* n_2 - X^* n_1 - X_1 n n_1$, and $\phi_2 = -\phi - \phi_1$, where $c = N_D/N_A^*$ is the compensation ratio. The g-r coefficients were obtained by a spatially homogeneous single-particle MC simulation [34–37]. The MC data were fitted by a smooth analytical function and substituted into (8.2–8.5). With these g-r parameters the static current density vs field characteristic typically shows an S-shaped curve, given in Fig. 8.4. To simulate the spatiotemporal behaviors, (8.2)–(8.5) have been solved with the method of particles [34], with cyclic transverse boundary conditions (periodic boundary conditions), which may be a possible approximation of the boundary conditions for the sample with planar-

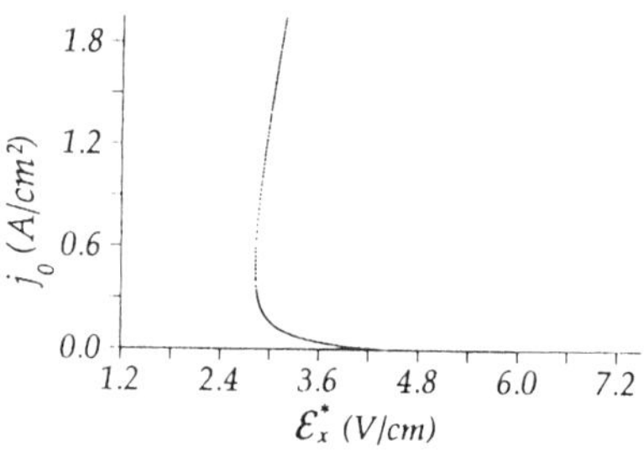

Figure 8.4 Static current density-field characteristic calculated with $N_A = 10^{14}$ cm^{-3}, $N_D = 5 \times 10^{12}$ cm^{-3}, $\epsilon = 16\,\epsilon_0$, $T_L = 4$ K, $\mu = \mu_\ell = 10^5$ cm^2/V s, $\tau_M = 10^{-12}$ s, $L_D = 5.6 \times 10^{-6}$ cm, $E_0 = 60.8$ V/cm, $X^* = 10^{-15}$, $T^* = 7.21 \times 10^{-5}$, $X_1 = 1.4 \times 10^{-6}$, $\alpha = 60.8(1 + \mu^2 B^2)^{-1/2}$, $X_1(E) = 7.85 \times 10^{-4} \times \exp[-11.3(\alpha E)^{-0.745}]$, $X_1^*(E) = 4.18 \times 10^{-2} \times \exp[-3.72(\alpha E)^{-0.66}]$, $T_1^s(E) = -1.2 \times 10^{-3} \times \exp[-0.2(-0.254 + \alpha E)^2] + 1.73 \times 10^{-3}(0.421 + \alpha E)^{-0.837}$ [14].

type ohmic contacts. After some initial transients the filament travels transversally with a constant velocity, as shown in Fig. 8.5. Since the Lorentz force drives the filament into the transverse direction, the traveling velocity v increases with the magnetic field intensity at the low-field regime in Fig. 8.6, but it saturates at the high-field regime with $B > 50$ mT. The traveling velocity takes a smaller value ($\approx 10^2$–10^3 cm/s) than the drift velocity of the carriers ($\approx 10^5$–10^6 cm/s), in agreement with experiments [23].

More adequate boundary conditions are the Dirichlet boundary conditions $E_z(0,t) = E_z(W,t)$ for a sample with finite width W and with planar-type contacts. With the Dirichlet boundary conditions the filament travels to the boundary and is pinned there [14]. With triangular contacts (Fig. 8.7), which may serve as a model for nonideal planar-type ohmic contacts, the spatiotemporal behaviors show irregular electronic turbulence with the following scenario [14]. Traveling of the filament is confined within the

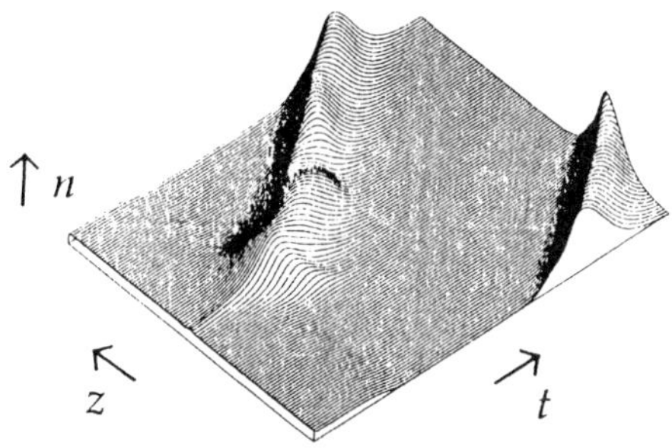

Figure 8.5 Spatiotemporal dynamics of the carrier density $n(z,t)$ in the presence of an externally applied magnetic field under the periodic boundary conditions for $j_0 = 34$ mA/cm^2, $B = 10$ mT with planar contacts. The total time is 8.1×10^{-6}s, the sample width is $W = 90$ μm [14].

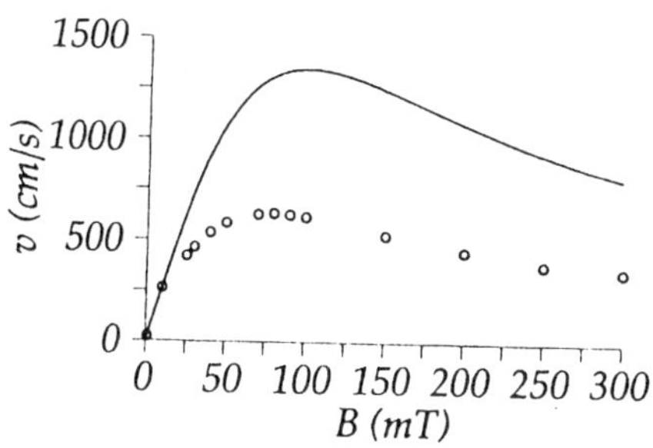

Figure 8.6 Transverse velocity of the traveling filaments as a function of the magnetic field strength (dots) for $j_0 = 93.76$ mA/cm^2, $W = 120$ μm. The full line shows the velocity obtained by the linear stability analysis for the same control parameters [14].

effective sample width $L(z)(< W)$ because the longitudinal electric field at the sample boundary is so small that the filament cannot travel to and survive at the sample boundaries. Qualitatively, the triangular contacts yield the following scenario: the free carriers inside the moving filament will be spread over an increasing effective sample length $L(z)$, which reduces the maximum carrier density in the filament. In addition, the drift field and therefore the generation rate decrease [14]. Both effects tend to destroy the filament. As a result, the total current through the sample decreases after the destruction, which increases the sample voltage and thus enhances filamentation again due to increased generation rate [14]. The oscillatory regime in Fig. 8.7b is the traveling and destruction process of the filament, which leads to chaotic voltage and current oscillations (Fig. 8.7b) and has been found in experiments on p-type Ge [23].

In the model simulations the time-scales for these processes are given by the time needed by the filament to travel across the sample, the g-r time and the dielectric relaxation time of the electric field. Depending on the ratios of these time-scales and the geometry of the contacts, different behaviors of the filament are possible [14]. For small magnetic fields and comparable time-scales the filament rests in the middle of the sample and shows an asymmetric breathing of the filament walls, as observed in n-type GaAs [15, 16]. As a function of magnetic field, Hüpper *et al.* have found different states of the filament: asymmetric breathing oscillations, small regular oscillations due to periodic nascence and destruction of traveling filaments, chaotic oscillations due to intermittent nascence and destruction, and pinning of the filament at the transverse boundary after some transient traveling sequences [14].

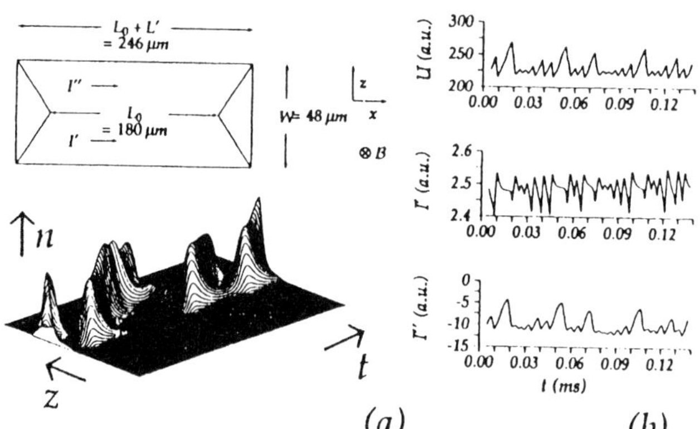

Figure 8.7 Traveling filaments for $j_0 = 0.85$ mA/cm^2, $B = 20$ mT with triangular contacts. The inset shows the sample geometry. (a) Spatiotemporal dynamics of n(z,t). The total time is 2×10^{-5}s. (b) Time series of the sample voltage U(t) and currents I', I'' (see inset), al in arbitary units [14].

8.1.3 Nucleation process of a current filament

Kunz *et al.* [38] have developed their model simulations into the problems of longitudinal breakdown modes and breakdown in 2D domains, which provides information for understanding nucleation process of a current filament. The model equations are essentially equivalent to (8.2)–(8.5) [14, 38] with $B = 0$, but extended to the 2D degrees of freedom. They studied the nucleation process of a current filament for p-type Ge, where they strongly support that the SNDC can be explained by standard g-r kinetics, including impact ionization, only if at least the ground and first excited levels are taken into account, namely, they support the multilevel impact-ionization model [20]. Gaa *et al.* [39] have also studied numerically the nucleation process of the filamentary current in n-type GaAs, where the numerical procedures are identical to those used by Kunz *et al.* [38]. Comparison of the numerical results between p-type and n-type semiconductor materials may be meaningful. Here the numerical procedures are introduced briefly only for p-type Ge. The spatiotemporal evolution of free hole density $n_{\rm h}$ is governed by the continuity equation, similarly to (8.2):

$$\dot{n}_{\rm h} = -\frac{1}{e}\nabla \cdot \mathbf{j}_{\rm h} + \phi(n_{\rm h}, \mathbf{n}_t, |\nabla u|), \tag{8.6}$$

$$\phi = X_1^{\rm s} n_{t2} - T_1^{\rm s} n_{\rm h} p_t + X_1 n_{\rm h} n_{t1} + X_1^* n_{\rm h} n_{t2}, \tag{8.7}$$

where the dot denotes the partial derivative with respect to time, e is the electron charge, u is the electrostatic potential,

$$E = \sqrt{E_x^2 + E_y^2} = |\nabla u|,$$

and $\mathbf{n}_t$ are the hole densities trapped at the ground and excited levels of acceptor formally given by the vector presentation: $\mathbf{n}_t = (n_{t1},\ n_{t2})$. In (8.7) $X_1^{\rm s}$ is the thermal ionization coefficient, $T_1^{\rm s}$ is the hole capture coefficient, X_1, X_1^* are the hole impact-ionization coefficients from the ground and excited level, respectively, $p_t = N_{\rm A} - n_{t1} - n_{t2}$ is the density of ionized acceptors, and $N_{\rm A}$ is the total density of acceptors. Similarly, the rate equations for the trapped holes at the ground and excited levels are written as:

$$\dot{\mathbf{n}}_t = \boldsymbol{\phi}_t(n_{\rm h},\ \mathbf{n}_t,\ |\nabla_u|), \tag{8.8}$$

with

$$\phi_{t1} = T^* n_{t2} - X^* n_{t1} - X_1 n_{\rm h} n_{t1}, \quad \phi_{t2} = -\phi - \phi_{t1}, \tag{8.9}$$

where $\boldsymbol{\phi}_t = (\phi_{t1},\ \phi_{t2})$, and X^*, T^* denote the transition coefficients from the ground level to the excited level and vice versa, respectively [38]. Within the drift-diffusion approximation the current density $\mathbf{j}_{\rm h}$ is expressed by:

$$\mathbf{j}_{\rm h} = -e(D_{\rm h}\nabla n_{\rm h} + \mu_{\rm h} n_{\rm h} \nabla u), \tag{8.10}$$

where μ_h is the hole mobility and D_h is the hole diffusion constant given by the Einstein relation $D_h = \mu_h k_B T_L / e$ (T_L: lattice temperature). Besides the continuity equation (8.6) the electrostatic potential u must satisfy the Poisson equation:

$$-\nabla \cdot (\epsilon \nabla u) = e(n_h + n_{t1} + n_{t2}) - N_A^*), \qquad (8.11)$$

where ϵ is the dielectric constant and $N_A^* = N_A - N_D$. The Poisson equation is equivalent to the charge conservation equation:

$$\nabla \cdot \mathbf{J} = 0, \qquad \text{with} \quad \mathbf{J} = -\epsilon \nabla \dot{u} + \mathbf{j}_h, \qquad (8.12)$$

where $\mathbf{J}$ is the total current density composed of the displacement current density $\epsilon \dot{\mathbf{E}}$ and the conduction current density $\mathbf{j}_h$. Equation (8.12) is obtained by taking the time derivatives of (8.11), subsequent elimination of n_h, n_{t1} and n_{t2} using (8.6) and (8.8) and taking into account the carrier conservation character of the g-r process $\phi + \sum_i \phi_{ti} = 0$.

The g-r coefficients as a function of electric field are obtained by the spatially homogeneous MC data [34–37]. The numerical parameters used for the g-r processes X^*, T^*, X_1^s are calculated on the basis of the Lax cascade-capture model [40, 41], using detailed balance [20]. The parameters for p-type Ge used by Kunz *et al.* [38] and n-type GaAs used by Gaa *et al.* [42] are given in Tables 8.1 and 8.2, respectively. The MC transport data are fitted by smooth analytic functions of the electric field [35–39, 42]. An example of the analytic functions for n-type GaAs is given in Fig. 8.8. For the MC simulation, optical phonon scattering is neglected as usual because of low lattice temperature. The relevant intraband scattering processes especially considered for n-GaAs are elastic ionized impurity scattering (by means of the Conwell–Weisskopf approximation) and inelastic acoustic deformation

Table 8.1 Material parameters for p-Ge at 4 K, obtained from Monte Carlo simulation

$N_A = 10^{14}$ cm^{-3}, $N_D = 5 \times 10^{12}$ cm^{-3}, $\varepsilon = 16\varepsilon_0$, $T_L = 4$ K, $\mu_h = 10^5$ cm^2 V^{-1} s^{-1}
$X^* = 10^{-3} s^{-1}$, $T^* = 7.21 \times 10^7$ s^{-1}, $X_1^S = 1.4 \times 10^6$ s^{-1}

$X_1(\varepsilon) = x_1 \exp[-x_2(\alpha\varepsilon)^{x_3^*}]$
$X_1^*(\varepsilon) = x_1^* \exp[-x_2^*(\alpha\varepsilon)^{x_3^*}]$
$T_1^S(\varepsilon) = t_1 \exp[-t_2(t_3 + \alpha\varepsilon)^2] + t_4(t_5 + \alpha\varepsilon)^{t_6}$
with

$x_1 = 8.26 \times 10^{-6}$ cm^3 s^{-1}	$x_1^* = 4.40 \times 10^{-4}$ cm^3 s^{-1}	$t_1 = -1.26 \times 10^{-5}$ cm^3 s^{-1}
$x_2 = 11.3$	$x_2^* = 3.72$	$t_2 = 0.2$
$x_3 = -0.745$	$x_3^* = -0.66$	$t_3 = -0.254$
$\alpha = 1.0$ cm V^{-1}		$t_4 = 1.82 \times 10^{-5}$ cm^3 s^{-1}
$t_5 = 0.421$		$t_6 = -0.887$

Table 8.2 Material parameters for n-GaAs

Parameter	Symbol	Simulation	Experiment
Donor concentration	N_D	7.0×10^{15} cm^{-3}	9.0×10^{13} cm^{-3}
Acceptor concentration	N_A	2.0×10^{15} cm^{-3}	7.0×10^{13} cm^{-3}
Lattice temperature	T_L	4.2 K	1.8 K
Mobility	μ	1.0×10^5 cm^2/V s	1.6×10^5 cm^2/V s (77 K)
Diffusion constant	D	36.2 cm^2/s	
Dielectric constant	ϵ	12.5	
GR coefficients	X_1^s	1.17×10^6 s^{-1}	
	X^*	3.36×10^3 s^{-1}	
	T^*	4.10×10^7 s^{-1}	
Sample dimensions		0.2×0.2 mm^2	3.6×3.6 mm^2
Distance of contacts		0.2 mm	4 mm
Sample thickness		14 μm	14 μm

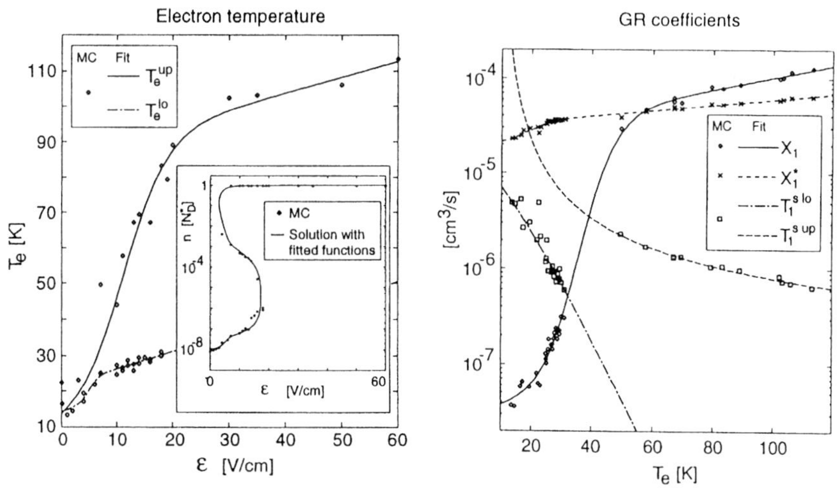

Figure 8.8 (a) Electron temperature T_e as a function of the electric field E: Monte Carlo (MC) data and fitted analytic representation. Material parameters: $N_D = 7 \times 10^{15}$ cm^{-3}, $N_A = 2 \times 10^{15}$ cm^{-3}, $\mu = 10^5$ cm^2/Vs, $T_L = 4.2$ K. Inset: stationary n(E) characteristic with g-r parameters given in (b). (b) g-r coefficients as a function of the electron temperature T_e: MC data and fitted analytical representations. $X_1^s = 1.17 \times 10^6$ s^{-1}, $X^* = 3.36 \times 10^3$ s^{-1}, $T^* = 4.1 \times 10^7$ s^{-1}, $\tau_M = 7.83 \times 10^{-14}$ s (momentum relaxation time) [39].

potential scattering. As a result of the MC simulation for n-GaAs, the impact-ionization coefficients X_1 and X_1^* as well as the capture coefficient T_1 depend not only on the electric field E, but also on the electron temperature T_e and the electron concentration n [34–37, 42]. For p-Ge it is noted that the g-r coefficients have been given also by analytic functions fitting the MC data, but merely as a function of the electric field [38] (see Table 8.1). The resulting stationary (homogeneous) carrier density-field characteristic $n^*(E)$ shows the SNDC in Fig. 8.9 for p-Ge (for n-GaAs see inset of Fig. 8.8a), which is obtained by $\phi = 0$, $\phi_{t1} = \phi_{t2} = 0$ in (8.7), (8.8) and assuming charge neutrality.

For p-Ge the charge neutrality is given by $\rho^* = e(-N_A + N_D + n_{t1}(n_h, E) + n_{t2}(n_h, E) + n_h) = 0$. Solving $\phi = 0$ in (8.7) and $\phi_{t1} = 0$ in (8.9) with $\rho^* = 0$ with respect to n_h, results in more than one solution $n_h(E)$ in a certain range, which means the bistability in the SNDC (Fig. 8.9). The bistability can be explained qualitatively as follows. For low electric fields (in the prebreakdown regime) almost all holes are trapped in the ground state of the acceptors. When the electric field increases, the few holes gain energy until they are able to impact-ionize carriers from the ground level (note: the excited levels are not populated [38]). The impurity breakdown occurs in the field regime that corresponds to the higher energy needed to ionize carriers from the ground level [38]. Thus, the abrupt transition occurs at the breakdown field from the lower branch to the upper branch of the S-shaped n^*–E characteristic. When the electric field decreases in the upper branch, free holes in the valence band will be trapped in the excited level. Since the transition time to the ground state is very long, the population of

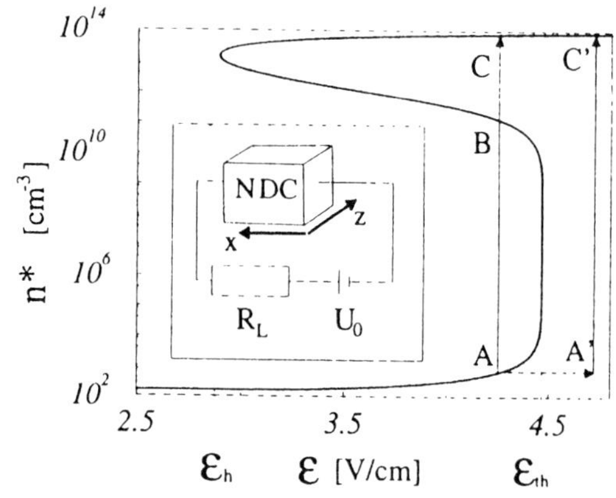

Figure 8.9 S-shaped carrier density–field relation for p-Ge at 4 K [from the model of eqns (8.7), (8.9) and the parameter values of Table 8.1]. For details of the captions see Ref. [38]. The inset shows the typical experimental set-up: the NDC sample is connected to a bias voltage U_0 and a load resistor R_L [38].

the excited level will be high until the energy of free holes is too small to ionize the carriers in the excited level [38]. With the lower electric field, at a sustained field, and abrupt transition from the upper to the lower branches occurs via the freeze-out, resulting in a hysteresis under voltage-controlled conditions.

In order to compute the spatiotemporal dynamics, a set of (PDEs) given by (8.6), (8.8), (8.12) must be solved with the initial conditions

$$u(0, \mathbf{x}) = u_0(\mathbf{x}), \; n_{\mathrm{h}}(0, \mathbf{x}) = n_{\mathrm{h},0}(\mathbf{x}), \; n_t(0, \mathbf{x}) = n_{t,0}(\mathbf{x}). \tag{8.13}$$

Kunz *et al.* used the semiconductor device simulator, ToSCA [38], which was originally designed to describe bipolar carrier transport in silicon-based devices. It has been modified to include g-r processes of band carriers with shallow impurities, where two-level impact-ionization processes are considered as usual. They investigated longitudinal breakdown modes and also breakdown in 2D domains [38] based on the two-level impact-ionization model [20]. Leaving the details of the algorithm [38], only the numerical results are introduced here.

For the longitudinal breakdown modes only longitudinal current flow between ohmic contacts is considered, neglecting transverse current components. Such an approach via the longitudinal modes is complementary to the investigations on the transverse breakdown modes [14, 30], where the dynamics of current filaments were analyzed under the assumption of pure transverse spatial modes of carrier densities and electric field. For simulations of the longitudinal breakdown modes, the g-r parameters in Table 8.1 are used, where sample length between ohmic contacts is $L_x = 0.56$ cm and bias voltage $U_0 = 3$ V, which corresponds to an average field $E_x \sim 5.4$ V/cm, is applied between ohmic contacts quasi-instantaneously within 1 ps of the initial simulation time. Figure 8.10 shows the evolution of the longitudinal profiles of the carrier densities n_{h} (Fig. 8.10a) and n_{t2} (Fig. 8.10b), where x is the longitudinal coordinate and t is the simulation time. Figure 8.11 shows the corresponding electric field distribution. In Fig. 8.10a the dark regions indicate contact zones, where inclusion of the contact zones is designed to model carrier injection into the bulk more realistically than by just applying Dirichlet boundary conditions for n_{h} [38]. That is, hole transport takes place by drift and diffusion only and the g-r processes are assumed to be inactive in the contact zones. Hence, the contact regions have been designed to serve as hole reservoirs. Also note that the effective doping density and hole mobility in the contact zones are chosen to be equal to those in the bulk [38]. As an initial condition, the bulk of p-Ge is assumed to be in thermal and chemical equilibrium with vanishing currents.

Initially (at $t = 0$), the carrier concentration at contact zones is kept at $n_{\mathrm{h}} \approx 9.5 \times 10^{13}$ cm^{-3} as hole reservoirs, while the carrier concentration in the bulk ($n_{\mathrm{h}} \approx 1 \times 10^{3}$ cm^{-3}) is many orders of magnitude smaller than the value at the contacts, as shown in Fig. 8.10a. Under the assumed voltage

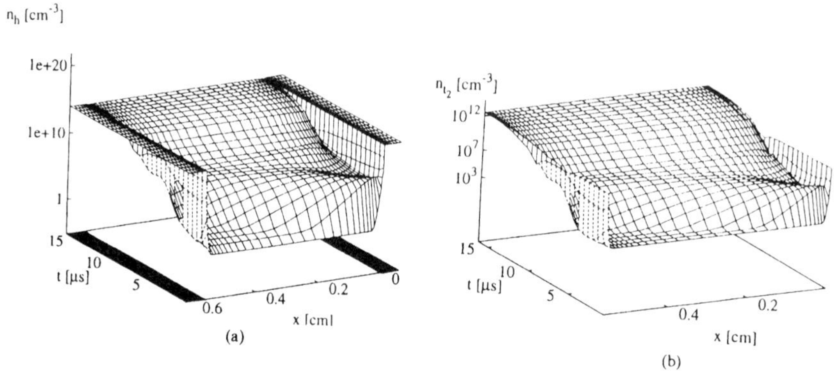

Figure 8.10 Carrier density (*a*) n_{h} and (b) n_{t2} in terms of the longitudinal coordinate x and time t. In (a) the contact regions are marked dark, in (b) they are omitted since the carrier concentrations n_{t1} and n_{t2} are not defined there. $U_0 = 3$ V, $L_x = 0.56$ cm, $R_{\mathrm{L}} = 0$, other parameters as in Table 8.1 [38].

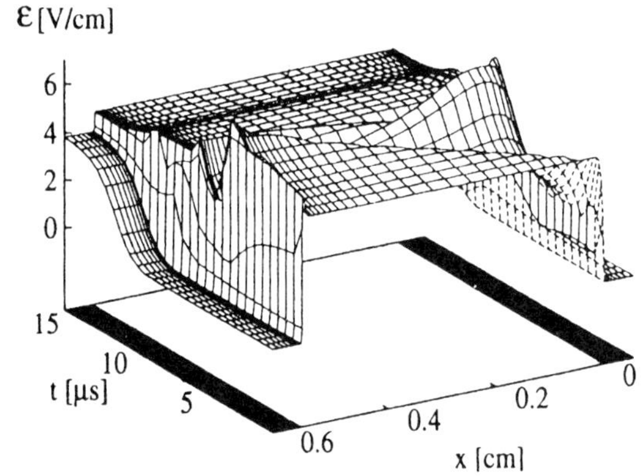

Figure 8.11 Electric field E_x in terms of the longitudinal coordinate x and time t (same parameters as in Fig. 8.10). The threshold field E_{th} is $E_{\mathrm{th}} = 4.44$ V cm^{-1} [38].

control, the electric field E_x (Fig. 8.11) reacts quasi-instantaneously, where the applied electric field is settled to be larger than the threshold field for the impact-ionization avalanche [38]. Subsequently, the carriers are injected into the bulk from the ohmic contact at $x = 0$ (anode) by drift and diffusion, whereby the impact-ionization avalanche of shallow acceptors is initiated in the anode region. In Fig. 8.10a such an initiation is clearly seen in the vicinity of anode at the initial stage within 5 μs, followed by an increased carrier concentration $n_{\mathrm{h}} \approx 10^{10}$ cm^{-3}. Corresponding to Fig. 8.10a, n_{t2} (carrier concentration at the excited level of acceptors) grows almost simultaneously, owing to enhanced trapping into the excited impurity state, as shown in Fig. 8.10b. Since $n_{t1} \gg n_{t2}$ (see Fig. 8.10b), n_{t1} hardly changes.

This initial stage of the impact-ionization avalanche evolves by front propagation from anode to cathode (Fig. 8.10). Kunz *et al.* [38] stressed that the front propagation is the result of collective g-r phenomena (impact ionization) propagating like a water-wave front, and not of the drift of individual carriers, and the front velocity estimated from the kink in E_x (Fig. 8.11) is less than 10^5 cm/s, compared with an expected drift velocity of $v_x = \mu E_x > 5 \times 10^5$ cm/s.

Since the voltage drop occurs predominantly at the high ohmic downstream of the front, the electric field there increases, whereas the field upstream decreases (see Fig. 8.11). Hence, impact ionization downstream is encouraged, whereas further creation of holes upstream of the front is inhibited [38]. As seen in Fig. 8.10, the front becomes less pronounced in the sample when it reaches the cathode. As soon as the front reaches the cathode the slight increase in $n_{\rm h}$ at the anode, resulting from a delayed reaction of the electric field there (Fig. 8.11), becomes relevant. Looking in detail in Fig. 8.11, the anode region has a higher resistance than the cathode region at the stage of 5–7 μs, leading to a larger voltage drop and, thus, the creation of a high-field domain there [38]. This domain enhances again the impact ionization, resulting in a further increase in $n_{\rm h}$. As a consequence of the same mechanism as before, the high-field domain propagates to the cathode. Such a seesaw-like mode determines the dynamics of this second stage of the breakdown process and decreases in amplitude (see Fig. 8.11), until at $t \approx 9$ μs the carrier concentration is almost homogeneous with the value $n_{\rm h} \approx 10^{10}$ cm^{-3}, which is still four orders of magnitude less than the steady-state value $n_{\rm h} \approx 10^{14}$ cm^{-3}. Subsequently, further impact-ionization processes drive the system to the steady state within $4 - 5$ μs ($t > 10$ μs in Fig. 8.10).

From the simulations, Kunz *et al.* found three stages of longitudinal impact-ionization breakdown: a stage of front creation and propagation (stage I) is followed by a stage dominated by a seesaw-like mode (stage II) and one in which the impact-ionization process completes over the whole sample (stage III). In Fig. 8.12a the total current density J_x in the sample is plotted as a function of time (by dots), which is compared with a spatially homogeneous model calculation (solid line), each for a bias voltage $U_0 = 3$ V and with the same length $L_x = 0.56$ cm between ohmic contacts, where the applied electric field is larger than the threshold field by $E = 1.2E_{\rm th}$ ($\delta E/E_{\rm th} = 0.2$). With the spatially homogeneous model, the total current density increases exponentially with time. For the longitudinal breakdown modes, there appears a stagnation period in stage II, as seen in Fig. 8.10, resulting in a delay time $\tau_{\rm D}$ in completing the current filamentation ($\tau_{\rm D} = 15$ μs in Fig. 8.12a). Here, the delay time is defined as the time interval between applying a bias voltage U_0 (quasi-instantaneously) and the rise of the sample current [38]. Figure 8.12b shows the delay times of the longitudinal model (dark squares) for different values of the relative excess

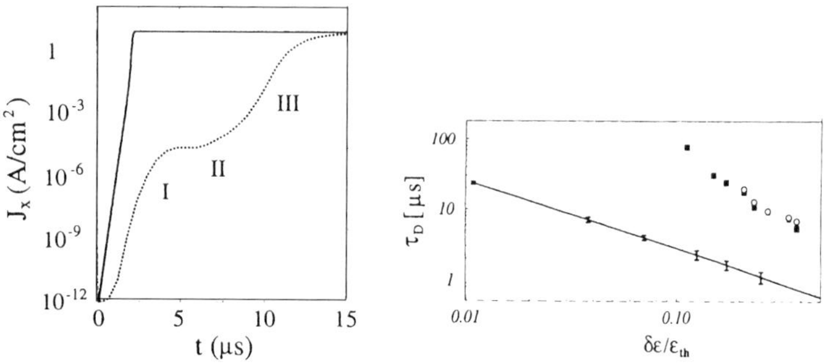

Figure 8.12 (a) Total current density J_x in the sample as a function of time t for the homogeneous model (solid line) and the longitudinal 1D model (dots) corresponding to a relative excess field $\delta E/E_{\text{th}} = 0.2$ (same parameters as in Fig. 8.11) [38]. (b) Delay times of the longitudinal model (solid squares) for different values of the relative excess field $\delta E/E_{\text{th}} = U_0/(L_x E_{\text{th}}) - 1$ in comparison with the values of the homogeneous model (thick line and thick error bars) and the 2D simultations (open circles) [38].

field $\delta E/E_{\text{th}} = U_0/(L_x E_{\text{th}}) - 1$ in comparison with the values of the homogeneous model (solid line with error bars). It is found that the delay times provided by the longitudinal model are six to 10 times larger than those of the homogeneous model.

Kunz *et al.* [38] further studied breakdown in 2D domains in p-Ge, which demonstrates pattern formation of a current filament in 2D space. They chose a square sample geometry with $L_x = L_z = 0.56$ cm. Ohmic contacts are modeled by applying Dirichlet boundary conditions to two regions of length $L_c = 0.028$ cm at the centers of the sample edges directed parallel to the z-axis, as indicated in Fig. 8.13a. At the contacts n_h is fixed to a value $n_h = 10^{15}\text{cm}^{-3} \gg N_A$ to model ohmic contacts [38]. Initial values of n_{t1}, n_{t2} are chosen such that initial distribution of n_h, $n_{h,0}(x)$ is in thermal equilibrium corresponding to typical values on the high resistivity branch of the S-shaped carrier density-field characteristic. Similarly to the longitudinal model, the potential at the anode $(x = 0)$ is applied to $U_0 =$ 3 V quasi-instantaneously within 1 ps. In Fig. 8.13b, c, pattern formation of a current filament is clearly seen, where the current filamentation completes just before 20 μs. The delay times before completing the current filamentation are also plotted in Fig. 8.12b by open circles, the values of which are slightly larger than those from the longitudinal model. Since the 2D model simulations have been done with point contacts, the filament simulated in Fig. 8.13 is induced by the dipole-like electric-field distribution [15, 16] and hence is "field-organized". In contrast, the electric-field distribution is uniform, with planar contact geometries [9–13], and thus current filament(s) can be created in a "self-organized" manner, which arises

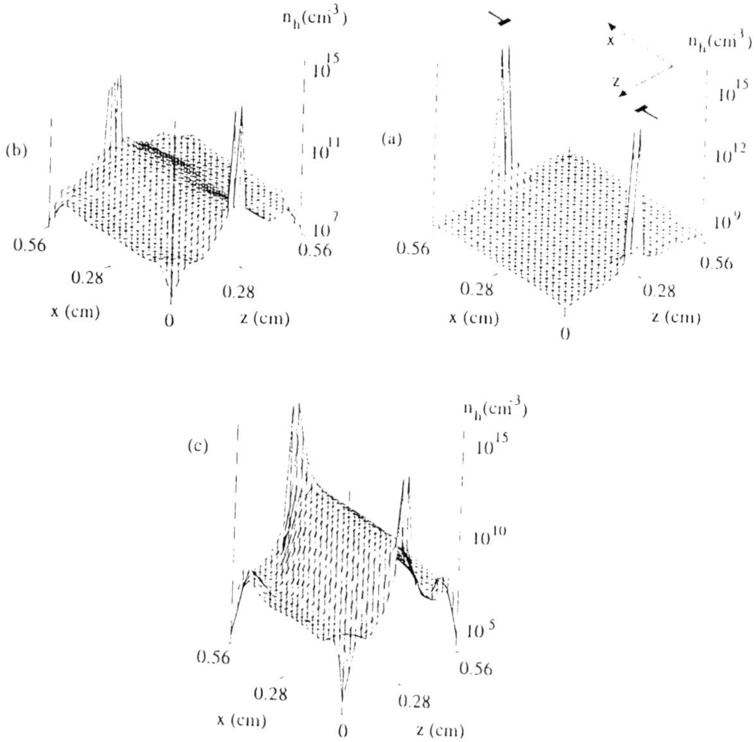

Figure 8.13 Carrier density distribution n_h as a function of x and z for several times after applying the bias voltage $U_0 = 3$ V: (a) $t = 5$ μs, (b) 10 μs, (c) 20 μs. In (a) the sample configuration with point contacts is shown. Numerical parameters as in Table 8.1 [38].

as dissipative structures as a result of a bifurcation from the thermodynamic branch of the current density–field characteristic in Fig. 8.9 [38].

Two-dimensional model simulations [24, 35–39, 42, 43] may provide almost complete numerical and theoretical knowledge for pattern formation of a current filament induced by the impact-ionization avalanche. Three-dimensional model simulations might be complete, but they may be time-exhaustive and provide little additional information since measurement techniques are restricted to 1D or 2D visualization of current filaments [8, 10–13, 15–18]. Gaa *et al.* [36, 37, 39] and Kunihiro *et al.* [35] also successfully simulated the spatiotemporal dynamics of filament formation in n-GaAs. An example by Gaa *et al.* [39] is shown in Fig. 8.14, where a square sample geometry was chosen with side length $L_x = L_z = 0.02$ cm and, similarly to the case for p-Ge [38], Dirichlet boundary conditions were applied to point contacts which locate at two opposite regions of length $L_c = 8 \times 10^{-4}$ cm at the centers of the sample edges. At the contacts n is

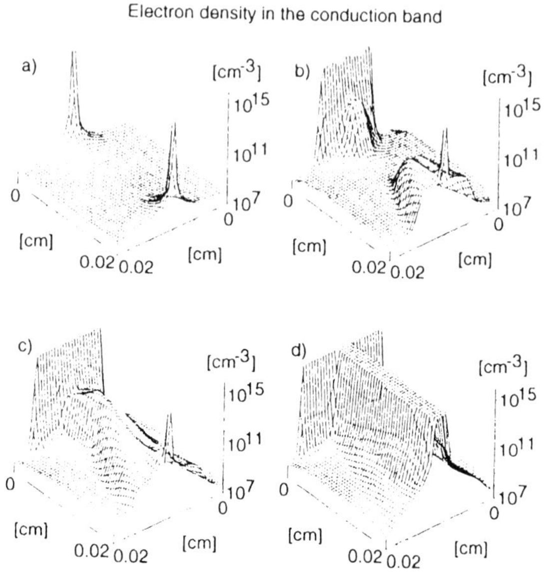

Figure 8.14 Temporal evolution of the electron density n: (a) $t = 0$ ns, $U = 0$ V, (b) $t = 0.5$ ns, $U = 0.48$ V, (c) $t = 1.0$ ns, $U = 0.48$ V, (d) $t = 2.5$ ns, $U = 0.48$ V. The applied voltage corresponds to $E = 24$ V/cm in the homogeneous $n(E)$ characteristic of the inset in Fig. 8.8 [39].

fixed to a value $n_D = 5 \times 10^{15} \text{cm}^{-3}$ to model ohmic contacts. All other boundaries are treated as insulating, where the components of the current density **j** and the electric field **E** perpendicular to the boundaries vanish [39]. Within 1 ps the voltage is linearly increased from $U = 0$ V to $U = 0.48$ V in Fig. 8.14. At an initial stage ($t = 0$ ns, Fig. 8.14a), the electron density is enhanced locally only at small regions in the vicinity of the contacts because of the ohmic nature of the contacts. Owing to the assumed voltage control, the electric field E reacts quasi-instantaneously, giving a dipole-like field distribution and enlarged areas of increased electron density at both the cathode and the anode. By the subsequent impact-ionization avalanche at the cathode a front moves toward the anode at $t \sim 0.5$ ns, as seen in Fig. 8.14b. The propagation of the front is accompanied by a high-field domain in which the electron temperature is very high [39]. The mechanism of the front propagation is quite similar to the case in p-Ge [38]. When the front meets the region of increased carrier density around the anode, a rudimentary filament is formed, as in Fig. 8.14c, where $t = 1.0$ ns. Finally, the rudimentary filament grows by the subsequent impact-ionization process, reaching the mature state in Fig. 8.14d, where $t = 2.5$ ns.

As shown by the transient total current $I(t)$ in Fig. 8.15, Gaa *et al.* [39] also found three stages of impact-ionization breakdown: a stage of front

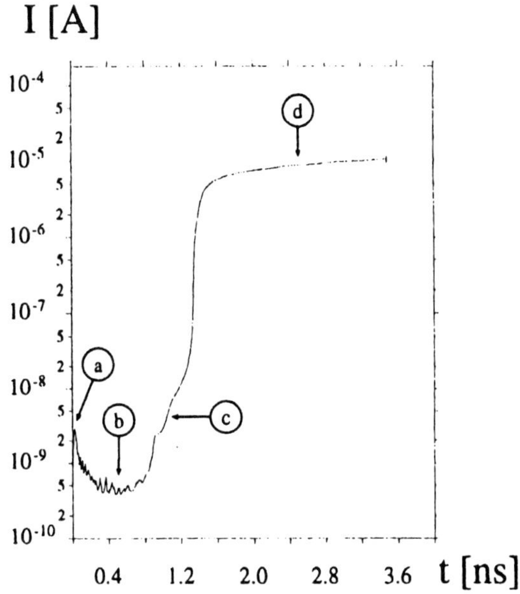

Figure 8.15 Total current $I(t)$ during the formation of a filament [39].

creation and propagation (stage I; (b) in Fig. 8.15) followed by a stage of stagnation after the front has reached the anode (stage II; (c) in Fig. 8.15) and one in which the rudimentary filament grows to a mature filament (stage III; (d) in Fig. 8.15).

Kunihiro *et al.* [35] investigated pattern formation of current filaments in n-type GaAs under crossed electric and magnetic fields. In experiments [44], it has been well known that a filament formed under a small bias current is bent by the Lorentz force under the crossed electric and magnetic fields and that, for an intermediate and relatively large bias current, significant asymmetry of the pattern image at opposite filament boundaries can be measured by scanning laser microscopy [15, 44]. The model simulations by Kunihiro *et al.* [35] and Gaa *et al.* [42] have explained satisfactorily an essential part of the experimental results [15, 44, 45]. The basic equations are similar to those in (8.6–8.12), which are available for p-type semiconductors, but taking into account the current density $\mathbf{j}(\mathbf{B})$ in the presence of a transverse magnetic field $\mathbf{B}$ normal to the current flow. The temporal evolution of the system is governed by the rate equations [35]:

$$\frac{\partial n}{\partial t} = \frac{1}{e}\nabla \cdot \mathbf{j} + X_1^s n_2 - T_1^s n p_t + X_1 n n_1 + X_1^* n n_2, \qquad (8.14)$$

$$\frac{\partial n_1}{\partial t} = T^* n_2 - X^* n_1 - X_1 n n_1, \tag{8.15}$$

$$\frac{\partial n_2}{\partial t} = -X_1^s n_2 + T_1^s n p_t - X_1^* n n_2 - T^* n_2 + X^* n_1 + X_1 n n_1, \tag{8.16}$$

where $\mathbf{j}$ is the current density (i.e., $\mathbf{j}(\mathbf{B})$, e is the electron charge, $p_t = N_D - n_1 - n_2$ is the density of ionized donors, and N_D is the total density of donors. The g-r coefficients are defined as follows: X_1^s is the thermal ionization coefficient of the excited donor level, T_1^s is its recombination coefficient, X_1, X_1^* are the impact-ionization coefficients from the ground and excited level, respectively, X^*, T^* are the transition coefficients from the ground level to the excited level and vice versa, respectively. These g-r coefficients are obtained from spatially homogeneous MC simulations, as usual [38, 39, 42].

The current density $\mathbf{j}(\mathbf{B})$ under the transverse magnetic field is written as:

$$\mathbf{j}(\mathbf{B}) = \frac{\mathbf{j}(0) + \mu\mathbf{B} \times \mathbf{j}(0)}{1 + |\mu\mathbf{B}|^2}, \tag{8.17}$$

where μ is the electron mobility and $\mathbf{j}(0)$ is the current density for $B = 0$, which is expressed with the drift-diffusion approximation as:

$$\mathbf{j}(0) = e(n\mu\mathbf{E} + D\nabla n), \tag{8.18}$$

with the diffusion constant D. It is often advantageous to express the continuity equation in terms of the total current density $\mathbf{J} = \epsilon\dot{\mathbf{E}} + \mathbf{j}$ by:

$$\nabla \cdot \mathbf{J} = 0, \tag{8.19}$$

where $\mathbf{J}$ is composed of the displacement current density $\epsilon\dot{\mathbf{E}}$ and the conduction current density $\mathbf{j}(\mathbf{B})$. (8.14)–(8.19) are solved numerically [35]. A homogeneously doped n-GaAs sample at liquid helium temperature ($T_L =$ 4.2K) has been considered. The simulated sample represents a square epitaxial GaAs layer with side lengths $L_x = L_z = 0.02$ cm and a thickness of $L_y = 1.4 \times 10^{-4}$ cm. Point contacts have been modeled by applying Dirichlet boundary conditions in the center of two opposite sample edges as usual, where the electron concentration is fixed to $n = 5 \times 10^{15}\text{cm}^{-3}$ and the widths of the contacts are $L_c = 8 \times 10^{-4}$ cm [35]. The material parameters are given in Table 8.3.

A transverse magnetic field of 5 mT has been applied. The nucleation and propagation mechanisms in Fig. 8.16 are almost the same as those under zero magnetic field presented by Gaa *et al.* [39]. In Fig. 8.16d, it is clearly shown that a boundary of the current filament is bent due to the Lorentz force, resulting in the increased width of the filament compared with the case of zero magnetic field, where the magnetic field is pointing downwards and

Table 8.3 Parameter values for n-GaAs

Parameter	Symbol	Value
Donor concentration	N_D	7.0×10^{15} cm^{-3}
Acceptor concentration	N_A	2.0×10^{15} cm^{-3}
Lattice temperature	T_L	$4.2K$
Mobility	μ	1.0×10^5 cm^2/V s
Dielectric constant	ϵ	$10.9\epsilon_0$

the current flows from anode ($x = 0.0$ cm) to cathode ($x = 0.02$ cm). In Fig. 8.16a, the nucleated front at the cathode propagates to the anode (see Fig. 8.16a, b). The spatial distribution of the electric field strength at $t = 2.5$ ns, i.e., when a steady state is practically reached, is shown in Fig. 8.17, where

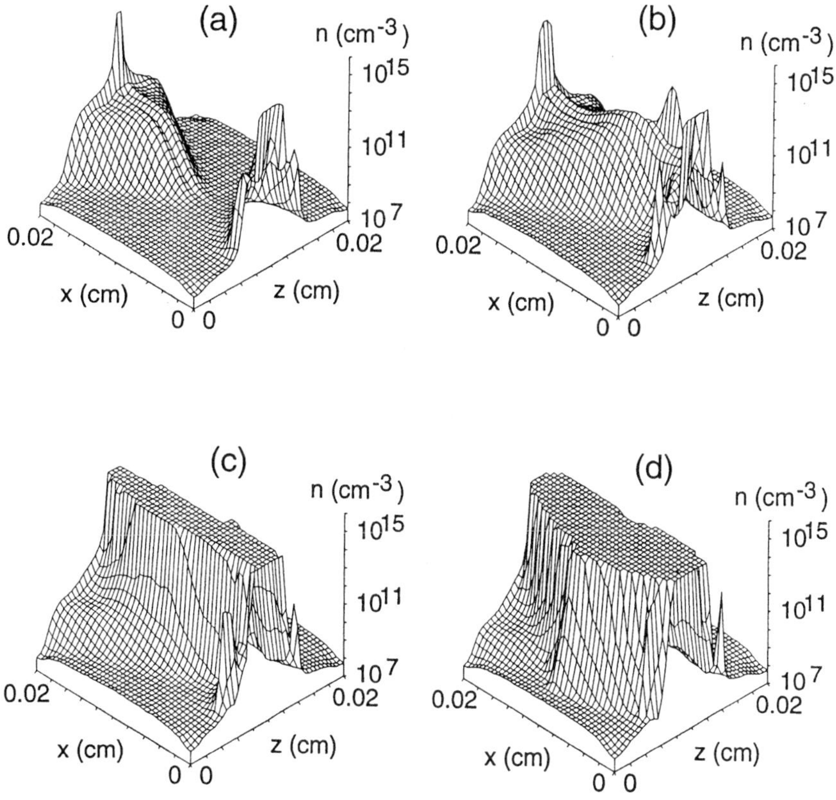

Figure 8.16 Temporal evolution of the electron density n as a function of the spatial coordinates x and z for a square sample with two point contacts at $B = 5$ mT. (a) $t = 0.5$ ns, (b) $t = 1.0$ ns, (c) $t = 1.6$ ns, (d) $t = 2.5$ ns. The applied voltage $V = 0.48$ V corresponds to an average field $E_{\text{av}} = 24$ V/cm [35].

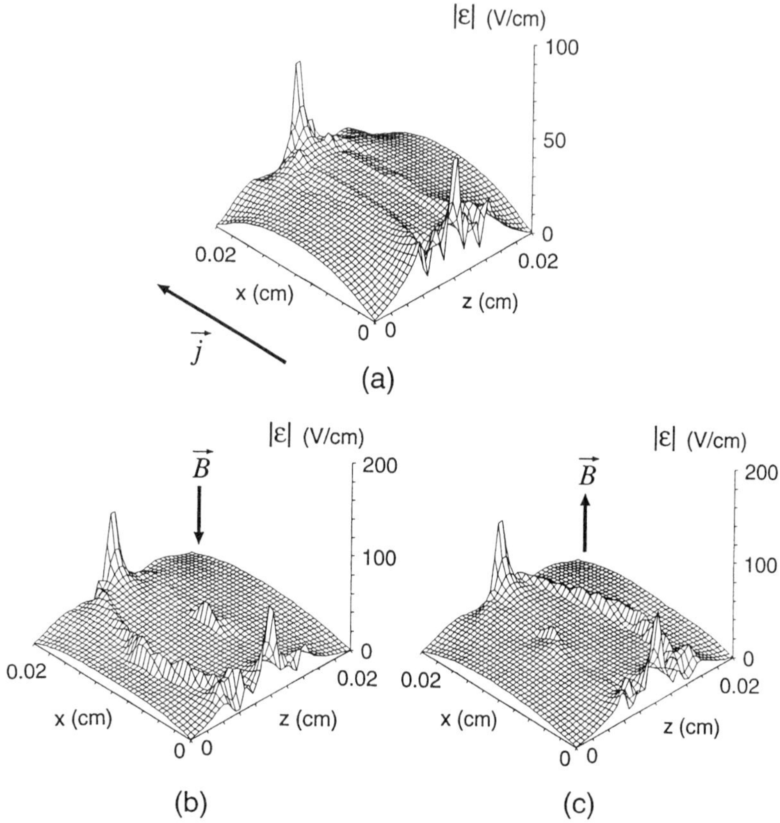

Figure 8.17 Spatial distribution of electric field strength $E(x, z)$ for (a) $B = 0$, (b) $B = 5$ mT, and (c) $B = -5$ mT at the quasi-steady state ($t \sim 2.5$ ns), $V = 0.48$ V. A magnetic field pointing downwards is defined as positive and upwards as negative [35].

directions of the magnetic fields and the current density are indicated by arrows. Compared with the symmetric distribution of the electric field for $B = 0$ (Fig. 8.17a), the field exhibits strong asymmetry at opposite filament boundaries for $B = \pm 5$ mT (Fig. 8.17b, c). The electric fields at the filament borders pointing in the direction of the Lorentz force are much stronger than at the opposite borders [35]. These asymmetric fields are in good agreement with the asymmetric ridge-like structure observed by scanning laser microscopy under a transverse magnetic field [15,44] (see later).

Transient dynamics of the total current during the formation of current filament are shown in Fig. 8.18 for different magnetic fields. The transient dynamics of the total current are described by the same mechanism as given by Gaa *et al.* [39]. The dynamics consist of four stages corresponding to (a)–

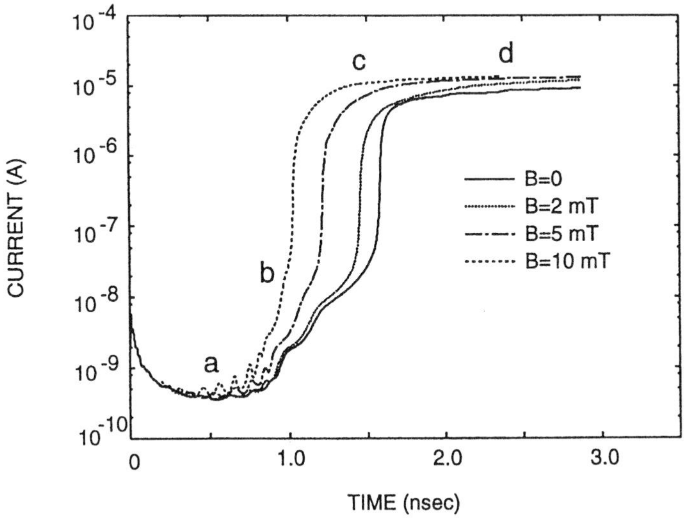

Figure 8.18 Total current $I(t)$ during the formation of a filament for different magnetic fields. The letters (a)–(d) corresponds to the stages of the filament formation shown in (a)–(d) of Fig. 8.16 [35].

(d) in Fig. 8.18: (a) low-current state corresponding to a stage of carrier front creation and propagation from the cathode; (b) slight increase and stagnation in the current corresponding to the formation of a rudimentary filament followed by the backward motion of a high-field domain; (c) strong increase in the total current accompanied by the establishment of a homogeneous straight filament; and (d) almost saturated current with a filament bent by the Lorentz force [39]. These basic processes have never been affected by applying magnetic fields with different strengths; however, note in Fig. 8.18 that higher magnetic fields accelerate the transition from almost insulating to highly conducting states (mature state of the filament) [35].

Other remarkable features in the model simulations by Kunihiro *et al.* [35] are given by the transverse profiles of the electric field strength $E(x, z) = (E_x + E_z)^{1/2}$ at $x = 0.01$ cm in Fig. 8.19 and the lateral electric field $E_z(x, z)$ at $x = 0.01$ cm in Fig. 8.20, where the shaded areas in Figs. 8.19 and 8.20 are the transverse cross-sections of electron concentration. For $B = 0$, the electric field profile in Fig. 8.19a is symmetric with respect to the center of the sample ($z = 0.01$ cm) during all stages of filament formation. In the first stage of the nucleation ($t = 0.5$ ns in Fig. 8.19a) the electric field shows a gentle peak at the middle of the sample due to the dipole-like electric field distribution between the point contacts [35]. After the formation of rudimentary filament, the field profile exhibits two maxima which arise

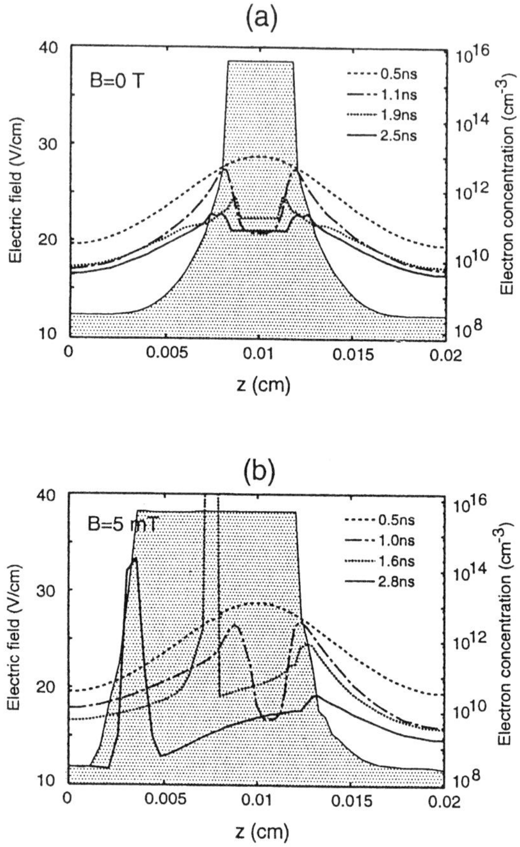

Figure 8.19 Cross-sections of the electric field strength $E(x, z)$ in the middle between the two point contacts ($x = 0.01$ cm) for (a) $B = 0$ and (b) $B = 5$ mT. The different times correspond to the stage of filament formation shown in (a)–(d) of Fig. 8.16. Shaded areas indicate the electron concentration in the quasi-steady state ($V = 0.48$ V) [35].

from the lateral field (see Fig. 8.20a) localized at the filament borders due to the breakdown of the charge neutrality. At the filament borders electrons tend to diffuse towards the outside regions from the rudimentary filament, leading to the weak carrier accumulation at the outside edge and the slight depletion at the inside edge because immobile ionized donors are left inside the filament borders, causing local breakdown of the charge neutrality. Hence, the lateral field is created only at the filament borders (Fig. 8.20a) owing to the local space-charge dipole, which prevents the electrons from diffusing outwards and stabilizes the filament.

For $B = 5$ mT, the electric field in Fig. 8.19b also retains an almost symmetric profile until the stage of a rudimentary filament ($t \sim 1.0$ ns; see

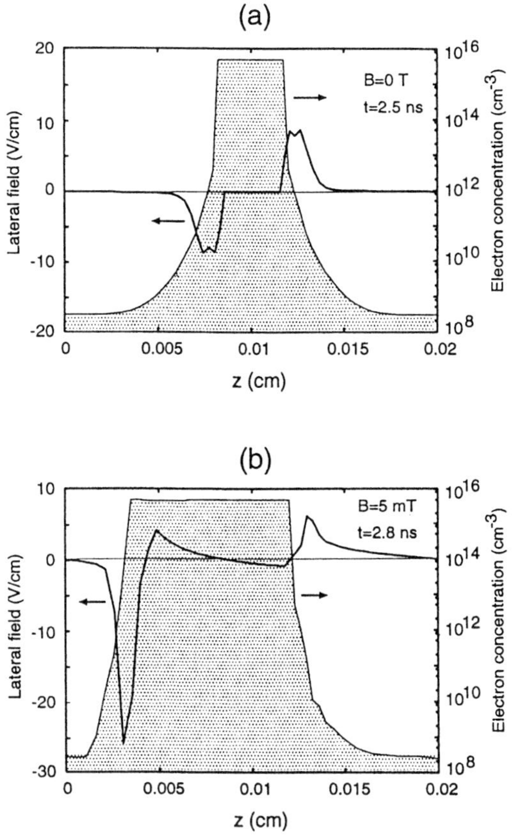

Figure 8.20 Cross-sections of the lateral electric field $E_z(x, z)$ at $x = 0.01$ cm for (a) $B = 0$ and (b) $B = 5$ mT in the quasi-steady state ($V = 0.48$ V) [35].

Fig. 8.18). However, once the impactionization avalanche sets in to form the rudimentary filament, the electric field exhibits strong asymmetry, as shown in Fig. 8.19b. On the filament border pointing in the direction of the Lorentz force, the diffusion of the electrons out of the filament is enhanced [35]. As a result, more negative and positive space charges are generated by mobile electrons and immobile ionized donors, respectively, at this border, giving rise to the strong lateral field in Fig. 8.20b. The lateral field serves as an attractive force which compensates the Lorentz and diffusion forces, stabilizing the filament border under a transverse magnetic field. On the other filament border, electrons diffusing outerwards are suppressed by the Lorentz force pointing in the opposite direction, thereby less space charge and a weaker lateral field are created on this border than for $B = 0$, as shown in Fig. 8.19b and 8.20b. This significant asymmetric profile of the lateral field simulated by Kunihiro *et al.* [35] explains the asymmetric signals observed by laser scanning microscopy [15, 44], where the signal of one filament border grows whereas the signal of the other border decays with increasing magnetic field. Kunihiro *et al.* [35] suggested that this asymmetric lateral field is also the source of the finite Hall voltage across the sample found in the experiments [16] and it may play a crucial role in inducing chaotic oscillations [16].

8.1.4 Spatiotemporal behaviors of localized current filaments in silicon p–n–p–n diodes

Spatiotemporal behaviors of the localized current filaments in silicon p–n–p–n diodes have been also studied extensively by Niedernostheide *et al.* [24, 46–50]. Here, the theoretical and numerical aspects [47, 50] of the experiments [24, 46–50] by Niedernostheide *et al.* will be introduced. Their works have been relative to studies on spatiotemporal pattern formations in glow discharge systems [51–53], yielding a strong similarity between spatiotemporal behaviors of the current filaments in the glow discharge systems and the semiconductor diode systems.

Let us start with the device structure and the model equations [47, 50]. The S-shaped I–V characteristic for the p^+–n^+–{\curr p}–n^- diode is shown in Fig. 8.21 [47], where a dc voltage supplier V_s and a load resistance R_0 are in series with the diode (see the inset figure). The device structure is

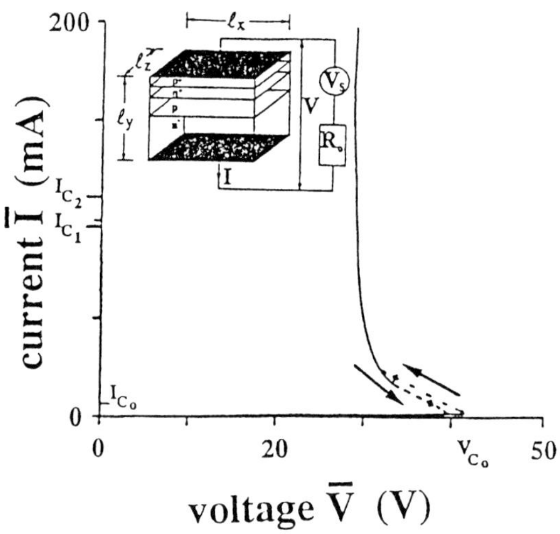

Figure 8.21 Current–voltage characteristic $I(V)$ of the p^+–n^+–p–n^- device. The inset shows the schematic structure of the device with the geometrical dimensions $\ell_x = 2.6$ mm, $\ell_y = 0.88$ mm and $\ell_z = 0.7$ mm [47].

composed of two subsystems; one is a p^+–n^+–p transistor (subsystem I) and the other a forward-biased p–n^- junction (subsystem II), as shown in Fig. 8.22. The n^- layer is weakly doped with a donor concentration of about 1.35×10^{13} cm^{-3} and very close to an intrinsic material, so that the injection of electrons from the n^- layer to the p layer and therefore into subsystem I can be neglected [47]. In subsystem I, the outermost p^+–n^+ junction is forward biased, while the inner part of the n^+–p junction is reverse biased (see Fig. 8.22) and blocks the current up to the voltage at which avalanche breakdown becomes important. Therefore, the most important part of the p^+–n^+–p–n^- diode is the reverse-biased n^+–p junction at which nucleation of a current filament starts via the avalanche breakdown. Together with a positive-feedback mechanism by the outer p^+–n^+ junction, this subsystem I (the p^+–n^+–p transistor) exhibits an S-shaped *I*–*V* characteristic in Fig. 8.21; i.e., the p^+–n^+–p transistor has a current-controlled negative resistance [47].

In Fig. 8.22, Niedernostheide *et al.* [47] have restricted their considerations to only two space dimensions; the two parts of the device, namely the p^+–n^+–p transistor and the p–n^- junction, are coupled as follows:

$$j_{\mathrm{pnp}}(x, y = 0, t) = j_{\mathrm{pn}}(x, y = 0, t), \tag{8.20}$$

$$V(t) = V_{\mathrm{i}}(x, t) + V_{\mathrm{p}}(x, t), \tag{8.21}$$

where the two subsystems are divided at the center of the p layer ($y = 0$) in Fig. 8.22; $j_{\mathrm{pnp}}(x, y = 0, t)$ and $j_{\mathrm{pn}}(x, y = 0, t)$ denote the current densities

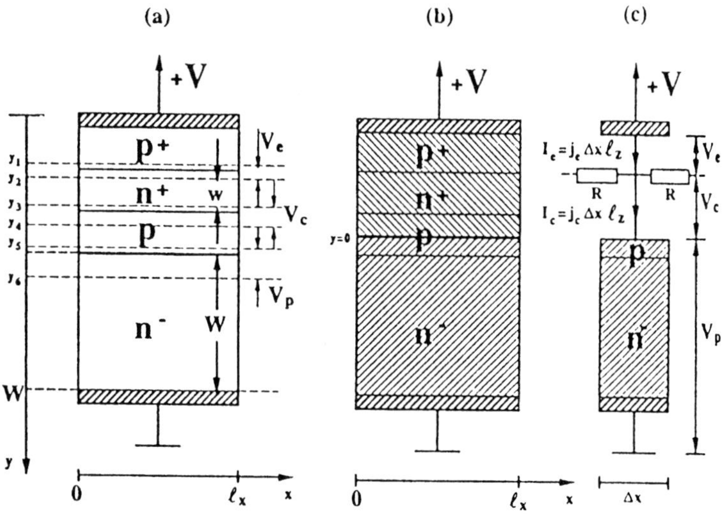

Figure 8.22 (a) Whole p^+–n^+–p–n^- device and (b) the proposed two-layer arrangement. The spreading of the emitter current (c) is determined by the base resistivity $R = \Delta x/(\sigma_b \ell_z w)$. y_1–y_6 mark the boundaries of the space-charge regions of the *p*–*n* junctions [47].

through the p^+–n^+–p transistor and the p–n^- junction at the interface $y = 0$ between the two subsystems. In (8.21) $V_i(t)$, $V_p(t)$ and $V(t)$ are the voltage drops across the p^+–n^+–p structure, the p–n^- junction and the whole device, respectively [47]. The model equations are derived by considering basic equations of the p^+–n^+–p transistor (section 8.1.4.1) and basic equations of the p–n^- junction (section 8.1.4.2).

8.1.4.1 *Description of the p^+–n^+–p transistor*

In the p^+–n^+–p structure, an equation for the emitter voltage V_e arising from the emitter and collector currents can be derived as follows. The emitter current density $j_e(x,t) = j_e(x, y = y_2, t)$ is composed of a displacement current due to the emitter capacity C_e, a diffusion current by injected carriers and a recombination current:

$$j_e(x,t) = C_e \frac{\partial V_e(x,t)}{\partial t} + j_s \left[\exp\left[\frac{V_e(x,t)}{V_T}\right] - 1 \right] + j_r \left[\exp\left[\frac{V_e(x,t)}{2V_T}\right] - 1 \right], \tag{8.22}$$

where j_s and j_r denote saturation current densities of the forward-biased emitter-base diode, and the thermal voltage V_T is defined by $V_T = k_B T/e$, with the elementary charge e, the Boltzmann constant k_B and the temperature T. For the explicit expression of j_s and j_r, see Ref. [47]. As is well known in standard books, the collector current density $j_c(x,t) = j_c(x, y = y_3, t)$ can be written as the sum of a displacement current with the collector capacity C_c, the collector saturation current, the transfer current of holes injected from the emitter and a leakage current:

$$j_c(x,t) = C_c \frac{\partial V_c(x,t)}{\partial t} + M j_{sc} + M\beta j_s \left[\exp\left[\frac{V_e(x,t)}{V_T}\right] - 1 \right] + \frac{V_c(x,t)}{\rho_L}, \tag{8.23}$$

where V_c is the collector voltage drop (Fig. 8.22c), M is the multiplication factor given by:

$$M = 1/[1 - (V_c/V_b)^m] \tag{8.24}$$

(V_b: the common base breakdown voltage; m: a constant number), j_{sc} describes the saturation current density of the reverse-biased collector, β is the base transport factor, and ρ_L the resistivity of the leakage. Taking into account the spreading of the current in the n^+ base with conductivity σ_b and width w, the emitter and collector current densities are connected by:

$$j_e(x,t) = j_c(x,t) + w\sigma_b \frac{\partial^2 V_e(x,t)}{\partial x^2}, \tag{8.25}$$

with the boundary conditions

$$\frac{\partial V_e(x=0,t)}{\partial x} = 0, \qquad \frac{\partial V_e(x=\ell,t)}{\partial x} = 0. \tag{8.26}$$

Equations (8.22)–(8.25), together with the boundary conditions (8.26) describe completely the behavior of the p^+–n^+–p transistor [47].

The reverse-biased collector has a rather small depletion layer capacity, leading to a fast relaxation of the collector voltage drop V_c. Therefore, by neglecting C_c in (8.23) and inserting (8.22) and (8.23) into (8.25), one has [47]:

$$\begin{aligned} C_e \frac{\partial V_e}{\partial t} = {} & w\sigma_b \frac{\partial^2 V_e}{\partial x^2} - (1-\beta M) j_s [\exp[V_e/V_T] - 1] + M j_{sc} \\ & + (V_i - V_e)/\rho_L - j_r[\exp[\mathrm{V_e}/2\mathrm{V_T}] - 1], \end{aligned} \tag{8.27}$$

with $V_i(x,t) = V_e(x,t) + V_c(x,t)$ (see Fig. 8.22c).

8.1.4.2 *Description of the p–n^- junction*

In their considerations, Niedernostheide *et al.* [47] made six assumptions for a simple description of the p–n^- junction. For better interpretation, a schematic arrangement of the p^+–n^+–p–n^- device in which a filamentary current flows through the p^+–n^+–p transistor is shown in Fig. 8.23 [47]. The assumptions are as follows.

(i) The p–n^- junction is an ideal Schockley diode.
(ii) The n^- is weakly doped, as described already, and thereby the injection of electrons from the n^- layer to the p layer can be neglected.
(iii) The width of the space-charge region and the width of the p region are neglected, i.e., $y_5 = y_6 = 0$ in Fig. 8.22. The hole concentration at the center of the p–n^- junction is therefore, together with (i), given by:

$$p(x, y=0, t) = p_{n0} \exp[V_p(x, y=0, t)/V_T], \tag{8.28}$$

where p_{n0} is the equilibrium concentration of holes in the n^- layer and V_p is the voltage drop across the p–n^- junction.
(iv) At the metallic contact $y = W$, surface recombination of holes with a surface recombination velocity s is associated, so that the following boundary condition is valid:

$$\frac{\partial p(x, y=W, t)}{\partial y} = -s \frac{p(x, y=W, t) - p_{ns0}}{D_p} \tag{8.29}$$

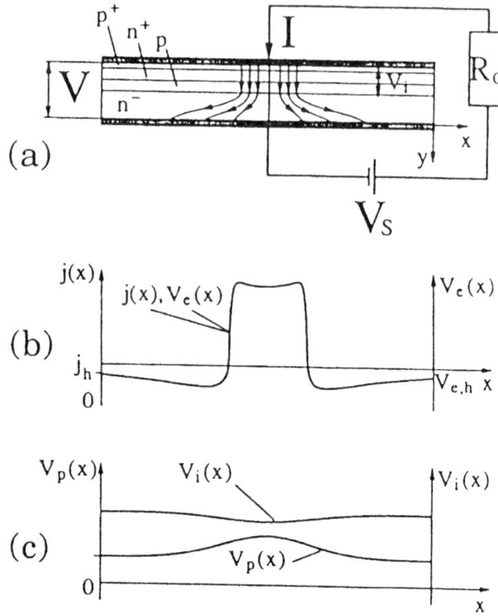

Figure 8.23 (a) Schematic arrangement of the p^+–n^+–p–n^- device with (b) a localized current filament that can develop because of the spreading of holes in the n^- layer. (c) The spreading leads to the long-range distribution of the voltages V_p and $V_i = V - V_p$ [47].

where D_p is the diffusion constant of holes and p_{ns0} is the hole equilibrium concentration at the metallic contact $y = W$.

(v) Low injection of holes in the n^- layer is assumed.

(vi) At the boundaries $x = 0$ and ℓ_x, homogeneous Neumann boundary conditions are assumed:

$$\partial p(x = 0, y, t)/\partial x = 0, \quad \partial p(x = l_x, y, t)/\partial x = 0. \tag{8.30}$$

With these assumptions, the spatiotemporal behaviors of $p(x, y, t)$ for $0 < y < W$ are determined from the following diffusion equation:

$$\frac{\partial p(x, y, t)}{\partial t} = D_p \left[\frac{\partial^2}{\partial x^2} + \frac{\partial^2}{\partial y^2}\right] p(x, y, t) - \frac{p(x, y, t) - p_{n0}}{\tau_p}, \tag{8.31}$$

where D_p is the diffusion constant of holes and τ_p is the lifetime of the holes in the n^- layer. In Fig. 8.23, note that the current flow in the p–n^- layer is only by the diffusion current:

$$\mathbf{j}_p(x, y, t) = -eD_p \nabla p(x, y, t), \quad 0 < y < W, \quad 0 < x < \ell_x, \tag{8.32}$$

where, especially at the interface at $y = 0$,

$$j(x, y = 0, t) = -eD_{\rm p}[\partial p(x, y = 0, t)/\partial y]. \tag{8.33}$$

Since the main interest here concerns the transverse distribution of holes, the model equation (8.31) is simplified by integrating it over the y-axis $\bar{p}(x, t) = (1/W)\int_0^W p(x, y, t)\mathrm{d}y$. Here, the surface recombination rate $1/\tau_{\rm s}$ is defined as:

$$\frac{1}{\tau_{\rm s}} = -\frac{D_{\rm p}[\partial p(x, y = W, t)/\partial y]}{\int_0^W [p(x, y, t) - p_{\rm n0}]\mathrm{d}y} = s\frac{p(x, y = W, t) - p_{\rm ns0}}{\int_0^W [p(x, y, t) - p_{\rm n0}]\mathrm{d}y}. \tag{8.34}$$

Then the boundary condition (8.29) becomes:

$$\frac{\partial p(x, y = W, t)}{\partial y} = -\frac{\bar{p}(x, t) - p_{\rm n0}}{D_{\rm p}\tau_{\rm s}}W. \tag{8.35}$$

Integrating (8.31) and using the boundary conditions (8.33), (8.35), one obtains:

$$\frac{\partial \bar{p}(x, t)}{\partial t} = D_{\rm p}\frac{\partial^2 \bar{p}(x, t)}{\partial x^2} + \frac{j(x, y = 0, t)}{eW} - \frac{\bar{p}(x, t) - p_{\rm n0}}{\tau}, \tag{8.36}$$

where an effective lifetime τ is defined as $1/\tau = (1/\tau_{\rm s} + 1/\tau_{\rm p})$ [47]. Finally, the description for the p–n$^-$ junction (subsystem II) is completed by replacing the Neumann boundary condition (8.30) by

$$\partial \bar{p}(x = 0, t)/\partial x = 0, \qquad \partial \bar{p}(x = \ell_x, t)/\partial x = 0. \tag{8.37}$$

8.1.4.3 *Basic model equations*

The basic model equations are (8.27) for the region of the p$^+$–n$^+$–p–n$^-$ transistor (subsystem I) and (8.36) for the p–n^- region (subsystem II). The following replacements are useful in order to discuss the fundamental dynamical behaviors expected from the model equations [47]:

$$q(V_{\rm e}, V_{\rm i}) = (1 - \beta M)j_{\rm s}\rho_{\rm c}[\exp(V_{\rm e}/V_{\rm T}) - 1] + j_{\rm r}\rho_{\rm c}[\exp(V_{\rm e}/2V_{\rm T}) - 1] \\ - Mj_{\rm sc}\rho_{\rm c} - (V_{\rm i} - V_{\rm c})(\rho_{\rm c}/\rho_{\rm L}), \tag{8.38a}$$

$$Q(V_{\rm e}, V_{\rm i}, V) = \frac{\tau}{eW}\{Mj_{\rm sc} + \beta Mj_{\rm s}[\exp(V_{\rm e}/V_{\rm T}) - 1] + (V_{\rm i} - V_{\rm e})/\rho_{\rm L}\} \\ - [\bar{p}(V_{\rm i}, V) - p_{\rm n0}], \tag{8.38b}$$

where $\rho_{\rm c} = \tau_{\rm e}/C_{\rm e}$, $V_{\rm i}(x, t)$ is the voltage drop across the p$^+$–n$^+$–p transistor and, therefore, M can be written as $M = M(V_{\rm i} - V_{\rm e})$. By putting $L = (D_{\rm p}\tau)^{1/2}$ and $\ell = (w\sigma_{\rm b}\rho_{\rm c})^{1/2}$, the model equations can be rewritten in the following form:

$$\tau_e \frac{\partial V_e(x,t)}{\partial t} = \ell^2 \frac{\partial^2 V_e(x,t)}{\partial x^2} - q(V_e, V_i), \tag{8.39a}$$

$$\tau \frac{\partial \bar{p}[V_i(x,t)]}{\partial t} = L^2 \frac{\partial^2 \bar{p}[V_i(x,t)]}{\partial x^2} + Q(V_e, V_i, V), \tag{8.39b}$$

where $\bar{p}[V_i]$ is given by the algebraic function (8.28) with $V_i = V - V_p$. The emitter voltage V_e is essentially determined by the function $q(V_e, V_i)$. Since $-q(V_e, V_i)$ in (8.38a) is an increasing function of V_e, namely $-\partial q(V_e, V_i)/\partial V_e > 0$ for a certain range of the voltage, the injection of holes from the p^+ emitter leads to a positive feedback and to an autocatalytic increase in V_e, which is due to the multiplication phenomena of the collector current density j_c with the multiplication factor M in the collector region [47]. Therefore, the autocatalytic increase in V_e and hence in j_c is physically based on the fact that holes injected from the emitter are multiplied in the collector region [47]. In the operating mode of the avalanche transistor, the current density of the base is negative and the inequality $(1 - \beta M) < 0$ holds. If the voltage drop V_i across the transistor part is assumed to be constant, this inequality means that the holes multiplied in the collector cause an electron base current from the collector which is large enough to induce an additional injection of holes from the emitter, and so on. The avalanche breakdown in the collector region may lead to pattern formation of a current filament, as illustrated schematically in Fig. 8.23a. Thus, the transistor part of the p^+–n^+–p–n^- diode contains an autocatalytic process if V_i is constant [47]. For this reason, Niedernostheide *et al.* [47, 50] treated the variable V_e as an activator. A set of equations (8.39a,b) describes typically the activator–inhibitor model with short-range activation and long-range inhibition ($\ell < L$), where the inhibitor is the averaged hole concentration $\bar{p}$.

In an actual problem of the p^+–n^+–p–n^- diode, the value of V_i is not constant because of direct interaction of the transistor part with the p–n^- junction via the current conservation. The holes injected in the n^- layer increase the hole concentration $\bar{p}(x,t)$ and therefore the value of the voltage drop V_p across the p–n^- junction due to (8.28). Since $V_i(x,t) = V - V_p(x,t)$, V_i decreases by an increase in V_p, which limits the autocatalytic process. Thus, V_i may be treated as an inhibitor [47, 50] (note: $V_i = V_e + V_c$ in Fig. 8.22). In (8.38b), (8.39b), it can be seen that for each positive value of V_i and V_e the relation $\partial Q/\partial V_i > 0$ is fulfilled, indicating that the voltage V_i works as an inhibitor [47]. By considering the external circuit, the third equation for the voltage drop $V(t)$ across the whole device is obtained as:

$$R_0 C_{ex} \frac{\partial V(t)}{\partial t} = V_s - V - R_0 \int_0^{\ell_x} j_c(V_e, V_i)\,dx, \tag{8.40}$$

where V_S is the value of the applied voltage, C_{ex} is the parasitic capacity parallel to the device, and R_0 the load resistance (see Fig. 8.21). The three equations (8.39a)–(8.40) describe the model equation for spatiotemporal behaviors of the p^+–n^+–p–n^- diode, where each equation has its characteristic time constant or relaxation time, $\tau_e, \tau, R_0 C_{ex}$, respectively. For a special case of $R_0 = 0$ and/or $R_0 C_{ex} \ll \min(\tau_e, \tau)$, one has $V = V_s$ from (8.40) and hence the model equations are reduced to (8.39a) and (8.39b) with $V = V_s$ (=constant). In these equations one also has the characteristic length ℓ and the diffusion length L, respectively. The value of the ratio ℓ/L is rather small ($\ell/L < 1$), because $\ell = (w\sigma_b\rho_c)^{1/2}$ is determined by the spatial spread of the base current in the thin n^+ base of the transistor part, while L is the diffusion length which characterizes the spreading of holes in the thick n^- layer due to the ambipolar diffusion of electrons and holes [47]. The ratio of the time constant τ_e/τ may also be less than 1, because the effective lifetime τ of holes in the n^- layer can be much larger than the relaxation time τ_e in the transistor. With these values of the ratios ($\tau_e/\tau < 1$, $\ell/L < 1$), a qualitative analysis of this system of equations is possible using the theory by proposed Kerner and Osipov [54], if for a certain range of the parameter V the conditions

$$\partial Q/\partial V_i > 0, \quad \partial q/\partial V_e < 0, \tag{8.41}$$

$$\frac{\partial q}{\partial V_e}\frac{\partial Q}{\partial V_i} - \frac{\partial q}{\partial V_i}\frac{\partial Q}{\partial V_e} > 0 \tag{8.42}$$

are fulfilled.

8.1.4.4 *Standing localized current filaments*

As seen in Fig. 8.21, the S-shaped I–V characteristic is quite similar to those in bulk semiconductors. In the p^+–n^+–p–n^- structure, a standing localized current filament (stable filament) can be generated by the avalanche breakdown in the reverse-biased n^+–p region, where transition from homogeneous current density distribution to the standing localized current filament occurs by switching from the lower to the upper branches of the S-shaped I–V characteristic (Fig. 8.21). In the model equations (8.39a), (8.39b) with the values of the various parameters listed in Table 8.4, it is possible to explain qualitatively how the standing localized current filament can be formed [47]. The homogeneous solutions of (8.39a) and (8.39b) can be obtained by putting $q(V_e, V_i) = 0$ and $Q(V_e, V_i, V) = 0$. From $q(V_e, V_i) = 0$ (see (8.38a)), V_i can be determined numerically as a function of V_e. The relation $V_i(V_e)$ exhibits an N-shaped curve, as shown in Fig. 8.24a, where the highly nonlinear character is due to the avalanche breakdown with the multiplication factor $M(V_i - V_e)$. The curve $Q(V_e, V_i, V_s) = 0$ for the voltage-bias condition ($R_0 = 0$) has a small negative slope in Fig. 8.24a [47].

Table 8.4 Parameter values for the $p^+ - n^+ - p - n^-$ diode

Parameter	Parameter meaning and manner of determination	Standard parameter value (typical parameter range)
C_e	capacity per unit square of the $p^+ - n^+$ junction; estimated by the doping profile, see Ref. [9]	1×10^{-4} F/cm^2 $10^{-2} - 10^{-6}$ F/cm^2
D_p	diffusion coefficient of holes in n^- layer; determined from literature, see, e.g., Ref. [25]	10 cm^2/s 5 – 20 cm^2/s
j_r	saturation current density of the recombination current in the $p^+ - n^+$ junction; determined by fitting the experimental to the numerical $I(V)$ characteristic	3×10^{-7} A/cm^2 $10^{-5} - 10^{-8}$ A/cm^2
j_s	saturation current density of the diffusion current in the $p^+ - n^+$ junction; determined by fitting the experimental to the numerical $I(V)$ characteristic	1.5×10^{-11} A/cm^2 10^{-11}–10^{-14} A/cm^2
j_{SC}	saturation current density ofthe $n^+ - p$ collector; determined by fitting the experimental to the numerical $I(V)$ characteristic	2×10^{-8} A/cm^2 $10^{-8} - 10^{-14}$ A/cm^2
ℓ_x	sample width in x direction; measured after finishing preparation procedure	0.5 cm 0.1 – 1.0 cm
ℓ_z	sample width in z direction; measured after finishing preparation procedure	0.05 cm 0.03 – 0.1 cm
L	effective diffusion length of holes in the n^- region; defined by $L = (D_p\tau)^{1/2}$	0.02 cm 0.01 – 0.05 cm
m	coefficient determining the dependence of the multiplication factor on the collector voltage; determined from the literature, e.g., Ref. [25]	3 2 – 5
p_{n0}	equilibrium concentration of holes in the n^- layer; determined by the doping profile, see Ref. [9]	1×10^7 cm^{-3} $10^7 - 3 \times 10^7$ cm^{-3}
R_0	external load resistor	1 Ω 0.1 – 10^6 Ω
T	temperature	300 K 77 – 370 K
V_b	breakdown voltage of the $n^+ - p$ collector; determined by fitting the experimental to the numerical I (V) characteristic	42 V 35 – 45 V
V_T	thermal voltage determined by kT/e , with Boltzmann constant k and elementary charge e	0.026 V 0.0065 – 0.032 V
w	thickness of the n^+ base; determined by the doping profile, see Ref. [9]; the influence of a drift field in the base may be taken into account by an effective vase width smaller than the real one	0.0003 cm $10^{-4} - 10^{-3}$ cm
W	thickness of the n^- substrate; determined by doping profile, see Ref. [9]	0.06 cm 0.04 – 0.08 cm
β	base transport factor; determined by fitting the experimental to the numerical I (V) characteristic	0.6 0.5 – 0.9
ρ_L	leakage resistance of the collector; determined by fitting the experimental to the numerical I (V) characteristic	40 000 Ω cm^2 $10^4 - 10^5$Ω cm^2
σ_b	average conductivity of the n^+ base; determined by doping profile, see Ref. [9]	10 (Ω cm)$^{-1}$ 1 – 100 (Ω cm)$^{-1}$
τ	effective hole lifetime; defined by the hole lifetime in the n^- region and the surface recombination rate at the cathode	10 μs 10 – 300 μs

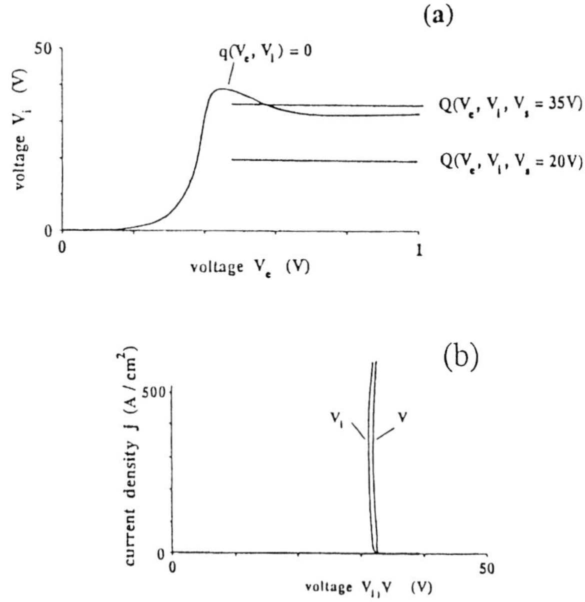

Figure 8.24 (a) Dependence of the transistor voltage V_i on the emitter voltage V_e for $g(V_\mathrm{e}, V_\mathrm{i}) = 0$ and $Q(V_\mathrm{e}, V_\mathrm{i}, V_\mathrm{s})$. The calculated current–voltage characteristics in (b) refer to the transistor part $[j(V_\mathrm{i})]$ and the whole p^+–n^+–p–n^- device $[j(V)]$ for a homogeneous current distibution [47].

The intersection points of the null clines $q(V_\mathrm{e}, V_\mathrm{i}) = 0$ and $Q(V_\mathrm{e}, V_\mathrm{i}, V) = 0$ correspond to homogeneous stationary solutions of the system. Since the emitter current density j_e is given by a monotonic function of V_e by (8.22), a set of the intersection points obtained as a function of V_s gives the S-shaped j–V curve in Fig. 8.24b. After the avalanche breakdown the voltage drop V_i across the transistor part decreases as a result of the negative differential resistance in the collector part.

Returning to the original model equations (8.39), a standing localized current filament can be formed owing to the diffusion terms, where the width ℓ_s of the localized filament in Fig. 8.23b is approximately close to the value of the diffusion length L of the holes in the n^- layer. In Fig. 8.23a, the filamentary current density in the transistor part is many times larger than in the vicinity of the filament, and holes diffuse not only to the cathode but also along the x-direction in the thick n^- layer [47]. This broadening of the spatial hole distribution with high hole concentration in the n^- layer causes a corresponding broad distribution of the voltage drop $V_\mathrm{p}(x, t)$ across the p–n^- junction, as shown in Fig. 8.23c. Accordingly, the voltage drop $V_\mathrm{i}(x, t)$ across the transistor exhibits a dip structure, which suppresses further growing of the localized filament in Fig. 8.23b. The dip structure extends to a certain distance from the center of the current filament, and V_i reaches a value corresponding to a homogeneous state outside the filament. This distance is determined by the value of L, which in turn is determined by the diffusion length of holes in the n^- region, as found in (8.39b). Thus, the characteristic width of a localized current filament is determined by L. Niedernostheide *et al.* [47, 50] estimated that the lifetime τ_p of holes in the n^- layer of the device is about 100 μs. The value of $L = (D_p\tau)^{1/2}$ is therefore about 350 μm for $D_p = 12.5$ cm^2/s if τ is approximated by τ_p, and is of the order of the observed width of the filament.

Summarizing the qualitative nature of a standing localized current filament, the following description [47] can be true. The existence of the localized current filament is linked with the fact that the value of the voltage drop V_i across the transistor locally decreases and suppresses a further increase in the current density in the filament. Therefore, the maximal current density in the filament is limited. A localized filament exists, if $L > \ell$, owing to the competition of the positive feedback (avalanche breakdown) in the transistor and the negative feedback followed by a less pronounced decrease in V_i due to the hole diffusion in the n^- layer of the p–n^- junction. The boundaries ("walls") of the filament are characterized by sharply changing current densities (Fig. 8.23b) resulting from the significant positive feedback in the transistor part.

8.1.4.5 Spontaneous transition from standing to pulsating, rocking and traveling localized current filaments

Using the model equations with the values of the various parameters listed in Table 8.4, Niedernostheide *et al.* have explained numerically [50, 55] their

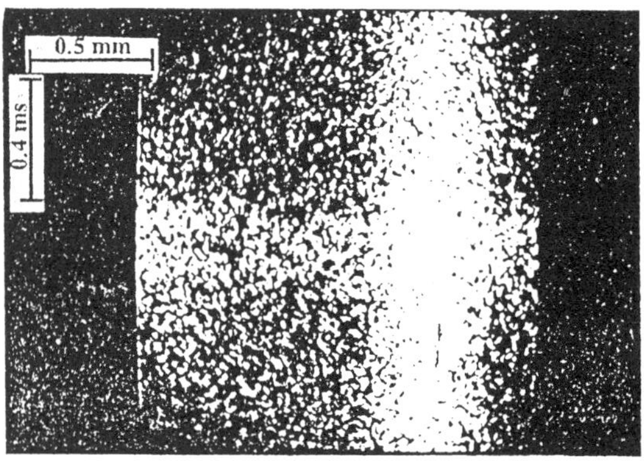

Figure 8.25 Streak camera record of the emitted light density of a standing current filament for the current $I = 91.3$ mA [47].

experimental results of the spontaneous transition from a standing localized current filament to the rocking localized current filament [46, 47, 50], transition from the locking to the traveling filament [46, 47, 49, 50], and also the spatiotemporal behavior of the period doubling [48]. An example of the standing localized current filament observed in the experiments is shown in Fig. 8.25 [47].

Spontaneous transition from standing to pulsating localized current filaments

Stability of the standing localized current filament can be argued by adding the fluctuations δj at the filament walls. Qualitatively, the fluctuations may be symmetric (δj_0) as in Fig. 8.26b, or antisymmetric (δj_1) as in Fig. 8.26c.

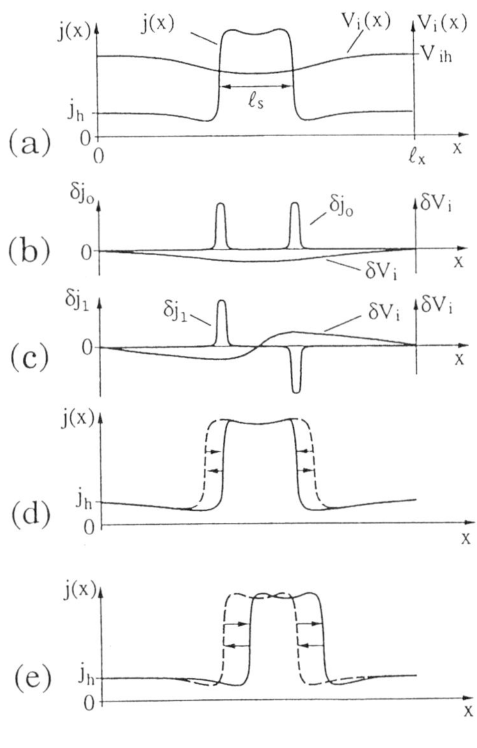

Figure 8.26 (a) Spatial stable stationary distribution of the current density (activator) and the voltage V_i (inhibitor). (b) Symmetric and (c) Antisymmetric critical fluctuations δj_0 and δj_1 of the current density with the corresponding fluctuations δV_i of the transistor voltage lead to (d) pulsating and (e) locking filaments, respectively. The solid and dashed lines mark the two extreme positions of the filament [47].

Spontaneous transition from standing to pulsating current filaments is a kind of bifurcation from a stationary state to a dynamical motion of the localized filament, in such a way that the symmetric fluctuations grow at the bifurcation point. For the symmetric fluctuations δj_0 at the filament walls, the filament tends to grow in the x-direction, since the fluctuations δj_0 and thus the fluctuations δV_{e0} of the activator induce the local avalanche breakdown at the walls of the filament. If $\tau \gg \tau_e$, a corresponding alternation δV_i (Fig. 8.26b) continues after some time delay, which in turn has a damping influence on the growth of the fluctuation δj. That is, the critical fluctuation δj_0 of the activator cannot instantly be suppressed by the inhibitor if $\tau_e/\tau \ll 1$. Hence, the repetitive behaviors of the instant activation and the delayed response of the inhibitor induce the pulsating localized current filament with the oscillating frequency

$$\omega_c = \left(\frac{\ell}{L\tau_e\tau}\right)^{1/2}$$

(breathing filament, see Fig. 8.26d) after the bifurcation. However, the pulsating localized current filament can only be observed if the load resistor R_0 in the external circuit is sufficiently small or even zero [47], because, for a constant dc bias V_s, it is only possible through a voltage-controlled bias condition that the total current can oscillate by the breathing filament.

Spontaneous transition from standing to rocking localized current filament

If the resistor in the external circuit is relatively large, or even for $R_0 = \infty$ (current-controlled bias condition), the symmetric fluctuation δj_0 may die out, since the fluctuations of the total current cannot be allowed for the current-controlled bias condition, as expected from the j–V curve in Fig. 8.24b. Instead, the antisymmetric fluctuation δj_1 can grow, keeping the constant total current. In this case, the fluctuation δV_i of the inhibitor is also antisymmetric (Fig. 8.26c) and the resultant dynamical behavior is characterized by a rocking localized current filament [46, 47, 50] in Fig. 8.26e, where the oscillating frequency of the localized current filament has been estimated as $\omega_{c1} \approx \omega_c$ [47].

Since any value of the total current (Fig. 8.24b) corresponds to a definite value of the filament width ℓ_s, the spontaneous transition from a standing to a rocking current filament occurs if the total current I exceeds a certain value I_c corresponding to the critical width ℓ_{sc}, where the total current must be constant during the rocking motion. Thus, a rocking localized current filament appears instead of a pulsating one if the external resistor as well as the total current exceed a certain critical value [47].

As in many cases for the bifurcation phenomena, two types of spontaneous transition from standing to rocking current filaments can be

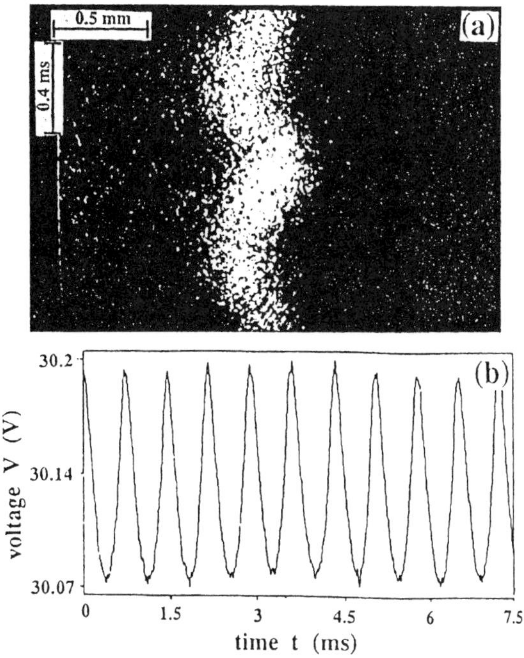

Figure 8.27 (a) Streak camera record of the emitted light density of a traveling current filament and (b) time series of the voltage drop V across the sample for the average current $I = 106.5$ mA [47].

distinguished. The first type is characterized by a transition from a standing to a rocking current filament connected with a discontinuous rise in the amplitude of the rocking current filament. The frequency of such a rocking current filament may be much less than ω_c because the oscillation may be the relaxation type with two characteristic times of the activator (V_e or j_e) and the inhibitor (V_i) For the second type, the spatial amplitude of the rocking current filament increases smoothly from zero after the bifurcation at $I = I_{c1}$, where the frequency is near ω_c. This type of transition has been observed experimentally by Niedernostheide *et al.* [46, 47, 50, 55]. An example of the rocking localized current filament observed in the experiments is shown in Fig. 8.27 [47].

Traveling localized current filament

In experiments, the transition from a rocking to a traveling current filament has been found [46, 47, 49, 50], when the total current exceeds the critical current I_{c2}, as can be seen in Fig. 8.28. The traveling current filament is moving with a constant velocity that depends only on the parameters of the

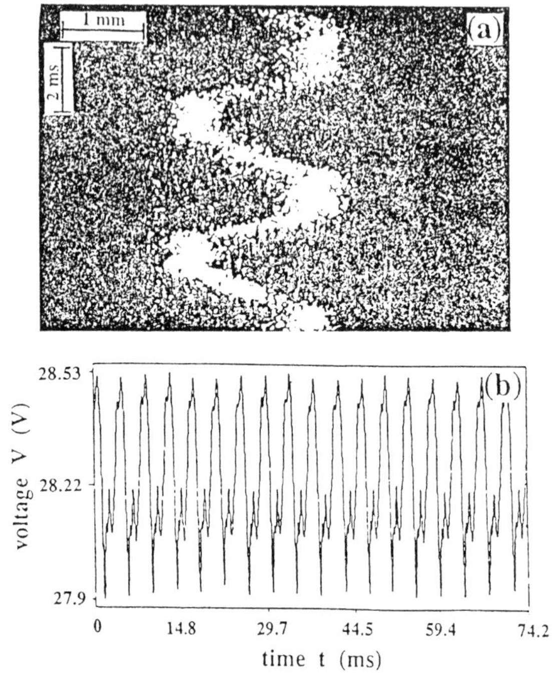

Figure 8.28 (a) Streak camera record of the emitted light density of a traveling current filament and (b) time series of the voltage drop V across the sample for the average current $I = 146.6$ mA [47].

device [47]. Consequently, the filament has to approach one of the semiconductor surfaces of the sample. Niedernostheide *et al.* have observed experimentally the repelling of the filament at the surfaces [47, 49]. The behavior is quite similar to the reflection of the solitary wave at a boundary. By the repelling of the filament at the surfaces, the filament oscillates between the two edges by the traveling mode, where the velocity of this localized current filament is almost constant except near the surfaces. Only if the filament reaches one of the surfaces can a sharp change in the velocity v to zero and a change of the sign of the velocity be observed [47]. During the repelling period, the filament characterized experimentally by the lightening region (Fig. 8.28a) does not exactly reach the semiconductor surfaces. This experimental observation has been explained by using the general properties of traveling autosolitons [56] in active systems for which the inequality

$$\tau_e/\tau < \ell/L \ll 1 \tag{8.43}$$

is valid [47].

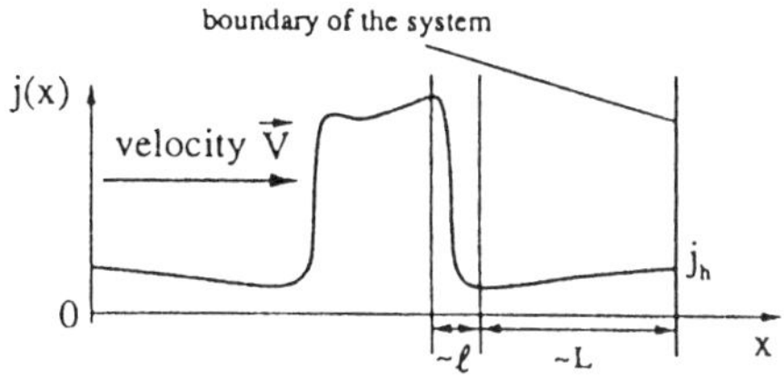

Figure 8.29 Schematic distribution of a traveling current filament near the boundary [47].

From the theory of traveling autosolitons in infinite extended systems, described by two nonlinear differential equations for an activator and an inhibitor without global inhibition as in (8.39) [47], it is known [54] that, in front of the first wall of the traveling autosoliton, a diffusion precursor of the inhibiting component δV_i exists because of the relatively broad distribution of $\delta V_i(x, t)$ (e.g., see Fig. 8.26c). As explained schematically in Fig. 8.29, the diffusion precursor leads to a refractory region of order L in size [47]. In this refractory region the current density changes smoothly from j_{min} to j_h in Fig. 8.29. Thus, the diffusion precursor in front of the traveling filament acts as a feeler of the filament. By this feeler, the filament is able to notice the semiconductor surface if it has approached the surface to a distance L [47]. The velocity of the filament begins to decrease and the repelling mechanism starts. Niedernostheide *et al.* [47] also noted that the repelling mechanism is of the same nature as the repelling due to collisions of current filaments in extended systems observed, e.g., in gas-discharge system [52] and that the repelling effect of the traveling current filament at the boundary of the system (p^+–n^+–p–n^- diode system) or at another current filament (in a gas-discharge system [52]) is quite different from the well-known behavior of traveling in many active systems described by models of the FitzHugh–Nagumo type [57, 58]. For the latter case, it is expected that the collision of the current filament leads to an annihilation.

8.1.4.6 Numerical results and comparison with experiments

Niedernostheide *et al.* [50] carried out model simulations under two different boundary conditions. Note that only the Neumann boundary conditions (8.26), (8.37) are considered in [47], in which only the qualitative arguments of the spatiotemporal behaviors have been made. From their detailed numerical investigations [50], the spatiotemporal dynamical behaviors are different when using different boundary conditions, which are homogeneous Neumann boundary conditions (case A) and mixed boundary conditions (case B) [50].

Case A: Homogeneous Neumann boundary conditions for V_e and p:

$$\frac{\partial V_e(x=0,t)}{\partial x} = 0, \qquad \frac{\partial V_e(x=\ell_x,t)}{\partial x} = 0, \tag{8.44a}$$

$$\frac{\partial \bar{p}(x=0,t)}{\partial x} = 0, \qquad \frac{\partial \bar{p}(x=\ell_x,t)}{\partial x} = 0. \tag{8.44b}$$

Case B: Mixed boundary conditions, i.e., Dirichlet boundary conditions for V_e and homogeneous Neumann boundary conditions for $\bar{p}$:

$$V_e(x=0,t) = V_{e0}, \qquad V_e(x=\ell_x,t) = V_{e0}, \tag{8.45a}$$

$$\frac{\partial \bar{p}(x=0,t)}{\partial x} = 0, \qquad \frac{\partial \bar{p}(x=\ell_x,t)}{\partial x} = 0. \tag{8.45b}$$

In case A, (8.44a) means that there is no lateral current in the n^+ base at the sample boundaries, and (8.44b) that there is no lateral diffusion at the sample boundaries in the n^- region. In case B, the Dirichlet boundary condition for V_e may be useful if the emitter voltage is limited or fixed, e.g., by surface recombination [47, 50]. A set of the model equations (8.39a), (8.39b) and

$$V = V_s - R_0\ell_z \int_0^{\ell_x} j(x,t)\mathrm{d}x \tag{8.46}$$

have been numerically integrated by using the finite-difference method, where in particular the Crank–Nicolson method has been applied for discretization in time [50]. In (8.46) $j(x,t)$ is the current density in the device chosen at an arbitrary cross-section parallel to the metallic contacts. Parameter values in Table 8.4 have been used. The control parameter is the applied source voltage V_s or equivalently the averaged total current $\bar{I}$. The numerical results are as follows, according to the boundary conditions, case A and case B, respectively.

Case A

For the calculations using homogeneous Neumann boundary conditions (case A), Niedernostheide *et al.* [50] chose homogeneous distributions for the initial values of V_e and $\bar{p}$. In order to simulate statistical fluctuations, which are unavoidable in the real systems, a pseudorandom variable of about 0.01% of the corresponding values is added to both distributions in each time step. The value of the load resistor has typically been chosen to be $R_0 = 10\ \mathrm{k\Omega}$.

As a function of the control parameter V_s, Niedernostheide *et al.* [50, 55] obtained: (i) transition from a homogeneous current–density distribution to a static localized filament; (ii) transition from a static to a breathing localized filament; (iii) period-doubling cascade of an oscillating localized filament; and finally (iv) transition to a traveling localized filament.

Transition from the homogeneous state to a static localized current filament can be seen when a driving voltage V_s is applied such that the total current lies on the upper branch of an S-shaped I–V curve, which is always possible when the avalanche breakdown occurs in the p^+–n^+–p transistor part, where $\beta M \gg 1$. With the homogeneous Neumann boundary conditions, it is remarkable that the static localized filament always develops near one of the system boundaries. The boundary at which the filament rises is selected by pseudorandom fluctuations of the system variables prepared in the numerical calculations [50]. An example of the localized filament is shown in Fig. 8.30a. The dark areas in Fig. 8.30 mark those parts of the distribution in which a certain threshold value, approximately two-thirds of the maximum current density, is exceeded [50]. Thus, these regions correspond to high-current regions and indicate the spatial position of the current filament.

The static filament is stable only up to a certain total current I_{c1}, corresponding to a critical width of the filament. Above the critical value with increasing V_s, the wall of the filament becomes unstable and a breathing filament develops, as shown in Fig. 8.30 for different values of the applied driving voltage. For $V_s = 33.890$ V (Fig. 8.30a) the distribution is stable, whereas for $V_s = 33.895$ V (Fig. 8.30b) a small modulation of the filament wall can be seen [50, 55]. The amplitude of this wall oscillation increases continuously with increasing V_s as shown in Fig. 8.30c, where $V_s = 33.90$ V. The wall oscillations are accompanied by oscillations in the total current $I(t)$ and in the voltage drop $V(t)$ across the device via the bias condition with $V(t) = V_s - R_0 I(t)$.

A detailed analysis of the breathing motion is possible if one investigates the spatial distributions $j_c(x, t)$ and $V_p(x, t)$ at different stages of the oscillation period. The sample voltage $V(t)$ oscillates as in Fig. 8.31a. For different times t_1–t_6 during an oscillation period, the spatial distributions for

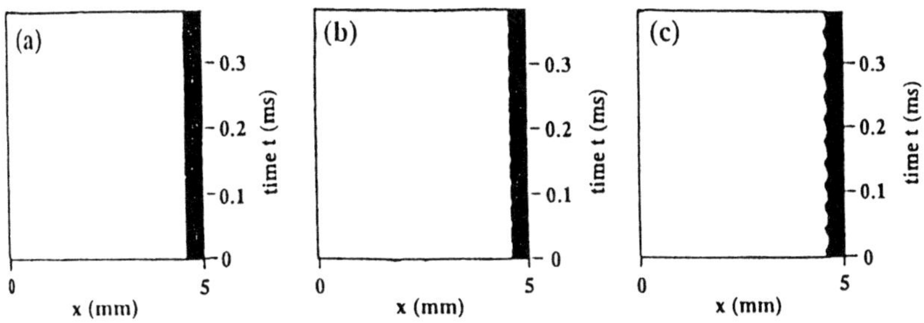

Figure 8.30 Numerically calculated spatiotemporal evolution of $j_c(x, t)$ for different values of V_s; parameters as in Table 8.4 and (a) $(V_s, \bar{I}) = (33.890$ V, 1.445 A), (b) (33.895 V, 1.450 A), (c) (33.900 V, 1.4555 A) [50].

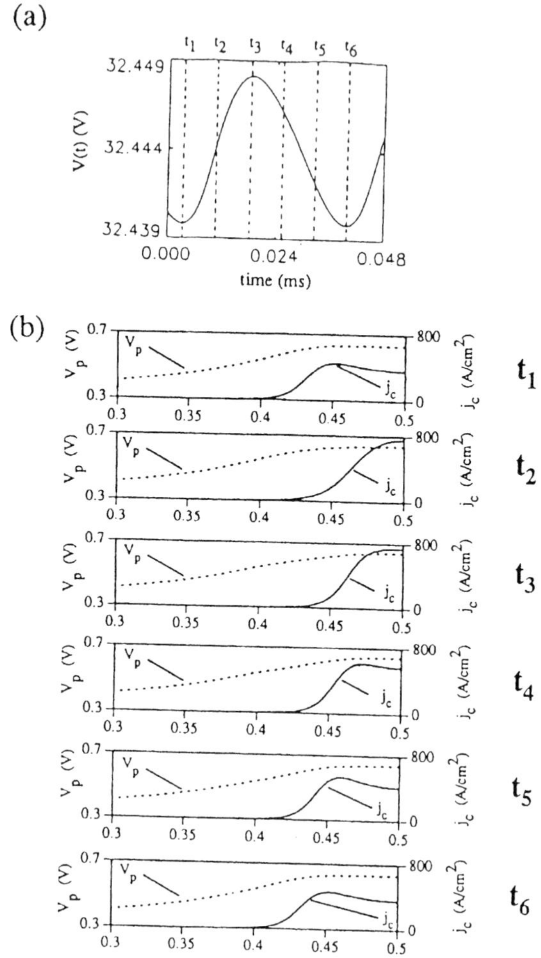

Figure 8.31 (a) Numerically calculated time trace V(t) of the device voltage and (b) distributions $V_p(x)$ and $j_c(x)$ for different times during an oscillation period; parameters as in Table 8.4 and $V_s = 33.94$ V, $I = 1.4896$ A, $t_1 = 2$ μs, $t_2 = 9.5$ μs, $t_3 = 17$ μs, $t_4 = 24.5\mu$s, $t_5 = 32\mu$s, $t_6 = 39.5$ μs with respect to the time zero in (a) [50].

j_c and V_p are shown in Fig. 8.31b. Here it is noted that the maximum elongation (or minimum contraction) of the wall in the x-direction of Fig. 8.30b or Fig. 8.30c corresponds to a maximum (minimum) total current $I(t)$, and thus to a minimum (maximum) voltage $V(t)$ via $V(t) = V_s - R_0 I(t)$. Keeping this relation in mind, the minimum and maximum values of $V(t)$ in Fig. 8.31a at t_1 and t_3 correspond to the spatial distribution of $j_c(x, t)$ at t_1 and t_3 in Fig. 8.31b, respectively. During the breathing mode of the filament, the potential distribution V_p representing the inhibitor via the relation $V_p = V - V_i$ follows behind the current–density distribution $j_e(x, t)$, which represents the activator. In Fig. 8.31b, the variation in V_p indicated by the dotted lines is quite small and, therefore, the role of the inhibitor cannot be well resolved in the figure. For larger average currents $\bar{I}$ (or larger applied voltage V_s), the spatiotemporal motion undergoes a period-doubling bifurcation to chaos [50, 55]. The spatiotemporal evolution is shown in Fig. 8.32: (a) a period 2 motion, (b) a period 4 motion, and (c) a chaotic (or turbulent) motion for values of $V_s = 33.9450$, 33.9495 and 33.9505 V corresponding to an average current of $\bar{I} = 1.501$, 1.510 and 1.511 A, respectively [50]. Although the localized filament is attracted to the boundary of the sample in Fig. 8.32, for the chaotic motion in Fig. 8.32c there are some instants when the filament is able to separate from the boundary.

Transition to a traveling localized filament can be seen with further increase in the control parameter, as shown in Fig. 8.33, where $V_s = 33.954$ V and $I = 1.512$ A. In Fig. 8.33b, the filament oscillates for a certain period of time near one of the system boundaries. During this period, the amplitude of the wall oscillations increases in the course of time, and if the amplitude exceeds a threshold value, the filament is detached from the boundary and travels with a constant velocity of about $v = 60$ μm/μs through the sample [50]. When it reaches the opposite boundary the filament is reattached there, oscillates for another few periods with increasing breathing amplitude, and again travels back to the right boundary. For this traveling localized filament the voltage $V(t)$ is shown in Fig. 8.33a, and the spatial distributions

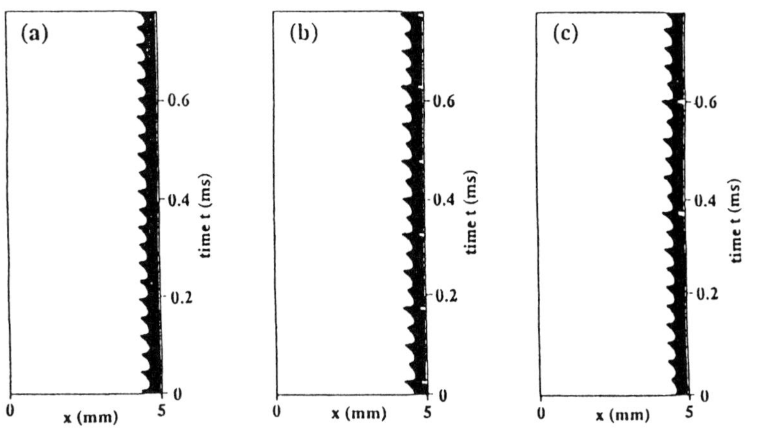

Figure 8.32 Numerically calculated spatiotemporal evolution of $j_c(x, t)$ or different values of V_s; parameters as in Table 8.4 and (a) $(V_s, \bar{I}) = (33.9450$ V, 1.501 A), (b) (33.9495 V, 1.510 A), (c) (33.9505 V, 1.511 A) [50].

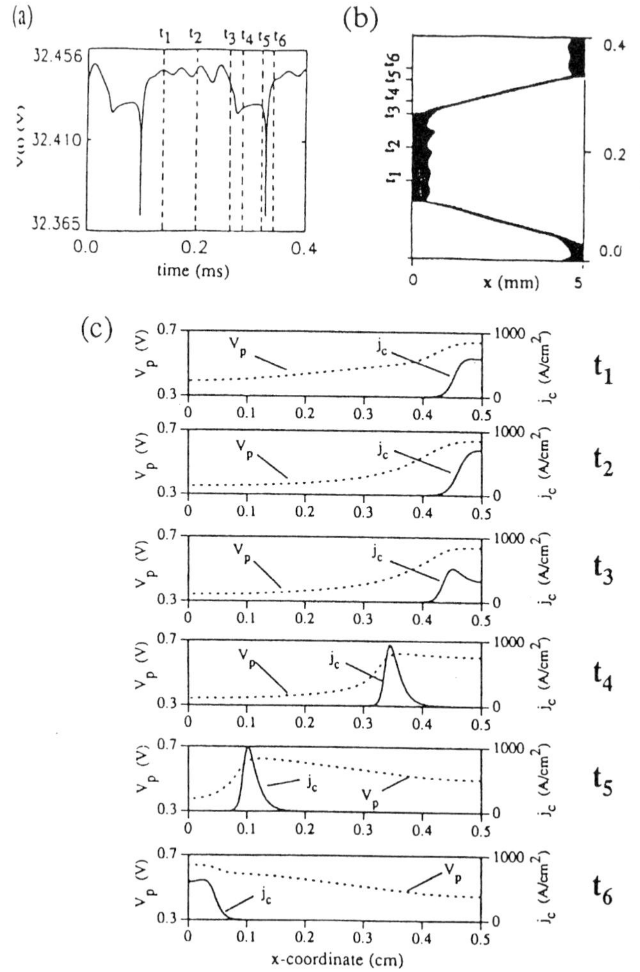

Figure 8.33 (a) Numerically calculated time trace $V(t)$, (b) spatiotemporal evolution $j_c(x, t)$, and (c) distributions $V_p(x)$ and $j_c(x)$ for different times during the travel from one boundary to the other; parameters as in Table 8.4 and $V_s = 33.954$, $\bar{I} = 1.512$ A, $t_1 = 140$ μs, $t_2 = 200$ μs, $t_3 = 260$ μs, $t_4 = 280$ μs, $t_5 = 300$ μs, $t_6 = 340$ μs with respect to the time zero in (a) [50].

$j_c(x)$ and $V_p(x)$ are shown in Fig. 8.33c for different times marked in the figure of $V(t)$. The relations among Fig. 8.33a–c seem to be rather complicated compared with Fig. 8.31 [50]. The relatively small local minima in the trace of $V(t)$ are accompanied by maximum filament elongation during the breathing oscillation (Fig. 8.33b). The decrease in the minima of $V(t)$ during the time between t_1 and t_2 marked in Fig. 8.33a corresponds to an increase in local maxima in the current trace $I(t) = [V_s - V(t)]/R_0$, indicating the growing breathing mode in Fig. 8.33b. It is clearly shown that the global minima of $V(t)$ (e.g., between t_5 and t_6 in Fig. 8.33a coincide with those moments at which the traveling filament reaches one of the system boundaries. The transition from the boundary oscillation to the traveling filament and the reverse trapping can be explained in Fig. 8.33c. In Fig. 8.33c, the widely spread inhibitor $V_p(x)$ moves with a certain delay behind the activator j_c, which can be clearly seen at the stage t_4 at which the filament is already in the traveling mode. Interplay of the diffusion precursor [47], [50] of the inhibitor distribution in front of the leading activator wall with the boundary of the sample is essential for the reflection of the filament when it approaches the boundary. The Neumann boundary acts as a mirror for the filament and the inhibitor precursor of the imaginary filament interacts with the traveling filament, resulting in a reflection of the filament [47, 49, 50]. An immediate reflection of the filament, however, is prevented by the widely distributed inhibitor tail behind the activator. For more details see Ref. [50].

An increase in V_s leads to a more complex behavior of breathing oscillation of the filament near the sample boundary. With further increasing V_s, the durations of the breathing oscillations become smaller, and finally, when $I > 1.86$ A, the breathing oscillations vanish completely and a well-defined traveling motion between the sample boundaries can be seen [50].

Case B

By using the mixed boundary conditions in which Dirichlet boundary conditions are applied to V_e in (8.45a), transition from a static to a rocking localized filament and transition from a rocking to a traveling localized current filament can be seen [46, 47, 49, 50]. The Dirichlet boundary conditions (8.45a) impose fixed values for V_e (e.g., V_e $(x = 0, \ell_x) = 0.63$V) at the sample boundaries. For those values $V_e(x = 0, \ell_x)$ that are smaller than the value $V_{e,h}$ of the corresponding spatially homogeneous distribution determined by $q(V_e, V_i) = 0$ and $Q(V_e, V_i, V) = 0$, a localized current filament does not develop directly at the boundary but can be formed only in a certain distance from the sample boundaries. For sufficiently small values of $V_e(x = 0, \ell_x)$, the boundaries are exclusively repulsive, which leads to the formation of a symmetric current filament at the center of the sample [50].

An example of a locking localized current filament induced after the transition from a static current filament is shown in Fig. 8.34, where V_e $(x = 0, \ell_x) = 0.63$ V for the Dirichlet boundary conditions [50]. For such a locking motion, the fundamental frequency of the locking oscillation rises at the bifurcation point discontinuously and decreases with increasing $\bar{I}$. The

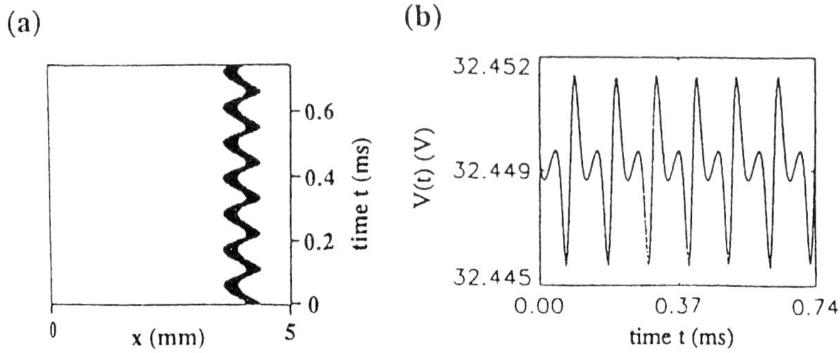

Figure 8.34 (a) Numerically clculated spatiotemporal evolution of j_c and (b) time trace of the device voltage $V(t)$ in the case of a locking current filament; parameters as in Table 8.4 and $V_s = 32.9$ V, $\bar{I} = 450$ mA, $V_e(x = 0,\ \ell_x) = 0.63$ V [50].

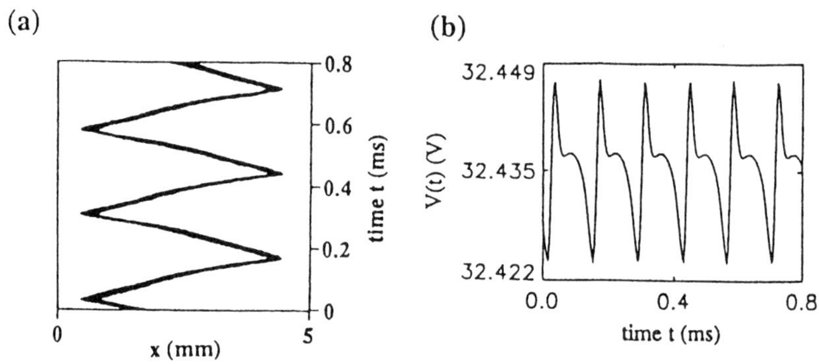

Figure 8.35 (a) Numerically calculated the evolution of the spatial distribution j_c and (b) time trace of the device voltage $V(t)$ in the case of a traveling current filament; parameters as in Table 8.4 and $V_s = 33.05$, $\bar{I} = 61$ mA, $V_3(x = 0,\ \ell_x) = 0.63$ V [50].

maxima of $V(t)$ in Fig. 8.34b, corresponding to minima in $I(t)$, appear when the filament turns back (see Fig. 8.34a); the small local maxima correspond to the inversion points located in the interior of the sample, and the global ones correlate with those moments at which the filament reaches near the sample boundary. The voltage minima correspond to the maximal filament velocity. For detailed discussions see Ref. [50]. An example of the transition from a locking to a traveling localized filament is shown in Fig. 8.35.

Many experimental observations by the streak camera image have been presented for the spatiotemporal behaviors of the localized current filaments in the p^+–n^+–p–n^- diode [46–49, 55], and successfully analyzed by the model equations [47, 50], as described above. When the applied source voltage V_s is varied as a control parameter, period doubling bifurcation can be observed [48] as typically shown in Fig. 8.36, where self-generated oscillations of the voltage drop across the device occur in the range of a few kHz and the amplitudes of the oscillations are in the range of a few tenths of percents of the dc part of the device voltage. In Fig. 8.36, the waveforms of the device voltage (first column) are analyzed by the 2D phase portrait by projecting the phase space trajectories onto the $V(t_{n+5}) - V(t_n)$ phase plane (second column), where $V(t_n)$ denotes the nth value of the discrete recorded

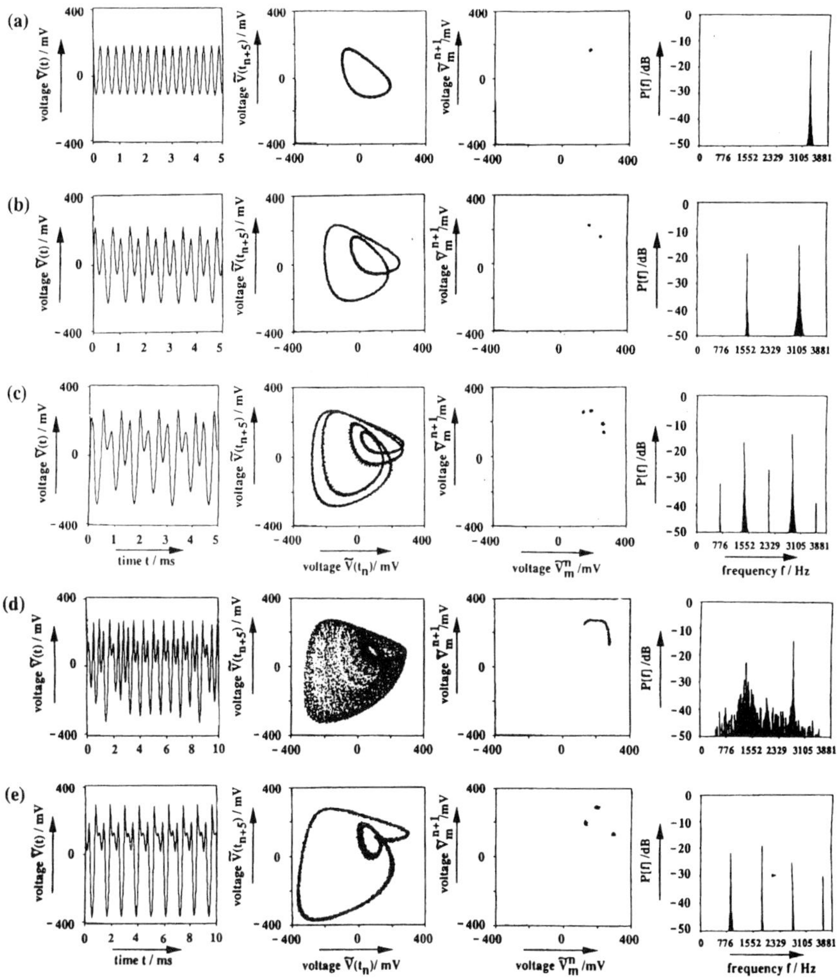

Figure 8.36 Time trace $V(t)$, projection of the phase space onto $V(t_{n+5}) - V(t_n)$, return map obtained by plotting the $(n+1)$th maximum V_m^{n+1} and power spectrum P(f) for different values of the dc driving voltage V_s. Parameters: sample ST2.25, $t_{n+5} - t_n = 100\ \mu s$, room temperature, $R_0 = 1\ k\Omega$, $V_s =$ (a) 168 V, (b) 174 V, (c) 180 V, (d) 182 V and (e) 186 V [48].

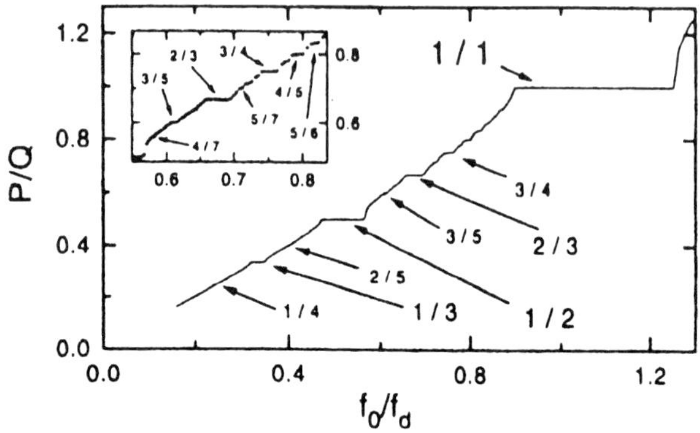

Figure 8.37 Locking ratio P/Q as a function of the frequency ratio f_0/f_d. Parameters are $V_{s0} = 49.5$ V, $V_d = 352$ mV, $R_0 = 510\ \Omega$, $T = 314.8$ K and $f_0 = 4717$ Hz, for sample S89 [55].

time series $V(t)$ [50]. The return maps constructed from successive local voltage maxima V_m of $V(t)$ (third column) and the power spectra of $V(t)$ (fourth column) clearly exhibit well-resolved period doubling bifurcation. The spatiotemporal dynamical behaviors in the regime of period doubling are due to the rocking motion of the localized current filament, while it is noted that, in the model simulations, breathing of the localized filament can cause the period doubling (Fig. 8.32).

A universality of the circle-map behavior [59] is also found [55]. Niedernostheide *et al.* [55] investigated the nonlinear response of the localized filament by applying a dc + ac bias with the driving frequency f_d, where the ac bias is superimposed on the dc bias V_s in a sinusoidal form of $\tilde{V}_s \sin(2\pi f_d t)$. They observed the complicated behaviors of frequency-locked, quasi-periodic and chaotic motions [55]. In Fig. 8.37, the winding number $W = f_i/f_d = P/Q$ vs the frequency ratio $\Omega = f_0/f_d$ is drawn for fixed values of V_s and $\tilde{V}_s$, where f_0 is the fundamental frequency or the natural frequency of self-generated oscillation of the localized filament under the dc bias and f_i is the fundamental response frequency under the dc + ac bias. Apparently, Fig. 8.37 reveals a well-defined devil's staircase. The global feature of the bifurcations in the $f_d - \tilde{V}_s$ control parameter space has shown Arnold tongues [59].

Spatial patterns of the localized current filaments for both the p^+–n^+–p–n^- and n^+–p–n–p^- diodes have also been studied [60] by scanning electron microscopy [11, 13]. For the spatiotemporal behaviors in the p–i–n diode, see Ref. [61]. For the general theory of pattern formation of localized current filaments (or autosolitons), see Kerner [56] and by Schöll [20].

8.1.5 Experimental aspects of pattern dynamics in bulk semiconductors – part I

Experimental studies on the spatial and/or spatiotemporal behaviors of the current filaments in bulk semiconductors have been extensively reported by many authors using various methods: low-temperature scanning electron microscopy [11, 13], scanning laser microscopy [15, 16] and measurements of photoluminescence quenching [45] to reconstruct the spatial patterns of the filaments. In the Si p–i–n diodes, the potential-probe method and spatially resolved methods of optical absorption and recombination radiation [61] have been used. In the Si p^+–n^+–p–n^- diode, as introduced in Chapter 3 and section 8.1.4, a method using a streak camera [46–50] has proved to be excellent for studying the spatiotemporal pattern dynamics of the localized current filaments. Major differences between the measurements of the filaments in the bulk semiconductors and the device structures p^+–n^+–p–n^- diode and p–i–n diode) are as follows. First, in the device structures, a current filament can be formed at room temperature [46–50, 60, 61], while it can only be measured at low temperatures (4.2 K) for the bulk semiconductors, which are doped with shallow impurities [9–11, 13, 15, 16, 20]. For this reason, the potential-probe method [61], which is a noninvasive technique, is generally difficult to handle. Secondly, in contrast to the measurements in the device structures [46–50], recombination radiation is not followed by current filamentation in the bulk semiconductors, so that external perturbations such as electron beam [10, 11, 13] and laser beam [15, 16, 45] must be used to visualize the current filaments. Since the filament boundaries are extremely sensitive to such perturbations, many measurement techniques are invasive ones which may more or less alter or perturb the original dynamical behaviors of the filaments under the dark condition.

Among many observations, Prettl's group have extensively investigated the image patterns of a current filament and its dynamical behaviors in n-type GaAs using scanning laser microscopy [15, 16] and measurements of photoluminescence quenching [45]. For measurements by low-temperature scanning electron microscopy in GaAs and Ge, the author refers to the books cited in Refs [10, 20]. Many fundamental properties of the current filament in n-GaAs under longitudinal magnetic fields have been clarified by Prettl's group [15, 16, 28, 45, 63–65]. Their elaborate experimental studies are introduced here, along with recent experimental results obtained in the author's laboratory.

Using scanning laser microscopy Brandl *et al.* [15, 16], Novàk and Prettl [65] and Spangler *et al.* [44, 64] have investigated the static pattern images and also the dynamical behaviors of the current filaments in n-GaAs under longitudinal magnetic fields. Epitaxial layers of n-type GaAs with a thickness of 16 μm [16, 62–64] to 29 μm [15, 44, 65] have been used.

Table 8.5 Geometry and material parameters of the sample

Sample dimensions	–	4.1×3.7 mm
Contact distance	c	3.5 mm
Contact radius	R	< 0.5 mm
Layer thickness	–	29 μm
Donor density	N_D	11.9×10^{14} cm^{-3}
Dompensation ratio	N_A/N_D	0.91
Free electron density at 77K	n	9.6×10^{13} cm^{-3}
Electron mobility at 77 K	μ_n	4.5×10^4 cm^2 V^{-1} s^{-1}

Depending on the material parameters the dynamical behaviors are found to differ. In particular, the compensation ratio N_A/N_D(77% in [16, 62–64] to 92% in [15, 44, 65]) and the electron mobility (8.9×10^4 cm^2/Vs [16, 62–64] to 4.5×10^4 cm^2/Vs [15, 44, 65] at 77 K) are important material parameters in order to observe self-sustained chaotic oscillations under the magnetic fields. The donor concentration and electron density are 5.7×10^{14}/cm^3 and 1.3×10^{13}/cm^3, respectively, for material of low compensation ratio [16, 64–66], while they are 1.19×10^{15}/cm^3 and 9.6×10^{13}/cm^3, respectively, for material of high compensation ratio [15, 62–64] (see Table 8.5). Ohmic point contacts have been used in their investigations [15, 16, 62, 65], in which a single current filament has been observed. In contrast to the planar contacts, by which the field distribution is uniform and the multifilaments can be formed by **self-organization**, the single filament can only be formed by **field organization** in the case of the ohmic point contacts, where the field distribution is given by the dipole-like potential distribution [65]. Brandl *et al.* [16, 28, 66, 67] and Spangler and Prettl [64] have observed a **Ruelle–Takens–Newhouse (RTN)** route to chaos for the dynamical behaviors of the current filament under a magnetic field. As described below, they attribute

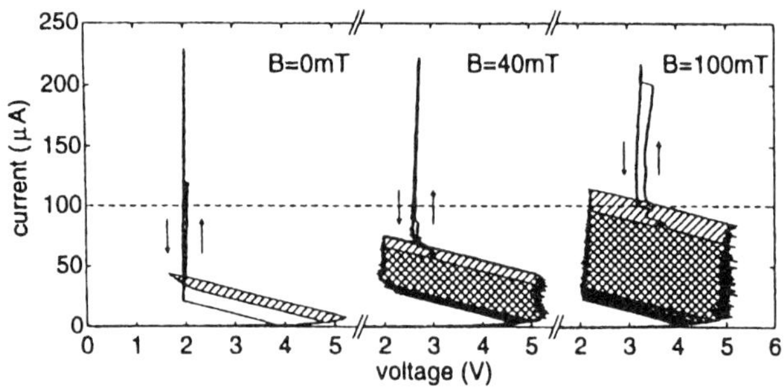

Figure 8.38 Current–voltage characteristics of a n-GaAs epitaxial layer for various magnetic field strengths at $T = 1.8$ K. Hatched regions show the extent of self-sustained oscillation. Arrows, for stable current flow, and right and left inclined hatching in the oscillatory regions indicate increasing and decreasing currents, respectively [44].

the occurrence of chaos via the RTN scenario to the simultaneous appearance of the fundamental electric-field-dependent generation and recombination cycle, the Hall-field instability at the filament boundaries and the Hall voltage collapse [16].

Using a GaAs sample with a relatively low electron mobility (4.5×10^4 cm^2/Vs) and a high compensation ratio ($N_A/N_D = 0.9$), the self-sustained oscillation is quite regular under transverse magnetic fields [65]. In Fig. 8.38, the hatched areas indicate the extent of oscillations in the I–V plane. The regular oscillation of the filamentary current is relaxation-type, which is caused by the repetitive cycle of ignition and extinction of the filament. The I–V characteristics show hysteresis; in particular, the oscillatory regimes are different for increasing and decreasing current, as indicated in Fig. 8.38 by arrows and by right and left inclined hatching, respectively. The hatched area increases by increasing the magnetic field intensity. Thus, a stable filamentary current flow at zero magnetic field is destabilized by applying a magnetic field yielding regular oscillation. At a constant bias current of, e.g., 100 μA the limit of stability is approached at a magnetic field of about $B = 100$ mT.

For these samples of high compensation ratio, Brandl *et al.* [15] and Spangler *et al.* [44] have investigated image patterns of a filament in the stable postbreakdown regime [44]. Reconstruction of a current filament can be done by scanning a focused He–Ne laser beam of 15 μm diameter [15], or a semiconductor laser pulse, line by line across the sample surface of the n-GaAs epitaxial layer and by monitoring the photocurrent through the sample as a function of the position of the focal spot (see Chapter 3), where the power of the incident beam is typically less than 1 mW [15]. An example of the image patterns at high current levels under the various magnetic field intensities measured by scanning laser microscopy is shown in Fig. 8.39 [44], where Spangler *et al.* have used a pulsed GaAlAs semiconductor laser of 25 ns long pulses, and the photosignals are synchronously recorded as a function of focal positions. In Fig. 8.39b, the image pattern at $B = 0$ mT reveals two symmetrical ridges corresponding to the filament boundaries [44], very similar to those observed in p-Ge on applying the low-temperature electron microscope technique [10, 11, 13]. The ridge-like structure appears due to the local impurity breakdown induced by the laser irradiation at the filament boundaries. In Fig. 8.39c–f, measurements at 30 mT and 90 mT are plotted. For each magnetic field strength two recordings are shown, differing in the orientation of the magnetic field. For the right column the field points upwards in the sense of Fig. 8.39a, and in the left column the field points downwards. By applying the magnetic fields, the spatial distribution of the photoresponse changes drastically, i.e., the symmetry of both ridges is broken. By increasing the magnetic field the response of one ridge on the side of the filament, which points in the direction of the Lorentz force ($\mathbf{J} \times \mathbf{B}$), decays, whereas the signal in the other ridge grows rapidly [15,

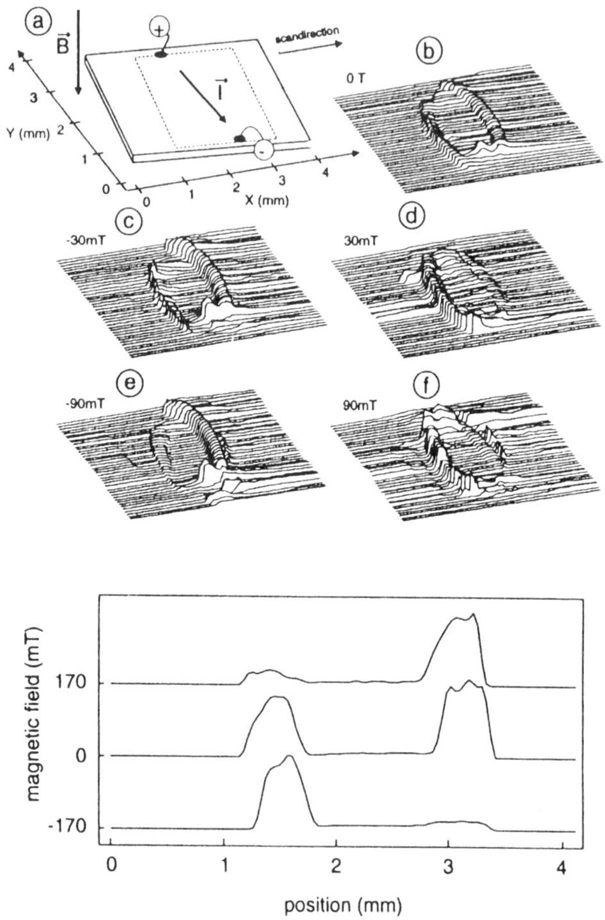

Figure 8.39 Top: reconstruction of a current filament at $I = 1$ mA for various magnetic fields. The direction of the current and the orientation of the magnetic field are indicated in (a), which shows the sample geometry. Bottom: photosignal along a single line in the center of the filament for various positive (pointing upwards) and negative (pointing downwards) magnetic fields [44].

44, 65, 67]. Inverting the magnetic field mirrors the spatial pattern at the axis of symmetry of the $B = 0$ structure.

The appearance of such an asymmetric pattern of the ridge structure under the magnetic fields is tightly related to the formation of the Hall field as described below. At $B = 0$, the ridge structure appears as a result of local impurity breakdown induced by the photoirradiation at the filament boundaries. The charge neutrality is locally broken [20], because the impact-ionized carriers at a boundary tend to diffuse outwards, whereas the

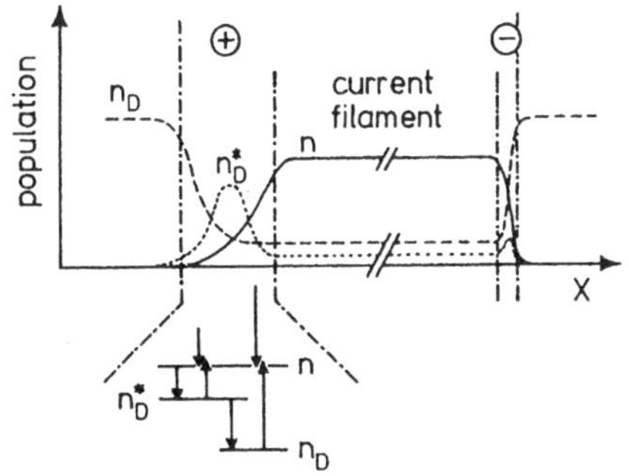

Figure 8.40 Schematic concentration profile of free carrier n, excited donors n_D^* and donors in the ground state n_D along a linescan across a current filament. The insert shows the most important impact ionization generation and relaxation process occuring in the transition zones marked with ⊕ and ⊖ according to the formation of the Hall field in a transverse magnetic field [15].

ionized impurities are immobile and fixed, whereby the electrical double layers are formed along the filament boundaries. The electrical double layer of the boundary (the transition zone of the filament) is totally charge neutral, but very sensitive and unstable to external perturbations. Applying a magnetic field causes redistribution of the electrons due to the Lorentz force at the boundaries. Figure 8.40 shows schematically the concentration profiles of free carriers by n, the donor ground state n_D and the excited state n_D^* [15], where the Lorentz force is pointing from left to right. By the Lorentz force the concentration profile of free electrons tends to move to the left side; in the figure, transition zones at the left and right sides are marked with plus and minus, respectively. That is, charge neutrality is lifted in the filament boundaries by depletion and accumulation of electrons on the left and right side of the filament, respectively, and hence the left transition zone is positively charged and the right one negatively charged, forming the Hall field in a transverse magnetic field [15, 16]. The combined effect of the Lorentz force, the Hall field and the well-known $1/B$ dependence of the electron diffusion length on the magnetic field intensity narrows the right boundary and widens the left boundary of the filament, respectively.

Brandl *et al.* [15] strongly suggest that a large population of excited donor states is established at the left boundary (see Fig. 8.40). The asymmetric ridge structures may be understood qualitatively by the multilevel impact-ionization model [20], as illustrated in Fig. 8.40 by an energy diagram indicating the relevant electron generation and recombination processes. Interband irradiation will generate additional free electrons during a laser linescan. This local increase in the free carrier concentration and an increased population of excited shallow donors in the transition zone will trigger an impactionization avalanche. The resultant situation is not a steady state, so re-establishment of the former situation is achieved through relaxation of the electrons into the donor states. By cyclic repetition of this process, an autonomous current fluctuation is obtained by irradiating the transition zone [15]. Such fluctuations have usually been observed as current spikes at the filament boundaries [15], but the induced photocurrent is averaged for each linescan, yielding an increased smooth photosignal at the left transition zone in Fig. 8.39. For the right narrow transition zone the population of excited donor states is smaller, so multiplication of interband generated electrons will be less effective [15]. As a consequence, the peak response in the photocurrent and the width are smaller for the right transition zone than for the left one. This is clearly observed in the recordings shown in the bottom diagram of Fig. 8.39, where the signal is plotted along a single line in the filament center for $B = 0$ and ± 170 mT [44]. The asymmetric ridge structure may be explained by the asymmetric spatial structure of the transverse electric field under the magnetic field, as investigated numerically by Kunihiro *et al.* [35].

In contrast to the above-mentioned situation at high currents, bending of the filament due to the Lorentz force has been found at small currents (e.g., at 100 μA in the I–V curve of Fig. 8.38) for samples of high compensation ratio [44]. The situation is clearly shown in Fig. 8.41, where the current flow in Fig. 8.41a is from top to bottom with the magnetic field pointing upwards for $B > 0$ (in contrast to Fig. 8.40a). The situation is similar to the case of a current-carrying elastic wire fixed at the electrical contacts [44]. By inverting the current flow and the magnetic field, as indicated by double arrows in Fig. 8.41, identical structures are obtained, which are expected from the symmetry of the Lorentz force. In the bottom of Fig. 8.41 a photosignal along a single line in the center between the contacts is plotted for various magnetic field strengths up to ± 40 mT, showing the displacement of the center of the filament as a function of the magnetic field strength. The measurements were carried at 100 μA along the broken line in the I–V characteristic in Fig. 8.38. The recordings demonstrate the symmetry of the filament displacement with respect to inversion of the orientation of the magnetic field.

Keeping the magnetic field strength constant and increasing the bias current, the displacement of the center of the filament decreases as a function of the current, as shown in Fig. 8.42, where $B = 40$ mT. The displacement decreases slowly with rising current and approaches zero at about 0.8 mA (not shown in the figure, but clearly seen in Fig. 8.39). From these observations Spangler *et al.* considered an analogy between the filament and a string being fixed at the two point contacts which is subjected to a constant force field [44]. They attributed an internal tension to a filament and equated the restoring force to the Lorentz force. For small elongations the internal tension is given by $\sigma = (jB/8w)L^2$, where j is the current density in the filament, w the amplitude of the filament center and L

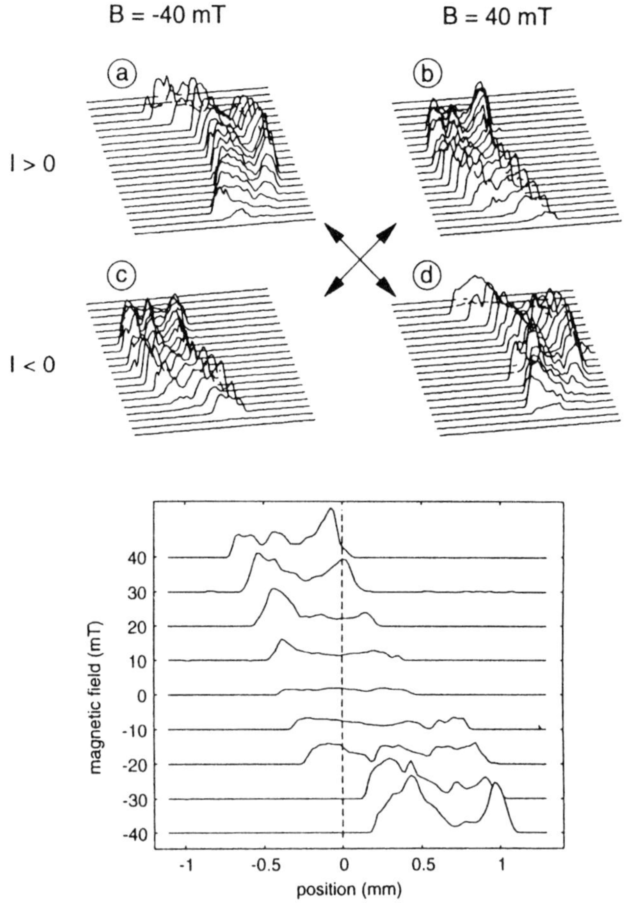

Figure 8.41 Top: reconstruction of a current filament at $B = 40$ mT and $I = 100$ μA. Spatially resolved linescans are shown for the magnetic field pointing upwards ($B > 0$) and downwards ($B < 0$) and for positive and negative currents. The double arrow in the center indicates the symmetry of the filament if both the current and magnetic field are inverted. Bottom: photosignal along a single line in the center of the filament at $I = 100$ μA for various positive (pointing upwards) and negative (pointing downwards) magnetic fields. The positive current flows from top to bottom [44].

the length of the filament. Taking the x from the measurements, Spangler *et al.* have determined and plotted in Fig. 8.42 for $B = 40$ mT [44].

Spangler *et al.* [44] discussed the stiffening of the filament observed with increasing the bias current, as follows. At small currents, just when a filament is formed, the electron temperature and mobility are not very different inside and outside the filament. The Lorentz force accumulates charges predominantly at the borders of the sample, causing a lateral electric

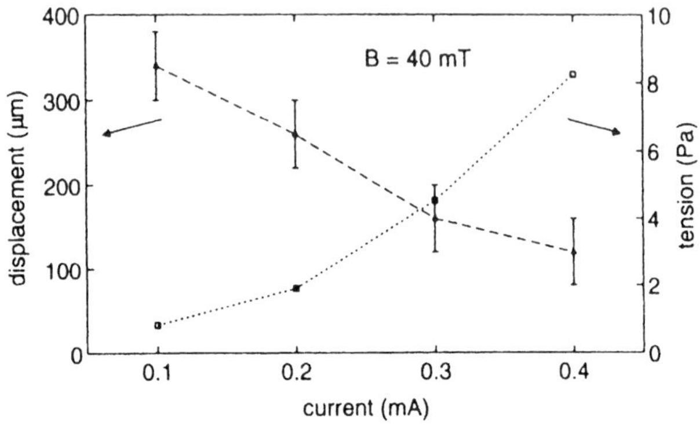

Figure 8.42 Lateral displacement (left ordinate) and tension (right ordinate) of a filament at $B = 40$ mT as a function of the bias current [44].

potential with constant slope $\phi = -E_{\mathrm{H}}x$ and the Hall field $E_{\mathrm{H}} = EB\,\mu$, where μ is the low-field mobility and E the longitudinal driving electric field. The free carrier distribution is not affected by the Hall field because the Poisson equation, $\mathrm{d}^2\phi/\mathrm{d}x^2 = 0$, remains unchanged. If E is homogeneous, as in the case of the planar-type contacts, the filament may drift freely in the direction of the Lorentz force. Such transverse motions of the filament have been observed in p-Ge [23, 68] and have been numerically simulated assuming the mobility to be spatially constant [14]. The point contacts in the present case generate a dipole-like electric field distribution in the semiconductor film [65]. With rising magnetic field the lateral displacement of the filament increases more and more until the driving electric field drops below the critical electric field and the filament is torn [44]. Now the current through the sample decreases and the voltage across the contacts increases, forming a new filament along the shortest distance between the contacts. This process is periodically repeated, yielding regular oscillations [44, 64]. For a moderate current, a bent filament can be observed in Fig. 8.41, without showing any spontaneous oscillations.

For high currents, in the order of 1 mA, the mobility in the filament μ_{h} is much larger than that of the low conductivity outside the filament μ_ℓ, where Spangler *et al.* [44] have found $\mu_{\mathrm{h}} \simeq 4 \times 10^4$ cm^2/V s, which is about one order of magnitude larger than μ_ℓ. Therefore, charges are now accumulated mainly at the filament borders. The surface charge of the filament $EB(\mu_{\mathrm{h}} - \mu_\ell) \simeq EB\,\mu_{\mathrm{h}}$ is practically equal to the total charge separated by the Hall effect. Thus, the potential due to the Hall field drops completely across the filament, keeping the inhomogeneous electron distribution (the filament) fixed in space. In their interpretation, the Hall field generated across the sample borders is negligible compared to the Hall field generated by the filament borders.

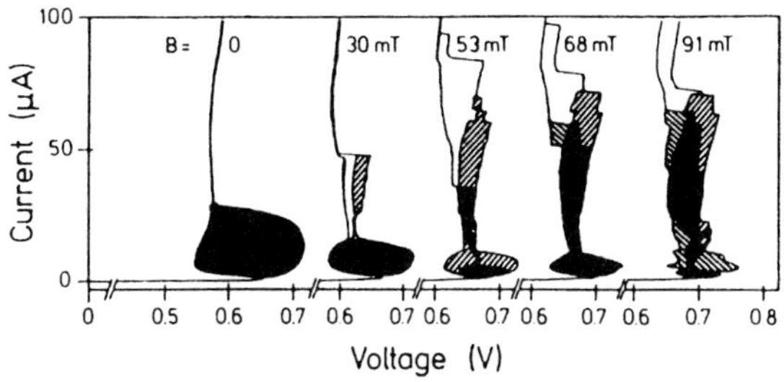

Figure 8.43 Current–voltage characteristics for various magnetic fields B (load resistor $R_L = 1$ MΩ and bath temperature $T = 4.2$ K) recorded by a method marking the extent of the oscillatory regime (hatching inclined from bottom left to top right denotes recording direction to increasing current; hatching inclined from top left to bottom right denotes decreasing current) [16].

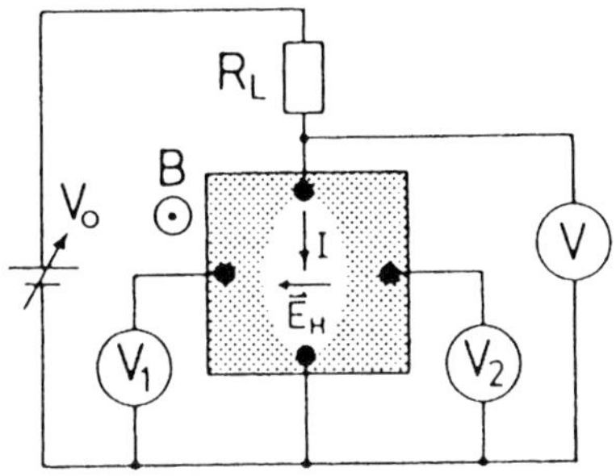

Figure 8.44 Schematic drawing of the experimental arrangement and the orientation of magnetic field B, current I and Hall field E_H. The contacts are in a quadratic arrangement on a 3×3 mm^2 square opposite contacts being 2 mm apart [16].

For n-GaAs epitaxial layers with high electron mobility ($\mu = 8.9 \times 10^4 \mathrm{cm}^2/\mathrm{Vs}$ at 77 K) and relatively low compensation ratio (77%), Brandl *et al.* [28, 66, 67] and Spangler and Prettl [64] found a RTN route to chaos in the postbreakdown regime under the transverse magnetic field. An example of the I–V characteristics for various magnetic fields B is shown in Fig. 8.43, where a load resistor of $R_L = 1$ MΩ is used. For $B = 0$, large-amplitude regular relaxation oscillations occur just above the threshold of impact-ionization breakdown, where the oscillations rise up to 5 MHz with increasing average current [16]. By applying the magnetic field a sharp transition to small-amplitude fluctuations of complex temporal behavior is observed, as shown by the hatched region in Fig. 8.43. In this latter regime, quasi-periodic, chaotic and intermittent switching current behaviors are all observed. For a transition to chaos according to the RTN scenario, Brandl *et al.* [16] have investigated the Hall effect of the filament, and found Hall voltage collapse. The Hall voltage collapse is considered as a physical origin for the third mechanism, namely, the appearance of the third fundamental frequency f_3, which is responsible for the emergence of chaos after the quasi-periodic motion.

The measurements for the Hall voltage were carried out using the experimental arrangement shown in Fig. 8.44, where $R_L = 200$ kΩ, the magnetic field B is normal to the [100] sample surface and four Ohmic point contacts are formed in a rectangular arrangement on the sample surface [16]. As a probe of the current through the sample the voltage drop V is recorded across the longitudinal contacts. The voltages V_1 and V_2 at the lateral contacts are determined with respect to one of the longitudinal contacts, yielding the Hall voltage defined by $V_H = V_2 - V_1$. All voltages are measured as a function of time, applying cryogenic field-effect-transistor impedance transformers soldered directly on the sample contacts [16]. In Fig. 8.45, the voltage drops are plotted as a function of applied voltage V_0, where $B = 0.5$ T. In the prebreakdown regime ($V_0 < 1.1$V) a usual Hall voltage V_H is generated following the rule of conventional Hall effect. Above the threshold of the breakdown in the regime of regular relaxation oscillations, a synchronous cross-talk of the current is observed in both lateral voltages V_1 and V_2, yielding smaller amplitudes in V_H [16]. This behavior is again explained by the standard Hall effect. In the geometrical orientation of the magnetic field and the current flow in Fig. 8.44, V_1 and V_2 correspond to the potentials at the negatively and positively charged sides of the filament due to the Lorentz force by $\mathbf{J} \times \mathbf{B}$, respectively.

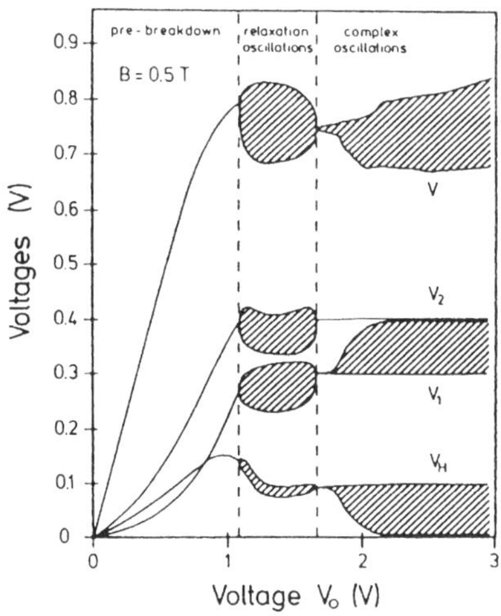

Figure 8.45 Potentials V, V_1, V_2 and Hall voltage V_H as a function of the external voltage V_0 for an applied magnetic field $B = 0.5$ T. The extent of fluctuations is marked by hatched areas [16].

Drastic changes in the Hall effect appear in the regime of complex multifrequency oscillations which is observed just above the regime of the regular relaxation oscillations. In spite of the fact that with increasing V_0 a sequence of quasi-periodic, mode-locking and chaotic behavior is observed in this regime, the potential V_2 at the positively charged side of the filament remains constant in time, as shown in Fig. 8.45. The potential V_1 at the negatively charged side is also constant in this region up to a critical value V_{0c} depending on the magnetic field strength. Increasing V_0 above V_{0c}, the voltage exhibits positive pulses in an intermittent manner. With further increases in V_0, some of the intermittent pulses happen to rise to the level of V_2, yielding the Hall-voltage collapse, where the dynamic ranges of the intermittent pulses in V_1 and the Hall voltages are given by hatching in Fig. 8.45. In Fig. 8.46, typical Hall-voltage bursts are shown simultaneously with the spontaneous oscillation of the longitudinal voltage V. Surprisingly, bursts in V_H appear to be triggered by a decrease in the oscillation amplitude in V and an increase in the average oscillating voltage V [16].

The essential microscopic mechanism causing all of the observed dynamic effects seems to be rather complicated, as interpreted by Brandl *et al.* [16]. For the occurrence of the quasi-periodic and frequency-locked oscillations, the longitudinal relaxation oscillation with the fundamental frequency f_1 may be coupled with the magnetic-field-induced transverse relaxation oscillation with the fundamental frequency f_2. Further coupling with the Hall-voltage collapse (the third mechanism with the fundamental frequency f_3) may initiate the chaotic behavior as in Fig. 8.46.

The details of the microscopic mechanism given by Brandl *et al.* [16] are as follows. Regular oscillations are simply due to oscillations of the longitudinal electric field and thus the impact-ionization probability around the critical values of breakdown, which repetitively causes ignition and extinction of the current filament (flushing filament). At a sufficiently large average current a stable filament is formed at $B = 0$ T, as expected from an I–V curve for $B = 0$ T in Fig. 8.43. Applying a magnetic field, redistribution of the free electrons at the filament borders occurs due to the

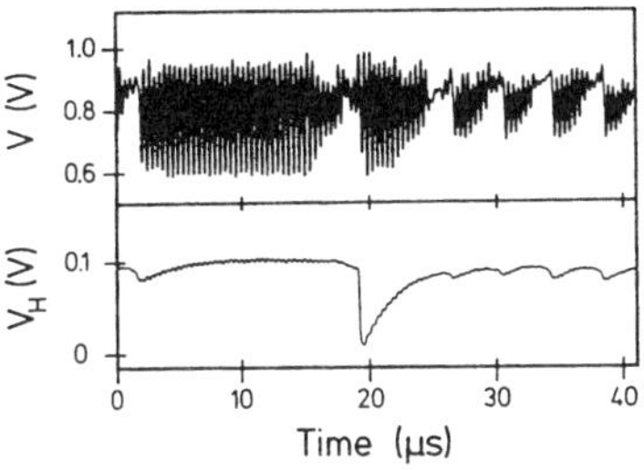

Figure 8.46 Temporal characteristics in the longitudinal voltage and the Hall voltage V_H for an applied magnetic field $B = 0.28$ T and $V_0 = 3.6$ V[16].

Lorentz force. As shown in [15], the impact-ionization probability is much larger in the positively charged border than in the negatively charged border (see Fig. 8.40). Thus, under the magnetic fields, the conditions of the impact-ionization instability at the positively charged border are met at a lower longitudinal voltage than at the negative side. The observed dynamical processes may be understood in the following way [16]. Above a critical voltage in the presence of a current filament, free carriers are generated at the positively charged border by impact ionization, mainly from populated excited donor states [15, 16], leading to a lateral excursion of the filament. This causes an oscillatory instability at this filament boundary by the same mechanism as for the ignition–extinction cycle of the total filament. Coupling to dielectric relaxation [16] generates the quasi-periodic and frequency-locked current behavior. Obviously, in Fig. 8.40, free electrons penetrate only partly into the positively charged region, leaving the external voltage V_2 practically constant (Fig. 8.45). The microscopic mechanism for the occurrence of chaos may be described qualitatively as follows [16]. Continuous relaxation of electrons into the donor ground state in the fluctuating filament border reduces the amplitude of oscillation. Thus, the average longitudinal voltage V increases (see Fig. 8.46) and eventually reaches the point where an impact-ionization avalanche may also be initiated in the opposite negatively charged boundary, driving the system into chaos. A pulse of large current density on the negative side of the filament occurs. Assuming that pinching [16, 28] contracts the current flow towards the center of the filament, ionized donors are left behind. This might cause the observed rapid rise in the potential V_1, leading to the collapse of the Hall voltage (Figs. 8.45 and 8.46). Again, the increased average current lowers the longitudinal voltage V, terminating this process and restabilizing the current flow. Decay times of the voltage bursts (see bursts in V_H of Fig. 8.46) are determined by the sample resistivity and internal and external capacities [16].

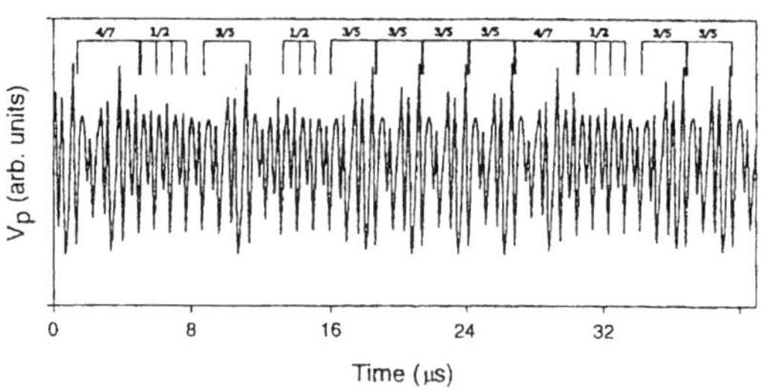

Figure 8.47 Time series for $f_d = 1.097$ MHz and $V_d = 0.82$ V above the critical overlapping and mixing of resonances in an erratic way with frequency ratios 1/2, 3/5 and 4/7 [64].

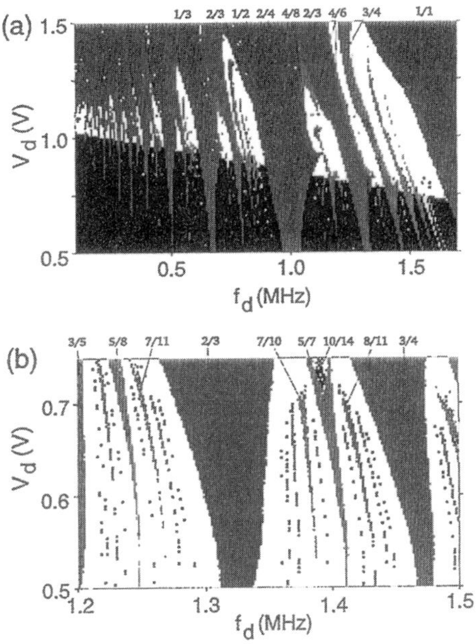

Figure 8.48 Experimental phase diagram for drive amplitude V_d and frequency f_d. The maximum resolved number P was 14 and it was not distinguished between quasi-periodic and chaotic regimes. Coloring is as follows: color (P): blue (1), green (2), red (3), lightblue (4), pink (5), yellow (6). The critical line in (a) (upper diagram) was plotted from the measured devil's staircase in Fig. 8.49 by hand and then the quasi-periodic regime below was colored in black. Diagram (b) is part of (a) but recorded with a higher resolution [64].

Using the same n-GaAs epitaxial sample which exhibits the chaotic behavior through the RTN scenario, Spangler and Prettl [64] tested universal circle-map behavior (Chapter 6) by applying the dc + ac bias and with zero magnetic field. For the sample, only regular relaxation spontaneous oscillations with the fundamental frequency f_0 have been observed under the zero magnetic field, which are strictly periodic. Thus, by driving the self-generated regular oscillations by an external ac bias ($V_d \sin(2\pi f_d t)$), the circle-map behavior may be well resolved. Above the critical line of the phase diagram, chaos appears as a direct result of overlapping of Arnold tongues. An example of the chaotic behavior for the sample voltage is shown in Fig. 8.47, in which mixing of the frequency-locked states appears in an erratic way. The global phase diagram is given in Fig. 8.48, where quasi-periodic regimes are colored black, chaotic regimes white, and Arnold tongues numbered by the rational number P/Q are in different colors. Figure 8.48b is a part of Fig. 8.48a but recorded with a higher resolution. The sequence of the frequency-locked states follow the

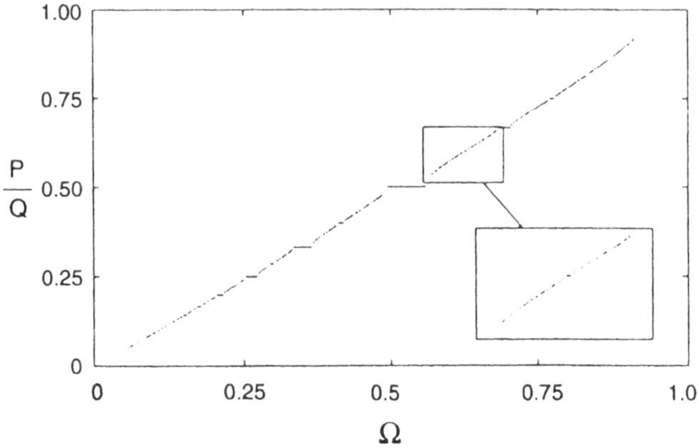

Figure 8.49 Experimental devil's staircase with 312 resolved mode steps. The inset shows the self-similarity [64].

Farey tree. Apparently, a critical line exists which divides the diagram into two sections between the black and white regions in Fig. 8.48a. Below the critical line the Arnold's tongues are embedded in the quasi-periodic sea. Above the line the Arnold's tongues begin to overlap. Spangler and Prettl also showed a break-up of the invariant two-torus at the critical line, and period doubling cascades at the right borders of the Arnold's tongues above the critical line [64]. They accurately determined the critical line which is given as an appearance of overlapping mode-locked resonance. On the line, the devil's staircase can be seen, as observed in Fig. 8.49, where they have resolved 312 mode-locking steps. For more detailed discussions see Ref. [64]. For the intermittent breakdown of current-oscillation tori, see Ref. [63].

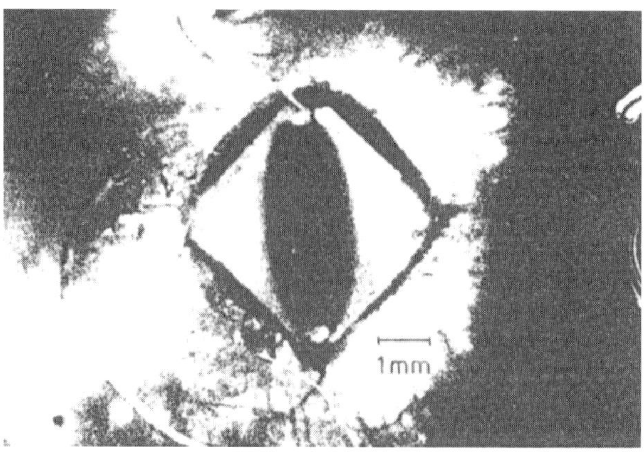

Figure 8.50 Photographic picture of dot-contacted sample A on IR-sensitive film (Kodak Infrared High Speed, f/5.6, exposure 10 s) at a current of 0.7 mA [45].

Among various measurement techniques [11, 15, 46–50, 60, 61] recently developed is a visualization method of current filaments by photoluminescence quenching [45]. This method is based on the quenching of photoluminescence inside current filaments and allows the direct photography of the filament structures, as shown by Eberle *et al.* [45] (see Fig. 8.50). The principle is quite simple: the presence of free electrons in current filaments reduces the intensity of impurity and excitonic luminescence due to impact ionization [69, 70]. It is especially effective for n-GaAs in which bound-exciton emission lines are predominant and electronic levels of free and bound excitons are close to donor ground states, giving rise to a sharp contrast of the photographic picture of photoluminescence quenching in the postbreakdown regime (Fig. 8.50). Photoluminescence was excited by the diffusive radiation from four red light-emitting diodes (emission peak at $\lambda \sim 630$ nm). Eberle *et al.* measured the filament structure in high-purity n-GaAs epitaxial layers with point ohmic contacts, using a linear photodiode array which consists of 1024 diodes, each 25 μm wide. Diode heights are restricted to 100 μm by a slit aperture and the horizontal array can be vertically translated by a stepper motor to record a full image of the sample. The photodiode signals are recorded with 12-bit resolution and digitally processed to give an image [45].

From their measurements several new findings are seen in Fig. 8.51. An inhomogeneous electron distribution is already present well below V_{th} Fig. 8.51a), which must be due to the electric field distribution between the point contacts. In the negative differential resistance (NDR) region the voltage across the sample oscillates with frequencies in the range of hundreds of kHz (gray rectangular region in Fig. 8.51). As the exposure time (1–4 s) of the recording is much longer than the oscillation period, the image is an average over many periods, giving the broad and diffuse distribution of quenched luminescence (Fig. 8.51b). For the broad image, Eberle *et al.* have strongly suggested lateral oscillations of the current flow by spatial oscillation mode-like breathing filaments. Moving up the current in the upper branch of the I–V characteristic, the lateral borders of the hot-electron distribution sharpen, which indicates current filamentation (Fig. 8.51c, d). These experimental findings are in good agreement with theoretical calculations done by Gaa *et al.* [42]. From the photoluminescence quenching measurements, images of multifilaments in a Corbino-disc sample have been also measured [45, 71]. For the instability in the Corbino-disc sample see Ref. [72].

8.1.6 Experimental aspects of pattern dynamics in bulk semiconductors – part II

Recent measurements in high-purity n-GaAs epitaxial layers, carried out by the author with technical assistance by students, are now presented. For the

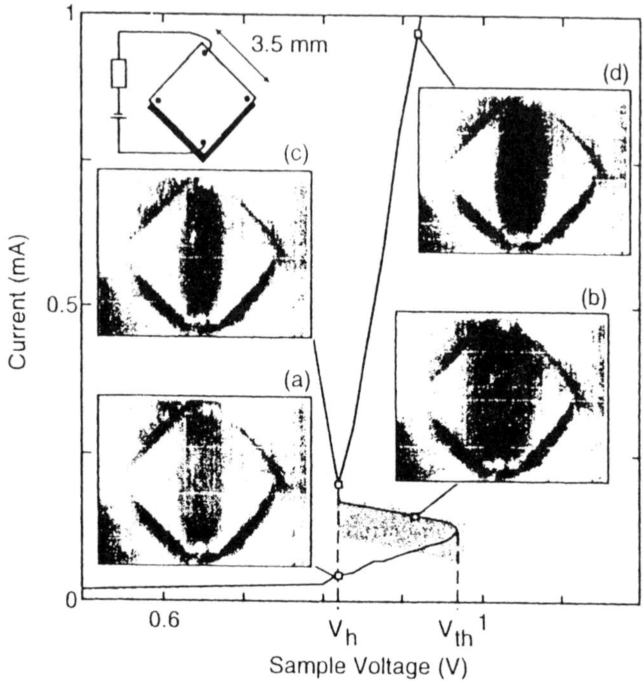

Figure 8.51 Spatially resolved photoluminescence from sample A at indicated points along the I–V curve. Dark areas in the images correspond to areas of hot electron quenched luminescence [45].

measurements by scanning laser microscopy and photoluminescence quenching, great appreciation is given to the original developments of the measurement techniques by Prettl's group, as mentioned in section 8.1.5 [15, 44, 45, 65–67, 71]. The main interests here include (i) the nucleation process of current filaments, (ii) the pattern dynamics of a periodically driven current filament, and (iii) the pattern dynamics under magnetic fields.

Two kinds of n-GaAs epitaxial layer are used: (i) $n = 2 \times 10^{14}$ cm^{-3}, $\mu_e \sim 8 \times 10^3$cm^2/V s at 300 K, thickness of the epitaxial layer $w \sim 12$ μm, and (ii) $N_D = 2.7 \times 10^{14}$ cm^{-3}, $n = 6.55 \times 10^{13}$ cm^{-3}, $\mu_e = 7.8 \times 10^3$ cm^2/V s at 300 K, $w = 20.8$ μm. For the latter sample the carrier concentration and the Hall mobility at 77 K are $n = 6.23 \times 10^{13}$ cm^{-3} and $\mu_e = 1.23 \times 10^5$ cm^2/V s. The compensation ratios are not accurate but at least 50% estimated from the μ_e vs n relation for both samples. In this section measurements are made for the latter sample, while the former sample was used in the previous investigations [9, 32].

Planar-type ohmic contacts and quasi-point contacts are formed by evaporating In on the sample surfaces and alloying it at 430°C in a vacuum

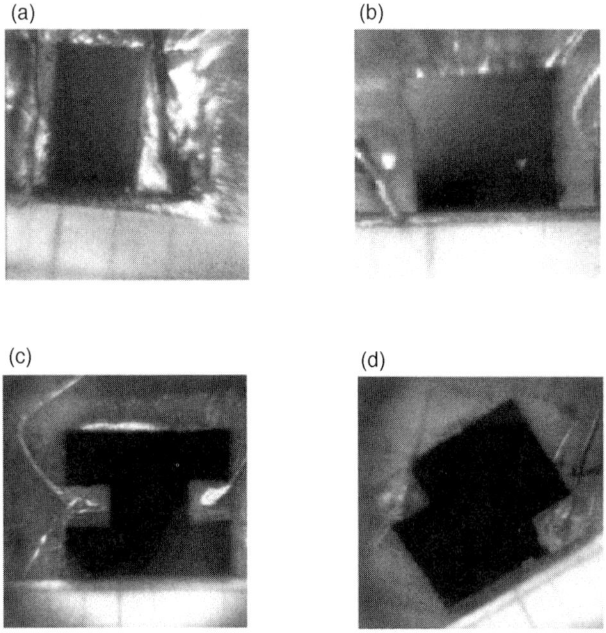

Figure 8.52 Photographs of sample geometries with planar-type ohmic contacts (a) #1213A, (b) #0205B, and squared quasi-point contacts (c) #0108C and (d) #0124D.

for 10 min, where two kinds of squared quasi-point contact are formed (Fig. 8.52c, d). For the samples with quasi-point contacts only a single current filament can be formed by the field organization because of the inhomogeneous field distribution as in a dipole-like field distribution [65]. The samples were glued, using silver paste, onto the holder of a rectangular plate (8×3.5 mm^2) which was attached at the end of a pipe of stainless steel (diameter 2.5 mm, length 120 cm). All measurements were taken at 4.2 K, by immersing the GaAs sample into liquid helium in a glass dewar. A dc bias or a dc + ac bias was applied to the sample in series with the load resistor ($R_L = 450\ \Omega$). Experimental arrangements for the scanning laser microscope and for measurements of photoluminescence quenching are shown schematically in Fig. 8.53. The measurement technique is simpler than that used by Brandl *et al.* [15, 67], without using a polygon mirror. The laser beam was linescanned by a scanning manipulator which can be driven by a computer-controlled 3D (xyz) stepping motor, where a 10 mW He–Ne laser is attached at a top stage of the manipulator. A spatial filter with a pinhole of 30 μm diameter is also attached to the laser head (Fig. 8.54). The scanning speed in the y-direction (see Fig. 8.53a) is adjustable, and normally

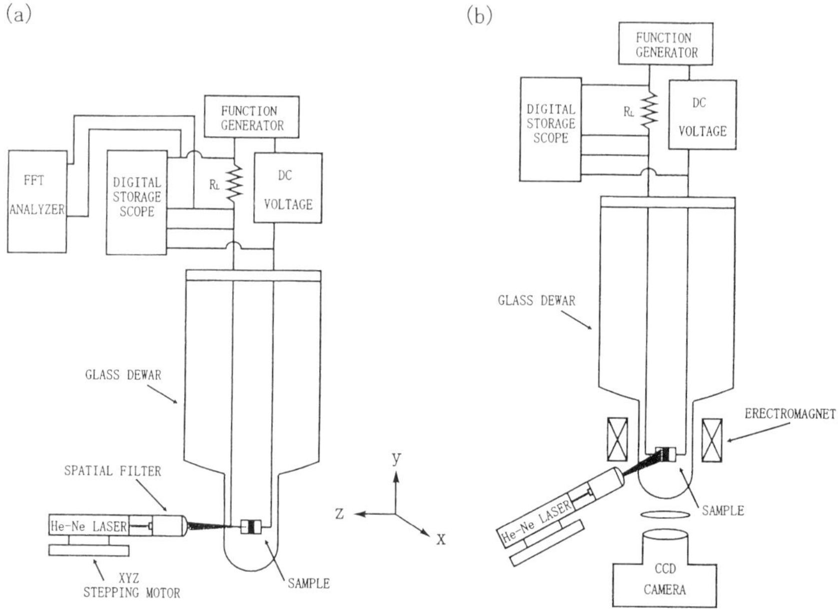

Figure 8.53 Experimental arrangements for scanning laser microscopy (left) and the measurements of photoluminescence quenching (right).

fixed at 40 s per linescan. The laser beam is attenuated using neutral density filters and focused on the sample surfaces with the minimum diameter of 30 μm. During each linescan the induced photosignals $\delta I(y)$ are monitored by a conventional fast fourier transform (FFT) analyzer and accumulated on a floppy disk (1024 points for each linescan). During the linescans the photosignals exhibit sharp spikes, especially at the filament borders, due to beam-induced local impurity breakdown. In order to reconstruct a smooth image pattern of the filament, successive $2m+1$ points (typically 11 points, $m=5$) of the signals are averaged by computing and $\langle \delta I(y) \rangle$ is plotted as a function of $y_i (i = m+1, m+2, \ldots, 1024-m)$ for each line scan in the y-direction, where

$$\langle \delta I(y) \rangle = \sum_{k=-m}^{m} \delta I(y_{1+k})/(2m+1).$$

Measurements of photoluminescence quenching can be taken by illuminating the defocused laser beam over the whole sample surface and detecting the photoluminescence pattern using a charge-coupled device (CCD) camera. The interference pattern of the laser beam is unavoidable, but there are no serious problems with analysis of the image pattern. A long-pass filter is used at the entrance of the CCD camera to block any visible and

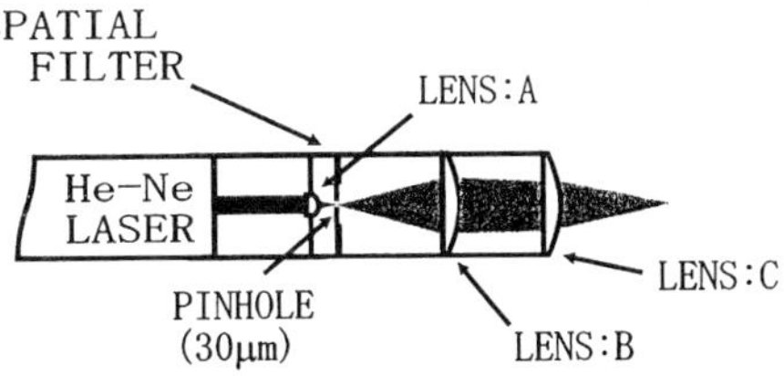

Figure 8.54 Construction of a spatial filter.

laser excitation lights. The exposure time of the CCD camera is adjusted and is typically 3–10 s; the exposure time is kept constant during a series of measurements. The photoluminescence quench patterns were also measured under transverse magnetic fields, using a standard electromagnet. With the electromagnet (Fig. 8.53) the transverse fields were applied by rotating the sample holder at $\theta \sim 45°(B_\perp = B \sin\theta)$. As a result, the image patterns are fairly contracted (see Fig. 8.63). I–V_0 characteristics under dark conditions are shown in Fig. 8.55 for the samples with different contact geometries in Fig. 8.52, where V_0 is the applied dc bias. Dimensions of the sample surfaces, contact spacing and breakdown fields E_b under dark conditions, and E_b^* under photoexcitation by the defocused laser beam are given in Table 8.6. By illuminating the defocused laser beam over the whole sample surface, the hysteresis regions of the I–V_0 characteristics become narrow by a considerable decrease in the breakdown voltages (Fig. 8.56). In Fig. 8.56 photographs of the photoluminescence quench patterns are shown at various applied voltages. For a sample with the quasi-point contacts (Fig. 8.52d) details of the pattern images are shown in Fig. 8.57 as a function of the applied voltage. In the prebreakdown regime (Fig. 8.57a), an inhomogeneous electron distribution already exists at two opposite sides of the squared point contacts, in agreement with the observations by Eberle *et al.* [45, 71]. The filament width increases sublinearly by increasing the applied voltage (Fig. 8.57b–f), the result of which is plotted by solid circles in Fig. 8.58, where the widths are measured at the center between the

Table 8.6 Dimensions of the samples and the breakdown fields in n-GaAs

Sample No.	#1213A	#0205B	#0108C	#0124D
Surface area (mm^2)	5.64 × 2.49	5.89 × 2.44	4.96 × 4.23	4.55 × 4.09
ℓ (mm)	1.55	2.52	2.38	2.84
E_b (V/cm)	5.16	4.68	5.05	6.16
E_b^* (V/cm)	4.52	2.77	3.15	3.70

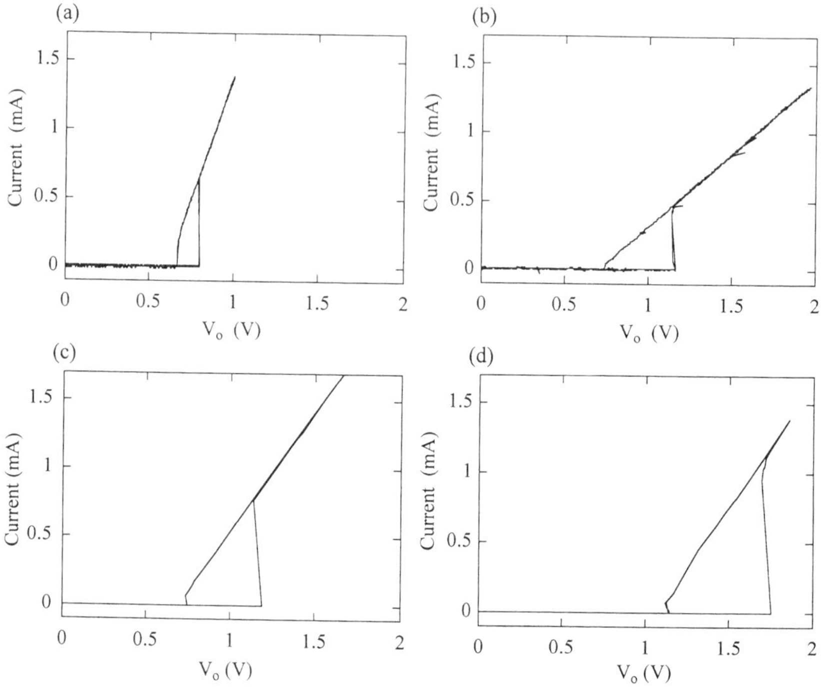

Figure 8.55 Current–voltage (I–V_0) characteristics observed under the dark conditions for the samples in Fig. 8.52. $T_L = 4.2$ K.

contacts. The sublinear increase in the filament width for $V_0 > 1.7$ V is in agreement with the experimental and numerical results obtained for the circular point contacts with a dipolar field distribution [65]. The widths measured by the scanning laser microscope (crosses) are slightly smaller than those by the photoluminescence quench patterns. This may be due to the photoexcitation density used for scanning laser microscopy being much lower than that used in the photoluminescence quenching measurements. For a sample with planar ohmic contacts the filament width increases linearly with V_0 [73], while it tends to saturate at higher voltages in Fig. 8.58.

Using these samples in Table 8.6 the nucleation process, dynamical behaviors of a periodically driven current filament and behaviors under magnetic fields are described below.

8.1.6.1 Nucleation process

The nucleation process [35–39, 42] of a current filament can be measured by photoluminescence quenching. For the measurements a pulse voltage $V_p = 0.855$ V with width T_p and a repetition frequency of 179 kHz (Fig. 8.59) was

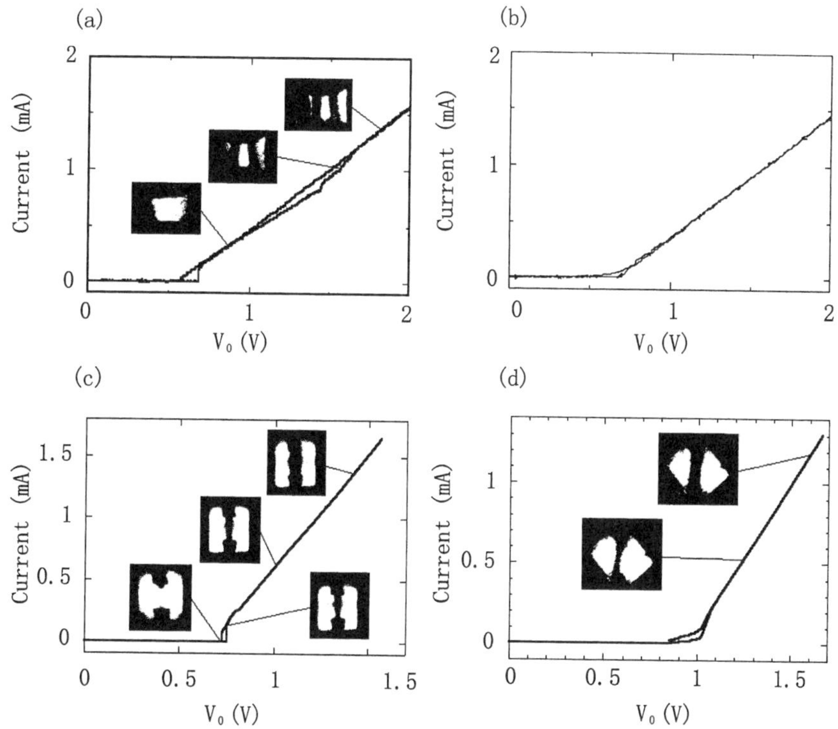

Figure 8.56 Current–voltage (I–V_0) characteristics measured by illuminating defocused laser light over the whole sample surfaces, where $T_L = 4.2$ K. A 10 mW He–Ne laser light, attenuated by a neutral density filter, was used as the light source. Photographs of the photoluminescence quench patterns are shown at various applied voltages V_0.

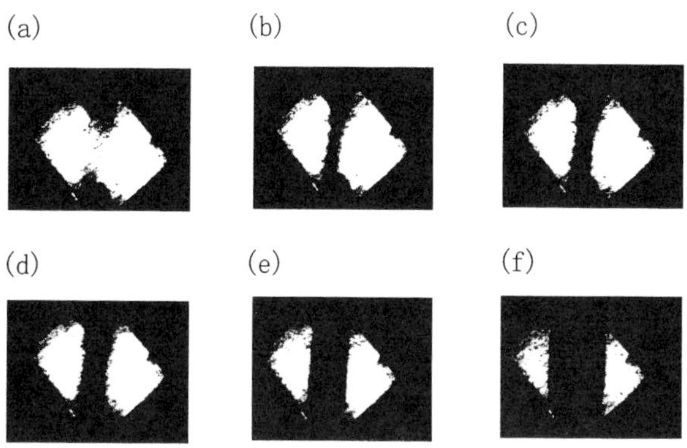

Figure 8.57 Quench patterns of the photoluminescence for sample #0124D at various applied voltages V_0: (a) 0.9 V, (b) 1.02 V, (c) 1.20 V, (d) 1.6 V, (e) 2.0 V, and (f) 2.7 V.

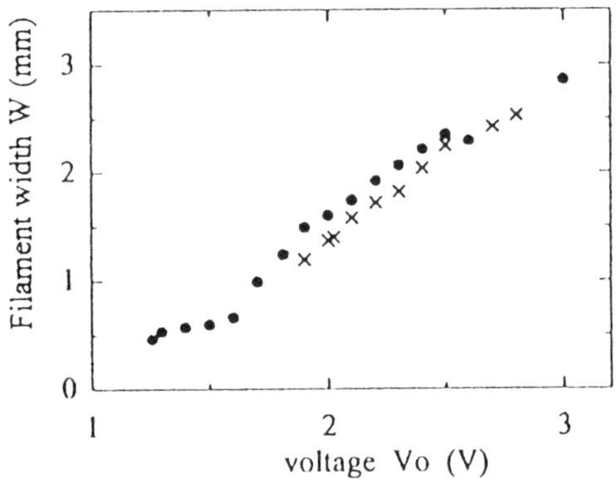

Figure 8.58 Filament widths measured as a function of applied voltage V_0 for the sample #0124D. Solid circles denote the widths measured by the quench patterns of photoluminescence, and crosses by scanning laser spectroscopy. $T_L = 4.2$ K.

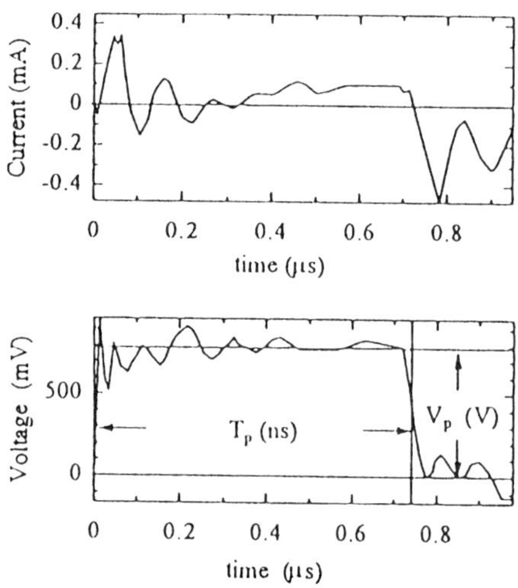

Figure 8.59 Bottom: applied pulse voltage with the repetition frequency of 179 kHz, pulse width T_p and pulse height $V_p = 0.855$ V. Top: the induced current through the sample #0108C. $T_L = 4.2$ K.

applied for the sample #0108C, where the current response is shown at the top of Fig. 8.59. In Fig. 8.60 the quenched patterns are shown for various pulse widths, where the photographs are produced by the color-imaging process. The nucleation of the filament is found to start at both the cathode (top contact in Fig. 8.60) and anode (bottom one) with $T_p > 50.5$ ns. The nucleation at the cathode is apparent because the electrons injected from the cathode have sufficient energy to impact-ionize the shallow donors, with the local field there being larger than the breakdown value. The carrier density

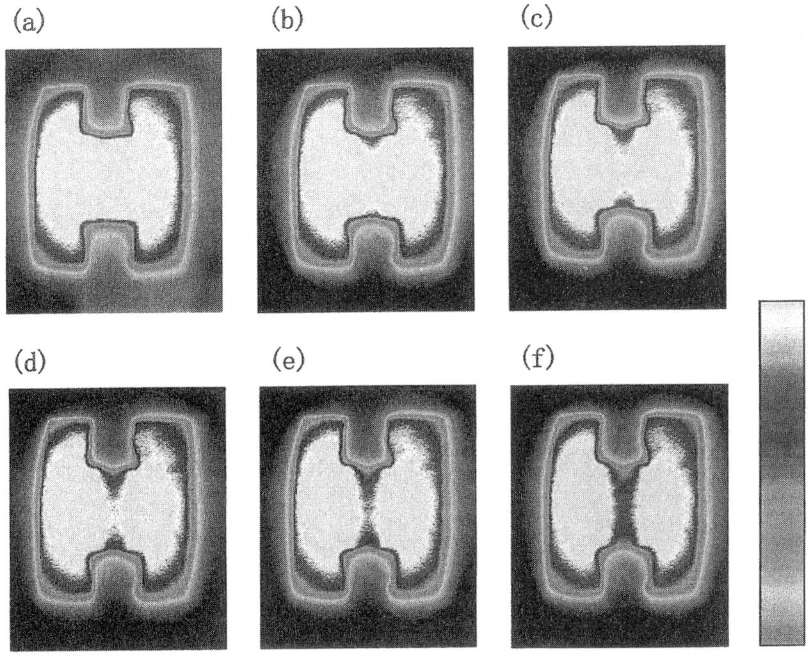

Figure 8.60 Quenched patterns of photoluminescence for various pulse widths T_p. The photographs were produced by the color-imaging process. $T_L = 4.2$ K. (a) $T_p = 5.5$ ns, (b) $T_p = 197$ ns, (c) $T_p = 337$ ns, (d) $T_p = 390$ ns, (e) $T_p = 434$ ns, and (f) $T_p = 667$ ns [213].

in the vicinity of the noninjecting anode is also increased by weak impact ionization, since a relatively large fraction of the applied voltage drops at the junction between the n-type anode and the insulating bulk in a manner similar to a reversely biased p–n junction, leading to a higher electric field here [35]. Fronts created at both contacts propagate (Fig. 8.60b, c) and reach each other at a pulse width of $T_p = 434$ ns in Fig. 8.60e, followed by a rudimentary filament [35, 39]. Kunihiro *et al.* explained the front propagation from the cathode by a local decrease in the electric field inside the nucleated region [35]; once the nucleation starts, the local field decreases as a result of the dielectric relaxation inside the nucleated region, forming a high-field domain at the front. Hence, the electric field ahead of the carrier front becomes higher than behind it, encouraging the impact ionization downstream. No impact ionization upstream occurs since the density of the impact-ionized carriers is very high inside the nucleated region, keeping the impact-ionization avalanche towards the outer region. A stage of stagnation [39] may exist, but has not been confirmed experimentally. Finally, a mature filament can be seen with $T_p \gtrsim 667$ ns (Fig. 8.60f).

Front velocity $v_{\rm front}$ and front mobility $\mu_{\rm front}$ can be estimated by measuring each location of the fronts attained by the pulse widths $T_{\rm p1}$ and $T_{\rm p2}(T_{\rm p1} > T_{\rm p2})$, where the location was determined as the distance between the anode edge and the center of the green-colored region of the front (Fig. 8.60). From the two image patterns, the difference $\delta\ell$ of the front locations propagated from the anode is measured and divided by the time difference $\Delta T = T_{\rm p1} - T_{\rm p2}$, yielding $v_{\rm front} = \delta\ell/\Delta T$. Since the electric field E is known for an applied voltage, then $\mu_{\rm front} = v_{\rm front}/E$. By this procedure $v_{\rm front}$ and $\mu_{\rm front}$ are plotted as a function of the time defined by $t = (T_{\rm p1} + T_{\rm p2})/2$ in Fig. 8.61 These values are found to be constant at the initial stage with $t \lesssim 280$ ns, followed by a sharp rise with $t \gtrsim 370$ ns. The attainable maximum values in Fig. 8.61 are $v_{\rm front} = 1.27 \times 10^6$ cm/s and $\mu_{\rm front} = 3.54 \times 10^5 \rm cm^2/Vs$, respectively.

Several differences are found between the experiments and the numerical investigations [35, 39]. In the numerical simulations [39] a mature filament can be formed within 2 ns, while it is observed only after several hundred ns in the experiments (667 ns in Fig. 8.60f). For the numerical results by Gaa *et al.* [39], $v_{\rm front} \sim 1.0 \times 10^7$ cm/s and $\mu_{\rm front} \sim 4 \times 10^5 \rm cm^2/$ Vs can be estimated. The estimated value of $v_{\rm front}$ is close to the saturated velocity at high fields, which is not attainable by low-field impact ionization. Such an overestimate in the model simulation is probably caused by using a fast dielectric relaxation time, and hence a much shorter time has been predicted for the formation of the mature filament [39].

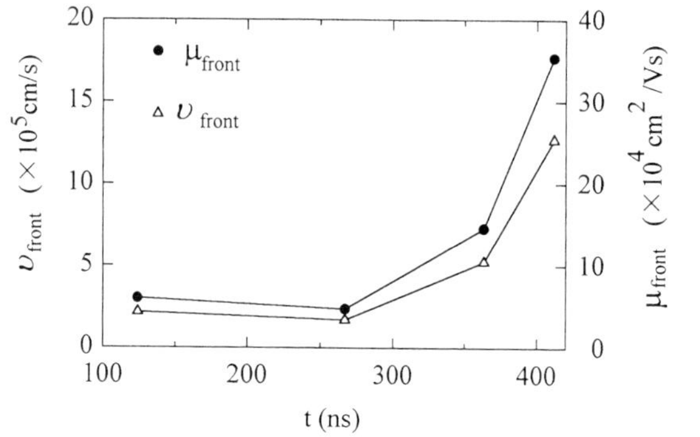

Figure 8.61 Front velocity and front mobility as a function of the time t defined by $t = (T_{\rm p1} + T_{\rm p2})/2$. The front velocity was obtained by measuring the difference of the front locations between the two adjacent photographs in Fig. 8.60, taken with the pulse widths $T_{\rm p1}$ and $T_{\rm p2}$ $(T_{\rm p1} > T_{\rm p2})$ and dividing by the time difference; $v_{\rm front} = \delta\ell/\Delta T$, where $\Delta T = T_{\rm p1} - T_{\rm p2}$. The front mobility was defined as $\mu_{\rm front} = v_{\rm front}/E$. Redrawn from [213].

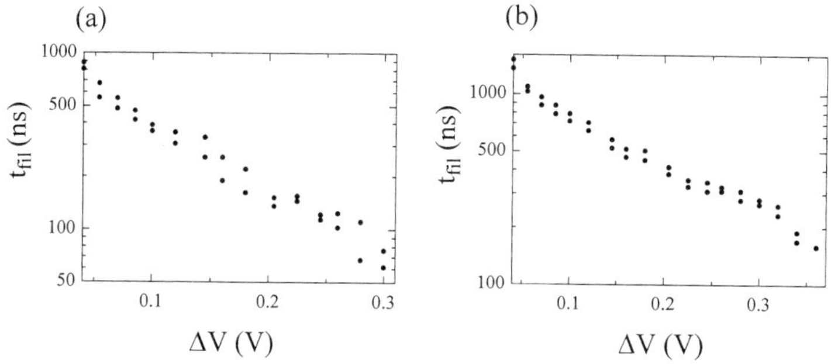

Figure 8.62 (a) Pulse width t_{fil} as a function of V under zero magnetic field ($B = 0$), where t_{fil} is the time necessary for the formation of a rudimentary filament and $\Delta V = V_0 - V_b$, and (b) t_{fil} vs ΔV measured under the transverse magnetic field, $B = 0.14$ T. Redrawn from [213].

The pulse width $t_{fil}(= T_p)$ necessary for the rudimentary filament (e.g., Fig. 8.60e) to be formed decreases with increasing applied pulse voltage V_0. In Fig. 8.62a t_{fil} are plotted as a function of ΔV on a semilogarithmic scale, where t_{fil} was determined from the colored-image pattern as a time at which the green-colored regions of the fronts propagating from both contacts begin to meet, and $\Delta V = V_0 - V_b$ (the breakdown voltage $V_b = 0.75$ V, see Fig. 8.56c). Under the transverse magnetic field ($B = 0.14$ T) the time t_{fil} becomes longer, as shown in Fig. 8.62b. This experimental fact disagrees with the numerical results by Kunihiro *et al.* [35] (see Fig. 8.18) that the application of the magnetic field accelerates the impact ionization. Their suggestion is that a finite lateral Hall field E_z is generated inside the filament and hence the local electric field should be always higher for $B \neq 0$ than for $B = 0$, because $E(B) = \sqrt{E_x^2 + E_z^2} > E_x$, where the longitudinal field E_x is expected to be essentially the same with and without a magnetic field [35]. While their interpretation is quite true, the experimental result in Fig. 8.62b strongly indicates that the magnetoresistance effect [74], rather than the local electric field, may dominate the nucleation process of the filament, where the magnetoresistance (Hall mobility) may increase (decrease) by applying the transverse magnetic field.

The image patterns under the transverse magnetic field are shown for various pulse widths in Fig. 8.63, where the pulse voltage is 0.91 V and $B = 0.14$ T. The nucleation at the anode takes place at a local position of the contact edge shifted from its center due to the Lorentz force, and is suppressed remarkably, as shown in Fig. 8.63b, c. The time t_{fil} necessary for the rudimentary filament (~ 412 ns, Fig. 8.63d) is approximately twice as long as that for a zero magnetic field ($t_{fil} \sim 210$ ns) at $V_p = 0.91$ V ($\Delta V = 0.16$ V). Finally, the rudimentary filament grows into the mature filament (Fig. 8.63e) after 725 ns.

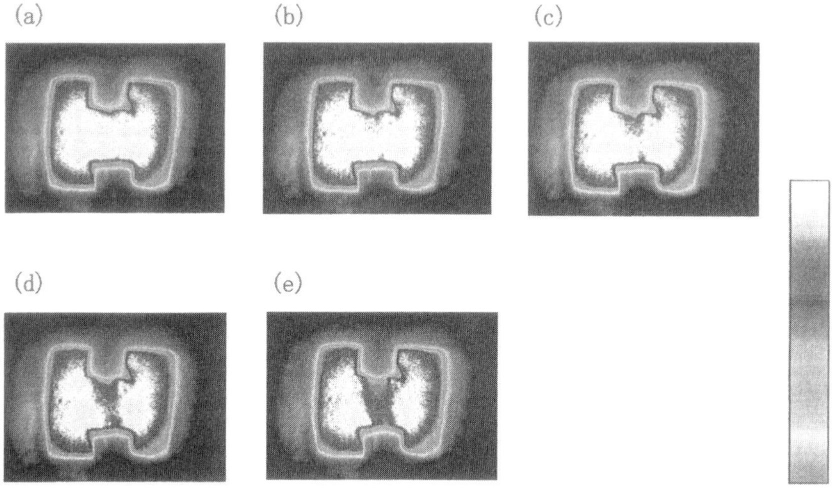

Figure 8.63 Quenched patterns of photoluminescence for various pulse widths under magnetic fields, $B = 0.14$ T. (a) $T_p = 348$ ns, (b) $T_p = 378$ ns, (c) $T_p = 412$ ns, (d) $T_p = 481$ ns, and $(e) T_p = 725$ ns. [213]

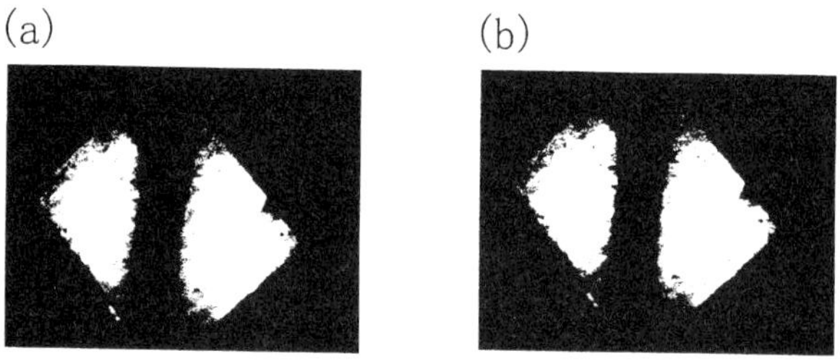

Figure 8.64 Quenched patterns of (a) the photoluminescence under a dc bias with $V_0 = 1.71$ V (a), and (b) the dc + ac bias with $V_0 = 1.71$ V, $V_{ac} = 50.1$ mV and $f_0 = 1.9$ MHz.

8.1.6.2 Pattern dynamics of a periodically driven current filament

The pattern dynamics of the filament in the epitaxial layers of bulk semiconductors have not been well known, and are far from being fully understood. This is due to the difficulty in taking snapshots of spatio-temporal patterns for the oscillatory filament in the MHz range, and also because the measurement techniques are invasive [65]. In order to understand the turbulent pattern, more sophisticated measurement techniques are required in the future. Here, the author's recent measurements are presented of the periodically driven current filament by means of photoluminescence quenching and scanning laser microscopy.

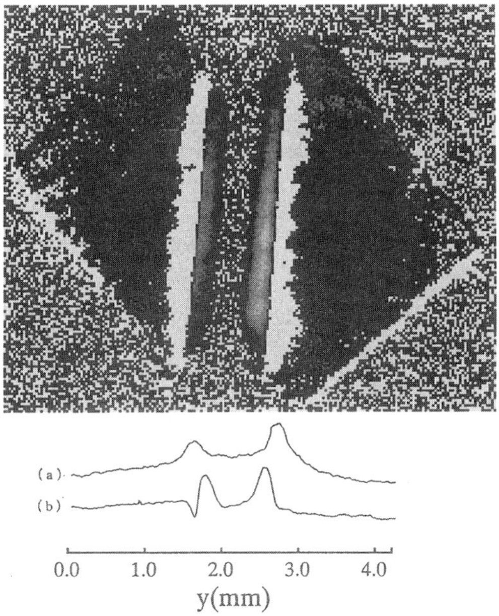

Figure 8.65 Top: a contrast pattern by color-imaging, obtained by subtracting Fig. 8.64b from Fig. 8.64a. Bottom: transversal spatial pattern of the photosignals observed by scanning laser microscopy, where the linescan has been done from left to right at the center of the sample surface between the contacts, where traces (a) and (b) are the photosignals measured by applying the dc bias with $V_0 = 1.71$ V and the dc + ac bias with $V_0 = 1.71$ V, $V_{ac} = 50.1$ mV and $f_0 = 1.9$ MHz, respectively. Redrawn from [213].

By applying a dc + ac bias in the form of $V_0 + V_{ac} \sin(2\pi f_0 t)$, one can typically observe period doubling bifurcation when V_0 (or V_{ac}) is decreased (or increased) as a control parameter in the postbreakdown regime [9, 29]. For the stable oscillation of period 1 in the postbreakdown regime, the filament may exhibit an enforced breathing mode [73]. Figure 8.64b shows the time-averaged quenching pattern of the photoluminescence for sample #0124D under the dc + ac bias, where $V_0 = 1.71$ V, $V_{ac} = 50.1$ mV and $f_0 = 1.9$ MHz. For reference, Fig. 8.64a is the image pattern under the dc bias. No remarkable differences can be seen with and without the ac bias in Fig. 8.64. However, by subtracting the brightness of the modulated pattern (Fig. 8.64b) from that of the image pattern under the dc bias (Fig. 8.64a), a clear contrast of the two patterns by color-imaging can be resolved along the filament boundaries, as shown in Fig. 8.65. Along each of the filament boundaries two stripes appear, where the inner ones, colored purple, have relatively thin widths. The symmetric structure of the stripes around the center of the filament strongly suggests the enforced breathing mode of the filament [73]. In the bottom part of Fig. 8.65 the photosignals observed by

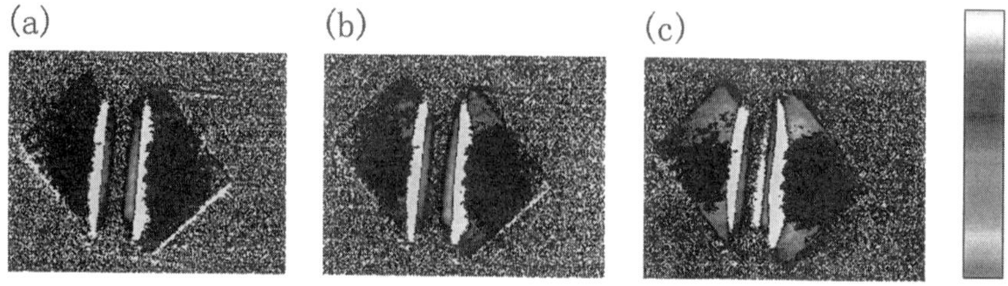

Figure 8.66 Contrast patterns by color-imaging, obtained by subtracting the modulated quenched pattern of photoluminescence under a dc + ac bias with: (a) $V_0 = 1.71$ V, $V_{ac} = 50.1$ mV, $f_0 = 1.9$ MHz, (b) $V_0 = 1.71$V, $V_{ac} = 70$ mV, $f_0 = 1.9$ MHz, and (c) $V_0 = 1.71$ V, $V_{ac} = 108$ mV, $f_0 = 1.9$ Hz from the static pattern of Fig. 8.64a under the dc bias with $V_0 = 1.71$ V.

scanning laser microscopy are shown, where the laser scans were made from left to right in the center between the contacts, and the curves (a), (b) are the photosignals obtained by applying the dc bias and the dc + ac bias, respectively. It is well known that the two ridges of the photosignals under the dc bias lie on the filament boundaries [15, 44, 65, 67], which was confirmed exactly by comparing Fig. 8.64a with curve (a) of Fig. 8.65. The ridge structure becomes sharp and narrow in width by applying the dc + ac bias, as shown in curve (b) of Fig. 8.65, where the width corresponds to the distance between two inner purple stripes of the contrast pattern. Hence, from Fig. 8.65, it is found that the filament breathes periodically, taking its minimum width 0.7 mm and maximum width 1.5 mm. Previously, it has been shown that the photosignals exhibit a double-peak structure at each of the filament boundaries [73], indicating the dynamic range of the breathing filament. For the sample (#0124D) in Fig. 8.65, the peaks of the photosignals corresponding to the outer stripes are absent in curve (b); instead, sharp dips are seen just outside the inner peaks. Increasing the amplitude of the ac voltage, the dynamic range of the breathing filament increases slightly, as shown in Fig. 8.66, where (a) $V_{ac} = 50$ mV, (b) $V_{ac} = 70$ mV, and (c) $V_{ac} = 108$ mV. Especially for Fig. 8.66c, two inner stripes can be seen at the right boundary. Note also the regions of contrast colored purple around the squared contacts, which are attributed to the difference pattern of the weakly quenched photoluminescence by the impact ionization of bound excitons and impurities. The brightness of the contrast around the contact edges increases with increasing driving amplitude. These regions are, however, discriminated from the filamentary region.

The bifurcation diagram in Fig. 8.67, which was measured under dark conditions, shows an appearance of period doubling by increasing V_{ac}, where $V_0 = 1.71$ V, $f_0 = 1.9$ MHz and ΔI is the amplitude of the current, measured from the minimum to the maximum at each cycle of the current oscillations. Period 1 (P1) bifurcates to period 2 (P2) at $V_{ac} = 190$ mV. The

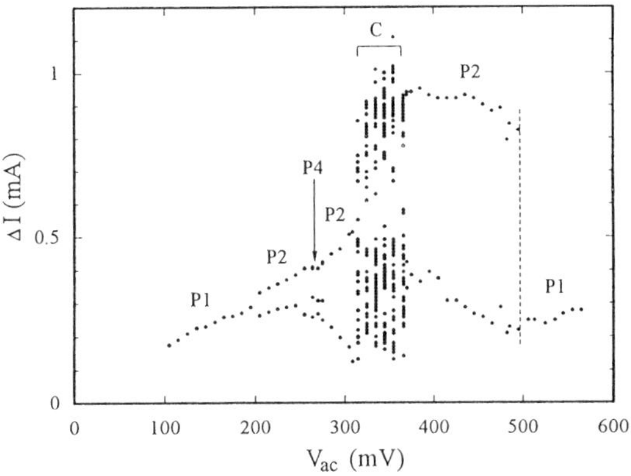

Figure 8.67 Bifurcation diagram measured under the dark condition, where V_{ac} is the control parameter. $V_0 = 1.71$ V, $f_0 = 1.9$ MHz. ΔI is the amplitude of the current, plotted by measuring from the minimum to maximum at each cycle of the current oscillations. P_j denotes period j ($j = 1, 2, 4$) and C implies chaos. Redrawn from [213].

period doubling bifurcation is not a usual one in the sense that period 4 (P4) appears at $V_{ac} = 264$–285 mV. Finally, chaos appears at $V_{ac} = 315$–366 mV. Corresponding to the bifurcation diagram, the currents and the phase portraits are shown in Fig. 8.68.

The transverse profiles of the photosignals were investigated with various driving amplitudes, where laser scanning was done horizontally along the center line between the contacts. In Fig. 8.69, (a)–(g) indicate the photosignals for period 1, (h) for the bifurcation point, (i) and (j) for period 2, (k) for period 4, (l)–(n) for period 2, (o) and (p) for chaos, (q) for period 2, and (r) for period 1. With increasing V_{ac}, in Fig. 8.69a–g, the separation between the two photocurrent peaks, which is identical to the separation between the inner stripes in Fig. 8.65, decreases gradually, indicating an increase in the breathing amplitudes. Simultaneously, a dip structure appears just outside each peak in Fig. 8.69a–g. During the laser scan (40 s/linescan), the laser spot perturbs the filamentary current, especially when it traverses across the boundaries and distorts the phase portraits in Fig. 8.68, and hence acts as a spatial control parameter. Distortions of the phase portraits and the simultaneous changes in the offset current by the laser spot cause the negative and positive photosignals in Fig. 8.69a–g when they are detected and averaged by the FFT analyzer. At the bifurcation point in Fig. 8.69h the profile of the photosignals becomes remarkable, showing sharp negative dips at the boundaries and even at the center of the filament. Such violence can always be observed at the bifurcation point [73], where the offset level of the phase portrait decreases at the negative dips. When the laser spot traverses the center of the filament,

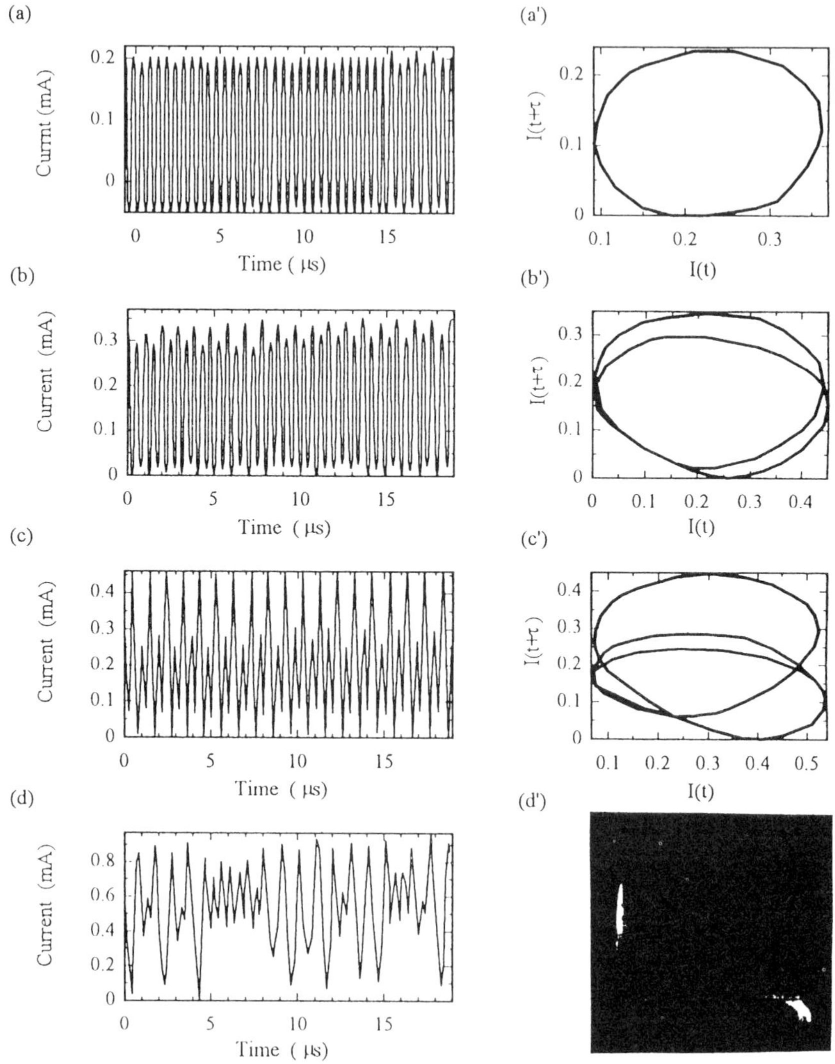

Figure 8.68 (a–d) Current-time characteristics and $(a' - d')$ the corresponding phase portraits under the dark condition, where $V_0 = 1.71$ V and $f_0 = 1.9$ MHz (a, a') $V_{ac} = 177$ mV, (b, b') $V_{ac} = 265$ mV, (c, c') $V_{ac} = 285$ mV, and (d, d') $V_{ac} = 333$ mV.

random switching between periods 1 and 2 occurs in the phase portraits, corresponding to Fig. 6.69h. For the transverse distributions Fig. 8.69ℓ–p, the structures exhibit more complicated situations, especially in the chaotic regime, yielding no definite physical picture. However, it is also noticed that a complex distribution (e.g., in Fig. 8.69ℓ) can be reproduced during many

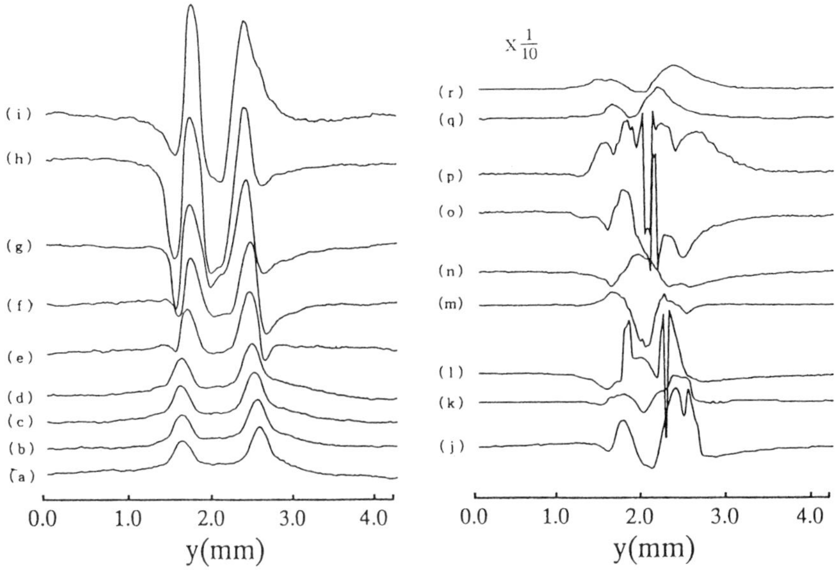

Figure 8.69 Transverse profiles of the photosignals measured by scanning laser microscopy under the dc + ac bias with various driving amplitudes. The transversal linescan was done at the center between the contacts. $V_0 = 1.7$ V, $f_0 = 1.9$ MHz. The driving amplitudes V_{ac} are (a) 105 mV, (b) 115 mV, (c) 125 mV, (d) 135 mV, (e) 165 mV, (f) 175 mV, (g) 185 mV, (h) 190 mV, (i) 195 mV, (j) 205 mV, (k) 255 mV, (l) 285 mV, (m) 295 mV, (n) 305 mV, (o) 315 mV, (p) 325 mV, (q) 425 mV, and (r) 520 mV.

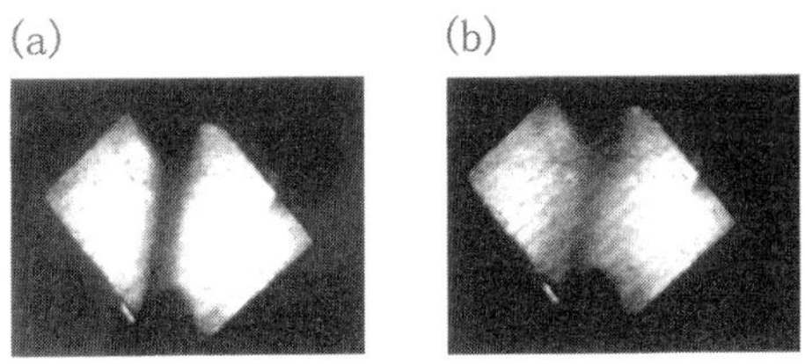

Figure 8.70 Quench patterns of photoluminescence (a) under a dc bias with $V_0 = 1.36$ V, and (b) under the dc + ac bias with $V_0 = 1.36$ V, $V_{ac} = 158$ mV, $f_0 =$ 1.9 MHz.

linescans for a given driven amplitude V_{ac}. Thus, by means of scanning laser microscopy, a physically meaningful picture can only be argued for the breathing filament of period 1 and its bifurcation point (Fig. 8.69a–h). From the measurements of photoluminescence quenching, it was found that a transition from the enforced breathing mode to the bulk-dominated oscillating mode [73] occurs after the first period doubling.

As an example of bulk-dominated oscillation, Fig. 8.70b shows the quench pattern of the photoluminescence observed with $V_0 = 1.36$ V, $V_{ac} =$

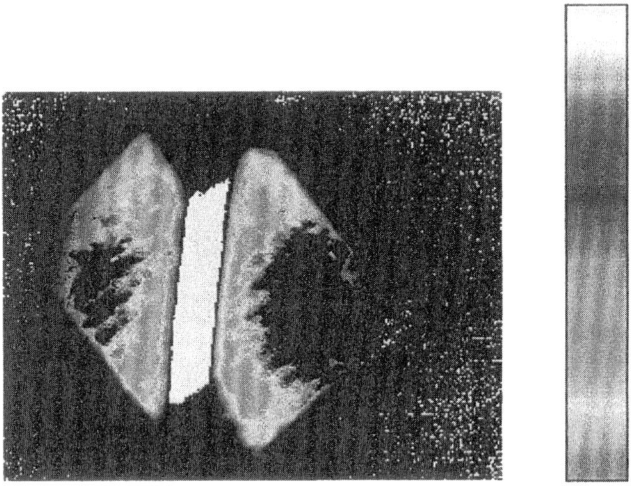

Figure 8.71 Contrast pattern by color-imaging, obtained by subtracting Fig. 8.70a from Fig. 8.70b.

158 mV and $f_0 = 1.9$ MHz, where the oscillating state is period 3. Apparently, the quench pattern in Fig. 8.70b is broad and vague, in contrast to the definite pattern of the filament under the dc bias (Fig. 8.70a). The filament channel is not definitely confirmed and the darkness is less prominent. From the contrast pattern in Fig. 8.71, which is obtained by subtracting the brightness of pattern (a) from (b), it is found that the band colored light blue corresponds to the filamentary channel under the dc bias (Fig. 8.70a), where the intensity of the subtracted contrast is negative in this band. Thus, the bulk of the filament is destabilized by the ac drive. This is the bulk-dominated oscillating mode [20, 73] of the filament. The bulk-dominated oscillating mode is related to the phase space trajectories (I, V) entering the unstable NDR region, while for the enforced breathing mode the trajectory of period 1 lies stably on the upper branch of the S-shaped I–V curve. The vague pattern in Fig. 8.70b is due to the time-average of the quenched photoluminescence pattern, and hence the average pattern of the repetitive cycle (period 3) of the ignition and extinction of the filament, but there is no temporal resolution by the CCD camera. Such a bulk-dominated oscillating mode can typically be seen after the first period doubling [73].

8.1.6.3 Bendable filaments and spontaneous oscillation under magnetic fields

As described previously, bendable filaments and the spontaneous instabilities induced under the crossed electric and magnetic fields have been extensively studied in n-GaAs epitaxial layers by Brandl *et al.* [16, 28], Kunz

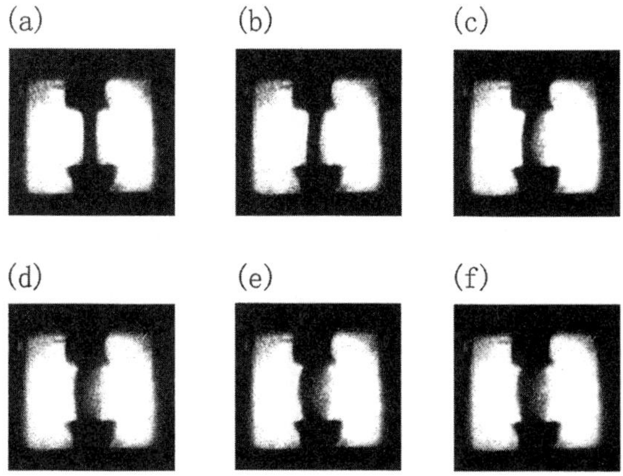

Figure 8.72 Image patterns of a current filament, by means of photoluminescence quenching, under various magnetic field strengths: (a) $B = 0$, (b) $B = 0.048$ T, (c) $B = 0.095$ T, (d) $B = 0.119$ T, (e) $B = 0.158$ T. $V_0 = 0.95$ V.

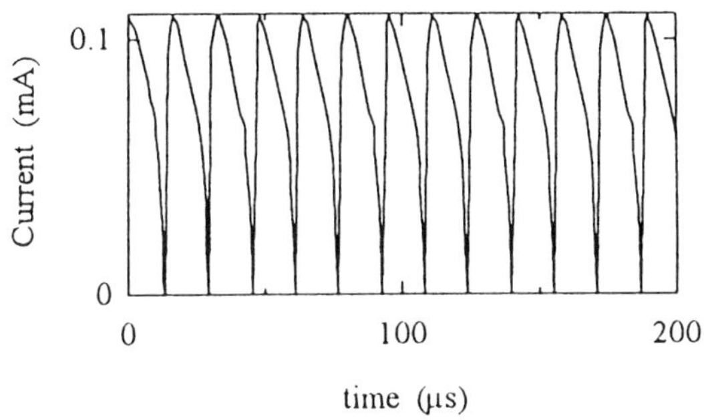

Figure 8.73 Spontaneous current oscillation observed with $B = 0.13$ T and $V_0 = 0.94$ V. The oscillation frequency is 64 kHz.

et al. [38], Kunihiro *et al.* [35] and Spangler *et al.* [44]. The results for the samples in Table 8.6 are now presented.

The current filaments are bendable by applying transverse magnetic fields which are perpendicular to the sample surface [44]. Figure 8.72 shows the pattern images of the filament under various magnetic field strengths, where the transverse magnetic fields were applied by rotating the sample holder by $\theta = 46.5°(B_\perp = |\mathbf{B}| \sin\theta,\ B_\parallel = |\mathbf{B}| \cos\theta)$, the transverse components are pointing normal to the sample surface of the photographs and the filamentary current flows from bottom (anode) to top (cathode) contacts. Note that, first, with the experimental arrangement in Fig. 8.53, the transverse components can be applied only by rotating the sample holder (one can also apply purely transverse magnetic fields with $\theta = 90°$, but no photoexcitation is possible in this case). Secondly, the photographs in Fig.

8.72 have been rotated clockwise by 90° from the original configuration of Fig. 8.52c, so that the currents flow horizontally in the actual arrangement of Fig. 8.53. Bending of the filament due to the Lorentz force becomes remarkable by increasing the magnetic field and is clearly seen in Fig. 8.72, where $V_0 = 0.94$ V, and the magnetic field strengths are (a) $B_\perp = 0$ T, (b) $B_\perp = 0.048$ T, (c) $B_\perp = 0.095$ T, (d) $B_\perp = 0.119$ T, and (e) $B_\perp = 0.158$ T. By inverting the polarity of the applied voltage, the filament bends in the opposite direction [44]. With $B_\perp > 0.09$ T, the filamentary current begins to show spatiotemporal spiking (~ 60 kHz). The spontaneous current oscillation exhibits a sharp rise and subsequent slow decay, as shown in Fig. 8.73, where $B_\perp = 0.13$ T and the fundamental oscillation frequency is 64 kHz. In Fig. 8.72c–f, light gray parts indicate the shadow of the oscillatory regions. The spontaneous spatiotemporal spiking extends from the positively charged right boundary towards the outer transverse region in Fig. 8.72c.

By taking the color imaging in Fig. 8.74a, b, the nonoscillating filamentary region was found to be colored red and the width increased slightly by increasing the magnetic field from 0.119 T (Fig. 8.74a) to 0.158 T (Fig. 8.74b), where patterns (a) and (b) in Fig. 8.74 correspond to Fig. 8.72d, f respectively. It is also found by the color imaging that the impact-ionization avalanche is most significant at both contacts, where the cusps of

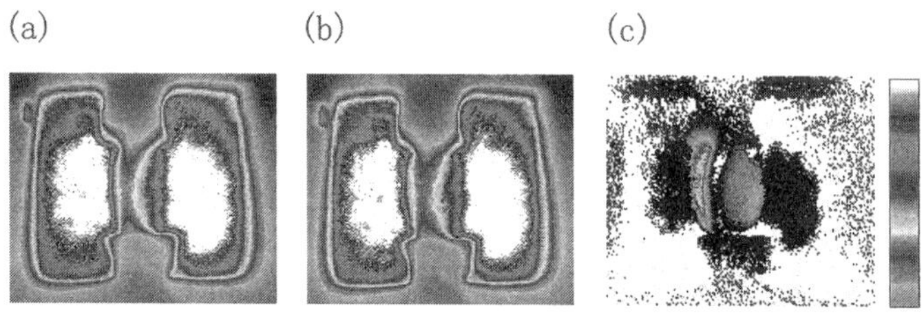

Figure 8.74 (a) Color-imaging of the current filament at $B = 0.119$ T for Fig. 8.73d, (b) color-imaging at $B = 0.158$ T for Fig. 8.74f, and (c) the contrast pattern by the subtraction $(a) - (b)$. $V_0 = 0.94$ V.

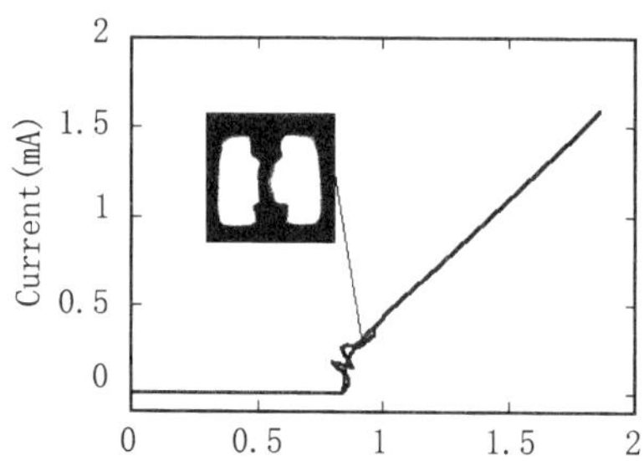

Figure 8.75 Current–voltage characteristic under the transverse magnetic field with $B = 0.13$ T.

dark blue that formed at the contacts indicate the lowest level of brightness. The shadows of the oscillatory region in Fig. 8.72d, f are shown more clearly by the isoclines with different colors at the right boundary in Fig. 8.74a, b. In Fig. 8.74c the contrast (difference) pattern is obtained by the subtraction (a) – (b). An island of the isocline on the left-hand side is the contrast of the left boundaries of the bent filaments in (a) and (b), and the broad area of contrast on the right (colored dark blue) indicates the oscillatory region. Between them, the dark region represents the center of the bent filament.

The spontaneous oscillations are not due to the repetition cycles of ignition and extinction of the filament, as expected from Figs. 8.72 and 8.74. Instead, the I–V characteristic in Fig. 8.75 indicates clearly the filamentary current never relaxes to the lower branch of the I–V curve during the oscillatory motions, where the current oscillation can be observed with the applied voltages in the range $0.85\ \mathrm{V} < V_0 < 1.0\ \mathrm{V}$. So far, two different types of fundamental current instabilities have been observed under transverse magnetic fields by Brandl *et al.* [15] and Spangler *et al.* [44, 62], as introduced in section 8.1.5. The instability observed by Brandl *et al.* is caused by photoexcitation during the laser scanning across the filament boundaries. In this case, the current instability is remarkable, especially when the focused laser beam excites the positively charged filament boundary [15] (see section 8.1.5, Fig. 8.40). Another type of spontaneous oscillation is due to the repetition cycle of ignition and extinction of the filament observed for the samples with point contacts [44, 62] (see section 8.1.5, Fig. 8.38). Under the small current bias, the current filament is bent by the Lorentz force. With increasing magnetic field strength the lateral displacement of the filament increases more and more until the driving electric field generated by the dipole-like field distribution drops below the critical field and the filament is torn [44]. This occurs in the center of the filament where the electric field is smallest [44]. Immediately after the extinction the voltage across the contacts increases again owing to the dielectric relaxation response, forming a new filament along the shortest distance between the contacts [44]. This process is repeated periodically, resulting in the regular oscillation. In the present case of Fig. 8.72 the spontaneous oscillations are similar to those observed by Brandl *et al.* [15]. However, the spontaneous oscillations observed by Brandl *et al.* [15] occur during the laser scanning around the transition region of the positively charged boundary, while the oscillatory region in Fig. 8.72 extends far beyond the transition region. Therefore, it is strongly suggested that the transversal front of the impact-ionization avalanche along the transition region of the filament propagates towards the outer insulating region and then relaxes to its original boundary of the bent filament due to the Lorentz force, yielding a repetitive process of asymmetric breathing mode under the transverse magnetic field.

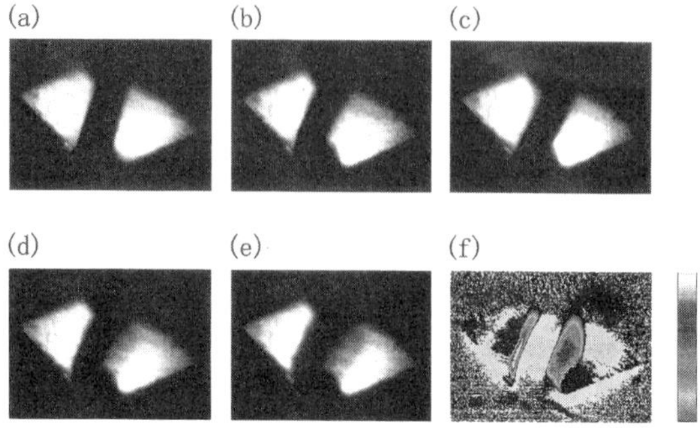

Figure 8.76 Image patterns of the current filament observed with $V_0 = 1.76$ V at various magnetic field strengths: (a) $B = 0$, (b) $B = 0.055$ T, (c) $B = -0.055$ T, (d) $B = 0.11$ T, and (e) $B = 0.165$ T. The top contact is the cathode for (a), (b), (d) and (f). Pattern (c) was obtained by inverting the polarity of the applied bias, and simply noted as $B = -0.05$ T. The contrast pattern (f) by color-imaging was obtained by the subtraction (b)–(e).

Similar behaviors have been confirmed for sample #0124D. By increasing the applied bias above the critical voltage the displacement of the filament due to the Lorentz force becomes less significant, without showing any spontaneous oscillations, owing to an increased internal tension of the fat filament [44]. This is typically true for sample #0108C (see Fig. 8.75). Depending on the contact geometry, spontaneous spatiotemporal spiking can be observed even with the applied bias far above the breakdown voltage, as found in sample #0124D. Figure 8.76 shows the image patterns observed with an applied voltage $V_0 = 1.76$ V at various magnetic field strengths, where (a) $B_\perp = 0$, (b) $B_\perp = 0.055$ T, (d) $B_\perp = 0.11$ T, and (e) $B_\perp = 0.165$ T; the top contact is the cathode for (a–e) but (c) was obtained by inverting the polarity of the applied voltage with $B_\perp = -0.055$ T (top contact: anode). The magnetic field $B_\perp$ is pointing downwards, perpendicular to the sample surface. Bending of the filament is not significant in Fig. 8.76b, but the difference can be seen clearly between Fig. 8.76a and b. By inverting the polarity (Fig. 8.76c) the bend is seen slightly in the opposite direction. With increasing magnetic field strength the filament shifts towards the left sides at each of the square contacts in Fig. 8.76d, e, and the bend is pronounced near the anode. The slide of the filament along the contact edges due to the Lorentz force was found to be more significant for the samples with planar contacts. In Fig. 8.76d, e light gray regions on the right side of the filaments indicate the areas of spatiotemporal spiking. The contrast pattern is given in Fig. 8.76f by subtracting (e) from (b). Again, an isocline island on the right is the region of spatiotemporal spiking. Between two isocline islands a shadow of the original filament in Fig. 8.76b is colored light blue. Such spatiotemporal spiking is found to be possible for this sample geometry even with a relatively large bias current ($V_0 = 1.76$ V), as follows. An insulated region along the right boundary of the filament is left behind by the displacement of the filament, at which the electric field is relatively high, and it is especially verified by the fact that owing to the displacement the squared contact at the anode begins to reveal its shadow in Fig. 8.76d, e. Thus, under photoexcitation, an impact-ionization avalanche may be facilitated along the transition region of the positively charged right boundary, and further the charge carrier front at the right boundary may propagate laterally. Finally, spontaneous oscillation may result, owing to the same mechanism of asymmetric breathing as described above for sample #0108c (Fig. 8.74).

The spatial distribution of the filament can also be investigated by scanning laser microscopy [15, 65], the results of which are shown in Fig. 8.77a, b, corresponding to the quench patterns in Fig. 8.76b, c, respectively,

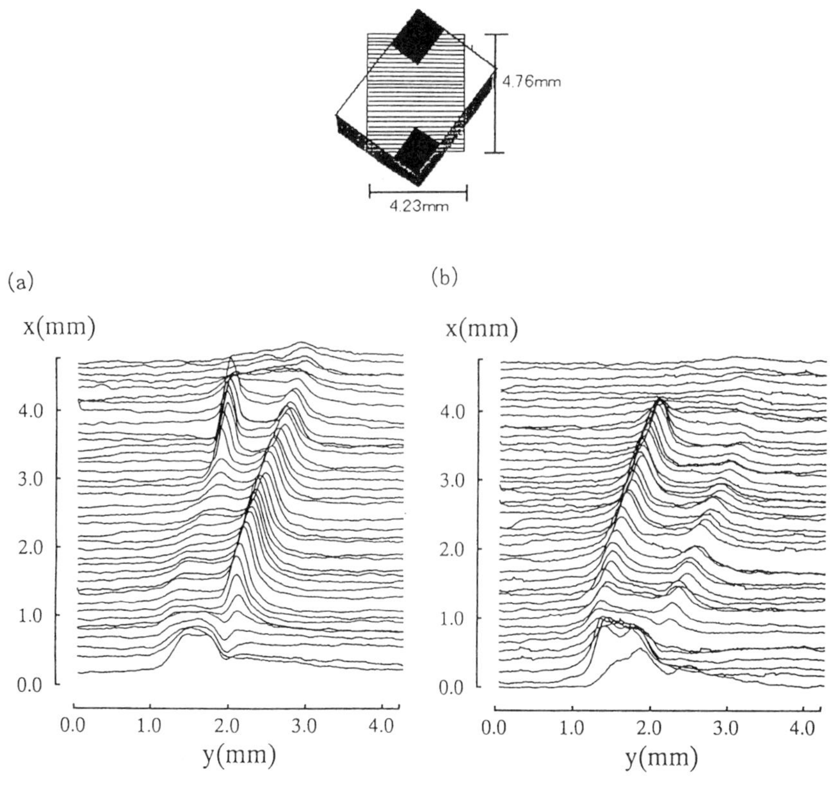

Figure 8.77 Spatial pattern images of the current filament by scanning laser microscopy, measured with (a) $B = -0.055$ T, and (b) $B = -0.055$ T, corresponding to Fig. 8.76b, c. Pattern (b) was obtained by inverting the polarity of the applied bias, and simply noted as $B = -0.05$ T. Top: sample geometry of #0124D; the area of the linescans is hatched.

where $B_\perp = 0.055$ T, the directions of the magnetic field and the current are the same as described in Fig. 8.76b, c, and the area of the linescans is indicated by hatching on the schematic sample geometry. Here, the results simply confirm the previously reported experimental results of Spangler *et al.* [44], showing the effectiveness of the measurement technique. At the positively charged boundary the photosignals are enhanced, showing a significant peak, while they are depressed at the negatively charged boundary [15]. The widths of the transitions region are approximately 300–400 μm from Fig. 8.77. By taking color-imaging of the quench pattern in Fig. 8.76b corresponding to Fig. 8.77a, the widths are found to be 200 μm at the negatively charged left boundary and have slightly larger values of 270–290 μm at the positively charged right boundary, in agreement with the proposal by Brandl *et al.* [15] (see Fig. 8.40).

For the samples with planar-type contacts, multifilaments can be formed with a relatively high voltage (Fig. 8.56a). Under the transverse magnetic field the filaments exhibit a considerable displacement as a result of remarkable sliding along one of the contact edges, as well as showing the bending behavior. Finally, the oscillatory mechanism of the current filament under transverse magnetic fields has been clarified by recent measurements [75, 76].

8.2 Towards a second new paradigm

As described in Chapter 5, universal behaviors of low-dimensional chaos have been strongly confirmed and evidenced in the nonlinear carrier transport arising from the SNDC (in the case of current filamentation) and the NNDC (in the case of high-field domain) in semiconductors. In the SNDC type the temporal behaviors have been shown to be related to the breakdown of perfect delay in the cusp catastrophe (Chapter 4); at the turning points of the S-shaped characteristic the critical slowing down is an important factor in the response of the phase space trajectory to reveal low-dimensional chaos. Under a periodically driven condition and even for the spontaneous oscillatory regime, the impact-ionization avalanche does not proceed instantaneously because of the nature of the critical slowing down when the electric field rises above the critical point. As a result, there is interplay between the upper and lower branches of the S-shaped characteristic. Roughly speaking, firing and misfiring occur in an erratic but deterministic manner, yielding low-dimensional chaos of spatially extended current filament. However, the relation between low-dimensional chaos and pattern dynamics is generally unclear. Low-dimensional chaos is only part of the pattern dynamics. Towards a second new paradigm, future objectives will be to clarify the pattern dynamics which exhibit low-dimensional chaos in semiconductor systems and also to study more general features of the pattern dynamics, as stated below.

The strategy of chaos theory was to show the universality underlying many dynamical systems by using truncated models (such as the logistic map) neglecting the spatial degrees of freedom. It is also true, however, that the pattern dynamics does not necessarily reveal such a universal scenario of low-dimensional chaos. Rather, the low-dimensional chaos theory is far from a panacea to explain all of the complex dynamical behaviors. Many degrees of freedom can also make for complicated motions that are effectively random (e.g., molecular chaos) [77]. A typical example of low-dimensional chaos followed by well-defined pattern dynamics is the Rayleigh–Bénard convection, showing a high quality of spatiotemporal behaviors via the RTN scenario [78]. In the BZ reaction–diffusion system [79, 80], low-dimensional dynamical behaviors have normally been investigated by stirring the reactant tank to prevent autocatalytic pattern formation. The pattern dynamics of the Zhabotinsky system is far from being understood purely in terms of the low-dimensional chaos theory. Examples showing spatiotemporal patterns during period doubling seem to be few. In that sense, the semiconductor chaos system may provide the model case for studying the pattern dynamics.

Since the impact-ionization process is autocatalytic, followed by the carrier diffusion, the semiconductor system has an analogy with the reaction–diffusion systems in chemistry [8]. An approach towards finding common features among different systems has been taken at the Workshop "Self-Organization in Activator-Inhibitor Systems: Semiconductors, Gas-Discharge and Chemical Active Media" [8]. Schöll has made a general remark on the spatiotemporal pattern formation in semiconductors, relating to the activator–inhibitor system as modeled by the nonlinear reaction-diffusion equations [8]:

$$\tau_a \frac{\partial a(x,t)}{\partial t} = f(a,u) + \ell^2 \frac{\partial^2 a}{\partial x^2} \tag{8.47}$$

$$\tau_u \frac{\partial u(x,t)}{\partial t} = g(a,u) + L^2 \frac{\partial^2 u}{\partial x^2}, \tag{8.48}$$

where a and u are the local activator and the long-range inhibitor, respectively, τ_a, τ_u are the characteristic time constants, ℓ, L are the characteristic diffusion lengths, and f and g denote appropriate (generally nonlinear) functions which describe the homogeneous autocatalytic model. In semiconductor transport instability the characteristic nonlinear function is the current density j vs electric field E (or carrier density vs electric field). As shown in Chapter 5 and present chapter, this characteristic is non-monotonic and displays negative differential conductivity (SNDC or NNDC). The activator is generally an internal variable which triggers the instability via an autocatalytic process [8].

In semiconductor systems, the activator variable is provided by the carrier density n for the low-temperature impurity breakdown with the basic current continuity equation:

$$\dot{\rho} + \operatorname{div} \mathbf{j} = 0, \tag{8.49}$$

$$\mathbf{j} = e(n\mu\mathbf{E} + D\nabla n), \tag{8.50}$$

where $\rho = e(N_D - N_A - N_D^0 - n)$ for n-type semiconductors and $\mathbf{j}$ is the total current density which is composed of the drift current density $en\,\mu\mathbf{E}$ with $\mathbf{E} = (E_\parallel, E_\perp)$ and the diffusion current density $eD\,\nabla n$. (8.49) with (8.50) gives the rate equation of the space-dependent g-r process for the impact-ionized carriers (note: $\dot{\rho} = -e(\dot{N}_D^0 + \dot{n})$). For another case of the instability, the activator variable is provided by the interface charge density ρ_s in semiconductor heterostructures and low-dimensional structures (for the oscillatory mechanism via real space transfer in AlGaAs/GaAs heterostructure see section 8.3 [8, 81]. The inhibitor is generally associated with the electric field E according to the dielectric relaxation:

$$\epsilon\dot{\mathbf{E}} = \mathbf{J} - en\,\mu\mathbf{E}, \tag{8.51}$$

where the total current density $\mathbf{J}$ is composed of the displacement current density $\epsilon\dot{\mathbf{E}}$ and the drift current density with the carrier mobility μ, and ϵ is the permitivity. For the total current density $\mathbf{J}$ one may add the diffusion current $eD\nabla n$, so that $\mathbf{J} = \epsilon\dot{\mathbf{E}} + \mathbf{j}$. For the longitudinal field (8.51) holds, neglecting longitudinal diffusion current as in [35–37]. According to (8.50), the dielectric relaxation acts as an inhibitory process which reduces the internal space charge field [8]. The realistic form of the inhibitor (sample voltage V) includes the circuitry condition via a series load resistor R and a parallel external capacitor C_{ext}, and an applied bias voltage U_0 [8]:

$$C\dot{U} = (U_0 - U)/R - I(a, U), \tag{8.52}$$

where $C = \epsilon A/L + C_{\text{ext}}$ is the sum of the internal and external capacitance. For the planar-type ohmic contacts $E_\parallel = U/\ell_s$ (ℓ_s: spacing between the contacts), and then (8.52) is also the rate equation for the longitudinal electric field. For the inhomogeneous current densities such as the current filament, the total current $I(a, U)$ is obtained by integrating the conduction current density over the sample cross-section:

$$I = \int_A en(x, y, t)\mu(x, y, t)\,\mathrm{d}x\,\mathrm{d}y.$$

Thus, (8.52) represents a global coupling. It is noted that the basic rate equations for the low-temperature impurity breakdown (8.2, 8.3; [35–37]) are different from the typical reaction–diffusion equations (8.47), (8.48). However, as described in section 8.1.4, the current instability of a filament in the p^+–n^+–p–n^- diode [47, 50] has been characterized typically by the reaction–diffusion model.

It will be important to survey the pattern dynamics investigated in other fields, such as in chemical reaction systems [8], since the semiconductor chaos system traditionally has less theoretical background. A typical example will be given by the reaction scheme of the "Brusselator" [4]:

$$\begin{aligned} A &\longrightarrow X \\ X + B &\longrightarrow Y \\ 2X + Y &\longrightarrow 3X \\ X &\longrightarrow E \end{aligned} \tag{8.53}$$

where A and B are the initial reactants, X and Y are intermediate products, and E is the final product. For the BZ reaction X and Y correspond to $[Ce^{3+}]$ and $[Ce^{4+}]$, respectively. Putting the reaction rate as unity in each part of (8.53), the reaction–diffusion equations in one dimension are given by:

$$\dot{X} = A - (B+1)X + X^2 Y + D_X \frac{\partial^2 X}{\partial x^2}, \tag{8.54}$$

$$\dot{Y} = BX - X^2 Y + D_Y \frac{\partial^2 Y}{\partial x^2}, \tag{8.55}$$

where D_X and D_Y are the diffusion constants. From the null-clines $f(X, Y) = A - (B+1)X + X^2 Y = 0$ in (8.54) and $g(X, Y) = BX - X^2 Y = 0$ in (8.55), the homogeneous steady-state solution is $(X_0 = A, Y_0 = B/A)$. For the homogeneous model equations $\dot{X} = f(X, Y)$, $\dot{Y} = g(X, Y)$ the linear stability analysis can be done easily by putting $X = X_0 + \xi_x \exp(\lambda t)$ and $Y = Y_0 + \xi_Y \exp(\lambda t)$, yielding

$$\lambda = \frac{a + d \pm \sqrt{(a-d)^2 + 4bc}}{2}, \tag{8.56}$$

$$a = B - 1, \quad b = A^2, \quad c = -B, \quad d = -A^2. \tag{8.57}$$

Let us take B as a control parameter. For $B \geq 1 + A^2 \equiv B_H$, $a + d \geq 0$ is fulfilled. Furthermore, with $B \sim B_H, (a-d)^2 + 4bc \sim -4A^2 < 0$. Thus, if the control parameter B increases above B_H, the system begins to oscillate around the equilibrium point (X_0, Y_0), forming a limit cycle. That is, **Hopf bifurcation** occurs at $B = B_H$. The spatiotemporal behaviors can be analyzed in (8.54), (8.55) by putting $X = X_0 + \xi_x \exp(iqx + \lambda t)$ and $Y = Y_0 +$

$\xi_Y \exp(iqx + \lambda t)$. From the linear stability analysis the eigenvalue satisfies

$$\begin{vmatrix} a - D_X q^2 - \lambda & b \\ c & d - D_Y q^2 - \lambda \end{vmatrix} = 0, \tag{8.58}$$

then

$$\lambda^2 + \alpha(q)\lambda + \beta(q) = 0, \tag{8.59}$$

with

$$\begin{aligned} \alpha(q) &= 1 + A^2 - B + (D_X + D_Y)q^2, \\ \beta(q) &= A^2 + [A^2 D_X + (1 - B)D_Y]q^2 + D_X D_Y q^4. \end{aligned} \tag{8.60}$$

By increasing B as a control parameter two types of bifurcation are possible at the wave numbers $q = 0$ and $q = q_c$, respectively. In (8.60) $\alpha(q)$ vs q^2 is linear. At $q = 0, \alpha(0) = 1 + A^2 - B$. Then this result gives Hopf bifurcation with $B = B_H$ as in the homogeneous case. The bifurcation scenario is called the **Hopf scenario** [82]. From (8.59), $\lambda = +i\sqrt{\beta(0)} = \pm iA$, which means growth of the homogeneous oscillation mode with $q = 0$. However, in the

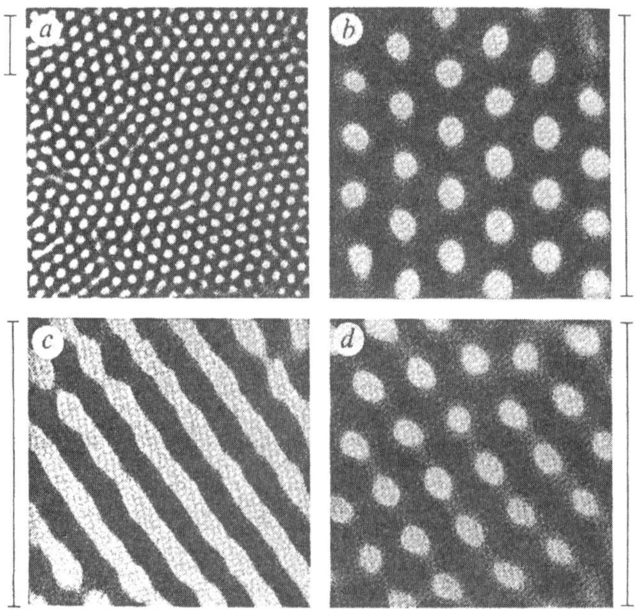

Figure 8.78 Stationary chemical patterns formed in a continuously fed reactor of the chlorite-iodide-malonic acid reaction. (a, b) hexagons; (c) stripes; (d) mixed states. the bars represent 1 mm. For details see Ref. [84].

actual system, the bifurcation may occur at a most feasible local region since the equilibrium point (X_0, Y_0) may show a small spatial fluctuation around its value $(A, B/A)$. Then the oscillatory behavior may propagate outside from the local region.

The second type is **Turing bifurcation** [8, 82]. In (8.60), $\beta(q)$ vs q^2 is parabolic. At

$$q_c = (A/\sqrt{D_X D_Y})^{1/2} \quad \text{and} \quad B = B_{\mathrm{T}} = (1 + A\sqrt{(D_X/D_Y)})^2,$$

one obtains $\beta(q_c) = 0$, yielding $\lambda^2 + \alpha(q_c)\lambda = 0$ and thus $\lambda = 0, -\alpha(q_c)$, where $\alpha(q_c) > 0$. At $B = B_T$, this yields the soft-mode bifurcation with $\lambda = 0$, followed by the spatially periodic pattern $\propto \cos(q_c x)$ with the wave number q_c. If $B_{\mathrm{H}} > B_{\mathrm{T}}$, the bifurcation scenario from the homogeneous state is called the **Turing scenario** [8, 83]. It is stressed that the periodic pattern is stationary. The Brusselator model is quite excellent because of the easy analysis and the universality of its result. Many experimental reaction–diffusion systems have exhibited such universality [8, 84–87]. An example of Turing-like patterns is given by Ouyang and Swinney, who studied the chemical patterns (Fig. 8.78) in a thin disk-shaped gel in contact with a reservoir of reagents of the chlorite-iodide-malonic acid reaction [84]. The reaction medium is a 2.0-mm-thick polyacrylamide gel disk (24.5 mm diameter), which is sandwiched between two 0.4-mm-thick porous glass disks. The bifurcation occurs from a spatially uniform state to a patterned state, as a control parameter is varied. Figure 8.78 shows an example of the hexagonal, striped and mixed patterns [84]. Turing-like patterns have been also observed experimentally in a new chemical reaction–diffusion system containing methylene blue, sulfide, sulfite and oxygen embedded in a polyacrylamide gel matrix [87]. The hexagon pattern has been confirmed from the spatial Fourier transform. Turing bifurcation has also been observed, in gas-discharge systems [88, 89], as described later, but not in semiconductor systems.

Concerning the exceptional absence of the Turing instability in the experiments on semiconductor systems, the following qualitative argument may be helpful. Consider the one-level impact-ionization model ([90, 91]; see this book). For the spatially extended model [90, 91], breakdown of the local charge neutrality and hence the transverse electric field play important roles in the pattern formation, as discussed later. In order to apply the above-mentioned argument to the Turing bifurcation the transverse electric field is neglected and local charge neutrality is assumed, i.e., $\rho = e(N_{\mathrm{D}} - N_{\mathrm{A}} - n - N_{\mathrm{D}}^0) = 0$. Generation–recombination processes at the excited donor states are neglected in the one-level impact-ionization model [90, 91]. The rate equation of the energy relaxation in [90, 91] is adiabatically eliminated, so that the electron mobility may be expressed as a function of the electron density and the electric field; $\mu(n, E)$. The argument is qualitative. The one-level impact–ionization model [90, 91] can be transformed into a reaction–

diffusion model:

$$\partial n/\partial t = f_{\mathrm{gr}} + D_{\mathrm{n}} \frac{\partial^2 n}{\partial x^2} \tag{8.61}$$

$$\partial E/\partial t = -\gamma_2(E - E_0 + Cn\mu E) + D_{\mathrm{E}} \frac{\partial^2 E}{\partial x^2} \tag{8.62}$$

where n denotes the carrier density, E the longitudinal electric field, f_{gr} the g-r rate, D_{n} and D_{E} ($\gg D_{\mathrm{n}}$) the diffusion constants, x the transverse coordinate, E_0 the applied electric field, C the slope of the load line, and μ the electron mobility. The diffusion term in (8.62) is formally expressed as an approximation for the global coupling (8.52). This approximation may be valid by taking $D_{\mathrm{E}} \to \infty$. The g-r rate is given by:

$$f_{\mathrm{gr}} = n(N_{\mathrm{D}} - N_{\mathrm{A}} - n)A_{\mathrm{I}} + (N_{\mathrm{D}} - N_{\mathrm{A}} - n)A_{\mathrm{T}} - n(n + N_{\mathrm{A}})B_{\mathrm{T}}, \tag{8.63}$$

where $N_{\mathrm{D}} - N_{\mathrm{A}}$ is the effective doping concentration, A_{I} is the impact-ionization rate, A_{T} is the thermal ionization rate, and B_{T} is the thermal recombination rate. The homogeneous solutions of (8.61), (8.62) exhibit an S-shaped n–E characteristic, where the electron mobility plays an important role via the screening effect of the ionized impurity scattering for the occurrence of the first-order phase transition [9, 29].

For the voltage-controlled bias condition with $C = 0$, the vertical load line intersects at three equilibrium points $(n_j,\ E_0)$ $(j = 1, 2, 3)$ in the S-shaped region, where the homogeneous rate equation for the carrier density behaves as $\dot{n} = f_{\mathrm{gr}} = -\gamma_1(n - n_1)(n - n_2)(n - n_3)$ and $n_1 < n_2 < n_3$. The middle point $(n_2,\ E_0)$ is a saddle point, and others are typically stable focus. From the linear stability analysis the characteristic exponents satisfy

$$(\lambda - a)(\lambda - d) - bc = 0, \tag{8.64}$$

where

$$\begin{aligned} a &= \frac{\partial f_{\mathrm{gr}}}{\partial n}\bigg|, \qquad b = \frac{\partial f_{\mathrm{gr}}}{\partial E}\bigg| > 0, \\ c &= -\gamma_2 C\bar{E}\left(\frac{\partial \mu}{\partial n}\bigg|\bar{n} + \bar{\mu}\right) < 0, \\ d &= -\gamma_2\left(1 + C\bar{\mu}\bar{n} + C\bar{n}\bar{E}\frac{\partial \mu}{\partial E}\bigg|\right) < 0. \end{aligned} \tag{8.65}$$

In (8.65), $|$ and $^{-}$ denote the equilibrium values. For $C = 0$ one has $c = 0$ and $d = -\gamma_2$ from (8.65) and hence $\lambda = a,\ d$, where $a = -\gamma_1(n_2 - n_1)(n_2 - n_3) > 0$ at the saddle point, and $a < 0$ at the stable focus since $a = -\gamma_1(n_1 - n_2)(n_1 - n_3)$ for $(n_1,\ E_0)$ and $a = -\gamma_1(n_3 - n_1)(n_3 - n_2)$ for $(n_3,\ E_0)$. At the breakdown field E_{b}, where $n_1^* = n_2^*$, the rate equation for n behaves as $\dot{n} = -\gamma_1(n - n_1^*)^2(n - n_3)$. Thus, $a = 0$ and $\gamma = 0$, d (< 0) at the critical point $(n_1^*,\ E_{\mathrm{b}})$, leading to the first-order phase transition to

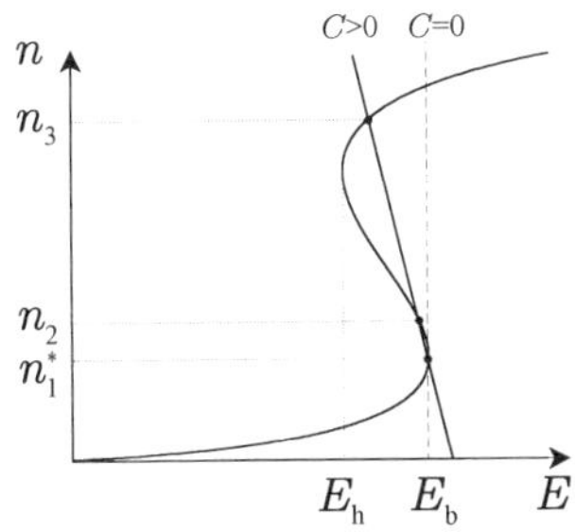

Figure 8.79 Schematic S-shaped n–E curve. The broken line indicates the load line with $C = 0$, while the solid line indicates the inclined bias condition.

$(n_3,\ E_b)$. For the case of the slightly inclined load line $(C \neq 0)$ with fixing (n_1^*, E_b), there are three equilibrium points $(n_1^*,\ E_b), (n_2,\ E_2)$ and $(n_3,\ E_3)$. The middle one (n_2, E_2) can still be a saddle point and the third one (n_3, E_3) a stable focus for a small value of C. (n_1^*, E_b) can be a stable focus in Fig. 8.79 since $\lambda = \{d \pm \sqrt{d^2 + 4bc}\}/2$ with $a = 0, d < 0$ and $bc < 0$. With a slight increase in the applied bias E_0, the first and middle equilibrium points come close to each other. Note that (n_1, E_1) is still a stable focus, keeping its stability character with $a > 0$ and $a + d < 0$ (this may be possible from the linear stability analysis, e.g., in section 4.3). The present argument will be confined to such a condition for the equilibrium point (n_1, E_1). In (8.61) and (8.62) the linear stability analysis around (n_1, E_1) can be done by putting $n = n_1 + \xi_n \exp(iqx + \lambda t)$, $E = E_1 + \xi_E \exp(iqx + \lambda t)$, where $a > 0$, $a + d < 0$, and $ad > bc$ from the requirement in (8.65). Then, the coefficients $\alpha(q),\ \beta(q)$ corresponding to (8.60) are given as:

$$\alpha(q) = (D_n + D_E)q^2 - (a + d), \tag{8.66}$$

$$\beta(q) = D_n D_E q^4 - (aD_E + dD_n)q^2 + ad - bc. \tag{8.67}$$

The eigenvalues $\lambda_\pm$ are given from (8.59) and (8.66), (8.67) as:

$$\lambda_\pm = \frac{-\alpha(q) \pm \sqrt{\alpha(q)^2 - 4\beta(q)}}{2}, \tag{8.68}$$

where $-\alpha(q) < 0$ from (8.66) since $a + d < 0$. Thus, the soft-mode phase transition is possible if it happens that $\beta(q) = 0$ at $(n_1,\ E_1)$, where $\lambda = 0, -\alpha(q)\ (< 0)$ from $\lambda(\lambda + \alpha(q)) = 0$. The Turing bifurcation may occur if the following conditions are met for $\beta(q) = 0$ with $q_c^2 > 0$:

$$q_c^2 = (aD_E + dD_n)/2D_n D_E > 0, \tag{8.69}$$

Since $a > 0,\ d < 0$ the condition for $q_c^2 > 0$ in (8.69) is:

$$aD_E > -dD_n. \tag{8.70}$$

Note that $\beta(q)$ is a parabolic function of q^2, where $\beta(0) = ad - bc > 0$ and a local minimum $\beta(q_c)$:

$$\beta(q_c) = -\frac{1}{4}\left(\frac{D_E}{D_n}a + \frac{D_n}{D_E}d\right)^2 + (ad - bc). \qquad (8.71)$$

For the equilibrium point (n_1^*, E_b), $\beta(q_c) > 0$ because $a = 0$ and $D_E \gg D_n$, yielding a spatially homogeneous solution. By increasing the applied bias E_0 slightly, $\beta(q_c)$ becomes negative because of the term $(D_E/D_n)a$ in (8.71), whereby the Turing bifurcation may occur with the wave number q_c. The model (8.61), (8.62) is typically the activator–inhibitor model with the activator n and inhibitor E, characterized by the long-range inhibition and the local activation ($D_E \gg D_n$) with $\gamma_1 \gg \gamma_2$.

In semiconductor systems the Turing bifurcation has not been observed, as already pointed out. Several points may be considered regarding this fact. First, the Turing bifurcation predicted from the linear stability analysis is limited only around the equilibrium point (n_1^*, E_b). Since the autocatalytic process in (8.61), (8.62) shows the first-order phase transition, the bifurcation may be subcritical. Also the Turing regime locates close to the breakdown field E_b, in a quite narrow range of the negative differential branch. Thus, after the bifurcation the Turing-like pattern might coalesce into a single filament under the influence of the external noise. Secondly, for the inclined load line the phase transition occurs at an electric field $E \sim E_b$ and simultaneously the Turing bifurcation may proceed with too small a characteristic wave number $q_c \sim a/2D_n$ with $D_E \to \infty$, where $a \sim 0$.

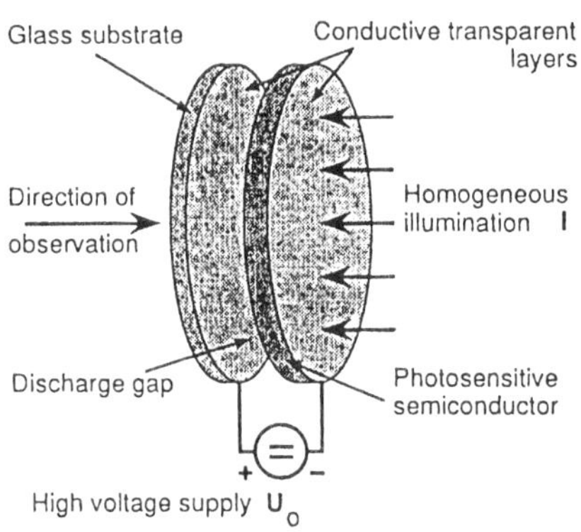

Figure 8.80 Experimental gas discharge device. The discharge gap is filled with nitrogen at a pressure of $p = 100$ hPa. Crystalline silicon doped with gold or zinc is used as a spatially distributed resistive load for the discharge. The discharge gap has a dimension of approximately 1 mm, and the lateral dimension of the active circular area is 20 mm in diameter. The two-layer structure is bounded by electrodes which are transparent to visible light [89].

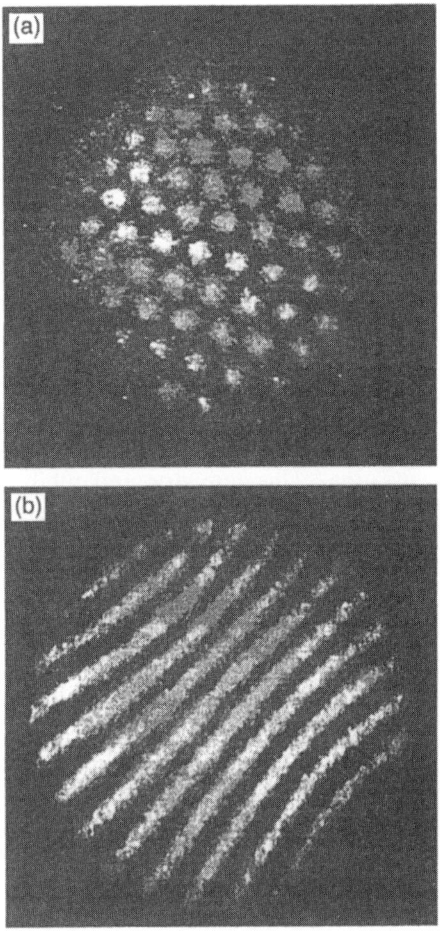

Figure 8.81 Stationary patterns in spatial distributions of the gas-discharge electric current. (a) Hexagon, and (b) stripes. The voltage drops at the gas-discharge system and the current densities averaged over the current channel are: (a) 1815 V, 9.3 $\mu A/cm^2$, and (b) 1935 V, 18.5 $\mu A/cm^2$ [89].

In the gas-discharge system the experimental geometry comprises a high resistive silicon plate, a planar discharge layer and a transparent indium tin oxide layer on a glass substrate, as shown in Fig. 8.80 [88, 89]. With the geometry Ammelt *et al.* [88] and Astrov *et al.* [89] have shown a subcritical bifurcation followed by a hexagonal pattern and a supercritical Turing bifurcation, as given in Fig. 8.81. They have also shown the spatiotemporal zigzag motion of the interacting filaments [88, 89]. The spatial (Turing bifurcation) and the spatiotemporal behaviors have been well explained by a two-layer model, composed of an ohmic resistive layer (resistive silicon plate in Fig. 8.80) and a nonlinear layer (a gas layer) with bistable properties [88, 89]. The interaction between the local voltage drop across the ohmic layer and the current density within the nonlinear layer is essential to produce a complex inhomogeneous distribution [88, 89].

An activator–inhibitor model for biological pattern formation has been given typically by [92–94]:

$$\frac{\partial a}{\partial t} = \rho_0 + \frac{a^2}{h} - \mu a + D_a \frac{\partial^2 a}{\partial x^2}, \tag{8.72}$$

$$\frac{\partial h}{\partial t} = \rho_1 + a^2 - \nu h + D_h \frac{\partial^2 h}{\partial x^2}, \tag{8.73}$$

where a and h are the activator and the inhibitor, respectively. In (8.72) the self-enhancing production rate (a^2) is slowed down by the influence of the inhibitor ($1/h$) [93]. In (8.73) the inhibitor production rate depends only on the activator concentration. The activator and inhibitor have the characteristic inverse relaxation times μ and ν, respectively. From the linear stability analysis [92, 93] the condition for the stable pattern formation is that the activator diffuses much more slowly than the inhibitor and has a longer time constant, i.e.,

$$D_a < D_h \quad \text{and} \quad \mu < \nu. \tag{8.74}$$

This means local activation and long-range inhibition. The patterns that can be generated by this type of interaction can be observed in many biological systems.

(a) (b)

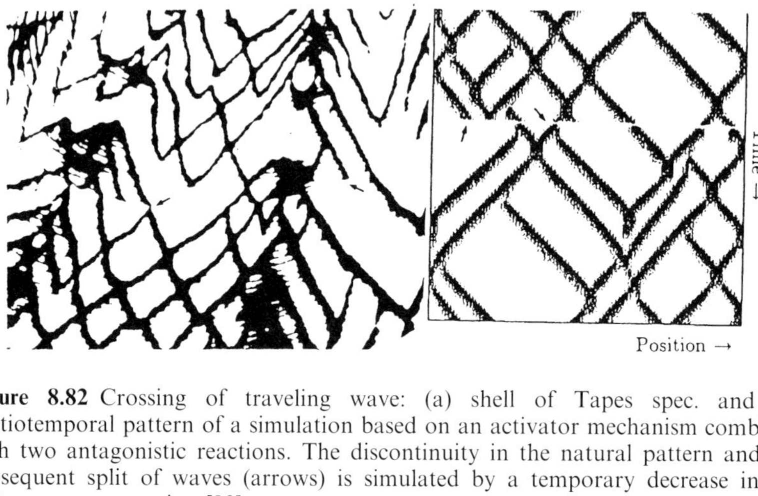

Figure 8.82 Crossing of traveling wave: (a) shell of Tapes spec. and (b) spatiotemporal pattern of a simulation based on an activator mechanism combined with two antagonistic reactions. The discontinuity in the natural pattern and the subsequent split of waves (arrows) is simulated by a temporary decrease in the activator concentration [93].

In contrast, Meinhardt has also shown that if

$$D_a > D_h \quad \text{and} \quad \mu > \nu, \tag{8.75}$$

oscillations and traveling waves can result [93]. Figure 8.82 shows an example of this mode by the pigmentation patterns on shells of mollusks, where the pigmentation pattern is shown on the left (a) and the model simulation on the right (b) [93]. The formation of patterns proceeds in a strictly linear manner since new pattern elements are added only along a narrow zone at the growing edge of the shell [93]. The second dimension is a time record of what has happened at the growing edge, showing the complete history of their formation as the time records of the traveling wave (Fig. 8.82a). The traveling waves in Fig. 8.82a can penetrate each other like collisions of solitons [93], the properties of which have been well reproduced by the model simulations in Fig. 8.82b. Meinhardt concludes that the reaction by two antagonists of cell local inhibitor (h) and long-range activator (a) with $D_a > D_h$ and $\mu > \nu$ is essential. As well as Turing instability these complex spatiotemporal behaviors may be universal among different physical systems. These subjects may also be highly interesting in semiconductor dynamical systems.

Period doubling bifurcation is the most well-known bifurcation scenario (Feigenbaum scenario) in the low-dimensional chaos theory. It is highly important to explore the relation between the low-dimensional deterministic nature and the spatiotemporal behaviors. In a p^+–n^+–p–n^- diode the spatiotemporal evolutions of a locking current filament during period doubling have been investigated numerically [50, 55] and experimentally [48], in the postbreakdown regime, high above the breakdown voltage of the I–V characteristic (see section 8.1.4). As mentioned in section 8.1.6, the pattern dynamics during the period doubling bifurcation in the bulk semiconductors have not been spatially resolved, except for period 1 and the first bifurcation point observed in a form of enforced breathing filament (see section 8.1.6). The spatiotemporal evolutions are expected to be quite different from those in the p^+–n^+–p–n^- diode because the chaotic motion occurs as a result of the interplay between the upper and lower branches of the S-shaped I–V curve and the oscillatory behaviors are due to the repetition cycle of the ignition and extinction of a current filament [15, 44]. Here, the spatiotemporal behaviors during the period doubling are supplemented by the numerical results [90, 91]. The difficulty in the experiments is in measuring the snapshots, but this will be done in the future.

The simulations were done using the one-level impact-ionization model (Chap. 3, [29, 95, 96]). They consider: (i) a rate equation by the detailed energy balance; (ii) g-r processes of the impact-ionized carriers n and the neutral donors N_{D}^0, by the continuity equation $\dot{\rho} = -\mathrm{div}\,\mathbf{J}$, $\rho = e(N_{\mathrm{D}} - N_{\mathrm{A}} - N_{\mathrm{D}}^0 - n)$ and $\dot{N}_{\mathrm{D}}^0 = -f_{\mathrm{gr}}(n, N_D^0)$; and (iii) a rate equation for the dielectric relaxations of the transverse and longitudinal electric fields, respectively, where only the transverse diffusion of the carriers is considered by $\dot{E}_\perp = -e(n\mu E_\perp + D\partial n/\partial x)$ (i.e., the 1D transverse breakdown mode is considered). The system is driven periodically by a dc + ac bias. The longitudinal electric field is assumed to be homogeneous for the planar-type ohmic contacts and the rate equation is given by the global coupling (8.52) from Kirchhoff's law of the circuit. The ODEs after scale conversion are as follows [91]:

$$\dot{\varepsilon} = -\gamma_1(\sqrt{\varepsilon}(\varepsilon - 1) - (3\pi/32)\,\mu E_\parallel^2), \tag{8.76}$$

$$\dot{\nu} = -\frac{\partial^2 \nu}{\partial x^2} + \alpha\frac{\partial}{\partial x}(\nu\mu E_\perp) + f_{\mathrm{gr}}(\nu, \nu_{\mathrm{D}}), \tag{8.77}$$

$$\begin{aligned}\dot{\nu}_{\mathrm{D}} &= -f_{\mathrm{gr}}(\nu, \nu_{\mathrm{D}}) \\ &= -\gamma_2(\kappa_1\nu\,\nu_{\mathrm{D}} + \kappa_2\nu_{\mathrm{D}} - \kappa_3\nu(k/(k-1) - \nu_D)),\end{aligned} \tag{8.78}$$

$$\dot{E}_\perp = -\gamma_\perp\left(\frac{1}{\alpha}\frac{\partial\nu}{\partial x} + \nu\mu E_\perp\right), \tag{8.79}$$

$$\dot{E}_\parallel = -\nu_\parallel(E_\parallel - E_0 - E_{\mathrm{ac}}\sin(2\pi f_0 t) + cI), \tag{8.80}$$

where ε denotes the mean kinetic energy of free electrons, ν, ν_{D} the densities of the free electrons and the shallow donors, respectively, μ the carrier mobility which is described by the Brooks–Herring formula as given below,

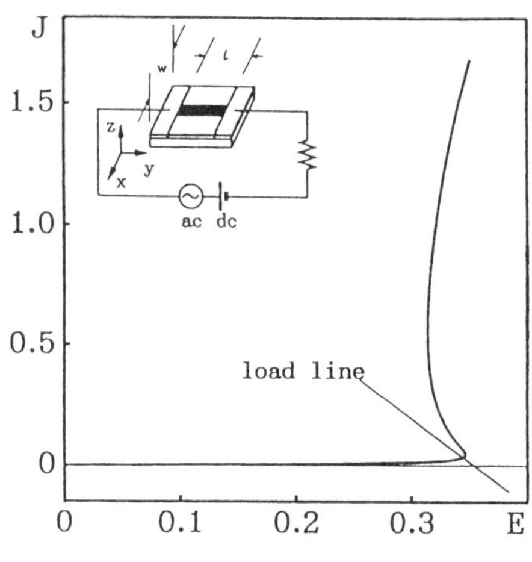

Figure 8.83 Static J–E characteristic obtained from the homogeneous ODEs. The inset shows the sample geometry [91].

κ_1 the impact-ionization rate, κ_2 the thermal ionization rate, κ_3 the thermal recombination rate, k the inverse of the compensation ratio N_D/N_A, $E_\perp$, $E_\parallel$ the transverse and longitudinal electric fields, respectively, γ_1, γ_2, $\gamma_\perp$, $\gamma_\parallel$ the inverse relaxation times in each of the rate equations, E_0 the dc bias, E_{ac} the amplitude of the ac bias, f_0 the driving frequency, c the inverse of the slope of the load line, and I the averaged current density $(1/w)\int_w en\,\mu E\,dx$ (w: width of the sample). The impact-ionization rate and the electron mobility are given by:

$$\kappa_1 = (1 + a\nu E_\parallel/\varepsilon)\sqrt{1/\varepsilon}\exp(18.75(1 - 1/\varepsilon)), \tag{8.81}$$

$$\mu = \varepsilon^{3/2}\frac{2/(k-1) + \nu_0}{2/(k-1) + \nu} \times \frac{\ln 51 - 50/51}{\ln u - (u-1)/u}, \tag{8.82}$$

where $u = 1 + 50(\nu_0/\nu)\varepsilon^2$. The transverse space coordinate x (see inset of Fig. 8.83) is divided into 50 grids, thereby one approximates $\partial^2\nu/\partial x^2 = \beta(\nu_{j+1} + \nu_{j-1} - 2\nu_j)$; $\partial\nu/\partial x = \sqrt{\beta}(\nu_{j+1} - \nu_j)$, where $j = 1, 2, \ldots, N$ and $\beta = (L_D/\Delta x)^2$ (L_D: Debye length). In (8.80) the average current density is given by

$$I = \frac{1}{N}\sum \nu_j\mu_j E_\parallel,$$

where $N = 50$. The Neumann boundary conditions are adopted: $\varepsilon_{N+1} = \varepsilon_N$, $\varepsilon_0 = \varepsilon_1$, etc. Corresponding to the reasonable values of the physical parameters (Chapter 5), we put $\gamma_1 = 100$, $\gamma_2 = 20.0$, $\gamma_\parallel = 1/c$ $(c = 0.28)$, $\gamma_\perp = 3.0$, $\kappa_2 = 0.05$, $\kappa_3 = 2.6$, $a = 0.11338$, $\alpha = 1.4$, $\beta = 30.0$, $k = 1.9$, $E_{ac} = 0.048$ and $f_0 = 100/350$ [91].

The homogeneous solution of the J–E curve for (8.76)–(8.82), is shown in Fig. 8.83, where J is the current density given by $\nu\mu E_\parallel$. Homogeneous

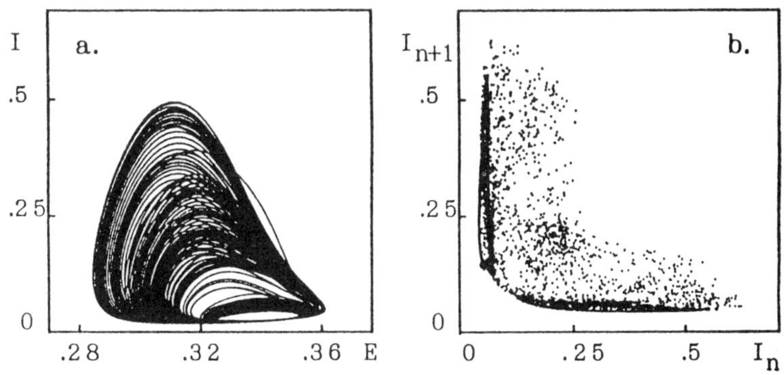

Figure 8.84 (a) Phase portrait I vs E, and (b) peak plots I_{n+1} vs I_n for the spatially extended system, where $E_0 = 0.353$ [91].

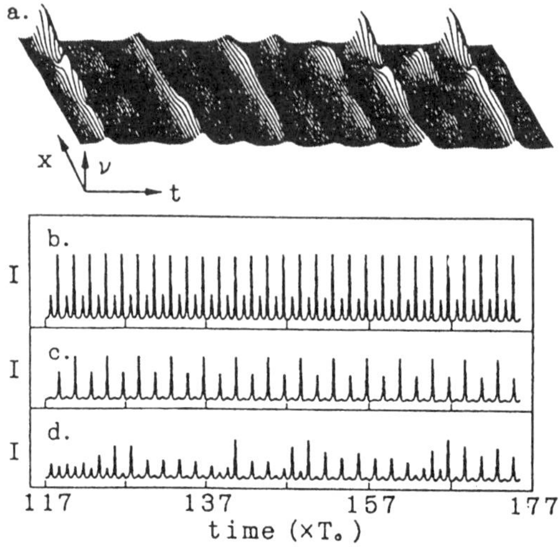

Figure 8.85 (a) Spatiotemporal pattern of electronic turbulence, and (b–d) averaged current as a function of time, where $T_0 = 1/f_0$. The dc bias was (a) $E_0 = 0.352$, (b) 0.368, (c) 0.365, and (d) 0.352. The waveform of period 1 has been omitted from the figure [91].

equations of the present model system exhibit chaotic behaviors under magnetic fields [95, 96]. Here, the load line $E_0 = E_{\|} + cJ$ is settled just below the breakdown field (see Fig. 8.83). When the load line is periodically driven for the homogeneous case, one can show easily period doubling route to chaos as a function of E_0. The spatially extended system (8.76)–(8.82) also shows period doubling bifurcation [90, 91]. The phase portrait (the average current density I vs electric field $E_{\|}$) and peak plots I_{n+1} vs I_n are shown for chaos in Fig. 8.84, where 6000 points are included in the peak plots. The phase portrait clearly indicates the deterministic nature of firing and misfiring of the filament around the breakdown field. The peak plots are similar to the λ-shaped return map (Chapter 5) obtained by the homogeneous equations of the approximated model B. The scatter of the data in the peak plots may be explained by an increased fractal dimension due to higher degrees of freedom in the spatially extended model system [91].

Figure 8.85 shows (a) the turbulent pattern (spatiotemporal chaos) of the carrier density ν and (b–d) the current waveforms for period 2, period 4 and chaos, respectively. The turbulent pattern in Fig. 8.85a indicates a chaotic ignition–extinction process among two or three current filaments, where ignition of a current filament at a local region tends to inhibit the firing at the next period at the same region [91]. The spatiotemporal behaviors seems to be ruled by the activator–inhibitor effect, where the activator is the carrier density and the inhibitor is the longitudinal electric field [91]. It is noted that the present model system is not given as a typical activator–inhibitor model system as in (8.61), (8.62). However, owing to the global coupling in (8.80), the integrated current of the filaments is fed back as an inhibitory effect, whereby the slow variable of the longitudinal electric field does not increase sufficiently to cause the impact-ionization avalanche at the next period and at the same region. As stated below, the transverse field participates more precisely in the pattern dynamics [91]. This situation occurs at the transition point of the *I–V* curve and is seen in the phase portrait of Fig. 8.84a.

The spatiotemporal patterns of periods 1 (a), 2 (b) and 4 (c) are shown as a function of E_0 in Fig. 8.86. The patterns of ε and μ are almost identical to the pattern of ν, which is a direct result of Haken's slaving principle with a relation $\gamma_1 \gg \gamma_2 \gg \gamma_\perp$ in (8.76)–(8.80). Since $\dot{\nu} = -(\alpha/\gamma_\perp)\partial \dot{E}_\perp/\partial x + f_{gr}$ from (8.77)–(8.79), the carrier density is slaved by the slow variable $E_\perp$. Therefore, the slow dynamics of the transverse field $E_\perp$ also plays a key role in the activator–inhibitor effect via the drift/diffusion terms in (8.77) and (8.79) [91]. In Fig. 8.87 the patterns of the carrier density and the transverse electric field are compared. The transverse field at the filament walls forms a sharp structure [91] ("wall field" [90, 91], and surprisingly it is retained even after the filaments died out. Further successive nucleations of the filaments take place at each of the local regions of the wall field. The formation and survival of the wall field are understood as follows. Once the nucleation of the current filament proceeds, the carrier diffusion takes place simultaneously. Since the ionized donors are spatially fixed the charge neutrality breaks down locally at the filament boundaries [20], which induces the transverse field (wall field), whereby the transverse drift current flows to counterbalance the diffusion current. Although the filament structure is smeared out after a half period by the ac voltage, the transverse field keeps its structure because it is not affected directly by the ac modulation in (8.79) and also because the transverse field responds slowly ($\gamma_\perp \ll \gamma_2$). Then, after the subsequent half period, the wall field acts as a triggering seed via the transverse drift current in (8.79). In the future experimental evidence may be hopefully obtained for the spatiotemporal patterns as in Fig. 8.86.

Besides the pure physical interests there are some challenges to apply the current filament dynamics to new devices. Kostial *et al.* have proposed using the well-controlled generation of multiple current filaments in a multistripe geometry of n-GaAs (up to 20 static current filaments in [97, 98]) as a "winner takes all" element in synergetic computer applications. In addition, there are more and more problems of complex dynamical behaviors in sophisticated device structures such as semiconductor heterostructures and superlattices, which will be described in section 8.3.2.

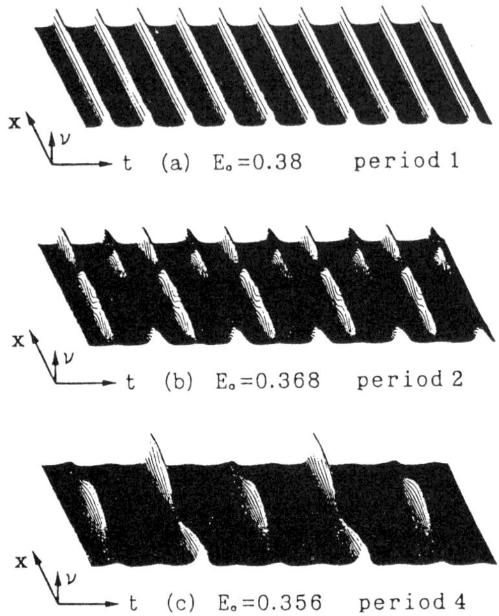

Figure 8.86 Spatiotemporal patterns of periods 1, 2 and 4 [91].

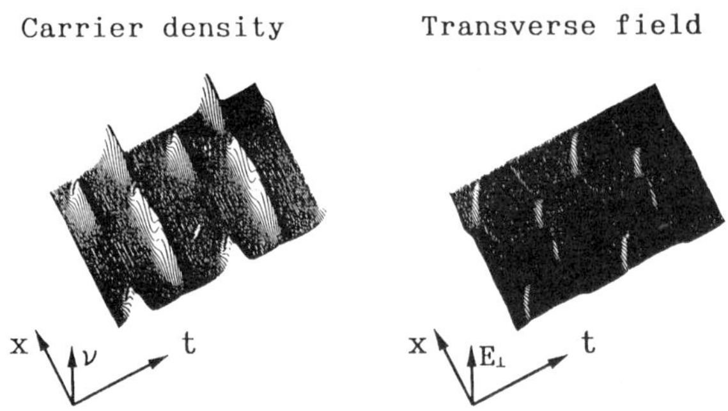

Figure 8.87 Comparison of the patterns of carrier density and transverse field, where $E_0 = 0.368$ [91].

8.3 Chaos in device structures and superlattices

In this final section the author introduces briefly chaos in electronic devices and circuits in section 8.3.1, and chaos in device structures and superlattices in section 8.3.2. Chaos in electronic devices and circuits is beyond the main interest of this book, but they are described briefly in section 8.3.1 because of the increasing interest in the application of chaos by synchronization [99] and controlling chaos [100], which are especially interesting to engineers. In section 8.3.2 the novel phenomenon of low-frequency oscillation (LFO) in the GaAs metal-semiconductor field-effect transistor (MESFET), the oscillatory behavior and chaos due to real space transfer [101] of 2D electron gas in heterostructures, and the oscillatory behaviors in the heterostructure hot-electron device are described. Also in section 8.3.2, multistability of the current–voltage characteristics, domain formation and the oscillatory behaviors due to resonant tunneling transport in GaAs/AlAs superlattices are introduced.

8.3.1 Chaos in electronic devices and circuits

8.3.1.1 Synchronization of chaos

At the beginning of 1990s several pioneering researchers began to find new directions to master intractable chaos: synchronization of chaos [99, 101] and controlling chaos [100, 103]. Since chaos is unpredictable over long periods in its deterministic time evolution, it was best for engineers to avoid it or to avoid operating devices in the chaotic regime. Since chaos shows sensitive dependence on the initial conditions and also exponential divergence of nearby orbits, two identical autonomous systems starting at nearly the same initial points in phase space have trajectories which quickly become uncorrelated, so that synchronization of chaos between the two subsystems seemed to be practically impossible. This synchronization was observed for the first time by Yamada and Fujisaka [104] in 1983. Independently, Pecora and Carroll [99] investigated synchronization and presented the idea that synchronization of chaos between two subsystems in the electronic chaos circuits can be applied to secure communications systems, exploiting the unique feature of chaotic signals.

Pecora and Carroll [99] examined their idea through several well-known systems (e.g., Rössler and Lorenz) with their strategy as stated below. For example, Rössler's attractor is solved by the following ODEs:

$$\begin{aligned}\dot{x} &= -y - z,\\ \dot{y} &= x + ay,\\ \dot{z} &= bx - (c - x)z.\end{aligned} \tag{8.83}$$

The system can be divided into two parts: the drive component (e.g., the variable y) and a subsystem (x, z). Consider another identical subsystem (x', z') which starts from quite different initial conditions and is driven by drive y. The ODEs for (x', z') are:

$$\begin{aligned}\dot{x}' &= -y - z',\\ \dot{z}' &= bx' - (c - x')z'.\end{aligned} \tag{8.84}$$

Then, can one expect synchronization of the response system (x', z') with

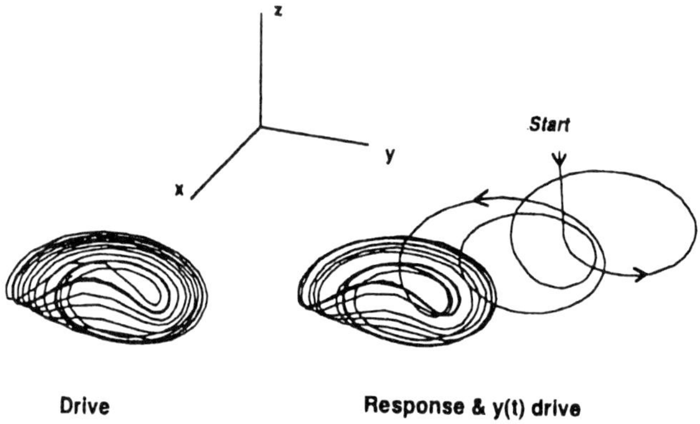

Figure 8.88 Attractors for the Rössler drive system and the (x', y') response system and y(t) drive variable [99].

(x, z)? Pecora and Carroll [99] showed that it was possible in Fig. 8.88, where after some transients the response of $x'(t)$ and $z'(t)$ approaches $x(t)$ and $z(t)$. That is, the subsystem (x', z') is completely slaved by drive y to synchronize with (x, z). The general idea for the synchronization is described as follows [99].

Consider an autonomous n-dimensional dynamical system,

$$\dot{\mathbf{u}} = \mathbf{f}(\mathbf{u}). \tag{8.85}$$

Divide the system, arbitrarily, into two subsystems $u = (v, w)$:

$$\dot{\mathbf{v}} = \mathbf{g}(\mathbf{v}, \mathbf{w}), \qquad \dot{\mathbf{w}} = \mathbf{h}(\mathbf{v}, \mathbf{w}), \tag{8.86}$$

where $\mathbf{v} = (u_1, \ldots, u_m)$, $\mathbf{g} = (f_1(u), \ldots, f_m(u))$, $\mathbf{w} = (u_{m+1}, \ldots, u_n)$, and $\mathbf{h} = (f_{m+1}(u), \ldots, f_n(u))$. Now generate a new subsystem w' identical to the w system, both of which are driven by the drive components v in (8.86). Then the total dynamical system which is composed of the drive component $\mathbf{v}$ and two subsystems $\mathbf{w}$ and $\mathbf{w}'$ can be described as [99]:

$$\dot{\mathbf{v}} = \mathbf{g}(\mathbf{v}, \mathbf{w}), \; \dot{\mathbf{w}} = \mathbf{h}(\mathbf{v}, \mathbf{w}), \; \dot{\mathbf{w}}' = \mathbf{h}(\mathbf{v}, \mathbf{w}'). \tag{8.87}$$

The subsystem $\mathbf{w}'$ will synchronize with $\mathbf{w}$ only if $\Delta\mathbf{w} = \mathbf{w}' - \mathbf{w} \to 0$ as $t \to \infty$. This requirement $\Delta\mathbf{w} \to 0$ for the synchronization leads to the variational equation:

$$\dot{\boldsymbol{\xi}} = D_w\mathbf{h}(\mathbf{v}(t), \mathbf{w}(t))\boldsymbol{\xi}, \tag{8.88}$$

where $D_w\mathbf{h}$ is the Jacobian of the $\mathbf{w}$ system vector field with respect to $\mathbf{w}$ only. The behavior of (8.88) depends on the Lyapunov exponents of the $\mathbf{w}$ system [99]. Pecora and Carroll [99] referred to these sub-Lyapunov exponents. The theorem of the synchronization has been given [99]: The subsystems $\mathbf{w}$ and $\mathbf{w}'$ will synchronize only if the sub-Lyapunov exponents are all negative.

In Table 8.7 the sub-Lyapunov exponents for the Rössler system and the Lorenz system are shown for various configurations of the drive and response [99]. For the Rössler system it is shown that only the y-drive configuration will synchronize with all the sub-Lyapunov components being negative (Fig. 8.88). Corresponding to Fig. 8.88 for the y-drive, the differences $\Delta x = x' - x$ and $\Delta z = z' - z$ diminish with time. However, for the x- or z-drive, the sub-Lyapunov exponents is positive, indicating failure of synchronization, so that one must choose the drive component adequately for a given system to be synchronized.

For the Lorenz system synchronization will occur for either x- or y-driving [99], as seen in Table 8.7. Figure 8.89a shows a plot of the time vs log of the differences $y' - y$ and $z' - z$ for the Lorenz attractor, where the drive component is x. The convergences to synchronization are consistent with the negative sub-Lyapunov exponents in Table 8.7. Figure 8.89b shows the results for the same situation, but with a slight change in the parameters of the response system. In this case the system partially synchronizes in that y' and z' stay within some neighborhood of y and z as they proceed around the attractor [99].

Pecora and Carroll [99, 102, 105] have examined the basic idea by using a modified version of an electronic chaotic circuit by Newcomb and Sathyan [106], realizing that the synchronized chaos circuit might be used for a secure communications system. As described by Ditto and Pecora [107], the proposed secure communications system should be composed of a transmitter and a receiver, each being located by an arbitrarily long distance and the signals buried in chaos being broadcasted from the transmitter to the receiver. The transmitter sends both the driving chaos signal $u_c(t)$ and the message buried in chaos $u_s(t) + u_c(t)$. Since the receiver is identical to the transmitter, the output signal in the receiver is completely synchronized by the driving signal $u_c(t)$. At the same time the receiver receives the private message by the encoded signal $u_s(t) + u_c(t)$. By subtracting the chaotic signal from the encoded signal, the original message can be obtained. To

Table 8.7

System	Drive	Response	Sub-Lyapunov exponents
Rössler	x	(y, z)	$(+0.2, -8.89)$
$a = 0.2, b = 0.2$	y	(x, z)	$(-0.056, -8.81)$
$c = 9.0$	z	(x, y)	$(+0.1, +0.1)$
Lorenz	x	(y, z)	$(-1.81, -1.86)$
$\sigma = 10, b = 8/3$	y	(x, z)	$(-2.67, -9.99)$
$r = 60.0$	z	(x, y)	$(+0.0108, -11.01)$

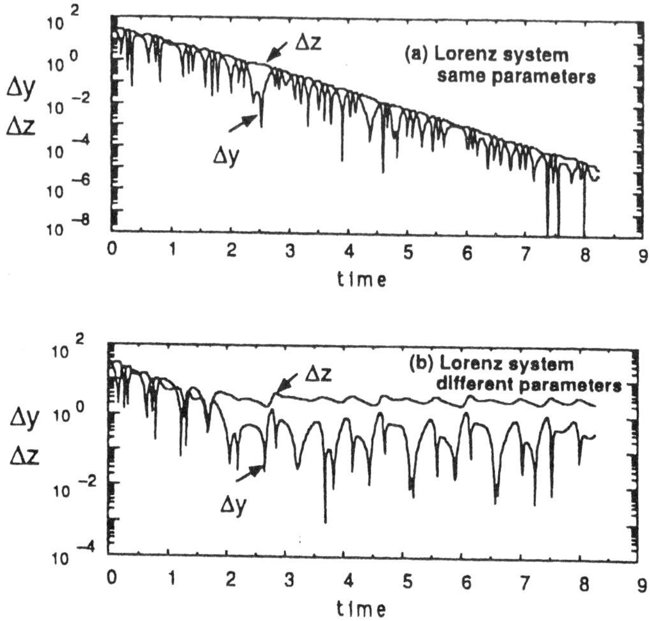

Figure 8.89 Differences $y' - y$ and $z' - z$ between the response variables and their drive counterparts for the Lorenz system for (a) when the parameters are the same for both systems, and (b) when the parameters differ by 5% [99].

build a practical secure communications system, however, engineers will be required to develop sophisticated schemes in which the encoding process involves more than just adding chaos to the message signal [107].

The idea of Pecora and Carroll [99, 102, 105] has prevailed and synchronization and various types of the communication systems [108, 109] have been investigated widely in the electronic chaos circuits of tunnel diode relaxation oscillators [110], Chua's circuit [111–118] and the Van der Pol-Duffing oscillator [119].

8.3.1.2 Controlling chaos

Another direction is controlling chaos. The theory of controlling chaos was first proposed by Ott, Grebogi and Yorke (OGY) in 1989 [100, 103, 120], and the experimental control of chaos in a parametrically driven magnetoelastic ribbon was examined by Ditto, Rauseo and Spano in 1990 [121]. Within a few years, many researchers had also verified the effectiveness of the OGY method and its modified version of controlling chaos in the experimental nonlinear dynamical systems of the thermal convection loop [122, 123], the diode resonator [124], the chaotic laser [125], the microcardial tissue [126] and the Belousov–Zhabotinsky reaction [127]. The OGY method can stabilize an arbitrary unstable periodic orbit by controlling chaos [100].

Chaos is a direct result of bifurcations from periodic orbits. Bifurcation means destabilization of an old periodic orbit, creating a new stable periodic orbit when a control parameter is changed. Thus, one has an unstable periodic orbit and a periodic orbit after the first bifurcation (e.g., the first period doubling). After the successive bifurcations, chaos (strange attractor in the phase space) emerges, where there is an infinite number of unstable periodic orbits. That is, an infinite number of unstable periodic orbits is embedded in the strange attractor with measure zero (see Chapter 1). The OGY method is a control technique by converting chaos to one of the unstable periodic orbits. Depending on the desired use, the system behavior could be changed by switching the temporal programming of the small parameter perturbations to stabilize one of the different periodic orbits [100]. No knowledge about dynamical system equations is required. Only chaotic data series $\{X(t)\}$, or embedding vectors in the phase space, are needed. The control can be made by analyzing the Poincaré surface of section.

For simplicity, assume that an experimental chaos system is three-dimensional, the dynamical equations of which are given by $d\mathbf{x}/\mathrm{d}t = \mathbf{F}(\mathbf{x}, \mathbf{p})$ with the control parameter $\mathbf{p}$ [100], where $\mathbf{x}$ is the 3D order parameter. It is not necessary to know about the dynamical equations. Now, the experimental time series of some scalar-dependent variable, say, $z(t)$, is measured. Then, as usual, phase space vector point is constructed by using delay coordinates with delay T [128]:

$$\mathbf{X}(t) = \{z(t),\ z(t-T),\ z(t-2T), \ldots,\ z(t-MT)\}. \tag{8.89}$$

A set of the vector points $\mathbf{X}(t)$ forms a strange attractor in the $(M+1)$-dimensional phase space when the system is uncontrolled. Let us investigate piercings of the orbits $\mathbf{X}(t)$ at a constant surface, where a set of the piercings forms the Poincaré surface of section. For our purposes, $M = D - 1$ is generally sufficient in (8.89) for the D-dimensional dynamical system, in contrast to the case of global embedding of the original phase space in the delay coordinate, for which $M + 1 = 2D + 1$ is required [128]). Hence, for the case considered ($D = 3$), the Poincaré surface of section is two-dimensional [100]. The surface of section can be constructed by taking piercing points at a constant surface, e.g., $z(t - MT) = c$.

Let $\boldsymbol{\xi}_1,\ \boldsymbol{\xi}_2, \ldots,\ \boldsymbol{\xi}_k, \ldots$ be the successive experimental points in the surface of section, where $\boldsymbol{\xi}_n$ denotes the coordinate at the nth piercing of the surface of section of the orbit $\mathbf{X}(t)$. Then, these points distribute very quickly over the whole surface of section of the strange attractor. Find all pairs $(\boldsymbol{\xi}_n,\ \boldsymbol{\xi}_{n+1})$ of closely located piercings. Then one can find an unstable fixed point (saddle point) $\boldsymbol{\xi}_F$ of period 1 about these pairs. Similarly, for the nearby

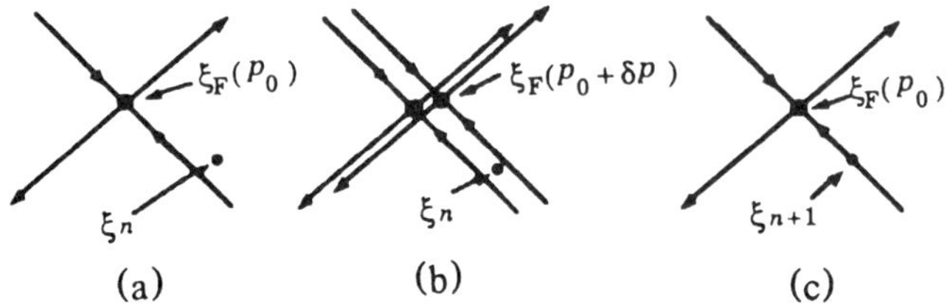

Figure 8.90 Schematic of the OGY control algorithm for a saddle fixed point: (a) the nth iterate $\boldsymbol{\xi}_n$ falls near the fixed point $\boldsymbol{\xi}_F(p_0)$; (b) turn on the perturbation of p to move the fixed point; (c) the next iterate is forced onto the stable manifold of $\boldsymbol{\xi}_F(p_0)$. Turn off the perturbation [12].

pairs ($\boldsymbol{\xi}_n$, $\boldsymbol{\xi}_{n+m}$) there is an unstable fixed point of period m around these pairs, where $\boldsymbol{\xi}_{n+j} \simeq \boldsymbol{\xi}_n$ only when $j = m$. From experimentally determined sequences it has been demonstrated that a large number (up to $m = 40$) of distinct unstable periodic orbits on a high-quality chaotic attractor can be determined [129]. For simplicity, concentrate on the case of period 1. Collecting these nearby pairs ($\boldsymbol{\xi}_n$, $\boldsymbol{\xi}_{n+1}$), the saddle point $\boldsymbol{\xi}_F$ can be determined with some accuracy. The saddle point has two main axes of stable and unstable manifolds with the eigenvalues λ_s and λ_u ($|\lambda_u| > 1 > |\lambda_s|$), respectively, where the unit vectors in each direction are denoted by $\mathbf{e}_s$ and $\mathbf{e}_u$. The approximate local linear map M is fitted to the collected experimental pairs ($\boldsymbol{\xi}_n$, $\boldsymbol{\xi}_{n+1}$):

$$\boldsymbol{\xi}_{n+1} - \boldsymbol{\xi}_F = \mathbf{M}(\boldsymbol{\xi}_n - \boldsymbol{\xi}_F), \tag{8.90}$$

where $\mathbf{M}$ is a 2×2 matrix [100, 121]. Knowing $\mathbf{M}$, (λ_s, λ_u) and $(\mathbf{e}_s, \mathbf{e}_u)$ could be experimentally accessible [100, 120, 121].

Having found these fundamental data, one can start to control chaos. Wait until a piercing point comes inside a small circle, $|\boldsymbol{\xi}_n - \boldsymbol{\xi}_F| < \delta\boldsymbol{\xi}^*$, and then control chaos by changing the parameter slightly from p_0 to $p_0 + \delta p$ [100, 120]. An attempt is made to confine the iterates of the map to a small neighborhood of the desired orbit $\boldsymbol{\xi}_F$. How is it possible? The illustration in Fig. 8.90 explains the OGY method [121]. When an iterate falls near the desired orbit (Fig. 8.90a), p is changed from its nominal value p_0 by δp, thereby changing the saddle point $\boldsymbol{\xi}_F(p_0)$ and its stable manifold (Fig. 8.90b), such that the next iterate will be forced back towards the stable manifold of the original desired orbit (Fig. 8.90c) for $p = p_0$ [121]. This is possible as follows. By a small perturbation δ_p the saddle point will shift slightly by $\delta\boldsymbol{\xi}_F$, and the new position $\boldsymbol{\xi}_F(p_0) + \delta\boldsymbol{\xi}_F$, is determined. For small δ_p one approximates $\mathbf{g} = \partial\boldsymbol{\xi}_F/\partial p|_{p=p_0} \simeq \delta\boldsymbol{\xi}_F/\delta p$, which allows an experimental determination of the vector $\mathbf{g}$ [100]. Knowing $\mathbf{g}$ the linear approximation of the map under the perturbation is given by [100]:

$$\boldsymbol{\xi}_{n+1} - \boldsymbol{\xi}_{\mathrm{F}}(p_0) = \delta p_n \mathbf{g} + [\lambda_{\mathrm{u}} \mathbf{e}_{\mathrm{u}} \cdot \mathbf{f}_{\mathrm{u}} + \lambda_{\mathrm{s}} \mathbf{e}_{\mathrm{s}} \cdot \mathbf{f}_{\mathrm{s}}][\boldsymbol{\xi}_n - \boldsymbol{\xi}_{\mathrm{F}}(p_0) - \delta p_{\mathrm{n}} \mathbf{g}], \quad (8.91)$$

where $\delta p_{\mathrm{n}} = \delta p$, and $\mathbf{f}_{\mathrm{u}}$ and $\mathbf{f}_{\mathrm{s}}$ are contravariant basis vectors defined by $\mathbf{f}_s \cdot \mathbf{e}_s = \mathbf{f}_u \cdot \mathbf{e}_u = 1$, $\mathbf{f}_{\mathrm{s}} \cdot \mathbf{e}_{\mathrm{u}} = \mathbf{f}_{\mathrm{u}} \cdot \mathbf{e}_{\mathrm{s}} = 0$. Just after the nth piercing the perturbation δp_{n}. Is determined so that $\boldsymbol{\xi}_{n+1}$ falls on the stable manifold of an unperturbed saddle point $\boldsymbol{\xi}_{\mathrm{F}}(p_0)$. δp_{n} can be chosen so that the scalar product $(\boldsymbol{\xi}_{n+1} - \boldsymbol{\xi}_{\mathrm{F}}) \cdot \mathbf{f}_{\mathrm{u}}$ of (8.91) behaves as a damping equation $(\boldsymbol{\xi}_{n+1} - \boldsymbol{\xi}_{\mathrm{F}}) \cdot \mathbf{f}_{\mathrm{u}} = \lambda_{\mathrm{s}}(\xi_{\mathrm{n}} - \xi_{\mathrm{F}} - \delta p_n \mathbf{g}) \cdot \mathbf{f}_{\mathrm{u}}$ (note: $|\lambda_{\mathrm{s}}| < 1$) :

$$\delta p_{\mathrm{n}} = \lambda_{\mathrm{u}}(\lambda_{\mathrm{u}} - 1)^{-1}(\boldsymbol{\xi}_n - \boldsymbol{\xi}_{\mathrm{F}}) \cdot \mathbf{f}_{\mathrm{u}}/(\mathbf{g} \cdot \mathbf{f}_{\mathrm{u}}), \quad (8.92)$$

where $(\boldsymbol{\xi}_n - \boldsymbol{\xi}_{\mathrm{F}}) \cdot \mathbf{f}_{\mathrm{u}}$ is the projected value of $\boldsymbol{\xi}_{\mathrm{n}} - \boldsymbol{\xi}_{\mathrm{F}}$ on the axis of unstable manifold. Equation (8.92) can be derived by subtracting the damping equation from (8.91). In this way one can capture the next piercing point on the axis of the stable manifold of $\boldsymbol{\xi}_{\mathrm{F}}(p_0)$. Once $\boldsymbol{\xi}_{n+1}$ falls on the stable manifold, the parameter perturbations can be set to zero, and the orbit for the subsequent time will approach $\boldsymbol{\xi}_{\mathrm{F}}(p_0)$ with the geometrical rate λ_{s} [100]. Since the iteration map in the actual systems may be nonlinear, slightly deviated from the linear approximation (8.90), it may not be possible to bring the orbit $(\boldsymbol{\xi}_{n+1})$ to the stable manifold. In addition, in the experimental systems, noise and errors prevent the correct determination of $\boldsymbol{\xi}_{\mathrm{F}}$, $\mathbf{f}_u$, $\mathbf{g}$ and λ_u, and thus prevent the next iterate being placed exactly on the stable manifold [100, 121]. In these cases one must keep on controlling; a new $\delta p = \delta p_{n+1}$ must be calculated for each iterate [121]. In order to achieve control, the parameter p can be varied in a small range: the parameter may be allowed to vary $-(p_0 + \delta p^*) < p < p_0 + \delta p^*$ [100]. Corresponding to this range, one can achieve control at the orbit which satisfies $|\boldsymbol{\xi}_n - \boldsymbol{\xi}_{\mathrm{F}}| < \boldsymbol{\xi}^*$ [121], where $\boldsymbol{\xi}^*$ is the allowed range for the control which is determined by the maximum perturbation δp^*. The condition $|\boldsymbol{\xi}_{\mathrm{n}} - \boldsymbol{\xi}_{\mathrm{F}}| < \boldsymbol{\xi}^*$ (a small circle) [100] may be replaced by a narrow stripe $(\boldsymbol{\xi}_n - \boldsymbol{\xi}_{\mathrm{F}}) \cdot \mathbf{f}_{\mathrm{u}} < \boldsymbol{\xi}^*$ [100, 130], where $\boldsymbol{\xi}^*$ is given from (8.92) by:

$$\boldsymbol{\xi}^* = \delta p^* |(1 - \lambda_{\mathrm{u}}^{-1})\mathbf{g} \cdot \mathbf{f}_{\mathrm{u}}|. \quad (8.93)$$

In an experiment of a parametrically driven magnetoelastic ribbon [106, 121, 130], the gravitationally buckled, amorphous magnetoelastic ribbon of $F_{81}B_{13.5}Si_{3.5}C_2$ is clamped at its base and periodically driven by a dc + ac magnetic bias, as shown schematically in Fig. 8.91. As the field varies, the Young's modulus alternatively increases and decreases, which acts as an important nonlinearity for revealing the chaotic buckling behavior [121, 130]. The voltage output $V(t)$ of the sensor is monotonically related to the ribbon's displacement from the vertical. For the experimental schematic diagram of controlling chaos, see Ref. [130]. The voltages were sampled at the drive period of the ac field by triggering a voltmeter off the ac signal, in a form of $X_n = V(t_n)$ where $t_n = n/f$; f is the driving frequency). Chaotic data

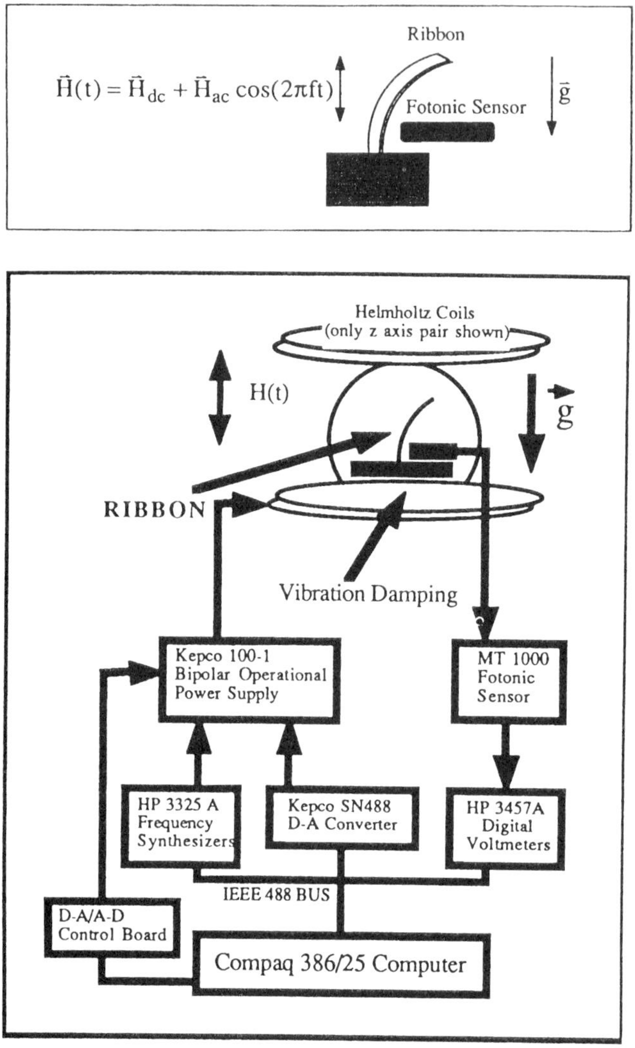

Figure 8.91 Experimental set-up for the control of a parametrically driven magnetoelastic ribbon (schematic). A triple axis set of Helmholtz coils serves to zero out the ambient field. In addition, the z-axis pair was used to gnerate the applied field [130].

are then given by a map of X_{n+1} vs X_n ($X_{n+1} = f(X_n)$), [121, 130], as shown in Fig. 8.92 [130]. In Fig. 8.92, the numbered points show the progression of the map after a close approach to the period 1 unstable fixed point, indicating that the orbit near the saddle point diverges very quickly over the whole attractor along the unstable manifold. In order to apply the OGY method to the 1D iterative map in Fig. 8.92, Ditto *et al.* [121] and Spano and Ditto [130] have defined the vector points,

$$\boldsymbol{\xi}_n = \begin{bmatrix} \mathbf{X}_{n+1} \\ \mathbf{X}_n \end{bmatrix}, \qquad \boldsymbol{\xi}_{\mathrm{F}} = \begin{bmatrix} \mathbf{X}_{\mathrm{F}} \\ \mathbf{X}_{\mathrm{F}} \end{bmatrix}, \tag{8.94}$$

where $\mathbf{X}_{\mathrm{F}}$ is the unstable fixed point in Fig. 8.92. Figure 8.93 shows the switching control between different periodic orbits [130]. Chaotic data are first controlled around a period 4 fixed point, and are then switched to period 1 control, followed by period 2 control, period 1 again, and finally back to period 4 control in Fig. 8.93. It takes a finite amount of time to acquire control when a switch is made. This is due to the fact that the OGY method specifies that one must wait for the system naturally to come within $\boldsymbol{\xi}^*$. For more details see Refs [121, 130]; the effect of noise on control is described in Ref. [130]. For more information see the book by Nayfeh and Balachandran [131].

Controlling chaos in electronic circuits has been investigated recently by many authors [124, 132–135]. Hunt [124, 132] has stabilized chaos in the diode resonator by 19 different orbits with periods as long as 23 drive cycles. For control in Chua's circuit see Ref. [111] and the references therein. A novel method of controlling chaos by a small time-continuous perturbation which does not require on-line computations has been proposed by Pyragas [136] and examined by the model simulations of (i) self-generated oscillations under crossed electric and magnetic fields in the regime of low-temperature impurity breakdown (dynamic Hall effect), and (ii) driven real-space transfer oscillations in modulation-doped heterostructures [137].

Chaos in circuits has been widely investigated based on the various motivations of the applications. The reports are quite diverse, so that the author cannot follow all of them, but they include applications of chaos in neural networks [138–141].

8.3.2 Chaos in device structures

8.3.2.1 GaAs metal-semiconductor field-effect transistor

This section discusses chaos in device structures which are not yet used as semiconductor devices of commercial basis but may be potentially developed as devices in the future. One of them is low-frequency oscillation (LFO) in the GaAs MESFET (metal-semiconductor field-effect transistor), related to the problem of semiconductor engineering [142–146], and another is the real-space transfer oscillations in GaAs/AlGaAs heterostructures [101, 147, 148]. By a few examples the author reminds semiconductor engineers that they should take precautions against the emergence of chaos in the fabrication of new devices.

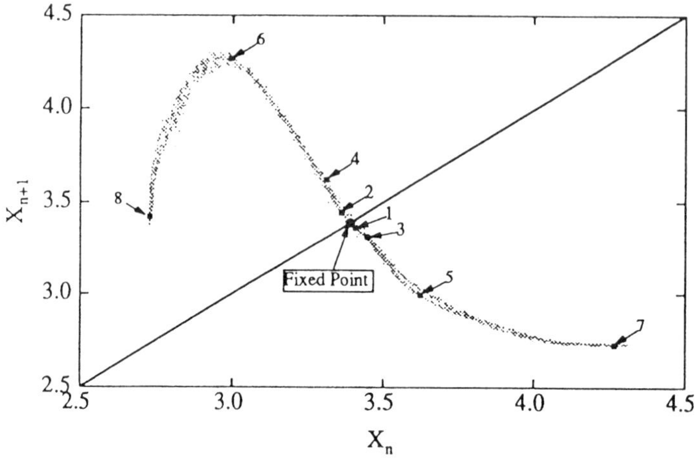

Figure 8.92 Typical attractor for the manetoelastic ribbon experiment. The numbered points show the progression of the map following a close approach to the period 1 unstable fixed point [130].

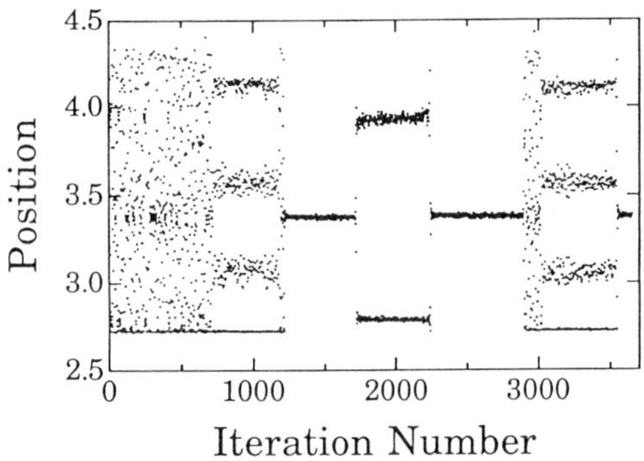

Figure 8.93 Switching between different periodic orbits by controlling. [130]; see also in Ref. [121].

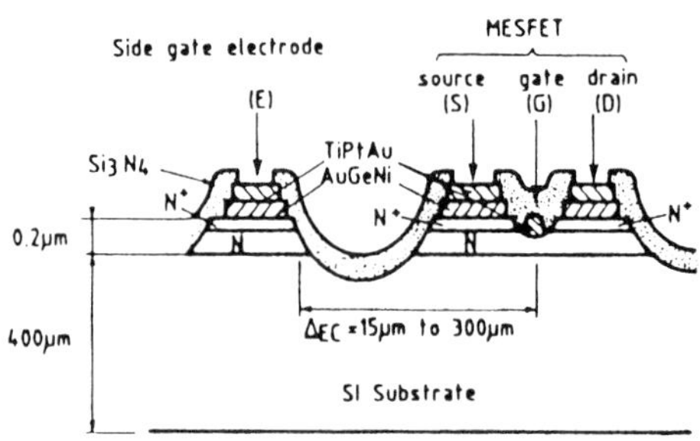

Figure 8.94 Cross-sectional view of a MESFET test-device structure with a side-gate electrode. [143].

In field-effect transistors (FETs), the drain current between source (S) and drain (D) is controlled by the gate voltage. GaAs MESFETs are expected to be used as high-speed, high-frequency transistors and integrated circuits (ICs). In the GaAs MESFETs fabricated on a semi-insulating substrate, however, the drain currents are modulated by a negative voltage applied on separated gate electrodes [142–146], a phenomenon called the "side-gating effect" or "back-gating effect" [142–146]. In a typical structure of the MESFET shown in Fig. 8.94 [143], the drain current is significantly affected and decreases with the application of a negative bias at a separate electrode, namely, the side-gate electrode ((E) in Fig. 8.94), showing the side-gating effect [143, 145, 146]. The back-gating effect is seen similarly when a bias is applied between a MESFET gate (G) and an additional electrode (back gate) formed at the bottom surface of a semi-insulating substrate (not shown in Fig. 8.94) [142, 144]. An example of the side-gating effect observed by Ohno and Goto [145] is shown in Fig. 8.95, where they have observed two types of side-gating effects, one with a threshold voltage for the drain current I_D at -5 V (Fig. 8.95a), and another without a threshold, in which the drain current begins to decrease from zero side-gate voltage (Fig. 8.95b). The side-gate effect is related to the leakage current I_{SG} in Fig. 8.95. Low-frequency oscillations (LFOs) are frequently followed by side-gating or back-gating effects [142–144, 146].

Such side-gating or back-gating effects as well as LFOs, are parasitic phenomena related to the leakage current through the SI substrate, which are detrimental effects in GaAs integrated circuits due to the interaction between closely spaced devices [142]. Many investigations have been done to clarify the parasitic phenomena [142–146]. The origins of the effect have been considered to be due to electron trapping on the Cr^{2+} and EL2 levels at the semi-insulating substrate–channel interface region [142] combined with the modulation of space charge by hot carrier transport at the channel–substrate interface region [143, 144]. The effects of the surface and hole injection on the side-gating effect have also been investigated in detail [143].

Many models have been proposed for the LFOs in semi-insulating GaAs [149–152]. Although the exact mechanism is not fully understood, it is generally accepted that the LFOs result from the field-enhanced capture of electrons by deep traps. The charge domains formed by trapping move with a velocity dependent on the current density and the time constant of the trap involved [149, 150]. The frequency of the current oscillation is determined by the transit time of the high-field domain traveling between an electron-injecting cathode and an anode on the semi-insulating GaAs. It has been also shown in an experiment [153] that two closely spaced deep levels (EL2 and EL0) cause the LFOs.

For various GaAs FET structures LFOs have been also observed [142–146, 154–156]. In some measurements, the LFOs under side-gating conditions have been related in a general way to oscillations occurring in

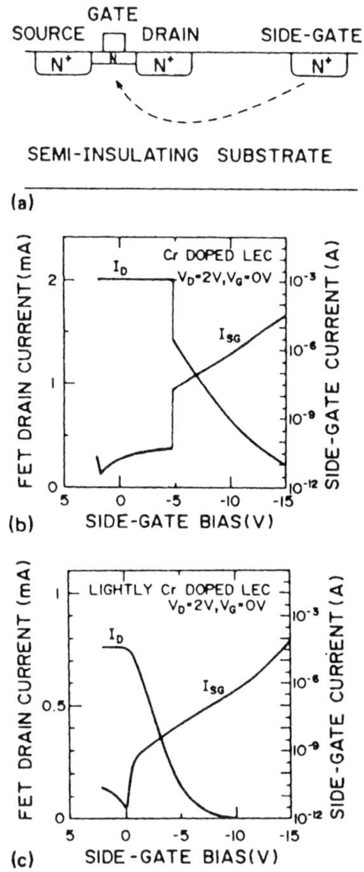

Figure 8.95 Two types of experimental side-gating effects. (a) Device configuration of GaAs MESFET with a side-gate. The distance between the side-gate and the transistor is 15μm. Negative bias voltages are applied on the side-gate electrode. (b) Side-gating effect with a threshold measured on the MESFET in a normal Cr-doped substrate. (c) Without a threshold measured on the MESFET in a lightly Cr-doped substrate [145].

semi-insulating substrates [155]. It has been also shown that the oscillations in the leakage current in a semi-insulating substrate directly result in oscillations in GaAs MESFET [144]. More physically, it has been shown [144] that the LFOs result from the modulation of the space-charge region at the active layer (channel)–substrate interface. Reviewing the various interpretations above, many problems remain unresolved before the exact mechanism can be resolved for LFOs in semi-insulating substrates and also in GaAs MESFET structures.

For the LFOs of GaAs MESFET under the side-gating condition, Yano *et al.* have shown that these oscillations reveal complex chaotic behaviors [146]. In their experiments the side-gating characteristics in Fig. 8.96 exhibit an S-shaped negative differential resistance (SNDR) under a constant current source measurement. In general, it is noted that the side-gate characteristics are roughly categorized into two groups [143]: one is a "gradual decrease" type without a threshold (Fig. 8.95c) and the other is a "step decrease" type with a threshold (Fig. 8.95b) [145, 146]. The S-shaped characteristic in Fig. 8.96 belongs to the latter type. The top part of Fig. 8.96 shows a characteristic of the drain current (I_D) vs side-gate voltage (in $-V_{SG}$), and the bottom part indicates the corresponding characteristic of side-gate current vs side-gate voltage. The GaAs MESFET circuit is shown in the inset of Fig. 8.96, where $V_G = 0V$, $V_S = 0V$, and $V_D = 0.5$ V. In Fig. 8.96, V_{SO} is the switchover voltage (breakdown voltage) and V_H is the

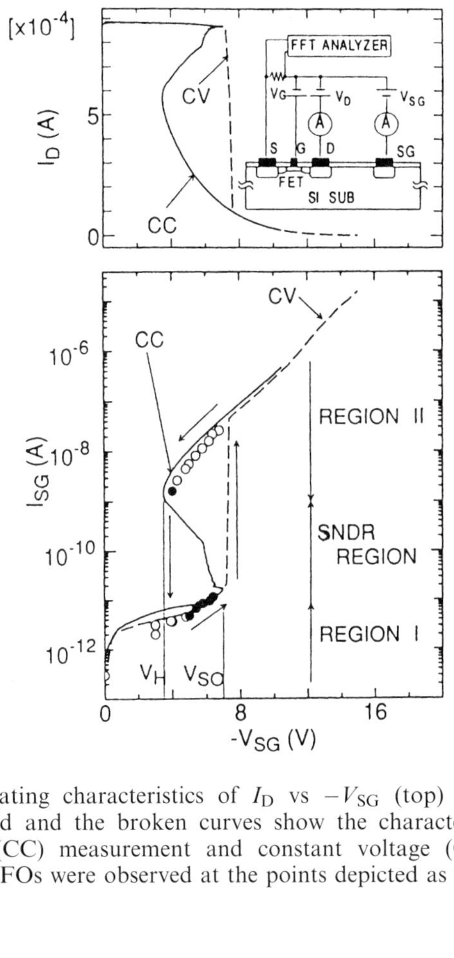

Figure 8.96 Side-gating characteristics of I_D vs $-V_{SG}$ (top) and I_{SG} vs $-V_{SG}$ (bottom). The solid and the broken curves show the characteristic obtained by constant current (CC) measurement and constant voltage (CV) measurement, respectively. The LFOs were observed at the points depicted as full circles [146].

holding voltage. The I–V curve is separated into three regions, region I, the SNDR region and region II.

LFOs have been observed for the "step decrease" type of FETs but not for the "gradual decrease" type. As the side-gate voltage V_{SG} changes from V_H to a more negative value, LFOs have been observed. In Fig. 8.96 solid circles indicate the points where the LFOs were observed, while open circle are the points where no oscillations were observed. LFOs have been observed in the lower branch of SNDR (solid circles in Fig. 8.96). When V_{SG} exceeds V_{SO}, the side-gate current increases sharply (see arrow in Fig. 8.96), while the drain current decreases sharply. After the phase transition no current oscillation can be observed [146]. By turning the side-gate voltage change in the reverse direction on the upper branch of the SNDR, no LFOs were observed (open circles) as low as V_H. When V_{SG} reached V_H the drain current oscillates again. The fluctuations in the drain current I_D were relatively small, 0.1–1% of the total drain current [146]. Under room light, Yano *et al.* found that the SNDR and the LFOs disappeared and that the side-gate characteristics became those of the "gradual decrease" type [146].

Under the SNDR condition Yano *et al.* [146] have observed complex oscillating behaviors of the drain current (Fig. 8.97). Oscillating waveforms depend largely on the lattice temperatures T_L. At $V_{SG} = -8.7$ V, fundamental frequencies for 320.1 K and 360.2 K are 9 and 0.6 Hz, respectively.

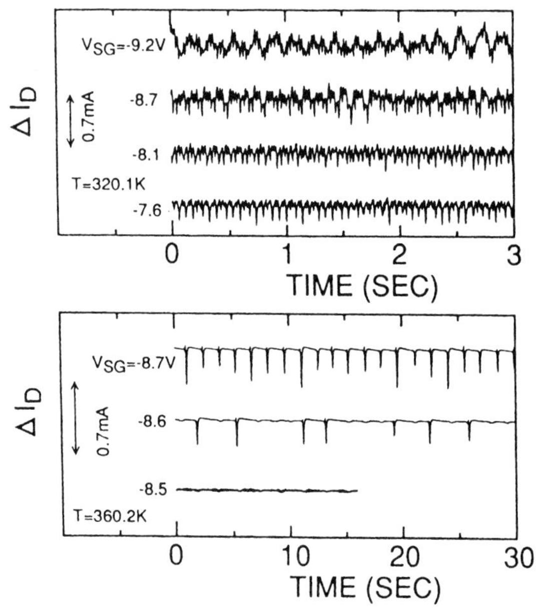

Figure 8.97 Waveforms or the LFOs at $T = 320.1$ K and $T = 360.2$ K [146].

For $T_L = 320.1$ K the waveforms are highly complex depending on the side-gate voltage, suggesting chaotic fluctuations. In the GaAs MESFET circuit, however, no clear bifurcations have been shown. In order to confirm the chaotic behaviors the correlation dimensions were analyzed by constructing the phase space attractor using a delay-time method. In Fig. 8.98 the correlation dimensions $\bar{\nu}$ are plotted as a function of the embedding dimension *m*. Yano *et al.* [146] obtained the saturated value $\bar{\nu}$, the lower bound of the fractal dimension, which was 1.7, 3.2, 3.2 and 3.6 for the waveforms at $V_{SG} = -7.3$, -8.1, -8.7 and -9.4 V, respectively. These saturated values indicate that the system behaviors could be deterministic and chaotic. For more details an adequate model simulation may be required.

For the SNDR Yano *et al.* [146] have interpreted the mechanism in Fig. 8.99. In region I (see Fig. 8.96), the side-gate current increases linearly as V_{SG} increases. Region I has been confirmed to be the Ohm's law region, the current of which is determined by the semi-insulating substrate resistance [146]. In this Ohm's law region, the potential distribution between the channel and the side-gate should be like n^+–n–n^+, as shown in Fig. 8.99b. In region I the activation energy, estimated from the temperature dependence of the side-gate leakage current, was found to be 0.7 eV, which is very close to the deep levels of the EL2 center of Cr^{2+} [142]. In region II light emission has been observed around the FET, probably at the channel–substrate interface [146]. The characteristic temperature in the 4000–5000 K range has been extraporated in the case of a Maxwellian distribution, assuming the emitted photon energy distribution to be directly related to the electron distribution [146]. These two results suggest the occurrence of the avalanche multiplication at the channel–substrate interface. In that case, holes generated by the avalanche multiplication may be collected in the vicinity of the side-gate n-region (Fig. 8.99c). This hole collection causes electron injection into the substrate from the side-gate [146]. As a result, double carrier injection occurs in the hole accumulation region of the semi-insulating substrate. The potential distribution between the channel and the side-gate should be like n^+–p–n^+, as shown in Fig. 8.99c. For the SNDR mechanism, impact ionization of deep traps in semi-insulating substrates has been also proposed [157].

Yano *et al.* [146] concluded that the decrease in the drain current is caused by the potential drop at the interface. In their interpretation, similar to Ref. [144], the LFOs reflect the variation in the channel–substrate junction depletion region thickness. They considered that the variation is caused by a combination of three mechanisms: local hole injection or generation, hole recombination in the substrate, and charge trapping–detrapping at the *n-i* junctions, both for the channel–substrate vicinity and for the side-gate vicinity [146]. Their main conclusion is that the LFO occurs as a result of the failure in balance between local hole injection and recombination in the

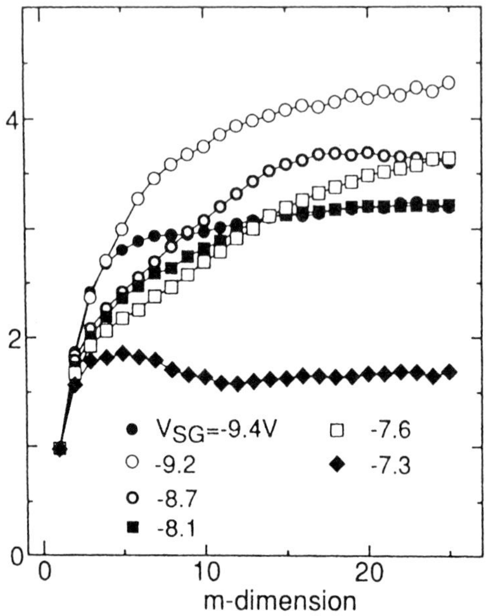

Figure 8.98 Critical exponents for the correlation dimension obtained for the various waveforms in Fig. 8.97, where $T = 320.1$ K. The obtained fractal dimensions are 1.7–3.6 [146].

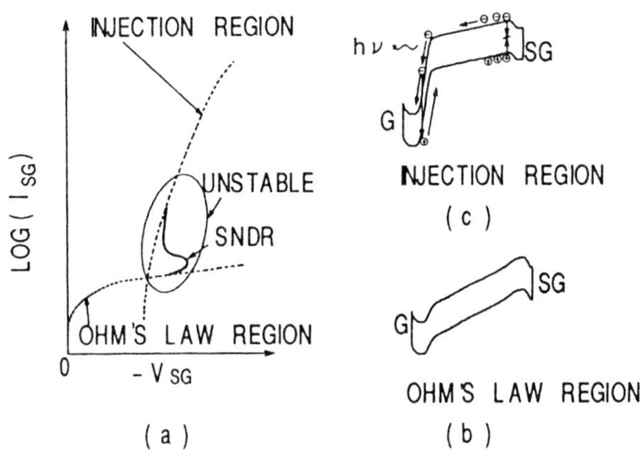

Figure 8.99 Diagram summarizing SNDR and LFO characteristics. (a) SNDR and LFOs appear at the cross-point of the Ohm's law curve and the injection curve; (b) potential distributions between the gate and the side-gate in the Ohm's law region; (c) potential distribution in the injection region. Avalanche breakdown occurs at high-field junction between channel and substrate. Generated holes are injected into a semi-insulating substrate [146].

semi-insulating substrates. Concerning these problems of the LFOs, no self-consistent model simulations have yet been presented. Because of the highly nonlinear and complex characters of the MESFET side-gating system, chaotic behaviors may not be so surprising. However, to the author's knowledge, the observation by Yano *et al.* [146] is the first one to explore chaos among a number of observations of LFOs in the MESFET side-gating effect.

8.3.2.2 *GaAs/AlGaAs heterostructures–real space transfer*

With the development of heterostructures and superlattices, the subject of the nonlinear carrier transport of low-dimensional carrier gas is growing and becoming more and more important [158, 159]. This section summarizes the numerical studies on the oscillatory instabilities and chaos for the real space transfer in heterostructures [101, 147, 148, 160–171] and also for the heterostructure hot-electron diode [172–175]. Although experimental studies in the heterostructures which show the N-shaped negative differential conductivity (NNDC) and oscillatory behaviors in the related problems of real space transfer are few, the numerical studies [160–170] may provide semiconductor engineers with precautions against the occurrence of chaos in modern semiconductor devices.

Basic ideas for the novel phenomenon of NNDC due to electron transfer in real space in a modulation-doped GaAs/Al_x $Ga_{1-x}As$ heterostructure have been proposed theoretically [101, 147, 148, 160–171] and experimentally [176–180]. Motivated by these investigations, and especially by Refs [176, 177, 181, 182], Aoki *et al.* [101] and Schöll and Aoki [160] have investigated numerically the oscillatory behaviors and chaos and have suggested the possibility of a high-frequency real space transfer oscillator, facilitating the numerical model simulations [161–170]. In the NNDC regime current oscillations of 2–25 MHz have been observed under the application of an ac-modulated bias [177], while in the numerical simulations [101, 160] intrinsic self-generated oscillations have been obtained in the 20–80 GHz range under dc bias. Figure 8.100 shows the conduction band diagram for the GaAs/AlGaAs heterostructure [160], where ΔE_c (denoted here by $\Delta\varepsilon_c$) is the conduction band discontinuity (interfacial band offset) and ϕ_B is the space-charge potential barrier. The AlGaAs layer is heavily n-doped with donor density N_D, while the GaAs layer is undoped. Under the thermal equilibrium the electrons fall into the GaAs quantum well, forming 2D electron gas (2DEG) at the interface. The electron mobility μ_1 of 2DEG is considered to be quite high owing to the reduced ionized impurity scattering, while mobility μ_2 in the AlGaAs layer may be much lower owing to strongly enhanced impurity scattering [147]. By applying a longitudinal electric field E with a few kV/cm, parallel to the layers, 2DEG are significantly heated. As a result, these electrons are thermionically emitted to the AlGaAs layer, where the electron mobility is quite low. Such a transfer is called "real space transfer", in analogy with the intervalley transfer in momentum space in the Gunn effect. Thus, with an electric field in the order of 2 kV/cm [176], the real space transfer from the high-mobility layer to the low-mobility layer causes NNDC, and oscillatory behaviors when the load line is inclined adequately [101, 160].

The model equations can be given by the hydrodynamic balance equation approach [101, 160]. A set of differential equations is described for the spatially averaged electron density n_i and the mean energy $\varepsilon_i (i = 1, 2)$, supplemented by the dielectric relaxation equation for the parallel electric field E and the rate equation for the interface potential barrier ϕ_B due to the space charge in the AlGaAs layer, by the following ODEs:

$$\frac{\mathrm{d}}{\mathrm{d}t} n_1 = \frac{J_{1\to 2} - J_{2\to 1}}{eL_1} \tag{8.95}$$

$$\frac{\mathrm{d}}{\mathrm{d}t}(n_1\varepsilon_1) = n_1 \left\langle \frac{\partial \varepsilon_1}{\partial t} \right\rangle_p + n_1 e\,\mu_1 E^2 + \frac{[J_{1\to 2}(\Delta\varepsilon_c + 2kT_1) - J_{2\to 1}(\Delta\varepsilon_c + 2kT_2)]}{eL_1} \tag{8.96}$$

$$\frac{\mathrm{d}}{\mathrm{d}t}(n_2\varepsilon_2) = n_2 \left\langle \frac{\partial \varepsilon_2}{\partial t} \right\rangle_p + n_2 e\,\mu_2 E^2 - \frac{[J_{1\to 2}(\phi_B + 2kT_1) - J_{2\to 1}(\phi_B + 2kT_2)]}{eL_1} \tag{8.97}$$

$$\left(1 + \frac{C}{C_i}\right) \varepsilon \frac{\mathrm{d}}{\mathrm{d}t} E = -\,\sigma_L (E - E_0) - \frac{en_1\,\mu_1 L_1 + en_2\,\mu_2 L_2}{L_1 + L_2} E \tag{8.98}$$

$$\frac{\mathrm{d}}{\mathrm{d}t}\phi_B = -e\,\mu_2 N_D (\phi_B - \phi_B^*) + \mu_2 \frac{e^3 L_1^2}{2\epsilon}(n_1^2 - n_1^{*2}) - e^2 L_1 L_2 \frac{\mathrm{d}}{\mathrm{d}t} n_1, \tag{8.99}$$

where

$$n_1 = \frac{1}{L_1}\int_{-L_1}^{0} n(x)\mathrm{d}x, \qquad n_2 = \frac{1}{L_2}\int_{0}^{L_2} n(x)\mathrm{d}x \tag{8.100}$$

are the spatially averaged electron densities in the two layers, $\varepsilon_i = \frac{3}{2}kT_i$ $(i = 1, 2)$ are the mean kinetic carrier energies, and T_i are the electron temperatures. The interface potential barrier in (8.99) is given by:

$$\phi_B = -e\int_0^{L_2} E_\perp \mathrm{d}x, \tag{8.101}$$

where $E_\perp$ is the perpendicular electric field in the AlGaAs layer which arises from the Poisson equation $\epsilon\ \partial E_\perp/\partial x = e(N_\mathrm{D} - n(x))$ due to the space charge [160]. Note that conservation of the total number of carriers holds:

$$n_1 L_1 + n_2 L_2 = N_\mathrm{D} L_2. \tag{8.102}$$

The thermionic emission current densities in (8.95), (8.96) are given by Bethe's theory [101]:

$$\begin{aligned} J_{1\to 2} &= -\, en_1(\varepsilon_1/3\pi m_1^*)^{1/2}\ \exp(-3\Delta\varepsilon_\mathrm{c}/2\varepsilon_1), \\ J_{2\to 1} &= -\, en_2(\varepsilon_2/3\pi m_2^*)^{1/2}\ \exp(-3\phi_\mathrm{B}/2\varepsilon_2), \end{aligned} \tag{8.103}$$

where m_i^* are the effective mass. Further, $\langle\partial\varepsilon_i/\partial t\rangle_\mathrm{p}$ are the energy loss rates due to polar optical scattering [160], $U_0 = E_0 d$ is the applied bias voltage, $C_\mathrm{i} = \epsilon h(L_1 + L_2)/d$ is the internal capacitance for sample dimensions h, d, and $L_1 + L_2, C$ is the external (package, contact) capacitance, ϵ is the permitivity, $\sigma_\mathrm{L} = [(h/d)(L_1 + L_2)R_\mathrm{L}]^{-1}$ with load resistance R_L, and * in (8.99) denotes steady-state values. Equations (8.95)–(8.97) can be derived from a moment expansion of the Boltzmann equation [20, 183], and represent particle continuity and energy balance. In particular, eqs (8.96), (8.97) include in the r.h.s. energy dissipation (first term), Joule heating (energy gain, second term), and energy flux due to convection, heat conduction and pressure-induced energy flow [101] (third term). Equations (8.98), (8.99) describe the dielectric relaxation of the parallel and perpendicular electric fields, respectively [160].

The physical origin of (8.99) is explained as follows. The perpendicular electric field is described by Poisson's equation:

$$\epsilon\ \partial E_\perp/\partial x = e[N_\mathrm{D} - n(x)], \tag{8.104}$$

and the continuity equation

$$\epsilon\ \partial E_\perp/\partial t = -e(\mu_2 n E_\perp + D_2(\partial n/\partial x)) + eL_1(\partial n_1/\partial t), \tag{8.105}$$

supplemented with the boundary conditions

$$\begin{aligned} \phi(0) &= -\ \phi_B, &\quad \partial\phi(0)/\partial x &= -eE_{\perp 0}, \\ \phi(L_2) &= 0, &\quad \partial\phi(L_2)/\partial x &= 0, \end{aligned} \tag{8.106}$$

where $\partial\phi/\partial x = -eE_\perp$, D_2 is the diffusion constant. Integrating (8.104) with aid of (8.99), (8.102), (8.106), one finds:

$$E_{\perp 0} = -e(N_\mathrm{D} - n_2)L_2/\epsilon = -en_1 L_1/\epsilon. \tag{8.107}$$

From (8.105) one obtains by integrating over $\int_0^{L_2} dx$ using (8.100) and

$n(x) = N_D - (\epsilon/e)(\partial E_\perp/\partial x)$, $\Delta n \equiv n(L_2) - n(0)$, and (8.107)

$$d\phi_B/dt = (e/\epsilon)[-\mu_2 N_D\ \phi_B + \mu_2 e^2 L_1^2 n_1^2/2\epsilon + eD_2\Delta n - eL_1L_2(dn_1/dt)]. \tag{8.108}$$

By putting (8.108) in the steady state and subtracting it, one obtains (8.99), where the approximation $\Delta n^* = \Delta n$ has been made [160]. For the numerical simulations the dimensionless variables are used: $\nu_i = n_i/N_D$, $\tilde{E} = E(\mu_1/v_{ds})$, $\tilde{\varepsilon}_i = \varepsilon_i/\ \Delta\varepsilon_c$, $\eta \equiv \phi_B/\ \Delta\varepsilon_c$, $\tilde{t} = t/\ \tau_r$, $\gamma_1 = \tau_r/\tau_0$, $\gamma_2 = \tau_r/\tau_\varepsilon$, $\gamma_3 = \tau_r/\tau_L$, $\ell = L_1/L_2$, $\ell_1 = L_1/L_D$, $\ell_2 = L_2/L_D$, where v_{ds} is the drift saturation velocity, $\tau_r \equiv \epsilon/(e\mu_2 N_D)$ is a dielectric relaxation time, $L_D = \sqrt{kT_L\ \mu_2\ \tau_r/e}$ is a Debye length (T_L: lattice temperature), $\tau_0 = \sqrt{3\pi m_1^*/\Delta\varepsilon_c}\ L_1 \exp(1.5)$ is an effective thermionic emission time, τ_ε is an effective energy relaxation time, and $\tau_L = \varepsilon(1 + C/C_i)/\sigma_L$ [101, 160]. The physical parameters are; $L_1 = 186$ Å, $L_2 = 1550$ Å, $h = 1$ mm, $d = 50\ \mu$m, $\Delta\varepsilon_c = 250$ meV ($x = 0.25$ for $Al_xGa_{1-x}As$), $\phi_B^* = 25$ meV, $N_D = 10^{16}/cm^3$, $m_1^* = 0.067\ m_0$, $m_2^* = 0.087175\ m_0$, $\mu_1 = 38\,000\ cm^2/Vs$, $\mu_2 = 152\ cm^2/Vs$, $v_{ds} = 2 \times 10^7$ cm/s, $T_L = 300$ K, $\epsilon = 12.5\ \epsilon_0$, $R_L = 283\ \Omega$, and $C = 1.57 \times 10^{-14}$ F.

The steady-state solutions of (8.95)–(8.97) yield N-shaped current–density field characteristics [101] which exhibit NDC for $E^* > 1.6$ kV/cm. Time-dependent simulations show the oscillatory instabilities with self-generated oscillations of the carrier densities, energy, field and potential barrier in a range of applied dc bias in the NDC regime [101, 160]. Figure 8.101 shows three types of oscillation which occur subsequently by increasing the applied bias field: (a) $E_0 = 11.749$ (6.18kV/cm), (b) 11.8063 (6.21 kV/cm), and (c) 11.914 (6.27 kV/cm). The phase portraits in Fig. 8.101 indicate anticlockwise limit cycles which change from low frequency ($\sim$ 25 GHz) with large amplitude (Fig. 8.101a) to high frequency ($\sim$ 40 GHz) with small amplitude (Fig. 8.101c) via a period-doubled state (Fig. 8.101b). The bifurcation has been investigated in the control parameter space (L_1, E_0). Complex nonlinear behaviors including intermittent spiking, boundary crisis and stationary chaos have been found in a limited narrow range of the control parameter space [160]. For more details see Ref. [160].

Reznik and Schöll have extended their model simulations to the oscillatory instabilities by real space transfer in double heterostructures [147, 163, 168], for the schematic structure given in Fig. 8.102a, where the GaAs quantum well layer is sandwiched by two identical AlGaAs layers [147]. Note that the previous works [101, 160] were done on the single heterostructure. They have shown that quantum size effects and the coupled real space transfer into both AlGaAs barriers lead to new nonlinear dynamical features, where classical thermionic emission and nonresonant tunneling from the 2D subbands $\varepsilon = \varepsilon_i + \hbar^2(k_y^2 + K_z^2)/2m_1^*$ $(i = 1, 2)$ of the GaAs quantum well with tunneling coefficients D_1, D_2 are incorporated. In

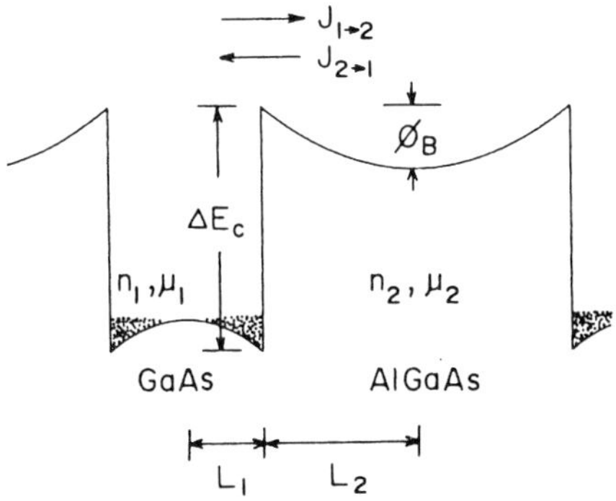

Figure 8.100 Schematic energy-band diagram of a GaAs/AlGaAs heterostructure $\Delta\varepsilon_c$ is the conduction band discontinuity, ϕ_B is the space-charge potential barrier [160]. In this tect, notation ΔE_c for the energy in the figure should be replaced by $\Delta\varepsilon_c$.

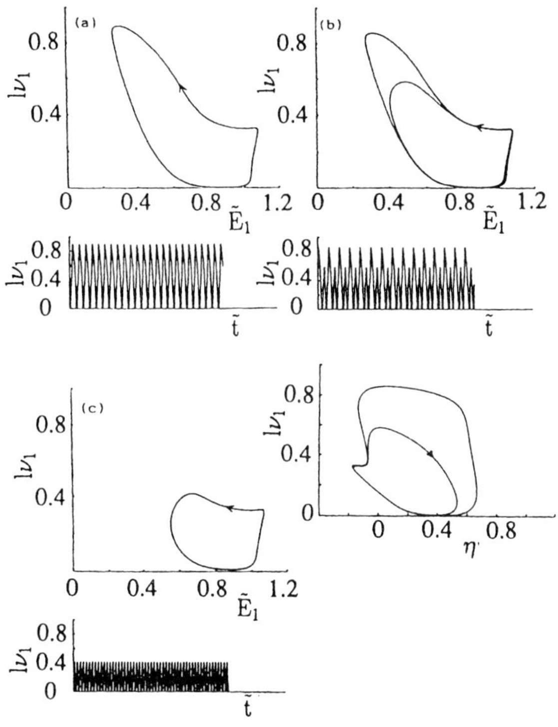

Figure 8.101 Simulation of self-generated oscillations for applied bias fields (a) $E_0 = 11.749$ (6.18kV/cm), (b) 11.8063 (6.21 kV/cm), and (c) 11.9114 (6.27 kV/cm). Phase portraits of $\ell\nu_1$ (electron density n_1 in units of $N_D L_2/L_1$ vs $\tilde{\varepsilon}_1$ (mean energy ε_1 in units of $\Delta\varepsilon_c$), and time series of n_1 vs time in units of τ_r are shown. In (b) the phase portrait of $\ell\nu_1$ vs η (potential barrier ϕ_B in units of $\Delta\varepsilon_c$) is also shown [160]. In this text, the notation $\tilde{E}_1$ for energy in the figures should be replaced $\tilde{\varepsilon}_1$.

the static current-field characteristic NDC is quenched as a result of the modified 2D density of states, but oscillatory instabilities still occur [147, 163]. In this case oscillatory real space transfer occurs in various directions, as shown in Fig. 8.102b. By increasing the applied bias U_0 one first observes a Hopf bifurcation of a nonsymmetric oscillation mode (NA) and then of a symmetric self-generated oscillation mode (SA), in which the electrons cycle synchronously or asynchronously, respectively, between the two barriers and the well [147, 163] (see the arrows in Fig. 8.102b during one periodic cycle). There exists bistability of these symmetric or asymmetric oscillation modes. SA or NA can be selected by the initial distributions of the electrons over the three layers. With further increase in U_0, periodic SA bifurcates to period 2 and then chaos via quasi-periodicity [147,163]. Reznik and Schöll have found many other interesting phenomena, such as a chaotic repeller.

Utecht and Schöll [147, 162] have further extended their model to coupled quantum wires. Oscillatory behaviors in two coupled modulation-doped wires A and B in Fig. 8.103 have been investigated, where wires can be realized by mesa etching. Each wire represents an identical real-space transfer oscillator. In the model simulations the quasi-1D electronic structure and density of states of the laterally confined electron gas in the GaAs layer have been taken into account. Coupling between two oscillators can be achieved via the load of the external circuit. The two quantum wires are coupled via the dielectric relaxation equation analogous to (8.98) [162]. As a consequence, for symmetric (identical) quantum wires it has been shown that only the symmetric oscillation mode is stable (Fig. 8.103a) while the nonsymmetric oscillation is transient, representing a periodic repeller (Fig. 8.103c, [162]). The stability character of the symmetric and non-symmetric mode can be interchanged by connecting an external capacitor C_{ext} parallel to the quantum wires (Fig. 8.103b). In this case, Utecht and Schöll have shown that the phase lag of the two wires leads to frequency-doubled current oscillations [147, 163].

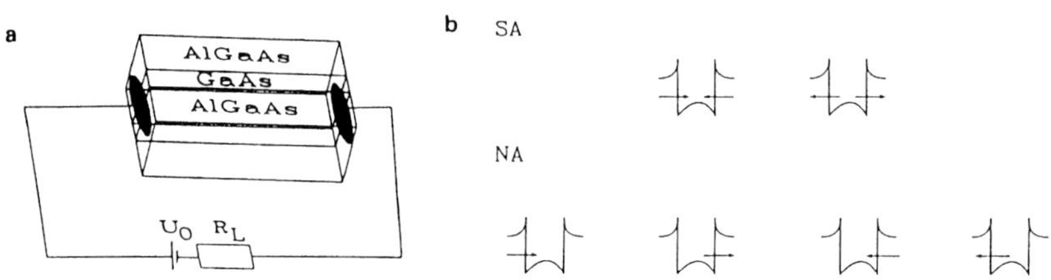

Figure 8.102 (a) Double heterostructure (schematic). (b) Symmetric (SA) and nonsymmetric (NA) oscillation mode of the double heterostructure (schematic). The arrows show different phases of the electron motion during one oscillation cycle [147].

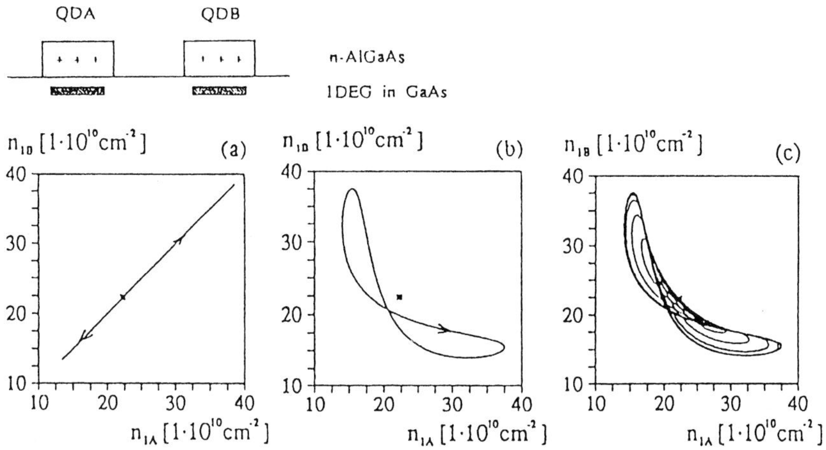

Figure 8.103 Phase portraits of carrier density in quantum wire B vs that in wire A (see top): (a) symmetric, (b) nonsymmetric, and (c) nonsymmetric transient oscillations for two symmetric quantum wires of width 50 Å. After Refs [147, 162].

Optically induced real space transfer has been investigated by Kunz and Schöll [184], who considered the real space transfer induced by intersubband infrared (IR) excitation of carriers in the GaAs quantum well and by subsequent tunneling into the adjacent AlGaAs layer, instead of field-induced carrier heating and thermionic emission under a relatively small electric field. The main part of their numerical results is the appearance of optoelectronic bistability without any oscillatory behaviors. For detailed discussions see Refs [147, 184].

8.3.2.3 GaAs/AlGaAs heterostructures–heterostructure hot electron diode

Another example is vertical hot-electron transport in the heterostructure hot-electron diode (HHED), which is a two-terminal device proposed by Hess *et al.* [185] and examined by Kolodzey *et al.* [186] and Higman *et al.* [187]. Motivated by the proposition of the new device [185], Wacker and Schöll have investigated numerically the self-generated oscillations in the HHED [172–175]. Originally, Hess *et al.* [185] proposed a new ultrafast switching device by a HEED which should work as quickly as 200 fs. Soon later Kolodzey *et al.* [186] have examined microwave frequency operation at 77 K in the HEED, which has a structure of three periods of $Al_{0.45}Ga_{0.55}$. As barriers alternating with GaAs quantum wells, and observed self-generated spontaneous oscillations with the harmonic spectral lines up to 10.5 GHz. (Note that resonant tunneling devices oscillate at frequencies as high as 200 GHz [187]; the high electron mobility transistor (HEMT) can produce amplification at frequencies as high as 94 MHz [188].) Higman *et al.* [187] have also shown room-temperature switching and S-shaped negative differential resistance in the HHED of a single barrier wide-gap AlAs/GaAs heterostructure.

The originally proposed mechanism for the HEED is shown in Fig. 8.104. The undoped AlGaAs layer ($\sim$ 2000 Å) is sandwiched by a thin undoped GaAs:n$^-$ layer ($\sim$ 2000 Å) and a thick GaAs:n$^+$ substrate [185]. By applying a negative bias to the left contact, vertical carrier transport occurs in two different modes of tunneling current and thermionic emission current. For lower electric fields in the lightly doped (nominally undoped) region, electrons will stay close to the conduction band minimum and drift towards the barrier. In this regime the current is due to tunneling through the barrier. For higher electric fields, the electron gas in the lightly doped layer will be heated to energies of typically 500 meV for a field of 10^4V, and the thermionic emission current over the barrier will dominate (Fig. 8.104a). By the mechanisms of these two modes, the *I–V* characteristic is the S-type (Fig. 8.104b), as evidenced experimentally under the pulsed operation [185].

Wacker and Schöll have modeled the novel HHED device under the current-controlled conditions, and have shown the appearance of voltage oscillations, stationary filament and periodic spiking of the filament [147, 168, 175]. In their model two states for the carrier transport are given by the conduction band diagram in Fig. 8.105, where the electric fields in layers 1 and 2 are denoted E_1 and E_2, and the current densities j_1 and j_2, respectively. These two layers are sandwiched between two n^+-GaAs buffer layers which act as good reservoirs of free electrons. At the interface the surface charge ρ_s (or 2DEG) accumulates, which is supplied by the lightly doped donors (N_D) in the GaAs layer. Electrons injected from a n^+–GaAs layer pass through the thin GaAs layer almost ballistically and reach the interface where there is a barrier $\Delta\varepsilon_c$ due to the interfacial band offset. The current density j_1 is transferred into the AlGaAs layer by tunneling at low electric fields (left of Fig. 8.105) and by thermionic emission at high electric fields (right of Fig. 8.105). Thus, one of the major descriptions in the model has been devoted to the carrier transport in the GaAs layer. In Ref [172] a displaced heated Maxwellian was used for the electron distribution in the GaAs layer. The appearance of ballistic peaks in the distribution function was later verified in MC simulations [189], and Wacker and Schöll have discussed pure ballistic transport instead of the displaced heated Maxwellian, whereby they have considered only polar optical scattering and electron transfer from Γ- to L-valleys as the dominant scattering mechanisms at high electric fields [174, 175].

The current density j_1 was first determined as follows [175]. Let the z-axis be perpendicular to the heterojunction. Then the electrical potential energy $-e\varphi(z)$ is given as in Fig. 8.105, where $\varphi(0) = 0$ and L_1 and L_2 are the lengths of the two layers. After the electrons are injected from the reservoir

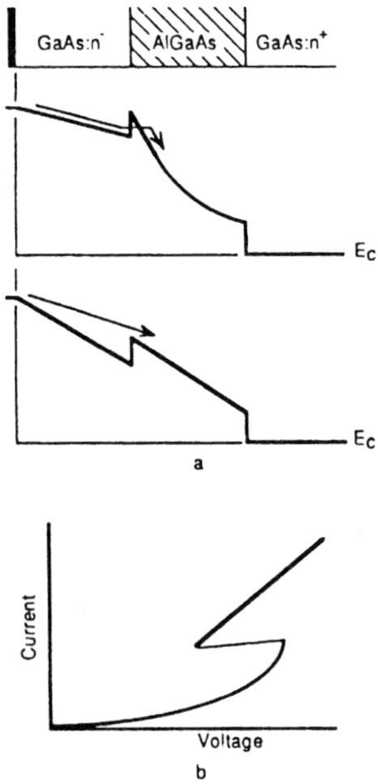

Figure 8.104 (a) Energy-band diagram of heterostructure hot-electron diode showing the two path electrons; tunneling current at low electric fields and thermionic emission current at high electric fields. (b) The resultant current–voltage characteristic [185].

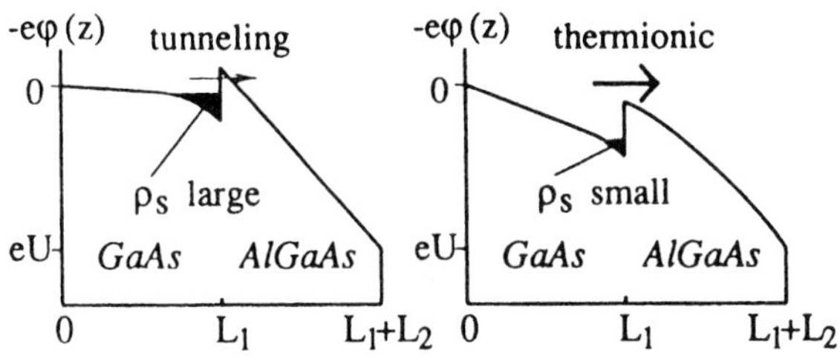

Figure 8.105 Schematic energy diagram of the HHED showing bistability between a low-current tunneling state and high-current state dominated by thermionic emission across the barrier [168, 175].

at $z = 0$, the electrons may have the ballistic energy $\varepsilon_{\mathrm{bal}}(z) = e\varphi(z)$, as found from Fig. 8.105. If $\varepsilon_{\mathrm{bal}}(z) < \Delta\varepsilon_{\mathrm{L}}$ (band offset), polar optical scattering is the main scattering mechanism. However, if $\varepsilon_{\mathrm{bal}}(z) \gg \hbar\,\omega_0$, which is justified for high electric fields, the velocity v of the electrons is nearly unchanged for small scattering angles [175]. Therefore, it is assumed that $m_\Gamma v(z)^2/2 = e\varphi(z)$ holds for $e\varphi(z) < \Delta\varepsilon_{\mathrm{L}}$. If $e\varphi(z) = \Delta\varepsilon_{\mathrm{L}}$ at some point z_{L} in the GaAs layer, the effective mass approximation breaks down. In this case, the maximum velocity in the Γ-valley is set as $v(z) = \sqrt{2\Delta\varepsilon_L/m_\Gamma}$ [190]. Furthermore, electron transfer from Γ- to L-valleys occurs with a constant intervalley scattering rate. Taking the scattering length $L_s \sim 1000$ Å, as indicated by MC simulations [191], the current density by the Γ-valley contribution is $j_{\Gamma 1}(z) \approx j_1[1 - (z - z_L/L_s)]$ and by the L-valley contribution $j_{L1} = j_1 - j_{\Gamma 1}(z)$. Owing to the larger effective mass and the effective deformation-potential scattering, the electron velocity v_L is much smaller in the L-valley. Here, $v_L = 10^7$ cm s^{-1} is used. Since

$$n_1 = n_{\Gamma 1} + n_{L1} \quad \text{and} \quad j_1 = en_{\Gamma 1}v_{\Gamma 1} + en_{L1}v_{L1}, \tag{8.109}$$

the ratio $j_{\Gamma 1}(L_1)/j_{L1}(L_1) \sim L_s/(L_1 - z_L)$ at the interface may read as $n_{\Gamma 1}v_{\Gamma 1}/n_{L1}v_{L1}(v_{\Gamma 1}$ and v_{L1} are known from the above arguments). From the ratio one can find $n_{\Gamma 1}$ and n_{L1}, and thus j_1 from (8.109), where n_1 is dependent on the electric field E_1, for instance $n_1 = N_D + \epsilon_1(-E_1)/(2eL_1)$ in Ref. [172]. In combination with the Poisson equation it is possible to derive the explicit expressions $U_1(E_1)$ and $j_1(E_1)$, where $U_1 = -\varphi(L_1)$ and $E_1 = -\varphi'(L_1)$ [175].

The current density is obtained by calculating the thermionic emission current density and tunneling current density, where the tunneling current density of the Γ-valley contribution is given by $j_{\Gamma 2} = j_{\Gamma 1}(L_1)\int d\varepsilon f_\Gamma(\varepsilon)D_\Gamma(\varepsilon)$ where $D_\Gamma(\varepsilon)$ is the WKB tunneling coefficient for the barrier at given energy ε. The distribution function $f_\Gamma(\varepsilon)$ exhibits a pronounced peak at the ballistic energy $\varepsilon = -eU_1$, but broadens owing to the energy loss by optical phonon scattering. The transmission current by the L-valley contribution is calculated similarly. The reflected electrons at the interface are considered to remain in the GaAs layer and reach the 2DEG interface. Such electrons participate again in the tunneling current, which is denoted j_T. The total current density in the AlGaAs layer is then given as $j_2(E_1, E_2) = j_{\Gamma 2} + j_{L2} + j_T$. In the GaAlAs layer, the electron transfer from the GaAs layer generates

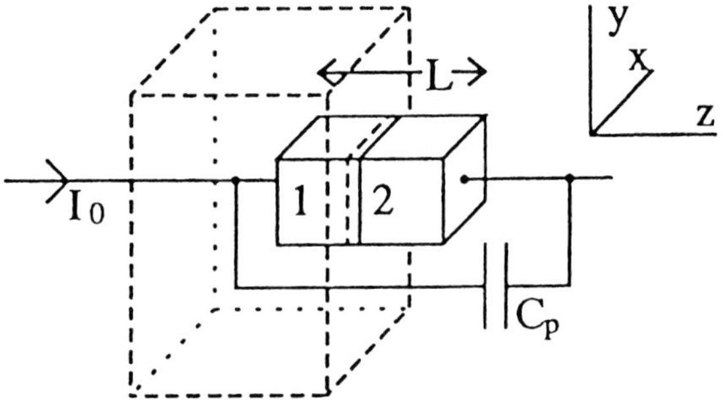

Figure 8.106 Sample geometry. Surface of intgration (dotted) used for determining the dynamics of U [168].

the charge density $\rho_2 = j_2/v_s (v_s = 10^7 \text{cm/s})$, which influences the voltage drop $U_2: \; U_2 = L_2 E_2 + \rho_2 L_2^2/2\epsilon_2$.

From these considerations the spatiotemporal dynamics under the current-controlled conditions are modeled for the rate equations of the charge density ρ_s at the interface and the potential drop $U = U_1 + U_2$ across the sample with the sample geometry in Fig. 8.106 [147, 168, 173, 175]:

$$\frac{\partial \rho_s(x, y, t)}{\partial t} = j_1 - j_2 - \left(\frac{\partial j_x^{\|}}{\partial x} + \frac{\partial j_y^{\|}}{\partial y} \right), \tag{8.110}$$

$$\frac{\mathrm{d}U}{\mathrm{d}t} = \frac{1}{C + C_p} \left\{ I_0 - \int_A \mathrm{d}x \, \mathrm{d}y \left(\frac{C}{C_1} j_1 + \frac{C}{C_2} j_2 \right) \right\}, \tag{8.111}$$

where $j_x^{\|}, j_y^{\|}$ are the current densities which are parallel to the interface, I_0 is the external current, $C_1 = 4A \; \epsilon_1/3L_1$, $C_2 = A \; \epsilon_2/L_2$, and $C^{-1} = C_1^{-1} + C_2^{-1}$. Equation (8.110) is the rate equation for the charge density which is derived from the continuity equation $\dot{\rho} + \text{div} \, \mathbf{j} = 0$. (8.111) is derived from Ampere's law, div $(\epsilon \dot{\mathbf{E}} + \mathbf{j}) = 0$. For the derivations see Refs [147, 168]. Wacker and Schöll have shown a Hopf bifurcation from the nonoscillatory stationary state to the oscillatory behaviors, as shown in Fig. 8.107 [147, 172]. Qualitative descriptions for the oscillation cycle, as visualized in the

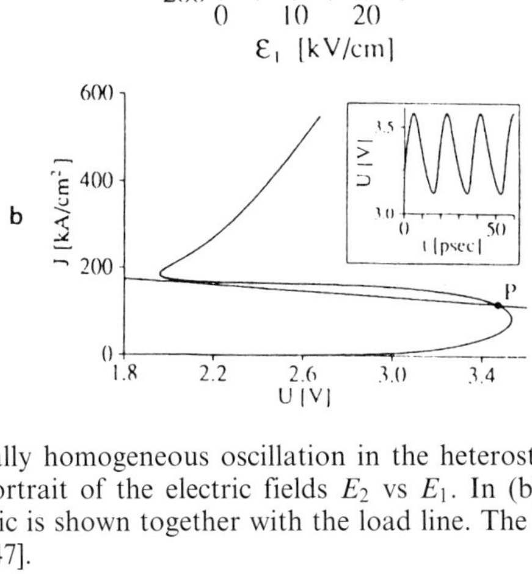

Figure 8.107 Spatially homogeneous oscillation in the heterostructure hot-electron diode. (a) Phase portrait of the electric fields E_2 vs E_1. In (b) the static current–voltage characteristic is shown together with the load line. The inset depicts voltage oscillations U(t) [147].

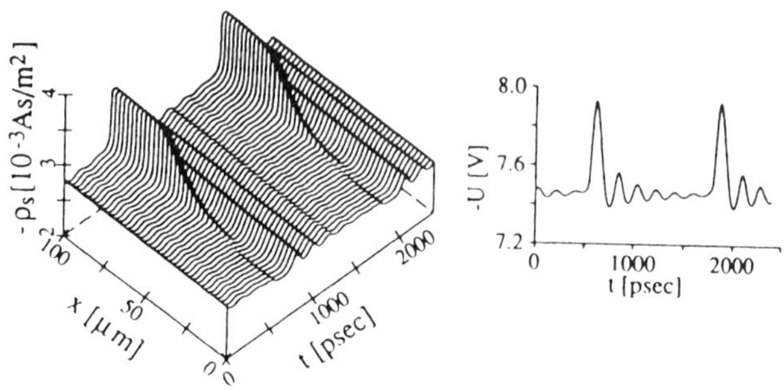

Figure 8.108 Spiking behavior in the HHED for the current-controled bias condition with $J_0 = 65.2\ \mathrm{kA/cm^2}$ [175].

(E_1, E_2) phase portrait, can be given as follows [147]. For small electric fields E_1, E_2 the tunneling current is small $(j_2 < j_1)$. Hence ρ_s increases from (8.110) and consequently E_2 increases while E_1 remains almost constant. For large E_2 the tunneling current becomes large due to a reduced barrier width (see Fig. 8.105), and thus $j_2 > j_1$, which decreases ρ_s and increases E_1, enhancing thermionic emission over the barrier. Subsequently, E_2 decreases until $j_2 < j_1$ is reached again, which completes the oscillatory cycle [147]. Full descriptions of the spatiotemporal behaviors describe the formation of a filament and the spiking behaviors [175]. Figure 8.108 shows the spiking of the transverse profiles of the charge density ρ_s (in the x-direction) under the current-controlled condition $J_0 = 65.2\ \mathrm{kA/cm^2}$.

For a better understanding of the complex phenomenon, Wacker and Schöll have investigated the oscillatory behaviors by introducing a very simple "generic" model [192]. With this model they have shown homogeneous steady state, stationary filament, homogeneous oscillation, periodic spiking and chaotic spiking. For details see Refs [168, 192].

8.3.2.4 Self-generated oscillations of domains in superlattices

Wide investigations into carrier transport in superlattices have been conducted [158, 159]. One of the variety of phenomena is the formation of electric-field domains [193–197] and oscillatory behaviors [198, 199]. The electric-field domain in GaAs–AlAs superlattices was first reported by Esaki and Chang in 1974 [193]. However, it is only recently that many researchers have begun to pay attention to the detailed mechanisms for domain formation and oscillatory behaviors [194–199]. Here, the recent experimental results are introduced followed by description of some of the theoretical approaches.

In the GaAs–AlAs superlattices, tunneling resonances occur with several subbands of energy ε_i $(i = 1, 2, \ldots)$ between adjacent GaAs quantum wells if

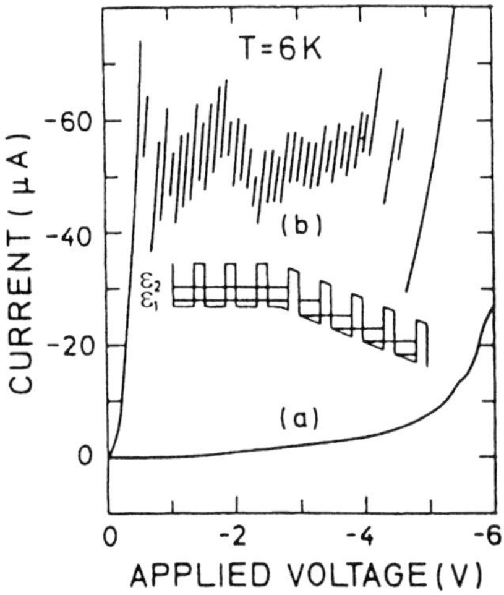

Figure 8.109 Dark current vs applied voltage, for (a) n^+–i–n^+ diode structure and (b) n^+–n–n^+ diode structure of SL. The inset shows a schematic diagram of the voltage distribution in the presence of domains [200].

an electric field $E_i = (\varepsilon_i - \varepsilon_1)/ed$ is applied perpendicular to the layers, where d is the period of the superlattices. If the GaAs quantum wells are n-doped at the large carrier densities, the applied voltage breaks up into at least two domains with two different electric fields E_1, E_2, where the domains are separated by a domain boundary (see below). Very often, $E_1 \sim 0$ and $E_2 = (\varepsilon_2 - \varepsilon_1)/ed$ takes a high value. Current is limited by the domain boundary because the transport by tunneling is nonresonant at the boundary. Typically, the superlattice which exhibits the domain boundary has the well width $d_W \sim 9$ nm (GaAs, Si doped at $3 \times 10^{17}/\mathrm{cm}^3$) and a barrier $d_B \sim 4$ nm (AlAs), and hence $d = d_W + d_B \sim 13$ nm with $N = 40$ periods [200]. The superlattice is sandwiched between two highly doped layers of AlGaAs which act as the electron reservoirs. Thus, the transport system is similar to the n^+–n–n^+ diode. Figure 8.109 shows the typical I–V characteristic for the n^+–n–n^+ configuration of the doped superlattice in (b), and the n^+–i–n^+ configuration of the undoped superlattice in (a), measured by Grahn *et al.*, where the superlattice periods are $N = 40$ [200]. The characteristic for the n^+–n–n^+ configuration exhibits 35 regularly spaced discontinuities, where jumps in the current can be understood as a movement of the boundary by one superlattice period. The schematic figure for the domains is shown in the inset of Fig. 8.109. In the plateau region of the I–V curve, the two domain field strengths $E_1 (\simeq 0$ V/cm) and E_2 are related to the resonant tunneling between the lowest subbands ($1 \rightarrow 1$) are resonant tunneling between the lowest and second subbands ($1 \rightarrow 2$) in the adjacent quantum wells, respectively (see the schematic illustration in

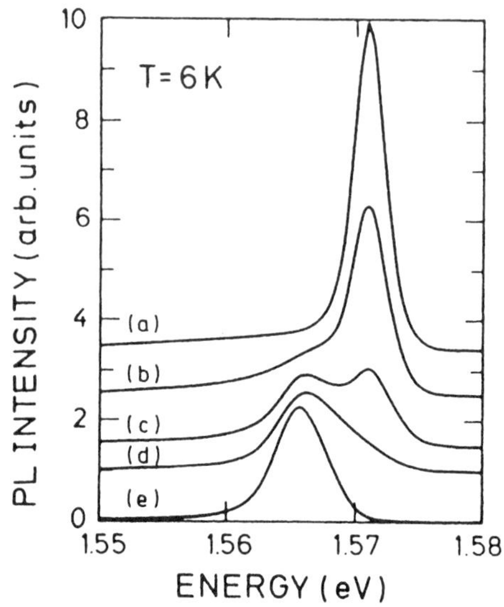

Figure 8.110 Photoluminescence spectra of the n^+–n–n^+ diode structure at different applied voltages: (a) −1 V, (b) −2 V, (c) −3 V, (d) −4 V, and (e) −5 V. The spectra are shifted vertically for clarity [201].

Fig. 8.109). The applied field is calculated by $E_A = V_0/Nd$, where V_0 is the applied voltage [200]. The voltage drop per period is simply V_0/N. For $V_0 = -4.6$ V, dividing by the number of jumps in Fig. 8.109b, the voltage drop per period is found to be 131 mV, in excellent agreement with the spacing between the first and second conduction subbands [200].

Direct evidence for the electric-field domains has been shown by Grahn *et al.*, who measured the photoluminescence spectra as a function of applied voltages [200]. By the quantum-confined Stark effect (QCSE) [195, 197], the photoluminescence spectra show a red shift and a direct identification for the well-defined high-field domain [200] (see the illustration in Fig. 8.109). The photoluminescence spectra in Fig. 8.110 show such evidence, where the excitation density is weak and well below the doping level in GaAs quantum wells. By increasing the applied voltage two distinct photoluminescence lines are observed, where the peaks do not shift in energy. The emission line at higher photon energy is due to the band-gap transition without QCSE which can be observed at low electric field in the regime of low-field domain, while the emission line at lower photon energy is caused by a red shift due to the QCSE. By comparing the energy positions of these lines with the Stark shift of the band-gap transitions, Grahn *et al.* [200] have identified the electric-field strengths of the domains as E_1 and E_2, in agreement with the indirect conclusion drawn from the I–V curve in Fig. 8.109. From these measurements it is quite reasonable to consider that the region of the high-field domain grows period by period with increasing applied voltage. The absence of the two-peak photoluminescence structure in the n^+–i–n^+ configuration of the superlattice is in agreement with the smooth *I–V* curve in Fig. 8.109, and hence this means a homogeneous electric field distribution without domain boundary [200].

Kastrup *et al.* [201] have shown hysteretic behaviors of the *I–V* characteristics of sharp current jumps. Figure 8.111 shows a typical example of the hysteretic behavior observed for negative bias voltages sweeping from 0 to -5 V (sweep-up) and from -5 to 0 V (sweep-down), where the superlattice consists of 40 periods of 9 nm GaAs wells and 4 nm AlAs barriers and the central 5 nm of each GaAs well is Si-doped at $3 \times 10^{17}/\text{cm}^3$. In Fig. 8.111, there are fewer than 40 branches (fewer than the superlattice periods, $N = 40$), suggesting that the domain boundary jumps over more than one period at the beginning and at the end of superlattice [201]. Between 0 and -2 V the branches clearly overlap for the two sweep directions, and are disconnected between the sweep directions for voltages between -2 and -4.8 V.

The hysteretic (bistable) behaviors are enlarged in Fig. 8.112. At the domain boundary the local transport is due to nonresonant tunneling, and the boundary between different field domains is connected with a 2D charge accumulation with the density $n_{2D} = \epsilon_0 \epsilon_r \Delta E/e$, where ΔE is the change in the electric field at the domain boundary and $\epsilon_0 \epsilon_r$ is the dielectric constant. By sweeping up the bias voltage, accumulated charge moves from a certain well *i* (domain boundary) to the adjacent well at *i*–1 via discharging at the well *i* and charging at *i*–1, whereby an upper branch may switch to a different branch, as indicated by the right arrows in Fig. 8.112b. Turning the voltage sweep towards lower voltages by the sweep-down, for instance at -3.5 V in Fig. 8.112b, the branch keeps its new state of the boundary locating at the $i - 1$ well, leading to the hysteretic behavior. Sweeping down the voltage further, the lower branch shows a jump to a state of the domain boundary locating at the well *i*. Therefore, each lower branch for the sweep-down is identical (connected) to the upper branch for the sweep-up, as indicated by the broken lines in Fig. 8.112b. Thus, the bistability occurs because for each current branch the domain boundary is located in a different well [201]. In Fig. 8.112a, the branches partly overlap, but the behaviors are similarly argued. Kastrup *et al.* [201] have also shown tristability or higher multistability for a superlattice sample with a Si-doping of $1 \times 10^{18}/\text{cm}^3$. For the multistability branches, the charge boundary may be located in one of three or more adjacent wells. By turning the voltage sweep carefully, any of the multistability branches can be realized [201]. For these measurements Kastrup *et al.* have used the model calculations.

Under laser excitation, the domains become unstable. Merlin *et al.* found damped current oscillations in the MHz range by using a steplike laser

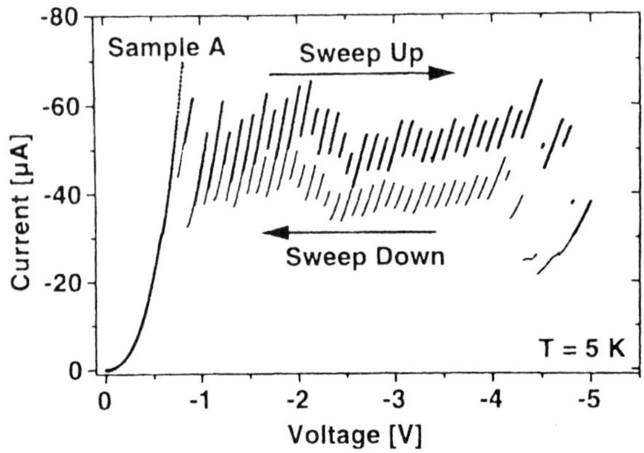

Figure 8.111 Typical current–voltage characteristic at 5 K of a doped superlattice. The sweep-up was taken from 0 to -5 V, and the sweep-down from -5 to 0 V. After Ref. [201].

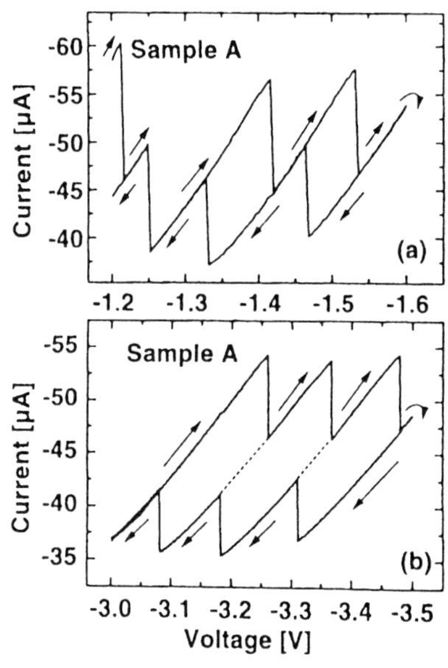

Figure 8.112 Two enlarged sections of the *I–V* characteristic at 5 K as shown in Fig. 8.111. In (a) a bistable region is shown, while in (b) the tristable regime is presented. The dashed lines in (b) have been added to indicate that the current branches are connected [201].

excitation [198]. Damped Gunn oscillations have been also found in strongly coupled superlattices by using intense picosecond light pulse [199]. Recently, Kastrup *et al.* observed self-generated stationary oscillations in doped GaAs–AlAs superlattices using an intensity stabilized Ti:sapphire laser

[202]. Under the intense laser excitation, sharp jumps in the current (cf. Fig. 8.109) smear out. Depending on the applied voltage the oscillating frequency varies between 250 kHz at low voltages and 20 MHz at high voltages [202]. A typical example of the oscillation is shown in Fig. 8.113, from the trace of the time-resolved current and the Fourier power spectra. The calculation revealed that the current oscillations are due to a motion of the domain boundary over several periods [202].

For these measurements of the high-field domains, Bonilla *et al.* [203, 204], Kastrup *et al.* [201], Wacker *et al.* [205], Prengel *et al.* [206] and Schöll and Wacker [168] have conducted intense investigations using their theoretical models. Theoretical approaches on the instability of the homogeneous field distribution have been taken by Laikhtman [207] and Laikhtman *et al.* [208]. Here, some essential parts of the theoretical works are described, after Prengel *et al.* [206], Wacker *et al.* [205] and Schöll and Wacker [168].

A GaAs/AlAs supperlattice with a period $d = L + B$ is considered, where L is the width of the GaAs well and B the width of the AlAs barrier. A weakly coupled quantum well system is considered, so that the width of the minibands calculated by means of the Kronig–Penny model is less than 1 meV [206]. For the weakly coupled quantum well system, the carrier transport can be modeled by the transition rates, as illustrated in the schematic conduction band diagram of Fig. 8.114 [168]. For the field domains the main transition rates may be due to the resonant tunneling between the first subbands (denoted by $1 \rightarrow 1$) and between the first and second subbands ($1 \rightarrow 2$). In order to describe the model adequately, various transition rates are further taken into account in Fig. 8.114, where X and Y are the rate coefficients for resonant tunneling $1 \rightarrow 2$ and $2 \rightarrow 1$, respectively, R_i is the rate of electrons crossing the barrier between two equivalent subbands i, and $A^{(k)}$ is the intersubband transition rate. Let $E^{(k)}$ be the averaged electric field between the wells with the index $(k-1)$ and (k). By denoting the 2D electron density by $n_1^{(k)}$ for the first subband in the kth well ($k = 1, \ldots, N$), and $n_2^{(k)}$ for the second subband, the rate equations for the electron densities are simply given by:

$$\begin{aligned}\dot{n}_1^{(k)} &= A^{(k)} + n_1^{(k-1)} R_1(E^{(k)}) + n_2^{(k+1)} Y(-E^{(k)}) + n_2^{(k+1)} Y(E^{(k+1)}) \\ &\quad - n_1^{(k)} R_1(E^{(k+1)}) - n_1^{(k)} (X(-E^{(k)}) + X(E^{(k+1)})),\end{aligned} \tag{8.112}$$

$$\begin{aligned}\dot{n}_2^{(k)} &= -A^{(k)} + n_2^{(k-1)} R_2(E^{(k)}) + n_1^{(k-1)} X(E^{(k)}) + n_1^{(k+1)} X(-E^{(k+1)}) \\ &\quad - n_2^{(k)} R_2(E^{(k+1)}) - n_2^{(k)} (Y(-E^{(k)}) + Y(E^{(k+1)})),\end{aligned} \tag{8.113}$$

where dots indicate the time derivatives. For the transition rates X and Y, Prengel *et al.* [206] have solved the Schrödinger equation by means of the perturbation theory. First, they calculated the first and second eigenstates $|1\rangle$ and $|2'\rangle$ of the single quantum well, where $|2'\rangle$ is the second eigenstate of

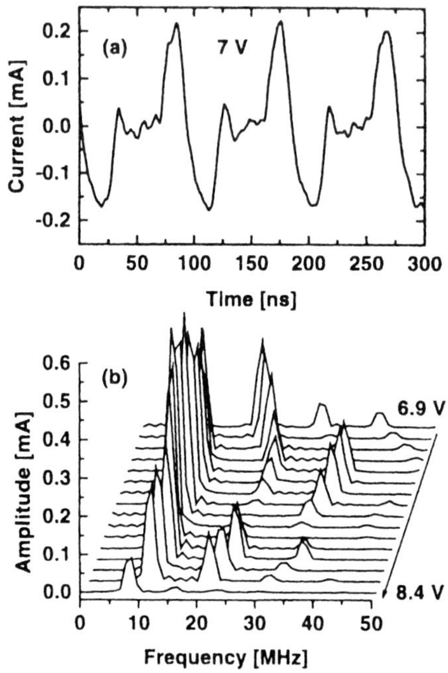

Figure 8.113 (a) Typical trace of the time-resolved current at 7 V with a dc average of 0.8 mA at 5 K. (b) Fourier transforms of the dark current oscillations vs frequency for different applied voltages (0.1 V steps) [202].

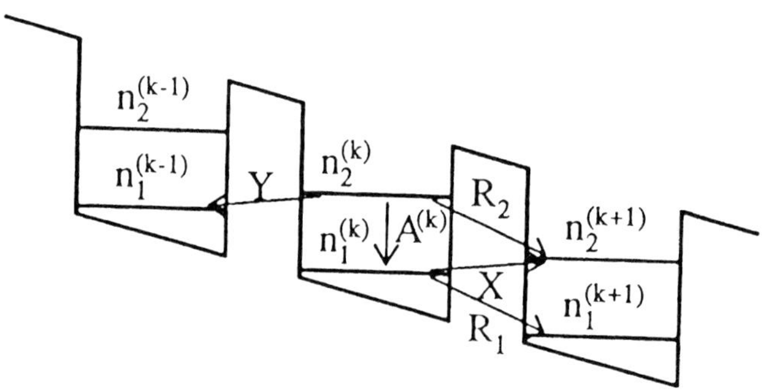

Figure 8.114 Conduction band vs position of the superlattice with an applied electric field and various transition rates (schematic). After Ref. [168]; see also Fig. 1 of Ref. [206].

the subband of the adjacent well. Each well is first treated as being isolated, but the eigenstate is solved under a small perturbation by an electric field (see Fig. 1 in [206]). These wave functions $\psi_1(x)$, $\psi_2(x)$ are called old wave functions. Then the coherent superposition of the old wave functions

$\psi_1(x) = \langle x|1\rangle$ and $\psi_2(x) = \langle x|2'\rangle$ gives rise to new, nonlocalized wave functions $\psi_+(x) = \langle x|+\rangle$, $\psi_-(x) = \langle x|-\rangle$ with corresponding energies ε_+, ε_-, approximately satisfying the Schrödinger equation,

$$H|\pm\rangle = (\varepsilon_\pm - V_0)|\pm\rangle \tag{8.114}$$

with the new Hamiltonian

$$H = T + V_1 + D_1 + V_2 + D_2, \tag{8.115}$$

where T is the kinetic energy of the electrons, V_1 and V_2 are the potential profiles for the two adjacent wells, and D_1 and D_2 are the perturbation potentials of the two wells under the applied voltage. The solution of (8.114) gives the mixed states between the time-dependent wave function given by [206]:

$$\Psi_\pm(x,t) = \exp\,[i(\varepsilon_\pm - V_0)t/\hbar]\psi_\pm(x). \tag{8.116}$$

From these treatments, time dependences of $\psi_1(x)$, $\psi_2'(x)$ are found. Starting with an electron in the state $|1\rangle$ at $t = 0$, its time-dependent wave function $|\Psi(t)\rangle$ oscillates in time with the frequency $\nu = (\varepsilon_+ - \varepsilon_-)/2\pi\hbar$ [168, 206]. The probability of finding the electron in level $2'$ at time $t > T_0$ is given by $|\langle 2'|1(t)\rangle|^2$, with $\psi_1(x,t) = \langle x|1(t)\rangle$. After $t = \pi\hbar/\Delta\varepsilon$ (half the oscillation period), $|\langle 2'|1(t)\rangle|^2$ becomes maximum in time, which is denoted by $W_{12'}$. The transition rate from $|1\rangle$ to $|2'\rangle$ can now be obtained by [168, 206]:

$$X(E) = W_{12'}\frac{\varepsilon_+ - \varepsilon_-}{2\pi\hbar} \tag{8.117}$$

Y is calculated analogously. In (8.112), (8.113) inverse tunneling processes are denoted by $X(-E)$ and $Y(-E)$, which are very small in general. The rate R_i $(i = 1,2)$ of electrons crossing the barrier between two equivalent subbands (eqs (8.112), (8.113)) can be given by using the expressions for the transport in a miniband $\varepsilon_i(q) = \varepsilon_i - \Delta_i \cos(qd)$:

$$R_i = \frac{\Delta_i}{\hbar}\sqrt{\frac{\tau_\mathrm{m}}{\tau_\mathrm{e}}}\frac{E/\tilde{E}}{1 + (E/\tilde{E})^2}, \tag{8.118}$$

where $\tilde{E} = \hbar(ed\sqrt{\tau_\mathrm{m}\,\tau_\mathrm{e}})$, and τ_m and τ_e are the scattering times for momentum and energy relaxations, respectively. This expression has been derived in Ref. [209]. For the intersubband transition rate $A^{(k)}$

$$A^{(k)} = n_2^{(k)}/\tau_{21} \tag{8.119}$$

has been used, where $\tau_{21} = 1$ ps is the intersubband scattering time [168]. If the energy difference $\varepsilon_2 - \varepsilon_1$ is greater than the energy $\hbar\omega_0$ of LO phonons, the main intersubband relaxation mechanism is polar optical scattering with $\tau_{21} = 1$ ps [206]. For the numerical simulations several fundamental equations are required in addition to (8.112), (8.113). From Gauss' law

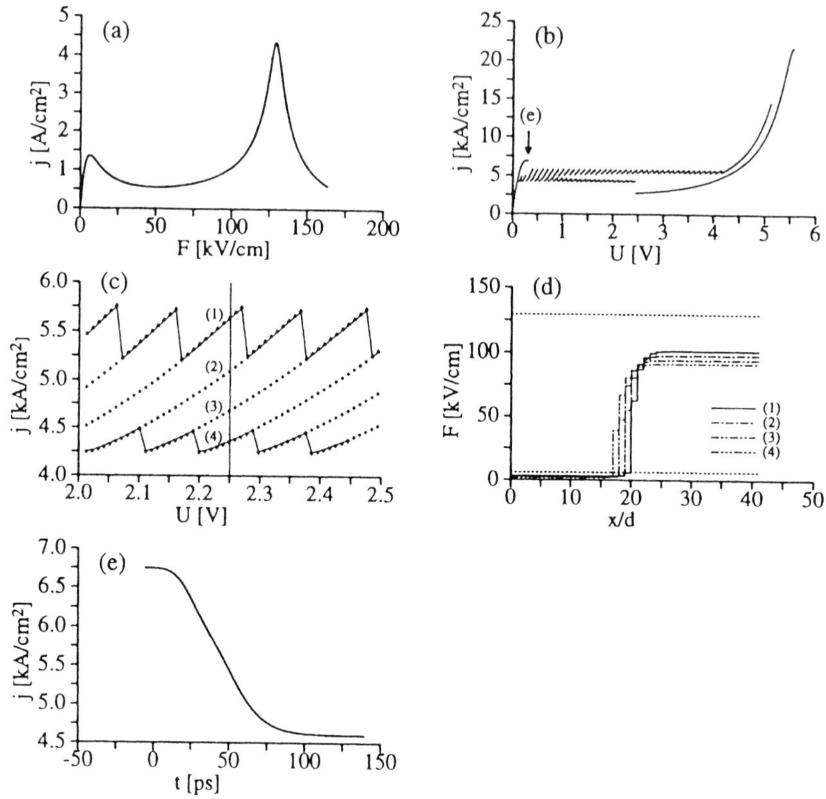

Figure 8.115 (a) Homogeneous steady state of j(F) for $N_D = 1.34 \times 10^{14}/\text{cm}^3$. (b) Current–voltage characteristic for $N_D = 6.7 \times 10^{17}/\text{cm}^3$ for sweep-up (upper line) and sweep-down (lower line) of the bias U. (c) Multistability for $2\ \text{V} < U < 2.5\ \text{V}$, where all stable branches are shown. Only the solid lines are found for sweep-up and sweep-down of the voltages. (d) Field distribution for the four stable states at $U = 2.1$ V, corresponding to the numbered points in (c). The dotted lines correspond to the maxima of $j(F)$ in (a). (e) Time dependence of the current if the bias is switched from $U = 0.295$ V to $U = 0.315$ V at $t = 0$ (see arrow in (b)). After Ref. [205].

one has:

$$e(n_1^{(k)} + n_2^{(k)} - N_D) = \epsilon_w \epsilon_0 \frac{E_{k+1} - E_k}{L}, \quad k = 1, \ldots, N, \tag{8.120}$$

where $\epsilon_w \epsilon_0$ is the dielectric constant in the wells. The applied bias voltage U is the sum of the voltage drops across each period:

$$d \sum_{k=1}^{N+1} E_k = U. \tag{8.121}$$

An important problem for modeling lies in the treatment of the contacts. Satisfactory descriptions of electrons tunneling from a 3D electron gas in the contact layer into different 2D electron populations inside the first quantum well of the superlattice and vice versa on the opposite electrode (anode) seem to be difficult. Prengel *et al.* have replaced the contact layers by additional quantum wells labeled 0 and $N+1$ and set the boundary conditions [206]:

$$n_1^{(0)} = n_1^{(1)}, \; n_2^{(0)} = n_2^{(1)}, \tag{8.122}$$

$$n_1^{(N+1)} = n_1^{(N)}, \; n_2^{(N+1)} = n_2^{(N)}, \tag{8.123}$$

Using (8.120), (8.121) the electric field E_k can be expressed by the electron densities $n_i^{(k)}$ and the applied voltage U (external parameter), thus it is possible to eliminate them from the rate equations. In this way, one obtains a set of $2N$ dynamic equations corresponding to (8.112), (8.113):

$$\dot{n}_1^{(k)} = f_1(n_1^{(1)}, \; n_2^{(1)}, \ldots, n_1^{(N)}, \; n_2^{(N)}; U), \tag{8.124}$$

$$\dot{n}_2^{(k)} = f_2(n_1^{(1)}, \; n_2^{(1)}, \ldots, n_1^{(N)}, \; n_2^{(N)}; U). \tag{8.125}$$

After solving (8.124), (8.125) by the Runge–Kutta procedure, the total current density from quantum well $k-1$ to k is given by [8, 205]:

$$\begin{aligned} j_{k-1\to k} \equiv j(E_k) = L[& n_1^{(k-1)}\{R_1(E_k) + X(E_k)\} - n_1^{(k)} X(-E_k) \\ &+ n_2^{(k-1)}\{R_2(E_k) + Y(-E_k)\} - n_2^{(k)} Y(E_k)]. \end{aligned} \tag{8.126}$$

For more details see Ref. [206].

For the uniform field distribution (homogeneous steady-state solution), $n_1^{(k)} = n_1, n_2^{(k)} = n_2$ $(n_1 + n_2 = N_\mathrm{D})$, $\dot{n}_1^{(k)} = \dot{n}_2^{(k)} = 0$ and $E_k = U/(N+1)d$ $(k = 1, \ldots, N+1)$. For the numerical parameters Prengel *et al.* [206] used those in Table 8.8. Figure 8.115a shows the I–V characteristic for the uniform field distribution [205], where $N_\mathrm{D} = 1.34 \times 10^{14}/\mathrm{cm}^3$ for a weakly doped sample [200]. The small peak at low field distribution is due to miniband conduction, and the large peak reflects resonant tunneling $1 \to 2'$. In general cases, the uniform distribution is not realized $(E_k = U/(N+1)d)$, especially for highly doped samples [200], followed by the domain formation. Figure 8.115b [205] shows such an example, where many current jumps (as large as the number of superlattice periods) and the hysteretic behavior [200, 201] are well explained. Multistability of the different current branches is shown in Fig. 8.115c. In particular, corresponding to current branches (1)–(4), it is indicated in Fig. 8.115d that the different branches are associated with different sizes of field domain [205]. The boundary between the domains is found to be of the order of two periods. Figure 8.115e shows the temporal switching of the current at the jump marked in Fig. 8.115b, showing the switching time in the order of 50 ps, which is quite small compared with the experimental value of 1 ns [210]. The difference may result from the large barrier width in the experiment.

Table 8.8 Numerical parameters in the simulations

Symbols	Values	Meanings of the symbols
L	90 Å	Thickness of GaAs layers
B	15 Å	Thickness of AlAs layers
d	105 Å	Superlattice period
V_0	982 meV	Conduction band discontinuity
m_1^*	$0.067m_e$	Effective electron mass in GaAs layer
m_2^*	$0.15m_e$	Effective electron mass in AlAs layer
$\in_W$	13.18	Relative permittivity of GaAs
T_L	5 K	Lattice temperature
T_{sc}	1 ps	Intersubband scattering time
τ_e	2 ps	Energy relaxation time
τ_m	0.005 ps	Momentum relaxation time
N	20	Number of quantum wells
E_1	56.2 meV	Ground state energy level (calculated)
E_2	221.5 meV	First excited state energy level (calculated)
λ_1	3.04×10^{-4}	Overlap integrals (calculated; see text)
λ_2	1.72×10^{-4}	Overlap integrals (calculated; see text)
n_c	1.34×10^{16} cm^{-3}	3D degeneracy concentration
N_D	5.35×10^{16} cm^{-3}	Donor concentration, $N_D = 4n_c$

In the model simulations by Wacker *et al.* [205], their main conclusions were that switching between different domain sizes occurs in the highly doped case only (as seen by a number of current jumps in Fig. 8.115b), while no domain formation is possible for the weakly doped case (Fig. 8.115a), in agreement with the experimental results [200, 201]. Furthermore, they found the oscillatory behaviors for intermediate doping [205]. Figure 8.116a shows the current oscillations for two different doping densities when the bias is switched from 0.01 V to 0.5 V at $t = 0$, where the lower trace is a transient damping oscillation for low doping density $N_D = 1.34 \times 10^{16}/\text{cm}^3$. Corresponding to persistent oscillations in Fig. 8.116a, the potential distribution is shown as a function of time in Fig. 8.116b. As seen clearly in Fig. 8.116b, the oscillation is due to a repetitive cycle of formation and annihilation of a pair of low-field domain ($x/d = 0$–30) and high-field domain ($x/d = 30$–40). It is noted that these oscillations are quite different from the oscillations in the Gunn diode, in which a spatially localized domain travels from cathode to anode repeatedly. Similar oscillations have been found experimentally by Grahn *et al.* [210], who studied transient oscillatory behavior using time-

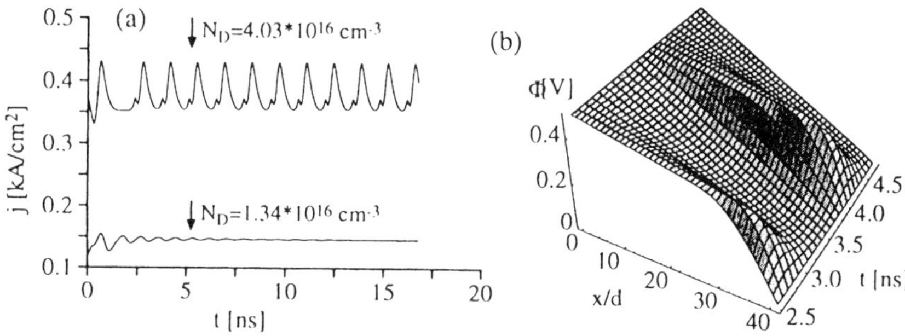

Figure 8.116 (a) Current oscillations for two different doping densities when the bias is switched from 0.01 V to 0.5 V at $t = 0$. (b) Time dependence of the potential distribution for $N_D = 4.03 \times 10^{16}/\mathrm{cm}^3$ [205].

resolved photoluminescence spectroscopy. Domain formation and oscillatory behaviors have been found to be quite sensitive to the growth-related disorder of superlattices. Wacker *et al.* [211] have shown numerically that the oscillating frequency increases in the presence of the disorder. For the numerical studies on transient oscillatory behaviors under laser excitation, see Bonilla *et al.* [203, 204].

For the heterostructure hot-electron diode, in a very recent reference [212] experimental oscillatory behaviors and MC simulations have been compared.

References

[1] S. Chandrasekhar, *Liquid Crystals* (Cambridge University Press, Cambridge, 1977); D. Demus and L. Richter, *Textures of Liquid Crystals* (Verlag Chemie, Weinkim, 1978).

[2] F.H. Busse, in *Hydrodynamic Instabilities and the Transition to Turbulence,* 2nd ed., eds. H.L. Swinney and J.P. Gollub (Springer, New York, 1985) pp. 97–137.

[3] R.H. Kraichnan and D. Montgomery, Rep. on Prog. in Phys. **43** (1980) 547–619.

[4] G. Nicolis and I. Prigogine, *Exploring Complexity – An introduction* (R. Piper, Munchen, 1989); M.L. Smoes, in *Dynamics of Synagertic Systems,* ed. H. Haken (Springer, Berlin, 1980) pp. 80–96.

[5] V.L. Bonch-Bruevich, I.P. Zvyagin and A.G. Mironov, *Domain Electrical Instabilities in Semiconductors* (Consultants Bureau, New York, 1975).

[6] J. Pozhela, *Plasma and Current Instabilities in Semiconductors* (Pergamon, Oxford, 1981).

[7] K. Kaneko, *Phenomenology and Characterization of Spatiotemporal Chaos in Dynamical Systems and Singular Phenomena,* ed. G. Ikegami (World Scientific, Singapore, 1987).

[8] H. Engel, F.-J. Niedernostheide, H.-G. Purwins and E. Schöll eds., *Self-Organization in Activator–Inhibitor Systems: Semiconductors, Gas-Discharge and Chemical Active Media* (Wissenschaft & Technik Verlag, Berlin, 1996).

[9] K. Aoki and K. Yamamoto, Appl. Phys. **A48** (1989) 111.

[10] J. Peinke, J. Parisi, O.E. Rössler and R. Stoop, *Encounter with Chaos – Self-Organized Hierarchical Complexity in Semiconductor Experiments* (Springer, Berlin, 1992).

[11] K.M. Mayer, R. Gross, J. Parisi, J. Peinke and R.P. Huebener, Solid State Commun. **63** (1987) 55.

[12] K. Aoki, U. Rau, J. Peinke, J. Parisi and R.P. Huebener, J. Phys. Soc. Jpn **59** (1990) 420; U. Rau, K. Aoki, J. Peinke, J. Parisi, W. Clauss and R.P. Huebener, Z. Phys. B–Condensed Matter **81** (1990) 53.

[13] K.M. Mayer, J. Parisi, J. Peinke and R.P. Huebener, Physica **D32** (1988) 306.

[14] G. Hüpper, K. Pyragas and E. Schöll, Phys. Rev. **B48** (1993) 17633.

[15] A. Brandl, M. Völcker and W. Prettl, Appl. Phys. Lett. 55 (1989) 238; Solid State Commun. **72** (1989) 847.

[16] A. Brandl, W. Kröninger, W. Prettl and G. Obermair, Phys. Rev. Lett. **64** (1990) 212.

[17] K.M. Mayer, J. Parisi, U. Rau and R.P. Huebener, Proc. 19th Int. Conf. on Phys. Semicond., ed. W. Zawadzki (Institute of Physics, Polish Academy of Sciences, 1988) p. 1411.

[18] D. Jäger, H. Baumann and R. Symanczyk, Phys. Lett **A117** (1986) 141.

[19] E. Schöll, in *Nonlinear Dynamics in Solids,* ed. H. Thomas (Springer, Berlin, 1992) p. 31; see also Proc. 19th Int. Conf. on Phys. Semicond., ed. W. Zawadzki (Institute of Physics, Polish Academy of Sciences, 1988) p. 1407; Phys. Rev. **B34** (1986) 1395; Solid-State Electron. **32** (1989) 1129; Solid-State Electron. **29** (1986) 687. Figures 8.1 and 8.2 are taken from E. Schöll and D. Drasdo, Z. Phys. B–Condensed Matter **81** (1990) 183.

[20] E. Schöll, *Nonequilibrium Phase Transitions in Semiconductors,* (Springer, Berlin, 1987).

[21] E. Kunz and E. Schöll, Z. Phys. B–Condensed Matter **89** (1992) 289.

[22] K.M. Mayer, J. Parisi and R.P. Huebener, Z. Phys. **B71** (1988) 171.

[23] W. Clauss, U. Rau, J. Peinke, J. Parisi, A. Kittel, M. Bayerbach and R.P. Huebener, J. Appl. Phys. **70** (1991) 232.

[24] F.-J. Niedernostheide, R. Dohmen, H. Willebrand, H.-J. Schulze and H.-G. Purwins, in *Nonlinear with Disorder,* Springer Proc. Phys. Vol. **67**, eds. F. Abdullaev, A.R. Bishop and S. Pnevmatikos (Springer, Berlin, 1992) p. 282.

[25] F.-J. Niedernostheide, M. Or-Guil and M. Kleinkes, in *Self-Organization in Activator–Inhibitor Systems: Semiconductors, Gas-Discharge and Chemical Active Media,* eds. H. Engel, F.-J. Niedernostheide, H.-G. Purwins and E. Schöll (Wisserschaft & Technik Verlag, Berlin, 1996) p. 16.

[26] F.-J. Niedernostheide ed., in *Nonlinear Dynamics and Pattern Formation in Semiconductors and Devices,* Springer Proc. Phys. Vol. **79** (Springer, Berlin, 1995) p. 168.

[27] A. Wacker and E. Schöll, Semicond. Sci. Technol. **7** (1992) 1456.

[28] A. Brandl and W. Prettl, Phys. Rev. Lett. **66** (1991) 3044.

[29] K. Aoki and N. Mugibayashi, Appl. Phys. **A48** (1989) 161.

[30] G. Hüpper and E. Schöll, Phys. Rev. Lett. **66** (1991) 2372.

[31] J. Peinke, U. Rau, W. Clauss, R. Richter and J. Parisi, Europhys. Lett. **9** (1989) 743.

[32] K. Aoki, Y. Kawase, K. Yamamoto and N. Mugibayashi, J. Phys. Soc. Jpn **59** (1990) 20.

[33] X.N. Song, D.G. Seiler and M.R. Loloee, Appl. Phys. **A48** (1989) 137.
[34] A. Cenys, G. Lasiene and K. Pyragas, Solid-State Electron. **35** (192) 975; B. Kehrer, W. Quade and E. Schöll, Phys. Rev. **B51** (1995) 7725.
[35] K. Kunihiro, M. Gaa and E. Schöll, Phys. Rev. **B55** (1997) 2207; see also Proc. 23rd Int. Conf. on Phys. Semicond., eds. M. Scheffler and R. Zimmermann (World Scientific, Singapore, 1996) p. 137.
[36] M. Gaa, K. Kunihiro and E. Schöll, in *Self-Organization in Activator–Inhibitor Systems: Semiconductors, Gas-Discharge and Chemical Active Media,* eds. H. Engel, F.-J. Niedernostheide, H.-G. Purwins and E. Schöll (Wissenschaft & Technik Verlag, Berlin, 1996) p. 38.
[37] M. Gaa, R.E. Kunz and E. Schöll, Phys. Rev. **B53** (1996) 15971.
[38] R.E. Kunz, E. Schöll, R. Nürnberg and H. Gajewski, Solid-State Electron. **39** (1996) 1155.
[39] M. Gaa, R.E. Kunz and E. Schöll, Proc. 9th Int. Conf. on Hot Carriers in Semicond., ed. J.P. Leburton (Plenumm Press, New York, 1996) 347; see also Phys. Rev. **B53** (1996) 15 971.
[40] M. Lax, Phys. Rev. **119** (1960) 1502.
[41] L. Reggiani and V.V. Mitin, Rev. Nuov. Cim. **12** (1989) 1.
[42] M. Gaa, R.E. Kunz, E. Schöll, W. Eberle, J. Hirschinger and W. Prettl, Semicond. Sci. Technol. (1996) in press.
[43] E. Schöll, in *Self-Organization in Activator – Inhibitor Systems: Semiconductors, Gas-Discharge and Chemical Active Media,* eds. H. Engel, F.-J. Niedernestheide, H.-G. Purwins and E. Schöll (Wissenschaft & Technik Verlag, Berlin, 1996) p. 10.
[44] J. Spangler, B. Finger, C. Wimmer, W. Eberle and W. Prettl, Semicond. Sci. Technol. **9** (1994) 373.
[45] W. Eberle, J. Hirschinger, U. Margull and W. Prettl, Appl. Phys. Lett. **68** (1996) 3329.
[46] F.-J. Niedernostheide, A. Arps, R. Dohmen, H. Willebrand and H.-G. Purwins, Phys. Stat. Sol. (B) **172** (1992) 249.
[47] F.-J. Niedernostheide, B.S. Kerner and H.-G. Purwins, Phys. Rev. **B46** (1992) 7559.
[48] F.-J. Niedernostheide, M. Kreimer, H.-J. Schulze and H.-G. Purwins, Phys. Lett. **A180** (1993) 113.
[49] F.-J. Niedernostheide, M. Kreimer, B. Kukuk, H.-J. Schulze and H.-G. Purwins, Phys. Lett. **A191** (1994) 285.
[50] F.-J. Niedernostheide, M. Ardes, M. Or-Guil and H.-G. Purwins, Phys. Rev. **B49** (1994) 7370.
[51] Yu. A. Astrov, L.M. Portsel, S.P. Teperick, H. Willebrand and H.-G. Purwins, J. Appl. Phys. **74** (1993) 2159.
[52] H. Willebrand, T. Hunteler, F.-J. Niedernostheide, R. Dohmen and H.-G. Purwins, Phys. Rev. **A45** (1992) 8766.
[53] E. Ammelt, Yu. A. Astrov and H.-G. Purwins, in *Nonlinear with Disorder,* Springer Proc. Phys. Vol. 67, eds. F. Abdullaev, A.R. Bishop and S. Pneumatikos (Springer, Berlin, 1992) p. 22.
[54] B.S. Kerner and V.V. Osipov, Sov. Phys. Usp. **32** (1989) 101.
[55] F.-J. Niedernostheide, C. Brillert, B. Kukuk and H.-G. Purwins, Phys. Rev. **B54** (1996) 14 012.
[56] B.S. Kerner, in *Nonlinear Dynamics and Pattern Formation in Semiconductors and Devices,* Springer Proc. Phys. Vol. 79, ed. F.-J. Niedernestheide (Springer, Berlin, 1995) p. 70.
[57] R. FitzHugh, Biophys. J. **1** (1961) 445.

[58] I. Nagumo, S. Arimoto and S. Yoshizawa, Proc. IRE **50** (1962) 2061.
[59] H.G. Schuster, *Deterministic Chaos–An Introduction,* 2nd rev. ed. (VCH, Weinheim, 1988) p. 155.
[60] A. Wierschem, F.-J. Niedernostheide, A. Gorbatyuk and H.-G. Purwins, Scanning **17** (1995) 106.
[61] D. Jäger and R. Symanczyk, in *Nonlinear Dynamics in Solids,* ed. H. Thomas (Springer, Berlin, 1992) p. 68; see also R. Symanczyk, in *Nonlinear Dynamics and Pattern Formation in Semiconductors and Devices,* Springer Proc. Phys. Vol. 79, ed. F.-J. Niedernestheide (Springer, Berlin, 1995) p. 200.
[62] J. Spangler, U. Margull and W. Prettl, Phys. Rev. **B45** (1992) 12 137.
[63] U. Margull, J. Spangler and W. Prettl, Phys. Rev. **B50** (1994) 14 166.
[64] J. Spangler and W. Prettl, Physica Scripta **T55** (1994) 25.
[65] V. Novàk and W. Prettl, in *Nonlinear Dynamics and Pattern Formation in Semiconductors and Devices,* Springer Proc. Phys. Vol. 79, ed. F.-J. Niedernostheide (Springer, Berlin, 1995) p. 144.
[66] A. Brandl, T. Geisel and W. Prettl, Europhys. Lett. **3** (1987) 401.
[67] A. Brandl and W. Prettl, In *Festkorperprobleme* (Solid State Physics), Vol. **30**, ed. U. Rossler (Vieweg, Braunschweig, 1990) p. 371.
[68] A. Kittel, M. Hirsch and J. Parisi, in *Self-Organization in Activator–Inhibitor Systems: Semiconductors, Gas-Discharge and Chemical Active Media,* eds. H. Engel, F.-J. Niedernostheide, H.-G. Purwins and E. Schöll (Wissenschaft & Technik Verlag, Berlin, 1996) p. 33.
[69] W. Bludau and E. Wagner, Phys. Rev. **B18** (1976) 5410.
[70] K. Aoki, T. Kobayashi and K. Yamamoto, Phys. Lett. **A79** (1980) 445.
[71] W. Eberle, in *Self-Organization in Activator–Inhibitor Systems: Semiconductors, Gas-Discharge and Chemical Active Media,* eds. H. Engel, F.-J. Niedernostheide, H.-G. Purwins and E. Schöll (Wissenschaft & Technik Verlag, Berlin, 1996) p. 44.
[72] U. Margull, in *Self-Organization in Activator–Inhibitor Systems: Semiconductors, Gas-Discharge and Chemical Active Media,* eds. H. Engel, F.-J. Niedernostheide, H.-G. Purwins and E. Schöll (Wissenschaft & Technik Verlag, Berlin, 1996) p. 50.
[73] K. Aoki, Proc. 23rd Int. Conf. on Phys. Semicond., eds. M. Scheffler and R. Zimmermann (World Scientific, Singapore, 1996) p. 2195.
[74] K. Seeger, *Semiconductor Physics – An Introduction* 3rd ed. (Springer, Berlin, 1985) pp. 61–72.
[75] J. Hirschinger, H. Kostial and W. Prettl, Solid State Commun. **106** (1998) 187.
[76] K. Aoki, Proc. 24th Int. Conf. Semicond. Ed. Gershoni (World Scientific, Singapore, 1998). Appeared in CD-ROM format.
[77] J.P. Crutchfield, J.D. Farmer, N.H. Packard and R.S. Shaw, Sci. Amer. **254** (1986) 46.
[78] J. Stavans, Phys. Rev. **A35** (1987) 4314.
[79] R.A. Schmitz, K.G. Graziani and J.L. Hudson, J. Chem. Phys. **67** (1977) 3040; J.L. Hudson, M. Hart and D.J. Marinko, J. Chem. Phys. **71** (1979) 1601.
[80] F. Argoul, A. Arneodo, P. Richetti and J.C. Roux, Chem. Res. **20** (1987) 436.
[81] A. Wacker and E. Schöll, Z Phys. **B39** (1994) 431.
[82] For Hopf bifurcation see J. Guckenheimer and P. Holmes, *Nonlinear Oscillations, Dynamical Systems, and Bifurcations of Vector Fields* (Springer, Berlin, 1983); for codimension-two Turing–Hopf bifurcations, see M. Meixner, A. De Wit, S. Bose and E. Schöll, Phys. Rev. **E55** (1997) 6690 and references therein.
[83] A.M. Turing, Phil. Trans. R. Soc. London **B327** (1952) 37.

[84] Q. Ouyang and H.L. Swinney, Nature **352** (1991) 610.
[85] V. Castets, E. Dulos and P. de Kepper, Phys. Rev. Lett. **64** (1990) 2953.
[86] P. de Kepper, V. Castets, E. Dulos and J. Boissonade, Physica **D49** (1991) 161.
[87] A.F. Münster and M. Watzl, in *Self-Organization in Activator – Inhibitor Systems: Semiconductors, Gas-Discharge and Chemical Active Media,* eds. H. Engel, F.-J. Niedernostheide, H.-G. Purwins and E. Schöll (Wissenschaft & Technik Verlag, Berlin, 1996) p. 164.
[88] E. Ammelt, Yu. Astrov and H.-G. Purwins, in *Self-Organization in Activator–Inhibitor Systems: Semiconductors, Gas-Discharge and Chemical Active Media,* eds. H. Engel, F.-J. Niedernostheide, H.-G. Purwins and E. Schöll (Wissenschaft & Technik Verlag, Berlin, 1996) p. 22.
[89] Yu. Astrov, E. Ammelt, S. Teperick and H.-G. Purwins, Phys. Lett. **A211** (1996) 184.
[90] K. Aoki, Proc. 22nd Int. Conf. on Phys. Semicond., ed. D.J. Lockwood (World Scientific, Singapore, 1995) p. 65.
[91] K. Aoki, Int. J. Bifur. Chaos **7** (1997) in press.
[92] A. Gierer and H. Meinhardt, Kybernetik **12** (1972) 30.
[93] H. Meinhardt, in *Self-Organization in Activator–Inhibitor Systems: Semiconductors, Gas-Discharge and Chemical Active Media,* eds. H. Engel, F.-J. Niedernostheide, H.-G. Purwins and E. Schöll (Wissenschaft & Technik Verlag, Berlin, 1996) p. 172.
[94] H. Meinhardt and M. Klinger, J. Theor. Biol. **126** (1987) 63.
[95] K. Aoki, Solid State Commun. **77** (1991) 87.
[96] K. Aoki, Phys. Lett. **A152** (1991) 485.
[97] H. Kostial, T. Ihn, M. Asche, R. Hey, K. Ploog and F. Koch, Jpn. J. Appl. Phys. **32** (1993) 491.
[98] H. Kostial, K. Ploog, R. Hey and F.G. Boebel, J. Appl. Phys. **78** (1995) 4560.
[99] L.M. Pecora and T.L. Carroll, Phys. Rev. Lett. **64** (1990) 821.
[100] E. Otto, C. Grebogi and J.A. Yorke, Phys. Rev. Lett. **64** (1990) 1196.
[101] K. Aoki, K. Yamamoto, N. Mugibayashi and E. Schöll, Solid State Electron. **32** (1989) 1149.
[102] L.M. Pecora and T.L. Carroll, Proc. 1st Experimental Chaos Conf., eds. S. Vohra, M. Spano, N. Shlesinger, L. Pecora and W. Ditto (World Scientific, Singapore, 1992) p. 121.
[103] E. Otto, C. Grebogi and J.A. York, in *Chaos: Soviet–American Perspectives on Nonlinear Science,* ed. D.K. Campbell (American Institute of Physics, New York, 1990).
[104] T. Yamada and H. Fujisaka, Prog. Theor. Phys. **69** (1983) 32.
[105] L.M. Pecora and T.L. Carroll, Phys. Rev. **A44** (1991) 2374.
[106] R.W. Newcomb and S. Sathyan, IEEE Trans. Circ. Syst. **30** (1983) 54.
[107] W.L. Ditto and L.M. Pecora, Sci. Amer. (Aug. 1993) 62.
[108] J. Gullicksen, M. de Sousa Vieira, M.A. Liberman, R. Sherman, A.J. Lichtenberg, J.Y. Huang and W. Wonchoba, Proc. 1st Experimental Chaos Conf., eds. S. Vohra, M. Spano, N. Shlesinger, L. Pecora and W. Ditto (World Scientific, Singapore, 1992) p. 137.
[109] V.S. Anishchenko, T.E. Vadivasova, D.E. Postnov and M.A. Safonova, Int. J. Bifur. Chaos **2** (1992) 633; Y.-C. Lai and C. Grebogi, Phys. Rev. **E47** (1993) 2357; T. Ushio, Phys. Lett **A198** (1995) 14.
[110] J.M. Perez, Z. Yu, J.M. Kowalski, G. Albert, C.L. Littler and X.N. Song, Proc 1st Experimental Chaos Conf., eds. S. Vohra, M. Spano, N. Shlesinger, L. Pecora and W. Ditto (World Scientific, Singapore, 1992) p. 327.

[111] L.O. Chua, IEICE Trans. Fundamentals **E77-A** (1994) 1811 and references therein.
[112] C.W. Wu and L.O. Chua, Int. J. Bifur. Chaos **3** (1993) 1619.
[113] Lj. Kocarev, K.S. Halle, K. Eckert, L.O. Chua and U. Parliz, Int. J. Bifur. Chaos **2** (1992) 709.
[114] U. Parliz, L.O. Chua, Lj. Kocarev, K.S. Halle and A. Shang, Int. J. Bifur. Chaos **2** (1992) 973.
[115] K.S. Halle, C.W. Wu, M. Itoh and L.O. Chua, Int. J. Bifur. Chaos **3** (1993) 469.
[116] H. Dedieu, M.P. Kennedy and M. Hasler, IEEE Trans. Circ. Syst.–II: Analog and Digital Signal Processing **40** (1993) 634; Special issue on Chaos in Nonlinear Electronic Circuits, Part C.
[117] M. Itoh, H. Murakami and L.O. Chua, IEICE Trans. Fundamentals **E77-A** (1994) 1000.
[118] M.J. Ogorzalek, IEEE Trans. Circ. Syst. – I: Fundamental Theory and Applications **40** (1993) 829; Special issue on Chaos in Nonlinear Electronic Circuits, Part B.
[119] K. Murali and M. Lakshmanan, Phys. Rev. **E48** (1993) 1624.
[120] T. Shinbrot, C. Grebogi, E. Otto and J.A. Yorke, Nature **363** (1993) 411.
[121] W.L. Ditto, S.N. Rauseo and H.L. Spano, Phys. Rev. Lett. **68** (1990) 3211.
[122] J. Singer, Y.-Z. Wang and H.H. Bau, Phys. Rev. Lett. **66** (1991) 1123.
[123] J. Singer and H.H. Bau, Phys. Fluids **A3** (1991) 2859.
[124] E.R. Hunt, Phys. Rev. Lett. **67** (1991) 1953.
[125] R. Roy, T.W. Murphy, T.D. Maier, Z. Gills and E.R. Hunt, Phys. Rev. Lett. **68** (1992) 1259; Z. Gills, C. Iwata, R. Roy, I.B. Schwartz and I. Triandaf, Phys. Rev. Lett. **69** (1992) 3169.
[126] A. Garfinkel, M.L. Spano, W.L. Ditto and J.N. Weiss, Science **257** (1992) 1230.
[127] V. Petrov, V. Gaspar, J. Masere and K. Showalter, Nature **361** (1993) 240.
[128] F. Takens, in *Dynamical Systems and Turbulence,* eds. D. Rand and L.S. Young (Springer, Berlin, 1981) p. 230; N.H. Packard *et al.,* Phys. Rev. Lett. **45** (1980) 712.
[129] G.H. Gunarante, P.S. Linsey and M. Vinson, Phys. Rev. Lett. **63** (1989) 1.
[130] M.L. Spano and W.L. Ditto, Proc. 1st Experimental Chaos Conf., eds. S. Vohra, M. Spano, N. Shlesinger, L. Pecora and W. Ditto (World Scientific, Singapore, 1992) 107.
[131] A.H. Nayfeh and B. Balachandran, *Applied Nonlinear Dynamics – Analytical, Computational, and Experimental Methods* (Wiley & Sons, New York, 1995) pp. 571–588.
[132] E.R. Hunt, Proc. 1st Experimental Chaos Conf., eds. S. Vohra, M. Spano, N. Shlesinger, L. Pecora and W. Ditto (World Scientific, Singapore, 1992) p. 333.
[133] M.P. Kennedy, Proc. 1st Experimental Chaos Conf., eds. S. Vohra, M. Spano, N. Shlesinger, L. Pecora and W. Ditto (World Scientific, Singapore, 1992) p. 340.
[134] G.A. Johnson, T.E. Tigner and R.E. Hunt, J. Circ. Syst. Comput. **3** (1993) 109.
[135] M. Adachi, Proc. Int. Symp. on Nonlinear Theory and its Applications (NOLTA '95, Las Vegas) ed. T. Saito (Housei University, 1995) p. 989.
[136] K. Pyragas, Phys. Lett. **A170** (1992) 421.
[137] E. Schöll, K. Pyragas, D. Cooper and R. Döttling, Proc. 8th Int. Conf. on Hot Carriers in Semicond. (HCIS-8), eds. J.F. Ryan and A.C. Maciel, Semicond. Sci. Technol. **9** (1994) 559.
[138] K. Aihara, T. Takabe and H. Toyoda, Phys. Lett. **A144** (1990) 333.

[139] K. Aihara, in *Bifurcation Phenomena in Nonlinear Systems and Theory of Dynamical Systems,* ed. H. Kawakami (World Scientific, Singapore, 1990) p. 143.

[140] L. Chen and K. Aihara, INNS Neural Networks **8** (1995) 915.

[141] T. Irita, T. Tsujita, M. Fujisawa and K. Hoh, Technical Report of IEICE **ED96-211** (1997) 17 (in Japanese); see also Y. Yasuda and K. Hoh, IEICE Trans. Fundamentals **E76-A** (1996) 1126; K. Hoh and Y. Yasuda, Jpn. J. Appl. Phys. **33** (1994) 594; T. Irita, T. Tsujita, M. Fujishima and K. Hoh, Jpn. J. Appl. Phys. **34** (1995) 1409.

[142] C. Kocot and C.A. Stolte, IEEE Trans. Electron. Devices **ED-29** (1982) 1059.

[143] S. Makram-Ebeid and P. Minondo, IEEE Trans. Electron. Devices **ED-32** (1985) 632.

[144] D.J. Miller and M. Bujatti, IEEE Trans. Electron. Devices **ED-34** (1987) 1239.

[145] Y. Ohno and N. Goto, J. Appl. Phys. **66** (1989) 1217.

[146] H. Yano, N. Goto and Y. Ohno, Proc. 7th Int. Conf. on Hot Carriers in Semicond., eds. C. Hamaguchi and M. Inoue, Semicond. Sci. Technol. **7** (1992) B491.

[147] E. Schöll, in *Negative Differential Resistance and Instabilities in 2-D Semiconductors,* eds. N. Balkan, B.K. Ridley and A.J. Vickers, NATO ASI Series B: Phys. Vol. **307** (Plenum Press, New York, 1993) p. 37.

[148] R. Döttling and E. Schöll, Phys. Rev. **B45** (1992) 1935.

[149] B.K. Ridley, Br. J. Appl. Phys. **17** (1966) 595.

[150] H.K. Sacks and A.G. Milnes, Int. J. Electron. **28** (1970) 565.

[151] H.K. Sacks and A.G. Milnes, Int. J. Electron. **30** (1971) 49.

[152] B.K. Ridley, J. Phys. C: Solid State Phys. **7** (1974) 1169.

[153] G.N. Maracas and D.A. Johnson, Appl. Phys. Lett. **46** (1985) 305.

[154] P. Canfield and L. Forbes, IEEE Electron. Device Lett. **EDL-6** (1985) 227.

[155] S. Makram-Ebeid, A. Mitonneau and G. Laurence, Proc. 2nd Conf. Semi-Insulating III–V Materials, eds. S. Makram-Ebeid and B. Tuck (Shiva Ltd, UK, 1982) p. 336.

[156] D. Miller, M. Bujatti and D. Estreich, in *GaAs IC Symp. Tech. Dig.* (1985) 31.

[157] Z-M. Li, S.P. McAlister, W.G. Mcmullan, C.M. Hurd and D.J. Day, J. Appl. Phys. **67** (1990) 15.

[158] For instance, see related papers in *Negative Differential Resistance and Instabilities in 2-D Semiconductors,* eds. N. Balkan, B.K. Ridley and A.J. Vickers, NATO ASI Series B: Physics Vol. 307 (Plenum Press, New York, 1993).

[159] For instance, see related papers in Proc. 23rd Int. Conf. on Phys. Semicond., eds. M. Scheffler and R. Zimmermann (World Scientific, Singapore, 1996).

[160] E. Schöll and K. Aoki, Appl. Phys. Lett. **58** (1991) 1277; Proc. 20th Int. Conf. on Phys, Semicond., eds. E. Anastassakis and J.D. Joannopoulos (World Scientific, Singapore, 1990) p. 1125.

[161] R. Döttling, E. Schöll and D. Reznik, Proc. NATO ASI on Chaotic Dynamics: Theory and Practice, ed. T. Bountis (Plenum Press, New York, 1992) p. 221.

[162] C. Utecht and E. Schöll, J. Appl. Phys. **74** (1993) 3602.

[163] D. Reznik and E. Schöll, Z. Phys. **B91** (1993) 309.

[164] R. Döttling and E. Schöll, in *Negative Differential Resistance and Instabilities in 2D Semiconductors,* eds. N. Balkan, B.K. Ridley and A.J. Vickers, NATO ASI Series B: Physics Vol. **307** (Plenum Press, New York, 1993) p. 179.

[165] R. Döttling and E. Schöll, Physica **D67** (1993) 418.

[166] R. Döttling and E. Schöll, Proc. MSS6, Solid State Electron. **37** (1994) 685.

[167] R. Döttling, D. Rudzick, E. Schöll, A. Straw, A.J. Vickers, N. Balkan and A. Da Cunha, Proc. HCIS-8, Semicond. Sci. Technol. **9** (1994) 559.
[168] E. Schöll and A. Wacker, in *Nonlinear Dynamics and Pattern Formation in Semiconductors and Devices,* ed. F.J. Niedernostheide (Springer, Berlin, 1995) p. 21.
[169] S. Bose, A. Wacker and E. Schöll, Phys. Lett. **A195** (1994) 144.
[170] O. Rudzick and E. Schöll, Semicond. Sci. Technol. **10** (1995) 143.
[171] D.P. Cooper and E. Schöll, Z. Naturforsch. **50a** (1995) 117.
[172] A. Wacker and E. Schöll, Appl. Phys. Lett. **59** (1991) 1702.
[173] A. Wacker and E. Schöll, Semicond. Sci. Technol. **7** (1992) 1456.
[174] A. Wacker and E. Schöll, Proc. 8th Vilnius Symp. on Ultrafast Phenomenon in Semiconductors (Academia, Vilnius, 1993) p. 37.
[175] A. Wacker and E. Schöll, Proc. HCIS-8, Semicond. Sci. Technol. **9** (1994) 592; A. Wacker, PhD Thesis TU-Berlin (Koster, Berlin, 1993) pp. 26–29.
[176] M. Keever, H. Shichijo, K. Hess, S. Banerjee, L. Witkowski and H. Morkoc, Appl. Phys. Lett. **38** (1981) 36.
[177] D. Coleman, J. Freeman, H. Morkoc, K. Hess, B.G. Streetman and M. Keever, Appl. Phys. Lett. **40** (1982) 493.
[178] A.J. Vickers, A. Straw and J.S. Roberts, Semicond. Sci. Technol. **4** (1989) 743.
[179] N. Balkan and B.K. Ridley, Semicond. Sci. Technol. **3** (1989) 507; Superlatt. Microstruct. **5** (1989) 539.
[180] P. Hendriks, E.A.E. Zwaal, J.G.A. Dubois, F.A.P. Blom and J.H. Wolter, J. Appl. Phys. **69** (1991) 302.
[181] K. Hess, H. Morkoc, H. Shichijo and B.G. Streetman, Appl. Phys. Lett. **35** (1979) 469.
[182] R. Sakamoto, K. Akai and M. Inoue, IEEE Trans. Electron. Devices **36** (1989) 2344.
[183] E. Schöll, Solid State Electron. **32** (1989) 1129.
[184] R. Kunz and E. Schöll, Phys. Rev. **B47** (1993) 4337.
[185] K. Hess, Y.K. Higman, M.A. Emanuel and J.J. Coleman, J. Appl. Phys. **60** (1986) 3775.
[186] K. Kolodzey, J. Lasker, T.K. Higman, M.A. Emanuel and J.J. Coleman, IEEE Electron. Device Lett. **9** (1988) 272.
[187] T.K. Higman, L.M. Miller, M.E. Favaro, M.A. Emanuel, K. Hess and J.J. Coleman, Appl. Phys. Lett. **53** (1988) 1623.
[188] P.M. Smith, P.C. Chao, K.H.G. Duh, L.F. Lester and B.R. Lee, Electron Lett. **22** (1986) 780.
[189] A.M. Belyantsev, E.V. Demidov and A. Yu. Romanov, Proc. 8th Vilnius Symp. on Ultrafast Phenomena in Semiconductors (Academia, Vilnius, 1993) p. 31.
[190] Z.S. Cribnikov, Sov. Phys. Semicond. **6** (1973) 1204.
[191] J.Y. Tang and K. Hess, IEEE Trans. Electron. Devices **ED-29** (1982) 1906.
[192] A. Wacker and E. Schöll, Z. Phys. **B93** (1994) 431.
[193] L. Esaki and L.L. Chang, Phys. Rev. Lett. **33** (1974) 495.
[194] K.K. Choi, B.F. Levine, R.J. Malik, J. Walker and C. Bethea, Phys. Rev. **B35** (1987) 4172.
[195] H.T. Grahn, H. Schneider, and K. von Klitzing, Phys. Rev. **B41** (1990) 2890.
[196] H.T. Grahn, W. Müller, K. von Klitzing and K. Ploog, Surf. Sci. **267** (1992) 579.
[197] M. Helm, J.E. Golub and E. Colas, Appl. Phys. Lett. **56** (1990) 1356.
[198] R. Merlin, S.H. Kwok, T.B. Norris, H.T. Grahn, K. Ploog, L.L. Bonilla, J. Galán, J.A. Cuesta, F.C. Martínez and J. Molera, Proc. 22nd Int. Conf. on

Phys. Semicond., ed. D.J. Lockwood (World Scientific, Singapore, 1995) p. 1039.

[199] H. Le Person, C. Minot, L. Boni, J.F. Palmier and F. Mollot, Appl. Phys. Lett. **60** (1992) 2397.

[200] H.T. Grahn, R.J. Haug, W. Müller and K. Ploog, Phys. Rev. Lett. **67** (1991) 1618.

[201] J. Kastrup, H.T. Grahn, K. Ploog, F. Prengel, A. Wacker and E. Schöll, Appl. Phys. Lett. **65** (1994) 1808.

[202] J. Kastrup, R. Klann, H.T. Grahn, K. Ploog, L.L. Bonilla, J. Galán, M. Kindelan, M. Moscoso and R. Merlin, Phys. Rev. **B52** (1995) 13 761.

[203] L.L. Bonilla, J. Galán, J.A. Cuesta, F.C. Martínez, and J.M. Molera, Phys. Rev. **B50** (1994), 8644.

[204] L.L. Bonilla, in *Self-Organization in Activator–Inhibitor Systems: Semiconductors, Gas-Discharge and Chemical Active Media,* eds. H. Engel, F.-J. Niedernestheide, H.-G. Purwins and E. Schöll (Wisserschaft & Technik Verlag, Berlin, 1996) p. 1.

[205] A. Wacker, F. Prengel and E. Schöll, Proc. 22nd Int Conf. on Phys. Semicond., ed. D.J. Lockwood (World Scientific, Singapore, 1995) p. 1075.

[206] F. Prengel, A. Wacker and E. Schöll, Phys. Rev. **B50** (1994) 1705.

[207] B. Laikhtman, Phys. Rev. **B44** (1991) 11 260.

[208] B. Laikhtman and D. Miller, Phys. Rev. **B48** (1993) 5395.

[209] A.A. Ignatov, E.P. Dodin and V.I. Shashkin, Mod. Phys. Lett. **B5** (1991) 1087.

[210] H.T. Grahn, R. Klann and W.W. Rühle, Proc. 22nd Int. Conf. on Phys. Semicond., ed. D.J. Lockwood (World Scientific, Singapore, 1995) p. 1169.

[211] A. Wacker, G. Schwarz, F. Prengel, E. Schöll, J. Kastrup and H.T. Grahn, Phys. Rev. **B52** (1995) 13 788.

[212] A. Reklaitis, R. Stasch, M. Asche, R. Hey, A. Krotkus and E. Schöll, J. Appl. Phys. **82** (1997) 1706.

[213] K. Aoki and S. Fukui, J. Phys. Soc. Jpn **67** (1998) 1106.

[214] K. Aoki, Phys. Stat. Sol. (b) **204** (1997) 481.

Index

euber4f€

~~139,00€~~

€ 95,–

(£ 110,00 ≈ 177,00 € (m))